The Mizoroki–Heck Reaction

Edited by

MARTIN OESTREICH

Organisch-Chemisches Institut, Westfälische Wilhelms-Universität, Münster, Germany

WILEY

A John Wiley and Sons, Ltd., Publication

This edition first published 2009
© 2009 John Wiley & Sons, Ltd

Registered office
John Wiley & Sons, Ltd, The Atrium, Southern Gate, Chichester, West Sussex, PO19 8SQ, United Kingdom

For details of our global editorial offices, for customer services and for information about how to apply for permission to reuse the copyright material in this book please see our website at www.wiley.com.

Library of Congress Cataloging-in-Publication Data

Oestreich, Martin.
    The Mizoroki-Heck reaction / Martin Oestreich
        p.   cm.
    Includes bibliographical references and index.
    ISBN 978-0-470-03394-4 (cloth)
  1. Heck reaction.   2. Palladium catalysts.   3. Organic compounds–Synthesis.   I. Title.
    QD505.O34 2008
    547′.2–dc22

                                                                    2008029113

British Library Cataloguing in Publication Data

A catalogue record for this book is available from the British Library

ISBN 978-0-470-03394-4 (H/B)

Set in 10/12pt Times by Aptara Inc., New Delhi, India
Printed and bound in Great Britain by CPI Antony Rowe, Chippenham, Wiltshire

The Mizoroki–Heck Reaction

Contents

Foreword

It is a rare honour to be asked to write a preface to a book based on chemistry which you helped begin. My research in palladium chemistry began before 1968. Initial results were obtained by preparing arylpalladium salts by exchange reactions between $Pd(OAc)_2$ or $PdCl_2$ and aryl-mercurials, -lead or -tin compounds and reacting these with alkenes to form arylalkenes.

The reactions occurred well even at room temperature, but the impracticality of using stoichiometric amounts of palladium severely limited the use. Subsequently, we overcame this problem by forming the organo-palladium salts by the reaction of aryl iodides directly with finely divided palladium, a reaction which Mizoroki discovered independently at about the same time.

This variation, however, suffered from another problem, namely that bromides generally worked poorly and chlorides very poorly, if at all. We then found that the addition of a triarylphosphine greatly helped the reactions with bromides and made this a practical synthetic reaction. Even so, it has taken many years for organic chemists to accept the use of palladium as a general reagent for organic synthesis.

The more recent discovery by Fu, that the use of tri-*tert*-butylphosphine as a ligand in the reaction in place of triphenylphosphine caused arylchlorides to react efficiently, greatly expanded the ability and practicality of the reaction so that it has now become a 'standard organic reaction' that is widely used by synthetic organic chemists and it has formed significant applications in the commercial synthesis of fine chemicals.

The current state of the art in this area is clearly shown by the variety and content of articles included in this volume. Surely, this publication will spur further research in the area and lead to still more important applications.

Richard F. Heck
Kingston, Ontario

Preface

I encountered Mizoroki–Heck chemistry in a practical sense upon joining the laboratory of Professor Larry E. Overman as a postdoctoral research fellow. At that time, I knew that the Mizoroki–Heck reaction belongs to a large family of palladium(0)-catalysed C–C bond-forming processes, all of them highly relevant to the construction of carbon skeletons ('forwards') and highly influential on bond disconnections ('backwards'). With my focus on complex molecule synthesis and enantioselective formation of congested quaternary carbon atoms in particular, I understood that the Mizoroki–Heck reaction is remarkable, in that its intramolecular variant meets this formidable challenge with impressive efficacy. In fact, the Mizoroki–Heck reaction is also the classic example of a catalytic asymmetric C–C bond-forming transition-metal catalysis. As a practitioner, I had to learn that the road to a successful Mizoroki–Heck reaction is paved with countless experimental subtleties and, oftentimes, failure. Seemingly minor changes in the reaction conditions greatly influenced the outcome, if there was any at all. Attempts to rationalize these observations based on literature precedence were frequently unfruitful, yet they opened the door to the tremendous and diverse amount of data on Mizoroki–Heck chemistry.

When I was approached by Paul Deards, commissioning editor at Wiley, to do a book on the Mizoroki–Heck reaction, I had considered myself aware of the breadth of the field but I had not seen the end of it yet. I foolishly agreed. However, while assembling the tentative table of contents, it became clear to me that the Mizoroki–Heck reaction, one of the pivotal C–C bond-forming reactions in synthetic organic chemistry, has gone through an immense growth over recent decades. It was obvious that it would be difficult to grasp all facets of Mizoroki–Heck chemistry in a single volume: quite fascinating to have a book on just one reaction! Areas that one perceives as niches when flicking through journals have matured, now deserving to be mentioned in their own right, which finally led to the division of the book into three sections and 16 chapters. There are chapters – and I will only mention one – that, for the first time, provide a comprehensive summary of the allocated area: *Mechanisms of the Mizoroki–Heck Reaction* compiled by Professor Anny Jutand is a wonderful example of that and, in my opinion, stimulates the appetite for the entire book.

To me, the preface written by Professor Richard F. Heck is the icing on the cake of a book that will hopefully be a useful reference and enjoyable reading for everyone being exposed to Mizoroki–Heck chemistry.

It is my pleasure to thank the international selection of contributors, 34 scientists from eight countries, for accepting my invitation and making this comprehensive monograph become reality. I certainly do not take their dedication, effort, share of expertise and insight for granted. I feel indebted to Paul Deards at Wiley for setting the ball rolling and, of course, Richard Davies at Wiley for being such a supportive project editor. Last, but not least, I express my sincere gratitude to my former graduate students, Dr Axel B. Machotta and Dr Sebastian Rendler, for their invaluable support during the final editing of all manuscripts. I truly enjoyed working with all of you!

Martin Oestreich
Münster/Westfalen, Germany

Contributors

Lutz Ackermann Institut für Organische und Biomolekulare Chemie, Georg-August-Universität, Tammannstraße 2, D-37077 Göttingen, Germany

Werner Bonrath DSM Nutritional Products, PO Box 2676, CH-4002 Basel, Switzerland

Robert Born Institut für Organische und Biomolekulare Chemie, Georg-August-Universität, Tammannstraße 2, D-37077 Göttingen, Germany

Irina P. Beletskaya Department of Chemistry, M. V. Lomonosov Moscow State University, Leninskie Gory, 119992 Moscow, Russia

Stefan Bräse Institut für Organische Chemie, Universität Karlsruhe (TH), Fritz-Haber-Weg 6, D-76131 Karlsruhe, Germany

Andrei V. Cheprakov Department of Chemistry, M. V. Lomonosov Moscow State University, Leninskie Gory, 119992 Moscow, Russia

Anthony G. Coyne Centre for Synthesis and Chemical Biology, UCD School of Chemistry and Chemical Biology, University College Dublin, Belfield, Dublin 4, Ireland

Amy B. Dounay Neurosciences Chemistry, Pfizer Global Research and Development, Groton Laboratories USA, MS 8220-4236, Eastern Point Road, Groton, CT 06340, USA

Eric M. Ferreira Division of Chemistry and Chemical Engineering, MC 164–30, California Institute of Technology, 1200 East California Boulevard, Pasadena, CA 91125, USA

Martin O. Fitzpatrick   Centre for Synthesis and Chemical Biology, UCD School of Chemistry and Chemical Biology, University College Dublin, Belfield, Dublin 4, Ireland

Lukas Gooßen   Fachbereich Chemie – Organische Chemie, Technische Universität Kaiserslautern, Erwin-Schrödinger-Straße (Gebäude 54), D-67663 Kaiserslautern, Germany

Käthe Gooßen   Bayer Healthcare AG, Pharmaceuticals, Building 402, D-42096 Wuppertal, Germany

Patrick J. Guiry   Centre for Synthesis and Chemical Biology, UCD School of Chemistry and Chemical Biology, University College Dublin, Belfield, Dublin 4, Ireland

Kenichiro Itami   Department of Chemistry and Research Center for Materials Science, Nagoya University, Chikusa-ku, Nagoya 464-8602, Japan

Anny Jutand   Département de Chimie, Ecole Normale Supérieure, CNRS, 24 Rue Lhomond, F-75231 Paris Cedex 5, France

Mats Larhed   Department of Medicinal Chemistry, Organic Pharmaceutical Chemistry, Uppsala University, BMC, PO Box 574, SE-75123 Uppsala, Sweden

Ulla Létinois   DSM Nutritional Products, PO Box 2676, CH-4002 Basel, Switzerland

Laura M. Levy   Institut für Organische und Biomolekulare Chemie, Georg-August-Universität Göttingen, Tammannstraße 2, D-37077 Göttingen, Germany

James T. Link   Abbott Laboratories, Department 47F, Building AP10, 100 Abbott Park Road, Abbott Park, IL 60064-6098, USA

Axel B. Machotta   Organisch-Chemisches Institut, Westfälische Wilhelms-Universität Münster, Corrensstraße 40, D-48149 Münster, Germany

Thierry Muller   Institut für Organische Chemie, Universität Karlsruhe (TH), Fritz-Haber-Weg 6, D-76131 Karlsruhe, Germany

Thomas Netscher   DSM Nutritional Products, PO Box 2676, CH-4002 Basel, Switzerland

Peter Nilsson   Department of Medicinal Chemistry, Organic Pharmaceutical Chemistry, Uppsala University, BMC, PO Box 574, SE-75123 Uppsala, Sweden

Martin Oestreich   Organisch-Chemisches Institut, Westfälische Wilhelms-Universität Münster, Corrensstraße 40, D-48149 Münster, Germany

Takashi Ohshima Department of Chemistry, Graduate School of Engineering Science, Osaka University, Toyonaka, Osaka 560-8531, Japan

Kristofer Olofsson Department of Medicinal Chemistry, Organic Pharmaceutical Chemistry, Uppsala University, BMC, Uppsala, Sweden

Larry E. Overman Department of Chemistry, 1102 Natural Sciences II, University of California, Irvine, Irvine, CA 92697–2025, USA

Jan Schütz DSM Nutritional Products, PO Box 2676, CH-4002 Basel, Switzerland

Masakatsu Shibasaki Graduate School of Pharmaceutical Sciences, The University of Tokyo, Hongo, Bunkyo-ku, Tokyo 113-0033, Japan

Brian M. Stoltz Division of Chemistry and Chemical Engineering, MC 164–30, California Institute of Technology, 1200 East California Boulevard, Pasadena, CA 91125, USA

Lutz F. Tietze Institut für Organische und Biomolekulare Chemie, Georg-August-Universität Göttingen, Tammannstraße 2, D-37077 Göttingen, Germany

Carol K. Wada Abbott Laboratories, Department 47F, Building AP10, 100 Abbott Park Road, Abbott Park, IL 60064-6098, USA

Jun-ichi Yoshida Department of Synthetic Chemistry and Biological Chemistry, Graduate School of Engineering, Kyoto University, Nishikyo-ku, Kyoto 615-8510, Japan

Haiming Zhang Division of Chemistry and Chemical Engineering, MC 164–30, California Institute of Technology, 1200 East California Boulevard, Pasadena, CA 91125, USA

1

Mechanisms of the Mizoroki–Heck Reaction

Anny Jutand

Département de Chimie, Ecole Normale Supérieure, CNRS, 24 Rue Lhomond,
Paris Cedex 5, France

1.1 Introduction

The palladium-catalysed Mizoroki–Heck reaction is the most efficient route for the vinylation of aryl/vinyl halides or triflates. This reaction, in which a C—C bond is formed, proceeds in the presence of a base (Scheme 1.1) [1, 2]. Nonconjugated alkenes are formed in reactions involving cyclic alkenes (Scheme 1.2) [1e, 2a,c,e,g] or in intramolecular reactions (Scheme 1.3) [2b,d–g] with creation of stereogenic centres. Asymmetric Mizoroki–Heck reactions may be performed in the presence of a chiral ligand [2]. The Mizoroki–Heck reaction has been intensively developed from a synthetic and mechanistic point of view, as expressed by the impressive number of reviews and book chapters [1, 2].

In the late 1960s, Heck reported that arylated alkenes were formed in the reaction of alkenes with a stoichiometric amount of [Ar–Pd—Cl] or [Ar–Pd–OAc], generated *in situ* by reacting ArHgCl with $PdCl_2$ or ArHgOAc with $Pd(OAc)_2$ respectively [3]. A mechanism was proposed which involves a *syn* migratory insertion of the alkene into the Ar–Pd bond, followed by a *syn* β-hydride elimination of a hydridopalladium [HPdX] (X = Cl, OAc) (Scheme 1.4a). In the case of cyclic alkenes, in which no *syn* β-hydride is available, a *syn* β'-hydride elimination occurs, leading to a nonconjugated alkene (Scheme 1.4b). Isomerization of the new C=C bond may occur by a *syn* readdition of HPdX in the reverse direction, followed by a *syn* β''-hydride elimination (Scheme 1.4c) [3c].

The Mizoroki–Heck Reaction   Edited by Martin Oestreich
© 2009 John Wiley & Sons, Ltd

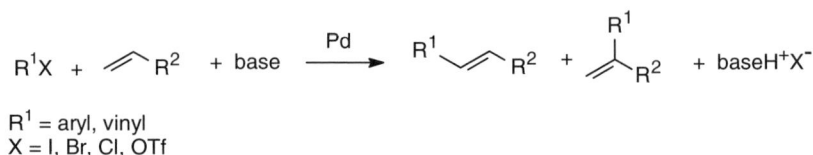

$$R^1X \;+\; \diagup\!\!\!\diagdown R^2 \;+\; \text{base} \;\xrightarrow{\;Pd\;}\; R^1\diagup\!\!\!\diagdown\!\!\diagup R^2 \;+\; \overset{R^1}{\diagdown\!\!\!\diagup}\!\!R^2 \;+\; \text{baseH}^+X^-$$

R^1 = aryl, vinyl
X = I, Br, Cl, OTf

Scheme 1.1

These pioneering studies by Heck have opened the way to a new reaction later called the Mizoroki–Heck reaction (Scheme 1.1). In 1971, Mizoroki *et al.* reported preliminary results on the PdCl$_2$-catalysed arylation of alkenes by iodobenzene in the presence of potassium acetate as base (Scheme 1.5) [4]. No new contribution to the mechanism was proposed, except that palladium particles, formed *in situ* in the reaction or deliberately added, were suggested to be the active catalyst [4].

In 1972, Heck and Nolley [5] improved the reactions by using Pd(OAc)$_2$ as catalyst and *n*-Bu$_3$N as base (Scheme 1.6). The reactions were performed without any solvent or in *N*-methylpyrrolidone (NMP) at 100 °C. More importantly, they proposed for the first time a full mechanism for the catalytic reactions. On the basis of what were at that time recent studies by Fitton and coworkers and Coulson on the formation of σ-ArPdXL$_2$ (L = PPh$_3$; X = I [6a,c], Br [6c], Cl [6b]) by oxidative addition of aryl halides to Pd^{0}L$_4$, Heck and Nolley proposed the formation of [ArPdI] in the oxidative addition of aryl iodide to palladium metal, generated *in situ* by reduction of Pd(OAc)$_2$ by the alkene. After reaction of [ArPdI] with the alkene as in Scheme 1.4a, the hydridopalladium [HPdI] decomposes to HI (quenched by the base) and palladium(0) available for another catalytic cycle [5].

In 1973, Mizoroki and coworkers [7] extended their preliminary work (Scheme 1.5) to aryl bromides; however, these were found to be considerably less reactive than aryl iodides: PhI > PhBr ≫ PhCl. Palladium black was identified to be more efficient than PdCl$_2$. The use of a phosphine ligand PPh$_3$ was mentioned as being slightly beneficial [7]. This was the last contribution of the Japanese group to those reactions.

In 1974, Dieck and Heck [8] developed the use of PPh$_3$ in association with Pd(OAc)$_2$ (Scheme 1.7). Aryl iodides were found to react faster than without PPh$_3$. More interestingly, the reaction was extended to aryl bromides at temperatures in the range 100–135 °C, but aryl chlorides were still unreactive [8].

In 1978, Heck and coworkers [1a, 9] introduced substituted triarylphosphines associated with Pd(OAc)$_2$. Among them, the tri-*o*-tolylphosphine, P(*o*-Tol)$_3$, was found to be more efficient than PPh$_3$ in reactions involving aryl bromides (experimental conditions of Scheme 1.7 at 75 °C). In 1983, Spencer improved the Mizoroki–Heck reactions catalysed by Pd(OAc)$_2$ associated with P(*o*-Tol)$_3$ upon using the polar solvent DMF and NaOAc as

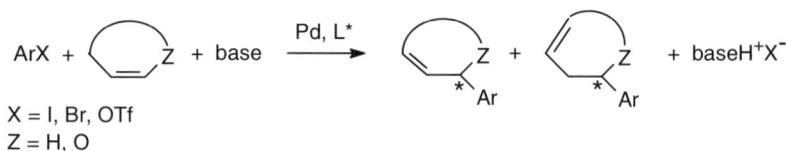

$$ArX \;+\; \bigcirc\!\!Z \;+\; \text{base} \;\xrightarrow{\;Pd,\,L^*\;}\; \bigcirc\!\!\underset{*}{Z}\!\!-\!Ar \;+\; \bigcirc\!\!\underset{*}{Z}\!\!-\!Ar \;+\; \text{baseH}^+X^-$$

X = I, Br, OTf
Z = H, O

Scheme 1.2

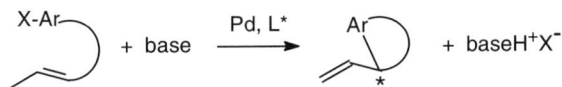

Scheme 1.3

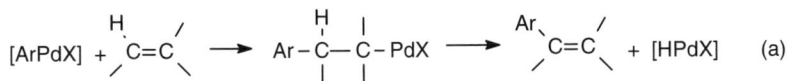

X = Cl, OAc

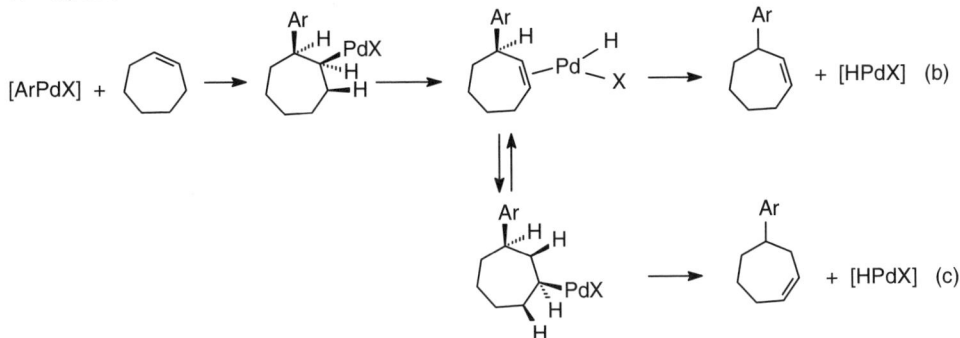

Scheme 1.4

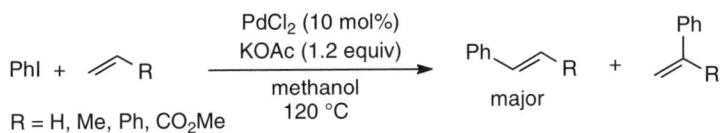

PhI + ═R

R = H, Me, Ph, CO$_2$Me

PdCl$_2$ (10 mol%)
KOAc (1.2 equiv)
methanol
120 °C

Ph ⌒R (major) + Ph ⌒R

Scheme 1.5

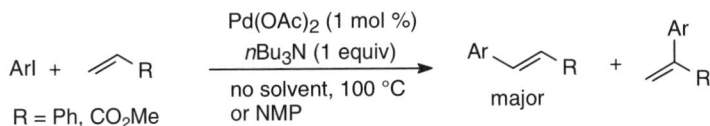

ArI + ═R

R = Ph, CO$_2$Me

Pd(OAc)$_2$ (1 mol %)
*n*Bu$_3$N (1 equiv)
no solvent, 100 °C
or NMP

Ar ⌒R (major) + Ar ⌒R

Scheme 1.6

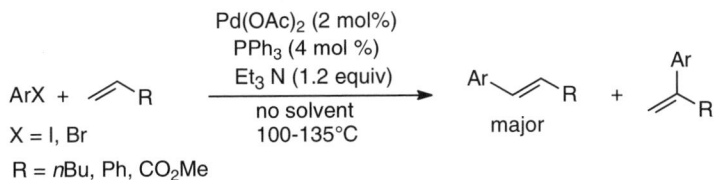

ArX + ═R

X = I, Br
R = *n*Bu, Ph, CO$_2$Me

Pd(OAc)$_2$ (2 mol%)
PPh$_3$ (4 mol %)
Et$_3$N (1.2 equiv)
no solvent
100-135°C

Ar ⌒R (major) + Ar ⌒R

Scheme 1.7

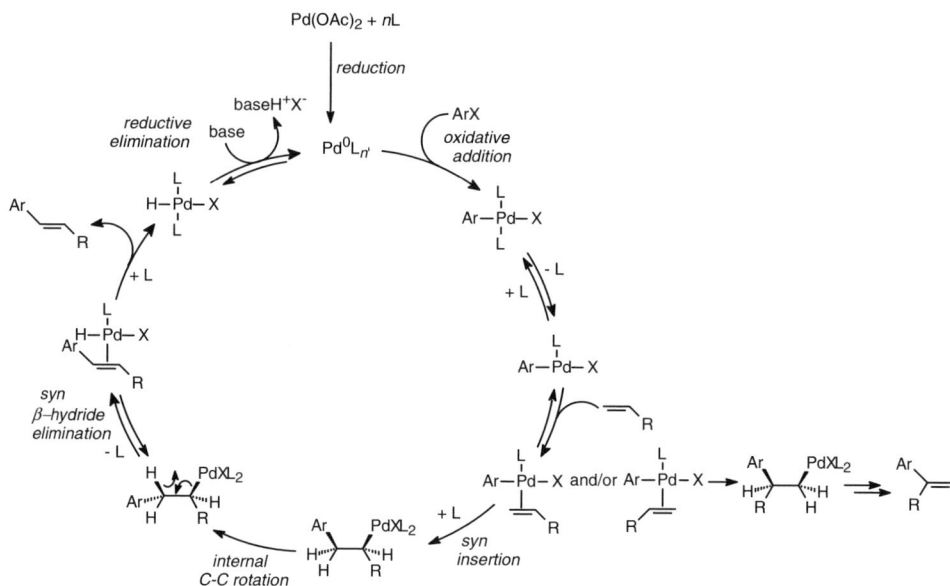

Scheme 1.8 *Mechanism proposed by Heck when the precursor is Pd(OAc)$_2$ associated with monophosphine ligands L.*

base [10a]. High turnover numbers (TONs) were thus achieved from aryl bromides (e.g. TON = 134 000 in the reaction of *para*-nitrophenyl bromide and ethyl acrylate at 130 °C) [10a]. Surprisingly, PPh$_3$ associated with Pd(OAc)$_2$ was more efficient than P(*o*-Tol)$_3$ (P/Pd = 4 in both cases) for the reaction of electron-withdrawing group (EWG)-substituted aryl chlorides which, however, exhibited low reactivity (21–50% yield after 6 h at 150 °C). Chlorobenzene was rather unreactive (4% yield) [10b].

In 1974, a mechanism was proposed by Dieck and Heck [8] for reactions catalysed by Pd(OAc)$_2$ associated with monophosphine ligands. Such a mechanism written by Heck as successive reactions [1a–c] is presented as the catalytic cycle in Scheme 1.8. After formation of a Pd(0) catalyst from the precursor Pd(OAc)$_2$ by a vaguely defined reduction process, the following steps of the catalytic cycle were proposed:

(i) The first step of the catalytic cycle is an *oxidative addition* of the aryl halide to a Pd(0) complex. Such a step is supported by oxidative additions of aryl halides to Pd^0(PPh$_3$)$_4$ reported by Fitton and Rick [6c] in 1971 with the reactivity order: ArI > ArBr ≫ ArCl and *p*-EWG-ArX > *p*-EDG-ArX (EDG = electron-donating group).

(ii) The oxidative addition gives a σ-aryl–palladium(II) halide, *trans*-ArPdXL$_2$ [6c], which first coordinates to the alkene after dissociation of one phosphine and then undergoes a *syn insertion* of the alkene, leading to a σ-alkyl–palladium(II) halide. The phosphine dissociation step is supported by the fact that the reaction of *trans*-PhPdBr(PPh$_3$)$_2$ with ethylene is inhibited by extra phosphine [9]. The reaction of ArPdXL$_2$ with an alkene, also referred to as *carbopalladation* (formation of a Pd—C bond), is at the origin of the regioselectivity of Mizoroki–Heck reactions [8, 9]. Indeed, two isomeric σ-alkyl–palladium(II) halide complexes may be formed in an α or β arylation of the alkene, leading to the branched or linear arylated alkene respectively (Scheme 1.8).

(iii) An internal C—C *bond rotation* in the σ-alkyl–palladium(II) halide brings an sp^3-bonded β-hydrogen in a *syn* position relative to the palladium atom. A *syn β-hydride elimination* gives a hydridopalladium(II) halide ligated to the arylated alkene. This reaction can be reversible (favoured in phosphine-free Mizoroki–Heck reactions), as proposed by Heck to explain some isomerization of the final arylated alkene by readdition of the hydridopalladium(II) halide onto its C=C bond with a reverse regioselectivity (see a similar process in Scheme 1.4b and c [8]).

(iv) After dissociation from the arylated alkene, the hydridopalladium(II) halide undergoes a reversible *reductive elimination* to regenerate the active Pd(0) complex. The base shifts this equilibrium towards the Pd(0) catalyst by quenching the hydrogen halide [9].

Under the same experimental conditions, same alkene, ligand and base, the reactivity order of aryl halides in Mizoroki–Heck reactions is usually: ArI > ArBr $\gg$ ArCl, suggesting that the *oxidative addition* is rate determining for the less reactive aryl halides. For the most reactive ones, the *complexation/insertion* of the alkene is considered as rate determining.

Besides the usual parameters of all reactions (temperature, solvent and concentration), other parameters may be varied (Pd precursors, ligands, bases, additives, etc.) to optimize Mizoroki–Heck reactions. Much work has been done in the last 30 years to perform Mizoroki–Heck reactions under mild conditions with high turnover numbers (TONs) [1v] and turnover frequencies (TOFs) [1, 2], to react aryl chlorides [1r,y], to improve the regioselectivity (α versus β arylation) [1e,g, 2g], to improve the enantioselectivity obtained with a chiral ligand when reacting cyclic alkenes [2a,c,e] or in intramolecular Mizoroki–Heck reactions [2b,d–f]. At the same time, the mechanism has been investigated to explain the high dependence of the efficiency and regioselectivity of Mizoroki–Heck reactions on the nature of the catalytic precursor, the base and the ligand.

It emerges that the main steps of the former textbook mechanism proposed by Heck have been confirmed (Scheme 1.8). However, the catalytic cycle may involve intermediate palladium complexes whose structures differ from those originally proposed, depending on the experimental conditions. One must also take into account the fact that new reagents (aryl triflates), new ligands (bidentate ligands, carbenes, bulky phosphines, etc.) and new precursors (palladacycles) have been introduced a long time after Heck's proposal.

The mechanisms of Mizoroki–Heck reactions performed from aryl derivatives are presented herein by highlighting how the catalytic precursors, the bases and the ligands may affect the structure and reactivity of intermediate palladium(0) or palladium(II) complexes in one or more steps of the catalytic cycle and, consequently, how they may affect the efficiency and regioselectivity of the catalytic reactions.

1.2 Mechanism of the Mizoroki–Heck Reaction when the Catalytic Precursor is Pd(OAc)$_2$ in the Absence of Ligand

As recalled in Section 1.1, Pd(OAc)$_2$ may be used as precursor without any phosphine ligand in Mizoroki–Heck reactions performed from aryl iodides [5]. However, Pd(OAc)$_2$

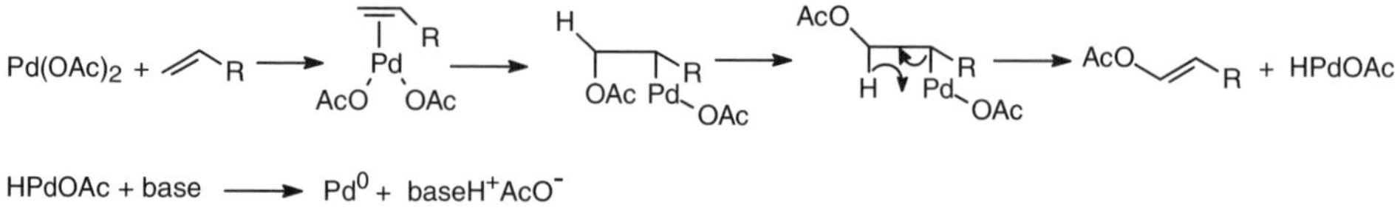

Pd(OAc)₂ + [alkene]→R → [Pd complex: AcO OAc with R] → [complex: OAc Pd OAc with R and H] → [complex: AcO H Pd OAc with R] → AcO—[alkene]→R + HPdOAc

HPdOAc + base ⟶ Pd⁰ + baseH⁺AcO⁻

Scheme 1.9

must be reduced *in situ* to a Pd(0) species which initiates the catalytic cycle by an oxidative addition to the aryl iodide.

Some reagents of Mizoroki–Heck reactions may play the role of reducing agents, such as alkenes proposed by Heck [3b], according to the mechanism depicted in Scheme 1.9: intramolecular nucleophilic attack of acetate onto the alkene coordinated to Pd(OAc)₂, followed by a β-hydride elimination leading to HPdOAc and subsequent formation of Pd(0) in the presence of a base [3b, 11, 12].

Amines used as bases in Mizoroki–Heck reactions have also been proposed as reducing agents. Indeed, a β-hydride elimination may take place in the amine coordinated to the Pd(II) centre, leading to HPdOAc and then to Pd(0) in the presence of the amine (Scheme 1.10) [13].

Pd(OAc)₂ + R¹R²N-CH₂R³ ⟶ [complex: OAc | H—Pd–OAc, R³CH—NR¹R²] ⟶ R³CH=N⁺R¹R² + HPdOAc

HPdOAc + amine ⟶ Pd⁰ + amineH⁺ AcO⁻

Scheme 1.10

Some additives, such as ammonium salts $R_4N^+X^-$ (X = Br, Cl, R = *n*-Bu), greatly improve Mizoroki–Heck reactions performed in the absence of phosphine ligands (Jeffery process [1h]). Reetz and Westermann [14a] have reported that thermolytic decomposition of Pd(OAc)₂ occurs at 100–130 °C in the presence of ammonium salts. The cleavage of the Pd–OAc bond generates $R_4N^+X^-$-stabilized Pd(0) nanoparticles (Scheme 1.11). Oxidative addition of PhI with those nanoparticles gives σ-bonded phenyl–Pd(II) species as PhPdI or $PhPdX_3^{2-}$ characterized by ¹H NMR spectroscopy (Scheme 1.11). Such Pd nanoparticles catalyse Mizoroki–Heck reactions [14]. Chlorobenzene can even react with styrene in the presence of Pd(OAc)₂ associated with the phosphonium salt $Ph_4P^+Cl^-$, yet at high temperatures (150 °C) in NMP [14b]. The homogeneous/heterogeneous character of the catalysis is under debate. De Vries *et al.* have developed so-called 'homeopathic' ligand-free Mizoroki–Heck reactions of aryl bromides and *n*-butyl acrylate, in the presence of low loading of Pd(OAc)₂ (0.05 mol%), NaOAc as base, in NMP at 130 °C [15a]. The lower the initial Pd(OAc)₂ concentration, the higher the TOF becomes. Pd(OAc)₂ is proposed to generate soluble clusters of palladium stabilized by Na^+X^- (X = AcO⁻ or/and Br⁻ in the course of the reaction) (Scheme 1.12). The latter inhibit the formation of inactive palladium black and deliver the active ligand-free Pd(0) into the catalytic cycle. Upon increasing Pd(OAc)₂ concentration, the two equilibria are shifted towards their

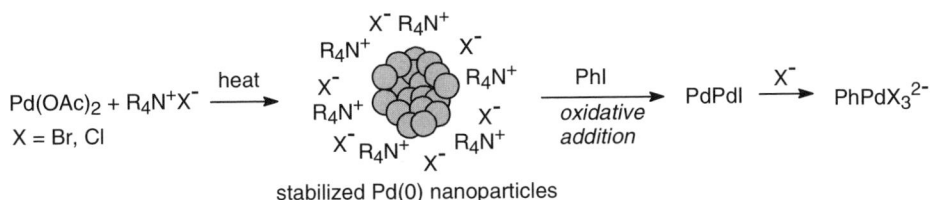

$$Pd(OAc)_2 + R_4N^+X^- \xrightarrow{\text{heat}} \text{stabilized Pd(0) nanoparticles} \xrightarrow[\text{oxidative addition}]{\text{PhI}} PdPdI \xrightarrow{X^-} PhPdX_3^{2-}$$

X = Br, Cl

Scheme 1.11

left-hand sides, namely towards the formation of inactive palladium black. However, the 'homeopathic' ligand-free palladium does not catalyse Mizoroki–Heck reactions with aryl chlorides.

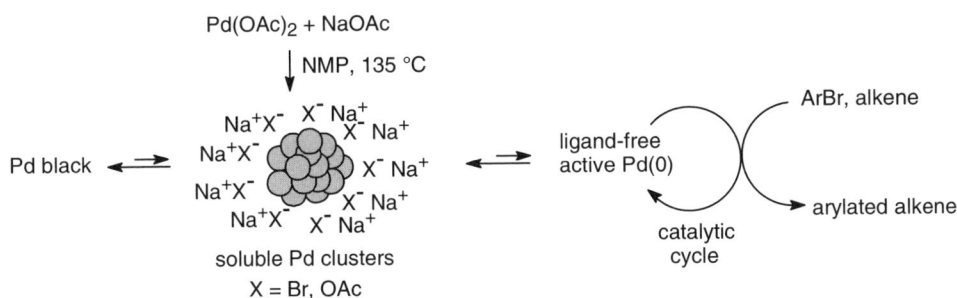

Pd(OAc)$_2$ + NaOAc

NMP, 135 °C

Pd black ⟷ soluble Pd clusters   X = Br, OAc ⟷ ligand-free active Pd(0)   catalytic cycle

ArBr, alkene → arylated alkene

Scheme 1.12

Mechanistic investigations of Mizoroki–Heck reactions performed in the absence of stabilizing ligands are rare due to some difficulty in characterizing intermediate palladium species and getting kinetic data [16]. Nevertheless, de Vries *et al.* [15b] have characterized anionic species by electrospray ionisation mass spectrometry: H_2O–$Pd^0(OAc)^-$ and $PhPd^{II}I_2^-$, in a Mizoroki–Heck reaction performed from PhI, butyl acrylate, NEt_3 as base, $Pd(OAc)_2$ as catalyst in NMP at 80 °C. Anionic species $Pd^{II}_2I_6^{2-}$, $PhPd^{II}I_2^-$ and $(\eta^2{-}CH_2{=}C(CH_3)CH_2OH)PhPd^{II}I_2^-$ have been characterized by quick-scanning extended X-ray absorption fine structure in the course of a catalytic reaction (PhI, $CH_2{=}C(CH_3)CH_2OH$, NEt_3, $Pd(OAc)_2$, in NMP) [15c].

1.3 Mechanism of the Mizoroki–Heck Reaction when the Catalytic Precursor is Pd(OAc)$_2$ Associated with Monophosphine Ligands

1.3.1 Pd(0) Formation from Pd(OAc)$_2$ in the Presence of a Monophosphine Ligand

The catalytic precursor $Pd^{II}(OAc)_2$ associated with a monophosphine such as PPh_3 is more efficient than $Pd^0(PPh_3)_4$ in Mizoroki–Heck reactions. Two problems arise: (i) how an active Pd(0) complex can be generated from $Pd^{II}(OAc)_2$ associated with PPh_3; (ii) why the latter precursor is more efficient than $Pd^0(PPh_3)_4$, whereas both are supposed to generate the same reactive species $Pd^0(PPh_3)_2$ in the oxidative addition to aryl halides [17].

$$Ph_3P \diagdown \diagup OAc$$
$$Pd^{II}$$
$$AcO \diagup \diagdown PPh_3$$

$$\xrightarrow[\text{DMF, 25 °C}]{k_{red} = 4 \times 10^{-4} \text{ s}^{-1}}$$

$$Pd^0(PPh_3)(OAc)^- + AcO\text{-}PPh_3^+$$

$$\downarrow PPh_3 \qquad\qquad \downarrow H_2O$$

$$Pd^0(PPh_3)_2(OAc)^- \qquad H^+ + (O)PPh_3 + AcOH$$

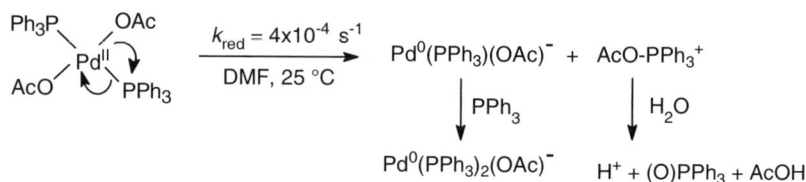

Overall reaction

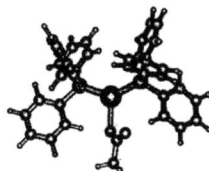

$$Pd(OAc)_2 + 3\,PPh_3 + H_2O \longrightarrow Pd^0(PPh_3)_2(OAc)^- + H^+ + (O)PPh_3 + AcOH$$

Pd–O = 2.369 Å

Scheme 1.13

In 1991, Jutand and coworkers discovered that a Pd(0) complex was generated *in situ* in tetrahydrofuran (THF) or dimethylformamide (DMF) at room temperature upon mixing $Pd(OAc)_2$ and n equivalents of PPh_3 ($n \geq 2$); that is, from the complex $Pd(OAc)_2(PPh_3)_2$ formed in the early stage [18]. In this process, PPh_3 is oxidized to the phosphine oxide, thereby attesting the reduction of Pd(II) to Pd(0) by the phosphine (Scheme 1.13) [18]. The rate of formation of the Pd(0) complex is not affected by the presence of alkene (decene) or amine (NEt_3) added in large excess to the initial mixture ($Pd(OAc)_2 + n$ equivalents of PPh_3 ($n \geq 2$)) [18]. Consequently, in the presence of a monophosphine ligand, the reduction of Pd(II) by the phosphine is much faster than that by the alkene or amine. Water does not modify the rate of formation of the Pd(0) complex [18]. The formation *in situ* of a Pd(0) complex from $Pd(OAc)_2$ and 5 equiv of PPh_3 in benzene in the presence of NEt_3 and water was confirmed shortly after by Osawa *et al.* [19].

Further studies by Amatore, Jutand *et al.* [20] on the kinetics of the reduction process have established that a Pd(0) complex is formed via an intramolecular reduction (reductive elimination) which takes place within the complex $Pd(OAc)_2(PPh_3)_2$ (first-order reaction for the palladium(II) and zero-order reaction for the phosphine) (Scheme 1.13). The rate constant of this rate-determining reduction process has been determined (k_{red} in Scheme 1.13). The reduction also delivers a phosphonium salt which is hydrolysed to phosphine oxide in a faster step (zero-order reaction for H_2O) (Scheme 1.13) [18].

The formation of a Pd(0) complex by reduction of a Pd(II) complex by a phosphine is quite general. It takes place as soon as a Pd^{II}–OR (R = H, $COCH_3$, $COCF_3$) bond is formed. Pd(0) complexes are indeed generated from: (i) $PdCl_2(PPh_3)_2$ upon addition of a base OH^- (via $PdCl(OH)(PPh_3)_2$), as established by Grushin and Alper [21]; (ii) $PdCl_2(PPh_3)_2$ after addition of acetate ions (via $PdCl(OAc)(PPh_3)_2$) [22]; (iii) the cationic complex $[Pd^{II}(PPh_3)_2]^{2+},2BF_4^-$ in the presence of water and PPh_3 (via $[Pd^{II}(OH)(H_2O)(PPh_3)_2]^+$) [23]; (iv) $Pd(OCOCF_3)_2 + nPPh_3$ [24].

The *in situ* formation of Pd(0) complexes takes place when Pd(OAc)$_2$ is associated with various phosphines: (i) aromatic phosphines (*p*-Z—C$_6$H$_4$)$_3$P (Z = EDG or EWG). The formation of the Pd(0) complex follows a Hammett correlation with a positive slope [20]. The more electron-deficient the phosphine, the faster the reduction process; this is in agreement with the intramolecular nucleophilic attack of the acetate onto the ligated phosphine as proposed in Scheme 1.13; (ii) aliphatic phosphines [20]; (iii) water-soluble phosphines, triphenylphosphine trisulfonate (trisodium salt) [25] and triphenylphosphinetricarboxylate (trilithium salt) [26]. One major exception is the tri-*o*-tolylphosphine P(*o*-Tol)$_3$, which cannot reduce Pd(OAc)$_2$ to a Pd(0) complex in DMF or THF. Instead, an activation of one C–H bond of the tolyl moieties by Pd(OAc)$_2$ takes place, leading to a dimeric *P,C*-palladacycle (see Section 1.5), as reported by Herrmann *et al.* in 1995 [27]. Such a Pd(II) *P,C*-palladacycle catalyses Mizoroki–Heck reactions [27]. It is, however, a reservoir of a Pd(0) complex, as recently established by d'Orlyé and Jutand [28] in 2005 (see Section 1.5).

The Pd(0) complex formed from Pd(OAc)$_2$ and 3 equiv PPh$_3$ is an anionic species Pd^0(PPh$_3$)$_2$(OAc)$^-$, where the Pd(0) is ligated by an acetate ion (Scheme 1.13) [29]. Further density functional theory (DFT) calculations by Shaik and coworkers [30] and Goossen *et al.* [31], support the formation of such anionic tri-coordinated Pd(0) complexes. The anionic 16-electron complex Pd^0(PPh$_3$)$_2$(OAc)$^-$ formed *in situ* from Pd(OAc)$_2$ and 3 equiv PPh$_3$ is, however, not that stable, yet it is more stable when generated in the presence of an amine (NEt$_3$) often used as a base in Mizoroki–Heck reactions. Pd^0(PPh$_3$)$_2$(OAc)$^-$ may indeed be destabilized by interaction of its acetate ligand with protons (generated in the hydrolysis of the phosphonium, Scheme 1.13) to give the unstable naked Pd^0(PPh$_3$)$_2$ complex. The capture of the protons by NEt$_3$ prevents this reaction and makes the anionic Pd^0(PPh$_3$)$_2$(OAc)$^-$ more stable (Scheme 1.14). This is the first unexpected role of the base [1k,m].

A Pd(0) complex is formed when 2 equiv of PPh$_3$ are added to Pd(OAc)$_2$. Since one PPh$_3$ is oxidized to (O)PPh$_3$, only one PPh$_3$ remains for the stabilization of the Pd(0) as in the formal complex [Pd^0(PPh$_3$)(OAc)$^-$]. The structure of the resulting Pd(0) complex must be more complicated, since two singlets of similar magnitude were observed in the ^{31}P NMR spectrum recorded in THF or DMF, suggesting the formation of *cis*- and *trans*-[Pd^0(OAc)(μ-OAc)(PPh$_3$)]$_2^{4-}$ [18]. These Pd(0) complexes are not stable; nevertheless, they react with iodobenzene, which explains why the catalytic precursor {Pd(OAc)$_2$ + 2PPh$_3$} is efficient in Mizoroki–Heck reactions [8, 9]. Addition of more than 3 equiv of PPh$_3$ to Pd(OAc)$_2$ leads to the formation of the saturated stable (18-electron) Pd^0(PPh$_3$)$_3$(OAc)$^-$ which is in equilibrium with Pd^0(PPh$_3$)$_2$(OAc)$^-$ [29]:

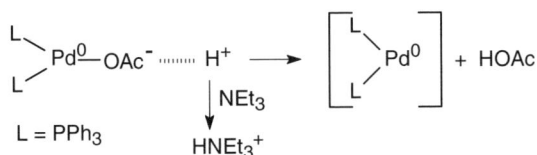

$$Pd^0(PPh_3)_3(OAc)^- \rightleftharpoons Pd^0(PPh_3)_2(OAc)^- + PPh_3$$

Scheme 1.14

1.3.2   Oxidative Addition

1.3.2.1   Oxidative Addition of Aryl Iodides

The 16-electron $Pd^0(PPh_3)_2(OAc)^-$ formed by reaction $Pd(OAc)_2$ and 3 equiv PPh_3 is found to be the only reactive species in the oxidative addition of iodobenzene [29]. It is more reactive than $Pd^0(PPh_3)_4$. Surprisingly, the expected *trans*-$PhPdI(PPh_3)_2$ is not produced in the oxidative addition; rather, a new complex *trans*-$PhPd(OAc)(PPh_3)_2$ is produced. The latter is in equilibrium with the cationic complex *trans*-$PhPdS(PPh_3)_2^+$ (S = DMF, THF) and AcO^- (Scheme 1.15) [18, 29, 32]. From the kinetics of the oxidative addition, it emerges that an intermediate complex is formed en route to *trans*-$PhPd(OAc)(PPh_3)_2$ in which the Pd(II) is still ligated by the iodide. Indeed, the kinetic curve for the release of iodide ions is S-shaped, proving that iodide is generated from an intermediate complex on the way to the final complex *trans*-$PhPd(OAc)(PPh_3)_2$ [18, 29]. The minimal structure for this intermediate was proposed as that of an anionic pentacoordinated complex $[PhPdI(OAc)(PPh_3)_2]^-$ (Scheme 1.15) [1m, 29, 33]. However, owing to its short life-time ($t_{1/2}$ = 30 s, DMF, 25 °C), this 18-electron complex does not play any role in the further step of Mizoroki–Heck reactions, being unable to be trapped by the slow reaction of the alkene before its evolution towards $PhPd(OAc)(PPh_3)_2$ [29].

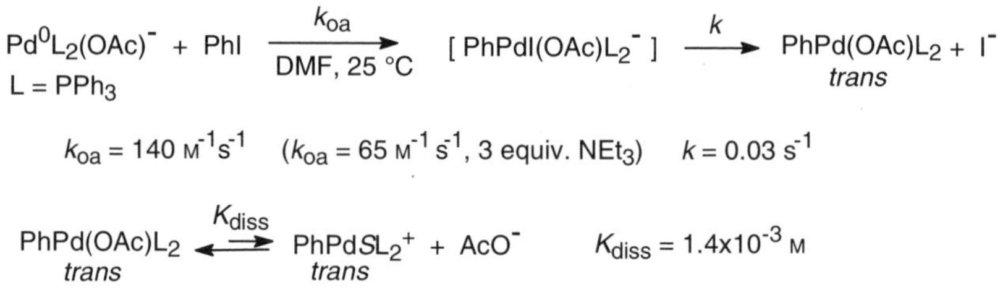

$$Pd^0L_2(OAc)^- + PhI \xrightarrow[\text{DMF, 25 °C}]{k_{oa}} [\,PhPdI(OAc)L_2^-\,] \xrightarrow{k} PhPd(OAc)L_2 + I^-$$

$$L = PPh_3 \qquad\qquad\qquad\qquad\qquad\qquad\qquad\qquad\qquad\qquad trans$$

$$k_{oa} = 140\ \text{M}^{-1}\text{s}^{-1} \quad (k_{oa} = 65\ \text{M}^{-1}\text{s}^{-1}, 3\ \text{equiv. NEt}_3) \quad k = 0.03\ \text{s}^{-1}$$

$$PhPd(OAc)L_2 \underset{}{\overset{K_{diss}}{\rightleftharpoons}} PhPdSL_2^+ + AcO^- \qquad K_{diss} = 1.4 \times 10^{-3}\ \text{M}$$

$$trans \qquad\qquad\quad trans$$

Scheme 1.15

Phosphine, Amines and Alkenes as Factors Affecting the Rate of the Oxidative Addition. Amatore, Jutand *et al.* [29] have established that excess PPh_3 slows down the oxidative addition by formation of the nonreactive $Pd^0(PPh_3)_3(OAc)^-$, thereby decreasing the concentration of the reactive $Pd^0(PPh_3)_2(OAc)^-$ by equilibrium with $Pd^0(PPh_3)_3(OAc)^-$.

The oxidative addition is also slower when performed in the presence of NEt_3, which stabilizes $Pd^0(PPh_3)_2(OAc)^-$ versus its decomposition by protons to the most reactive bent $Pd^0(PPh_3)_2$ (Scheme 1.14) [1m, 30]. This is the second unexpected role of the base: a decelerating effect on the oxidative addition.

The oxidative addition is also slower when performed in the presence of an alkene, one of the components of the Mizoroki–Heck reaction. Owing to the reversible complexation of the reactive $Pd^0(PPh_3)_2(OAc)^-$ by the alkene which generates the nonreactive complex $(\eta^2\text{–}CH_2\text{=}CHR)Pd^0(PPh_3)_2(OAc)^-$ (R = Ph, CO_2Me), the concentration of $Pd^0(PPh_3)_2(OAc)^-$ decreases, making the oxidative addition slower (Scheme 1.16) [34].

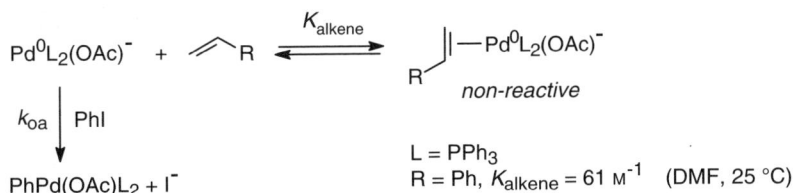

Scheme 1.16

1.3.2.2 Oxidative Addition of Aryl Triflates

Mosleh and Jutand have established that the oxidative addition of aryl triflates ($ArOSO_2CF_3$) to $Pd^0(PPh_3)_4$ gives in DMF the cationic complex *trans*-$ArPd(DMF)L_2^+$ as characterized by conductivity measurements (Scheme 1.17a) [35, 36]. When the same reaction is performed from $Pd^0(PPh_3)_2(OAc)^-$ generated from $Pd(OAc)_2$ and 3 equiv PPh_3, the neutral complex *trans*-$ArPd(OAc)L_2$ is formed in equilibrium with the cationic complex (Scheme 1.17b) [37]. This again emphasizes the important role of acetate ions delivered by the precursor $Pd(OAc)_2$.

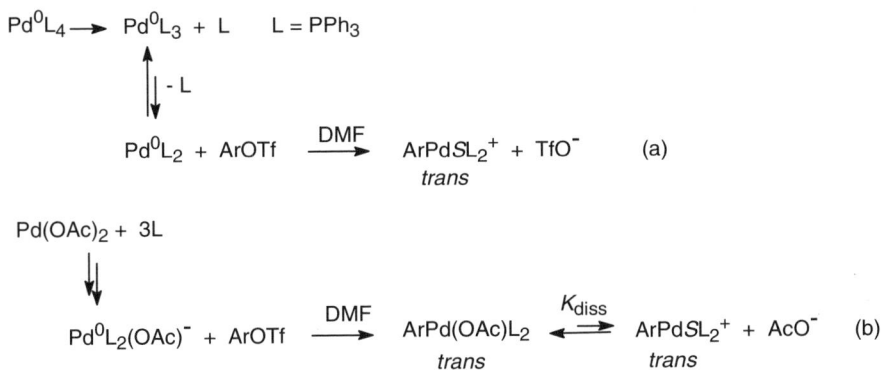

Scheme 1.17

1.3.3 Complexation/Insertion of the Alkene

In the mechanism postulated by Heck (Scheme 1.8), *trans*-$PhPdIL_2$ was proposed to be formed in the oxidative addition and to react with the alkene. However, such a complex is not generated in the oxidative addition when the precursor is $Pd(OAc)_2$ (Scheme 1.15). Moreover, one sees in Table 1.1, which presents the comparative reactivity of *trans*-$PhPdX(PPh_3)_2$ (X = I, OAc, BF_4) with styrene, that *trans*-$PhPdI(PPh_3)_2$ is inert towards styrene (100 equiv) in DMF at $20\,°C$. It is only after addition of AcO^- ions that (*E*)-stilbene is formed [29]. Indeed, acetate ions react with *trans*-$PhPdI(PPh_3)_2$ to generate *trans*-$PhPd(OAc)(PPh_3)_2$ (Scheme 1.18 [18, 29]) which reacts with styrene. *trans*-$PhPd(OAc)(PPh_3)_2$ generated in the oxidative addition of PhI to $Pd^0(PPh_3)_2(OAc)^-$ reacts with styrene to give (*E*)-stilbene (Table 1.1). The reaction is retarded by excess PPh_3 [29].

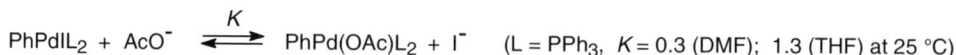

$$PhPdIL_2 + AcO^- \overset{K}{\rightleftharpoons} PhPd(OAc)L_2 + I^- \quad (L = PPh_3, \; K = 0.3 \; (DMF); \; 1.3 \; (THF) \; at \; 25\,°C)$$

Scheme 1.18

This is rationalized by the mechanisms depicted in Scheme 1.19 [1m]. The reaction of the alkene with *trans*-PhPdI(PPh$_3$)$_2$ is limited by the dissociation of PPh$_3$ (Scheme 1.19a), whereas the dissociation of one PPh$_3$ in *trans*-PhPd(OAc)(PPh$_3$)$_2$ is assisted by the bidentate character of the acetate ligand (Scheme 1.19b). This favours the approach of the alkene in a *cis* position relative to the Ph group.

The following reactivity order is observed in DMF at 20 °C (Table 1.1):

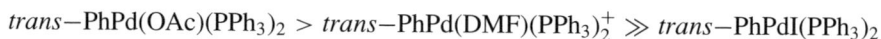

$$trans-PhPd(OAc)(PPh_3)_2 > trans-PhPd(DMF)(PPh_3)_2^+ \gg trans-PhPdI(PPh_3)_2$$

As expected, the cationic complex *trans*-PhPd(DMF)(PPh$_3$)$_2^+$ is more reactive towards styrene than *trans*-PhPdI(PPh$_3$)$_2$ but, surprisingly, is less reactive than PhPd(OAc)(PPh$_3$)$_2$ (Table 1.1) [29, 38]. The coordination of the cationic complex by the alkene gives a *trans* complex (Scheme 1.19c). Consequently, the insertion of the alkene is inhibited by the endergonic *trans/cis* isomerization of the alkene-ligated cationic complex, making route c slower than route b.

The amine NEt$_3$, which may be used as a base in Mizoroki–Heck reactions, does not play any role when the reaction of styrene is performed with isolated *trans*-PhPdIL$_2$ or *trans*-PhPdSL$_2^+$ (Table 1.1). The reaction is, however, accelerated by the amine when performed with *trans*-PhPd(OAc)L$_2$ generated in the oxidative addition of PhI to Pd^0(PPh$_3$)$_2$(OAc)$^-$ (formed *in situ* from Pd(OAc)$_2$ and 3 PPh$_3$) (Table 1.1). Protons are generated together with the Pd(0) complex (Scheme 1.13). Their interaction with the acetate ions shifts the equilibrium between PhPd(OAc)L$_2$ and PhPdSL$_2^+$ towards the latter, which is the less reactive one (Scheme 1.20). Addition of a base neutralizes the protons, increases the concentration of free acetate and, thus, that of the most reactive complex PhPd(OAc)L$_2$, making the overall carbopalladation step faster (Scheme 1.20) [1m]. Consequently, *trans*-PhPd(OAc)(PPh$_3$)$_2$ is a key intermediate in the carbopalladation step in Mizoroki–Heck reactions when the catalytic precursor is Pd(OAc)$_2$ associated with PPh$_3$.

Table 1.1 *Reaction of* trans-*PhPdX(PPh$_3$)$_2$ (X = I, OAc, BF$_4$) (2 mM) with styrene (0.2 M) in the presence or not of NEt$_3$ in DMF at 20 °C*

trans-PhPdX(PPh$_3$)$_2$	NEt$_3$ (equiv)	Time (h)	(E)-stilbene yield (%)
PhPd(OAc)L$_2$[a]	0	24	34
PhPd(OAc)L$_2$[a]	3	19	75
PhPdIL$_2$	0 or 3	24	0
PhPdIL$_2$ + 2 AcO$^-$	0 or 3	48	72
PhPdSL$_2^+$, BF$_4^-$	0 or 3	24	27

[a] *trans*-PhPd(OAc)(PPh$_3$)$_2$ is generated by oxidative addition of PhI (2 mM) to Pd(PPh$_3$)$_2$(OAc)$^-$ generated from Pd(OAc)$_2$ (2 mM) and 3 equiv PPh$_3$ in DMF.

Neutral mechanism   L = PPh$_3$

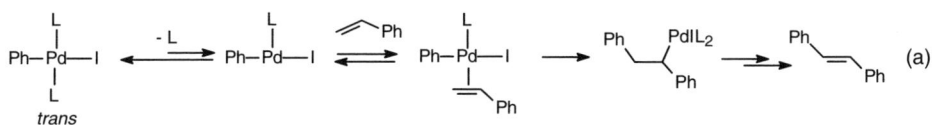

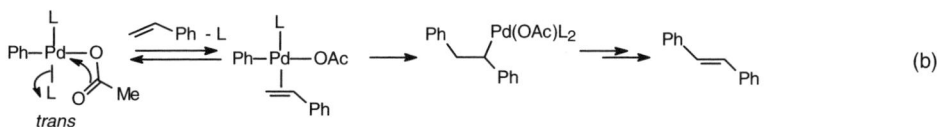

$$
\begin{array}{ccc}
\underset{\text{trans}}{\overset{L}{\underset{L}{\text{Ph--Pd--I}}}} & \xleftarrow{\text{-- L}} & \overset{L}{\underset{L}{\text{Ph--Pd--I}}} \rightleftharpoons \overset{L}{\underset{\underset{Ph}{\|}}{\text{Ph--Pd--I}}} \longrightarrow \underset{Ph}{\overset{Ph}{\diagdown}}\!\!\!\underset{Ph}{\overset{\text{PdIL}_2}{\diagup}} \longrightarrow \underset{Ph}{\overset{Ph}{=}}\quad\text{(a)}
\end{array}
$$

$$
\begin{array}{ccc}
\underset{\text{trans}}{\text{Ph--Pd--O}} & \xrightleftharpoons{\text{Ph -- L}} & \overset{L}{\underset{\underset{Ph}{\|}}{\text{Ph--Pd--OAc}}} \longrightarrow \underset{Ph}{\overset{Ph\;\;\text{Pd(OAc)L}_2}{}} \longrightarrow \underset{Ph}{\overset{Ph}{=}}\quad\text{(b)}
\end{array}
$$

Ionic mechanism

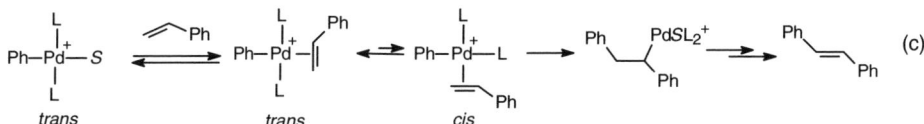

$$
\begin{array}{cccc}
\underset{\text{trans}}{\overset{L}{\underset{L}{\text{Ph--Pd--S}^+}}} & \rightleftharpoons & \underset{\text{trans}}{\overset{L}{\underset{L}{\text{Ph--Pd}^+}}} & \rightleftharpoons \underset{\text{cis}}{\overset{L}{\underset{\underset{Ph}{\|}}{\text{Ph--Pd--L}^+}}} \longrightarrow \underset{Ph}{\overset{Ph\;\;\text{PdSL}_2{}^+}{}} \longrightarrow \underset{Ph}{\overset{Ph}{=}}\quad\text{(c)}
\end{array}
$$

Scheme 1.19

1.3.4 Multiple Role of the Base

In summary, the base serves multiple purposes in Mizoroki–Heck reactions: (i) the base stabilizes $\text{Pd}^0(\text{PPh}_3)_2(\text{OAc})^-$ versus its decomposition to $\text{Pd}^0(\text{PPh}_3)_2$ by protons (Scheme 1.14); (ii) the base slows down the oxidative addition; (iii) the base accelerates the car-bopalladation step by increasing the concentration of the reactive $\text{PhPd}(\text{OAc})\text{L}_2$ (Scheme 1.20); (iv) the base favours the recycling of the Pd(0) complex from the hydridopalla-dium(II) (formed in the β-hydride elimination process) by shifting the reversible reductive elimination towards the Pd(0) complex. The formation of HPdXL_2 (X = I, L = PPh$_3$) has been proposed by Heck (Scheme 1.8). In the present case, where acetate ions play such an important role, it is not clear whether $\text{HPdI}(\text{PPh}_3)_2$ exists or not. In DMF, the cationic complex *trans*-$\text{HPdSL}_2{}^+$ may be formed with acetate (or iodide) as the counter anion. It has indeed been shown that the oxidative addition of Pd^0L_4 (L = PPh$_3$) with acetic acid is reversible (Scheme 1.21) and that the hydridopalladium complex formed in that reaction is cationic, $\text{HPd}(\text{DMF})(\text{PPh}_3)_2{}^+$ [36, 39]. One sees that the role of the base is more subtle than initially postulated. The consequences in terms of efficiency of the catalytic cycle are now presented.

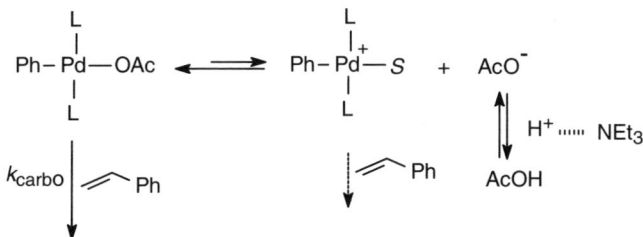

$$
\overset{L}{\underset{L}{\text{Ph--Pd--OAc}}} \rightleftharpoons \overset{L}{\underset{L}{\text{Ph--Pd--S}^+}} + \text{AcO}^-
$$

$$
k_{\text{carbo}} \downarrow \diagup\!\!\!\diagdown Ph \qquad\qquad \downarrow \diagup\!\!\!\diagdown Ph \qquad \updownarrow\; H^+ \cdots\! \text{NEt}_3 \qquad \text{AcOH}
$$

Scheme 1.20

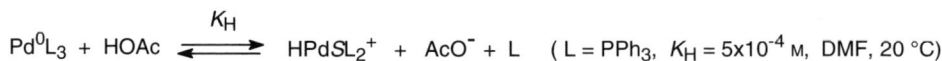

$$Pd^0L_3 + HOAc \xrightleftharpoons{K_H} HPdSL_2^+ + AcO^- + L \quad (L = PPh_3, \ K_H = 5 \times 10^{-4} \ \text{M, DMF, 20 °C})$$

Scheme 1.21

1.3.5 Catalytic Cycle

A mechanism is now proposed for Mizoroki–Heck reactions involving $Pd(OAc)_2$ as precursor associated with PPh_3 (Scheme 1.22). From the rate constants of the main steps given in Scheme 1.22, it appears that, for comparable iodobenzene and styrene concentrations, the overall carbopalladation (complexation/insertion of the alkene) from $PhPd(OAc)(PPh_3)_2$

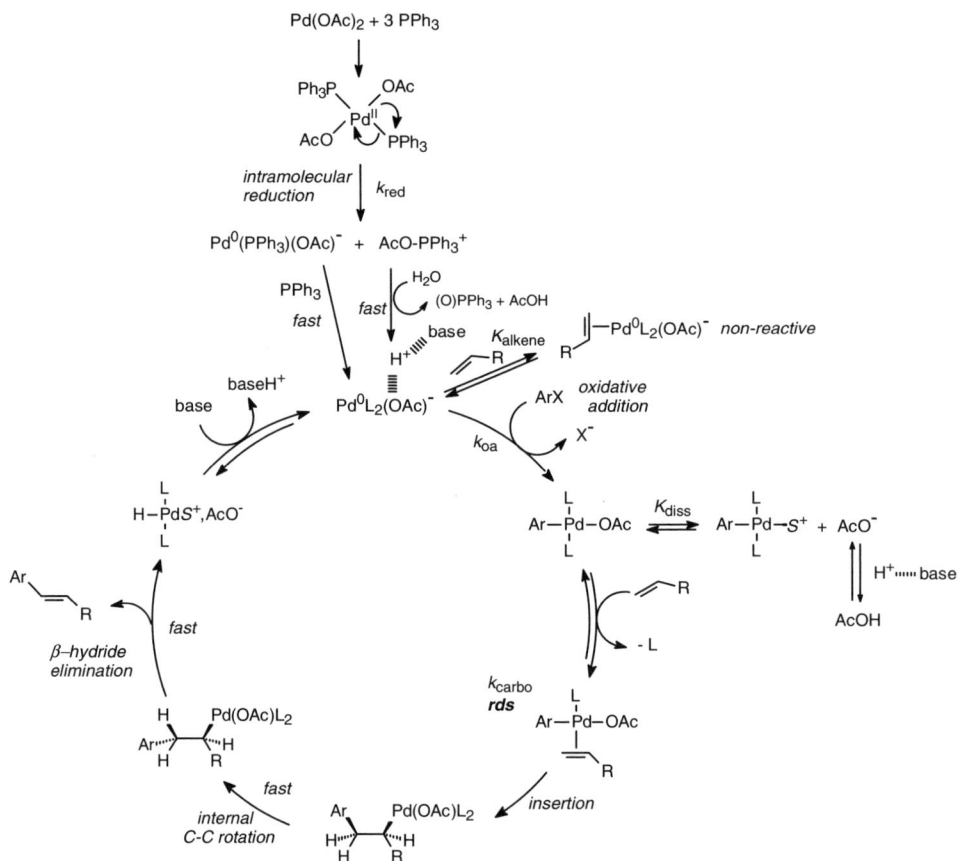

Scheme 1.22 *Mechanism of Mizoroki–Heck reactions with $Pd(OAc)_2$ as precursor associated with PPh_3. Rate and equilibrium constants in DMF at 25 °C when $ArX = PhI$, $R = Ph$ and base = NEt_3: $k_{red} = 4 \times 10^{-4} \ s^{-1}$, $K_{alkene} = 61 \ \text{M}^{-1}$, $k_{oa} = 140 \ \text{M}^{-1} s^{-1}$ ($65 \ \text{M}^{-1} s^{-1}$ in the presence of 3 equiv NEt_3), $K_{diss} = 1.4 \times 10^{-3} \ \text{M}$, $k_{carbo} = 5 \times 10^{-5} \ \text{M}^{-1} s^{-1}$ ($10^{-4} \ \text{M}^{-1} s^{-1}$ in the presence of 3 equiv NEt_3), k_{carbo} is the apparent rate constant of the overall carbopalladation process, which involves complexation and insertion of the alkene.*

is the slowest step of the catalytic cycle. This new mechanism highlights the crucial role of acetate ions which are delivered by the precursor $Pd(OAc)_2$. Indeed, AcO^- is a ligand of all the Pd(0) and Pd(II) complexes present throughout the catalytic cycle, principally in $PhPd(OAc)L_2$ involved in the rate-determining step.

1.3.5.1 Factors Controlling the Efficiency of a Catalytic Reaction

When the successive steps of a catalytic cycle are examined independently from each other, they have their own reaction rate. But when they are involved in a catalytic cycle, the effective rates of the successive steps are not independent from each other. Indeed, the rate of step i is $v_i = k_i[R_i][M_i]$, where k_i is the rate constant of step i, R_i is the reagent involved in step i and M_i is the catalytic species involved in step i, whose concentration $[M_i]$ is modulated, controlled by the rate of the previous reaction $(i-1)$ in which it is generated [1k]. All steps have the same reaction rate when the stationary regime is reached. It will be more easily reached if the intrinsic reaction rates of all elemental steps are as close as possible to each other. In other words, to increase the efficiency of a catalytic cycle one must accelerate the rate-determining step; that is, destabilize the stable intermediate species and also decelerate the fast reactions by stabilizing high-energy species [1k].

This is illustrated in the mechanism of the Mizoroki–Heck reaction depicted in Scheme 1.22. Indeed, three main factors contribute to slow down the fast oxidative addition of PhI: (i) *the anion* AcO^- delivered by the precursor $Pd(OAc)_2$, which stabilizes Pd^0L_2 as the less reactive $Pd^0L_2(OAc)^-$; (ii) *the base* (NEt_3) which indirectly stabilizes $Pd^0L_2(OAc)^-$ by preventing its decomposition by protons to the more reactive bent Pd^0L_2; (iii) *the alkene* by complexation of $Pd^0L_2(OAc)^-$ to form the nonreactive $(\eta^2\text{-}CH_2=CHR)Pd^0L_2(OAc)^-$. On the other hand, the slow carbopalladation is accelerated by the *base* and by the *acetate ions* which generate $ArPd(OAc)L_2$, which in turn is more reactive than the postulated $ArPdIL_2$. The base, the alkene and the acetate ions play, then, the same dual role in Mizoroki–Heck reactions: deceleration of the oxidative addition and acceleration of the slow carbopalladation step. Whenever the oxidative addition is fast (e.g. with aryl iodides or activated aryl bromides), this dual effect favours the efficiency of the catalytic reaction by bringing the rate of the oxidative addition closer to the rate of the carbopalladation [1m, 34].

The mechanism depicted in Scheme 1.22 is also valid for Mizoroki–Heck reactions performed with aryl triflates, since $ArPd(OAc)L_2$ complexes are formed in the oxidative addition (Scheme 1.17b) [37]. This mechanism is also applicable when the catalytic precursor is not $Pd(OAc)_2$ (e.g. $Pd^0(dba)_2$ and PPh_3, $PdCl_2(PPh_3)_2$ or $Pd^0(PPh_3)_4$ (dba = *trans,trans*-dibenzylideneacetone)), but when acetate ions are used as base. AcO^- is indeed capable of coordinating to Pd^0L_2 complexes to give $Pd^0L_2(OAc)^-$ [29] or react with $ArPdIL_2$ to generate the more reactive $ArPd(OAc)L_2$ [18].

The situation is problematic when considering less reactive aryl chlorides or deactivated aryl bromides involved in the rate-determining oxidative addition, since the alkene will also contribute to decelerate the slow oxidative addition by complexation of the reactive $Pd^0L_2(OAc)^-$ (Scheme 1.16). To solve this problem, one has to design a new ligand which will make the Pd(0) more reactive or introduce the alkene via a syringe pump, so that a low alkene concentration can be maintained throughout the catalytic reaction.

1.4   Mechanism of the Mizoroki–Heck Reaction when the Catalytic Precursor is Pd(OAc)$_2$ Associated with Bisphosphine Ligands

In 1990, Cabri *et al.* [40a] reported that the precursor Pd(OAc)$_2$ associated with a bidentate P^P ligand as dppp (1,3-bis-diphenylphosphinopropane) appeared to be more efficient than PPh$_3$ in Mizoroki–Heck reactions performed from aryl triflates and enol ethers (electron-rich alkenes); moreover, the regioselectivity in favour of the α-arylated alkenes was improved to 100%. Since that time, dppp associated with Pd(OAc)$_2$ has been used extensively to catalyse Mizoroki–Heck reactions and to investigate the factors that control the regioselectivity [1g, 40]. The chiral bidentate (*R*)-Binap (2,2′-bis(diphenylphosphino)-1,1-binaphthyl) associated with Pd(OAc)$_2$ has also been used by Shibasaki and coworkers [2b,d,41a] and Overman and Poon [41b] in intramolecular enantioselective Mizoroki–Heck reactions (also, see Link [2f] for an authorative review on the Overman–Shibasaki chemistry), as well as by Hayashi and coworkers [2a, 41c,d] to control the regioselectivity and enantioselectivity of intermolecular Mizoroki–Heck reactions performed from cyclic alkenes (see Schemes 1.3 and 1.2 (Z = O) respectively).

1.4.1   Pd(0) Formation from Precursor

As established by Amatore, Jutand *et al.* in 2001 [42], the reaction of Pd(OAc)$_2$ associated with 1 equiv of dppp does not generate an observable Pd(0) complex. This is a consequence of the reversibility of the reductive elimination which takes place within Pd(OAc)$_2$(dppp) formed in the early stage (Scheme 1.23). The intramolecular and, consequently, fast reverse reaction is an oxidative addition of the resulting Pd(0) complex to the phosphonium formed in the reductive elimination step but still ligated to the Pd(0) (Scheme 1.23). As a consequence, a Pd(0) complex is generated at low concentration in that endergonic equilibrium. However, if a second equivalent of dppp is added, then the anionic Pd(0) complex Pd^0(dppp)(OAc)$^-$ is formed by substitution of the mono-ligated P^P$^+$ (Scheme 1.23). The hydrolysis of the latter, which gives protons and the hemioxide dppp(O), contributes to the shift of the successive equilibria towards the formation of the anionic Pd(0) complex. The formation of the stable anionic Pd^0(dppp)(OAc)$^-$ complex is supported by DFT calculations [30]. As for Pd^0(PPh$_3$)$_2$(OAc)$^-$, Pd^0(dppp)(OAc)$^-$ is more stable in the presence of a base such as NEt$_3$. Indeed, Pd^0(dppp)(OAc)$^-$ may be decomposed into the unstable naked Pd^0(dppp) by interaction of its ligand acetate with protons formed together with

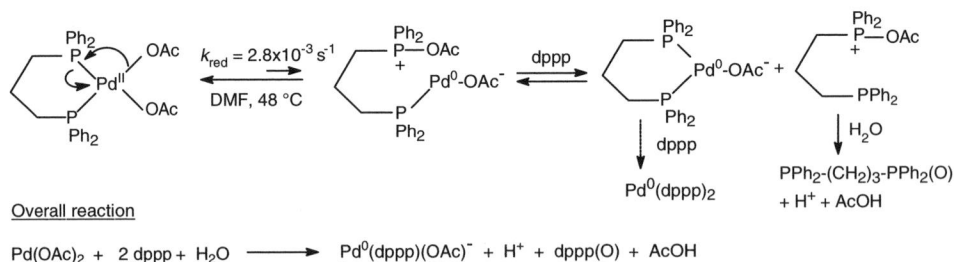

Overall reaction

Pd(OAc)$_2$ + 2 dppp + H$_2$O $\longrightarrow$ Pd^0(dppp)(OAc)$^-$ + H$^+$ + dppp(O) + AcOH

Scheme 1.23

$$Pd(OAc)_2 + H_2O + 3\ Binap + 2\ NEt_3 \longrightarrow Pd^0(Binap)_2 + Binap(O) + 2Et_3N\cdot HOAc$$

Scheme 1.24

the Pd(0) complex (process similar to that in Scheme 1.14 for PPh_3). Neutralization of the protons by NEt_3 stabilizes the anionic $Pd^0(dppp)(OAc)^-$ [42]. Addition of a third equivalent of dppp generates $Pd^0(dppp)_2$. The rate of formation of the Pd(0) complex from $Pd(OAc)_2$ and 2 equiv dppp (see k_{red} in Scheme 1.23) is slightly slower than that from $Pd(OAc)_2$ and 3 equiv PPh_3 in DMF at the same temperature [20, 42].

The formation of $Pd^0((R)\text{-Binap})_2$ and $(R)\text{-Binap}(O)$ from $Pd(OAc)_2$ and 3 equiv of (R)-Binap in benzene was reported earlier by Ozawa *et al.* in 1992 [19]. Owing to excess (R)-Binap, the overall reaction gives $Pd^0((R)\text{-Binap})_2$ (Scheme 1.24). The reaction is performed in the presence of NEt_3. Water is found to be a crucial additive, with the formation of the Pd(0) complex being faster in the presence of water [19]. The mechanism of formation of $Pd^0(Binap)_2$ must be similar to that involving dppp (Scheme 1.23). The intramolecular reduction process is reversible and the role of water is to shift the successive equilibria (similar to those in Scheme 1.23 with (R)-Binap instead of dppp) towards the formation of the final $Pd^0(Binap)_2$ by the irreversible hydrolysis of the phosphonium salt. ^{18}O-labelled $(R)\text{-Binap}(O)$ is indeed formed over time when using $^{18}OH_2$ [19]. Interestingly, the accelerating effect of water in the reduction process is specific to bidentate ligands. It is not observed with PPh_3 (Scheme 1.13) [18] because the intramolecular reduction process which delivers the Pd(0) complex is in that case rendered irreversible because of a much slower backward oxidative addition (intermolecular reaction).

1.4.2 Oxidative Addition

Owing to the presence of the hemioxide dppp(O) formed together with $Pd^0(dppp)(OAc)^-$, the oxidative addition of PhI does not form the expected PhPdI(dppp) complex; rather, the cationic complex $PhPd(dppp)(dppp(O))^+$, ligated by both the dppp ligand and the hemioxide which behaves as a monodentate ligand (Scheme 1.25) [42], is formed. The mechanism of the oxidative addition is quite complicated, involving dimeric anionic Pd(0) complex, and the overall reaction is slower (by a factor 300) than that performed from $Pd^0(PPh_3)_2(OAc)^-$ at identical concentrations of PhI and NEt_3 [42]. The complex $PhPd(dppp)(dppp(O))^+$ reacts with excess iodide or acetate ions to give PhPdI(dppp) and PhPd(OAc)(dppp)

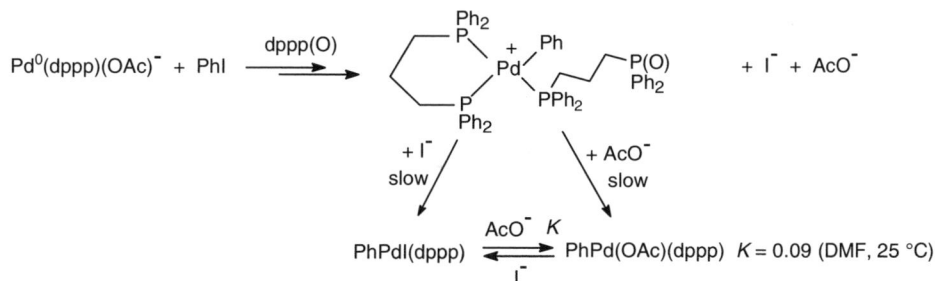

Scheme 1.25

respectively [42]. Moreover, those two complexes are able to exchange their anions in a reversible reaction (Scheme 1.25) [43].

1.4.3   Complexation/Insertion of the Alkene–Regioselectivity

Regioselectivity is one of the major problems of Mizoroki–Heck reactions. It is supposed to be affected by the type of mechanism: *ionic* versus *neutral*, when the palladium is ligated by bidentate P^P ligands. The ligand dppp has been taken as a model for the investigation of the regioselectivity. Cabri and Candiani [1g] have reported that a mixture of branched and linear products is formed in Pd^0(P^P)-catalysed Mizoroki–Heck reactions performed from electron-rich alkenes and aryl halides (Scheme 1.26a) or aryl triflates in the presence of halide ions (Scheme 1.26b). This was rationalized by the so-called *neutral* mechanism (Scheme 1.27). The neutral complex ArPdX(P^P) is formed in the oxidative addition of Pd^0(P^P) to aryl halides or to aryl triflates in the presence of halides. The carbopalladation proceeds from the neutral ArPdX(P^P) after dissociation of one phosphorus. The coordination of the alkene may proceed in two ways, leading to a mixture of linear and branched alkenes (Scheme 1.27). This mechanism involving *neutral* complexes is more sensitive to steric factors than to electronic factors with a preferential migration of the aryl group onto the less substituted carbon of the alkene, leading to the linear alkene (Scheme 1.27b). This mechanism was also proposed by Cabri and Candiani [1g] for electron-deficient alkenes which are even more reactive towards neutral complexes than electron-rich alkenes. Cabri and Candiani [1g] and Hayashi *et al.* [2a] have reported that branched alkenes are mainly produced from electron-rich alkenes in Pd^0(P^P)-catalysed Mizoroki–Heck reactions

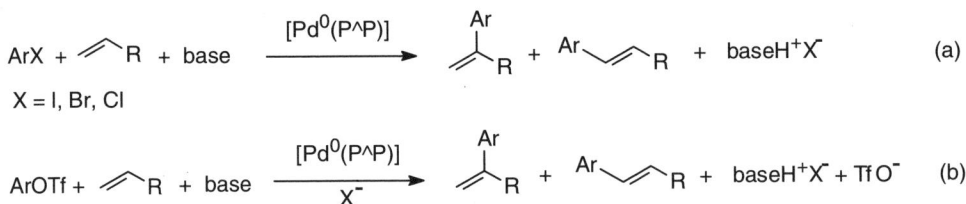

$$\text{ArX} + \diagup\text{R} + \text{base} \xrightarrow{[Pd^0(P^\wedge P)]} \overset{\text{Ar}}{\underset{\text{R}}{\diagup}} + \text{Ar}\diagup\text{R} + \text{baseH}^+\text{X}^- \qquad (a)$$

X = I, Br, Cl

$$\text{ArOTf} + \diagup\text{R} + \text{base} \xrightarrow[\text{X}^-]{[Pd^0(P^\wedge P)]} \overset{\text{Ar}}{\underset{\text{R}}{\diagup}} + \text{Ar}\diagup\text{R} + \text{baseH}^+\text{X}^- + \text{TfO}^- \qquad (b)$$

Scheme 1.26

Neutral mechanism

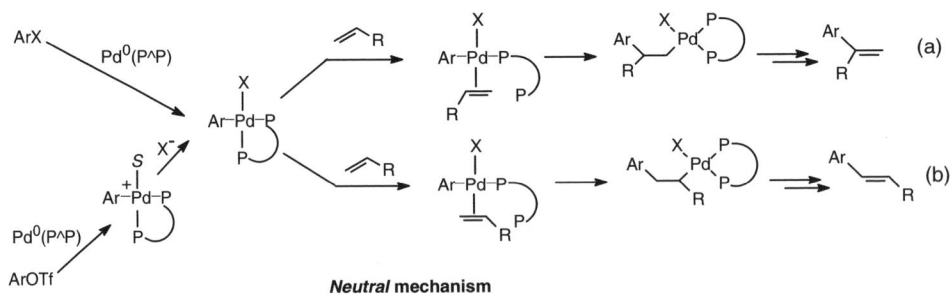

Scheme 1.27   Textbook neutral *mechanism for the regioselectivity of Mizoroki–Heck reactions (the C—C internal rotation is not shown).*

performed from aryl halides in the presence of a halide scavenger (Ag$^+$ [40c,d, 44a,b], Tl$^+$ [40c,d, 44c], K$^+$ in aqueous DMF [40j]) (Scheme 1.28a) or from aryl triflates (Scheme 1.28b) [40a,d,e,g, 41c,d]. Since cationic complexes ArPdS(P$^\wedge$P)$^+$ (S = solvent) may be generated by (i) dissociation of ArPdX(P$^\wedge$P) [45], (ii) abstraction of X in ArPdX(P$^\wedge$P) by a halide scavenger [46, 47] or (iii) by oxidative addition of aryl triflates to Pd^0(P$^\wedge$P) complexes [48], they have been proposed in the reaction with alkenes in the so-called *ionic* mechanism; that is, involving *cationic* Pd(II) complexes (Scheme 1.29). The carbopalladation proceeds by coordination of the alkene to the cationic ArPdS(P$^\wedge$P)$^+$. This *ionic* mechanism is more sensitive to electronic factors than to steric factors. When R = EDG, the coordination of the polarized alkene proceeds in one major pathway with a selective migration of the aryl moiety onto the charge-deficient α-carbon of the electron-rich alkene, leading to the branched alkene (α-arylation) (Scheme 1.29a) [1g, 49].

The *ionic* mechanism has been investigated from isolated cationic complexes ArPdS(P$^\wedge$P)$^+$ at low temperatures so that the σ-alkyl–PdS(P$^\wedge$P)$^+$ intermediates can be observed after reaction with alkenes CH$_2$=CHR. In the absence of a base, Brown and Hii [46] have characterized σ-ArCH$_2$CH(R)–Pd(THF)(dppf)$^+$ (Ar = Ph, R = CO$_2$Me; dppf = 1,1′-bis(diphenylphosphino)ferrocene) by ^{1}H and ^{31}P NMR spectroscopy. The β-hydride elimination takes place, leading to the linear alkene (E)-ArCH=CHR and HPd(THF)(dppf)$^+$. The latter cannot generate a Pd^0 complex in the absence of a base and reacts with the initial alkene (Scheme 1.30a). Similarly, when reacting PhPd(THF)(Binap)$^+$ with the electron-rich alkene 2,3-dihydrofuran, the intermediate σ-alkyl–Pd(THF)(Binap)$^+$ is observed and HPd(THF)(Binap)$^+$ is also found to react with the initial 2,3-dihydrofuran [50]. At $-60\,°$C, σ-ArCH$_2$CH(Ar′)–Pd(THF)(dppp)$^+$ complexes are stabilized by interaction of the Pd^{II} centre with the adjacent Ar′ group, which restricts the C—C internal rotation [51].

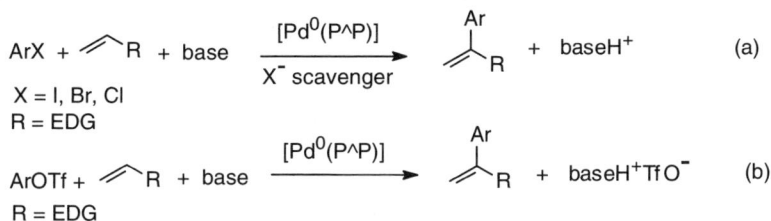

Scheme 1.28

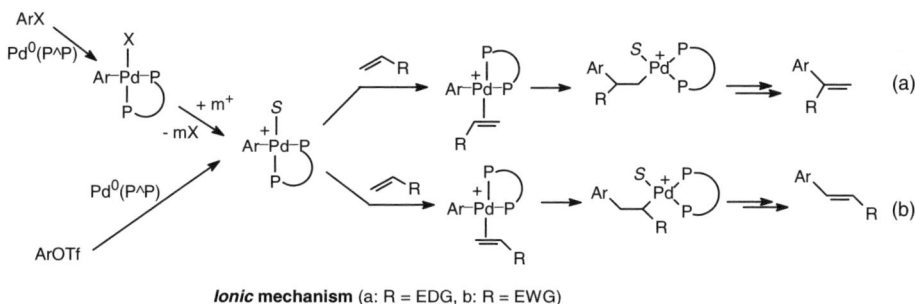

Ionic mechanism (a: R = EDG, b: R = EWG)

Scheme 1.29   *Textbook* ionic *mechanism for the regioselectivity of Mizoroki–Heck reactions.*

Ionic mechanism

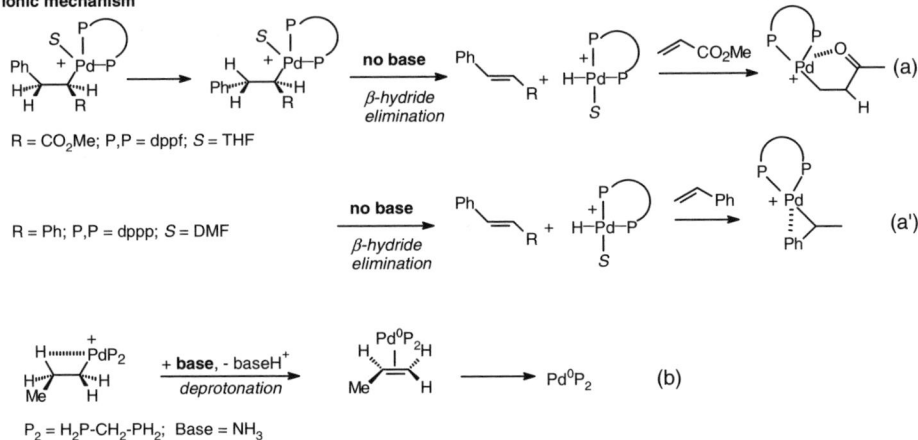

Scheme 1.30 Ionic mechanism: (a) Experimental work. (a′) Experimental work. The branched alkene $Ph_2C{=}CH_2$ is also formed. (b) DFT calculations.

Åkermark and coworkers [47] have characterized σ-PhCH$_2$CH(R)–Pd(DMF)(dppp)$^+$ (R = Ph) formed by reacting PhPd(dppp)$^+$BF$_4^-$ with styrene in DMF at $-20\,^\circ$C. This complex delivers (E)-stilbene and HPd(DMF)(dppp)$^+$ which reacts with styrene in the absence of a base (Scheme 1.30a′). Therefore, in the absence of a base the hydridopalladium is quenched by the starting alkene and no Pd(0) is formed. Supported by DFT calculations, Deeth *et al.* [52] have proposed that a Pd^0 complex is generated directly from the cationic σ-alkyl–Pd(II) adduct in the presence of a base, via the easier deprotonation of the agostic hydrogen (Scheme 1.30b); the energetically less favoured β-hydride elimination is therefore bypassed.

Amatore, Jutand *et al.* [42] have established that the oxidative addition of PhI to Pd^0(OAc)(dppp)$^-$ (generated from Pd(OAc)$_2$ and 2 equiv dppp) gives the cationic complex PhPd(dppp)(dppp(O))$^+$; this is followed by reaction of iodide ions released from PhI in the course of a catalytic reaction giving PhPdI(dppp) or/and PhPd(OAc)(dppp) whenever acetate ions are used as bases (Scheme 1.25). The reaction of PhPd(dppp)(dppp(O))$^+$ with alkenes (styrene, methyl acrylate) is so slow that this complex must be considered as a transient complex on the way to PhPdI(dppp) and/or PhPd(OAc)(dppp). These two complexes, which exchange their anions (Scheme 1.25), are in equilibrium with the common cationic complex PhPd(DMF)(dppp)$^+$ in DMF (Scheme 1.31) [43]. Consequently, two *neutral* phenyl–palladium(II) complexes are candidates, in addition to the *cationic* PhPdS(dppp)$^+$, for the reaction with alkenes. The kinetics of the reaction of isolated PhPdX(dppp) (X = I, OAc) with electron-deficient, neutral and electron-rich alkenes in the absence of a base has been followed by ^{31}P NMR spectroscopy in DMF. It emerges that PhPd(OAc)(dppp) reacts with styrene and methyl acrylate via PhPd(DMF)(dppp)$^+$; that

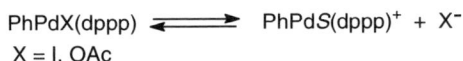

$$\text{PhPdX(dppp)} \rightleftharpoons \text{PhPd}S\text{(dppp)}^+ + \text{X}^-$$
$$\text{X = I, OAc}$$

Scheme 1.31

S = DMF   ***Ionic* mechanism**

$$PhPdX(dppp) \underset{k_{-1}}{\overset{k_1 \quad K_1}{\rightleftharpoons}} PhPdS(dppp)^+ + X^-$$

X = OAc
R = CO$_2$Me, Ph

X = I
R = Ph

K'_2

$$PhPd(dppp)(\eta^2\text{-}CH_2\text{=}CH\text{-}R)^+$$

k'_3

$$Ph\text{-}CH_2\text{-}CH(R)\text{-}PdS(dppp)^+$$

$$Ph \diagup\!\!\diagup R$$

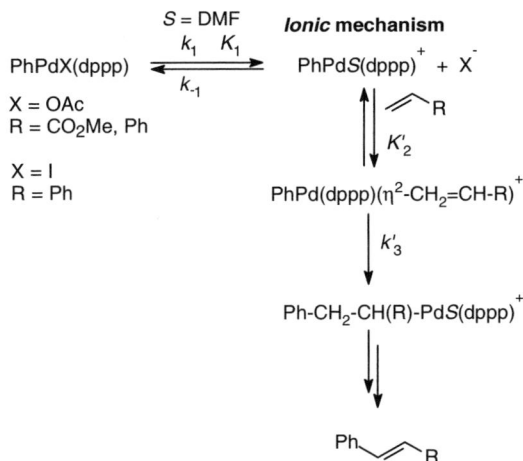

Scheme 1.32   *The reactive species with neutral and electron-deficient alkenes from kinetic data.*

is, via an *ionic* mechanism (Scheme 1.32) [43], as with isobutylvinyl ether (Scheme 1.33) [53]. The mechanism of the reaction of PhPdI(dppp) with alkenes is substrate dependent. PhPdI(dppp) reacts with styrene via PhPd(DMF)(dppp)$^+$ in the *ionic* mechanism (Scheme 1.32) [43], as with isobutylvinyl ether (Scheme 1.33) [53], whereas PhPdI(dppp) and PhPd(DMF)(dppp)$^+$ react in parallel with the more reactive methyl acrylate (*neutral and ionic* mechanisms) (Scheme 1.34) [43]. All reactions are retarded by coordinating anions (I$^-$ and AcO$^-$) at constant ionic strength, which is in agreement with the pure *ionic* mechanisms of Schemes 1.32 and 1.33 and the mixed mechanism of Scheme 1.34, with PhPd(DMF)(dppp)$^+$ being more reactive than PhPdI(dppp). All reactions are accelerated upon increasing the ionic strength. The higher the ionic strength, the higher the concentration of PhPd(DMF)(dppp)$^+$ is and the faster the overall reaction with the alkene. In all

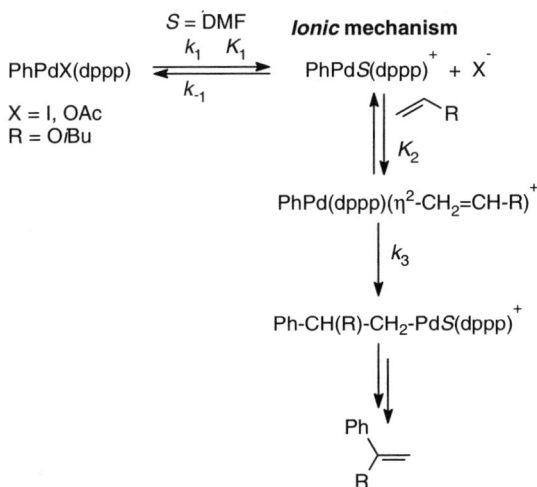

S = DMF   ***Ionic* mechanism**

$$PhPdX(dppp) \underset{k_{-1}}{\overset{k_1 \quad K_1}{\rightleftharpoons}} PhPdS(dppp)^+ + X^-$$

X = I, OAc
R = O*i*Bu

K_2

$$PhPd(dppp)(\eta^2\text{-}CH_2\text{=}CH\text{-}R)^+$$

k_3

$$Ph\text{-}CH(R)\text{-}CH_2\text{-}PdS(dppp)^+$$

$$Ph\diagdown{\Large=}$$
R

Scheme 1.33   *The reactive species with an electron-rich alkene from kinetic data.*

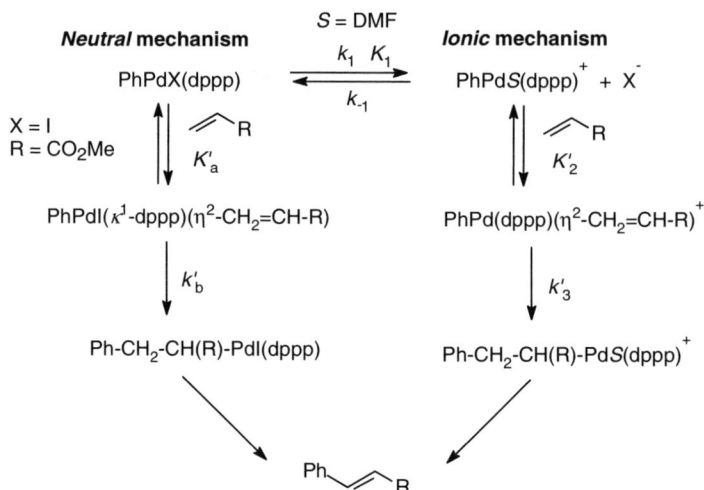

Scheme 1.34   *The reactive species with an electron-deficient alkene from kinetic data.*

cases, the linear product (E)-PhCH=CHR is formed as: (i) the unique product for R = CO_2Me from PhPdX(dppp) (X = I, OAc) [43]; (ii) the major product for R = Ph from PhPdI(dppp) (80% selectivity) and PhPd(OAc)(dppp) (82% selectivity) [47]. The branched product CH=CH(Ph)R is formed as the major product for R = OiBu from PhPdI(dppp) (90% selectivity) [53]. In a first approach, the slower formation of the minor product has not been considered and the alkene insertion step has been regarded as irreversible, as postulated in textbook mechanisms (Schemes 1.27 and 1.29). The equilibrium and rate constants for the formation of the major product have been determined in DMF (Table 1.2) with the following reactivity orders [43, 53]:

- whatever the complex PhPdX(dppp) (X = I, OAc)

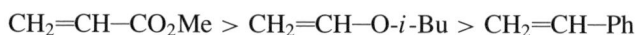

$$CH_2=CH-CO_2Me > CH_2=CH-O-i-Bu > CH_2=CH-Ph$$

- whatever the alkene

$$PhPd(DMF)(dppp)^+ \gg PhPdX(dppp) \quad (X = I, \ OAc)$$

Table 1.2   *Equilibrium and rate constants of the reaction of PhPdX(dppp) (X = I, OAc) with alkenes in DMF at 25 °C (Schemes 1.32–1.34) [43, 53]*

PhPdX(dppp)	⁒CO₂Me	⁒CO₂Me	⁒OiBu	⁒Ph
X	k_1 $(10^{-4}\,s^{-1})$	$k'_a k'_b/(k'_{-a} + k'_b)$ $(10^{-5}\,M^{-1}\,s^{-1})$	$K'_1 K'_2 k'_3$ $(10^{-6}\,s^{-1})$	$K_1 K_2 k_3$ $(10^{-7}\,s^{-1})$	$K'_1 K'_2 k'_3$ $(10^{-8}\,s^{-1})$
OAc	1.6(±0.1)	—	1.5(±0.1)	1.2(±0.1)	6.6(±0.1)
I	1.1(±0.1)	2.5	1.0(±0.1)	1.7(±0.1)	3.8(±0.1)

PhPd(OAc)(dppp) is more reactive than PhPdI(dppp) with styrene and methyl acrylate, in agreement with the fact that the dissociation of PhPd(OAc)(dppp) to the reactive PhPd(DMF)(dppp)$^+$ is more effective than that of PhPdI(dppp) (compare the respective values of k_1 in Table 1.2). Interestingly, an inversion of reactivity is observed with the electron-rich isobutylvinyl ether (Table 1.2).

Therefore, except for the reaction of methyl acrylate with PhPdI(dppp), all reactions exclusively proceed from the *cationic* complex PhPd(DMF)(dppp)$^+$. However, the simplified mechanisms established in Schemes 1.32 and 1.33 cannot explain the high dependence of the regioselectivity on experimental conditions. Indeed, Åkermark and coworkers [47] have reported reactions of isolated PhPdX(dppp) (X = I, OAc, BF$_4$) with styrene, focusing on the effect of X on the regioselectivity of the reaction. In DMF, the ratio PhCH=CHPh/CH$_2$=CPh$_2$ decreases in the order OAc > I > BF$_4$. According to the mechanism proposed in Scheme 1.32, which establishes the exclusive reactivity of styrene with the *cationic* complex PhPd(DMF)(dppp)$^+$ generated from PhPdX(dppp) (X = I, AcO), all complexes PhPdX(dppp) (X = I, AcO, BF$_4$) should afford the same regioselectivity, namely that given by the cationic complex PhPd(DMF)(dppp)$^+$BF$_4$$^-$, which is not observed experimentally [47].

In more recent studies by Xiao and coworkers [40m,n], Mizoroki–Heck reactions catalysed by Pd(OAc)$_2$ associated with dppp and performed from the electron-rich alkene (*n*-butylvinyl ether) and aryl halides (without any halide scavenger, i.e. under the conditions of the textbook *neutral* mechanism of Scheme 1.27 proposed by Cabri and Candiani [1g]) give a mixture of branched and linear products in DMF, whereas the branched product is exclusively produced in ionic liquids (in the absence of halide scavengers) in a faster reaction. Whatever the medium, the *cationic* complex ArPdS(dppp)$^+$ is always the sole reactive complex with electron-rich alkene (Scheme 1.33) [53]. Consequently, the regioselectivity should not vary with the experimental conditions.

A more elaborated mechanism was thus proposed by Jutand and coworkers [53] which rationalizes the regioselectivity of Mizoroki–Heck reactions performed in DMF with dppp as ligand (Scheme 1.35). The complexation of the alkene to the reactive cationic complex PhPd(DMF)(dppp)$^+$ (**1**$^+$) may generate the two isomers **2**$^+$ and **2**$'$$^+$ and then complexes **3**$^+$ and **3**$'$$^+$ in a reversible insertion step [54]. When considering electron-rich alkenes, the branched product is formed as a major product because $K_2K_3k_4 \gg K'_2K'_3k'_4$ (Scheme 1.35, route a). If anion X$^-$ (I$^-$, AcO$^-$) is present at high concentration, cationic complexes **3**$^+$ and **3**$'$$^+$ may be reversibly quenched by the anion as neutral complexes **4** and **4**$'$ respectively. The linear product will be formed as a major product whenever $K'_2K'_3K'_5k'_6[X^-] \gg K_2K_3k_4$ (route b$'$ faster than route a in Scheme 1.35) or $K'_2K'_3K'_5k'_6 \gg K_2K_3K_5k_6$ if $K_5k_6[X^-] \gg k_4$ (route b$'$ faster than route a$'$).

According to Scheme 1.35, the major branched product will be formed from an electron-rich alkene in a reaction involving either pure *cationic* PhPd(DMF)(dppp)$^+$ or *neutral* PhPdX(dppp) which reacts via the *cationic* complex, both in the absence of extra anions (I$^-$, AcO$^-$) (Scheme 1.35, route a). The linear product will be formed if the reaction is performed in the presence of a large excess of anions (I$^-$, AcO$^-$), as it often occurs in real catalytic reactions, but still via the *cationic* PhPd(DMF)(dppp)$^+$ (Scheme 1.35, route b$'$). The inversion of reactivity observed in the reaction of the electron-rich isobutylvinyl ether – PhPdI(dppp) being more reactive than PhPd(OAc)(dppp) (Table 1.2 [53]), although it is less dissociated to the cationic complex – may be understood as the contribution of the

Ionic mechanism

$$\text{PhPdX(dppp)} \underset{k_{-1}}{\overset{k_1 \; K_1}{\rightleftharpoons}} \text{PhPdS(dppp)}^+ \; + \; X^- \qquad S = \text{DMF}$$

X = I
X = OAc

$\mathbf{1^+}$

K_2 / K'_2

$$\underset{\mathbf{2^+}}{\text{Ph–Pd–P}} \qquad \underset{\mathbf{2'^+}}{\text{Ph–Pd–P}}$$

K_3 / K'_3

branched (a) ... k_4 ... $\mathbf{3^+}$ ... $\mathbf{3'^+}$ ... k'_4 ... **linear** (b)

if X = I, OAc
at high concentration

$X^- \; K_5$ / $X^- \; K'_5$

(a') ... k_6 ... $\mathbf{4}$ ... $\mathbf{4'}$ ... k'_6 ... (b')

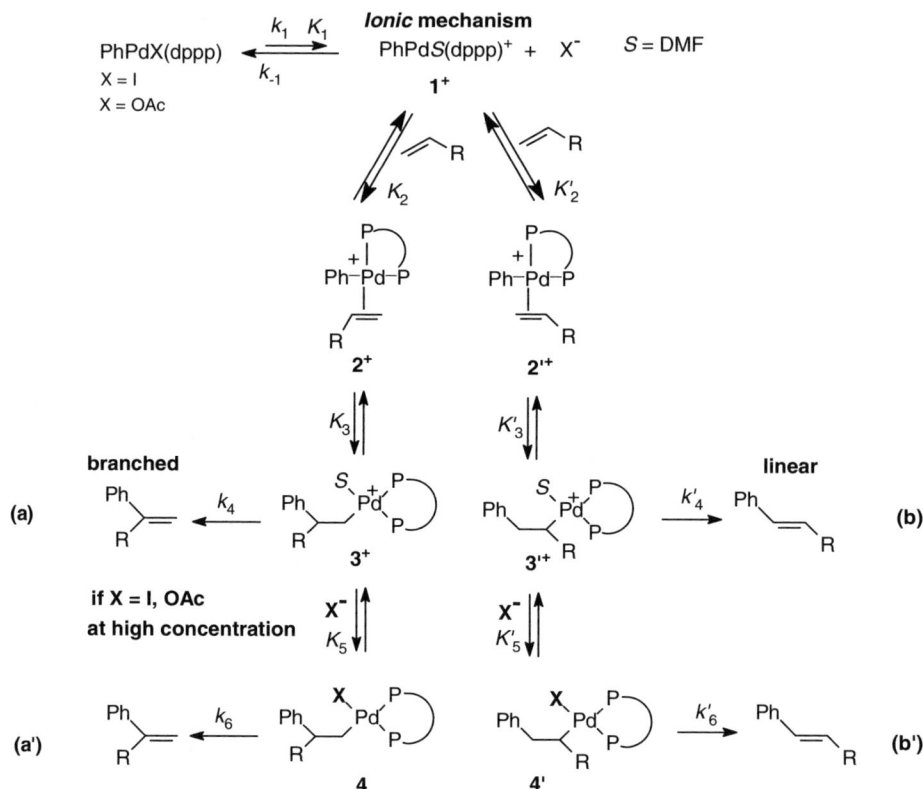

Scheme 1.35 *Mechanism which rationalizes the regioselectivity of Mizoroki–Heck reactions in DMF when $P^\wedge P = dppp$ (the C—C internal rotation in complexes $\mathbf{3^+}$, $\mathbf{3'^+}$, $\mathbf{4}$ and $\mathbf{4'}$ is omitted for more clarity).*

formation of the branched product via route a' in parallel to route a (Scheme 1.35). In that case, what would be determined is not $K_1K_2k_3$ (as proposed in Table 1.2 when the alkene insertion was considered as irreversible) but $K_1K_2K_3k_4 + K_1K_2K_3K_5k_6[X^-] = K_1K_2K_3(k_4 + K_5k_6[X^-])$. Since K_2K_3 does not depend on X, one has to compare the values of $K_1(k_4 + K_5k_6[X^-])$ for PhPd(OAc)(dppp) and PhPdI(dppp). One knows that PhPd(OAc)(dppp) is more dissociated than PhPdI(dppp) towards the cationic complex ($K_1^{OAc} > K_1^I$), but one does not know the relative values of K_5k_6 for AcO$^-$ or I$^-$. The affinity of AcO$^-$ for complex $\mathbf{3^+}$ must be lower than that of I$^-$ ($K_5^{OAc} < K_5^I$), suggesting an antagonist effect and thus an inversion of the reactivity.

As far as the less polarized styrene is concerned, routes a and b in Scheme 1.35 are in competition in the reaction of the isolated *cationic* PhPd(DMF)(dppp)$^+$ with styrene, leading to a mixture of linear *l* and branched *b* products ($b/l = 42/58$ [47]). The reaction of styrene with *neutral* PhPdX(dppp) (X = I, OAc), both reacting via the *cationic* complex, gives the major linear product ($b/l = 20/80$ and $18/82$ respectively [47]) because the faster reaction of I$^-$ and AcO$^-$ with $\mathbf{3'^+}$ favours route b' in Scheme 1.35.

In the catalytic reactions of Xiao and coworkers [40m,n] performed in DMF from aryl bromides and *n*-butylvinyl ether, a mixture of branched and linear products is formed because the halide ions released at high concentration in the course of the catalytic reaction react with the *cationic* complexes of type 3^+ and $3'^+$ to give the *neutral* complexes 4 and 4'; this accounts for the mixture of branched and linear products. At high ionic strength, such as in ionic liquids, the dissociation of ArPdX(dppp) towards the reactive *cationic* complex ArPdS(dppp)$^+$ is favoured, its concentration is increased and, consequently, the reaction must be faster, which is observed. But more importantly, the reaction of halide ions with the cationic complexes of type 3^+ and $3'^+$ is slowed down by the high ionic strength (consequently, no need of halides scavengers), inhibiting the formation of complexes 4 and 4' (lower part of Scheme 1.35). The regioselectivity of the reaction performed in ionic liquids (major branched product) is thus given by the route a of Scheme 1.35, via the major cationic complexes $2^+/3^+$.

1.4.4 Catalytic Cycles

As established above, the regioselectivity of Mizoroki–Heck reactions performed in DMF is sensitive to the presence of coordinating anions such as halide or acetate (Scheme 1.35). The carbopalladation step always proceeds from the more reactive *cationic* complex ArPdS(dppp)$^+$ (Schemes 1.35 and 1.36), not from *neutral* ArPdX(dppp), except for the reaction of ArPdI(dppp) with the most reactive methyl acrylate, performed in the absence of acetate ions (Schemes 1.34 and 1.37).

At identical concentrations of iodobenzene and alkenes CH$_2$=CHR (R = Ph, CO$_2$Et, OiBu), the oxidative addition is always faster [42] than the reaction of alkene with Ph-PdX(dppp), which is the rate-determining step (DMF, 25 °C) [43, 53].

1.5   Mechanism of the Mizoroki–Heck Reaction when the Catalytic Precursor is a *P,C*-Palladacycle

1.5.1   Pd(0) Formation from a *P,C*-Palladacycle

In contrast to PPh$_3$, P(*o*-Tol)$_3$ cannot reduce Pd^{II}(OAc)$_2$ to a Pd(0) complex, but a *P,C*-palladacycle, *trans*-di(μ-acetato)-bis[*o*-(di-*o*-tolylphosphino)benzyl]dipalladium (5) is formed via a cyclometallation [27, 55]. The palladacycle 5 is an efficient catalyst for Mizoroki–Heck reactions involving aryl bromides and activated aryl chlorides (i.e. substituted by EWGs) [1j,l,o,s–v, 27, 55]. When 5 is used as catalyst in C—N cross-coupling reactions, Louie and Hartwig [56] have established that the true catalyst is a Pd(0) complex, Pd^0{P(*o*-Tol)$_3$}$_2$ formed by reduction of the palladacycle by the nucleophile (a secondary amine as a hydride donor in the presence of a strong base).

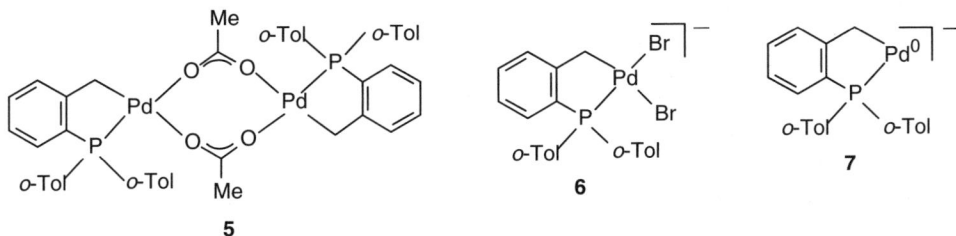

5

6

7

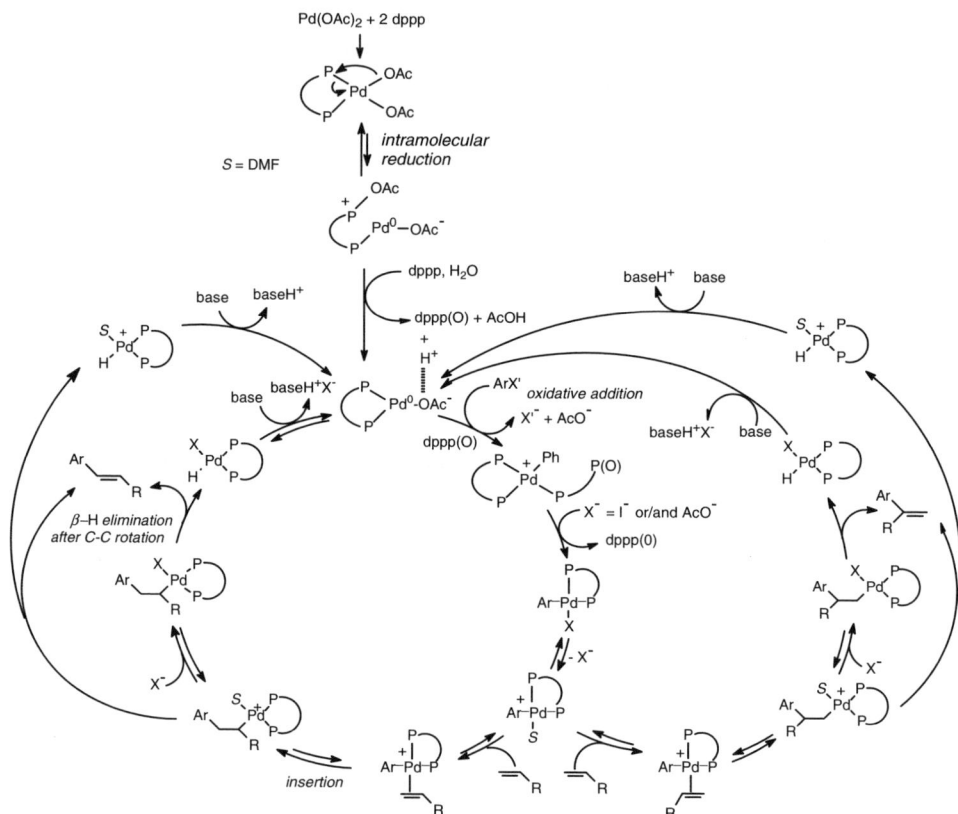

Scheme 1.36 *Mechanism of the Mizoroki–Heck reaction when the catalytic precursor is Pd(OAc)$_2$ associated with dppp: (i) when X' = X = I, R = Ph, O-i-Bu (the formation of HPdS(dppp)$^+$ may be by-passed if the base is strong enough to deprotonate the agostic H in the σ-alkyl–PdS(dppp)$^+$ complexes, see Scheme 1.37); (ii) when acetate is used as base with X' = I, X = OAc, R = Ph, O-i-Bu, CO$_2$Me.*

In Mizoroki–Heck reactions, in the absence of clearly identified reducing agent and due to the recovery of the monomeric *P,C*-palladacycle **6** at the end of a Mizoroki–Heck reaction performed from an aryl bromide, a catalytic cycle involving Pd^{II}/Pd^{IV} complexes was first proposed [27, 57]. DFT calculations have, however, established that the oxidative addition of iodobenzene with a Pd(II) complex is energetically not favoured at all [58]. Some Mizoroki–Heck reactions proceed with an induction period when the base is a tertiary amine [55b], but no induction period is observed when an acetate salt is used as a base [27, 55]. This is why the palladacycle **5** is often associated with an acetate salt used as base [27, 55a]. The induction period was explained by Beller and Riermeier [55b] as a slowly occurring reduction of the palladacycle **5** to give the active Pd(0) complex [Pd^0{P(*o*-Tol)$_3$}]. Even in the absence of any identified reducing agent, Böhm and Herrmann [59] have also proposed the reduction *in situ* of the palladacycle **5** to an anionic Pd(0) complex **7**, still ligated to the benzyl moiety of the ligand. The first step of the catalytic cycle would

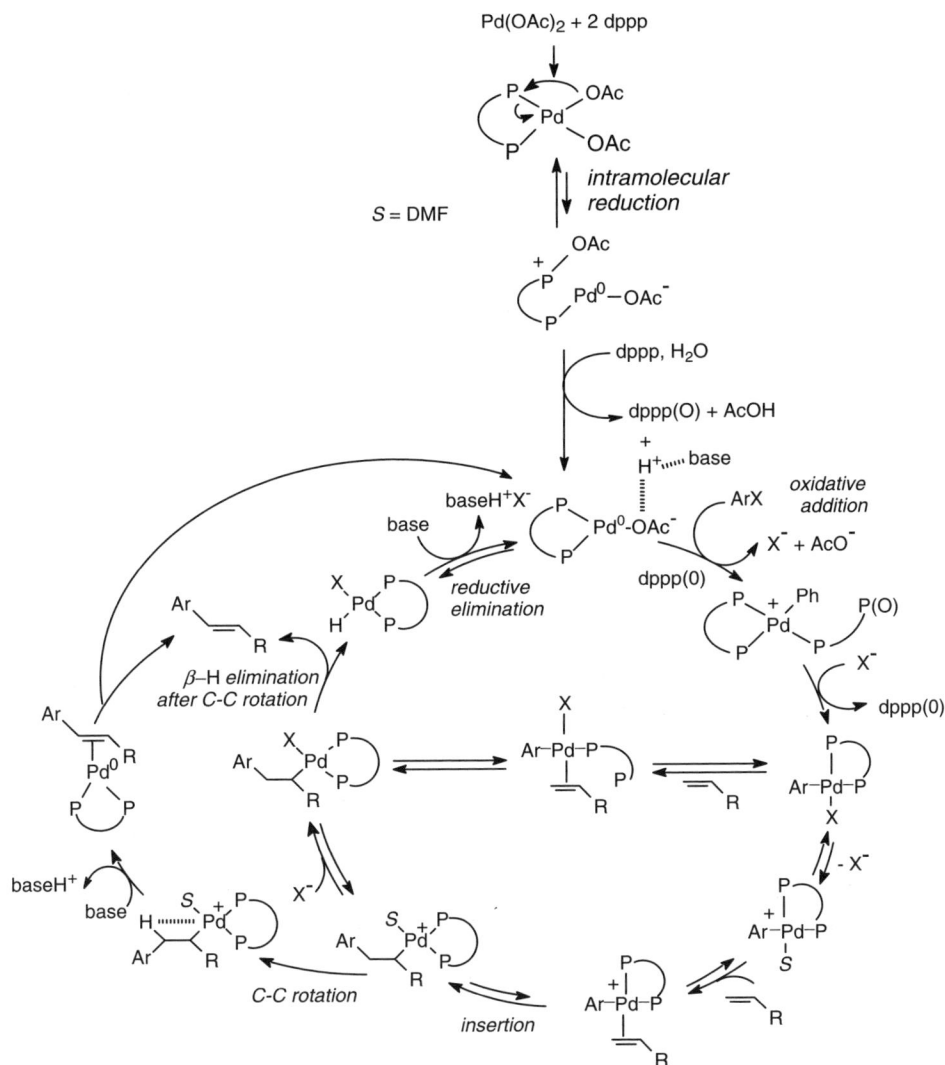

Scheme 1.37 *Mechanism of the Mizoroki–Heck reaction when the catalytic precursor is Pd(OAc)₂ associated with dppp and when the base is not an acetate salt (e.g. base = NEt₃): X = I, R = CO₂Me.*

then be the classical oxidative addition of an aryl halide to a Pd(0) complex, as in a classical catalytic cycle involving Pd^0/Pd^{II} complexes.

In 2005, upon investigation of the electrochemical properties of the palladacycle **5** in DMF, d'Orlyé and Jutand [28] showed that a Pd(0) complex (characterized by its oxidation peak at +0.2 V versus saturated calomel electrode (SCE)) is generated by the electrochemical reduction of **5** and its monomers in DMF (Scheme 1.38). However, the final Pd(0) species is not the electrogenerated complex **7** proposed by Böhm and Herrmann [59].

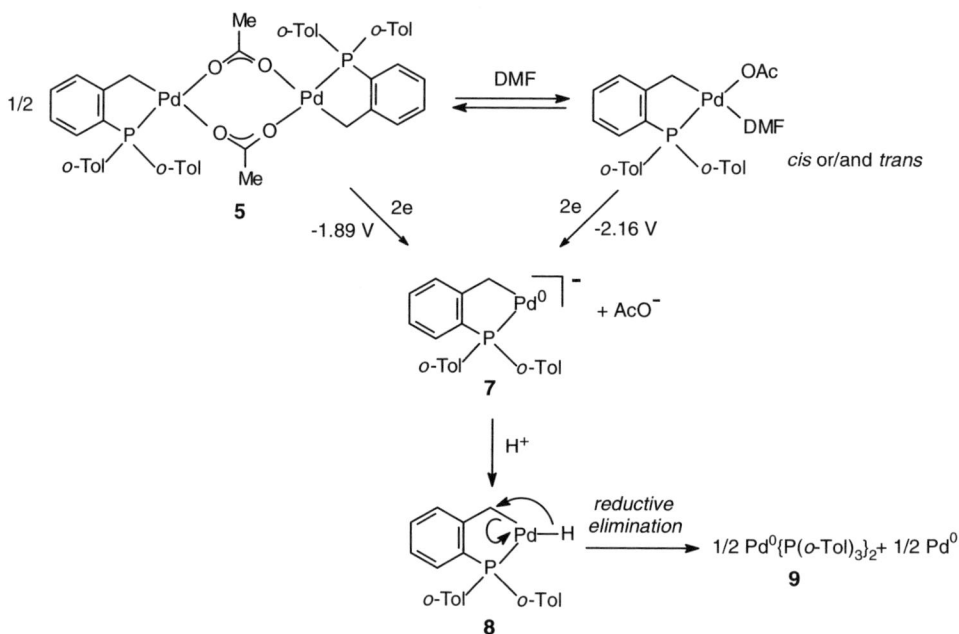

Scheme 1.38 *Electrochemical reduction of palladacycle 5 in equilibrium with monomers in DMF at 25 °C (peak potentials are measured versus SCE reference electrode).*

Instead, $Pd^0\{P(o\text{-}Tol)_3\}_2$ (**9**) is formed upon fast protonation of **7** followed by reductive elimination from complex **8** (Scheme 1.38) [28]. Therefore, in DMF, the palladacycle **5** is reduced to a Pd^0 complex at a rather high negative potential that could be reached by zinc powder (Scheme 1.38). Such a strong reducing agent is, however, never present in Mizoroki–Heck reactions. No oxidation peak was detected when the cyclic voltammetry of **5** was performed directly towards oxidation potentials, establishing that a Pd(0) complex is not generated spontaneously from the palladacycle **5** in DMF at 25 °C.

In 2005, d'Orlyé and Jutand [28] hypothesized that a Pd(0) complex might be generated *in situ* from the palladacycle **5** in an endergonic reductive elimination between the benzylic group attached to the Pd(II) centre and the *cis*-acetate ligand (Scheme 1.39). This reaction generates the monophosphine–Pd(0) complex **10** ligated by the new ligand **11** formed in the reductive elimination. The backward reaction in Scheme 1.39, an intramolecular oxidative

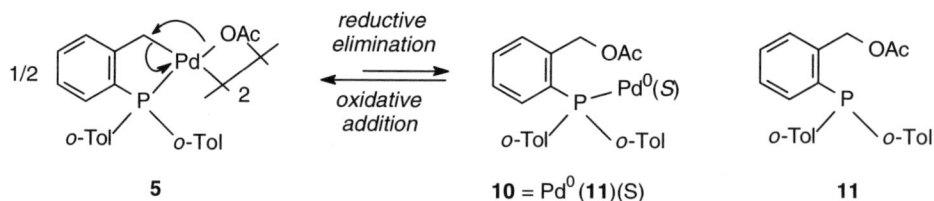

Scheme 1.39 *Formation of a Pd^0 complex by reductive elimination (S = solvent).*

addition of the Pd(0) with the C—O bond of the *o*-benzylic acetate in complex **10**, must be very fast due to its intramolecular character. This is why the equilibrium in Scheme 1.39 would lie in favour of the palladacycle **5** at 25 °C. Consequently, the Pd(0) complex generated at low concentration in the endergonic reductive elimination could be detected (e.g. by an oxidation peak), provided the equilibrium in Scheme 1.39 is shifted towards its right-hand side by trapping the low-ligated Pd(0) complex **10** by additional ligands: P(*o*-Tol)$_3$, dba or AcO$^-$. Those additives have been selected because they cannot reduce the palladacycle **5** to a Pd(0) complex.

D'Orlyé and Jutand [28] have indeed shown that a Pd(0) complex is generated *in situ* from the palladacycle **5** after addition of dba and P(*o*-Tol)$_3$ in large excess at 80 °C in DMF; that is, in the absence of any reducing agents. This Pd(0) complex has been characterized by an oxidation peak (+0.19 V versus SCE) which disappears after addition of PhI, confirming *a posteriori* the formation *in situ* of a Pd(0) from **5** or its monomeric form in DMF. The oxidation peak potential characterizes the complex Pd^0{P(*o*-Tol)$_3$}$_2$ (**9**) formed from complex **10**, due to the large excess of P(*o*-Tol)$_3$ (Scheme 1.40) [28].

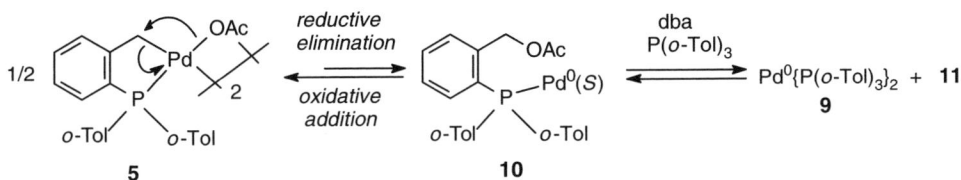

Scheme 1.40 *Formation of a detectable Pd(0) complex from **5** (S = DMF).*

As mentioned above, acetate salts are often used as base in palladacycle-catalysed Mizoroki–Heck reactions [1j,s, 27, 55]. D'Orlyé and Jutand have established that an anionic monopalladacycle (**12**) is formed upon addition of excess *n*-Bu$_4$NOAc to **5** in DMF (Scheme 1.41). After further addition of a stoichiometric amount of P(*o*-Tol)$_3$ (P/Pd = 1) and excess dba [60], an oxidation peak is detected after 1 h at 80 °C, at a slightly less positive potential (+0.14 V versus SCE) than that obtained in the absence of acetate ions (see above). This oxidation peak disappears when the solution is cooled to 20 °C and appears again upon increasing the temperature to 60–80° C. It definitively disappears after addition of PhI, confirming that a Pd(0) complex has been generated

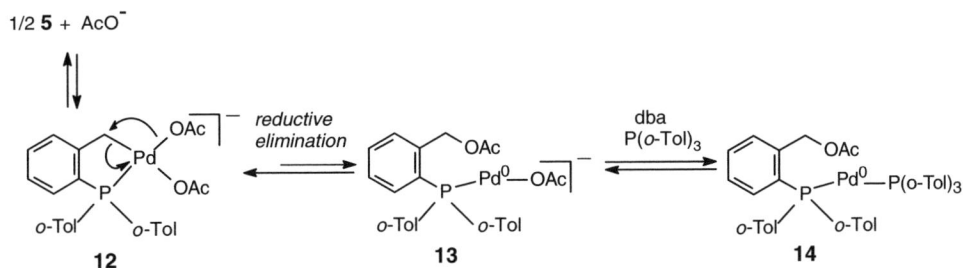

Scheme 1.41 *Formation of a detectable Pd(0) complex **14** from palladacycle **5** in the presence of acetate ions and P(o-Tol)$_3$ in DMF at 80 °C.*

in situ from the palladacycle **5** via the anionic complex **12** in the absence of any reducing agent (Scheme 1.41). The reversible reductive elimination has been shifted towards the formation of complex **14** by successive stabilization of the Pd(0) complex by acetate and P(*o*-Tol)$_3$ (Scheme 1.41) [28].

The oxidative addition of Pd^0(dba)$_2$ to PPh$_2$(*o*-benzyl acetate) (**15**) (a related but less-hindered ligand than **11**) generates a mononuclear *P,C*-palladacycle **16** (Scheme 1.42) [28, 61]. The cleavage of the benzyl–OAc bond of the ligand **15** by a Pd(0) complex by oxidative addition supports the idea that the formation of the Pd(0) complex **10** in Scheme 1.39 is, indeed, reversible [28].

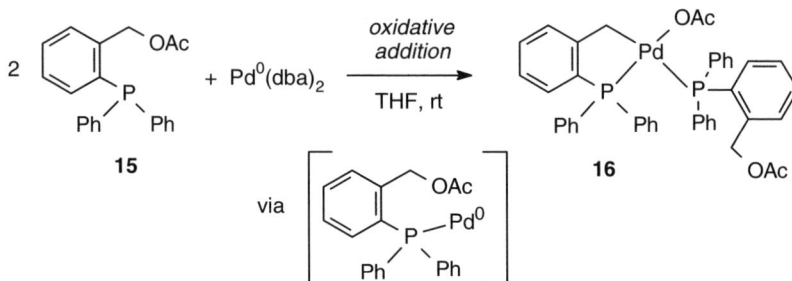

Scheme 1.42

Therefore, d'Orlyé and Jutand have established that the *P,C*-palladacycle **5** is a reservoir of a monophosphine–Pd(0) complex Pd^0{P(*o*-Tol)$_2$(*o*-benzyl–OAc)}(DMF) **10**, generated *in situ* by a reductive elimination between the OAc ligand and the *o*-benzyl moiety of the phosphine ligand in a reversible process. Such a reaction is favoured by acetate anions, often used as base in Mizoroki–Heck reactions, via the formation of an anionic monomeric *P,C*-palladacycle complex (**12**) ligated by acetate. This explains why no induction period is observed in Mizoroki–Heck reactions when NaOAc is used as base (see above) [27, 55]. This reversible formation of a Pd(0) complex ensures the stability of the *P,C*-palladacycle structure in Mizoroki–Heck reactions. Indeed, the anionic bromide-ligated monomeric palladacycle **6** has been observed by Herrmann *et al.* in the course of Mizoroki–Heck reactions performed from aryl bromides [27]. Such an anionic palladacycle is formed by reaction of **5** with bromide ions released in the catalytic reaction or voluntarily added. The complex **6** may undergo a reversible reductive elimination between the benzylic carbon and a bromide ligand to give the anionic Pd(0) complex **17** similar to **13** (Scheme 1.43).

1.5.2 Catalytic Cycle

A catalytic cycle is proposed for Mizoroki–Heck reactions involving a *P,C*-palladacycle precursor based on the fact that a monoligated Pd(0) complex is formed from *P,C*-palladacycle precursors (see above). The structure of the Pd(0) complex **10** is close to that of Pd^0{P(*o*-Tol)$_3$} generated from Pd^0{P(*o*-Tol)$_3$}$_2$ as the minor but active species in oxidative additions of aryl bromides, as reported by Hartwig and Paul [62a]. The oxidative addition gives the dimeric complex [ArPd(μ-Br){P(*o*-Tol)$_3$}]$_2$ in equilibrium with the former T-shaped complex ArPdBr{P(*o*-Tol)$_3$} prone to react with a nucleophile [62b,c]. Such a mechanism must be valid for the Pd(0) complexes **10** or **13** generated *in situ* from the

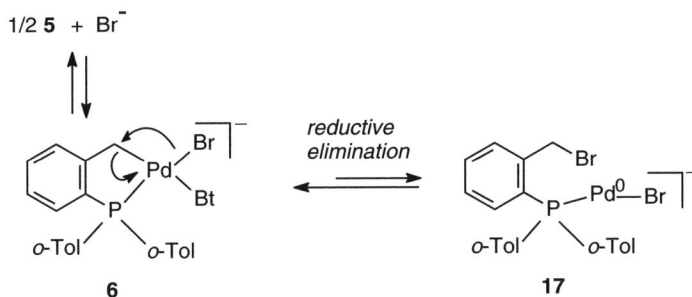

Scheme 1.43

P,C-palladacycle precursors and ligated to the monophosphine **11** (Scheme 1.44). Kinetic data are still missing for most steps of the catalytic cycle [63]. The oxidative addition of aryl bromides is probably not rate determining, since the rate of the overall reaction is highly dependent on the structure of the alkene, as evidenced by competitive reactions of two different alkenes with the same aryl bromide in the presence of **5** [55a]. However, the alkene might also favour the reductive elimination in the palladacycle **5**, which slowly delivers the active Pd(0) complex into the catalytic cycle [60].

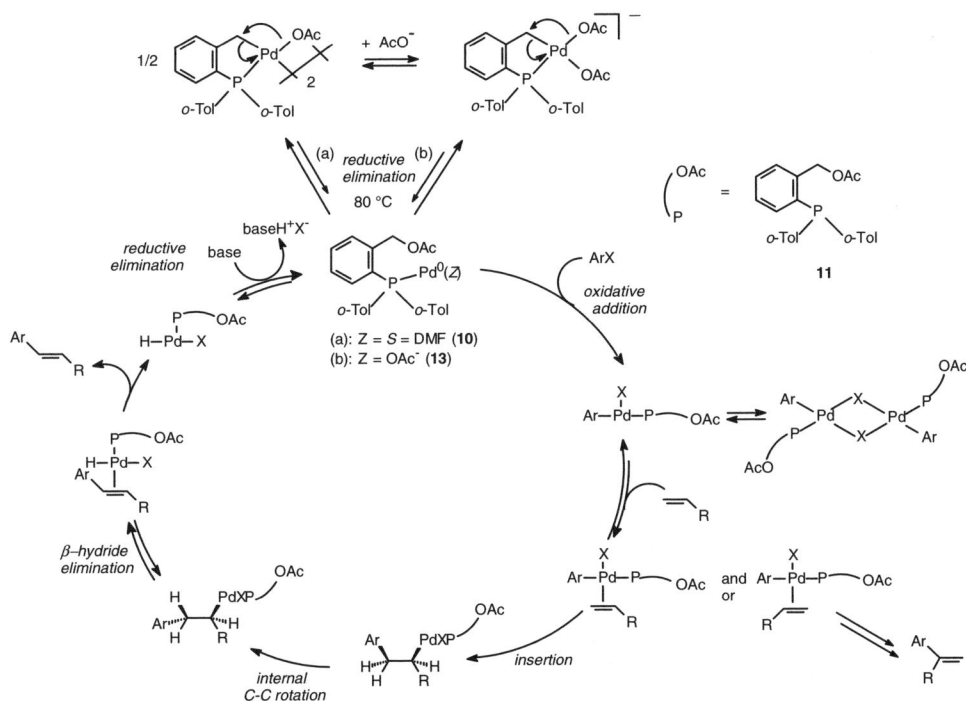

Scheme 1.44 *Mechanism for P,C-palladacycle-catalyzed Mizoroki–Heck reactions performed in DMF (route a) or in the presence of acetate ions as a base (route b).*

1.6   Mechanism of the Mizoroki–Heck Reaction when the Ligand is an *N*-Heterocyclic Carbene

N-Heterocyclic carbenes (NHCs, named Cb in the following, Chart 1.1) were introduced as ligands by Herrmann *et al.* [64] in their search to activate aryl chlorides or poorly reactive aryl bromides in Mizoroki–Heck reactions (for recent reviews, see [1p,q,s,t,w,x]). NHCs are indeed strong σ-donor and weak π-acceptor ligands [65] which make Pd(0) complexes more electron-rich and thus favour the oxidative addition of relatively unreactive aryl halides. C=C unsaturated NHCs have been introduced in Mizoroki–Heck reactions via a $PdI_2(Cb)_2$ precursor (R_1 = Me, R_2 = H in Chart 1.1, left) by Herrmann *et al.*, who observed an acceleration of the reactions upon addition of a reducing agent (hydrazine). This establishes that Pd(0) ligated by carbene(s) is the active species in the oxidative addition step [64].

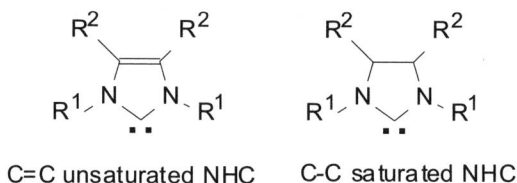

C=C unsaturated NHC   C–C saturated NHC

Chart 1.1

$PdX_2(Cb)_2$ (X = halide, acetate) precursors may be formed from a Pd(II) salt (e.g. $Pd(OAc)_2$) and *N*-heterocyclic azolium salts which are deprotonated into the NHC ligand [1p, 64, 66a–c]. They are also generated *in situ* when *N*-heterocyclic azolium salts are used as ionic liquid solvents [66d,e]. Isolated stable NHC-ligated Pd(0) complexes [67] are also used as catalysts in Mizoroki–Heck reactions [68].

1.6.1   Oxidative Addition

The oxidative addition of aryl halides to $Pd^0(Cb)_2$ complexes has been reported and the complexes *trans*-ArPdX(Cb)_2 (Cb = *cyclo*-C{NR^1CR^2}_2, X = I, R^1 = R^2 = Me; X = Cl, R^1 = *t*-Bu, R^2 = H; Chart 1.1, left) formed in the reaction have been isolated and characterized [68a, 69, 70]. The aryl–palladium(II) complexes are always ligated by two carbene ligands irrespective of their bulkiness.

Kinetics data on the oxidative addition are scarce. In 2003, Roland and coworkers [71] used $PdX_2(Cb^a)_2$ (X = I, Cl) as an efficient precursor for Mizoroki–Heck reactions performed from aryl bromides at moderate temperatures (Scheme 1.45). Since $Pd^0(Cb^a)_2$ could not be isolated, its reactivity with aryl halides was followed by cyclic voltammetry, the transient $Pd^0(Cb^a)_2$ being generated in the electrochemical reduction of the precursors $PdX_2(Cb^a)_2$ in DMF (Scheme 1.45). The rate constants k of the oxidative addition of aryl halides to $Pd^0(Cb^a)_2$ have been determined (Table 1.3) [71].

In the late 2003, Caddick and coworkers [70] reported that the isolated $Pd^0(Cb^b)_2$, where Cb^b is much more bulky than Cb^a, reacts with 4-CH_3–C_6H_4–Cl via $Pd^0(Cb^b)$ in a dissociative mechanism (Scheme 1.46). The reactivity of $Pd^0(Cb^b)$ in the dissociative

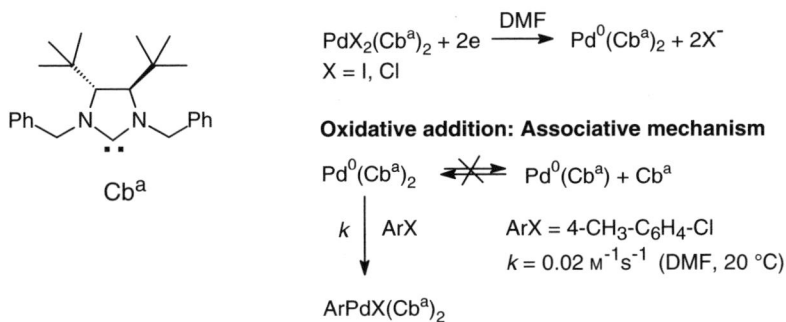

$$PdX_2(Cb^a)_2 + 2e \xrightarrow{\text{DMF}} Pd^0(Cb^a)_2 + 2X^-$$
$$X = I, Cl$$

Oxidative addition: Associative mechanism

$$Pd^0(Cb^a)_2 \xcancel{\rightleftharpoons} Pd^0(Cb^a) + Cb^a$$

$$k \downarrow ArX$$

$$ArX = 4\text{-}CH_3\text{-}C_6H_4\text{-}Cl$$
$$k = 0.02 \text{ M}^{-1}\text{s}^{-1} \text{ (DMF, 20 °C)}$$

$$ArPdX(Cb^a)_2$$

Scheme 1.45

mechanism is controlled by the value of its rate constant k and its concentration, which is very low due to the endergonic equilibrium with $Pd^0(Cb^b)_2$ (see the value of K in Scheme 1.46): rate $= k[Pd^0(Cb^b)][4\text{-chlorotoluene}] = kK[Pd^0(Cb^b)_2][4\text{-chlorotoluene}]/[Cb^b]$. As a result, the overall reaction is quite slow and could be followed by ^{1}H NMR ($t_{1/2} \approx 24\,$h) when $[Pd^0] = [4\text{-chlorotoluene}] = 0.2\,$M) [70].

DFT calculations reported by Green *et al.* [72] in 2005 on Pd(0) complexes ligated by two C=C unsaturated carbenes close to Cb^b support the dissociative mechanism with this assumption: the bulkier the substituent on the N atoms is, the lower the dissociation energy of the biscarbene–Pd(0) complex is.

In 2006, Jutand and coworkers [73] extended their former work of 2003 to the reactivity of the electrogenerated $Pd^0(Cb^a)_2$ with aryl chlorides. The reactions take place at 20 °C in DMF (Scheme 1.45, Table 1.3). As in all oxidative additions [6c], the following reactivity orders have been established:

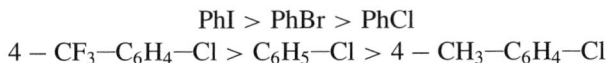

$$PhI > PhBr > PhCl$$
$$4 - CF_3-C_6H_4-Cl > C_6H_5-Cl > 4 - CH_3-C_6H_4-Cl$$

Jutand and coworkers [73] have established that $Pd^0(Cb^a)_2$ is the reactive species in an associative mechanism (Scheme 1.45); this is in contrast to $Pd^0(Cb^b)_2$, which reacts via $Pd^0(Cb^b)$ in a dissociative mechanism (Scheme 1.46) [70]. This is rationalized by steric factors. The stable Cb^b carbene is bulky and thus prompted to dissociation from $Pd^0(Cb^b)_2$ [72]. Conversely, the ligand Cb^a is less stable and less bulky than Cb^b, as evidenced by

Table 1.3 *Rate constants* k *of the oxidative addition of aryl halides to Pd(0) complexes electrogenerated from Pd(II) precursors in DMF at 20 °C (Scheme 1.45)*

Pd(0) complex	k (M^{-1} s^{-1})				
	PhI	PhBr	PhCl	4-CH$_3$—C$_6$H$_4$—Cl	4-CF$_3$—C$_6$H$_4$—Cl
$Pd^0(Cb^a)_2$	>1180	1180	0.13	0.02	0.35
$Pd^0(Cb^a)(PPh_3)$	830	2	n.d.	n.d.	n.d.

n.d.: not determined.

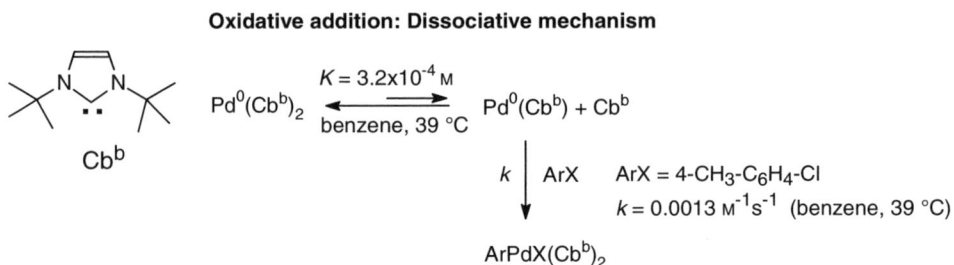

Oxidative addition: Dissociative mechanism

Cb^b

$$Pd^0(Cb^b)_2 \underset{\text{benzene, 39 °C}}{\overset{K = 3.2\times10^{-4} \text{ M}}{\rightleftharpoons}} Pd^0(Cb^b) + Cb^b$$

k | ArX   ArX = 4-CH$_3$-C$_6$H$_4$-Cl
          k = 0.0013 M^{-1}s^{-1} (benzene, 39 °C)

$ArPdX(Cb^b)_2$

Scheme 1.46

the X-ray structure of the parent complex PdI$_2$(Cb^a)$_2$ [71]. Therefore, Pd^0(Cb^a)$_2$ is less inclined to dissociation than Pd^0(Cb^b)$_2$. Interestingly, comparison of the reactivity of 4-chlorotoluene with Pd^0(Cb^a)$_2$ and Pd^0(Cb^b)$_2$ shows that Pd^0(Cb^a)$_2$, which reacts via the associative mechanism (Scheme 1.45), is more reactive than Pd^0(Cb^b)$_2$, which reacts via Pd^0(Cb^b) in the dissociative mechanism (Scheme 1.46). Moreover, when comparing their respective rate constants k, one sees that Pd^0(Cb^a)$_2$ (k = 0.02 M^{-1} s^{-1}, 20 °C, Scheme 1.45) is even more reactive than Pd^0(Cb^b) (k = 0.0013 M^{-1} s^{-1}, 39 °C, Scheme 1.46) at identical concentrations of Pd(0) and 4-chlorotoluene [73].

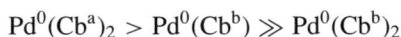

$$Pd^0(Cb^a)_2 > Pd^0(Cb^b) \gg Pd^0(Cb^b)_2$$

In other words, the involvement of a monoligated Pd^0(Cb) as the active species is not a guarantee for a fast oxidative addition, because Pd^0(Cb) is always generated at low concentration in its endergonic equilibrium with the nonreactive Pd^0(Cb)$_2$ complex.

Therefore, the structure of the reactive Pd(0) in oxidative addition (Pd^0(Cb) versus Pd^0(Cb)$_2$) is governed by the bulk of the carbene ligand; but the reactivity is not necessarily controlled by the structure of the reactive species, since a bis-ligated Pd^0(Cb)$_2$ (e.g. Cb = Cb^a) may be even more reactive than a monoligated Pd^0(Cb') (e.g. Cb' = Cb^b). Electronic factors must also be taken into consideration, C—C saturated carbenes (as Cb^a) being stronger σ-donors than C=C unsaturated carbenes (as Cb^b) [65g].

The mixed carbine–phosphine Pd^0(Cb^a)(PPh$_3$) generated in the electrochemical reduction of PdI$_2$(Cb^a)(PPh$_3$) also reacts in an associative mechanism and is less reactive than Pd^0(Cb^a)$_2$ (Table 1.3) [71, 73], showing that the carbene Cb^a is indeed much more electron donating than PPh$_3$. However, Pd^0(Cb^a)(PPh$_3$) is much more efficient than Pd^0(Cb^a)$_2$ in Mizoroki–Heck reactions performed from PhBr [71], suggesting that the oxidative addition is not rate determining.

1.6.2   Complexation/Insertion of the Alkene

Because of the lack of kinetic data, the mechanism of the insertion of the alkene is currently speculative. The alkene might react with *trans*-ArPdX(Cb)$_2$ either via the cationic complex ArPd(Cb)$_2^+$ (after dissociation of X$^-$) or via the neutral ArPdX(Cb) (after dissociation of one Cb). DFT calculations by Rösch and coworkers [74] in 1998, reinforced by experimental work by Cavell and coworkers [68a] in 1999 on the comparative reactivity of CH$_2$=CH—CO$_2$-n-Bu with isolated *trans*-ArPdI(Cb)$_2$ and ArPd(Cb)$_2^+$ (generated *in situ* by halide abstraction in dichloromethane), led to the conclusion that the alkene reacts faster with the cationic *trans*-ArPdS(Cb)$_2$$^+$ than with *trans*-ArPdI(Cb)$_2$, suggesting that the dissociation of the strong σ-donor Cb is thermodynamically less favoured (Scheme 1.47). However, the reaction involving the cationic complex gives rise to by-products formed by reductive elimination of the carbene ligand with the alkyl group or with the hydride, giving imidazoliums (Scheme 1.47) [68a].

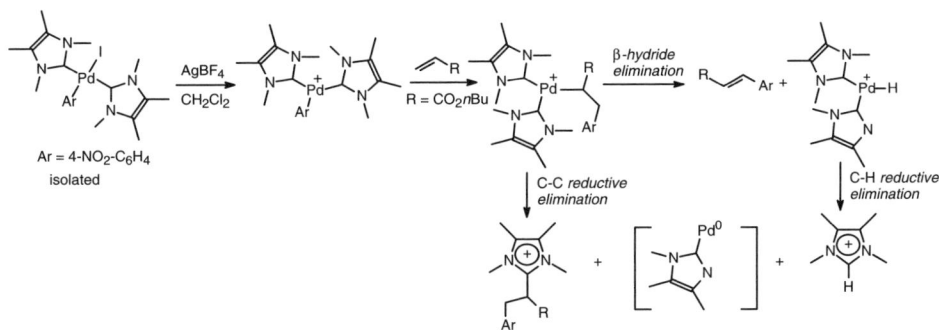

Scheme 1.47 *Reaction of an alkene with a cationic ArPd(Cb)$_2$$^+$ and subsequent by-reactions (the precomplexation of the alkene to the cationic ArPd(Cb)$_2$$^+$ is omitted).*

DFT calculations by Rösch and coworkers [74] showed that for PhPdCl(P$^\wedge$Cb) complexes, where the bidentate ligand P$^\wedge$Cb is a monophosphine linked to an NHC, the reaction with an alkene proceeds via dissociation of the more labile phosphine; that is, via a neutral Cb-linked ArPdX complex (Scheme 1.48). Later on, bidentate P$^\wedge$Cb ligands generated *in situ* from phosphine–imidazolium salts proved to be efficient in Mizoroki–Heck reactions employing aryl bromides – even deactivated ones, as pioneered by Nolan and coworkers [1q, 75]. Interestingly, the DFT calculations have paved the way to fruitful experiments.

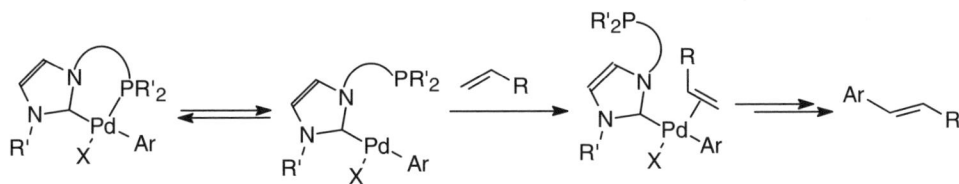

Scheme 1.48

It appears reasonable to predict that ArPdX(Cb)(PR$_3$) complexes generated in the oxidative addition of ArX to Pd^0(Cb)(PR$_3$) would dissociate to ArPdX(Cb), which reacts with the alkene. Such a dissociation of the phosphine must be even easier than the intramolecular dissociation of the phosphine in the bidentate P$^\wedge$Cb ligand proposed above. This is probably why mixed complexes Pd^0(Cb)(PR$_3$) are more efficient than Pd^0(Cb)$_2$ in Mizoroki–Heck reactions performed from aryl bromides [71, 76], even if they are less reactive than Pd^0(Cb)$_2$ in the oxidative addition. Indeed, the high stability of the Cb–Pd(II) bond combined with the easy dissociation of the phosphine in ArPdX(Cb)(PR$_3$) favours the complexation/insertion of the alkene.

1.6.3   Catalytic Cycles

The above considerations are summarized in three individual catalytic cycles: the *ionic* mechanism catalysed by a Pd(0) coordinated to one or two C—C saturated or C=C unsaturated NHC monocarbenes (Scheme 1.49); the *neutral* mechanism catalysed by mixed Pd(0) complexes coordinated to one C—C saturated or C=C unsaturated NHC monocarbene and one phosphine (Scheme 1.50); and the *neutral* mechanism catalysed by Pd(0) coordinated to a bidentate P^Cb ligand (Scheme 1.51).

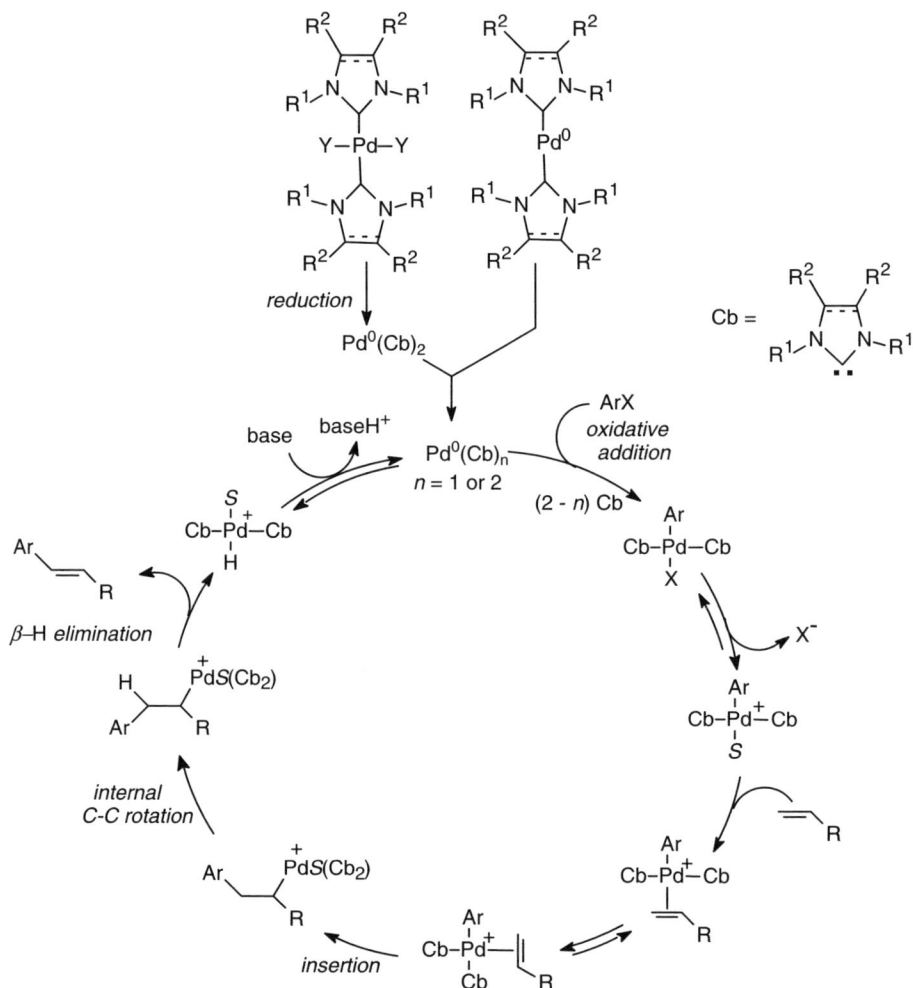

Scheme 1.49 *Ionic mechanism for Mizoroki–Heck reactions catalysed by a Pd(0) coordinated to one or two C—C saturated or C=C unsaturated N-heterocyclic monocarbenes (only one way for the coordination of the alkene is presented). The reactive species is $Pd^0(Cb)$ for a bulky carbene and $Pd^0(Cb)_2$ for a nonbulky carbene. The aryl–palladium complex formed in the oxidative addition is always ligated by two Cb ligands delivered by the Pd(0) or Pd(II) precursor even if $Pd^0(Cb)$ is the reactive species.*

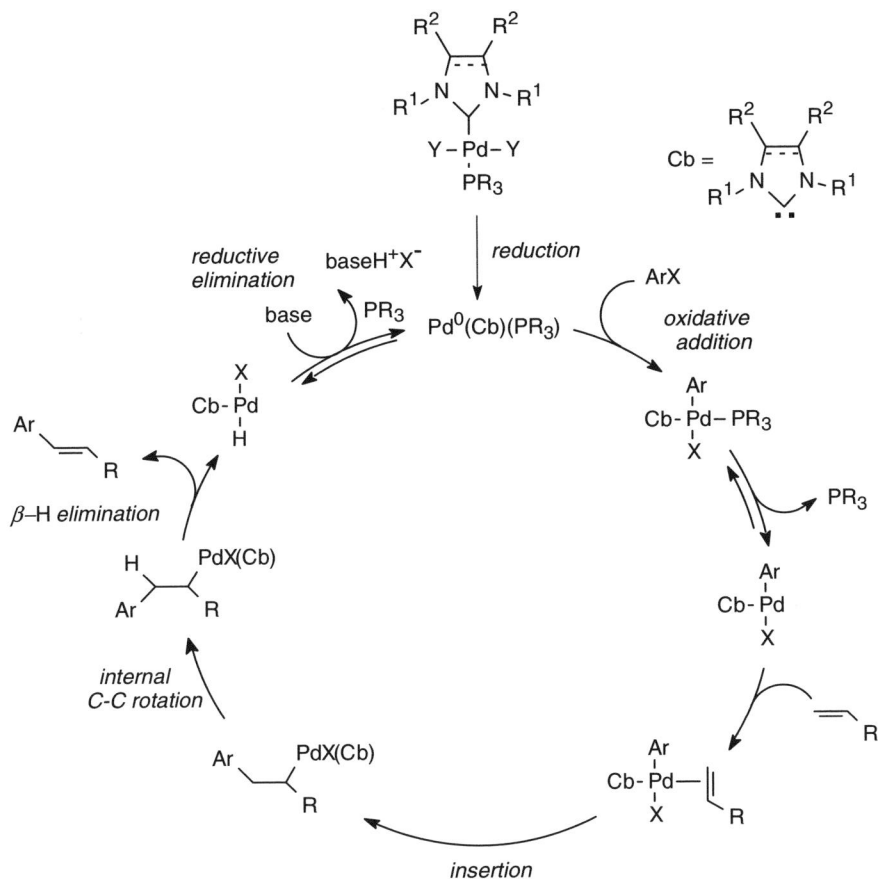

Scheme 1.50 Neutral *mechanism for Mizoroki–Heck reactions catalysed by mixed Pd(0) complexes coordinated to one C—C saturated or C=C unsaturated N-heterocyclic monocarbene and one phosphine (only one orientation of the alkene is presented).*

1.7 Mechanism of the Mizoroki–Heck Reaction when the Ligand is a Bulky and Electron-Rich Monophosphine

The introduction of *P,C*-palladacycles [27, 55a] or NHC-ligated palladium complexes by Herrmann *et al.* [64] has permitted the use of activated aryl chlorides (substituted by EWGs) in Mizoroki–Heck reactions. However, the reactions were performed at high temperatures (120–160 °C) and chlorobenzene or nonactivated aryl chlorides (substituted by EDGs) were not reactive [1r,y]. The further significant improvement was the introduction of bulky and electron-rich phosphines by Littke and Fu [77] in association with the precursor Pd$^0_2$(dba)$_3$. Among them, tri-*tert*-butylphosphine, P-*t*-Bu$_3$, is the best ligand for reactions of nonactivated aryl chlorides under mild conditions: P-*t*-Bu$_3$ ≫ PCy$_3$ (Cy = cyclohexyl) [77a]. The efficiency of the Mizoroki–Heck reactions involving P-*t*-Bu$_3$ is, however, very dependent on the base. By replacing Cs$_2$CO$_3$ by Cy$_2$NMe, the Mizoroki–Heck reactions

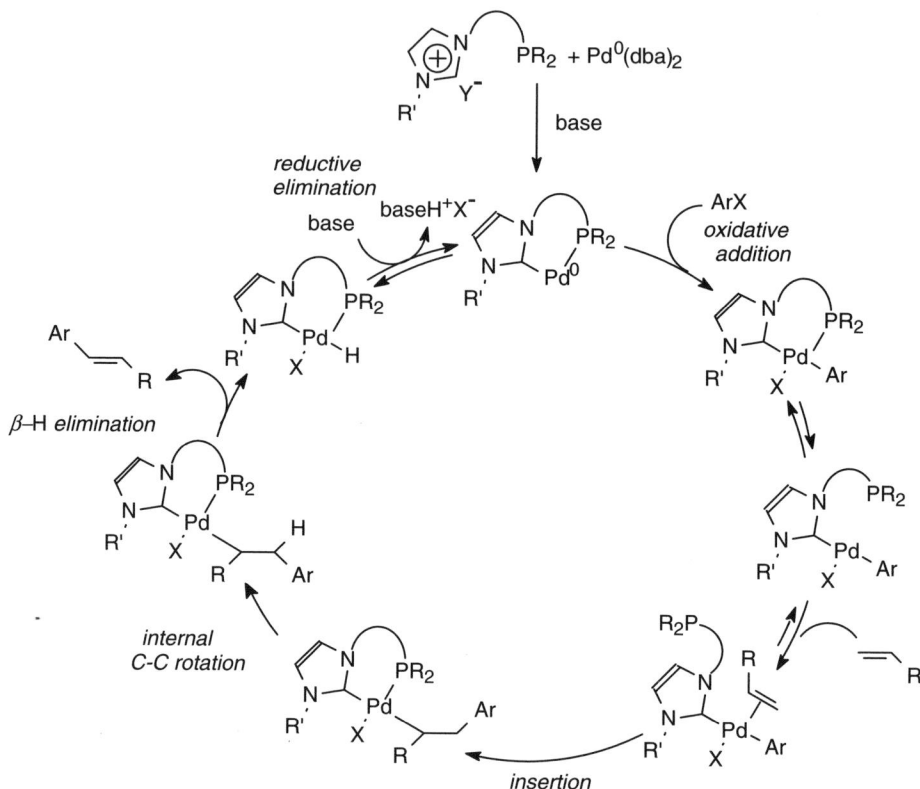

Scheme 1.51 Neutral *mechanism for Mizoroki–Heck reactions catalysed by Pd(0) coordinated to a bidentate P^Cb ligand.*

from activated aryl chlorides are performed at room temperature, whereas nonactivated or hindered aryl chlorides are converted at 100–120 °C [77b]. In the same phosphine series, changing one or two *t*-Bu groups, as in Fc′–P-*t*-Bu$_2$ (Fc′ = aryl-substituted ferrocenyl) or (Ad)$_2$P-*t*-Bu (Ad = adamantyl), namely increasing the bulk of the phosphine, allows Mizoroki–Heck reactions with nonactivated aryl chlorides (NaOAc as base at 110 °C [78] and K$_3$PO$_4$ as base at 110 °C [79] respectively).

1.7.1 Oxidative Addition

Barrios-Landeros and Hartwig [80a] have reported the mechanism of the oxidative addition of aryl bromides with Pd(0) complexes ligated by bulky electron-rich phosphines such as Pd^0(Fc′–P-*t*-Bu$_2$)$_2$ (Fc′ = aryl-substituted ferrocenyl). All Pd(0) complexes are found to react via a monophosphine complex Pd^{0}L in a dissociative mechanism, leading to the monophosphine T-shaped complex ArPdXL which may be stabilized by a weak agostic Pd–H bond (H from the ligand) [80b,c] (Scheme 1.52). The mechanism of the oxidative addition of PhBr to Pd^0(P-*t*-Bu$_3$)$_2$ (P-*t*-Bu$_3$: cone angle 182°, $pK_a = 11.4$) is not reported, but the T-shaped structure of the complex PhPdBr(P-*t*-Bu$_3$) formed in the oxidative addition suggests that the reaction proceeds via the dissociative mechanism of Scheme 1.52 [80c]. The reaction of aryl chlorides is more problematic. The oxidative addition of PhCl with Pd^0(Fc′–P-*t*-Bu$_2$)$_2$ proceeds by the dissociative mechanism of Scheme 1.52 [80a], whereas no reaction occurs with Pd^0(P-*t*-Bu$_3$)$_2$ [80c]. However, evidence for the reactivity of Pd^0(P-*t*-Bu$_3$) with PhCl has been established in oxidative additions performed in the presence of bases [80d]. This suggests that the concentration of Pd^0(P-*t*-Bu$_3$) in its equilibrium with Pd^0(P-*t*-Bu$_3$)$_2$ must be very low and that decreasing the ratio P-*t*-Bu$_3$/Pd should be beneficial for catalytic reactions. Indeed, Littke and Fu [77b] have observed that, in a Mizoroki–Heck reaction performed from 4-chloroacetophenone and styrene in the presence of Cy$_2$NMe as base (i.e. in conditions where the oxidative addition is rate limiting), the catalytic systems {1.5% Pd^0$_2$(dba)$_3$ + 3% P-*t*-Bu$_3$} or {1.5% Pd^0(P-*t*-Bu$_3$)$_2$ + 0.75% Pd^0$_2$(dba)$_3$} for which the ratio Pd/P = 1 give the same conversion of 30% after the same reaction time, whereas using 3% of Pd^0(P-*t*-Bu$_3$)$_2$ leads to only 2% conversion. When the precursor is Pd^0$_2$(dba)$_3$ associated with P-*t*-Bu$_3$ so that Pd/P = 1, the major complex Pd^0(dba)(P-*t*-Bu$_3$) must be formed, which dissociates to Pd^0(P-*t*-Bu$_3$) more easily than Pd^0(P-*t*-Bu$_3$)$_2$ does [81].

$$Pd^0L_2 \rightleftharpoons Pd^0L + L \qquad \begin{array}{c} L-Pd-X \\ | \\ Ar \end{array}$$

Pd^{0}L$_2$ ⇌ Pd^{0}L + L   →(ArX)→ ArPdXL

structure of isolated complexes

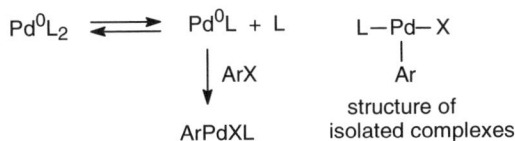

Scheme 1.52 *Dissociative mechanism for the oxidative addition.*

When less bulky ligands are involved, such as PCy$_3$ (cone angle 170°, $pK_a = 9.7$), the complex Pd^0(PCy$_3$)$_2$ reacts with aryl halides, including chlorobenzene, in an associative mechanism [82a, Brown and Jutand, in preparation] to give the bis-phosphine complex ArPdX(PCy$_3$)$_2$ [82] (Scheme 1.53). Therefore, the structure of the aryl–palladium(II) complex, ArPdXL versus ArPdXL$_2$, is controlled by steric factors rather than by electronic factors.

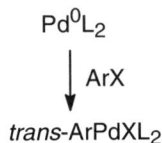

Pd^{0}L$_2$
↓ ArX
trans-ArPdXL$_2$

Scheme 1.53 *Associative mechanism for the oxidative addition (L = PCy$_3$, X = I, Br, Cl).*

1.7.2 Complexation/Insertion of the Alkene

The reactivity of alkenes with isolated ArPdXL (L = P-*t*-Bu$_3$) complexes has not been investigated. The coordination of such 14-electron complexes by the alkene should not be rate limiting. However, the phosphine and the halide sit in a *trans* position in isolated

ArPdXL complexes (Scheme 1.52) [80c]. If that structure is maintained in solution, then the coordinated alkene and the aryl group will be *trans* related. This does not favour the alkene insertion, which requires an isomerization prior to insertion (see 1.7.4).

1.7.3   Role of the Base in the Recycling of the Pd(0) Complex

Investigation of the hydridopalladium complex formed in the β-hydride elimination step by Hill and Fu [83] in 2004 provided mechanistic insights for the first time on its key involvement in the success of Mizoroki–Heck reactions performed from aryl chlorides. With Cs_2CO_3 as base, $HPdCl(P\text{-}t\text{-}Bu_3)_2$ was detected by ^{31}P NMR spectroscopy (first identification of a hydridopalladium in the course of a Mizoroki–Heck reaction); in contrast, only $Pd^0(P\text{-}t\text{-}Bu_3)_2$ was observed with Cy_2NMe as base, not the hydridopalladium complex. This explains why Cy_2NMe is more efficient than Cs_2CO_3 for Mizoroki–Heck reactions performed from aryl chlorides, because it is more effective for the regeneration of the Pd(0) catalyst from $HPdCl(P\text{-}t\text{-}Bu_3)_2$.

The kinetics of the reaction of *trans*-$HPdCl(P\text{-}t\text{-}Bu_3)_2$ with Cy_2NMe leading to $Pd^0(P\text{-}t\text{-}Bu_3)_2$ was investigated by Hill and Fu [83]. The overall reaction is made irreversible in the presence of a large excess of amine, as in catalytic reactions. The formation of the Pd(0) complex is inhibited by addition of $P\text{-}t\text{-}Bu_3$, which is consistent with an initial dissociation of one $P\text{-}t\text{-}Bu_3$ prior reductive elimination (Scheme 1.54) [83].

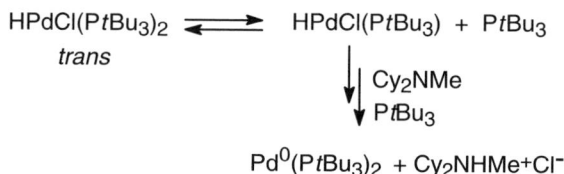

$$HPdCl(P\textit{t}Bu_3)_2 \rightleftharpoons HPdCl(P\textit{t}Bu_3) + P\textit{t}Bu_3$$
$$\textit{trans}$$

$$\downarrow \text{Cy}_2\text{NMe}$$
$$\downarrow \text{P}\textit{t}\text{Bu}_3$$

$$Pd^0(P\textit{t}Bu_3)_2 + Cy_2NHMe^+Cl^-$$

Scheme 1.54

Crystallization of $HPdCl(PR_3)_2$ (R = *t*-Bu or Cy) reveals that the P–Pd–P angle is $161°$ in the bent $HPdCl(P\text{-}t\text{-}Bu_3)_2$ but $180°$ in $HPdCl(PCy_3)_2$, which is thus less prone to reductive elimination [83]. This explains why, when using the same base, $P\text{-}t\text{-}Bu_3$ is more efficient than PCy_3 for Mizoroki–Heck reactions performed from aryl chlorides [77].

1.7.4   Catalytic Cycle

The efficiency of bulky and electron-rich phosphines in Mizoroki–Heck reactions seems to be due to their ability to generate monophosphine–Pd(0) or –Pd(II) complexes in each step of the catalytic cycle (Scheme 1.55). Steric factors are probably more important than electronic factors. One sees from Fu's studies that the last step of the catalytic cycle in which the Pd(0) complex is regenerated in the presence of a base may be rate determining. The role of this last step has been underestimated for a long time. Provided this step is favoured (e.g. with $P\text{-}t\text{-}Bu_3$ as ligand and Cy_2NMe as base), the oxidative addition of aryl chlorides would appear to be rate determining. However, Mizoroki–Heck reactions performed from the same aryl chloride with the same Pd(0) catalyst and same base but

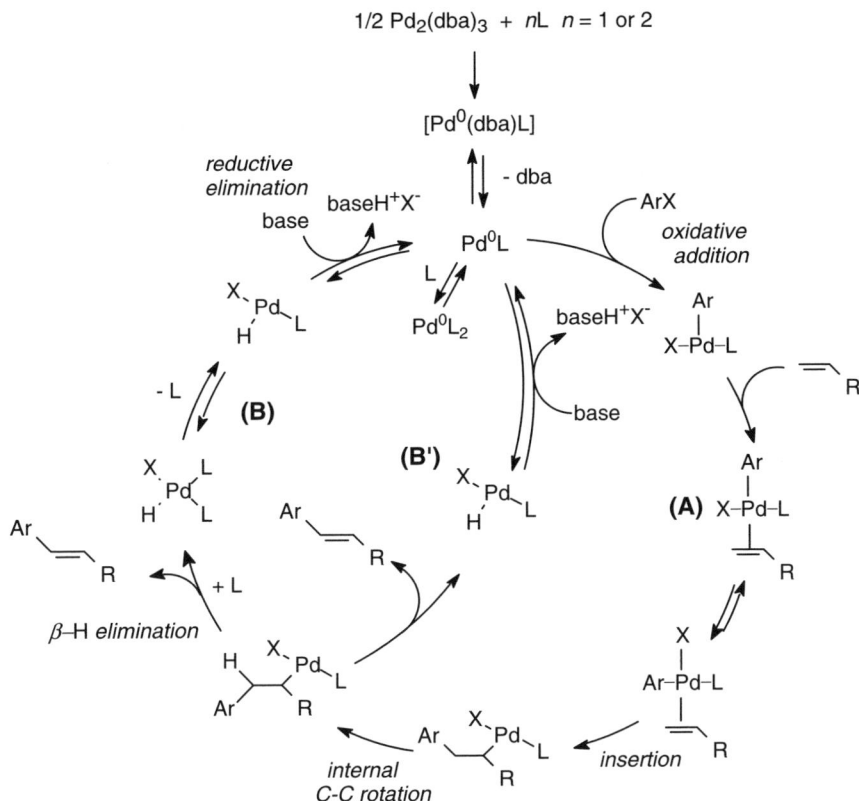

Scheme 1.55 *Mechanism of Mizoroki–Heck reactions performed from ArCl when L = P-t-Bu$_3$ (only one orientation of the alkene is presented). (A) + (B): catalytic cycle when n = 2; (A) + (B') catalytic cycle when n = 1.*

with different alkenes (e.g. styrene versus methyl acrylate) require different reaction times and temperatures incompatible with a rate-determining oxidative addition [77a].

1.8 Conclusion

The Mizoroki–Heck reaction is a subtle and complex reaction which involves a great variety of intermediate palladium complexes. The four main steps proposed by Heck (oxidative addition, alkene insertion, β-hydride elimination and reductive elimination) have been confirmed. However, they involved a considerable number of different Pd(0) and Pd(II) intermediates whose structure and reactivity depend on the experimental conditions, namely the catalytic precursor (Pd(0) complexes, Pd(OAc)$_2$, palladacycles), the ligand (mono- or bis-phosphines, carbenes, bulky monophosphines), the additives (halides, acetates), the aryl derivatives (ArX, ArOTf), the alkenes (electron-rich versus electron-deficient ones), which may also be ligands for Pd(0) complexes, and at least the base, which can play a multiple role as well. The efficiency and regioselectivity of Mizoroki–Heck reactions will only be optimal after finely balancing those parameters.

Depending on the experimental conditions, the catalytic cycles may involve: (i) Pd nanoparticles, anionic tri-ligated $Pd^0L_2(OAc)^-$ (L = monophosphine or L_2 = bidentate bisphosphine complexes), neutral Pd^0L_2 (L = carbene, monophosphine) or monoligated Pd^0L (L = bulky phosphine, bulky carbene); (ii) neutral $ArPd^{II}XL_2$ (X = halide or acetate) or cationic $ArPd^{II}SL_2^+$ complexes (L = monophosphine, monocarbene or L_2 = bidentate bisphosphine), T-shaped complex ArPdXL (L = bulky phosphines); (iii) $HPdClL_2$ or HPdClL (L = bulky monophosphine) complexes, and so on.

Kinetic data, however, are still missing for most steps which follow the oxidative addition for the precursors or ligands recently introduced in Mizoroki–Heck reactions (palladacycles, bulky phosphines and carbenes).

References

1. For reviews on the Mizoroki–Heck reactions, see: (a) Heck, R.F. (1978) New applications of palladium in organic synthesis. *Pure Appl. Chem.*, **50**, 691–701; (b) Heck, R.F. (1979) Palladium-catalyzed reactions of organic halides with olefins. *Acc. Chem. Res.*, **12**, 146–51; (c) Heck, R.F. (1982) Palladium-catalyzed vinylations of organic halides. *Org. React.*, **27**, 345–90; (d) Trost, B.M. and Verhoeven, T.R. (1982) Organopalladium compounds in organic synthesis and catalysis, in *Comprehensive Organometallic Chemistry*, Vol. **8** (eds G. Wilkinson, F.G.A. Stone and E.W. Abel), Pergamon Press, Oxford, pp. 854–83; (e) Daves, G.D. Jr and Hallberg, A. (1989) 1,2-Additions to hetero-substituted olefins by organopalladium reagents. *Chem. Rev.*, **89**, 1433–45; (f) de Meijere, A. and Meyer, F.E. (1994) Fine feathers make fine birds: the Heck reaction in modern garb. *Angew. Chem., Int. Ed. Engl.*, **33**, 2379–411; (g) Cabri, W. and Candiani, I. (1995) Recent developments and new perspectives in the Heck reaction. *Acc. Chem. Res.*, **28**, 2–7; (h) Jeffery, T. (1996) On the efficiency of tetraalkylammonium salts in Heck type reactions. *Tetrahedron*, **52**, 10113–30; (i) Crisp, G.T. (1998) Variations on a theme – recent developments on the mechanism of the Heck reaction and their implications for synthesis. *Chem. Soc. Rev.*, **27**, 427–36; (j) Herrmann, W.A., Böhm, V.P.W. and Reisinger, C.-P. (1999) Application of palladacycles in Heck type reactions. *J. Organomet. Chem.*, **576**, 23–41; (k) Amatore, C. and Jutand, A. (1999) Mechanistic and kinetic studies of palladium catalytic systems. *J. Organomet. Chem.*, **576**, 254–78; (l) Beletskaya, I.P. and Cheprakov, A.V. (2000) The Heck reaction as a sharpening stone of palladium catalysis. *Chem. Rev.*, **100**, 3009–66; (m) Amatore, C. and Jutand, A. (2000) Anionic Pd(0) and Pd(II) intermediates in palladium-catalyzed Heck and cross-coupling reactions. *Acc. Chem. Res.*, **33**, 314–21; (n) Bräse, S. and de Meijere, A. (2002) Palladium-catalyzed reactions involving carbopalladation, in *Handbook of Organopalladium Chemistry for Organic Synthesis*, Vol. **1** (ed. E.-i. Negishi), John Wiley & Sons, Inc., New York, pp. 1123–32; (o) Larhed, M. and Hallberg, A. (2002) The Heck reaction (alkene substitution via carbopalladation–dehydropalladation) and related carbopalladation reactions, in *Handbook of Organopalladium Chemistry for Organic Synthesis*, Vol. **1** (ed. E.-i. Negishi), John Wiley & Sons, Inc., New York, pp. 1133–78; (p) Herrmann, W.A. (2002) N-Heterocyclic carbenes as ligands for metal complexes – challenging phosphane ligands in homogeneous catalysis. *Angew. Chem. Int. Ed.*, **41**, 1290–309; (q) Hillier, A.C., Grasa, G.A., Viciu, M.S. *et al.* (2002) Catalytic cross-coupling reactions mediated by palladium/nucleophilic carbene systems. *J. Organomet. Chem.*, **653**, 69–82; (r) Littke, A.F. and Fu, G.C. (2002) Palladium-catalyzed coupling reactions of aryl chlorides. *Angew. Chem. Int. Ed.*, **41**, 4176–211; (s) Herrmann, W.A., Ölefe, K., Preysing, D.V. and Schneider, S.K. (2003)

Phospha-palladacycles and *N*-heterolytic carbene palladium complexes: efficient catalysts for CC-coupling reactions. *J. Organomet. Chem.*, **687**, 229–48; (t) Bedford, R.B. (2003) Palladacyclic catalysts in C—C and C–heteroatom bond-forming reactions. *Chem. Commun.*, 1787–96; (u) Beletskaya, I.P. and Cheprakov, A.V. (2004) Palladacycles in catalysis – a critical survey. *J. Organomet. Chem.*, **689**, 4055–82; (v) Farina, V. (2004) High-turnover palladium catalysts in cross-coupling and Heck chemistry: a critical overview. *Adv. Synth. Catal.*, **346**, 1553–82; (w) Cavell, K.J. and McGuinness, D.S. (2004) Redox process involving hydrocarbylmetal (*N*-heterocyclic carbene) complexes and associated imidazolium salts: ramifications for catalysis. *Coord. Chem. Rev.*, **248**, 671–81; (x) Alonso, F., Beletskaya, I.P. and Yus, M. (2005) Non-conventional methodologies for transition metal catalysed carbon–carbon coupling: a critical overview. Part 1: the Heck reaction. *Tetrahedron*, **61**, 11771–835; (y) Zapf, A. and Beller, M. (2005) The development of efficient catalysts for palladium-catalyzed coupling reactions of aryl halides. *Chem. Commun.*, 431–40; (z) Knowles, J.P. and Whiting, A. (2007) The Heck–Mizoroki cross-coupling reaction: a mechanistic perspective. *Org. Biomol. Chem.*, **5**, 31–44.

2. (a) Hayashi, T., Kubo, A. and Ozawa, F. (1992) Catalytic asymmetric arylation of olefins. *Pure Appl. Chem.*, **64**, 421–7; (b) Shibasaki, M., Boden, C.D.J. and Kojima, A. (1997) The asymmetric Heck reaction. *Tetrahedron*, **53**, 7371–95; (c) Loiseleur, O., Hayashi, M., Schmees, N. and Pfaltz, A. (1997) Enantioselective Heck reactions catalyzed by chiral phosphinooxazolidine–palladium complexes. *Synthesis*, 1338–45; (d) Shibasaki, M. and Vogl, E.M. (1999) The palladium-catalyzed arylation and vinylation of alkenes – enantioselective fashion. *J. Organomet. Chem.*, **576**, 1–15; (e) Loiseleur, O., Hayashi, M., Keenan, M. *et al.* (1999) Enantioselective Heck reactions using chiral P,N-ligands. *J. Organomet. Chem.*, **576**, 16–22; (f) Link, J.T. (2002) The intramolecular Heck reaction. *Org. React.*, **60**, 157–534; (g) Oestreich, M. (2005) Neighbouring-group effects in Heck reactions. *Eur. J. Org. Chem.*, 783–92.

3. (a) Heck, R.F. (1968) Arylation, methylation, and carboxyalkylation of olefins by Group VIII metal derivatives. *J. Am. Chem. Soc.*, **90**, 5518–26; (b) Heck, R.F. (1969) The mechanism of arylation and carbomethoxylation of olefins with organopalladium compouds. *J. Am. Chem. Soc.*, **91**, 6707–14; (c) Heck, R.F. (1971) Electronic and steric effects in the olefin arylation and carboalkoxylation reactions with organopalladium compounds. *J. Am. Chem. Soc.*, **93**, 6896–901.

4. Mizoroki, T., Mori, K. and Ozaki, A. (1971) Arylation of olefins with aryl iodide catalyzed by palladium. *Bull. Soc. Chem. Jpn.*, **44**, 581.

5. Heck, R.F. and Nolley, J.P. Jr (1972) Palladium-catalyzed vinylic hydrogen substitution reactions with aryl, benzyl, and styryl halides. *J. Org. Chem.*, **37**, 2320–2.

6. (a) Fitton, P., Johnson, M.P. and McKeon, J.E. (1968) Oxidative addition to palladium(0). *J. Chem. Soc., Chem. Commun.*, 6–7; (b) Coulson, D.R. (1968) Ready cleavage of triphenylphosphine. *J. Chem. Soc., Chem. Commun.*, 1530–1; (c) Fitton, P. and Rick, E.A. (1971) The addition of aryl halides to tetrakis(triphenylphosphine)palladium(0). *J. Organomet. Chem.*, **28**, 287–8.

7. Mori, K., Mizoroki, T. and Ozaki, A. (1973) Arylation of olefin with iodobenzene catalyzed by palladium. *Bull. Soc. Chem. Jpn.*, **46**, 1505–8.

8. Dieck, H.A. and Heck, R.F. (1974) Organophosphinepalladium complexes as catalysts for vinylic hydrogen substitution reactions. *J. Am. Chem. Soc.*, **96**, 1133–6.

9. Ziegler, C.B. and Heck, R.F. (1978) Palladium-catalyzed vinylic substitution with highly activated aryl halides. *J. Org. Chem.*, **43**, 2941–6.

10. (a) Spencer, A. (1983) A highly efficient version of the palladium-catalysed arylation of alkenes with aryl bromides. *J. Organomet. Chem.*, **258**, 101–8; (b) Spencer, A. (1984) Homogeneous palladium-catalysed arylation of activated alkenes with aryl chlorides. *J. Organomet. Chem.*, **270**, 115–20.

11. (a) Tsuji, J. (1980) *Organic Synthesis with Palladium Compounds*, Springer-Verlag, Berlin, pp. 16–7; (b) Tsuji, J. (1995) *Palladium Reagents in Organic Synthesis*, John Wiley & Sons, Ltd, Chichester.

12. Kitching, W., Rappoport, Z., Winstein, S. and Young, W.G. (1966) Allylic oxidation of olefins by palladium acetate. *J. Am. Chem. Soc.*, **88**, 2054–5.

13. Collman, J.P., Hegedus, L.S., Norton, J.R. and Finke, R.G. (1987) *Principles and Applications of Organotransition Metal Chemistry*, University Science Books, Mill Valley, CA, p. 725.

14. (a) Reetz, M.T. and Westermann, E. (2000) Phosphane-free palladium-catalyzed coupling reactions: the decisive role of Pd nanoparticles. *Angew. Chem. Int. Ed.*, **39**, 165–8; (b) Reetz, M.T., Lohmer, G. and Schwickardi, R. (1998) A new catalyst for the Heck reaction of unreactive aryl halides. *Angew. Chem. Int. Ed.*, **37**, 481–3.

15. (a) De Vries, A.H.M., Mulders, J.M.C.A., Mommers, J.H.M. *et al.* (2003) Homeopathic ligand-free palladium as a catalyst in the Heck reaction. A comparison with a palladacycle. *Org. Lett.*, **5**, 3285–8; (b) De Vries, A.H.M., Parlevliet, F.J., Schmieder-van de Vonderwoort, L. *et al.* (2002) A practical recycle of a ligand-free palladium catalyst for Heck reactions. *Adv. Synth. Catal.*, **344**, 996–1002; (c) Evans, J., O'Neill, L., Kambhampati, V.L. *et al.* (2002) Structural characterisation of solution species implicated in the palladium-catalysed Heck reaction by Pd K-edge X-ray absorption spectroscopy: palladium acetate as a catalytic precursor. *J. Chem. Soc., Dalton Trans.*, 2207–12.

16. (a) Schmidt, A.F., Al Halaiqa, A. and Smirnov, V.V. (2006) Interplays between reactions within and without catalytic cycle of the Heck reaction as a clue to the optimization of synthetic protocol. *Synthesis*, 2861–73; (b) Köhler, K., Kleist, W. and Pröckl, S.S. (2007) Genesis of coordinatively unsaturated palladium complexes dissolved from solid precursors during Heck coupling reactions and their role as catalytically active species. *Inorg. Chem.*, **46**, 1876–83.

17. Fauvarque, J.F., Pflüger, F. and Troupel, M. (1981) Kinetics of the oxidative addition of zerovalent palladium to aromatic iodides. *J. Organomet. Chem.*, **208**, 419–27.

18. Amatore, C., Jutand, A. and M'Barki, M.A. (1992) Evidence of the formation of zerovalent palladium from Pd(OAc)$_2$ and triphenylphosphine. *Organometallics*, **11**, 3009–13.

19. Ozawa, F., Kubo, A. and Hayashi, T. (1992) Generation of tertiary phosphine-coordinated Pd(0) species from Pd(OAc)$_2$ in the catalytic Heck reaction. *Chem. Lett.*, **21**, 2177–80.

20. Amatore, C., Carré, E., Jutand, A. *et al.* (1995) Rate and mechanism of the formation of zerovalent palladium complexes from mixtures of Pd(OAc)$_2$ and tertiary phosphines and their reactivity in oxidative additions. *Organometallics*, **14**, 1818–26.

21. Grushin, V.V. and Alper, H. (1993) Alkali-induced disproportionation of palladium(II) tertiary phosphine complexes [L$_2$PdCl$_2$], to LO and palladium(0). Key intermediates in the biphasic carbonylation of ArX catalyzed by [L$_2$PdCl$_2$]. *Organometallics*, **12**, 1890–901.

22. M'Barki, A.M. (1992) PhD thesis, ENS, University Paris VI, unpublished results.

23. Amatore, C., Jutand, A. and Medeiros, M.J. (1996) Formation of zerovalent palladium from the cationic Pd(PPh$_3$)$_2$(BF$_4$)$_2$ in the presence of PPh$_3$ and water in DMF. *New J. Chem.*, **20**, 1143–8.

24. Amatore, C., Jutand, A., Lemaître, F. *et al.* (2004) Formation of anionic palladium(0) complexes ligated by the trifluoroacetate ion and their reactivity in oxidative addition. *J. Organomet. Chem.*, **689**, 3728–34.

25. Amatore, C., Blart, E., Genêt, J.P. *et al.* (1995) New synthetic applications of water-soluble acetate Pd/TPPTS catalyst generated *in situ*. Evidence for a true Pd(0) species intermediate. *J. Org. Chem.*, **60**, 6829–39.

26. Gélin, E., Amengual, R., Michelet, V. *et al.* (2004) A novel water-soluble *m*-TPPTC ligand: steric and electronic features – recent developments in Pd- and Rh-catalyzed C—C bond formations. *Adv. Synth. Catal.*, **346**, 1733–41.

27. Herrmann, W.A., Brossmer, C., Öfele, K. *et al.* (1995) Palladacycles as structurally defined catalysts for the Heck olefination of chloro- and bromoarenes. *Angew. Chem., Int. Ed. Engl.*, **34**, 1844–8.

28. D'Orlyé, F. and Jutand, A. (2005) *In situ* formation of palladium(0) from a *P,C*-palladacycle. *Tetrahedron*, **61**, 9670–78.

29. Amatore, C., Carré, E., Jutand, A. *et al.* (1995) Evidence for the ligation of palladium(0) complexes by acetate ions: consequences on the mechanism of their oxidative addition with phenyl iodide and PhPd(OAc)(PPh$_3$)$_2$ as intermediate in the Heck reaction. *Organometallics*, **14**, 5605–14.

30. Kozuch, S., Shaik, S., Jutand, A. and Amatore, C. (2004) Active anionic zero-valent palladium catalysts: characterization by density functional calculations. *Chem. Eur. J.*, **10**, 3072–80.

31. For DFT calculations on a related anionic Pd(0) complex, Pd^0(PMe$_3$)$_2$(OAc)$^-$, see: (a) Goossen, L.J., Koley, D., Hermann, H.L. and Thiel, W. (2004) The mechanism of the oxidative addition of aryl halides to Pd-catalysts: a DFT investigation. *Chem. Commun.*, 2141–3; (b) Goossen, L.J., Koley, D., Hermann, H.L. and Thiel, W. (2006) Palladium monophosphine intermediates in catalytic cross-coupling reactions: a DFT study. *Organometallics*, **25**, 54–67.

32. Amatore, C., Carré, E. and Jutand, A. (1998) Evidence for an equilibrium between neutral and cationic arylpalladium(II) complexes in DMF. Mechanism of the reduction of cationic arylpalladium(II) complexes. *Acta Chem. Scand.*, **52**, 100–6.

33. From DFT calculations, the oxidative addition of PhI to the anionic Pd^0(PMe$_3$)$_2$(OAc)$^-$ gives PhPd(OAc)(PMe$_3$)$_2$ as the final compound via anionic [(η^2-PhI)Pd^0(OAc)(PMe$_3$)$_2$]$^-$, where the Pd(0) centre is ligated by one C=C bond of PhI; see: Goossen, L.J., Koley, D., Hermann, H.L. and Thiel, W. (2005) Mechanistic pathways for oxidative addition of aryl halides to palladium(0) complexes: a DFT study. *Organometallics*, **24**, 2398–410.

34. (a) Amatore, C., Carré, E., Jutand, A. and Medjour, Y. (2002) Decelerating effect of alkenes in the oxidative addition of phenyl iodide to palladium(0) complexes in Heck reactions. *Organometallics*, **21**, 4540–5; (b) Jutand, A. (2004) Dual role of nucleophiles in palladium-catalyzed Heck, Stille and Sonogashira reactions. *Pure Appl. Chem.*, **76**, 565–76.

35. Jutand, A. and Mosleh, A. (1995) Rate and mechanism of oxidative addition of aryl triflates to zerovalent palladium complexes. Evidence for the formation of cationic (σ-aryl)palladium complexes. *Organometallics*, **14**, 1810–7.

36. Jutand, A. (2003) The use of conductivity measurements for the characterization of cationic palladium(II) complexes and for the determination of kinetic and thermodynamic data in palladium-catalyzed reactions. *Eur. J. Inorg. Chem.*, 2017–40.

37. Jutand, A. and Mosleh, A. (1997) Nickel- and palladium-catalyzed homocoupling of aryl triflates. Scope, limitation and mechanistic aspects. *J. Org. Chem.*, **62**, 261–74.

38. The higher reactivity of alkenes with cationic *trans*-PhPdS(PMe$_3$)$_2$$^+$ than with neutral *trans*-PhPdBr(PMe$_3$)$_2$ has been reported: Kawataka, F., Shimizu, I. and Yamamoto, A. (1995) Synthesis and properties of cationic organopalladium complexes. remarkable rate enhancement in olefin insertion into the palladium-aryl bond by the generation of a cationic palladium complex from *trans*-[PdBr(Ph)(PMe$_3$)$_2$]. *Bull. Soc. Chem. Jpn.*, **68**, 654–60.

39. Amatore, C., Jutand, A., Meyer, G. *et al.* (2000) Reversible formation of a cationic palladium(II) hydride [HPd(PPh$_3$)$_2$]$^+$ in the oxidative addition of acetic acid or formic acid to palladium(0) in DMF. *Eur. J. Inorg. Chem.*, 1855–9.

40. For Heck reactions catalysed by Pd(OAc)$_2$ associated with dppp, see Ref. [1g] and: (a) Cabri, W., Candiani, I. and Bedeshi, A. (1990) Ligand-controlled α-regioselectivity in palladium-catalyzed arylation of butyl vinyl ether. *J. Org. Chem.*, **55**, 3654–5; (b) Cabri, W., Candiani, I., De-Bernardinis, S. *et al.* (1991) Heck reaction on anthraquinone derivatives: ligand, solvent, and salts effects. *J. Org. Chem.*, **56**, 5796–800; (c) Cabri, W., Candiani, I., Bedeshi, A. and Santi, R. (1991) Palladium-catalyzed α-arylation of vinyl butyl ether with aryl halides. *Tetrahedron Lett.*, **32**, 1753–6; (d) Cabri, W., Candiani, I., Bedeshi, A. *et al.* (1992) α-Regioselectivity in palladium-catalyzed arylation of acyclic enol ethers. *J. Org. Chem.*, **57**, 1481–6; (e) Cabri, W.,

Candiani, I., Bedeshi, A. and Santi, R. (1992) Palladium-catalyzed arylation of unsymmetrical olefins. Bidentate phosphine ligand controlled regioselectivity. *J. Org. Chem.*, **57**, 3558–63; (f) Larhed, M. and Hallberg, A. (1996) Microwave-promoted palladium-catalyzed coupling reactions. *J. Org. Chem.*, **61**, 9582–4; (g) Larhed, M. and Hallberg, A. (1997) Direct synthesis of cyclic ketals of acetophenones by palladium-catalyzed arylation of hydroxyalkyl vinyl ethers. *J. Org. Chem.*, **62**, 7858–62; (h) Vallin, K.S.A., Larhed, M., Johansson, K. and Hallberg, A. (2000) Highly selective palladium-catalyzed synthesis of protected α,β-unsaturated methyl ketones and 2-alkoxy-1,3-butadienes. High-speed chemistry by microwave flash heating. *J. Org. Chem.*, **65**, 4537–42; (i) Qadir, M., Möchel, T. and Hii, K.K. (2000) Examination of ligand effects in the Heck arylation reaction. *Tetrahedron*, **56**, 7975–9; (j) Vallin, K.S.A., Larhed, M. and Hallberg, A. (2001) Aqueous DMF–potassium carbonate as a substitute for thallium and silver additives in the palladium-catalyzed conversion of aryl bromides to acetyl arenes. *J. Org. Chem.*, **66**, 4340–3; (k) Xu, L.J., Chen, W.P., Ross, J. and Xiao, J.L. (2001) Palladium-catalyzed regioselective arylation of an electron-rich olefin by aryl halides in ionic liquids. *Org. Lett.*, **3**, 295–7; (l) Vallin, K.S.A., Zhang, Q.S., Larhed, M. *et al.* (2003) A new regioselective Heck vinylation with enamides. Synthesis and investigation of fluorous-tagged bidentate ligands for fast separation. *J. Org. Chem.*, **68**, 6639–45; (m) Mo, J., Xu, L. and Xiao, J. (2005) Ionic liquid-promoted, highly regioselective Heck arylation of electron-rich olefins by aryl halides. *J. Am. Chem. Soc.*, **127**, 751–60; (n) Mo, J. and Xiao, J. (2006) The Heck reaction of electron-rich olefins with regiocontrol by hydrogen-bond donors. *Angew. Chem. Int. Ed.*, **45**, 4152–7.

41. (a) For seminal works on intramolecular Heck reactions catalyzed by Pd(OAc)$_2$ associated with (*R*)-Binap, see: Sato, Y., Sodeoka, M. and Shibasaki, M. (1989) Catalytic asymmetric C—C bond formation: asymmetric synthesis of *cis*-decalin derivatives by palladium-catalyzed cyclization of prochiral alkenyl iodides. *J. Org. Chem.*, **54**, 4738–9; (b) Overman, L.E. and Poon, D.J. (1997) Asymmetric Heck reactions via neutral intermediates: enhanced enantioselectivity with halide additives gives mechanistic insights. *Angew. Chem., Int. Ed. Engl.*, **36**, 518–21; (c) For seminal works on intermolecular Heck reactions catalyzed by Pd(OAc)$_2$ associated with (*R*)-Binap, see: Ozawa, F., Kubo, A. and Hayashi, T. (1991) Catalytic asymmetric arylation of 2,3-dihydrofuran with aryl triflates. *J. Am. Chem. Soc.*, **113**, 1417–9; (d) Ozawa, F., Kubo, A., Matsumoto, Y. and Hayashi, T. (1993) Palladium-catalyzed asymmetric arylation of 2,3-dihydrofuran with phenyl triflate. A novel asymmetric catalysis involving a kinetic resolution process. *Organometallics*, **12**, 4188–96.

42. Amatore, C., Jutand, A. and Thuilliez, A. (2001) Formation of palladium(0) complexes from Pd(OAc)$_2$ and a bidentate phosphine ligand (dppp) and their reactivity in oxidative addition. *Organometallics*, **20**, 3241–9.

43. Amatore, C., Godin, B., Jutand, A. and Lemaitre, F. (2007) Rate and mechanism of the reaction of alkenes with aryl palladium complexes ligated by a bidentate *P,P* ligand in Heck reactions. *Chem. Eur. J.*, **13**, 2002–11.

44. (a) Karabelas, K., Westerlund, C. and Hallberg, A. (1985) The effect of added silver nitrate on the palladium-catalyzed arylation of allyltrimethylsilanes. *J. Org. Chem.*, **50**, 3896–900; (b) Karabelas, K. and Hallberg, A. (1986) Synthesis of (*E*)-(2-arylethenyl)silanes by palladium-catalyzed arylation of vinylsilanes in the presence of silver nitrate. *J. Org. Chem.*, **51**, 5286–90; (c) Grigg, R., Loganathan, V., Santhakumar, V. and Teasdale, A. (1991) Suppression of alkene isomerisation in products from intramolecular Heck reactions by addition of Tl(I) salts. *Tetrahedron Lett.*, **32**, 687–90.

45. Portnoy, M., Ben-David, Y., Rousso, I. and Milstein, D. (1994) Reactions of electron-rich arylpalladium complexes with olefins. Origin of the chelate effect in vinylation catalysis. *Organometallics*, **13**, 3465–79.

46. Brown, J.M. and Hii, K.K. (1996) Characterization of reactive intermediates in palladium-catalyzed arylation of methyl acrylate (Heck reaction). *Angew. Chem., Int. Ed. Engl.*, **35**, 657–9.

47. Ludwig, M., Strömberg, S., Svensson, M. and Åkermark, B. (1999) An exploratory study of regiocontrol in the Heck type reaction. Influence of solvent polarity and bisphosphine ligands. *Organometallics*, **18**, 970–75.

48. Jutand, A., Hii, K.K., Thornton-Pett, M. and Brown, J.M. (1999) Factors affecting the oxidative addition of aryl electrophiles to 1,1′-bis(diphenylphosphino)ferrocenepalladium(η^2-methylacrylate), an isolable Pd[0] alkene complex. *Organometallics*, **18**, 5367–74.

49. For DFT calculations on ionic versus neutral mechanism, see: (a) Von Schenck, H., Åkermark, B. and Svensson, M. (2003) Electronic control in the regiochemistry in the Heck reaction. *J. Am. Chem. Soc.*, **125**, 3503–8; (b) Deeth, R.J., Smith, A. and Brown, J.M. (2004) Electronic control in the regiochemistry in palladium-phosphine catalyzed intermolecular Heck reactions. *J. Am. Chem. Soc.*, **126**, 7144–51.

50. Hii, K.K., Claridge, T.D.W. and Brown, J.M. (1997) Intermediates in the intermolecular asymmetric Heck arylation of dihydrofurans. *Angew. Chem., Int. Ed. Engl.*, **36**, 984–7.

51. Hii, K.K., Claridge, T.D.W., Giernoth, R. and Brown, J.M. (2004) Conformationally restricted arene intermediates in the intermolecular asymmetric arylation of vinylarenes. *Adv. Synth. Catal.*, **346**, 983–8.

52. Deeth, R.J., Smith, A., Hii, K.K. and Brown, J.M. (1998) The Heck olefination reaction; a DFT study of the elimination pathway. *Tetrahedron Lett.*, **39**, 3229–32.

53. Amatore, C., Godin, B., Jutand, A. and Lemaître, F. (2007) Rate and mechanism of the heck reactions of arylpalladium complexes ligated by a bidentate *P,P* ligand with an electron-rich alkene (isobutyl vinyl ether). *Organometallics*, **26**, 1757–61.

54. For reversible alkene insertion, see: Catellani, M. and Fagnola, M.C. (1994) Palladacycles as intermediates for selective dialkylation of arenes and subsequent fragmentation. *Angew. Chem., Int. Ed. Engl.*, **33**, 2421–2.

55. For seminal works, see Ref. [27] and: (a) Herrmann, W.A., Brossmer, C., Reisinger, C.-P. *et al.* (1997) Palladacycles: efficient new catalysts for the Heck vinylation of aryl halides. *Chem. Eur. J.*, **3**, 1357–64; (b) Beller, M. and Riermeier, T.H. (1998) Phosphapalladacycle-catalyzed Heck reactions for efficient synthesis of trisubstituted olefins: evidence for palladium(0) intermediates. *Eur. J. Inorg. Chem.*, 29–35.

56. Louie, J. and Hartwig, J.F. (1996) A route to Pd⁰ from PdII metallacycles in amination and cross-coupling chemistry. *Angew. Chem., Int. Ed. Engl.*, **35**, 2359–61.

57. Shaw, B.L. (1998) Speculations on new mechanisms for Heck reactions. *New J. Chem.*, **22**, 77–9.

58. Sundermann, A., Uzan, O. and Martin, J.M.L. (2001) Computational study of a new Heck reaction mechanism catalyzed by palladium(II/IV) species. *Chem. Eur. J.*, **7**, 1703–11.

59. Böhm, V.P.W. and Herrmann, W.A. (2001) Mechanism of the Heck reaction using a phospha-palladacycle as the catalyst: classical versus palladium(IV) intermediates. *Chem. Eur. J.*, **7**, 4191–7.

60. The beneficial role of dba in the formation of the Pd(0) complex is not quite clear, but alkenes are known to favour reductive elimination. See: Giovannini, R. and Knochel, P. (1998) Ni(II)-catalyzed cross-coupling between polyfunctional arylzinc derivatives and primary alkyl iodides. *J. Am. Chem. Soc.*, **125**, 11186–7. The alkene in a Heck reaction may play this role and explain why **5** "does not react with PhBr until the olefin is added to the mixture", as observed by Herrmann *et al.* [55a].

61. A monomeric palladacycle is formed because the ligand **15** is less hindered than **11**. Dimeric palladacycles are indeed cleaved to monomeric palladacycles by less-hindered ligands as PPh₃, see: Cheney, A.J. and Shaw, B.L. (1972) Transition matal–Carbon bond. Part XXXL. Internal

metallations of palladium(II)-*t*-butyldi-*o*-tolylphosphine) and di-*t*-butyl-*o*-tolylphosphine complexes. *J. Chem. Soc., Dalton Trans.*, 860–5.

62. (a) Hartwig, J.F. and Paul, F. (1995) Oxidative addition of aryl bromide after dissociation of phosphine from a two-coordinate palladium(0) complex, bis(tri-*o*-tolylphosphine) palladium(0). *J. Am. Chem. Soc.*, **117**, 5373–4; (b) Paul, F., Patt, J. and Hartwig, J.F. (1995) Structural characterization and simple synthesis of {Pd[P(*o*-Tol)₃]₂}, dimeric palladium(II) complexes obtained by oxidative addition of aryl bromides, and corresponding monometallic amine complexes. *Organometallics*, **14**, 3030–9; (c) Paul, F., Patt, J. and Hartwig, J.F. (1994) Palladium-catalyzed formation of carbon–nitrogen bond. Reaction intermediates and catalyst improvements in the hetero cross-coupling of aryl halides and tin amides. *J. Am. Chem. Soc.*, **116**, 5969–70.

63. From kinetic data obtained by calorimetry in a Heck reaction performed from a dimeric *N,C*-palladacycle, Blackmond and coworkers propose the formation of a dimeric arylpalladium(II) [ArPd(μ-X)L]₂ in oxidative addition of an undefined [Pd⁰Lₙ] complex, generated *in situ* from the *N,C*-palladacycle. Such a dimeric complex is supposed to be in equilibrium with the reactive ArPdXL₂; see: Rosner, T., Le Bars, J., Pfaltz, A. and Blackmond, D.G. (2001) Kinetic studies of Heck coupling reactions using palladacycle catalysts and kinetic modeling of the role of dimer species. *J. Am. Chem. Soc.*, **123**, 1848–55.

64. Herrmann, W.A., Elison, M., Fischer, J. *et al.* (1995) Metal complexes of *N*-heterocyclic carbenes. A new structural principle for catalysts in homogeneous catalysis. *Angew. Chem., Int. Ed. Engl.*, **34**, 2371–4.

65. For properties of NHCs and ligated NHCs, see: (a) Herrmann, W.A. and Köcher, C. (1997) *N*-Heterocyclic carbenes. *Angew. Chem., Int. Ed. Engl.*, **36**, 2162–87; (b) Bourissou, D., Guerret, O., Gabbaï, F.P. and Bertrand, G. (2000) Stable carbenes. *Chem. Rev.*, **100**, 39–91; (c) Lee, M.-T. and Hu, C.-H. (2004) Density functional study of *N*-heterocyclic and diamino carbene complexes: comparison with phosphines. *Organometallics*, **23**, 976–83; (d) Dorta, R., Stevens, E.D., Scott, N.M. *et al.* (2005) Steric and electronic properties of *N*-heterocyclic carbenes (NHC): a detailed study on their interaction with Ni(CO)₄. *J. Am. Chem. Soc.*, **127**, 2485–95; (e) Cavallo, L., Correa, A., Costabile, C. and Jacobsen, H. (2005) Steric and electronic effects in the bonding of *N*-heterocyclic ligands to transition metals. *J. Organomet. Chem.*, **690**, 5407–13; (f) Scott, N.M. and Nolan, S.P. (2005) Stabilization of organometallic species achieved by the use of *N*-heterocyclic carbene (NHC) ligands. *Eur. J. Inorg. Chem.*, 1815–28; (g) Diez-Gonzalez, S. and Nolan, S.P. (2007) Stereoelectronic parameters associated with *N*-heterocyclic carbene (NHC) ligands: a quest for understanding. *Coord. Chem. Rev.*, **251**, 874–83.

66. (a) Herrmann, W.A., Elison, M., Fischer, J. *et al.* (1996) *N*-Heterocyclic carbenes: generation under mild conditions and formation of Group 8–10 transition metal complexes relevant to catalysis. *Chem. Eur. J.*, **2**, 772–80; (b) Andrus, M.B., Song, C. and Zhang, J. (2002) Palladium–imidazolium carbene catalyzed Mizoroki–Heck coupling with aryl diazonium ions. *Org. Lett.*, **4**, 2079–3082; (c) Huynh, H.V., Ho, J.H.H., Neo, T.C. and Koh, L.L. (2005) Solvent-controlled selective synthesis of a *trans*-configured benzimidazoline-2-ylidene palladium(II) complex and investigation of its Heck-type catalytic activity. *J. Organomet. Chem.*, **690**, 3854–60; (d) Xu, L., Chen, W. and Xiao, J. (2000) Heck reaction in ionic liquids and the *in situ* identification of *N*-heterocyclic carbene complexes of palladium. *Organometallics*, **19**, 1123–7; (e) Jin, C.-M., Twanley, B. and Shreeve, J.M. (2005) Low-melting dialkyl- and bis(polyfluoroalkyl)-substituted 1,1′-methylenebis(imidazolium) and 1,1′-methylenebis (1,2,4-triazolium) bis(trifluoromethanesulfonyl)amides: ionic liquids leading to bis(*N*-heterocyclic carbene) complexes of palladium. *Organometallics*, **24**, 3020–3.

67. For the synthesis of Pd⁰(carbene)₂ with C≡C unsaturated carbenes, see: (a) Arnold, P.L., Cloke, F.G.N., Geldbach, T. and Hitchcock, P.B. (1999) Metal vapor synthesis as a straightforward route to Group 10 homoleptic carbene complexes. *Organometallics*, **18**, 3228–33; (b) Titcomb, L.R., Caddick, S., Clocke, F.G.N. *et al.* (2001) Unexpected reactivity of two-coordinate

palladium–carbene complexes; synthetic and catalytic implications. *Chem. Commun.*, 1388–9; (c) For the synthesis of Pd⁰(carbene)₂ with a C—C saturated carbene, see Ref. [67b].

68. (a) Guinness, D.S., Cavell, K.J., Skelton, B.W. and White, A.H. (1999) Zerovalent palladium and nickel complexes of heterocyclic carbenes: oxidative addition of organic halides, carbon–carbon coupling processes, and the Heck reaction. *Organometallics*, **18**, 1596–605; (b) Selvakumar, K., Zapf, A. and Beller, M. (2002) New palladium carbene catalysts for the Heck reaction of aryl chlorides in ionic liquids. *Org. Lett.*, **4**, 3031–3.

69. Caddick, S., Cloke, F.G.N., Hitchcock, P.B. *et al.* (2002) The first example of simple oxidative addition of an aryl chloride to a discrete palladium *N*-heterocyclic carbene amination precatalyst. *Organometallics*, **21**, 4318–9.

70. Lewis, A.K.D., Caddick, S., Cloke, F.G.N. *et al.* (2003) Synthetic, structural, and mechanistic studies on the oxidative addition of aromatic chlorides to a palladium (*N*-heterocyclic carbene) complex: relevance to catalytic amination. *J. Am. Chem. Soc.*, **125**, 10066–73.

71. Pytkowicz, J., Roland, S., Mangeney, P. *et al.* (2003) Chiral diaminocarbene palladium(II) complexes: synthesis, reduction to Pd(0) and activity in the Mizoroki–Heck reaction as recyclable catalysts. *J. Organomet. Chem.*, **678**, 166–79.

72. Green, J.C., Herbert, B.J. and Lonsdale, R. (2005) Oxidative addition of aryl chlorides to palladium *N*-heterocyclic carbene complexes and their role in catalytic arylamination. *J. Organomet. Chem.*, **690**, 6054–67.

73. Roland, S., Mangeney, P. and Jutand, A. (2006) Reactivity of Pd(0)(NHC)₂ (NHC = *N*-heterocyclic carbene) in oxidative addition with aryl halides in Heck reactions. *Synlett*, 3088–94.

74. Albert, K., Gisdakis, P. and Rösch, N. (1998) On C—C coupling by carbene-stabilized palladium catalysts: a density functional study of the Heck reaction. *Organometallics*, **17**, 1608–16.

75. (a) Yang, C., Lee, H.M. and Nolan, S.P. (2001) Highly efficient Heck reactions of aryl bromides with *n*-butyl acrylate mediated by a palladium/phosphine-imidazolium salt system. *Org. Lett.*, **3**, 1511–4; (b) Wang, A.-E., Xie, J.-H., Wang, L.-X. and Zhou, Q.-L. (2005) Triaryl phosphine-functionalized *N*-heterocyclic carbene ligands for Heck reaction. *Tetrahedron*, **61**, 259–66.

76. Herrmann, W.A., Böhm, V.P.W., Gstöttmayr, C.W.K. *et al.* (2001) Synthesis, structure and catalytic application of palladium(II) complexes bearing *N*-heterocyclic carbenes and phosphines. *J. Organomet. Chem.*, **617–618**, 616–28.

77. (a) Littke, A.F. and Fu, G.C. (1999) Heck reactions in the presence of P(*t*-Bu)₃: expanded scope and milder reaction conditions for the coupling of aryl chlorides. *J. Org. Chem.*, **64**, 10–11; (b) Littke, A.F. and Fu, G.C. (2001) A versatile catalyst for Heck reactions of aryl chlorides and aryl bromides under mild conditions. *J. Am. Chem. Soc.*, **123**, 6989–7000.

78. Shaughnessy, K.H., Kim, P. and Hartwig, J.F. (1999) A fluorescence-based assay for high-throughput screening of coupling reactions. Applications to Heck chemistry. *J. Am. Chem. Soc.*, **121**, 2123–32.

79. Ehrentraut, A., Zapf, A. and Beller, M. (2000) New efficient palladium catalyst for Heck reactions of deactivated aryl chlorides. *Synlett*, 1589–92.

80. (a) Barrios-Landeros, F. and Hartwig, J.F. (2005) Distinct mechanisms for the oxidative addition of chloro-, bromo-, and iodoarenes to a bisphosphine palladium(0) complex with hindered ligands. *J. Am. Chem. Soc.*, **127**, 6944–5; (b) Stambuli, J.P., Bühl, M. and Hartwig, J.F. (2002) Synthesis, characterization, and reactivity of monomeric, aryl palladium complexes with a hindered phosphine as the only dative ligand. *J. Am. Chem. Soc.*, **124**, 9346–7; (c) Stambuli, J.P., Incarvito, C.D., Bühl, M. and Hartwig, J.F. (2004) Synthesis, structure, theoretical studies, and ligand exchange reactions of monomeric, T-shaped arylpalladium(II) halide complexes with an additional, weak agostic interaction. *J. Am. Chem. Soc.*, **126**, 1184–94; (d) Alcazar-Roman, L.M. and Hartwig, J.F. (2001) Mechanism of aryl chloride amination: base-induced oxidative addition. *J. Am. Chem. Soc.*, **123**, 12905–6.

81. (a) For the formation of major $Pd^0(dba)L_2$ complexes when L is not a bulky phosphine, see: Amatore, C. and Jutand, A. (1998) Role of dba in the reactivity of palladium(0) complexes generated *in situ* from mixtures of $Pd(dba)_2$ and phosphines. *Coord. Chem. Rev.*, **178–180**, 511–28; (b) For the formation of major $Pd^0(dba)L$ when L is a bulky ligand, see: Yin, J., Rainka, M.P., Zhang, X.X. and Buchwald, S.L. (2002) A highly active suzuki catalyst for the synthesis of sterically hindered biaryls: novel ligand coordination. *J. Am. Chem. Soc.*, **124**, 1162–3.

82. (a) Galardon, E., Ramdeehul, S., Brown, J.M. *et al.* (2002) Profound steric control of reactivity in aryl halide addition to bisphosphane palladium(0) complexes. *Angew. Chem. Int. Ed.*, **41**, 1760–3; (b) Huser, M., Youiniou, M.T. and Osborn, J.A. (1989) Chlorocarbon activation – catalytic carbonylation of dichloromethane and chlorobenzene. *Angew. Chem., Int. Ed. Engl.*, **28**, 1386–8.

83. Hills, I.D. and Fu, G.C. (2004) Elucidating reactivity differences in palladium-catalyzed coupling processes: the chemistry of palladium hydrides. *J. Am. Chem. Soc.*, **126**, 13178–9.

2

Focus on Catalyst Development and Ligand Design

Irina P. Beletskaya and Andrei V. Cheprakov

Department of Chemistry, M. V. Lomonosov Moscow State University, Leninskie Gory, Moscow, Russia

2.1 Introduction

The Mizoroki–Heck reaction was discovered more than 35 years ago [1, 2]; since then, it has been recognized as one of the most fundamental types of the organic processes catalysed by palladium complexes.

For any general transition metal-catalysed process, researchers seek to achieve protocols which:

  (i) possess a wide synthetic scope – that is, they have the ability to process all imaginable types of substrate;
 (ii) are, at the same time, highly selective;
(iii) rely on practically available catalysts; and
(iv) maximize the effectiveness of the catalyst, either by increasing the turnover rate (measured by turnover frequency (TOF) values) and/or catalyst longevity (as estimated by turnover numbers (TONs)) or, alternatively, by designing recyclable catalytic systems.

Even a very brief analysis of publications on Mizoroki–Heck reactions reveals immediately that the goals of researchers in this area fit well into this paradigm. However, it is rather apparent that these goals have not been achieved so far by any particular protocol, or by a compact group of protocols. Very impressive results on, say, catalytic efficiency are obtained for a very narrow subset of substrates (aryl iodides and reactive aryl bromides

The Mizoroki–Heck Reaction   Edited by Martin Oestreich
© 2009 John Wiley & Sons, Ltd

reacting with simple terminal alkenes, such as acrylates or styrene). Any step aside is punished by a loss of all effectiveness gained. Selectivity and broadening of scope to less reactive substrates are achieved by individual tailoring of custom-made ligands and precatalysts, and these parameters have reverse effects. More than any other catalytic process, the Mizoroki–Heck reaction deserves critical reassessment of the backgrounds of reasoning.

Here, we wish to show why the task of designing a common catalytic system for the Mizoroki–Heck reaction is doomed from the beginning, because there are several largely incompatible types of catalytic process unified under this general name. This chapter is a rather impudent effort to bring at least some order into the realm of mysterious cocktails [3] published in their scores each month as 'novel' catalytic systems for Mizoroki–Heck reactions. If there have already been so many, how could it be that problems still persist?

In many other important transition metal-catalysed organic reactions, the development of methods hinges upon development of the catalyst or precatalyst. This is, in turn, mainly dependent on the progress in the design of ligands. During recent decades, we have witnessed manifold progress in transition metal catalysis in general and palladium catalysis in particular, mainly due to the introduction of new generations of state-of-the-art ligands.

Although Mizoroki–Heck chemistry also followed the same ligand-driven development, the results may appear puzzling. In 2000 we speculated [4] that the Mizoroki–Heck reaction is the best model process (*a sharpening stone*) for the development of new catalytic systems, and that this development reciprocates in palladium catalysis as a whole. As is evident now, this metaphor should be read in a literal sense: The sharpening stone sharpens other instruments, but not itself! Palladium chemistry has indeed advanced dramatically since then. New effective catalysts have been described for almost all palladium-catalysed reactions except the Mizoroki–Heck reaction, which turns out to be stubbornly reluctant to keep pace with other classical palladium-catalysed chemistry.

First, we have to conclude that in Mizoroki–Heck reactions the object of design should not be confined to tuning ancillary ligands or precatalysts. The Mizoroki–Heck reaction differs from other palladium-catalysed transformation, even from closely related cross-coupling reactions. This might be rationalized by the coordination sphere of palladium, which cannot be controlled as precisely as in many other palladium-catalysed reactions. Mizoroki–Heck reactions are so strongly dependent on all factors (amount of precatalyst, nature of base, media, additives, reaction temperature and even such intimate details as the order of mixing) that we cannot reasonably just discuss the influence of the catalyst but should rather keep in mind the catalytic system as a whole. The latter should be regarded as the target of design.

2.2   General Considerations: Types of Catalytic System

No universal catalytic system for the Mizoroki–Heck reaction has so far been revealed, and it is hardly to be expected that any such system will be revealed. In principle, this is no different from other palladium-catalysed reactions in which specific subtasks are also more effectively performed not with a single universal catalyst but by a manifold of methods varying in ligand, solvent and ancillary reagents.

The search for new protocols has been evolving in the past decade at an increasing pace. What is the driving force for such interest and what has actually been revealed

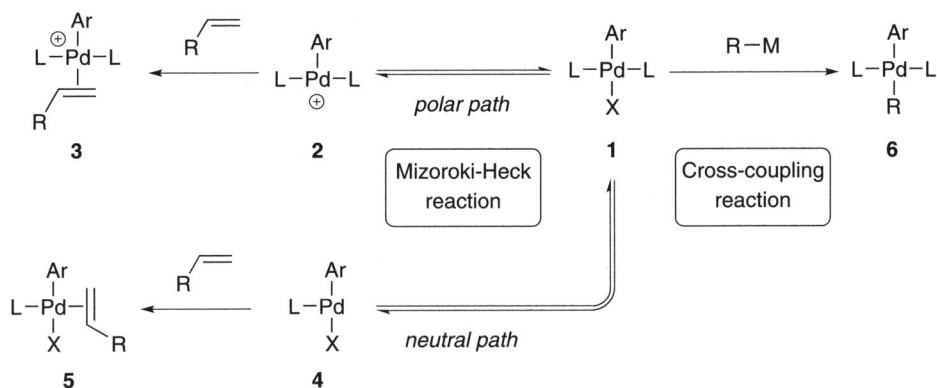

Scheme 2.1

along this path? This is not an easy task. Surprisingly, no attempts to classify the already published systems have been made; thus, the key tasks/challenges to be solved have not been identified yet. Still, in Mizoroki–Heck reactions, a unique situation has been created, because so many, often chaotic, efforts are the best basis for unbiased conclusions. Since hundreds of researchers independently developed catalytic systems involving a limited set of substrates, it is now quite clear which levels of catalytic activity correspond to each substrate class. This, in turn, allows one to discard ungrounded conclusions about the purported influence of various factors on catalytic activity. For example, in what concerns the effects of ancillary ligands, we see no use to consider *ligand effects* if a similar level of activity has been achieved independently in other studies lacking this particular ligand or a related one. The availability of a representative number of independent studies dealing with the same process imparts logical strength to such conclusions.

The Mizoroki–Heck reaction is often discussed and studied in parallel with palladium-catalysed cross-coupling reactions with explicit or implicit consideration that the regularities found for either one are relevant to the other. The history of both branches, however, shows that this basis of thinking is partially misleading. The difference between Mizoroki–Heck reactions and cross-couplings is profound and stems, most likely, from an important difference in the mechanisms (Scheme 2.1). Both processes are initiated by the same oxidative addition step and proceed through the common aryl palladium(II) intermediate **1**. The next steps, however, are different, particularly from the viewpoint of the coordination capabilities of the palladium(II) centre, which is well-known for bearing four ligands in the square planar configuration. In cross-coupling reactions, strong nucleophiles readily substitute a halide in the coordination sphere via a neutral pathway (**1** → **6**). In Mizoroki–Heck reactions, a weak neutral ligand, the alkene, cannot substitute a halide directly. It has to substitute either a neutral ancillary ligand through a neutral pathway (**1** → **5**) involving prior dissociation (**1** → **4**) or a halide indirectly via the cationic palladium(II) complex **2** (polar or cationic pathway, **1** → **3**). Thus, a cross-coupling allows for two monodentate ligands or one bidentate ligand to remain in the coordination shell. This opens up vast possibilities for the choice of such ligands to control precisely the behaviour of the catalytically active species, its reactivity and selectivity.

The task of controlling the catalytic cycle of the Mizoroki–Heck reaction turns out to be much more difficult. The neutral path, formally analogous to the aforementioned cross-coupling, tolerates only a single stable ligand in the coordination shell of palladium(II). This requirement almost excludes all bidentate ligands and puts strong limitations on the choice of monodentate ligands, which usually prefer to form diligated complexes. As far as the extant body of data shows, this situation is unambiguously realized only in a small number of ligand-accelerated systems.

The polar path is apparently more common for Mizoroki–Heck reactions. With bidentate ligands, it serves as the basis for various regio- and enantio-selective protocols. However, owing to some peculiarities in the coordination chemistry of palladium, it has little or no application in the version with two stable monodentate ligands. Still, this path dominates in common Mizoroki–Heck reactions, where all ancillary ligands are labile or hemilabile and the coordination shell of palladium is largely undefined.

It is important to stress that, in most other transition-metal-catalysed reactions, the control of catalytic processes is achieved through the choice, refinement and adjustment of stable ancillary ligands remaining in the coordination shell throughout all steps of the catalytic cycle. In those areas, the design of ligands is the essence of the art. The Mizoroki–Heck reaction, on the other hand, is very reluctant towards the control of catalytic activity via ancillary ligands.

In Mizoroki–Heck reactions, the developer cannot just rely on ligands. All components of the catalytic system strongly influence the nature of the catalytic species. The nature of the ligands in the coordination sphere of the catalytically active species often remains unknown, being speculative at the most. More often than not, Mizoroki–Heck reactions proceed through intermediates with coordination shells filled by undefined ancillaries. The art of the 'Mizoroki–Heck reaction designer' is thus the systematization and judicious choice of a multicomponent catalytic system involving substrates, reaction media, bases and additives. It should not be overlooked that the nature of the palladium complexes involved heavily depends on temperature.

Thus, the analysis of extant data has led us to the conclusion that as many as four distinct types of catalytic system can be roughly identified for common Mizoroki–Heck reactions. Their main features with respect to the scope of substrates, dependence on specific ancillary ligands, achievable catalytic efficiency and dependence on specific conditions are briefly summarized in Table 2.1.

This classification relies on the dependence on various parameters defining the operation of the individual catalytic system. *All* parameters are important. The only parameter which cannot be considered is reaction time, because representative information on the kinetics of various Mizoroki–Heck reactions is scarce. Normally, all Mizoroki–Heck reactions are performed with long exposures (12–48 h or more), and reaction time for particular substrates is almost never exactly established or optimized. A rough estimate of reaction kinetics is expressed through TONs and TOF values.

A very important aspect, which should be dealt with in the design of new – and the discussion of the known – protocols, is the *stability* or, more correctly, the *robustness* of the catalytic process itself. By this, we mean the dependence of a running process on the deviation of parameters: The more robust or deviation-proof a given process is, the more convenient it is in synthetic terms to be reproduced, scaled-up and extended in scope. The type 1 systems are the most robust. In fact, these processes might be regarded as the

Table 2.1 *Four major types of catalytic system occurring in intermolecular Mizoroki–Heck reactions*

	Type 1	Type 2	Type 3	Type 4
Typical substrates	Aryl iodides, activated aryl bromides	Unactivated aryl bromides, activated aryl chlorides	Unactivated aryl chlorides, unactivated aryl bromides	Aryl triflates, aryl halides in the presence of halide scavengers
Palladium precatalyst	Any form of Pd^0 or Pd^{II}	Any form of Pd^0 or Pd^{II}	Phosphine complexes of Pd	Diphosphine complexes of Pd
Obligatory ancillary ligands	Not needed	Not needed	Electron rich, bulky monophosphine	Chelating diphosphine
Main function of ligands	Reserving	Supporting	Activating	Restricting
Coordination shell of palladium	Undefined, contains only labile or hemilabile ligands	Undefined, contains only labile or hemilabile ligands	Contains one stable, monodentate ligand	Contains one stable, bidentate ligand
Achievable TON values	High to enormous	Modest to high	Low	Low to very low
Optimal palladium precatalyst loadings (mol%)	<0.1, can be as low as 10^{-5}	Optimal range 0.05–0.5; higher and lower values are detrimental	0.5–5	1–20
Temperatures	From r.t. to 180 °C and higher	120–180 °C and higher	80–120 °C for aryl chlorides, both higher and lower temperatures are detrimental; r.t. to 50°C for aryl bromides	80–120 °C, both higher and lower temperatures are detrimental
Solvents	Polar coordinating	Polar coordinating	Nonpolar, non-coordinating	Preferably non-coordinating
Palladium nanoparticles	Often observed	Almost always observed	None	None
Preparative and economic value	Good; limited by substrate cost	Poor; require chemical engineering, but may be attractive for industry	Good; limited by catalyst cost	Very good

stereotype of a broadly applicable process, as very wide variations of all parameters are feasible when using aryl iodides and activated aryl bromides as electrophilic components. The only exclusions are reactions performed with very low precatalyst loadings to achieve record-making TONs or TOF values. As these experiments are very sensitive, reproducibility is a serious concern. The type 2 systems are unstable by default. This is what separates such systems from type 1 systems, with which they share many parameters. The border between type 1 and type 2 with respect to composition of the reacting systems is not sharp. In many cases, it is only the extreme sensitivity towards the deviation of parameters from those optimized, as well as substrate sensitivity; this assists in unambiguous assignment to type 2. Type 3 and type 4 systems are perfectly stable and robust within the limits set by a given realization. The preparative appeal of these systems, even in spite of the high cost of catalysts, is determined by a high degree of versatility with respect to substrates. Once developed, a given type 3 or type 4 protocol might be applied to other substrate–alkene pairs with good hopes for success.

2.2.1   Substrate Dependence

A common ground that is explicitly or implicitly defended in the majority of studies on Mizoroki–Heck reactions is that the limiting stage for the whole cycle is the oxidative addition step. By this criterion, the most important substrates, aryl halides, are subdivided into very reactive (aryl iodides and electron-deficient aryl bromides), less reactive (all other aryl bromides and electron-deficient aryl chlorides) and very unreactive (all other aryl chlorides). As evident as this classification may seem, it is not based on any solid proof. Indeed, if it were really so important, the oxidative addition step should have been characterized by very strong dependence on substituent effects in these substrates. However, this has not been observed in either Mizoroki–Heck reactions or in any other palladium-catalysed reaction of aryl halides. The Hammett reaction constant values ρ, whenever measured, are rather modest in value [5]. Such values could hardly have accounted for the well-known enormous distance between the reactivity of, for example, a typical activated substrate **7** and a typical deactivated substrate **8** (Figure 2.1).

   We should stress that this difference only applies to processes in which palladium species have an undefined coordination shell (as in type 1 and type 2 systems). For true ligand-accelerated type 3 processes, which are controlled by a definite stable ligand in the coordination shell, the difference between electron-rich and electron-poor ('inactivated' versus 'activated') substrates is indeed not dramatic, which allows for the formulation of broad-scope robust protocols.

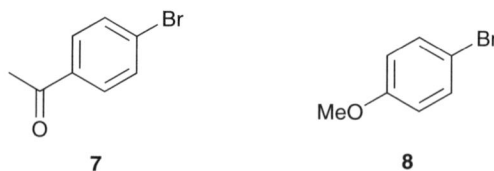

Figure 2.1   *Activated and deactivated substrates.*

Otherwise, we see two sorts of substrate dependence. The first is associated with halide dependence. In systems relying on undefined catalytic species (type 1 or type 2), iodides are always better than bromides, and both are much better than chlorides even if we compare electron-deficient aryl chlorides or bromides with electron-rich aryl bromides or iodides. On the other hand, iodides are generally unwelcome in type 3 processes. This strong influence of halide stems not only from the kinetics of oxidative addition, but also on the ligand properties of halide ions. Iodide is a strongly bonded ligand which markedly influences the nature and composition of palladium species, whereas chloride is a much more labile ligand. Therefore, the catalytic cycles in which aryl chlorides are involved are generally less stable, with the catalyst more prone to deactivation. This may account for the known paradox that aryl chlorides are reactive in many catalytic systems without special ligands but yields and TONs are unsatisfactory.

The other sort of substrate dependence is a large substituent effect on net results (yields, TONs and average TOF values) which somehow goes along with a very modest substituent effect in the assumed rate-determining step, which is the oxidative addition. This is directly associated with main principles of catalyst design in the Mizoroki–Heck reaction: only those factors matter which govern the overall integral performance (averaged over full reaction time) of a given process. In this context, any immediate differential values (initial or peak rates, differential TOF values measured at low conversion levels or any rate factors obtained by competitive reaction methods) are not relevant because they tell us nothing about the outcome of a given catalytic process.

2.2.2 Ancillary Ligands

The Mizoroki–Heck reaction involves an immense variety of ancillary ligands, since it has been a matter of common agreement that phosphines play a distinct role different from any other ancillary ligand. Heck himself introduced two major types of catalytic system: (1) phosphine-free for aryl iodides [2]; (2) using Ph$_3$P [6, 7] or (2-tolyl)$_3$P [8] for aryl bromides. This dichotomy has laid the basis for further division of all new protocols into those requiring phosphine ligands (phosphine assisted) and those that are *phosphine free* (quite often erroneously referred to as *ligand free*).

The analysis of hundreds of protocols published so far, however, prompts a conclusion that this dichotomy is rather misleading, as it does not reflect the real differences of catalytic systems. Thus, the analysis of the role of ligands should be more refined. The roles of ancillary ligands can be categorized into four major divisions:

(i) Support of zero-valent palladium as reactive monomeric complexes and prevention of clusterization eventually leading to inactive metal sediments (*supporting function*). This function is provided either by simple labile and hemilabile monodentate ligands, including monophosphines or/and by many other specifically or fortuitously added nonphosphine ligands (e.g. bases, anions or even coordinating solvents).

(ii) Binding palladium into a relatively stable complex, which slowly releases active palladium species during the reaction by means of thermal decomposition or reaction with components of the reaction system (*reserving function*). This function is ensured by all sorts of strongly binding mono-, di- or tri-dentate ligands; the complexes thus formed, *slow-release precatalysts* (SRPCs), are stable as such even at high temperatures but are more or less slowly attacked and cleaved by the components of the reaction mixture.

(iii) Tuning the reactivity and selectivity of catalytic species by increasing the rates of oxidative addition, particularly into less active bonds (*activating function*). This function is secured by bulky monodentate electron-rich ligands, almost exclusively phosphines, which are capable of the essential increase of reactivity of the palladium centre in the oxidative addition step. Owing to their own steric bulk, these restrict themselves from overcrowding the coordination sphere of palladium.

(iv) Tuning the selectivity of catalyst by intimate control of coordination sphere, which allows for selection of mechanism (e.g. choice between nonpolar and polar mechanisms) and suppresses alternative pathways of lower selectivity (*restricting function*). This function is provided by stable bidentate ligands, almost exclusively diphosphines, which form a stable coordination sphere required for highly regio- or stereo-selective pathways while restricting nonselective side-reactions at the same time.

2.2.3   Bases

Bases play multiple roles in Mizoroki–Heck reactions, involving not only the regeneration of palladium(0) from the palladium(II) hydride intermediate, but also taking part in (1) preactivation and reduction of palladium(II) precatalysts, (2) cleavage of SRPC and (3) stabilization of palladium(0) in the form of weak labile ligands thereby preventing palladium(0) compounds from deactivation. Common bases used in Mizoroki–Heck reactions comprise tertiary amines (Et_3N and n-Bu_3N are by far the most popular), which are usually used in nonpolar solvents and ligand-assisted protocols, and inorganic bases (acetates, carbonates, bicarbonates, phosphates as well as fluorides; precedents of the latter are few though [9]), which are more useful in polar, coordinating solvents in the presence of simple palladium salts or SRPC. No useful trends in the application of particular bases have been revealed. At some point in time, acetates were widely popular among practitioners, but all speculations about a pivotal role of the acetate ion in the support and activation of palladium(0) are not definitely supported by experimental evidence. There are a lot of cases when other bases perform better and, most probably, many other basic ions are able to perform similarly.

Non-nucleophilic bases (i-Pr_2NEt, Cy_2NEt and proton sponge) are widely used in type 3 and type 4 processes, where the base should perform only as a Brønsted base and where various side-effects associated with Lewis basicity, ligation and nucleophilicity are considered to be detrimental.

On the other hand, bases possessing high Lewis basicity might serve as ligands in phosphine-free protocols. For example, tetramethylguanidine (TMG) and 1,4-diazabicyclo [2.2.2]octane (DABCO) were shown to improve yields markedly in phosphine-free reactions of aryl iodides, bromides and activated chlorides, compared with identical system without additive. Representative protocols: $PdCl_2$ (0.1 mol%), TMG, NaOAc, N,N-dimethylacetamide (DMA), 140 °C or $Pd(OAc)_2$ (0.0001–5 mol%), DABCO, K_2CO_3, dimethylformamide (DMF), 120 °C.

It should be noted that palladium(II) hydride intermediates, by all evidence, are quite acidic. Consequently, strong bases are not needed for deprotonation, and weak bases (e.g. $NaHCO_3$) do quite well. In fact, it is well known that, in the absence of a base, the Mizoroki–Heck reaction does take place but is rapidly quenched after a few catalytic turnovers. From this, we might assume that it is not the deprotonation of 'Pd–H' itself,

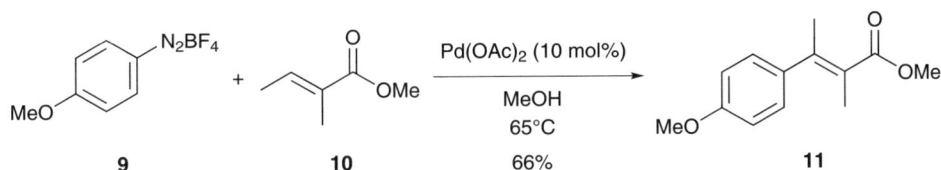

Scheme 2.2

but rather the prevention of accumulated protic acidity in reaction media that is the main function of base.

Still, there are base-free Mizoroki–Heck reactions, which are known to occur with arene diazonium salts. In this particular case, the absence of base is advantageous, as spontaneous decomposition of diazonium salts is known to be accelerated by nucleophiles and bases. High precatalyst loadings are required, as, for example, in the arylation of disubstited acrylates **10** (**9** → **11**, Scheme 2.2) [10]; a base-free reaction might only be sustained for a few turnovers.

The mechanism of action of these base-free Mizoroki–Heck reactions is an intriguing question. These reactions are performed under very mild conditions, so any uncontrolled thermal decomposition pathways might be ruled out. Evidently, palladium(II) hydride intermediates sometimes regenerate palladium(0) spontaneously, either by reductive elimination or because of enhanced acidity. While there is no well-documented evidence of reductive elimination, Milstein and coworkers [11] presented a system with enhanced acidity that is rendered by an electron-rich bulky phosphine **14** in the coordination sphere. These authors described base-free Mizoroki–Heck reactions in the presence of ligand **14** (**12** → **15**, Scheme 2.3), which was shown to increase the acidity of palladium(II) hydride

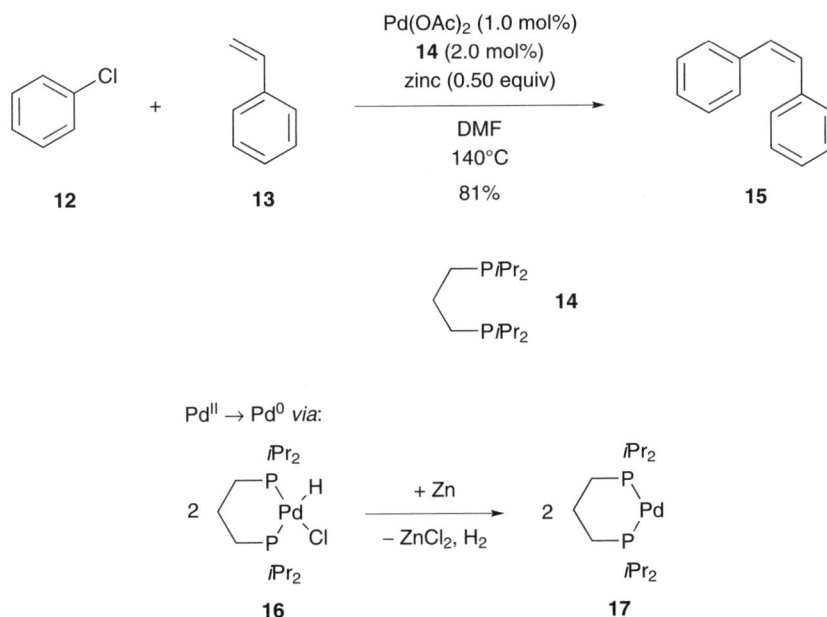

Scheme 2.3

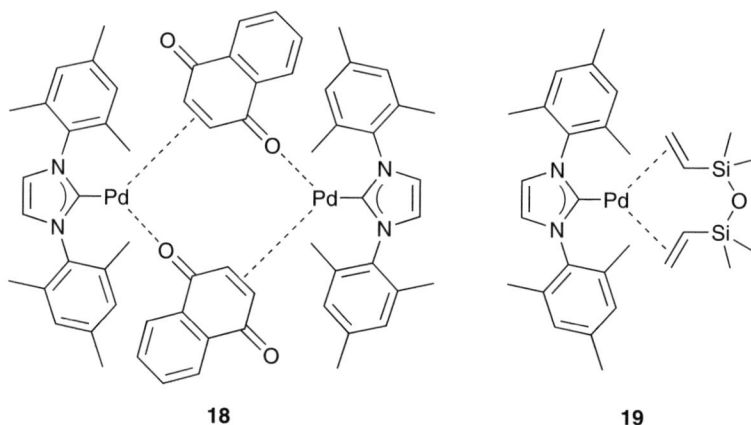

18   **19**

Figure 2.2 *Preformed palladium(0) carbene complexes.*

intermediate **16** almost to the level of a strong mineral acid; regeneration of palladium(0) species **17** was achieved in the presence of zinc as a reducing metal. So far, this ligand and the reported base-free protocol have not found any further application [12].

There is, however, a logical link between the Milstein–Portnoy system [11] and reports on the effective use of electron-rich carbene ligands for base-free Mizoroki–Heck reactions of diazonium salts independently reported by Andrus *et al.* [13] and Beller and coworkers [14]. An equimolar mixture of Pd(OAc)$_2$ and SIPr·HCl was shown to be able to support Mizoroki–Heck reactions of a vast series of arene diazonium tetrafluoroborates with styrenes and acrylates in high yield at room temperature in tetrahydrofuran (THF) [13]. On the other hand, preformed palladium(0) carbene complexes **18** and **19** of the IMes ligand could be used for the same purpose [14]. Slightly elevated temperatures (50–75 °C) were required though, apparently for predissociation of complexes to release active monocarbene palladium complexes (Figure 2.2). The systems are operative at 0.1–2 mol% catalyst loading, thus affording a decent number of base-free catalytic cycles. This performance might be rationalized by the electron-rich carbene ligands, which presumably account for the high acidity of the respective palladium(II) hydride intermediates. These are then able to regenerate palladium(0) spontaneously, even under the conditions of gradually decreasing pH value. In both systems, the presence of carbene ligands in the coordination shell of palladium is essential for proper modulation of the reactivity.

True base-free Mizoroki–Heck reactions should not be mistaken with base-free Mizoroki–Heck reactions of such special substrates as acid anhydrides, in which it is the leaving carboxylate group which serves as base [15]; this reaction benefits from halide additives [16].

2.2.4 Additives

Catalytic systems often contain additional components, which amplify the performance. The best example is the so-called Jeffery or Jeffery–Larock family of protocols [17–22] using simple palladium(II) salts as precatalysts in aqueous solvents in the presence of quaternary ammonium salts, usually *n*-Bu$_4$NBr. Apart from *n*-Bu$_4$NBr, related salts such as n-Bu$_4$NCl [17, 23, 24], Et$_4$NCl [25, 26], BnEt$_3$NBr [27] and LiCl [22, 28, 29], and even nonionic phase transfer agents such as polyethyleneglycols (PEGs) [30], are used.

There are two general trends in the application of salt additives: (1) reactions of aryl iodides under mild conditions in type 1 systems and (2) reactions of less reactive substrates under harsh conditions in type 2 systems. Among the probable functions performed by additives are:

(i) ligand exchange to replace the more strongly bound iodide by more labile chloride, bromide, acetate or related anionic ligands (retardation of the Mizoroki–Heck reaction by iodide was explicitly observed with pentafluorophenyl halides as substrates, of which C$_6$F$_5$I is by far less reactive than C$_6$F$_5$Br [9]);
(ii) formation of anionic complexes of palladium, which were shown to be more reactive in oxidative addition reactions [31, 32];
(iii) facilitation of the preactivation step, as there are indications that n-Bu$_4$NBr shortens the initial latent period [33], likely due to traces of tertiary amine present in such salts;
(iv) stabilization of small clusters of palladium(0) by tetraalkylammonium salts or PEGs, thereby preventing further growth of palladium metal particles;
(v) phase-transfer action of tetraalkylammonium salts or PEGs should not be neglected, at least for partially heterogeneous, aqueous systems or in those cases when poorly soluble inorganic bases are used.

These functions are likely to be in a complex interplay in real systems, so that a blend of multiple effects is seen as a result.

The crucial role of additives in the regulation of palladium deactivation by accumulation was demonstrated by Dupont and coworkers [34] for a system in which the reaction rate of oxidative addition is low. n-Bu$_4$NBr suppresses the nucleation and growth of metallic particles and, thus, allocates palladium(0) species more time to enter the catalytic cycle. The role of such additives is to balance the rate of palladium(0) formation, the rate of oxidative addition and the rate of nucleation of metal particles [35]. The suppression of inactivation is realized through both the formation of anionic complexes of palladium **22** and inhibition of further growth of palladium metal particles (Scheme 2.4). This mode of action requires the application of quaternary ammonium halides, which presumably protect small, newly generated and still very reactive clusters **20** of palladium from further aggregation to large, almost inactive sedimenting particles **21** [36].

The addition of n-Bu$_4$NBr or related additives facilitates the reactions in the presence of simple precatalysts to afford effective straightforward protocols. These are particularly useful for cases in which the use of supporting ligands is inefficient in order to avoid the likely overloading of the palladium coordination sphere. This comes into play when substrates themselves are capable of coordinating to palladium, such as in Mizoroki–Heck reactions of heteroaromatic substrates [37], aminoacrylates [24, 26, 27, 38–40], resin-supported aminoacrylates [41], aminoacrylate peptides [42] or both [43]. We note, though, that this is again not an absolute rule, and sometimes a classical protocol utilizing Ph$_3$P complexes works better for such substrates [44]. Mizoroki–Heck reactions in the presence of additives is quite effective even in demanding macrocyclizations (**23** → **24**, Scheme 2.5) [45].

The addition of quaternary ammonium salts [46–48] or LiCl [28, 49–51] is practised in phosphine-free reactions of aryl iodides and bromides with unsaturated alcohols, leading to

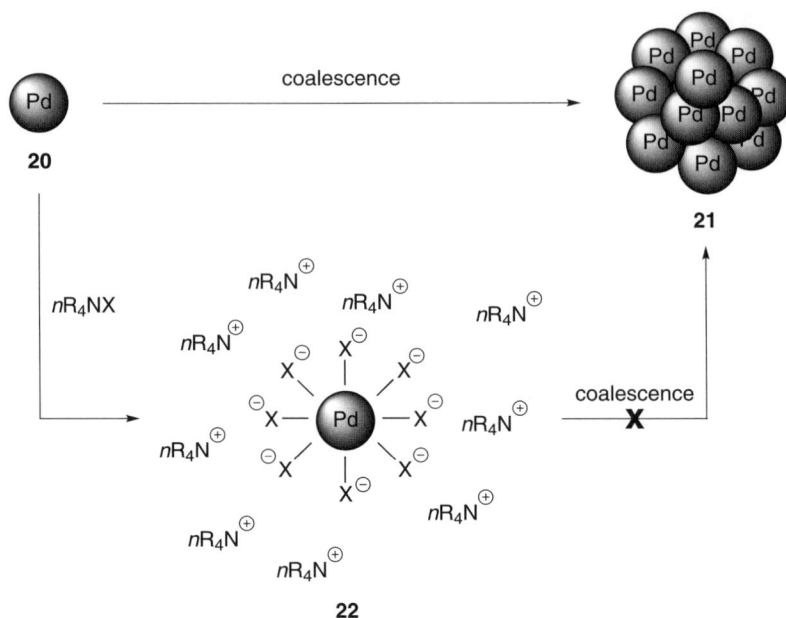

Scheme 2.4

arylalkanones. This protocol gives satisfactory results even in an unusual Mizoroki–Heck reaction of α-iodoenones **25** with allylic or homoallylic alcohols **26** (**25** $\rightarrow$ **27**, Scheme 2.6) [52].

The application of additives in type 1 systems to allow for performing the reactions at room temperature under very mild conditions is particularly important for thermally sensitive substrates, such as in Mizoroki–Heck reactions of vinylketones [29]. Also, the additives are considered as a prerequisite of successful scale-up of protocols developed at a micro- or semimicro-scale [53].

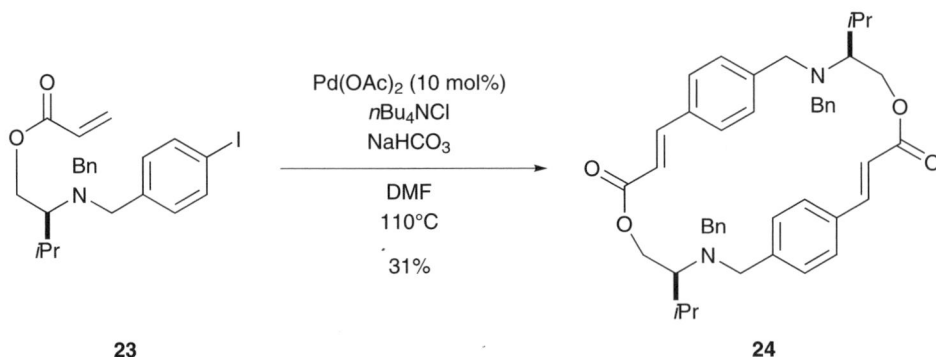

Scheme 2.5

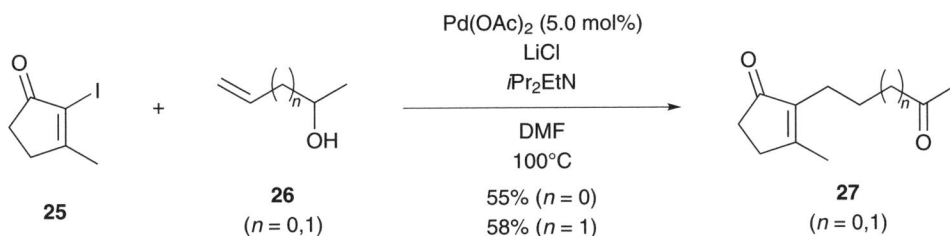

Scheme 2.6

Some finer effects caused by additives were reported. Thus, the use of the pair n-Bu$_4$NOAc–KCl was used to control the selectivity of reductive elimination in the arylation of acrolein acetals (Scheme 2.7). Interestingly, a more usual 'reverse' combination KOAc–n-Bu$_4$NCl gave worse results with respect to selectivity [54, 55], which clearly shows that the Mizoroki–Heck reaction is often extremely sensitive towards subtle details of formulation of the catalytic system. As proposed by Cacchi and coworkers [54, 68] for the n-Bu$_4$NOAc–KCl cocktail [55–57], the coordination sphere of palladium is saturated to form an anionic complex **28** (**28** → **29** + **30**), in which palladium is not capable of coordinating to the aryl π-system as in **31** (**31** → **30**, Scheme 2.7).

The other important class of additives is halide anion scavengers such as silver(I) or thallium(I) salts [58–61]. The effect of these salts is associated with abstraction of a halide anion, thereby leaving behind a free coordination site for alkene binding. This polar pathway might be useful if:

1. The reaction is performed in the presence of a conventional phosphine ligand in syntheses of complex molecules. This approach is quite popular in intramolecular Mizoroki–Heck chemistry, which is beyond the scope of this chapter (see Chapters 5, 6 and 16).

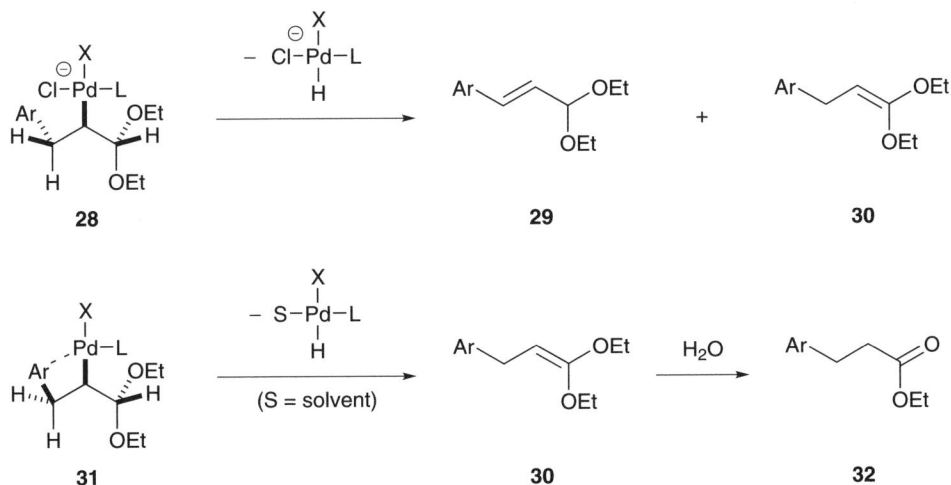

Scheme 2.7

2. It is desirable to enter the polar pathway in reactions of aryl halides in order to avoid the use of aryl triflates or arenediazonium salts, whose reactions are polar by default (cf. type 4 systems).

Using AgOAc is doubly beneficial, as it acts as a halide scavenger effect and as a base. This might be useful if the addition of stronger bases must be avoided to overcome side-reactions [62].

2.2.5   Media

It is often claimed that Mizoroki–Heck reactions can be performed in practically any media and, indeed, the variety published is enormous. However, the Mizoroki–Heck reaction cannot be regarded as media insensitive.

Judicious scrutiny of data clearly shows a dichotomy in the use of solvents. Some prefer solvents of low polarity and low Lewis basicity; others prefer media of high polarity and high coordination capability. We believe that this dichotomy is rooted in the distinction between the types of Mizoroki–Heck catalytic system.

As is seen from discussion above, the vast majority of Mizoroki–Heck reactions belong to either type 1 or type 2 processes, which involve palladium with undefined coordination shells. In the presence of halide ions (liberated during the reaction or derived from additives) and anionic bases, the coordination shell of palladium contains as many anionic ligands as it can possibly bear [32]. Thus, these catalyses do only pass through coordination equilibria involving charged reactants on both sides. Similar considerations apply to type 4 systems, which involve cationic intermediates.

Hence, the majority of Mizoroki–Heck reactions show a clear preference for polar media, which lends assistance to both deligation of halide and support of the palladium catalyst: (1) highly Lewis basic organic solvents such as DMF, DMA, *N*-methylpyrrolidone (NMP) and dimethylsulfoxide (DMSO); (2) aqueous solvents; (3) ionic liquids.

On the other hand, true ligand acceleration (type 3 processes) shows preference for solvents of low polarity and lower Lewis basicity (toluene, dioxane and THF) with soluble tertiary amines as bases. In this respect, these Mizoroki–Heck reactions resemble cross-coupling processes, which also display strong preference for these solvents. Reactions in nonpolar solvents (toluene or xylene) have been known since Heck's seminal articles [8]. The halide remains a crucial subject of concern: in reactions catalysed by phosphine complexes of palladium, aryl iodides prefer triarylphosphines and polar solvents, whereas reactions of aryl bromides and chlorides indeed prefer electron-rich trialkylphosphines and nonpolar solvents [63–65].

In the last decade, a lot of attention has been paid to environmental aspects. As to the Mizoroki–Heck reaction, environmentally benign media currently involved in the design of catalytic systems encompass supercritical carbon dioxide ($scCO_2$), fluorous systems, water and aqueous systems, solvent-free systems [66]. In this context, it should be noted that the so-called *solvent-free* reactions are actually not literally such, but are performed in media composed of substrates and often liquid amine. This was described as early as in 1972 by Heck himself [2, 8] (microwave heated version [53]). Amines are good coordinating solvents; during the reaction, the amine is transformed into amine salt, which, being a major constituent or reaction mixture in the absence of a true solvent, adds to the net media polarity.

2.2.6   Temperature

Temperature is a very important parameter in Mizoroki–Heck reactions, determining type and operation mode of catalytic systems. As with other parameters, there are controversial views on the role of temperature. It is sometimes believed to require high temperatures. This idea is generally used when new precatalysts having high thermal stability are designed. Thus, this property is implicitly or explicitly regarded as beneficial. *Which temperature is really needed in a general Mizoroki–Heck reaction?*

Indeed, there is ample data showing that none of the individual stages or the catalytic cycle as a whole require high temperatures to occur. Even the oxidative addition step proceeds with all substrates (including aryl chlorides) at room temperature or very modest temperatures (below 50–60 °C). High temperature is not necessarily needed and might even be detrimental to both type 3 and, particularly, type 4 processes because both rely on a defined coordination sphere, which cannot be maintained at high temperatures.

In type 1 and type 2 Mizoroki–Heck reactions, high temperature is required when SRPCs are involved as palladium sources. In type 2 reactions, the need to speed up the oxidative addition step is more important when the palladium catalyst has no particular activating ligands. In this case, a lower rate of oxidative addition would result in the build-up of unsupported palladium(0) with concomitant deactivation.

Room-temperature Mizoroki–Heck reactions deserve special comment. Aryl iodides are capable of affording Mizoroki–Heck products with reactive alkenes at room temperature or slightly elevated temperature, provided that the reaction time is sufficiently long (days or weeks). Even 'lazy' precatalysts, such as nanoparticles, do catalyse Mizoroki–Heck reactions at room temperature [67]. Mizoroki–Heck reactions at low temperatures are synthetically useful if the substrates cannot tolerate basic conditions at elevated temperatures; for example, acroleins [17, 68]. Selective monosubstitution in diodobenzenes is achieved at low temperature [69].

An ingenious system of this sort was developed by Dupont and coworkers involving a *room-temperature SRPC* [34, 70]: palladacycle **33**, which is in equilibrium with PdCl$_2$ and *N,N*-dimethylaminomethylphenylacetylene (**34**) in solution (Scheme 2.8). This complex

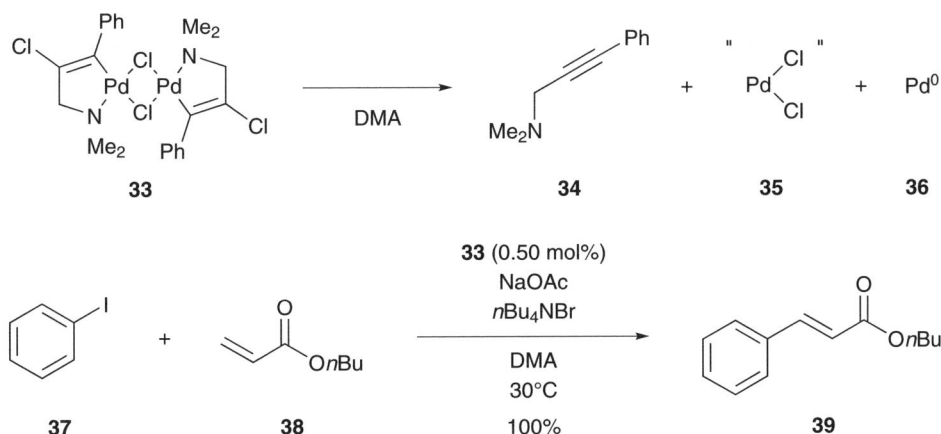

Scheme 2.8

effects room-temperature reactions of aryl iodide **37** with *n*-butyl acrylate (**38**) using 1.0 mol% of palladium (**37** → **39**) [70].

The room-temperature Mizoroki–Heck reaction can be extended to activated aryl bromides using electrochemically generated palladium(0) at high catalyst loading (30 mol%) [71].

2.3 Four Types of Intermolecular Mizoroki–Heck Catalytic System

2.3.1 The Type 1 Catalytic System

This type might be referred to as the classical Mizoroki–Heck type, because, as is now evident, it was these systems which were first described. This type requires substrates which are highly reactive in the oxidative addition, which means that any form of palladium(0) emerging within the catalytic system reacts with the aryl halide. Therefore, these reactions occur in the presence of any complexes having at least three labile or hemilabile ligands. These are generated from practically any palladium-containing precursor, including palladium metal itself in a reasonable dispersion state. Additional activation towards oxidative addition is achieved through coordination of simple anionic ligands, acetate or chloride, as seen in the influence of salt additives (see above). Reactions in the presence of such activating additives proceed relatively slow at room temperature. Phosphines or similar strongly coordinating ligands often retard the processes. Systems of this sort do not demand anaerobic conditions. Sometimes very low ('homeopathic') loadings of palladium precatalysts are needed, yielding very high TONs and TOF values.

It should be noted that extreme care should be exercised when designing catalytic systems for type 1 reactions and in interpreting the results. As soon as these reactions are effectively catalysed by trace amounts of almost any palladium precursor, the nature of catalytically active species cannot be reliably established. Hence, the effects and the real role of potential components of the catalytic system are in question. As a result, the suggestion of catalytic reactions in the absence of deliberately added precatalysts or in the presence of palladium compounds, which are well known not to release palladium, is illegitimate [72, 73].

From a preparative viewpoint, type 1 reactions are reliable, high yielding, allow for simple scale-up and are easily optimized. Application of these processes to other substrate–alkene pairs is usually successful as long as the aryl halide is sufficiently activated towards oxidative addition.

On demand, such processes achieve very high TONs or TOF values after optimization using various published SRPCs. We note here that any reported ultrahigh TONs and TOF values are difficult to reproduce.

An interesting realization of a type 1 system is the so-called recyclable protocols, in which only a minor portion of the initially added SRPC is consumed. The major portion remains intact and might be reused.

2.3.2 The Type 2 Catalytic System

Type 2 might be referred to as *Spencer's systems*. Spencer was the first who discovered that, after thorough optimization of the reaction conditions, even less reactive substrates (activated aryl chlorides or bromides) do react in a system, which utilizes a simple palladium precatalyst [74–77]. We believe that such catalytic processes should be regarded as a realization of a fundamentally different Mizoroki–Heck catalytic cycle, which requires a different approach to the design of its catalytic system. While Heck himself introduced two protocols for aryl iodides (using phosphine-free palladium(II) precursor) and aryl bromides (using phosphine complexes of palladium(0)), subsequent research starting from the seminal papers of Spencer [74–77] revealed that both aryl bromides and activated aryl chlorides undergo Mizoroki–Heck reactions in the presence Pd(OAc)$_2$ [78], (EtS)$_2$PdCl$_2$ [79], more sophisticated phosphine complexes and palladacycles [80, 81]. All these reactions are clearly similar in terms of their characteristics and features: (1) high reaction temperatures; (2) polar, high boiling-point solvents; (3) palladium loadings roughly in the range 0.05–0.5 mol%; (4) strong and very uneven dependence on both conditions and substrates used; very good performance is only seen for a limited set of substrates. Any deviation from optimized parameters – solvent, base, additives, temperature, precatalyst loading, purity of reagents and media components, air, and even the efficiency of stirring [82] – leads to poorer results.

The amount of palladium precursor must be individually tuned in each of the protocols. The effectiveness and yields generally decrease with loadings that are too high and too low. The former lead to deactivation of the catalyst through sedimentation of palladium black and the latter lead to unpractical low reaction rates and poor yields, TONs and TOF values.

Systems of this kind are highlighted in the studies of de Vries and coworkers, who showed that whatever sort of palladium precatalyst is employed, if conditions and initial loadings are optimized, Mizoroki–Heck reactions can be successfully performed with all sorts of aryl halides except unactivated aryl chlorides. Although there seems to be no difference between this kind and what we regard as type 1 systems, there is an apparent attribute revealing the intimate difference: type 2 reactions show a bell-like dependence on the initial palladium loading, since either higher or lower starting concentrations of precatalyst are strongly detrimental to the outcome (Figure 2.3). We have hypothesized [4, 83] and de Vries and coworkers [32, 36, 84] have confirmed that such paradoxical behaviour (more catalyst is worse than less catalyst) should be associated with deactivation of palladium due to formation of clusters, nanoparticles and eventually bulk metal (palladium black); this is facilitated at higher concentrations of precatalysts in environments lacking good or simply enough supporting ligands. Conversely, in type 1 systems, the reactive substrates themselves effectively scavenge palladium(0), which is why this phenomenon is not even observed.

Only in those systems in which the rate of oxidative addition is low does the build-up of palladium(0) become a serious side-reaction. This happens if preactivation processes supply more palladium(0) than can be consumed in the oxidative addition step. Ideally, palladium(0) should, once formed, immediately enter the catalytic cycle, which should regenerate the active catalyst after each turnover. In reality, a certain fraction of palladium exits this cycle due to various deactivation processes (absorption on surfaces or reoxidation to palladium(II)). In a pseudostationary state, the losses of palladium(0) from the cycle are compensated by 'feeding' from the precatalyst reservoir. Any deviation of this balance will rapidly lead to 'starvation' of the catalytic system. In summary, both build-up (more palladium(0) is formed than consumed by the cycle) and loss (more palladium is withdrawn from the cycle than provided by the precatalyst) will quench the catalytic system.

There are two possible realizations of type 2 systems. In the first, palladium is added as a simple salt or complex in the presence of stabilizing additives. In the second, palladium is added as an SRPC, either preformed or formed *in situ* from a simple palladium source

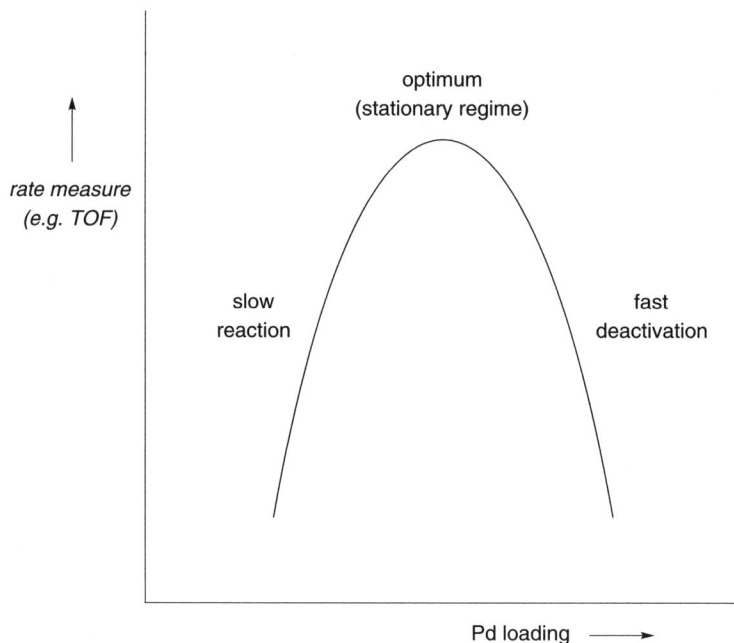

Figure 2.3   *Bell-like dependence of type 2 reactions on initial palladium loading.*

and a strongly coordinating ligand. In this case, the palladium loading is typically higher, as the SRPC is presumed to maintain a reservoir of palladium.

We have to state explicitly that there is no significant ligand acceleration effect revealed in type 2 systems. This even applies when the palladium precatalyst bears ligands which are otherwise well known for their high effectiveness in oxidation addition processes. Type 2 processes are performed under conditions where full cleavage (i.e. dissociation) of any ligand occurs prior to the catalysis itself. Thus, even though t-Bu$_3$P is known to impart high reactivity to palladium(0) complexes under mild conditions in cross-coupling reactions as well as type 3 Mizoroki–Heck protocols (see below), the performance of (t-Bu$_3$P)$_2$Pd under type 2 conditions (NaOAc, DMA, 140 °C) is roughly the same or worse than the performance of such typical SRPCs as palladacycles [85]. A phosphine complex is more readily activated to enter the catalytic process but it is also more rapidly deactivated because of a more labile coordination shell.

The type 2 Mizoroki–Heck reaction carries both a positive and a negative message. The positive one is that all sorts of substrates except electron-rich aryl chlorides and sterically hindered alkenes usually afford reasonable yields using rather unpretentious sources of palladium (simple salts or palladacycles). The negative message is that each process, each combination of substrates, the concentration of reagents and experimental realization (shape and materials of the reaction vessel, as well as order of mixing and heating) must be optimized separately and thoroughly.

To these also belong systems based on Doucet–Santelli tetraphosphine ligands **41–43**, with Tedicyp (**41**) being studied in depth (Figure 2.5) [90]. Ligand **41** was tested extensively in reactions of aryl bromides with unprecedented variety in the choice of the alkenes;

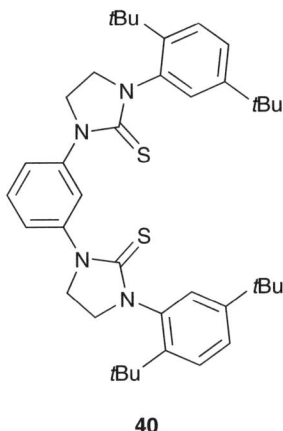

40

Figure 2.4   *Ligand* **40**.

aside from standard styrene and acrylates [89, 91, 92], simple terminal alkenes [93], terminal alkenoic acids [94], vinyl sulfides/sulfoxides/sulfones/sulfonates [95], allylic alcohols [96, 97], enol ethers [91, 98–101], vinylic ketones [102], and even allenes [103] were subjected to Mizoroki–Heck couplings. All these were performed under reaction conditions typical for the type 2 with K_2CO_3 or $NaHCO_3$ as base and high boiling-point polar solvents DMF or DMA at 110–150 °C. Precatalyst loadings were in the range 0.01–2 mol%, although in a few cases amounts as low as 0.001 mol% were possible giving record-making TONs (10^8) and TOF values ($5 \times 10^6 \, h^{-1}$) for selected activated aryl bromides and aryl iodides [104].

In spite of very impressive results obtained with ligand **41**, we cannot state for sure if this and similar ligands have any proven role in the reactivity of catalysts involving them. First, **41** has not been proven applicable to nonactivated aryl chlorides yet; and second, the reactions of activated aryl bromides were performed at temperatures exceeding 110 °C. Besides, the comparison of independently obtained data in the standard reaction of 4-bromoanisole (**8**) with acrylate (Table 2.2, entry 5) performed either with a

Table 2.2   *Examples of typical type 2 Mizoroki–Heck reactions with 4-bromoanisole (**8**) as a substrate*

Entry	Substrate	Precatalyst	Pd (mol%)	Base	Solvent	T (°C)	Yield (%)	TON	Ref.
1	t-Butyl acrylate	Pd(OAc)₂	0.05	NaOAc	NMP	135	91	1 820	[84]
2	Styrene	Pd(MeCN)₂Cl₂ + 6Ph₄PCl	0.01	NaOAc	NMP	130	97	9 800	[86]
3	n-Butyl acrylate	Pd(OAc)₂ + DABCO	0.001	K₂CO₃	DMF	120	70	70 000	[87]
4	Styrene	Pd(dba)₂ + **40**	0.5	NaOAc	NMP	160	75	150	[88]
5	n-Butyl acrylate	[Pd(allyl)Cl]₂ + **41**	0.001	K₂CO₃	DMF	130	82	82 000	[89]

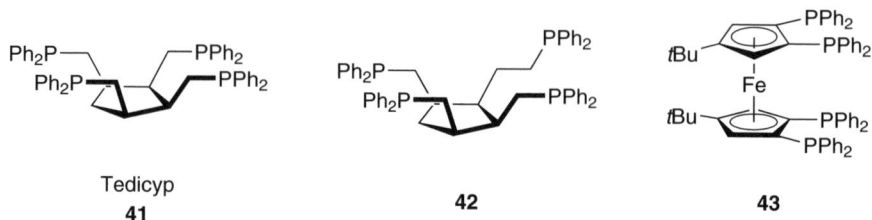

Figure 2.5   *Doucet–Santelli tetraphosphine ligands.*

Tedicyp-based system or with a simple phosphine-free system (entries 1 and 3) under otherwise almost identical conditions reveals a more or less matching performance. It seems very likely that tetrapodal phosphines serve as effective supporting ligands, thereby ensuring almost infinite stability of palladium(0) species.

Typically, type 2 catalytic systems are exploited at high temperatures starting approximately at 120 °C, but usually exceeding 140 °C; lower temperatures are sometimes possible after optimization. Thus, Gürtler and Buchwald [25] showed that, using their phosphine-free catalyst system, not only aryl iodides, but also aryl bromides **44**, including electron-rich substrates, afford high yields in the arylation of disubstituted alkenes **45** at temperatures as low as 85–100 °C (**44** → **46**, Scheme 2.9). Both base and additive in this nice protocol are poor ligands for palladium. Thus, this system involves well-exposed catalytically active centres of high reactivity, which are protected from fast deactivation by the anionic additive Et$_4$NCl.

As type 2 systems are maintained by rather high levels of soluble palladium concentrations in the reaction media, palladium from any support will be depleted rapidly. Recycling of a supported catalyst will thus be difficult. Aside from stable palladium complexes, there are a few supported systems which deliver reasonable type 2 performance. One of the most active systems was introduced by Molnár and Papp [105] with palladium(II) absorbed by ion-exchange onto montmorillonite clay. The precatalyst operates at high temperatures (150–160 °C) in the presence of Na$_2$CO$_3$ in NMP and delivers high yields of Mizoroki–Heck products with, for example, phenyl bromide, **8** and 4-chloroacetophenone (**47**) at 0.001–0.1 mol% catalyst loading (**47** → **48**, Scheme 2.10). With more reactive substrates (e.g. phenyl bromide) this precatalyst survives two or three reuses, which shows that palladium is extracted from the support lattice at a sparing rate.

Another supported system is Alper's silica-immobilized pincer complex **49**, which effects the reaction between electron-rich aryl bromides and standard alkenes using a 1 mol%

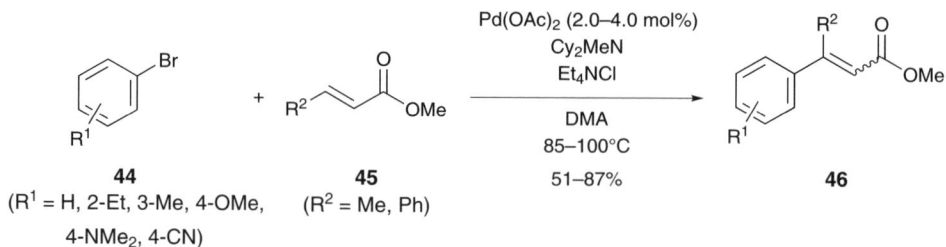

Scheme 2.9

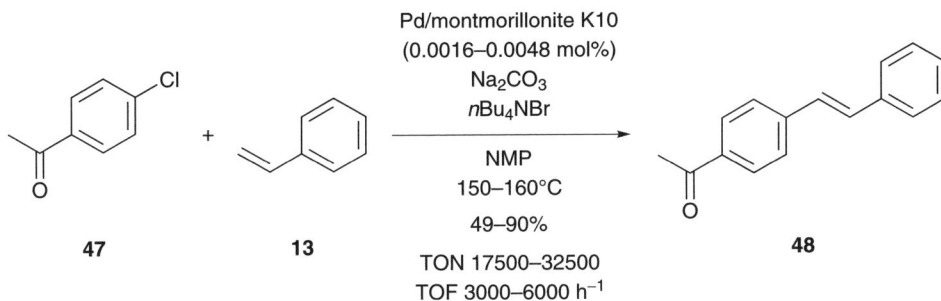

Scheme 2.10

palladium loading (Na$_2$CO$_3$, NMP, 140 °C) (Figure 2.6). The supported precatalyst **49** is recyclable three times. It withstands several days heating at 140 °C in the reaction media with only partial cleavage. It was shown that chemical bonding between silica surface and complex is essential for such stability, since neither a physically absorbed pincer complex nor a mechanical mixture survived such conditions [106].

Activated aryl chlorides have been reported many times to undergo high-yielding Mizoroki–Heck couplings under the same conditions as unactivated aryl bromides using palladium precatalysts such as palladacycles [80, 81, 107–109] and heterocyclic carbene complexes [33, 110–115], as well as systems involving phosphite ligands [116], phosphinous acids [117], dialkylsulfides [79], and palladium supported on montmorillonite K10 clay [105]. All these processes clearly show typical features of type 2 processes. A thorough investigation of representative systems of this sort performed by Zapf and Beller revealed several characteristic features; broad variation of several parameters (ligand, additive, bases and solvent) showed no comprehensible trends in their variation (**50** → **51**, Scheme 2.11, Table 2.3) [118]. Moreover, a change of any of these parameters (e.g. base) reshuffles the trends established for the variation of another (e.g. additive). Figure 2.7 shows compound **52** (part of Table 2.3).

The question at to whether type 2 catalytic systems might be adjusted to processing nonactivated aryl chlorides is of particular interest. Many investigations show that phenyl chloride and even electron-rich aryl chlorides are not totally unreactive, and poor to modest yields of Mizoroki–Heck products were obtained in many systems. This prompts the important conclusion that aryl chlorides are able to (1) enter catalytic cycles in the absence of a catalyst bearing special ligands (see below) and (2) successfully run through a few turnovers but rapid catalyst deactivation prevents reliable operation.

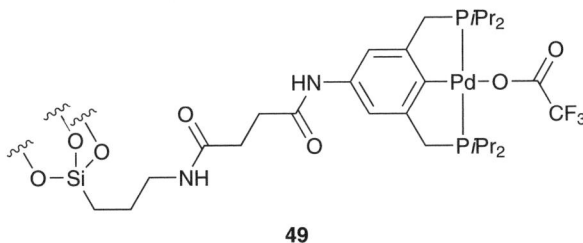

Figure 2.6 *Alper's silica-immobilized pincer complex.*

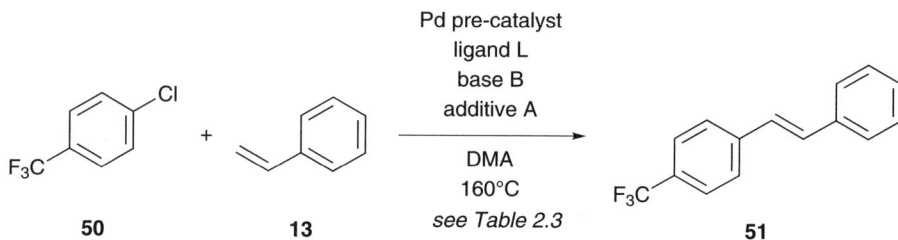

Scheme 2.11

The earliest observation that phenyl chloride is sufficiently reactive in a Mizoroki–Heck reaction was reported in 1984: Pd(OAc)$_2$ (2 mol%), 1,2-bis(diphenylphosphino)ethane (dppe) or Ph$_3$P, NaOAc, DMF–H$_2$O, 130 °C. In the reaction with styrene, stilbene was obtained in 45–53% yield [119]. Essentially identical results were disclosed simultaneously by Spencer [74]. Thus, it is not the activity but, rather, the longevity of the catalytic system limiting such processes. If the conditions are thoroughly optimized, then it could well be that the lifetime of the active catalytic species will be extended. The first – and one of the very few demonstrations of that – was given by Reetz *et al.* [86], who showed that, in the presence of the phosphonium salt Ph$_4$PCl (**56**), phenyl chloride (**55**) furnishes Mizoroki–Heck product **60** in almost quantitative yields (TON = 48): Pd(MeCN)$_2$Cl$_2$, Ph$_4$PCl (6 equiv based on palladium), NaOAc, NMP, 150 °C (Scheme 2.12). The role of phosphonium salt additive **56** is likely to serve in a conjugate Mizoroki–Heck cycle (Scheme 2.12, right), which helps to reactivate the palladium catalyst **54**. The key intermediate of the conjugate cycle **58** loses the neutral Ph$_3$P ligand at the high temperature employed, thereby providing a bypass route (**53** → **54** → **58** → **57**) to the key intermediate **57** of the main cycle (Scheme 2.12, left). This strategy, however, is not effective if applied to less reactive aryl chlorides, since, in this case, the conjugate becomes faster than the main cycle. This, in turn, creates a situation of competitive consumption and fast depletion of the phosphonium salt, which leads to scrambling of aryl groups from the aryl chloride and

Table 2.3 *Beller's systematic screening for aryl chloride activation*

	Conditions				Screen	
Entry	Pd	L	B	A	of	(Yield [%])
1	Pd(OAc)$_2$	Screen	Na$_2$CO$_3$	*n*-Bu$_4$NBr	L	*n*-Bu$_3$P (83) > Cy$_3$P (71) ≫ *t*-Bu$_3$P (31)
2	**52**	None	screen	*n*-Bu$_4$NBr	B	CaO (67) > NaOAc (64) > KOAc (61), Na$_2$CO$_3$ (61) ≫ K$_3$PO$_4$ (39) > *n*-Bu$_4$NOAc (31) > Cs$_2$CO$_3$ (25) ≫ Et$_3$N (12)
3	**52**	None	NaOAc	Screen	A	*n*-Bu$_4$NBr (64) ≫ *n*-Bu$_4$NCl(aq) (36) > *n*-Bu$_4$NI (35) > *n*-Bu$_4$NOAc (34) ≫ *n*-Bu$_4$NCl (21) > *n*-Bu$_4$NHSO$_4$ (16)
4	**52**	None	Na$_2$CO$_3$	Screen	A	*n*-Bu$_4$NCl(aq) (89) ≫ *n*-Bu$_4$NBr (61) > *n*-Bu$_4$NHSO$_4$ (58) ≫ *n*-Bu$_4$NOAc (43) ≫ *n*-Bu$_4$NCl (29)

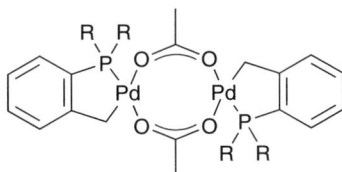

52 (R = 2-tolyl)

Figure 2.7 *Herrmann–Beller palladacycle.*

the phosphonium salt. On the other hand, with aryl bromides this system works effectively to provide one of the most effective published type 2 systems.

The use of a supported palladium complex in the Mizoroki–Heck coupling a phenyl chloride was demonstrated to provide high TON (23 000) and yield (89%) in the reaction with styrene [120].

Another spectacular result was obtained with phosphinite pincer palladacycle (**61**, Figure 2.8) [121]. While this precatalyst delivered typical type 1 performance in DMF (rather modest for this class), this complex is able to process even electron-rich aryl chlorides in high yields in the less polar solvent dioxane in the presence of the rarely used base CsOAc. Given i-Pr$_2$P groups as sidearms, a true ligand-accelerating effect could be suspected. However, the reaction is rather slow at 0.67 mol% palladium loading, requiring 5 days at 120 °C or 1 day at 180 °C. Apparently, the latter temperature is too high for any definitely ligated palladium species to dominate. Therefore, this system is more likely to represent a finely optimized type 2 process, which shows that the SRPCs are even applicable to the least reactive substrates provided there is an optimal match of release rate and catalytic cycle rate.

Thus, in order to afford good results, each type 2 reaction (for each pair of substrates) should be optimized anew and the optimal source of palladium must established. Type 2 reactions, therefore, justify the immense amount of work on various palladium

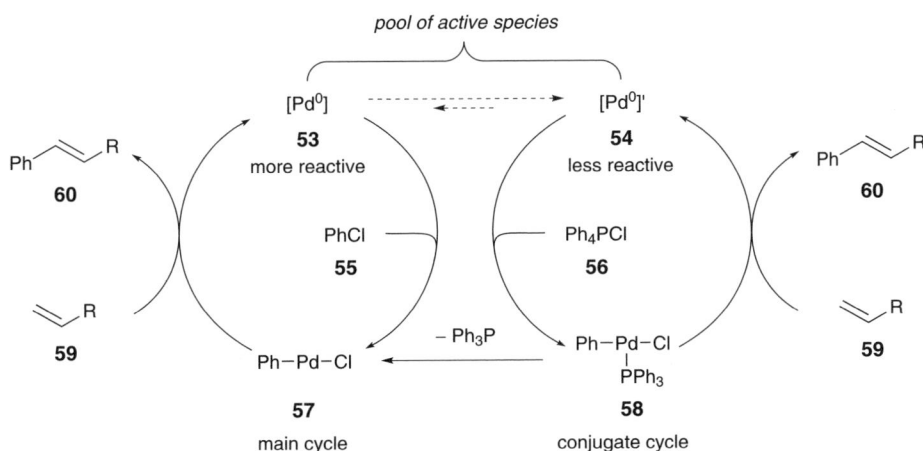

Scheme 2.12

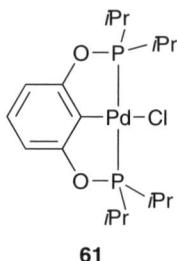

61

Figure 2.8   *Phosphinite pincer palladacycle.*

precatalysts, including innumerable palladacycles, carbene complexes, phosphine and nonphosphine complexes, because any of these might provide the optimal balance of activation–inactivation–catalysis rates.

2.3.3   The Type 3 Catalytic System

This type might be designated as the Hartwig–Fu version of the Mizoroki–Heck reaction, as these researchers' groups proposed the first unambiguous realizations of such catalytic cycles [122, 123]. Non-activated aryl chlorides cannot be reliably processed in reasonable yields using any of the type 1 or type 2 systems. The presence of purpose-built ligands facilitating the oxidative addition is obligatory in order to achieve preparatively useful yields. This type encompasses true *ligand-accelerated* [124] or *ligand-assisted* (phosphine-assisted [4]) processes.

Historically, this distinction was first made by Heck on the basis of two different systems, one based on a simple palladium salt for aryl iodides [2] and another based on the Ph₃P or (2-tolyl)₃P complex of palladium for aryl bromides [6–8]. However, as the performance of these initial systems was far from the levels achieved later, we cannot conclude for sure that a ligand-accelerating effect is observed in these protocols. We know today that aryl iodides and reactive aryl bromides are highly reactive practically with any form of labile palladium complex, so that the ligand-accelerating effect cannot be reliably established for these substrates in the many published protocols (type 1 and type 2 systems).

At least for aryl iodides, the application of phosphine complexes of palladium still continues to be practised in complex preparations where mild conditions are desirable to secure high selectivity. In these cases, halide scavenger additives are often used to enable the polar pathway, because phosphine ligands under mild conditions block the coordination places need for alkene binding. Owing to the low reactivity of those systems, high loadings of precatalyst are only tolerable in complex molecule synthesis. For example, this approach enabled clean vinylation of vinylic boronate **63** to afford complex diene **64** (**62 → 64**, Scheme 2.13) [62].

From the discussion of type 2 Mizoroki–Heck reactions, we have seen that aryl bromides and chlorides can be made reactive without the application of certain ligands. Thus, the diagnosis as to whether a given system might be considered as a true ligand-accelerated one is made based on a simple observation: if a given ligand catalysed the Mizoroki–Heck coupling under markedly milder conditions (that is, within a few hours or less and at

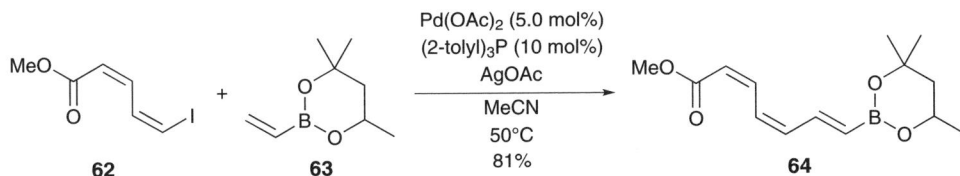

Scheme 2.13

temperatures below 100 °C), then we might tentatively assume that the ligand-accelerating effect is operating.

This assignment is easy for electron-rich aryl chlorides. So far, none of simpler catalytic systems has been found to be able to process donor-functionalized aryl chlorides with good to high yields reliably. A good benchmark substrate for this test is 4-chloroanisole. Thus, if a given catalyst provides reasonable yields for this substrate under mild conditions, it might be a clear indication of ligand acceleration. It should be noted that this classification is only based on the substantial extant body of data on reactions with standard alkenes, acrylates and styrenes. With a limited set of data available for nonstandard alkenes, this classification might become biased and inconclusive.

A tentative catalytic cycle for ligand-accelerated Mizoroki–Heck reactions is shown in Scheme 2.14. A pivotal feature of this mechanism is the requirement to have neutral complex **4** with a single ancillary ligand, as related coordinatively saturated species **1** or

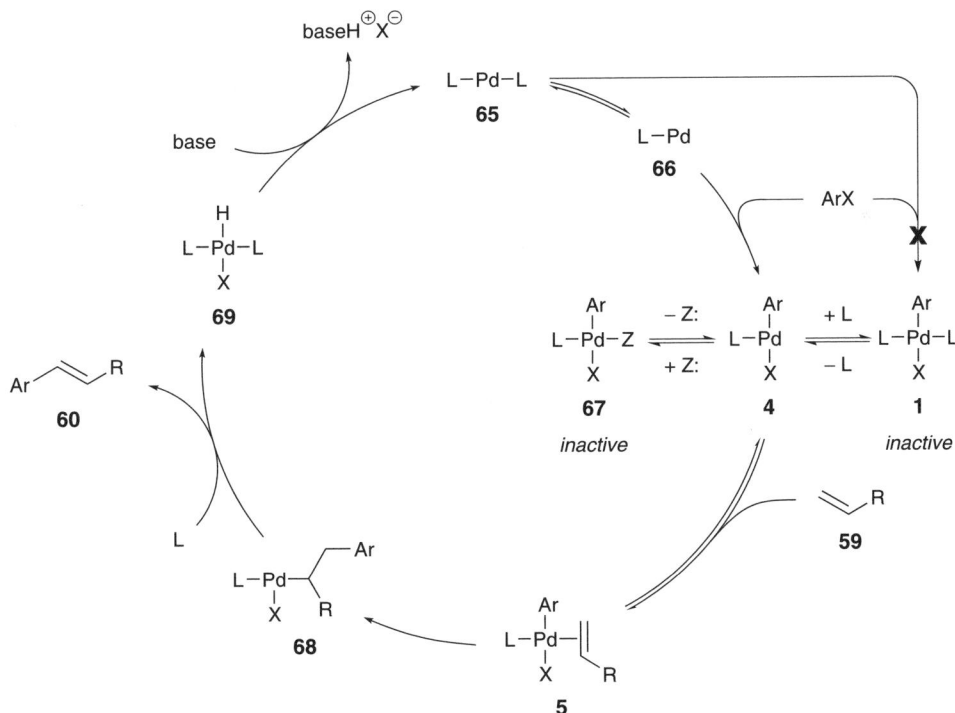

Scheme 2.14

67 otherwise do not offer a free coordination site for the alkene (**4** → **5**). It is thus obvious that this ancillary ligand must the accelerating ligand and that there is no space for useless spectator ligands. Complexes with two ancillary ligands **1** apparently cannot coordinate the alkene effectively and, therefore, are inactive. In this mechanism, any components which might serve as ligands for palladium should best be avoided, because the presence of such components diminishes the concentration of active species with a single ancillary ligand **4**. In ligand-accelerated protocols, the use of (a) polar coordinating solvents, (b) nucleophilic bases and (c) any other coordinating additives should be avoided. These protocols should also rely on both electron-rich and bulky ancillary ligands, because this combination is a prerequisite for the formation of highly active, coordinatively unsaturated palladium complex **66** [125]. Complexes of type **66** are activated towards oxidative addition into unreactive substrates.

Here lies one other ambiguous point. While type 1 systems do show enormous TONs and TOF values and type 2 systems are able to show high TONs (usually of the order of 10^3, but 10^5 cycles are feasible by very careful optimization), type 3 or ligand-accelerated systems usually require generous loadings of palladium (rarely less than 0.5 mol%). This might be regarded as a paradox – indeed, the *ligand-accelerated* systems are markedly slower and less effective than common type 2 or type 1 systems, in which there is no proven ligand-acceleration effect. This paradox is misleading, however. High loadings of palladium precatalyst in the type 3 processes do not mean that all palladium is able to participate in the catalysis. As soon as type 3 processes are conducted under milder conditions (lower temperatures), the preactivation (**65** → **66**, Scheme 2.15) is less effective and most of the palladium loading is likely to remain in the inactive form of **65**. If the temperature is raised further, then deligation of palladium complex **66** with the special ligand would lead to nonligated unsupported palladium(0) **36**, which is not suitable for reactions with less reactive substrates due to deactivation (**36** → **70**). This picture also shows that the borderline (between type 2 and type 3) processes are mechanistically imprecise. Therefore, we deduce this classification not from unrecognizable mechanistic, but on observable preparative criteria.

The first systematic attempt to solve the problem and to develop a rational approach towards Mizoroki–Heck reactions of unactivated aryl chlorides was performed by Milstein and coworkers [11, 126]. These authors described analogues of bidentate 1,3-bis(diphenylphosphino)propane (dppp) and 1,4-bis(diphenylphosphino)butane (dppb) ligands with *i*-Pr instead of the phenyl groups to make their ligands (e.g. **14**, Scheme 2.3,

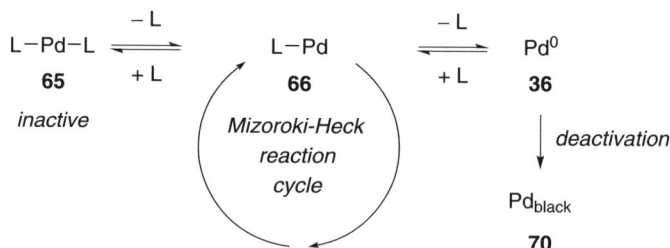

Scheme 2.15

see above) both bulky and electron rich. The palladium complexes of these ligands, in particular the dppb derivative [126], were indeed capable of smooth oxidative addition of aryl chlorides. However, as the ligand is bidentate, the polar pathway should be involved, which was ineffective in this particular case.

The problem of how to make aryl chlorides usable in palladium-catalysed reactions has particularly attracted considerable attention in the last decade [127–129]. The task to use aryl chlorides and to obtain preparatively useful yields and TONs for a wide selection of substrates might only be solved by application of monodentate ligands. These are actively involved in increasing the reactivity of palladium(0) species towards the oxidative addition to the unactivated $C(sp^2)$–Cl bond.

By making use of high-throughput screening, Hartwig and coworkers [123] discovered that less reactive substrates do participate in Mizoroki–Heck reactions in the presence of electron-rich bulky phosphine ligands $R(t\text{-Bu})_2P$ (R = t-Bu, Ph, Fc): $Pd(dba)_2$ (dba = E,E-dibenzylideneacetone; 3.0 mol%), $R(t\text{-Bu})_2P$ (6.0 mol%), NaOAc, DMF, 110 °C. Of these ligands, $t\text{-Bu}_3P$ has been identified as the most promising ancillary ligand for developing reliable protocols for Mizoroki–Heck reactions of aryl chlorides and bromides. Simultaneous with the efforts of Hartwig and coworkers [123], a similar protocol was disclosed by Littke and Fu [122]: $Pd_2(dba)_3$ (1.5 mol%), $t\text{-Bu}_3P$ (6.0 mol%), Cs_2CO_3, 1,4-dioxane, 100–120 °C. These protocols had their shortcomings, which were overcome using the sterically hindered tertiary amine Cy_2MeN as a base [130]. Unactivated aryl chlorides are now reactive towards both mono- and di-substituted alkenes (**71** → **72**, Scheme 2.16).

In this system, activated aryl chlorides and all kinds of aryl bromides are already reactive at room temperature. Instead of free $t\text{-Bu}_3P$, which is highly air sensitive, the air-stable salt $[t\text{-Bu}_3PH]BF_4$ (1.0–3.0 mol%) is applicable in combination with $Pd_2(dba)_3$ (0.50–1.5 mol%), Cy_2MeN, 1,4-dioxane, 120 °C or even at room temperature [131].

Again, in the true *ligand-accelerated* reactions, the temperature should be kept within optimal limits. An increase of temperature above 110–120 °C is not useful, since, owing to the ligand dissociation equilibria, phosphine ligands would dissociate, liberating nonphosphine palladium catalysts. These are then unreactive towards deactivated aryl chlorides under those conditions, which, in turn, do not resemble type 2 processes because these require polar coordinating solvents and coordinating bases (see above).

Of course, the remaining amounts of monophosphine complexes would even operate at higher temperatures and in the presence of coordinating solvents and bases, but such protocols are expected to be unstable (type 2 systems). For example, Indolese and coworkers [132, 133] have shown that their system, namely $PdCl_2$ (1.0 mol%), $t\text{-Bu}_3P$ (2.0 mol%) and

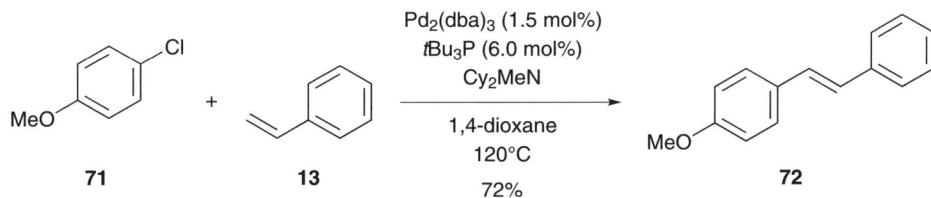

Scheme 2.16

Na$_2$CO$_3$, affords high yields in the reaction of 4-chloroanisole (**71**) with *n*-butyl acrylate (**38**) in DMA at 140 °C. This is likely to be the highest temperature that has appeared in literature for ligand-accelerated Mizoroki–Heck reactions of unreactive aryl chlorides. In the same protocol, several bulky secondary phosphines, namely bis(adamantyl)phosphine, bis(norbornyl)phosphine and *t*-Bu$_2$PH, were also operative yet inferior to *t*-Bu$_3$P [133]. Unlike the Hartwig–Fu protocol, this method has found no application so far, which is why its scope cannot be reliably assessed.

Quite recently, the arsenal of methods for aryl chlorides in the Mizoroki–Heck reaction was broadened by using the cheaper complex (Cy$_3$P)$_2$PdCl$_2$ with Cs$_2$CO$_3$ as the base in 1,4-dioxane at 120 °C [134]. The reported scope of substrates is broad, involving several nonstandard alkenes and many aryl chlorides; electron-rich aryl chlorides, except for 4-chlorotoluene and sterically hindered coupling partners, were excluded in this study.

On balance, the first steps have been made towards general type 2 and type 3 systems for processing aryl chlorides. The former is substrate specific and requires separate optimization for any pair of reagents and is, at best, applicable to aryl chlorides not bearing strongly electron-donating or bulky substituents; the latter is practically limited to the Hartwig–Fu protocol [122, 123, 130, 131], which is broad in scope at the expense of high catalyst and ligand loadings. Despite these fundamental advances, a general room-temperature protocol for Mizoroki–Heck reactions of aryl chlorides is still unavailable. It should be emphasized that, for many C–C and C–Het (Het = heteroatom) cross-coupling reactions, a number of effective (including room-temperature) protocols have been developed making aryl chlorides useful and convenient substrates. Apart from the need to use aryl chlorides in Mizoroki–Heck chemistry, type 3 systems solve the important task of using all kinds of aryl bromides under the very mild, reliable conditions often needed in complex syntheses.

Various variants of the Hartwig–Fu protocol have been used for the synthesis of optical materials, such as: pyrene-functionalized silsesquioxane for LED materials, Pd(*t*-Bu$_3$P)$_2$ (1.0 mol%), Cy$_2$MeN, toluene, 80 °C [135]; triphenylsilane-modified triarylamines **73** (previous conditions) [136]; diketone-modified styrene monomers **74**, Pd$_2$(dba)$_3$ (1.5 mol%), *t*-Bu$_3$P (3.0 mol%) (Cy$_2$MeN, 1,4-dioxane, r.t.) (Figure 2.9) [137, 138]. Other applications include stereospecific stepwise one-pot bis-arylation of vinylsulfide to form **75** [139] and tandem reactions involving Mizoroki–Heck reactions – Pd(OAc)$_2$ (10 mol%), [*t*-Bu$_3$PH]BF$_4$ (20 mol%), K$_2$CO$_3$, DMA, 130 °C [140]. The mild conditions of this method enabled the selective diarylation of vinylboronate **76** (**76** → **77**), which turned out to be a

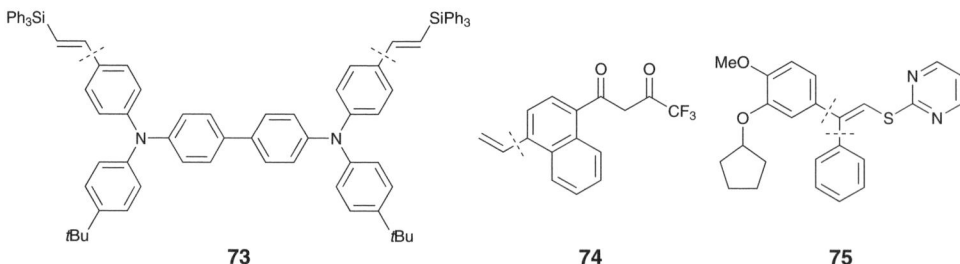

Figure 2.9 *Compounds made using the Hartwig–Fu protocol.*

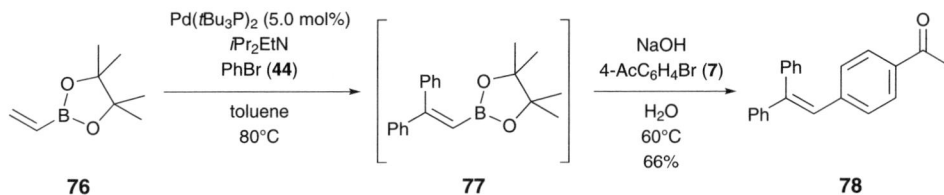

Scheme 2.17

good basis for the construction of various trisubstituted alkenes (**76** → **78**, Scheme 2.17) [141, 142].

Currently, it seems that, in what concerns the practical needs of synthesis, *ligand-accelerated* Mizoroki–Heck reactions depend completely on a single ligand, namely *t*-Bu$_3$P. Several related promising ligands identified by Hartwig and coworkers [123] in high-throughput screening tests have found no application in intermolecular Mizoroki–Heck chemistry.

TONs of 10 000 were achieved with an analogue of 1,1′-bis(diphenylphosphino)ferrocene (dppf; **79**, Figure 2.10) in the Mizoroki–Heck reaction of inactivated aryl bromides: [(η^3-allyl)PdCl]$_2$, **79**, K$_2$CO$_3$, DMF, 130 °C [92]. In fact, this is an excellent value for a type 3 ligand-accelerated reaction. However, the reaction conditions – polar solvent, coordinating base and rather high temperature – might suggest that this is yet another example of a type 2 process [92]. Several other potentially useful ligands belong to the family of 2-diorganophosphinobiphenyls, of which many representatives are highly popular in the cross-coupling reactions. In the Mizoroki–Heck reaction their performance so far is humble. Thus, sterically hindered ligands **80** showed good activity comparable to that of *t*-Bu$_3$P: Pd$_2$(dba)$_3$ (3.0 mol%), ligand **80** (R = Ph, 12 mol%), Cs$_2$CO$_3$, dioxane, 100 °C [143]. However, the true potential of this system cannot be reliably judged from a single test published. Another related phosphine, **81**, gave excellent results with highly electron-rich and sterically hindered aryl bromides: Pd$_2$(dba)$_3$ (1.0 mol%), ligand **81** (4.0 mol%), Cs$_2$CO$_3$, 1,4-dioxane, 150 °C [144]. A recent addition to this growing family of ligands is the very interesting electron-rich aminodiphosphines, which form an unusual four-membered chelate **82** with palladium [145]. These complexes were tested as precatalysts

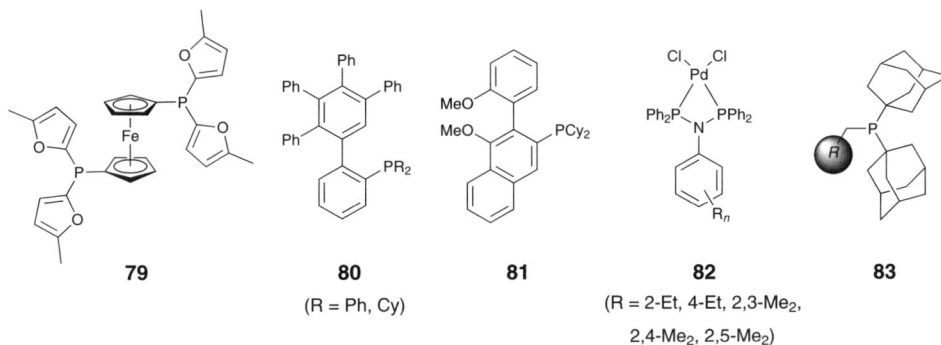

Figure 2.10   *Ligand motifs **79–83**.*

with aryl bromides, including both activated 4-bromoacetophenone (**7**), deactiviated 4-bromoanisole (**8**), and 4-bromotoluene; their impressive performance at 80 °C in 1,4-dioxane is also reflected in high chemical yields. Mild conditions and solvent unequivocally reveal a true *ligand-acceleration* effect: LPdCl$_2$ **82** (R = 2-Et or 2,5-Me$_2$, 0.001 mol%), K$_2$CO$_3$, 1,4-dioxane, 80 °C. We speculate that the strained four-membered ring is cleaved under these conditions; thus, the ligand behaves in a monodentate fashion.

As type 3 catalytic systems require high loadings of palladium and expensive phosphines, it is apparent that the development of economical protocols is highly desirable. However, these systems are not well suited for low-leaching recyclable systems. Bulky phosphines cannot form highly stable complexes, and deligation is a usual event. For example, it has been shown that bisphosphine complexes PdL$_2$ (**65**) with such ligands readily lose one ligand at as low a temperature as 70 °C [125], and deligation of the second ligand should require only marginally harsher conditions. Besides, such catalysts should be readily degradable by oxygen as well. Therefore, only a few attempts were undertaken and the results are modest. Diadamantylphosphine linked to linear polystyrene resin **83** (Figure 2.10) has been rather effective in the Mizoroki–Heck reaction of aryl bromides (including electron rich) with *n*-butyl acrylate: Pd(dba)$_2$ (0.5 mol%), ligand **83** (1.0 mol%), *i*-Pr$_2$NH, NMP, 100 °C [146]. Recyclability could not be achieved due to deterioration of the membrane by the solvent.

2.3.4   The Type 4 Catalytic System

This type of Mizoroki–Heck reaction is often referred to as the 'polar pathway' [60, 147, 148]. Bidentate phosphines are rarely used in nonasymmetric Mizoroki–Heck reactions, as against other C–C and C–Het cross-coupling reactions. However, in Mizoroki–Heck chemistry, a very important role is reserved for them to serve as ancillary ligands in various regio- and stereo-selective protocols. In such reactions, selectivity is governed by a well-defined coordination sphere of palladium, which requires the bidentate ligand to remain coordinated as chelate throughout. This polar mechanism was postulated to rationalize such catalytic cycles. Its main feature is the necessity to dissociate an anionic ligand in order to obtain a coordination site for the alkene coordination–migratory insertion event (**85** → **86** → **87**, Scheme 2.18).

Any pathways involving partial or full dissociation of bidentate ligands (**85** → **88** → **89** or **84** → **90** → **92**) should lead to the loss of configuration and, as a result, to nonselective side-reactions. Thus, the design of catalytic systems for type 4 reactions should, as a main goal, select those bidentate ligands which maintain a stable coordination environment under discrete reaction conditions, thereby suppressing unselective transformations. The most important function of ligands might be designated as being *restrictive*, to restrict all potential transformations to a single selective pathway (e.g. **84** → **87**, Scheme 2.18). The precatalysts or catalytic systems should also be designed in a way so that bis-chelation leading to inert complexes **91** is avoided (**84** → **90** → **91**) [149].

Type 4 Mizoroki–Heck reactions require mild conditions. Partial or full dissociation of bidentate ligands is not expected to require very high temperatures (cf. discussion of various SRPC), at which type 4 precatalysts would simply turn into yet another SRPC. Indeed, Shaw and Perera [149] showed that (L)PdCl$_2$ complexes with various bidentate phosphines (including dppf, dppe and dppp) make useful precatalysts for Mizoroki–Heck

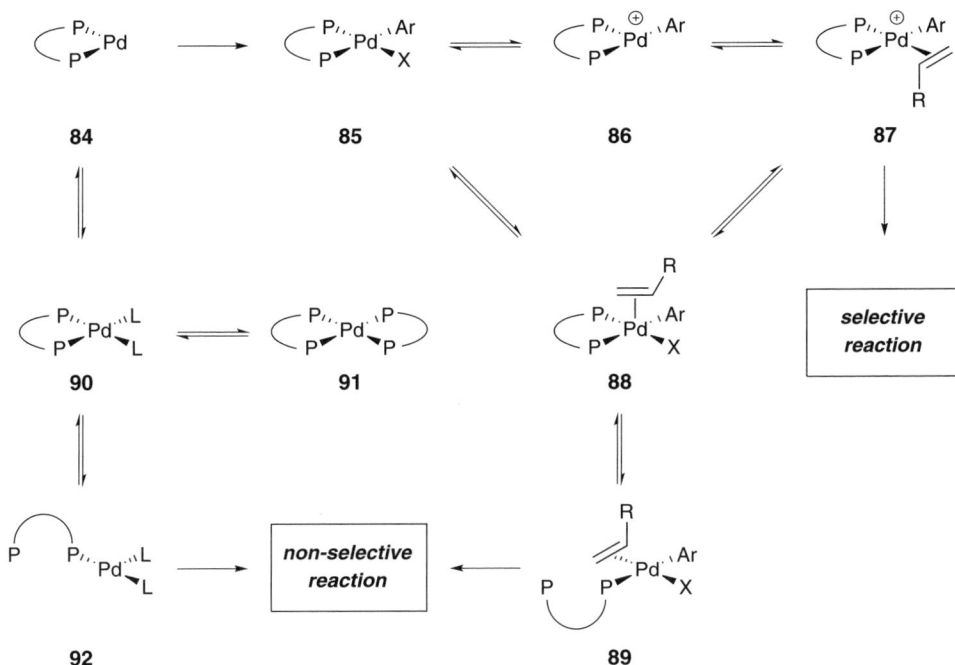

Scheme 2.18

reactions of aryl iodides and activated aryl bromides in type 1 reactions if used at elevated temperatures.

There are two major realizations of the polar pathway in intermolecular Mizoroki–Heck reactions: (1) enantioselective arylation of cyclic alkenes (Chapter 11) and (2) regioselective internal arylation of terminal alkenes (Chapter 3).

2.4 Palladium Precatalysts in Type 1 and Type 2 Mizoroki–Heck Reactions

2.4.1 SRPCs

As deduced from the discussion above, the true catalytically active species in type 1 or type 2 processes possesses undefined coordination shells independent (!) of the palladium precursor. Many of those precursors belong to what we prefer to call the SRPCs; that is, compounds which serve as a continuous reservoir of palladium throughout the process. Such compounds are either preformed complexes or a combination of a simple palladium source with an appropriate strongly binding (usually chelating) ligand. In the latter case, the ligand temporarily binds any excess of palladium as a relatively stable complex, the reservoir (Scheme 2.19). Systems in which the ligand is added in a stoichiometry (based on palladium) that is insufficient for quantitative formation of the most stable palladium–ligand pair are usually more effective. Monophosphines, depending on their steric bulk, readily form complexes with two to four ligand molecules per palladium. Thus, in the case of the preparatively often used 1 : 1 palladium : ligand ratio, a mixture of phosphine-free

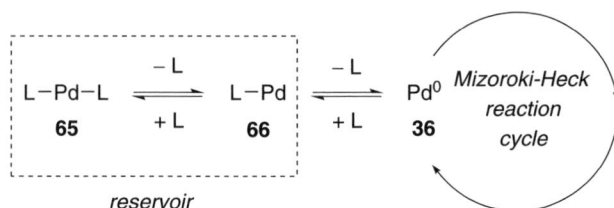

Scheme 2.19

catalyst **36** in equilibrium with mono- (**66**), di- (**65**), tri-, and tetra-phosphine complexes are obtained (Scheme 2.19). The more active phosphine-free palladium complexes **36** sustain the catalytic cycle, while any excess of palladium is stabilized in less active phosphine complexes as **65** or **66**, thereby preventing the formation of metal particles.

Classical monophosphines, such as Ph_3P and $(2\text{-tolyl})_3P$ introduced to the Mizoroki–Heck reaction by Heck and coworkers [6–8] and Spencer [77], are likely to belong to this class. Thus, in the nontrivial Mizoroki–Heck reaction of vinyl tosylate **93** with methyl acrylate (**94**), a mixture of $Pd(OAc)_2$ with Ph_3P (P/Pd = 3 : 1) produces a completely inactive system. Conversely, high yields of the desired product **95** are obtained either by portionwise addition of $Pd(OAc)_2$ or by the use of $Pd(OAc)_2$ and Ph_3P in a 1 : 1 or, even better, 1 : 0.9 ratio (Scheme 2.20) [150].

SRPCs fulfil different functions in type 1 and type 2 Mizoroki–Heck reactions. In type 1 reactions, which are generally very fast, the use of SRPCs is justified mainly by optional recycling. These reactions are effectively catalysed by very low amounts of soluble palladium. In principle, any form of palladium precursor will do; an SRPC simply facilitates handling. Another incentive is that losses of palladium at the preactivation step are avoided. This is particularly important in heterogeneous reaction systems, in which the overall catalytic efficiency might be limited not by the kinetics of the individual steps of the catalytic cycle, but by more subtle and poorly predictable transfer processes between separate phases. Again, this might result in instabilty of the catalytic process through local accumulation of palladium(0).

2.4.2   Nanoparticles

The involvement of very small particles of palladium metal ranging in size from a few to hundreds of nanometres in Mizoroki–Heck reactions, as well as in other palladium-catalysed reactions, is well-known.

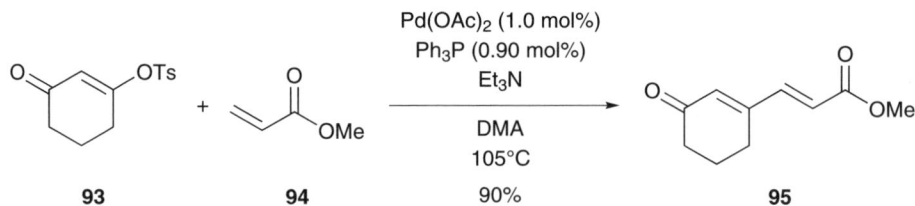

Scheme 2.20

At the early stages, palladium sols seemed very promising, which came from the belief that finely dispersed metals possess outstanding activity in various processes; for example, in the reduction of organic halides. Bearing in mind that Mizoroki–Heck reactions and related cross-coupling processes are largely limited by the oxidative addition of organic halides, high hopes for processing less reactive aryl bromides or even chlorides with palladium dispersions were not unfounded. Obviously, this would have been a bright perspective for Mizoroki–Heck chemistry, because such particles are readily recyclable and likely to be tunable by well-developed control of dispersity and size distribution of metal sols. However, we had already disputed this scenario in 2000 when only limited data were available [4]. Since then, a formidable number of publications dealing with either the use of dispersed palladium or the role of nanoparticles in catalysis have been published [151–153]. In this discussion, we should clearly differentiate between various possible scenarios of the involvement of multinuclear palladium(0) species in Mizoroki–Heck reactions:

Scenario 1. This is the most rigorous, using palladium nanoparticles as true heterogeneous catalysts. The catalytic process in this scenario should occur at metal surface or at special sites where palladium atoms possess free coordination sites required for catalytic action. Arguments in favour of this hypothesis are rather scarce and not convincing [4]. Pertinent to this scenario, two-dimensional palladium nanoparticles were obtained through encapsulation in graphite [154]. Apparently, such nanoparticles have no palladium atoms 'hidden inside', which immediately exposes all palladium content to the reaction media. However, the activity of the thus-prepared catalyst was rather low and it only supported the reaction of aryl iodides with styrene (K_2CO_3, n-Bu_4NBr, DMF, 100 °C). The authors reported that the encapsulated nanoparticles had stayed intact after the reaction because their initial morphology is not altered during the reaction. Arguments to the contrary are many:

(a) On the surface of palladium sols obtained under ordinary conditions – that is, without rigorous exclusion of oxidizing substances – all active sites are oxidized [155–157]. Therefore, these sites must be reactivated before entering Mizoroki–Heck catalytic cycle. This is certainly feasible but has not been realized so far. Moreover, it is well known that it is the saturation of free valencies of surface metal atoms by ligands (including oxidation) which actually protects the cluster (nanoparticle) from further aggregation and growth [157]. Thus, these phenomena – the activation of surface for catalytic activity by removal of all stabilizing ligands and the very existence of small clusters (nanoparticle) – are mutually incompatible. This has been excellently demonstrated in the fundamental studies of Schmid *et al.* [156] and Moiseev and Vargaftik [155].

(b) A Mizoroki–Heck reaction requires at least three free coordination sites at palladium. In palladium clusters, only vertex atoms satisfy this criterion. Vertex atoms are, however, rare. In sols with particle sizes larger than 8 nm, less than 1% of the total number of palladium atoms are at vertices [158, 159]. In larger palladium particles, catalytic activity is also associated with crystal lattice defects [160].

(c) Full dissolution of palladium sols and even palladium black sediments under the conditions of a typical Mizoroki–Heck reaction has been described. If this does happen once, then why should we suppose that in some other cases it does not?

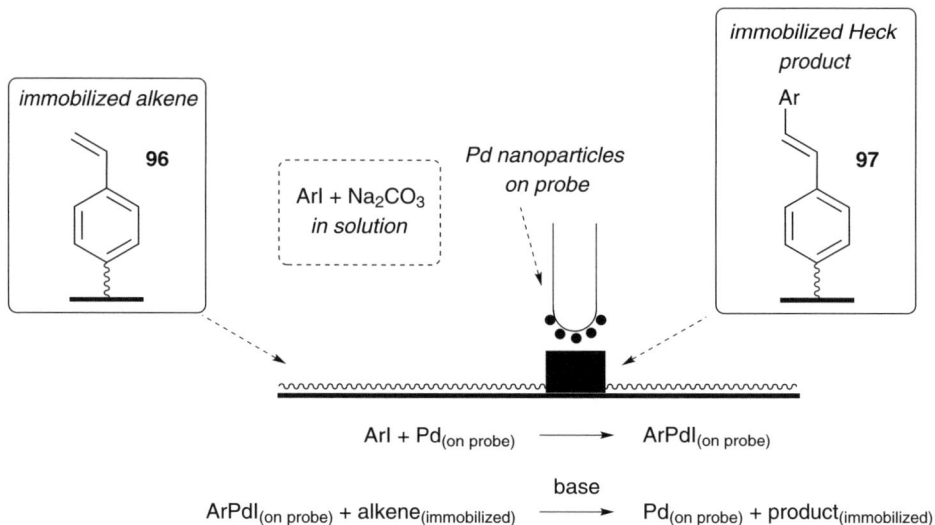

immobilized alkene

96

Arl + Na$_2$CO$_3$
in solution

Pd nanoparticles
on probe

immobilized Heck
product

Ar

97

$$Arl + Pd_{(on\ probe)} \longrightarrow ArPdl_{(on\ probe)}$$

$$ArPdl_{(on\ probe)} + alkene_{(immobilized)} \xrightarrow{\ base\ } Pd_{(on\ probe)} + product_{(immobilized)}$$

Scheme 2.21

(d) All investigations of palladium sols (nanoparticles) as Mizoroki–Heck catalysts have been performed with undemanding substrates under harsh reaction conditions. We know very well that these reactions would also proceed in the presence of trace amounts of palladium. This parameter cannot be reliably controlled.

Thus, this scenario might indeed be regarded as unrealistic for reactions on a preparative scale. However, it should not be discarded on micro- or submicro-scale. Actually, a very interesting implementation of it was demonstrated using atomic force microscopy (AFM) [161]. A nanometre-size probe carrying a layer of palladium nanoparticles leaves a clear trace of Mizoroki–Heck product **97** in the system, in which the alkene **96** is immobilized on the surface while the others are in solution (Scheme 2.21). The trace is drawn only where the probe is in contact with the surface, which makes possible the achieving of incredible resolutions of 10–15 nm, with only insignificant degree of blurring owing to escape of reacting system into the volume of the media. Indeed, if the Ar–Pd–I intermediate were to detach from the surface, then Mizoroki–Heck product would form at a distance from the probe and then resulting in blurring of the trace.

It should be noted that, in other catalytic reactions involving a fewer number of components and running under more tolerant conditions than in the Mizoroki–Heck reaction, nanoparticles and other finely organized nanostructures do by all means perform as well-defined catalysts providing catalytic sites involved in the nanostructure [162].

Scenario 2. The reaction proceeds in solution by leaching, but palladium is continuously resedimented on surface(s) for various reasons. It performs a shuttle service. This scenario (if operative) can be the basis of truly recyclable, net leaching-free systems. However, it is afflicted with a major shortcoming, namely the inevitable change of morphology of the initial metal particles due to Ostwald ripening [163]. Redistribution of matter among a multitude of small particles is governed by surface energy and entropy; thus, the initial fine

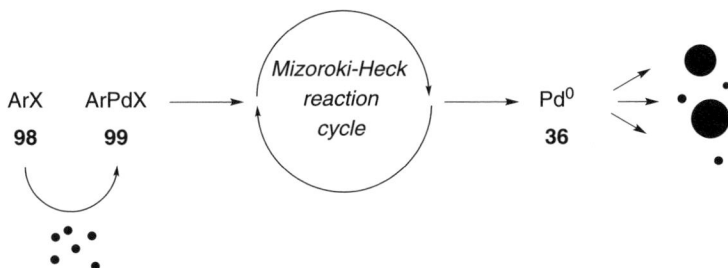

Scheme 2.22

dispersion containing many small nanoparticles turns into a polydisperse system, in which smaller particles tend to disappear through dissolution while larger particles grow larger until the activity decreases to inappropriate levels (Scheme 2.22).

Change of morphology accompanying the decrease of activity during the reaction was detected almost each time when this task was pursued by researchers [164–166]. Thus, in a system with uncontrolled exchange between nanophase (palladium nanoparticles dispersed in liquid media or supported over a surface), the Mizoroki–Heck reaction leads to degeneration of the nano-precatalyst. During such processes, the activity of palladium metal nanoparticles decreases dramatically due to an increase in polydispersity, accumulation of large crystallites, inclusion of impurities and other unidentified ageing processes [164–166].

It would thus be highly desirable to design systems in which forces would act against the degeneration to maintain monodispersity and the initial level of activity throughout. Not much data exists thus far in favour of this possibility. However, a solution could be sought in those systems where palladium nanoparticles function in organized nanoenvironments (porous supports or micellar aggregates). These environments might control size distribution of the metal dispersion throughout the catalytic process. Indeed, several papers show that various amphiphilic block copolymers might provide this service more or less effectively. These copolymers are built of hydrophobic and hydrophilic polymer chains, which form micellar aggregates is solution.

As early as 1997, Antonietti and coworkers [167] showed that palladium nanoparticles supported by polystyrene–polyvinylpyrrolidone block copolymer effect the Mizoroki–Heck reaction while retaining a monodisperse size distribution throughout. Very recently, a similar effect was demonstrated for *N*-vinylimidazole–*N*-vinylcaprolactam copolymer, in the presence of which a monodisperse palladium sol was generated directly in the reacting system [168].

Scenario 3. The reaction proceeds in liquid phase by leaching, the extent of which is very low. In this case, the heterogeneous catalyst performs a sacrificial service, but, owing to the slow leaching, the lifetime of such precatalysts is extended to afford high effective TONs (either in a single run or in several separate recyclings).

In the sense of this scenario, palladium dispersions (sols and nanoparticles) are just another group of SRPCs. The assortment of these nano-precatalysts described so far is broad and might be categorized into several distinct groups: (1) sols protected by small ions or molecules; (2) sols protected by low molecular weight surfactants; (3) sols protected by polymers; and (4) sols immobilized on surface or in pores of solid support. Additionally, by proper selection of protection shells, sols can be rendered lipophilic (dispersible in organic solvents of low or moderate polarity), hydrophilic (dispersible in water, aqueous media and other highly polar solvents) or bestowed with affinity to special media (fluorous solvents or scCO$_2$).

In general, the better protected these sols are, the lower their activity is. The history of palladium sols in Mizoroki–Heck chemistry began with the studies by Reetz and Lohmer [169], who showed that palladium sols generated by electrochemical reduction in the presence of propylene carbonate are active enough to be able to give good yields even with chlorobenzene (Na$_2$CO$_3$, 155 °C). This is likely to be the highest activity recorded in Mizoroki–Heck reactions with nano-precatalysts. The main problem with such sols is their poor longevity, which is directly connected with their high activity; a weak protection layer makes them unstable and not capable of long operation. High loadings of precatalysts were required even with reactive substrates. Two other seminal studies dealt with sols protected by tetraalkylammonium salts [170, 171], which gave moderately active precatalysts for type 1 reactions.

Scenario 4. Nanoparticles are formed during the reaction initiated by soluble palladium precatalysts. Our experience in phosphine-free (mostly aqueous) systems led us at an early stage to hypothesize that these protocols go hand in hand with formation of palladium sols [4, 83]. Since 2000, the observation of palladium sols in Mizoroki–Heck reactions has become ubiquitous. So, if one were to decide to compile a list of references that mention the formation of sols, that list would practically coincide with a list of references which deal with the development of new phosphine-free catalytic systems. In fact, the idea that it is the dispersed palladium metal which catalyses (or, more correctly, serves as precatalyst) the Mizoroki–Heck reaction should be traced back to the seminal publications by Heck in 1972 [2], who clearly attributed catalytic activity in a phosphine-free catalytic system to palladium metal formed by reduction of Pd(OAc)$_2$ by the alkene or the amine. This idea fell into oblivion until its rediscovery in the course of a general interest in nanoscale systems in the late 1990s.

Since then, statements that palladium nanoparticles or sols are true catalysts in a given catalytic system have become not infrequent. Indeed, in many cases it is clearly seen that, after the addition of precatalyst, a brownish-grey colour of palladium sol appears immediately and transmission electron microscope measurements reveal the nanoparticles of metal. From the discussion of various preformed sols in Mizoroki–Heck reactions (Scenarios 1–3), it becomes evident that palladium nanoparticles display rather limited activity with respect to scope, TONs and TOF values. On the other hand, in the discussion of various SRPCs, we see many examples of high activity and quite extended scope of substrates reaching to aryl chlorides.

In a given reacting system, usually not all palladium is engaged in the catalytic process. The amount engaged depends on kinetics and the concentrations of reactants. The excess of palladium exists either in palladium(II) (**102**) or as palladium(0) (**36**); both are unstable under the conditions of Mizoroki–Heck reactions and should be protected (Scheme 2.23): palladium(II) (**102**) is unstable towards reduction if not protected in the form of an SRPC (**102** → **103**), palladium(0) (**36**) is unstable towards clusterization and growth of metal particles (**36** → **101** → **70**) and is protected by additives or components of reaction media in the form of anionic complexes **100** [36, 84]. If we consider the reversibility of formation

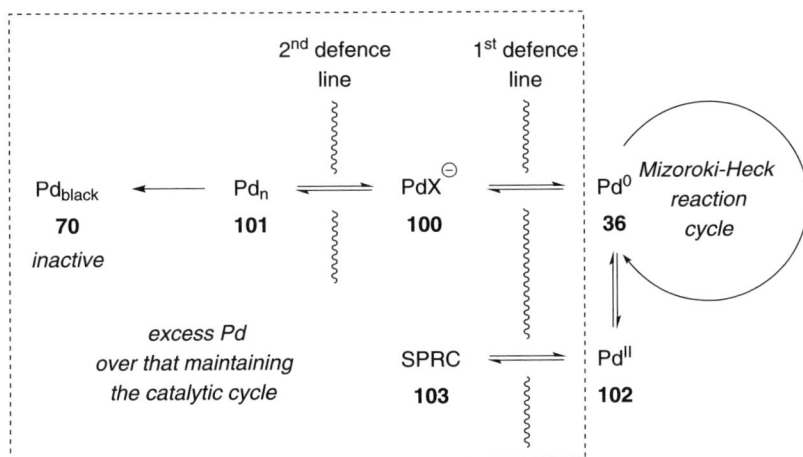

Scheme 2.23

of small clusters and nanoparticles **101**, these species should also be considered as a way to protect excessive palladium(0) from complete inactivation via the formation of large metal particles **70**.

Thus, the formation of palladium sols should be regarded as an indicator of a imbalance between the total amount of palladium in the system and the amount of palladium actually engaged in the catalytic process. The formation of palladium sols, therefore, is readily diagnosed by characteristic colours. In the reactions using supported precatalysts, the formation of sols is a token of inappropriate degree of leaching. The less the accumulation of palladium nanoparticles is, the better are the overall parameters of reacting system, because: (i) palladium in nanoparticles is mostly inactive (only surface atoms are engaged in solution–precipitation processes), which is detrimental to integral parameters such as TONs; (ii) palladium nanoparticles are seeds, which are prone to gathering all palladium in the system because their size steadily increases and palladium is thus being withdrawn from the cycle.

The apparent paradox that, in the systems in which palladium tends to form nanoparticles, TONs increase on decreasing the amount of palladium precatalyst had been noted by us in 1999 [83] and brilliantly and unambiguously established by Reetz and de Vries [36] in a series of studies devoted to the functioning of phosphine-free Mizoroki–Heck reactions. This countervailing trend between the palladium loading and TON certainly has its limitations: as soon as the balance between palladium engaged in the cycle and palladium in the reservoir is disturbed, accumulation of palladium nanoparticles will stop and TONs will decrease.

2.4.3   Supported Catalysts: Leaching versus Recycling

The idea that Mizoroki–Heck reactions in the presence of immobilized precatalysts are catalysed by homogeneous palladium species formed through leaching was declared by Shmidt and Mametova as early as 1996 [172] and further corroborated by subsequent studies [32]. Currently, we believe that no conclusive evidence of the opposite exists;

for discussion of nonleaching claims, see Ref. [173]. In some cases, the probability that immobilized catalyst might provide catalytic centres is rather high, yet still requiring additional solid proof.

The knowledge accumulated so far on supported catalysts in Mizoroki–Heck reactions leaves little room for doubt that these reactions involve leaching of palladium and are actually catalysed by homogeneous dissolved palladium complexes. Thus, supported systems provide another example of SRPCs, which, however, pursue a different goal than common nonsupported SRPCs.

Several types of supported catalyst are to be classified:

1. Palladium complexes, immobilized through appropriate linkers on various supports, might be regarded as a palladium complex precursor. In our opinion, from the viewpoint of reactivity and catalytic performance, there is no significant difference between these and other common precatalysts. All of them enter into the Mizoroki–Heck catalytic cycle via a preactivation procedure.
2. Palladium salts or complexes directly deposited on an appropriate support without a linker, but rather through surface groups native to the support material. These precatalysts are cheaper than the aforementioned systems. Although these possess a less understandable and more disordered structure, their activity in Mizoroki–Heck reactions (type 1 and, very rarely, type 2 systems) is comparable to precatalyst of more sophisticated design and architecture.
3. Palladium metal particles (nanoparticles) embedded into various supports. While such catalysts can be specially prepared, many investigations of suported precatalysts of other types show that sooner or later, during pretreatment or the reaction itself, palladium tends to be reduced and redeposited as clusters and nanoparticles [173]. Consequently, almost any supported catalyst, if not in the first run, turns into a nanoparticles-on-support material on recycling. Moreover, this kind of precatalyst is obtained directly *in situ* if the reaction is run first in the presence of a palladium salt and an appropriate polymer; for example, poly-*N*-vinylpyrrolidone or poly-*N*-vinylimidazole–*N*-vinylcaprolactam. The resulting supported nanoparticle precatalysts are further used in recycling runs [168].
4. Supported liquid-phase catalysts: More detailed consideration of supported and recyclable systems is beyond the scope of the present review and is well covered in a number of recent reviews [151, 174–178].

2.4.4   Carbene Complexes

2.4.4.1   *N*-Heterocyclic Carbene-based Complexes

Heterocyclic carbene ligands are apparently the most important nonphosphine ligand class. Bulky electron-rich carbenes effectively mimic the new generation of phosphine ligands while avoiding the disadvantageous properties of phosphines, such as high toxicity, flammability and high cost. In C–C and C–Het cross-coupling reactions, as well as in other important palladium-catalysed reactions, many new effective protocols involving carbene ligands and preformed carbene complexes have been introduced [179–181] (supported carbene complexes [176]).

Their role in the Mizoroki–Heck reaction is different. An analysis of published data shows that, in the intermolecular Mizoroki–Heck reaction, the use of carbene ligands is

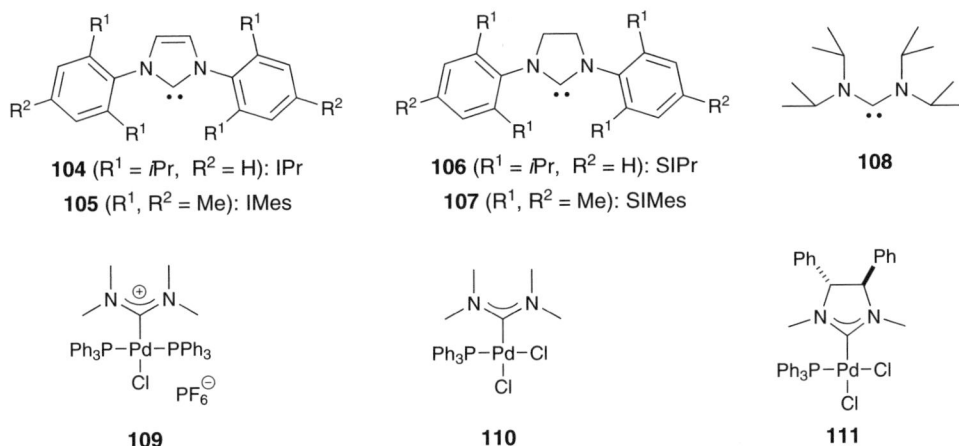

Figure 2.11   *Popular carbene ligands.*

practically restricted by a single function: the reservoir of palladium either in pre- or *in-situ*-formed SRPCs. Apparently, no unambiguous examples of activation of unreactive aryl chlorides (type 3 Mizoroki–Heck reactions) have been disclosed so far. Conversely, in cross-coupling reactions (e.g. Kumada reaction of Grignard reagents, Negishi reaction of organozinc reagents, Suzuki–Miyaura reaction of organoboron compounds, Buchwald–Hartwig amination reaction, as well as arylation of enolates), a number of effective protocols allowing for the use of unreactive aryl bromides and chlorides under very mild conditions (20–60 °C) using various carbene ligands are known. Popular ligands are IPr (**104**), IMes (**105**), SIPr (**106**) and SIMes (**107**, Figure 2.11).

Interestingly, the lead structures for the whole family of stable carbenes, the acyclic diaminocarbenes, have only been introduced and tested recently in palladium-catalysed reactions [182, 183]. One of the simplest carbenes **108** was shown to be active in Mizoroki–Heck reactions both as an added ligand (*n*-butyl acrylate, PhBr, Pd$_2$(dba)$_3$ (1 mol%), Cs$_2$CO$_3$, *n*-Bu$_4$NBr, DMA, 110 °C) [183] and as preformed complexes **109** or **110** (*n*-butyl acrylate, PhBr, **109** (1 mol%), Cs$_2$CO$_3$, NMP, 120 °C) [182], showing a rather typical performance for an electron-rich carbene ligand; high temperature and all other prerequisites of a type 2 process were required. No reliable features of a ligand-acceleration effect were noticed. With preformed complexes, no clear dependence on, for example, the important structural feature of the formal charge on palladium (cationic complex **109** versus neutral complex **110**) has been noted. Besides, all complexes of the types [(PPh$_3$)$_2$PdCl(L)]$^+$X$^-$ and [PPh$_3$PdCl$_2$(L)] with acyclic (e.g. **109** and **110**) or cyclic carbenes (**111**) showed comparable performance under the same conditions and in the same reaction [182]. Carbene **108** turned out to be an excellent ligand for room-temperature cross-coupling reactions (Suzuki and Sonogashira reactions) with electron-rich aryl bromides and chlorides, with the latter requiring slighty elevated temperatures (45 °C) [183]. In sharp contrast, Mizoroki–Heck reactions under essentially the same conditions required elevated temperatures (**44 → 112**, Scheme 2.24) [183]. This is yet another demonstration of the striking difference between the Mizoroki–Heck reaction and formally closely related cross-coupling processes, even

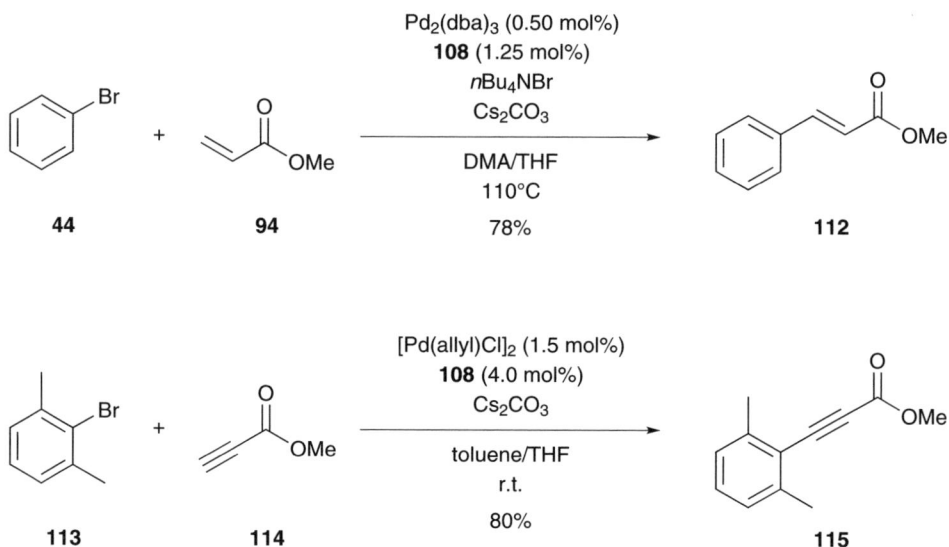

Scheme 2.24

the cross-coupling with acetylenes (**113** → **115**, Scheme 2.24) [183]. Its copper-free version is sometimes regarded as a special case of Mizoroki–Heck reactions ... apparently erroneously!

The reasons why carbene ligands are inferior to phosphines must be sought in their coordination properties. High-level theoretical calculations using Me_3P and dimethylimidazolydine as model reactants reveal similar reaction pathways for the respective PdL_2 complexes. With bis-phosphine complexes of palladium, these should be more reactive (lower activation barriers are predicted) in the oxidative addition than bis-carbene complexes [184]. However, for Mizoroki–Heck reactions, only a single strongly bonded ancillary ligand is tolerated in the coordination shell. An interesting insight into this behaviour was obtained by electrochemical studies with palladium complexes of two typical carbenes **116** and **117** (Figure 2.12) [185]. The following trends were established:

(a) The stability of carbenes and PdL_2 complexes depends on the bulkiness of the groups at the nitrogen atom. In the absence of bulky groups, neither carbenes themselves nor palladium complexes are stable.

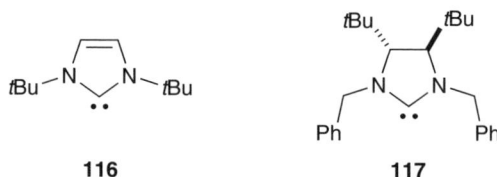

Figure 2.12   *Carbene ligands used in electrochemical investigations.*

(b) PdL_2 complexes are capable of oxidative addition to aryl chlorides at room temperature, and the addition to aryl bromides or iodides is extremely fast.
(c) Complex with saturated ligand **117** is more reactive than complex with unsaturated ligand **116**.
(d) Monocarbene complexes PdL, as well as mixed complexes Ph_3PPdL, are less active than PdL_2 in the oxidative addition.

These findings reveal that, although the reactivity of carbene complexes in oxidative addition is apparently very high and is not inferior to complexes of bulky electron-rich phosphines, the coordination properties clearly prefer bis-ligated complexes over mono-ligated complexes. This makes carbene complexes less effective, because the coordination sphere of palladium tends to be overcrowded and the PdL_2 intermediates are less labile. Even the most bulky carbene ligands are more strongly bonded to the metal centre than typical bulky phosphines. Therefore, carbene-based catalytic systems in Mizoroki–Heck reactions will only be effective at high temperatures.

Carbene ligands are used either as preformed palladium complexes or added as the respective salts. As soon as the complexes involve palladium(II), reductive preactivation is required, which is evidenced by rather long induction periods. Here lies one of many essential differences between carbene and phosphine systems: While phosphines are known to lend assistance in the reduction of palladium(II) to palladium(0), carbenes are not capable of this. This accounts for the need for harsh conditions, under which the preactivation processes (partial or full dissociation of complex and reduction by other components of reacting system) occur. The addition of reducing agents sometimes results in activation under milder conditions [186] or shortening of the induction period [33], but this is not common practice.

Reactions with carbene complexes as precatalysts are conveniently performed in ionic liquids [187]. Carbene ligands must be associated with an important subclass of ionic liquids, namely imidazolium salts. Carbenes are formed by deprotonation of imidazolium salts, yet this association is not as clear as it seems. While those used for ionic liquids usually bear simple alkyl groups, stable carbenes require bulky substituents. Still, we cannot exclude transient formation of carbenes and carbene complexes playing some role in the support and storage of palladium in systems realized in ionic liquids. In more complex systems, the interrelation between ionic liquid and the respective carbene ligand might be exploited by combining these features in a chelated carbene complex (Figure 2.13). Thus, **119** – when used in the reaction of aryl iodides with styrene or *n*-butyl acrylate in **118** as ionic liquid (**119** (2 mol%), Na_2CO_3, 120 °C) – could be recycled up to five times without deterioration by washing ionic liquid phase with both ether and water [188]. Neither organic waste nor salts were accumulated. In another example, hydroxyethylated

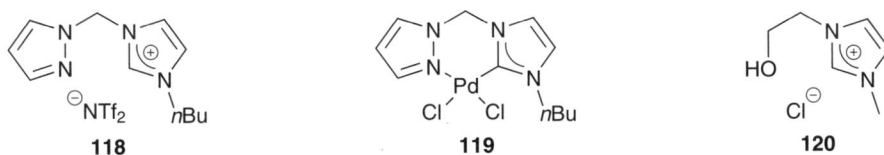

118   **119**   **120**

Figure 2.13 *Imidazolium ionic ligands and the respective carbene complex.*

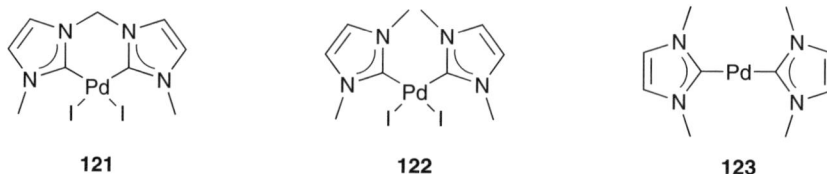

Figure 2.14 *Carbene complexes 121–123.*

ionic liquid **120** served well as a medium for standard Mizoroki–Heck reactions of aryl iodides and bromides with styrene and acrylates (Pd(OAc)$_2$ (2 mol%), K$_3$PO$_4$, 130 °C), allowing for recycling up to six times [189].

Considering carbene ligands and complexes in general, both monodentate and chelating varieties are known; both types were introduced by Herrmann and coworkers [33, 190]. These authors showed that those, either as preformed complexes or as mixtures of palladium and imidazolium salts, are useful for the Mizoroki–Heck reaction of several of aryl halides except unactivated aryl chlorides. Although a great number of papers followed [179], these two seminal publications summarize practically all essential achievements in this area. The reactions are performed in polar aprotic high boiling-point solvents (DMF, DMA or NMP) in the presence of inorganic salts as bases at high temperatures. The chelated bis-carbene complexe **121** is more thermally stable than the respective complex **122** with mono-carbene ligands (Figure 2.14). The performances of both types are comparable. The use of prereduced palladium complex **123** formed *in situ* affords faster reaction times, thus leading to much higher TONs. These systems are less stable and show inferior longevity; record-making TONs come along with modest yields and poor performance in scaled-up experiments. Moreover, in a separate later study, it was explicitly shown that the system based on carbene **121** and similar ligands show markedly inferior catalytic parameters in comparison with Pd(OAc)$_2$ with all parameters otherwise identical: Pd(OAc)$_2$ or **121** (0.05 mol%), NaOAc, NMP, 135 °C [191]. With both electron-rich 4-bromoanisole (**8**) and activated 4-bromoacetophenone (**7**), the expected trend was seen. In type 2 processes, the storage of palladium in an SRPC would lead to decrease of the average reaction rate due to a much lower effective concentration of active palladium species.

In the other palladium-catalysed reactions, the most essential progress since the introduction of heterocyclic carbene ligands in catalysis was associated with sterically hindered carbenes. These were initially developed by Grubbs and coworkers [192, 193] and Nolan and coworkers [194] for ruthenium-catalysed alkene metathesis. Only these carbenes reliably mimic electron-rich phosphines in cross-coupling chemistry. As we have already noted in the Mizoroki–Heck reaction, these ligands have so far failed to form catalysts active enough to enable type 3 Mizoroki–Heck reactions of unactivated chlorides. However, for aryl bromides, these ligands do show a performance clearly above the level of the typical type 2 systems. Thus, Yang and Nolan [195] showed that sterically hindered carbenes IMes (**105**) or SIPr (**106**) make good ligands for the reaction of various aryl bromides **44** with *n*-butyl acrylate (**38**) if the carbene is generated *in situ* in the presence of Pd(OAc)$_2$ (2 mol%), a carbonate base and a rather high temperature, namely 120 °C (palladium : ligand ratio = 1 : 2) (**44** → **39**, Scheme 2.25). This system

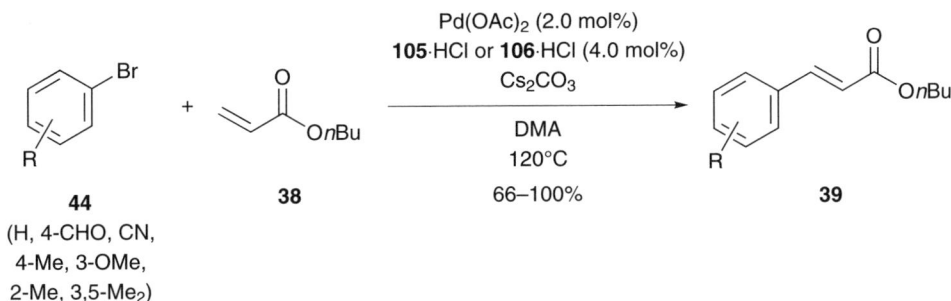

Scheme 2.25

is apparently at a borderline between type 2 and type 3. The activating influence of the carbene ligand is evidenced by a broad scope of bromides, which consistently give high yields using the same protocol. However, this system might also operate via a reservoir of palladium.

As we now see, carbene complexes might be used in both type 1 and type 2 processes. In the type 1 systems, these complexes are useful SRPCs which afford smooth reactions in the presence of very small catalyst loadings. Unfortunately, no consistent comparison between different types of carbene-based precatalysts is possible, as all studies published pursue different goals and almost never compare the newly introduced systems with the prototype Herrmann systems. In the published systems, a wide variation of all imaginable structural factors was permuted: (i) monodentate versus bidentate; (ii) monodentate *cis* versus *trans*; (iii) saturated versus unsaturated five-membered ring; (iv) size of chelating ring, even including such nontrivial structures as palladamacrocycle or palladabicycle; (v) strain; (vi) bulkiness; (vi) pre- versus *in-situ*-formed complex; (vii) monocarbene versus dicarbene complex; (viii) palladium(II) versus palladium(0) complexes (see figures and tables below).

The only (more or less) clear trend is that usually the less stable precatalyst performs better in a given model reaction. Stability is affected by geometric parameters, such as size of the chelate (five- and six-membered rings being more stable than larger rings) and steric bulk. The absence of trends is indicative of the SRPCs, for which the difference between various forms is relatively unimportant as soon as the conditions under which a given precatalyst releases active palladium species are realized.

Monodentate carbenes and their complexes have been obtained for almost all possible azolines and azolidines with a second heteroatom flanking the carbene-generating carbon centre. A selection of published data is shown in Table 2.4 and Figure 2.15.

The other indirect evidence in favour of the hypothesis that carbene ligands in Mizoroki–Heck reactions do not deliver any ligand acceleration might be seen in the following: simple 2-substituted imidazoles and imidazolines (both in a palladium : ligand ratio of 1 : 1 or as complex **141**, Figure 2.15) are incapable of forming carbenes but still serve as useful 'ligands' for Mizoroki–Heck reactions of aryl bromides; for example, 4-bromotoluene (Table 2.4, entry 19) [205]. Of all these heterocycles studied, the simplest, 2-methylimidazoline, showed the best performance, compared with bulkier 2-*t*-butyl- and 2-phenyl-substituted compounds, as well as 2-methylimidazole. The latter is not suprising,

Table 2.4 Monodentate carbene complexes

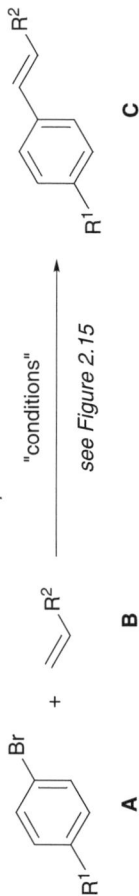

Entry	Pd-cat. (mol%)	A (R¹)	B (R²)	Base	Solvent	T (°C)	Yield (%)	TON	TOF (h^{-1})	Ref.
1	124 (4×10^{-4})	Ac	CO_2-n-Bu	NaOAc	DMA	120	29	7.4×10^4	24 000	[196]
2	125 (4×10^{-4})	Ac	CO_2-n-Bu	NaOAc	DMA	120	50	1.1×10^5	10 400	[196]
3	126 (2×10^{-3})	Ac	CO_2-n-Bu	NaOAc	DMA	120	n.d.	5.0×10^4	n.d.	[197]
4	127 (1×10^{-3})	Ac	CO_2-n-Bu	NaOAc	NMP	135	97	9.7×10^4	19 400	[198]
5	128 (0.05)	MeO	CO_2-n-Bu	NaOAc	NMP	135	87	1.7×10^3	348	[198]
6	129 (1×10^{-3})	Ac	CO_2-n-Bu	NaOAc	DMF	130	100	1.0×10^5	n.d.	[113, 199]
7	130 (0.05)	Ac	CO_2-t-Bu	NaOAc	DMF	100	95	n.d.	n.d.	[200]
8	130 (1.0)	OMe	CO_2-t-Bu	NaOAc	DMF	110	79	n.d.	n.d.	[200]
9	131 (1.0)	OMe	CO_2-t-Bu	NaOAc	DMF	110	53	n.d.	n.d.	[200]
10ᵃ	132 (1.0)	OMe	CO_2-t-Bu	$NaHCO_3$	DMF	140	94	n.d.	n.d.	[201]
11	133 (0.5)	Ac	CO_2Me	Cs_2CO_3	NMP	120	100	200	n.d.	[202]
12	134 (1.0)	Ac	H	Cs_2CO_3	1,4-Dioxane	80	97	n.d.	n.d.	[203]
13	135 (1.0)	Ac	H	Cs_2CO_3	1,4-Dioxane	80	94	n.d.	n.d.	[203]
14	136 (1.0)	OMe	H	NaOAc	DMA	130	92	n.d.	n.d.	[204]
15	137 (1.0)	OMe	H	NaOAc	DMA	130	98	n.d.	n.d.	[204]
16	138 (1.0)	OMe	H	NaOAc	DMA	130	62	n.d.	n.d.	[204]
17	139 (1.0)	OMe	H	NaOAc	DMA	130	83	n.d.	n.d.	[204]
18	140 (1.0)	OMe	CO_2nBu	Cs_2CO_3	NMP	120	90	n.d.	n.d.	[182]
19	141 (1.0)	Me	CO_2Me	K_2CO_3	DMF	120	79	n.d.	n.d.	[205]

ᵃWith addition of n-Bu₄NBr.

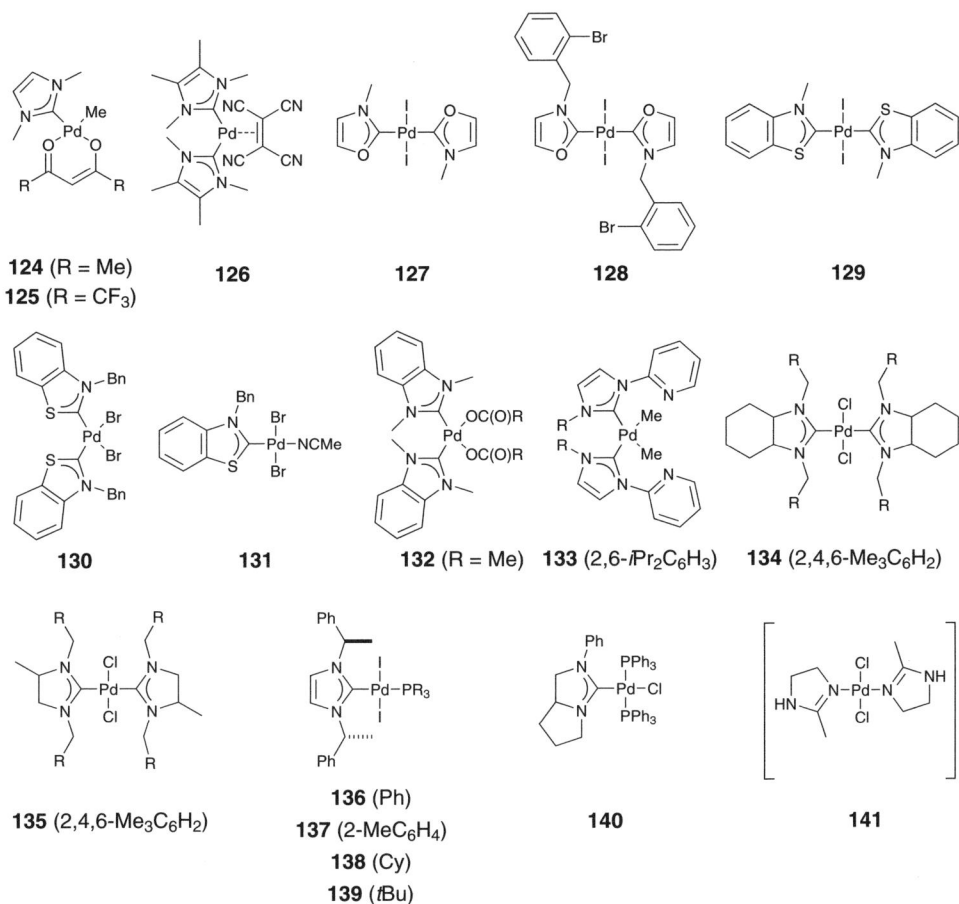

124 (R = Me)
125 (R = CF$_3$)

126

127

128

129

130

131

132 (R = Me) **133** (2,6-*i*Pr$_2$C$_6$H$_3$) **134** (2,4,6-Me$_3$C$_6$H$_2$)

135 (2,4,6-Me$_3$C$_6$H$_2$)

136 (Ph)
137 (2-MeC$_6$H$_4$)
138 (Cy)
139 (*t*Bu)

140

141

Figure 2.15 *Monodenate carbene complexes.*

since 2-methylimidazoline is more strongly bound due to lower steric strain and higher basicity [205, 206].

Even hybrid carbene–phosphine precatalysts such as **136–139** showed weak and uneven dependence on the phosphine ligand (Table 2.4, entries 14–17) [204]. The complexes with bulky electron-rich phosphines **138** and **139** were less reactive than those with conventional triarylphosphines **136** and **137**. This can hardly be regarded as surprising, as the deligation of the carbene is likely to be less probable than the deligation of the phosphine at the applied temperatures; activation of the precatalyst should inevitably occur by dissociation of the phosphine first. In this respect, the design of hybrid carbene–phosphine complexes for Mizoroki–Heck reactions is not promising.

Similar trends observed with more complex structural types of carbene ligands and their complexes is seen from the selected data presented in Tables 2.5 and 2.6 and Figures 2.16 and 2.17.

Table 2.5 *(Complexes of) chelating carbenes*

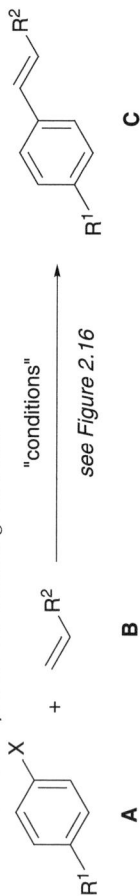

Entry	Pd-cat. (mol%)	A		B (R^2)	Base	Solvent	T (°C)	Yield (%)	TON	TOF (h^{-1})	Ref.
		R^1	X								
1	**142** (1×10^{-4})	Ac	Br	H	NaOAc	DMA	140	100	7.1×10^5	25 500	[207]
2	**142** (0.01)	MeO	Br	H	NaOAc	DMA	140	26	1.8×10^3	232	[207]
3	**143** + Pd(OAc)$_2$ (1.5)	Ac	Br	H	Cs$_2$CO$_3$	DMF/H$_2$O	80	92–97	n.d.	n.d.	[208]
4	**144** (1×10^{-3})	Ac	Br	CO$_2$-n-Bu	NaOAc	NMP	135	63	6.3×10^4	n.d.	[191]
5	**145** (1×10^{-3})	Ac	Br	CO$_2$-n-Bu	NaOAc	NMP	135	77	7.7×10^4	n.d.	[191]
6	**146** (4.0)	H	Br	CO$_2$-n-Bu	NEt$_3$	**152**	120	35	n.d.	n.d.	[209]
7	**147** (1×10^{-3})	Ac	Br	H	NaOAc	DMA	140	84–99	$(6.0–7.0) \times 10^4$	2000–2360	[210]
8	**148** (2×10^{-4})	H	Br	CO$_2$-n-Bu	NEt$_3$	DMF	140	79–84	4.0×10^5	n.d.	[211]
9	**149** (0.05)	MeO	Br	CO$_2$-n-Bu	NaOAc	NMP	135	85	1.7×10^3	n.d.	[198]
10^a	**150** (2×10^{-4})	H	I	CO$_2$-n-Bu	NEt$_3$	DMF	140	60	3.0×10^5	n.d.	[198]
11^a	**151** (1×10^{-3})	Ac	Br	CO$_2$-n-Bu	NaOAc	DMA	120	88	1.1×10^5	5240	[212]

^{a}With addition of n-Bu$_4$NBr.

Table 2.6 Complexes of chelating ligands with carbene and noncarbene binding sites

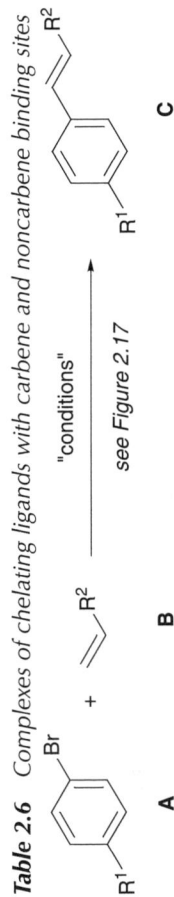

Entry	Pd-cat. (mol%)	A R^1	A X	B (R^2)	Base	Solvent	T (°C)	Yield (%)	TON	TOF (h^{-1})	Ref.
1	**152** (1.0)	Me	Br	CO$_2$-n-Bu	NaOAc	DMA	160	81	n.d.	n.d.	[213]
2	**153** (0.07)	Ac	Br	CO$_2$Me	Et$_3$N	NMP	140	60–66	850–950	n.d.	[202]
3	**154** (0.5)	Ac	Br	CO$_2$Me	Cs$_2$CO$_3$	DMA	120	100	200	n.d.	[202]
4	**155** (2 × 10^{-3})	H	I	H	Et$_3$N	DMA	110	96	4.8 × 10^4	2400	[214]
5	**156** (0.5)	Ac	Br	CO$_2$Me	NEt$_3$	DMA	120	67–92	134–185	n.d.	[202]
6	**157** (0.5)	Ac	Br	CO$_2$Me	NEt$_3$	DMA	120	100	200	n.d.	[202]

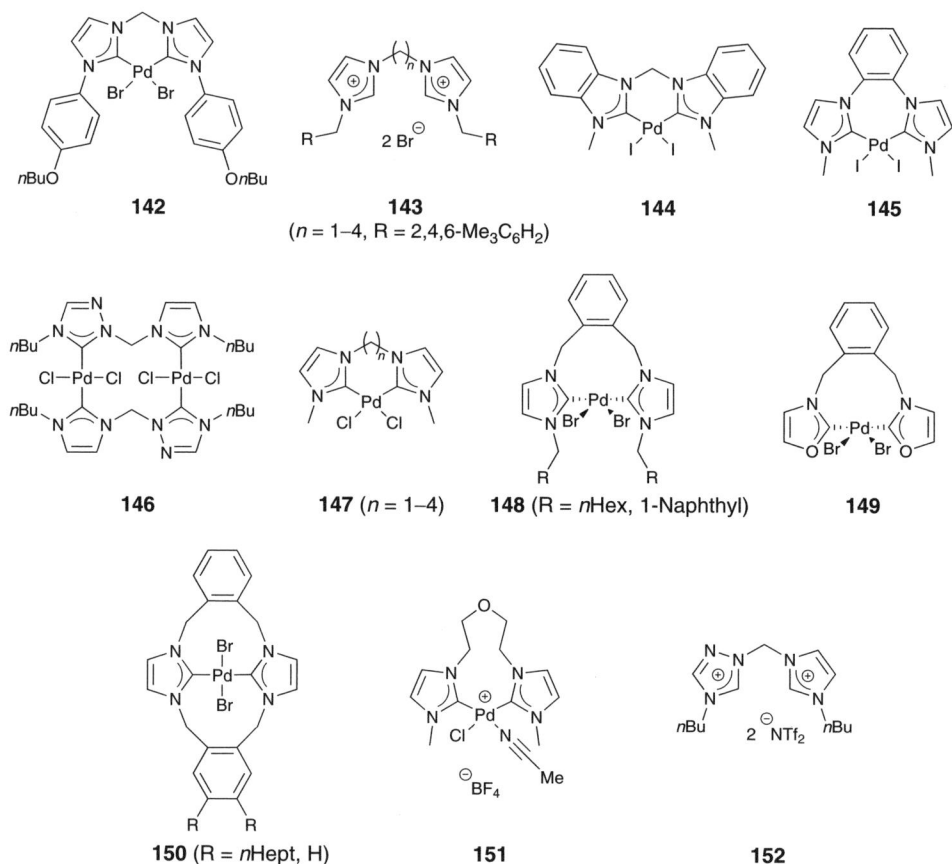

Figure 2.16   *(Complexes of) chelating carbenes.*

2.4.4.2   *N*-Heterocyclic Carbene-Based Pincer Complexes

A very interesting subclass is pincer complexes having carbene binding sites, **158–162** (Table 2.7, Figure 2.18). Very high robustness of pincer complexes makes them good candidates for SRPCs (type 1 processes), wheras for reactions with less active substrates the level of released palladium should be too low to enable effective turnover. There are a few examples of type 2 reactions using pincer complexes, but only using high temperatures above those normally used for type 2 reactions with less robust precatalysts [114]. Under otherwise identical conditions, the less robust CNC-type complex **161** is markedly more active than its more robust analogue **162** involving a CCC-palladabicyclic structure [114].

Where the comparison within a row of similar ligands can be drawn, common trends are observed. This is apparent in the case of σ-methyl palladium pincer complexes **159** (Figure 2.18) and their respective congeners, neutral chloropalladium complexes (not shown), in which the substituent at palladium has only a minor influence [216]. Not surprisingly, both precatalysts gave essentially the same results, since for catalytic activity

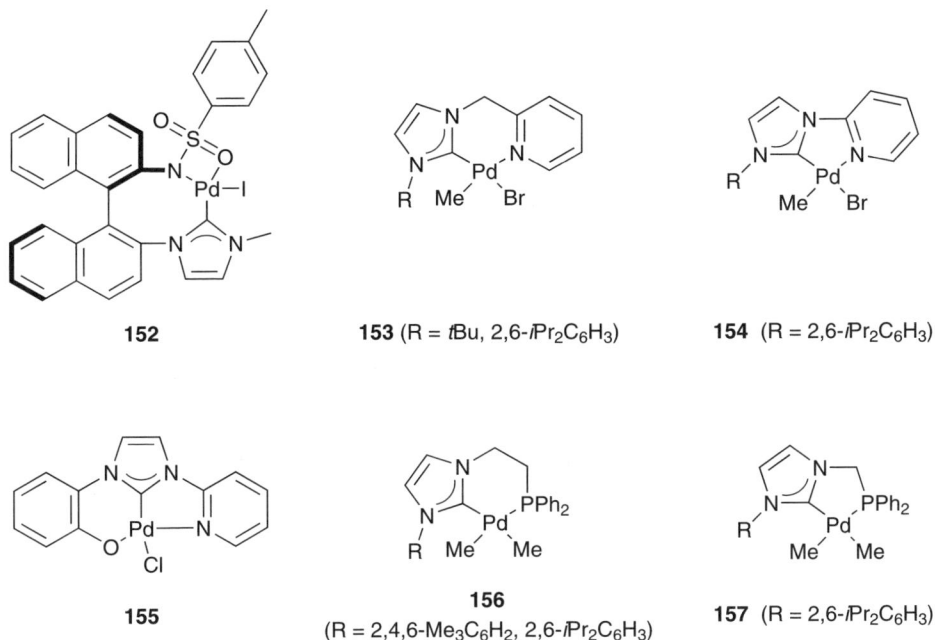

Figure 2.17 *Complexes of chelating ligands with carbene and noncarbene binding sites.*

the complex must be disassembled first. The same conclusion emerges from an apparent similarity of catalytic performance of complexes with different linkages **159** ($n = 0$) and **160** ($n = 1$) (Table 2.7, entries 3 and 4, Figure 2.18); the former should possess a more strained structure than the latter due to an unfavourable arrangement of the annulated rings at palladium, which force the ligands attached to palladium to deviate from the preferred planar configuration. In experiments, the activity of the more strained system **159** (R = mesityl) was higher than the activity of the otherwise identical less strained system **160** (R = mesityl); TONs dropped from 61 000 to 16 000 and TOF values decreased from 3000 to 800 h^{-1} [216]. On the other hand, for complexes with the same size of chelate rings, no regular significant trends in catalytic efficiency are observed on variation of the substituent in the carbene moiety from methyl to bulky mesityl or 2,6-diisopropylphenyl. Both these findings agree very well with the hypothesis that these pincer complexes are fully disassembled to deliver a catalytically active species; thus, only the factors of thermal stability matter, while all structural or electronic features in the ligand which might have had an influence on the reactivity of palladium complex exert an insignificant effect and are not involved in any clear trends.

A rather unusual confirmation of the fact that, under the conditions of Mizoroki–Heck reactions, only a small part of a robust SRPC is disassembled to afford a catalytically active species was shown in 'self-coupling' transformation of bis-carbene chelate **163** containing styrene residues in the reaction with active substrate 4-bromoacetophenone (**7**) (Scheme 2.26) [217]. A reasonable yield of 78% of doubly arylated complex **164** with conserved chelate structure was isolated.

Table 2.7 CCC- and CNC-type pincer complexes

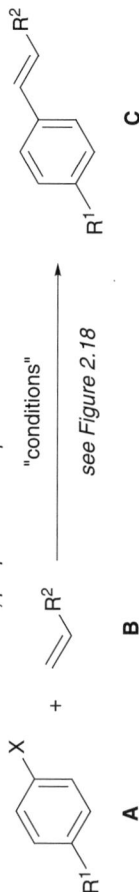

| Entry | Pd-cat. (mol%) | A | | B (R²) | Base | Solvent | T (°C) | Yield (%) | TON | TOF (h⁻¹) | Ref. |
		R¹	X								
1	**158** (1.0)	Ac	Br	H	NaOAc	DMA	110	97–98	n.d.	n.d.	[215]
2	**158** (1.0)	MeO	Br	H	NaOAc	DMA	110	46–50	n.d.	n.d.	[215]
3	**159** (1 × 10⁻³)	Ac	Br	CO₂-n-Bu	NaOAc	DMA	120	35–42	$(3.4\text{–}4.2) \times 10^4$	1700–3000	[216]
4	**160** (1 × 10⁻³)	Ac	Br	CO₂-n-Bu	NaOAc	DMA	120	16–73	$(1.6\text{–}7.2) \times 10^4$	800–3600	[216]
5	**161** (0.2)	Ac	Br	H	KOAc	DMA	165	97	485	n.d.	[114]
6	**161** (0.2)	MeO	Br	H	KOAc	DMA	165	97	485	n.d.	[114]
7[a]	**161** (0.2)	CHO	Cl	H	KOAc	DMA	165	90	450	n.d.	[114]
8[a]	**162** (0.2)	CHO	Cl	H	KOAc	DMA	165	70	350	n.d.	[114]

[a]With addition of n-Bu₄NBr.

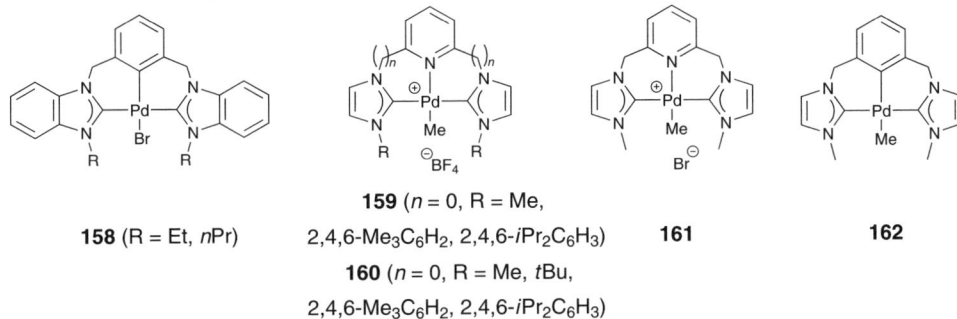

158 (R = Et, *n*Pr)

159 (*n* = 0, R = Me,
2,4,6-Me$_3$C$_6$H$_2$, 2,4,6-*i*Pr$_2$C$_6$H$_3$)
160 (*n* = 0, R = Me, *t*Bu,
2,4,6-Me$_3$C$_6$H$_2$, 2,4,6-*i*Pr$_2$C$_6$H$_3$)

161

162

Figure 2.18   *CCC- and CNC-type pincer complexes.*

2.4.4.3   Carbocyclic Carbene-based Complexes

The triumphal success of *N*-heterocyclic carbene ligands in transition metal-catalysed reactions stimulated the introduction of new ligands with a related binding motif, namely
carbocyclic carbenes. Apparently representing another class of SRPCs, these new ligands
show essentially the same behaviour in the Mizoroki–Heck reaction as the related derivatives of five-membered azoles. Possibly due to their lower stability and a higher tendency to
release catalytically active species, complexes **165–170** (Figure 2.19) show huge TONs in
the type 1 processes, as well as a reasonable level of performance in type 2 processes which
promise a good activity upon individual optimization (Table 2.8). The most remarkable of
the nonazole series are likely the carbenes, in which the vacant p-orbital is included in an
aromatic carbocycle, such as tropylidene, which has been shown by Herrmann *et al.* [218]
to form stable palladium complexes **165**. The latter serve as excellent precatalysts for type 1
Mizoroki–Heck reactions, giving very high TONs and TOF values (Table 2.8, entries 1
and 2). Cyclopropenylidene-derived congener **166** showed a comparable performance in
the same set of reactions, yet a much higher level of TOF/TON in type 2 Mizoroki–Heck
reactions was achieved at much lower yield (Table 2.8, entries 3 and 4) [219].

Carbenes derived from a benzene ring system were trapped in palladacycles **167** that are
formed from nitrones **171** (Scheme 2.27, three resonance structures of **167a–c** are shown)
[220]. Probably, **167** is more than just another SRPC. In fact, complexes **167** were shown
to be able to run Mizoroki–Heck reactions with deactivated aryl bromides at a catalyst
loading level of 10^{-4} mol%, giving very high TONs and TOF values along with high
yields of products (Table 2.8, entries 5 and 6) [220]. Could it be the long-awaited carbene
ligand with a true ligand-acceleration effect in Mizoroki–Heck reactions? Unfortunately,
this is unlikely because (i) the conditions of Mizoroki–Heck reactions are quite typical for
SRPCs and (ii) modest performance has been observed in the same system with activated

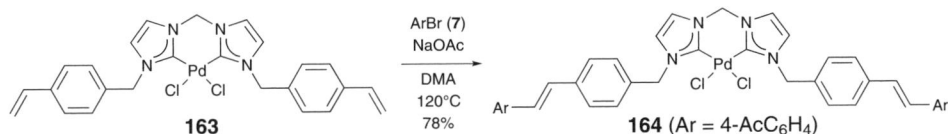

ArBr (**7**)
NaOAc
⟶
DMA
120°C
78%

163

164 (Ar = 4-AcC$_6$H$_4$)

Scheme 2.26

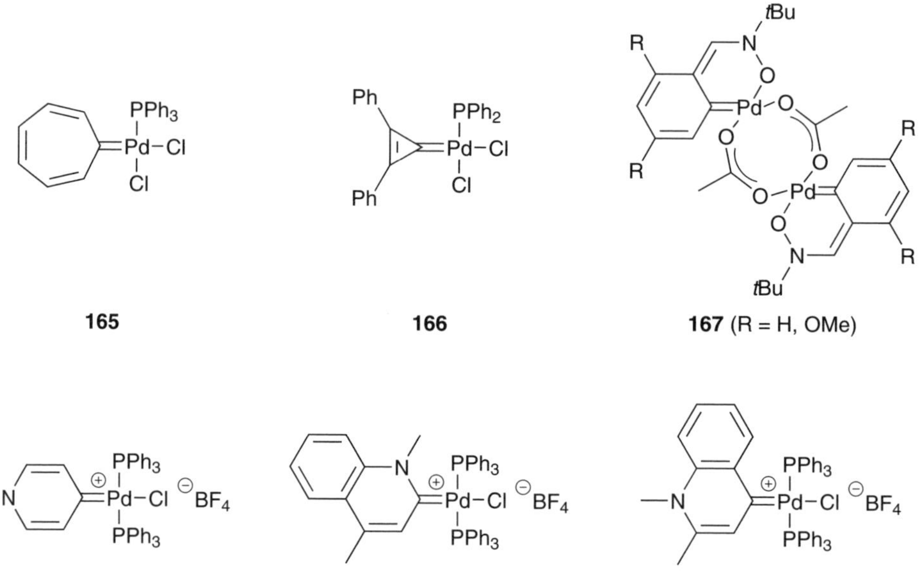

165   **166**   **167** (R = H, OMe)

168   **169**   **170**

Figure 2.19   *Carbocyclic carbenes and complexes thereof.*

aryl chlorides. Both these findings hardly fit into a type 3 paradigm. It is more likely that this system underlines that the performance of type 2 systems can be boosted by a good choice of precatalysts to the levels observable in typical type 1 systems. Having the technological perspective of the Mozoroki–Heck reaction in mind, this constitutes an encouraging insight.

On the other hand, a considerable number of carbene complexes based on derivatives of pyridinylidenes (**168**) and their analogues in the quinoline (**169** and **170**) and acridine series (not shown) were prepared and tested in Mizoroki–Heck reactions [221]. These ligands involve carbenes, which have a carbene core almost isostructural to the one in complex **167**. In the test reactions surveyed, an almost identical performance in type 1 and type 2 processes was observed, the latter being dramatically inferior to the performance of **167** (Table 2.8, entries 7–11) [221]. This, as well as the absence of any structural dependence on the performance, clearly indicates the slow-release operation.

2.4.5   Palladacycles

2.4.5.1   *C,P-, C,S-* and *C,N*-Palladacycles

Palladacycles are apparently the best studied subclass of the SRPCs. Moreover, the development of the concept itself originates from the invention and investigation of palladacycles, particularly the classical Herrmann–Beller palladacycle **52** (Figure 2.20, see below).

At the beginning, palladacyles were considered as a long-awaited class of palladium catalysts with a stable and very robust coordination sphere and were, thus, expected to open almost infinite possibilities for the development of so-called well-defined catalysts. In order to reconcile the coordination requirements of the classic Mizoroki–Heck reaction

Table 2.8 Performance of complexes of **165–170**

A + B $\xrightarrow[\text{see Figure 2.19}]{\text{"conditions"}}$ C

Entry	Pd-cat. (mol%)	A		B (R²)	Base	Solvent	T (°C)	Yield (%)	TON	TOF (h⁻¹)	Ref.
		R¹	X								
1	**165** (1×10^{-4})	Ac	Br	$CO_2\text{-}n\text{-Bu}$	NaOAc	DMA	145	100	1.0×10^6	n.d.	[218]
2	**165** (0.1)	MeO	Br	$CO_2\text{-}n\text{-Bu}$	NaOAc	DMA	145	42	420	n.d.	[218]
3	**166** (1×10^{-4})	Ac	Br	$CO_2\text{-}n\text{-Bu}$	NaOAc	DMA	145	94	9.4×10^5	n.d.	[219]
4	**166** (1×10^{-3})	MeO	Br	$CO_2\text{-}n\text{-Bu}$	NaOAc	DMA	145	54	5.4×10^4	n.d.	[219]
5	**167** (8×10^{-5})	H	Br	H	KOAc	DMA	140	89	1.1×10^6	n.d.	[220]
6	**167** (8×10^{-5})	MeO	Br	H	KOAc	DMA	140	75	9.4×10^5	n.d.	[220]
7	**168** (1×10^{-4})	Ac	Br	$CO_2\text{-}n\text{-Bu}$	NaOAc	DMA	145	56	5.6×10^5	n.d.	[221]
8	**169** (1×10^{-5})	Ac	Br	$CO_2\text{-}n\text{-Bu}$	NaOAc	DMA	145	55	5.5×10^5	n.d.	[221]
9	**169** (0.1)	MeO	Br	$CO_2\text{-}n\text{-Bu}$	NaOAc	DMA	145	86	860	n.d.	[221]
10	**170** (1×10^{-5})	Ac	Br	$CO_2\text{-}n\text{-Bu}$	NaOAc	DMA	145	62	6.2×10^6	n.d.	[221]
11[a]	**170** (0.5)	Ac	Cl	$CO_2\text{-}n\text{-Bu}$	NaOAc	DMA	150	43	86	n.d.	[221]

[a]With addition of $n\text{-Bu}_4\text{NBr}$.

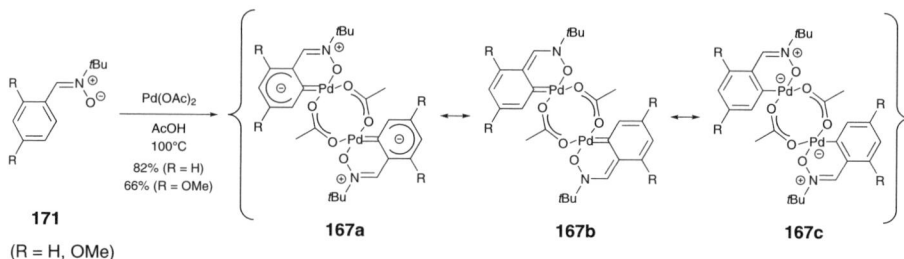

Scheme 2.27

mechanism with the fixed coordination shell of palladacycles, a new unusual palla-
dium(II)/palladium(IV) mechanism has been proposed [222, 223]. However, the data
gained before 2000 already offered enough evidence against both the hypothesis of an
alternative mechanistic pathway and well-defined catalysts for Mizoroki–Heck reactions
[4, 224].

Despite the thermally and chemically robust nature of palladacycles, their cleavage
under the conditions of the Mizoroki–Heck reaction or other palladium-catalysed reactions
apparently occurs. Various pathways for cleavage have been postulated that consist of ligand
exchange at labile coordination sites with subsequent reductive elimination at one of the
chelating arms of the palladacycle [225]. Exact details of these processes are not relevant
for the purpose of this discussion. What is relevant is the statement that palladacycles are
typical SRPCs. The rate of cleavage depends on the structure of the palladacycle, as well
as of the particular conditions used. Two variants are possible: (a) the cleavage is very slow
so that only a small amount of palladium is released – this is useful for type 1 processes,
particularly for those in which the precatalyst is recycled and reused; (b) the cleavage is
fast so that a considerable amount of the palladium is released during the reaction – this
is useful for type 2 processes. Apparently, palladacycles are by default not suitable for any
ligand-accelerated or ligand-selected process. By these means, they hardly can appear in a
true type 3 or type 4 context.

As is now evident, all Mizoroki–Heck chemistry of palladacyclic precatalysts can be
deduced from the behaviour of the first published representative of this class, namely the
Herrmann–Beller palladacycle **52** (Scheme 2.20) [107, 226]. This precatalyst is capable of
driving the type 1 reactions with activated aryl bromides, giving TON values up to 200 000
in common solvents (0.0005 mol% **52**, NaOAc, DMA, 135 °C) [107], which can be boosted
to 1 000 000 in ionic liquids [227]. With aryl iodides, the reactions can be run at as low

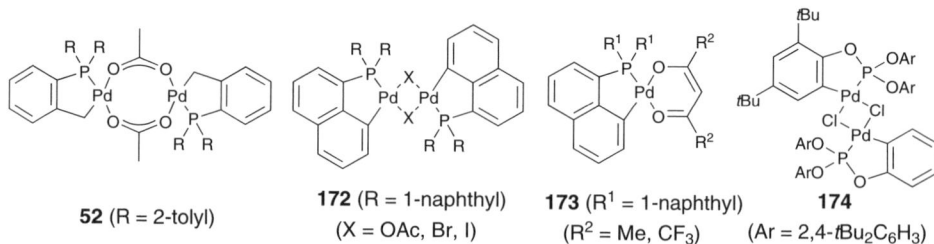

Figure 2.20   *Palladacycles **172–174** including Herrmann–Beller palladacycle **52**.*

as 85 °C [228]. In type 2 reactions, the performance of palladacycle **52** roughly matches many other similar SRPCs, and with electron-rich aryl bromides or activated aryl chlorides TONs up to 800 were achieved, preferentially in the presence of additives *n*-Bu₄NBr or LiBr [107].

The Herrmann–Beller palladacycle **52** is the only representative of this class of precatalysts, which has found real application in organic synthesis as a convenient stable source of palladium for various syntheses involving inter- and intra-molecular Mizoroki–Heck reactions [229].

Since their advent, a number of other palladacyclic precatalysts have been reported with broadly varied stability towards cleavage processes ranging from room-temperature release in the case of palladacycle **33** (cf. Scheme 2.8) [34, 70] to various PCP-pincer palladabicycles, which are stable up to temperatures of 150–160 °C (see Section 2.4.5.2). At the beginning, when the idea of palladacycles as well-defined palladium catalysts with fixed coordination shells prevailed, the leading design principle was to disclose sophisticated palladacyclic architectures, which presumably could solve complex synthetic tasks, including enantioselective processes. More stable and rigid structures were introduced, such as both dimeric and monomeric naphthalene-based palladacycles **172** and **173** [222]. However, as it turned out, the performance of the new structures closely matched the progenitor precatalyst **52**, showing very high TONs for reactions of aryl iodides.

An important step forward was achieved when the idea became accepted that the most valuable virtue of a palladacyclic precatalyst is not any sophisticated structure, but rather their ready availability and low cost in conjunction with simple handling and storage. In fact, the Herrmann–Beller palladacycle **52** already satisfies all these criteria par excellence; the only real or imaginary drawback is the relation of the latter with phosphines being regarded as toxic and expensive.

Encountering these disadvantages, a significant improvement was achieved with palladacycle **174** assembled from a readily available phosphite which was in turn derived from a commercially available sterically hindered phenol [230]. A direct comparison of **174** with the prototype palladacycle **52** is not viable, as the reactions with phosphite palladacycle **174** were generally conducted under more stringent conditions (temperatures of 160–180 °C). Yet, formidable TONs up to 1 000 000 in the reactions of *n*-butyl acrylate with activated substrates such as 4-bromoacetophenone (0.0001 mol% **174**, NaOAc, DMA, 180 °C) were accessible. Moreover, with a TON of 9800 for 4-bromoanisole (0.01 mol% **174**, K₂CO₃, DMA, 160 °C), **174** was quite reactive for type 2 reactions [230].

The palladacycle **174** virtually opened a new era in the design of palladacyclic precatalysts. The growing awareness of the fact that certainly all of the currently known precatalysts perform through the cleavage of palladacyclic shell to either a mono-phosphine complex [4] or even a manifold of complexes with undetermined coordination shells afforded a vast selection of various phosphorus-free palladacycles (Figures 2.21–2.23; see below). Indeed, if the palladacyclic scaffold is nonessential – at least from the viewpoint of ligand effect on activity (absence of ligand-acceleration effects) – then an idea for the choice of the compound to be transformed into a palladacyclic precatalyst as a simple disposable 'trash wrapper' materialized from this insight; the former should be a cheap and readily available compound directly from the shelf [228].

Palladacycles incorporating donating sulfur moieties ('CS-palladacycles') should be as stable as phosphorus-bearing precatalysts ('CP-palladacycles'); as a consequence, their

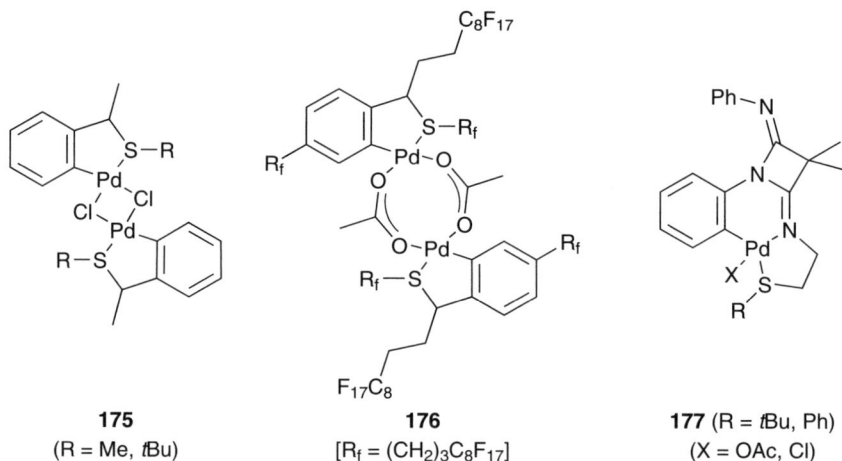

Figure 2.21   *Phosphorus-free palladacycles **175–177** (part 1).*

activation should necessitate rather harsh conditions. Such palladacycles are thus useful for type 1 processes, particularly within recyclable protocols, as well as type 2 reactions at high temperatures. Thus, a series of benzylsulfide-based palladacycles **175** are applicable to aryl iodides at 140 °C (0.002 mol% **175**, Et$_3$N, DMA) with TONs up to 50 000, while type 2 processes require 170 °C and *n*-Bu$_4$NBr additive to afford TON values up to 30 000 (Figure 2.21) [81].

CNS-palladabicycles **177** (1.0 mol% **177**, Cs$_2$CO$_3$, DMF, 135 °C) offer good activity for type 2 reactions to afford high yields in reactions of all sorts of aryl bromides with styrene [231]. At the same time, precatalysts **177** failed both in high TON type 1 reactions and with activated aryl chlorides [231].

A lot of trivial nitrogen-containing compounds (Schiff bases, oximes and benzylamines) readily form palladacycles. These are invariably useful precatalysts for Mizoroki–Heck

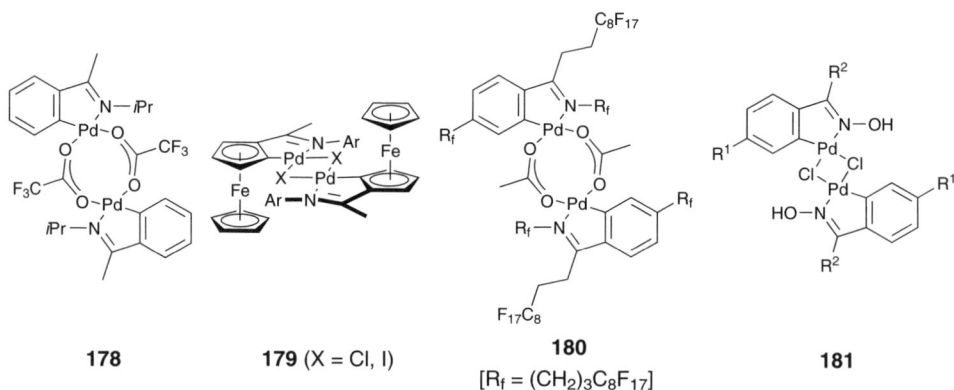

Figure 2.22   *Phosphorus-free palladacycles **178–181** (part 2).*

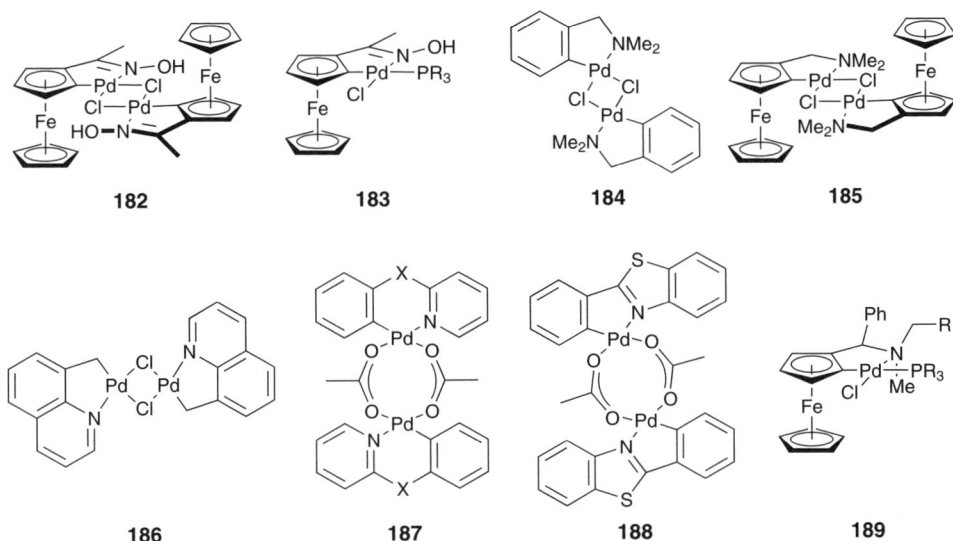

Figure 2.23 *'Phosphorus-free' palladacycles **182–189** (part 3).*

reactions and often show excellent performance in type 1 and type 2 reactions. This performance is often higher than that of prototype **52** [232]. The reason for this is obvious: phosphine-derived palladacycles are so robust that they fail to release all their palladium; an essential part of it is retained, which is useful for recycling experiments but effectively diminishes the TONs/TOF values, disallowing the obtaining of record-making results. On the other hand, less robust CN- and other related palladacycles are presumably cleaved quantitatively; in turn, any molecule released to the reaction vessel would work. This is good for record-making TON/TOF value studies, but poor for the design of recyclable systems.

Imine-derived palladacycles **178** were the earliest to be introduced to Mizoroki–Heck chemistry (Figure 2.22) [233]. Operating in a typical system (7×10^{-5} mol% **178**, Na_2CO_3 or Et_3N, NMP, 140 °C), palladacycles **178** not only effected the reaction of aryl iodides with acrylate, giving huge TONs approaching 1.5×10^6, but even performed almost as effectively with phenyl bromide, showing 132 900 cycles (TOF 1000 h^{-1}, a remarkable value). As with almost any other SRPC, the performance profile of these palladacycles is uneven: if styrene is subjected to turnover instead of acrylate, then **178** performs significantly worse [233]. Ferrocene-based imine palladacycle **179** was implemented using an unusual choice of reaction medium, namely a moderately polar system (Et_3N, 1,4-dioxane, 100 °C) generally used for true ligand-accelerated systems and not for type 1 or 2 processes [234]. In this system, aryl iodides reacted with acrylate to afford more than 7×10^6 catalytic cycles yet achieved over a rather long time of 339 h (average TOF value of 22 000 h^{-1}, which is rather modest value for an aryl iodide). The system was also highly alkene dependent and gave very modest results with activated aryl bromides [234]. It is the choice of polarity of the reaction media that most likely is responsible for such uneven performance.

Lower stability and faster palladium release are nicely seen in the behaviour of fluorous palladacycles **176** (Figure 2.21) and **180** (Figure 2.22) [235, 236]. These precatalysts show very high performance with 4-bromoanisole and methyl acrylate or styrene if the reaction is performed at high temperature (10^{-4} mol% **176** or **180**, Et$_3$N, DMF, 140 °C) [236]. The performance is not comparable at lower temperatures (80–100 °C), but both precatalysts allow recycling four times by extraction of the reaction media with fluorous solvent in order to remove the intact precatalyst [236]. Apparently, all precatalyst is cleaved at higher temperature and enters the reaction, thus affording excellent TONs, while at lower temperatures the extent of cleavage is diminished; released palladium lacks a fluorous shell, but the remaining catalytically inactive precatalyst can be extracted and recycled until complete depletion of the palladium reservoir is reached.

Another important class of palladacycles is derived from readily available oximes of benzaldehydes, benzophenones and related compounds with an oxime functionality facilitating *ortho*-palladation. The precatalysts **181** are reactive in type 1 reactions (0.0001 mol% **181**, K$_2$CO$_3$, NMP, 150 °C), as well as type 2 reactions of activated aryl chlorides, giving high TONs up to 70 000 (Figure 2.22) [237]. The palladacycles **181** clearly show that high TONs observed in type 1 reactions should not be considered as a measure of high catalytic activity. Indeed, while establishing an incredible record in the reactions with aryl iodides with TONs approaching 10^{10} (!), these precatalysts showed only modest performance for typical type 2 substrates, namely electron-rich aryl bromides [237, 238].

A separate line of oxime-derived palladacycles **182** and **183** is formed using the ferrocene platform, which showed a typical level of activity in type 1 reactions (0.01 mol% **182** or **183**, NaOAc, NMP, 140–150 °C) (Figure 2.23) [109].

A variety of related palladacycles **184–188** are obtained from various heterocyclic precursors or benzylamines, all showing the expected level of performance in type 1 reactions [228, 239, 240]. The precursors used for the preparation of such complexes are among the most readily available, thus approaching very closely the ideal of a 'trash wrapper' for palladium. These palladacycles are quite weak in comparison with CP- or CS-palladacycles, which makes them cleavable under milder conditions. Therefore, the reactions can be performed at temperatures as low as 70–85 °C although the addition of reducing agents is sometimes required to facilitate the cleavage of palladacycles [228, 240]. In addition, palladacycles **189** based on ferrocenylmethylamines were recently shown to serve well for reaction of phenyl bromide with styrene (0.1 mol% **189**, K$_2$CO$_3$, DMF, 140 °C) [241]. Among other examples, arylhydrazine palladacycles are of particular interest; these can serve as useful precatalysts in phase-transfer catalysis using water–substrate mixtures (K$_2$CO$_3$ or *i*-Pr$_2$NH, *n*-Bu$_4$NBr, water, 80 °C) [242].

2.4.5.2   Pincer Palladacycles

Hemilability of various pincer complexes in various reactions has been unambiguously established [243, 244].

Despite electron-rich bulky side-arms as in phosphine pincers **190**, **191** [245] or **192** [246] (Figure 2.24), these complexes behave strikingly different from their respective dialkyl or trialkylphosphine palladium complexes; the latter complexes show type 3 activtity (cf. Hartwig–Fu protocol; see above). PCP-pincer complexes **190–192**, however, are typical SRPCs exclusively suitable for type 1 reactions of aryl iodides and activated aryl bromides (Table 2.9, entries 1–6). Ligand-acceleration effects are not observed, which unequivocally underlines that the cleavage of these pincer complexes under the reacation conditions occurs to release nonphosphine palladium complexes with indeterminate coordination shell.

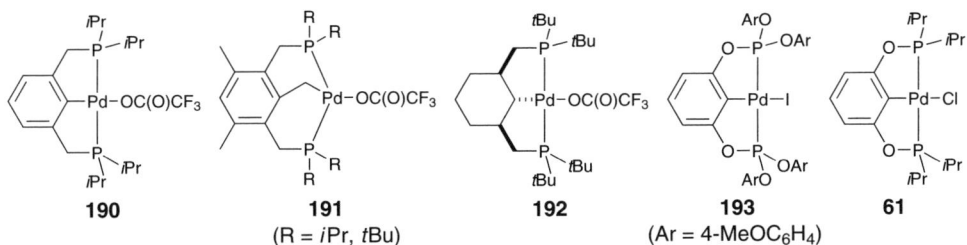

Figure 2.24 *PCP-pincer palladacycles.*

Similar reactivity is observed for pincer palladabicycles with phosphite sidearms **193** and **61** (Table 2.9, entries 7–10), although in these cases the optimization of conditions to the upper limits of activity appears promising. One of the record-making results was obtained by Shibasaki and coworkers [247] for pincer complex **193**, which in standard Mizoroki–Heck reactions of aryl iodides with acrylates showed TONs approaching 9 000 000 and TOF values approaching 150 000 h^{-1} (Table 2.9, entries 7–9); these values will be difficult to reproduce, since the palladium loading becomes comparable to trace elements contained in common reagents or on the walls of reaction vessels [249]. It should be noted that the records were achieved at very high temperatures uncommon in this area (up to 180 °C) and in the presence of the unusual previously undisclosed additive hydroquinone, the function of which remains unknown [247]. This work is very important for further development of Mizoroki–Heck systems as it clearly shows that reactive substrates, particularly aryl iodides, can be processed in Mizoroki–Heck reactions with common alkenes by almost negligible amounts of palladium. Thus, it seems hardly worthwhile to design sophisticated systems with complex coordination shells and intricate ligands. Another record maker is pincer complex **61** (Table 2.9, entry 10) [121, 248]. Besides the effective processing of aryl chlorides [121], this precatalyst, moreover, enables type 1 and type 2 operation with less reactive *gem*-disubstituted alkenes [248].

Pincer complexes with other chelating donors other than phosphorus are also useful as SRPCs. NCN- or SCS-structures are likely to be less stable than PCP-pincer complexes. Therefore, catalytic turnover using these precatalysts is possible at lower reaction temperatures. Generally, precatalysts **194–198** were tested in type 1 reactions with activated substrates (Table 2.10, Figure 2.25) [250–256]. Attachment of pincer complexes to various polymeric supports enabled the design of recyclable systems for type 1 reactions [250]. Moreover, lower stability of NCN- or SCS-pincer complexes ensures higher effective concentrations of catalytically active species; precatalyst **195** was shown to serve well in reactions with activated aryl chlorides (Table 2.10, entry 3), typical substrates for type 2 processes [252].

2.4.5.3 Palladacycle–Phosphine Complexes

One of the design ideas was to develop a good source of monophosphine palladium complexes formed from some convenient precursor. As soon as palladacycles were shown to release palladium, it seemed promising to design 1 : 1 palladacycle–phosphine hybrid complexes that were supposed to serve as a source of monophosphine species upon cleavage.

Table 2.9 *PCP-pincer palladacycles*

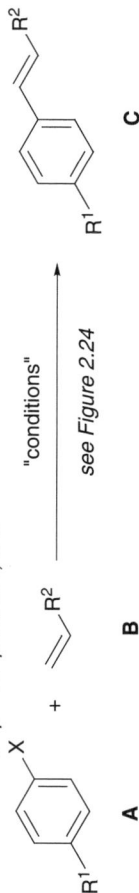

Entry	Pd-cat. (mol%)	A R^1	A X	B (R^2)	Base	Solvent	T (°C)	Yield (%)	TON	TOF (h^{-1})	Ref.
1	**190** (1.8×10^{-4})	H	I	CO$_2$Me	Na$_2$CO$_3$	NMP	140	91	5.2×10^5	n.d.	[245]
2	**191** (1.8×10^{-4})	H	I	CO$_2$Me	Na$_2$CO$_3$	NMP	140	91–100	$(1.4–5.3) \times 10^5$	n.d.	[245]
3	**191** (7.0×10^{-4})	CHO	Br	CO$_2$Me	Na$_2$CO$_3$	NMP	140	79	1.1×10^5	n.d.	[245]
4	**191** (7.0×10^{-4})	MeO	I	CO$_2$Me	Na$_2$CO$_3$	NMP	140	100	1.4×10^5	n.d.	[245]
5	**192** (1.0×10^{-3})	H	I	CO$_2$Me	Na$_2$CO$_3$	NMP	140	91	1.0×10^5	7140	[246]
6	**192** (6.0×10^{-4})	CHO	Br	CO$_2$Me	Na$_2$CO$_3$	NMP	160	87	1.5×10^5	3040	[246]
7	**193** (1.0×10^{-4})	H	I	CO$_2$-n-Bu	Na$_2$CO$_3$	NMP	180	95	9.5×10^5	53 000	[247]
8	**193** (1.0×10^{-5})	H	I	CO$_2$-n-Bu	Na$_2$CO$_3$	NMP	180	60	6.0×10^6	150 000	[247]
9	**193** (1.0×10^{-4})	MeO	I	CO$_2$-n-Bu	Na$_2$CO$_3$	NMP	140	43	4.3×10^5	4500	[247]
10	**61** (3.5×10^{-5})	H	I	CO$_2$Me	Na$_2$CO$_3$	DMF	—[a]	97	1.4×10^5	n.d.	[248]

[a]No temperature given in Ref. [248].

Table 2.10 NCN- and SCS-pincer palladacycles

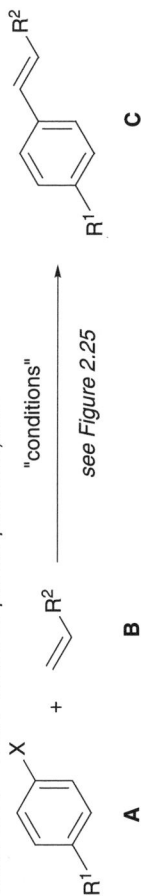

| Entry | Pd-cat. (mol%) | A | | B (R^2) | Base | Solvent | T (°C) | Yield (%) | TON | TOF (h^{-1}) | Ref. |
		R^1	X								
1	**194** (1.0×10^{-3})	MeO	I	CO_2Me	Et_3N	DMF	110	99	n.d	n.d.	[251]
2	**194** (0.1)	CHO	Br	CO_2Me	Et_3N	DMF	110	35	n.d	n.d.	[251]
3^a	**195** (0.5)	Me	Cl	CO_2-n-Bu	Na_2CO_3	DMA	140	43	n.d	n.d.	[252]
4	**196** (0.1)	OMe	I	Ph	Na_2CO_3	DMF	110	93	1000	n.d.	[253, 254]
5	**197** (1.0×10^{-3})	Me	I	Ph	NaOAc	DMA	165	n.d.	7.0×10^4	n.d.	[255]
6	**198** (0.1)	Ac	Br	CO_2Me	K_2CO_3	DMF	140	86	n.d.	n.d.	[256]

^{a} Addition of nBu$_4$NBr.

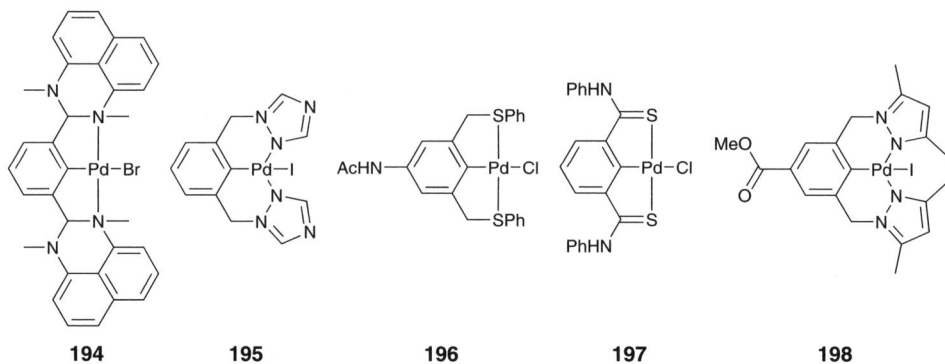

Figure 2.25 *NCN- and SCS-pincer palladacycles.*

This design idea, however, has not led to clear improvement yet. Indeed, if this principle were operative, then a hybrid complex would perform superior to the respective homoleptic phosphine complex or a mixture of a simple palladium source and a phosphine. In fact, this design idea implies that during the preactivation stage the hybrid complex **199** loses palladacyclic ligand while keeping phosphine to furnish monophosphine palladium species **200** (Scheme 2.28). Given the relative stability of the chelate palladacycle ring as opposed to a single phosphine ligand, this scenario does not seem very plausible.

Indeed, whenever such hybrid complexes were explored in parallel with the parent complexes (palladacycles without additional phosphine ligand) their activity was comparable. As an example, hybrid oxime palladacycles **183** showed comparable performances with respect to parent palladacycles **182** (see Figure 2.23) [109].

Indolese and coworkers [132, 133] have shown that simple CN-palladacycles **203** or **204** can be used in equimolar mixtures with bulky secondary or tertiary phosphines at 0.5 mol% loading (Na$_2$CO$_3$, DMA, 140 °C) in order to process deactivated aryl chloride **201** in good to high yields (**201** → **202**, Table 2.11, Figure 2.26). Systems using **203** or **204** (0.5 mol%) as palladium source were superior to the systems based on Pd(OAc)$_2$ – all other parameters were kept identical. If, however, PdCl$_2$ was taken as palladium source at twice as high a loading (1.0 mol%), consistently better results were achieved [132, 133]. The other interesting observation is that the difference between ligands is largely masked if palladacycles are taken as palladium source, so that the best ligand cannot capitalize the full potential. This is likely because the palladium released from palladacycle is the limiting factor. As soon as palladium is released slowly, there is always an effective excess of phosphine ligand over palladium in the system, and, as a result, the effective concentration of catalytically active monophosphine species is lower than in the case when an immediate source of palladium is used.

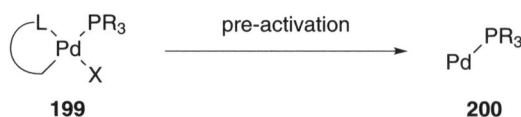

Scheme 2.28

Table 2.11   *Ligands for the Mizorok—Heck reaction of a deactivated aryl chloride*

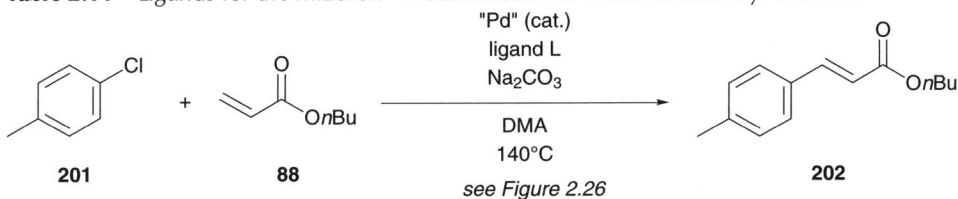

		Palladium source 'Pd'/yield (%)			
Entry	Ligand L (L : Pd = 1 : 1)	PdCl$_2$ (1.0 mol%)	Pd(OAc)$_2$ (0.5 mol%)	**203** (0.5 mol%)	**204** (0.5 mol%)
1	t-Bu$_3$P	98	49	70	71
2	t-Bu$_2$PH	77	53	71	77
3	Ad$_2$PH	78	52	74	65

Imine palladacycle **205** bearing tricyclohexylphosphine showed a good level of performance (0.1 mol% **205**, Et$_3$N, DMF, 140 °C), but only in type 1 reactions with reactive aryl bromides (Figure 2.26) [257]. The yields decreased to modest or poor levels with less reactive aryl bromides and aryl chlorides. This indicates the absence of a ligand-acceleration effect. The performance of hybrid complex **205** is evidently inferior to that of (Cy$_3$P)$_2$PdCl$_2$ as a precatalyst; the latter is capable of promoting a true ligand-accelerated reaction with aryl chlorides under milder conditions [134].

2.4.6   Nonphosphine Complexes

The rich coordination chemistry of palladium provides a plethora of more or less readily available complexes, almost all applicable in type 1 or type 2 Mizoroki–Heck reactions. Only a limited series of truly robust and almost indefinitely stable ('dead') palladium complexes, such as palladium porphyrinate or phthalocyanine complexes, as well as salen-type chelates with Schiff bases, are known not to release palladium even under the harshest conditions used in Mizoroki–Heck reactions. Results from the investigation of reactivity of

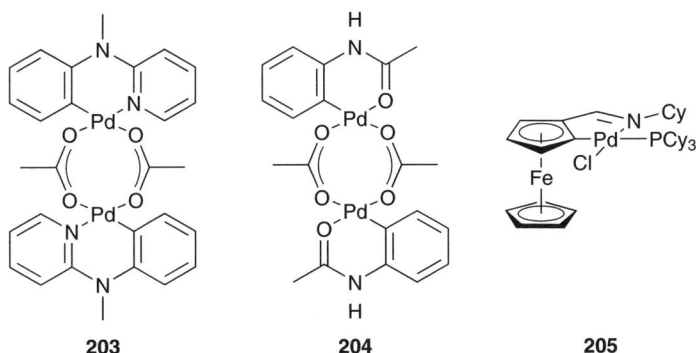

Figure 2.26   *'Phosphorus-free' palladacycles **203–205** (part 4).*

Table 2.12 (Complexes of) nonphosphine ligands

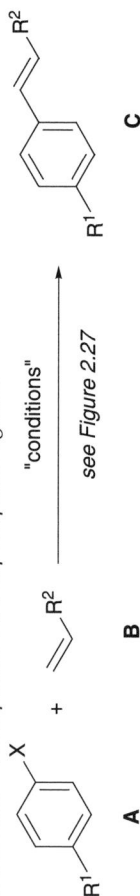

R^1-C$_6$H$_4$-X (**A**) + R^2 (**B**) →["conditions" see Figure 2.27] R^1-C$_6$H$_4$-CH=CH-R^2 (**C**)

Entry	Pd-cat. (mol%)	A R^1	A X	B (R^2)	Base	Solvent	T (°C)	Yield (%)	TON	TOF (h^{-1})	Ref.
1	**206** (2.5 × 10^{-4})	MeO	Br	Ph	K$_2$CO$_3$	H$_2$O	100	93	3.7 × 10^5	62 000	[242]
2[a]	**207** + PdCl$_2$(MeCN)$_2$ (5.0 × 10^{-4})	Me	I	CO$_2$-n-Bu	K$_3$PO$_4$	DMF	130	84	1.7 × 10^5	70 000	[258]
3	**207** + PdCl$_2$(MeCN)$_2$ (5.0)	MeO	Br	CO$_2$-n-Bu	K$_3$PO$_4$	NMP	80	52	n.d.	n.d.	[259]
4	**208** + PdCl$_2$ (1.0)	Me	Br	CO$_2$Me	K$_2$CO$_3$	DMF	120	42–84	n.d.	n.d.	[260]
5[a]	**40** + Pd(dba)$_2$ (0.1)	Me	Br	CO$_2$-n-Bu	K$_2$CO$_3$	NMP	130	75	n.d.	n.d.	[88]

[a] Addition of n-Bu$_4$NBr.

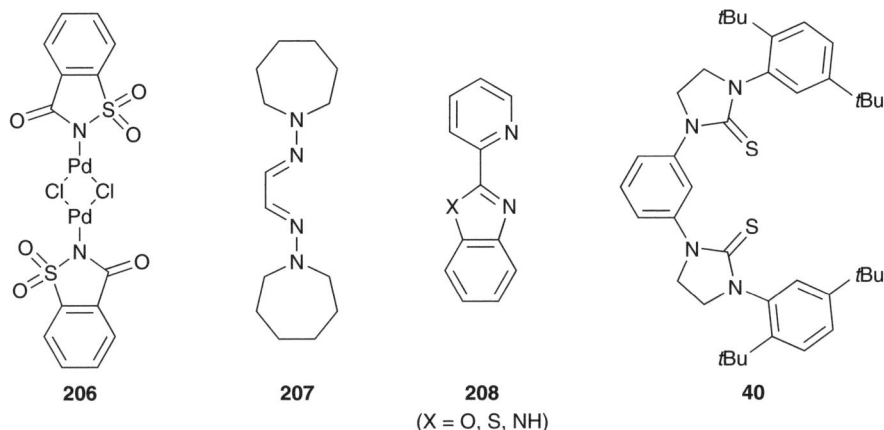

Figure 2.27   *Nonphosphine bidentate ligands and respective SRPC.*

such 'dead' complexes are somewhat questionable, since these complexes nearly always contain traces of undefined palladium compounds, which are well capable of triggering type 1 reactivity – not to be mistaken with the activity of the highly stable complex itself.

In turn, the majority of palladium complexes containing nitrogen, oxygen or similar donors as part of mono- or bi-dentate ligands are quite weak and labile, which makes them useful as SRPCs. Weaker ligands which can afford higher levels of active palladium species are useful in type 2 processes, thus opening appealing vistas for the search of practical protocols. The other incentive behind such complexes is their ready availability from inexpensive starting materials, along with convenient handling and a nontoxic profile.

Occasionally, rather unusual complexes regarding their utmost simplicity were chosen and behaved nicely. One of the brightest examples is a recent study on the use of palladium complex **206** with a trivial sulfonamides ligand, namely saccharine (Table 2.12, entry 1, Figure 2.27) [242]. Complex **206** is water soluble and, therefore, is applicable in a phase-transfer protocol in the presence of n-Bu$_4$NBr with water as the only solvent and the substrates as organic phase [242]. This effective realization of a type 2 protocol is characterized by one of the highest TONs/TOF values reported so far for electron-rich aryl bromides, which is particularly appealing in the light of the quite mild conditions. Moreover, this protocol can be extended to other alkenes, such as allylic alcohols [242].

Several other reported complexes showed less spectacular results, yet are still valuable for Mizoroki–Heck reactions with less reactive bromides. Typical systems most commonly consist of bidentate ligands. Recent reports included bidentate dinitrogen ligands derived from glyoxal bishydrazones or structurally similar monohydrazones of 2-pyridinecarbaldehyde; both showed good performance in standard Mizoroki–Heck reactions of aryl iodides and bromides (including electron-rich ones), as well as activated aryl chlorides with acrylates [258]. The azepane derivative **207** performed best in this study (Table 2.12, entries 2 and 3) [258]. Other simple bidentate NN-ligands derived from 2-(2-pyridyl)benzazoles were reported, the best being the derivative of benzimidazole **208** (Table 2.12, entry 4) [259].

Besides NN-ligands, SS-ligands show comparable properties, though limited data are available. Being apparently more stable, these precatalysts usually require higher temperatures to afford appropriate levels of activity. Bis-thiolate complexes (dppf)Pd(SR)$_2$ provide

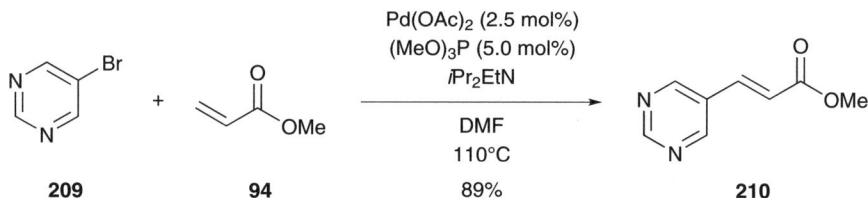

Scheme 2.29

sufficient catalytic activity to enable the reaction of phenyl bromide [260]. Bis(thiourea) ligand **40** showed considerable performance not only with aryl iodides, but also with inactivated aryl bromides, including 4-bromoanisole (Table 2.12, entry 5) [88]. More complex structures with several different chelation sites, such as NSO-ligands based on thiosemicarbazone, were also active in type 2 reactions but failed to give good yields with deactivated aryl bromides [261].

A number of bidentate ligands containing pyridine and other azine binding sites were described, involving five-membered [262], six-membered [263] and macrocyclic [262] chelate rings. All these ligands afford SRPCs active in type 1 Mizoroki–Heck reactions. Similar activity was recorded for phenanthroline-derived palladium complexes [264].

Phosphite ligands were introduced in Mizoroki–Heck reactions as early as in Spencer's studies [77], but found scarce application due to modest performance even in type 1 reactions. In some special applications, such as the reaction of heteroaryl bromide **209** with methyl acrylate (**94**), simple phosphite $(MeO)_3P$ performed better than phosphines, apparently because the substrates themselves are reasonable ligands for palladium, and a saturation of the coordination sphere may prevent alkene coordination (**209** → **210**, Scheme 2.29) [265].

More sophisticated phosphites have been introduced to the Mizoroki–Heck reactions, usually as their respective palladacycles or pincer complexes (see above). The same is true for phosphinites, a potentially very active ligand class that should display ligand-acceleration effects due to the electron-rich phosphorus centre. Simple phosphinites $Ph_2P(OR)$ and phosphonites $(PhO)_2P(OH)$ were studied in an interesting solvent-free system (0.2 mol% L_2PdCl_2, n-Bu_4NBr, $NaHCO_3$, 140 °C) in the reaction of phenyl bromide with n-butyl acrylate, thus showing a borderline type 1 or 2 performance [266]. Despite a highly electron-rich environment at palladium, bis(phosphinite) chelate complex **211** showed only modest activity for aryl iodides and bromides in Mizoroki–Heck reactions with acrylates or styrenes (Table 2.13, entry 1, Figure 2.28) [267]. The precatalyst was practically inactive towards aryl chlorides.

Quite representative is the group of ligands involving phosphine binding sites along with an N, O, or S chelating donor. The design behind such ligands apparently relied on the possibility of ligand acceleration to be effected by phosphine; therefore, bulky dialkylphosphine side-arms are not uncommon in the published structures. Not surprisingly, the data on catalytic activity of such precatalysts fit well into the slow-release paradigm, and all such catalysts deliver type 1 or type 2 Mizoroki–Heck reaction performance under typical conditions.

Iminophosphines capable of forming N,P-chelate rings with palladium appeared in several contributions, showing their applicability for type 2 processes. Both the monomeric

Table 2.13 *Performance of complexes of **211–215***

$$A + B \xrightarrow[\text{see Figure 2.28}]{\text{"conditions"}} C$$

Entry	Pd-cat. (mol%)	A R^1	A X	B (R^2)	Base	Solvent	T (°C)	Yield (%)	TON	TOF (h^{-1})	Ref.
1	**211** (5.0)	Ac	Br	CO$_2$Me	n-Bu$_3$N	DMF	120	86	n.d.	n.d.	[267]
2	**212** + Pd(OAc)$_2$ (2.0)	MeO	Br	Ph	—	n-Bu$_3$N/AcOH	130	85	n.d.	n.d.	[268]
3	**213** + Pd(OAc)$_2$ (2.0)	MeO	Br	Ph	—	n-Bu$_3$N/AcOH	130	92	n.d.	n.d.	[268]
4	**214** (1.0)	H	Br	Ph	Cs$_2$CO$_3$	MeO(CH$_2$)$_2$OH	120	89	n.d.	n.d.	[269]
5	**215** (1.0 × 10^{-3})	H	Br	Ph	NaOAc	DMA/H$_2$O	140	80–90	(8.5–9.6) × 10^5	n.d.	[270]

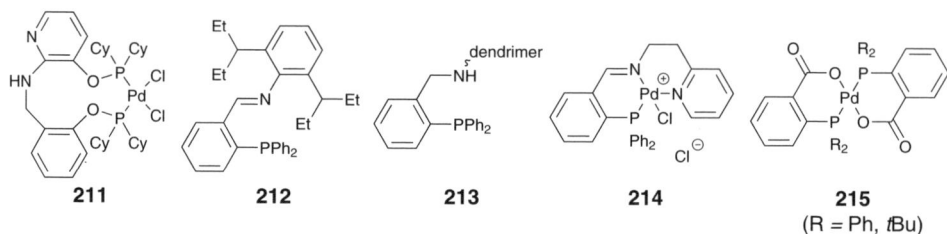

Figure 2.28 *Ligand used in catalyses of Table 2.13.*

ligand **212** and the same motif attached to diaminobutane (DAB) dendrimer **213** were capable of processing 4-bromoanisole in high yields, with all parameters fitting into the type 2 paradigm (Table 2.13, entries 2 and 3, Figure 2.28) [268]. Dendritic precatalysts showed no advantages over monomeric ligand, particularly if the 'waste' of ligand sites is taken into consideration: in the catalytic systems based on **212**, one atom of palladium was taken per two iminophosphine units, while in the dendrimer particle based on **213** there are 32 such sites. Given the molecular weight of dendrimer, such an approach could not have been regarded as practical even if dendrimeric catalyst were to show an unusual performance.

In the same row is the cationic complex **214** with a tridentate PNN-ligand (Table 2.13, entry 4) [269]. Among PO-ligands and their chelates, complexes **215** formed by 2-phosphinobenzoic acids are worth mentioning [270]. These complexes deliver type 2 activity, with PhBr showing high TONs and TOF values in aqueous solvent (Table 2.13, entry 5).

References

1. Mizoroki, T., Mori, K. and Ozaki, A. (1971) Arylation of olefin with aryl iodide catalyzed by palladium. *Bull. Chem. Soc. Jpn.*, **44**, 581.
2. Heck, R.F. and Nolley, J.P. Jr (1972) Palladium-catalyzed vinylic hydrogen substitution reactions with aryl, benzyl, and styryl halides. *J. Org. Chem.*, **14**, 2320–22.
3. Bräse, S. and de Meijere, A. (1998) Palladium-catalyzed coupling of organyl halides to alkenes – the Heck reaction, in *Metal-Catalyzed Cross-Coupling Reactions* (eds F. Diederich and P.J. Stang), Wiley-VCH Verlag GmbH, Weinheim, pp. 99–166.
4. Beletskaya, I.P. and Cheprakov, A.V. (2000) The Heck reaction as a sharpening stone of palladium catalysis. *Chem. Rev.*, **100**, 3009–66.
5. Böhm, V.P.W. and Herrmann, W.A. (2001) Mechanism of the Heck reaction using a phospha-palladacycle as the catalyst: classical versus palladium(IV) intermediates. *Chem. Eur. J.*, **7**, 4191–97.
6. Dieck, H.A. and Heck, R.F. (1974) Organophosphinepalladium complexes as catalysts for vinylic hydrogen substitution reactions. *J. Am. Chem. Soc.*, **96**, 1133–6.
7. Dieck, H.A. and Heck, R.F. (1975) A palladium-catalyzed conjugated diene synthesis from vinylic halides and olefinic compounds *J. Org. Chem.*, **40**, 1083–90.
8. Patel, B.A., Ziegler, C.B., Cortese, N.A. *et al.* (1977) Palladium-catalyzed vinylic substitution reactions with carboxylic acid derivatives. *J. Org. Chem.*, **42**, 3903–7.
9. Albeniz, A.C., Espinet, P., Martin-Ruiz, B. and Milstein, D. (2005) Catalytic system for the Heck reaction of fluorinated haloaryls. *Organometallics*, **24**, 3679–84.

10. Pastre, J.C. and Correia, C.R.D. (2006) Efficient Heck arylations of cyclic and acyclic acrylate derivatives using arenediazonium tetrafluoroborates. A new synthesis of the antidepressant drug (±)-paroxetine. *Org. Lett.*, **8**, 1657–60.

11. Portnoy, M., Ben-David, Y. and Milstein, D. (1993) Clarification of a remarkable chelate effect leads to palladium-catalyzed base-free arylation. *Organometallics*, **12**, 4734–5.

12. Arvela, R.K., Pasquini, S. and Larhed, M. (2007) Highly regioselective internal Heck arylation of hydroxyalkyl vinyl ethers by aryl halides in water. *J. Org. Chem.*, **72**, 6390–6.

13. Andrus, M.B., Song, C. and Zhang, J.Q. (2002) Palladium-imidazolium carbene catalyzed Mizoroki–Heck coupling with aryl diazonium ions. *Org. Lett.*, **4**, 2079–2.

14. Selvakumar, K., Zapf, A., Spannenberg, A. and Beller, M. (2002) Synthesis of monocarbenepalladium(0) complexes and their catalytic behavior in cross-coupling reactions of aryldiazonium salts. *Chem. Eur. J.*, **8**, 3901–6.

15. Stephan, M.S., Teunissen, A.J.J.M., Verzijl, G.K.M., and Vries, J.G.D. (1998) Heck reactions without salt formation: aromatic carboxylic anhydride as arylating agents. *Angew. Chem. Int. Ed.*, **37**, 662–4.

16. Shmidt, A.F. and Smirnov, V.V. (2000) Use of aromatic acid anhydrides as arylation agents in the Heck reaction. *Kinet. Catal.*, **41**, 743–4.

17. Jeffery, T. (1984) Palladium-catalysed vinylation of organic halides under solid–liquid phase transfer conditions. *J. Chem. Soc., Chem. Commun.*, 1287–89.

18. Jeffery, T. (1985) Highly stereospecific palladium-catalysed vinylation of vinylic halides under solid-liquid phase transfer conditions. *Tetrahedron Lett.*, **26**, 2667–70.

19. Jeffery, T. (1987) Palladium-catalyzed vinylation of acetylenic iodides under solid-liquid phase-transfer conditions. *Synthesis*, 70–1.

20. Jeffery, T. and Galland, J.-C. (1994) Tetraalkylammonium salts in Heck-type reactions using an alkali metal hydrogen carbonate or an alkali metal acetate as the base. *Tetrahedron Lett.*, **35**, 4103–6.

21. Jeffery, T. (1994) Heck-type reactions in water. *Tetrahedron Lett.*, **35**, 3051–4.

22. Larock, R.C. and Babu, S. (1987) Synthesis of nitrogen heterocycles via palladium-catalyzed intramolecular cyclization. *Tetrahedron Lett.*, **28**, 5291–4.

23. Villemin, D. and Nechab, B. (2000) Use of a new hydrophilic phosphine: DPPPA. Rapid and convenient Heck reaction in aqueous medium under microwave irradiation. *J. Chem. Res. (S)*, 429–31.

24. Ritzén, A., Basu, B., Wållberg, A. and Frejd, T. (1998) Phenyltrisalanine: a new, C_3-symmetric, trifunctional amino acid. *Tetrahedron: Asymmetry*, **9**, 3491–6.

25. Gürtler, C. and Buchwald, S.L. (1999) A phosphane-free catalyst system for the Heck arylation of disubstituted alkenes: application to the synthesis of trisubstituted olefins. *Chem. Eur. J.*, **5**, 3107–12.

26. Crombie, A.L., Kane, J.L., Shea, K.M. and Danheiser, R.L. (2004) Ring expansion–annulation strategy for the synthesis of substituted azulenes and oligoazulenes. 2. Synthesis of azulenyl halides, sulfonates, and azulenylmetal compounds and their application in transition-metal-mediated coupling reactions. *J. Org. Chem.*, **69**, 8652–67.

27. Willans, C.E., Mulders, J., de Vries, J.G. and de Vries, A.H.M. (2003) Ligand-free palladium catalysed Heck reaction of methyl 2-acetamido acrylate and aryl bromides as key step in the synthesis of enantiopure substituted phenylalanines. *J. Organomet. Chem.*, **687**, 494–7.

28. Dyker, G., Grundt, P., Markwitz, H. and Henkel, G. (1998) Heck reaction and Robinson-type annulation: a versatile combination. *J. Org. Chem.*, **63**, 6043–7.

29. Brasholz, M., Luan, X.S. and Reissig, H.U. (2005) Towards the rubromycins: an efficient synthesis of a suitable isocoumarin precursor, its lactam analogue, and palladium-catalyzed couplings. *Synthesis*, 3571–80.

30. Sauvagnat, B., Lamaty, F., Lazaro, R. and Martinez, J. (2001) Highly selective intermolecular Heck reaction for the soluble polymer supported synthesis of glutamic acid analogues. *Tetrahedron*, **57**, 9711–8.

31. Amatore, C. and Jutand, A. (2000) Anionic Pd(0) and Pd(II) intermediates in palladium-catalyzed Heck and cross-coupling reactions. *Acc. Chem. Res.*, **33**, 314–21.

32. De Vries, J.G. (2006) A unifying mechanism for all high-temperature Heck reactions. The role of palladium colloids and anionic species. *Dalton Trans.*, 421–9.

33. Herrmann, W.A., Elison, M., Fischer, J. *et al.* (1995) Metal complexes of *N*-heterocyclic carbenes – a new structural principle for catalysts in homogeneous catalysis. *Angew. Chem. Int. Ed. Engl.*, **34**, 2371–4.

34. Consorti, C.S., Flores, F.R. and Dupont, J. (2005) Kinetics and mechanistic aspects of the Heck reaction promoted by a CN-palladacycle. *J. Am. Chem. Soc.*, **127**, 12054–65.

35. Shmidt, A.F., Khalaika, A. and Bylkova, V.G. (1998) Role of a base in the catalytic arylation of olefins. *Kinet. Catal.*, **39**, 194–9.

36. Reetz, M.T. and de Vries, J.G. (2004) Ligand-free Heck reactions using low Pd-loading. *Chem. Commun.*, 1559–63.

37. Deng, J., Wu, Y.M. and Chen, Q.Y. (2005) Cross-coupling reaction of iodo-1,2,3-triazoles catalyzed by palladium. *Synthesis*, 2730–8.

38. Ritzen, A. and Frejd, T. (2000) Chiral, polyionic dendrimers with complementary charges – synthesis and chiroptical properties. *Eur. J. Org. Chem.*, 3771–82.

39. Basu, B. and Frejd, T. (1996) Catalytic asymmetric synthesis of bis-armed aromatic amino acid derivatives. Problems related to the synthesis of enantiomerically pure bis-methyl ester of the (*S*,*S*)-pyridine-2,6-diyl bis-alanine. *Acta. Chem. Scand.*, **50**, 316–22.

40. Carlstrom, A.S. and Frejd, T. (1989) Palladium-catalyzed synthesis of didehydroamino acid derivatives. *Synthesis*, 414–8.

41. Doi, T., Fujimoto, N., Watanabe, J. and Takahashi, T. (2003) Palladium(0)-catalyzed Mizoroki-Heck reaction and Rh(I)-catalyzed asymmetric hydrogenation of polymer-supported dehydroalanine system. *Tetrahedron Lett.*, **44**, 2161–5.

42. Chattopadhyay, S.K., Pal, B.K. and Biswas, S. (2005) Pd(0)-catalyzed Heck-type arylation of didehydropeptides. *Synth. Commun.*, **35**, 1167–75.

43. Crestey, F., Collot, V., Stiebing, S. and Rault, S. (2006) Synthesis of new dehydro 2-azatryptophans and derivatives via Heck cross-coupling reactions of 3-iodoindazoles with methyl 2-(acetylamino)acrylate. *Synthesis*, 3506–14.

44. Aitken, D.J., Faure, S. and Roche, S. (2003) Synthetic approaches to the southern part of cyclotheonamide C. *Tetrahedron Lett.*, **44**, 8827–30.

45. Gibson, S.E., Lecci, C. and White, A.J.P. (2006) Application of the Heck reaction in the synthesis of macrocycles derived from amino alcohols. *Synlett*, 2929–34.

46. Taylor, E.C. and Liu, B. (1999) A simple and concise synthesis of LY231514 (MTA). *Tetrahedron Lett.*, **40**, 4023–6.

47. Taylor, E.C., Gillespie, P. and Patel, M. (1992) Novel 5-desmethylene analogs of 5,10-dideaza-5,6,7,8-tetrahydrofolic acid as potential anticancer agents. *J. Org. Chem.*, **57**, 3218–25.

48. Muzart, J. (2005) Palladium-catalysed reactions of alcohols. Part B: formation of C–C and C–N bonds from unsaturated alcohols. *Tetrahedron*, **61**, 4179–212.

49. Dyker, G., Kadzimirsz, D. and Henkel, G. (2003) Macrocycles from simple building blocks by a multifold Heck-type coupling reaction. *Tetrahedron Lett.*, **44**, 7905–7.

50. Dyker, G. and Kadzimirsz, D. (2003) Synthesis of benzo[*a*]pyren-6-yl-substituted carboxylic acids. *Eur. J. Org. Chem.*, 3167–72.

51. Dyker, G. and Markwitz, H. (1998) A palladium-catalyzed domino process to 1-benzazepines. *Synthesis*, 1750–4.

52. Dyker, G., Markwitz, H. and Henkel, G. (2001) A putatively unfeasible Heck reaction – from cyclopentenones to annulated ring systems. *Eur. J. Org. Chem.*, 2415–23.

53. Leadbeater, N.E., Williams, V.A., Barnard, T.M. and Collins, M.J. (2006) Solvent-free, open-vessel microwave-promoted Heck couplings: from the mmol to the mol scale. *Synlett*, 2953–8.

54. Battistuzzi, G., Cacchi, S., Fabrizi, G. and Bernini, R. (2003) 3-Arylpropanoate esters through the palladium-catalyzed reaction of aryl halides with acrolein diethyl acetal. *Synlett*, 1133–6.

55. Noël, S., Luo, C.H., Pinel, C. and Djakovitch, L. (2007) Efficient heterogeneously palladium-catalysed Heck arylation of acrolein diethyl acetal. Selective synthesis of cinnamaldehydes or 3-arylpropionic esters. *Adv. Synth. Catal.*, **349**, 1128–40.

56. Noël, S., Djakovitch, L. and Pinel, C. (2006) Influence of the catalytic conditions on the selectivity of the Pd-catalyzed Heck arylation of acrolein derivatives. *Tetrahedron Lett.*, **47**, 3839–42.

57. Noël, S., Pinel, C. and Djakovitch, L. (2006) Direct synthesis of tricyclic 5*H*-pyrido[3,2,1-*ij*]quinolin-3-one by domino palladium catalyzed reaction. *Org. Biomol. Chem.*, **4**, 3760–2.

58. Karabelas, K., Westerlund, C. and Hallberg, A. (1985) The effect of added silver nitrate on the palladium-catalyzed arylation of allyltrimethylsilanes. *J. Org. Chem.*, **50**, 3896–900.

59. Abelman, M.M. and Overman, L.E. (1988) Palladium-catalyzed polyene cyclizations of dienyl aryl iodides. *J. Am. Chem. Soc.*, **110**, 2238–9.

60. Sato, Y., Sodeoka, M. and Shibasaki, M. (1990) On the role of silver salts in asymmetric Heck-type reaction. A greatly imrpoved catalytic asymmetric synthesis of *cis*-decalin derivatives. *Chem. Lett.*, 1953–4.

61. Burns, B., Grigg, R., Santhkumar, V. *et al.* (1992) Palladium catalysed tandem cyclisation–anion capture and hydride ion capture by alkyl- and π-allyl-palladium species. *Tetrahedron*, **48**, 7297–320.

62. Batsanov, A.S., Knowles, J.P. and Whiting, A. (2007) Mechanistic studies on the Heck–Mizoroki cross-coupling reaction of a hindered vinylboronate ester as a key approach to developing a highly stereoselective synthesis of a C1–C7 *Z,Z,E*-triene synthon for viridenomycin. *J. Org. Chem.*, **72**, 2525–32.

63. Saffarzadeh-Matin, S., Chuck, C.J., Kerton, F.M. and Rayner, C.M. (2004) Poly(dimethyl-siloxane)-derived phosphine and phosphinite ligands: synthesis, characterization, solubility in supercritical carbon dioxide, and sequestration on silica. *Organometallics*, **23**, 5176–81.

64. Lee, J.K., Williamson, R.M., Holmes, A.B. *et al.* (2007) A study of the Heck reaction in non-polar hydrocarbon solvents and in supercritical carbon dioxide. *Aust. J. Chem.*, **60**, 566–71.

65. Montilla, F., Galindo, A., Andres, R. *et al.* (2006) Carbosilane dendrons as solubilizers of metal complexes in supercritical carbon dioxide. *Organometallics*, **25**, 4138–43.

66. Clark, J.H. and Tavener, S.J. (2007) Alternative solvents: shades of green. *Org. Proc. Res. Dev.*, **11**, 149–55.

67. Reetz, M.T., Breinbauer, R. and Wanninger, K. (1996) Suzuki and Heck reactions catalyzed by preformed palladium clusters and palladium/nickel bimetallic clusters. *Tetrahedron Lett.*, **37**, 4499–502.

68. Battistuzzi, G., Cacchi, S. and Fabrizi, G. (2003) An efficient palladium-catalyzed synthesis of cinnamaldehydes from acrolein diethyl acetal and aryl iodides and bromides. *Org. Lett.*, **5**, 777–80.

69. Gibson, S.E., Jones, J.O., Kalindjian, S.B. *et al.* (2004) Synthesis of meta- and paracyclophanes containing unsaturated amino acid residues. *Tetrahedron*, **60**, 6945–58.

70. Consorti, C.S., Zanini, M.L., Leal, S. *et al.* (2003) Chloropalladated propargyl amine: a highly efficient phosphine-free catalyst precursor for the Heck reaction. *Org. Lett.*, **5**, 983–6.

71. Tian, J. and Moeller, K.D. (2005) Electrochemically assisted Heck reactions. *Org. Lett.*, **7**, 5381–3.

72. Leadbeater, N.E. and Marco, M. (2003) Transition-metal-free Suzuki-type coupling reactions: scope and limitations of the methodology. *J. Org. Chem.*, **68**, 5660–7.
73. Kostas, I.D., Coutsolelos, A.G., Charalambidis, G. and Skondra, A. (2007) The first use of porphyrins as catalysts in cross-coupling reactions: a water-soluble palladium complex with a porphyrin ligand as an efficient catalyst precursor for the Suzuki–Miyaura reaction in aqueous media under aerobic conditions. *Tetrahedron Lett.*, **48**, 6688–91.
74. Spencer, A. (1984) Homogeneous palladium-catalysed arylation of activated alkenes with aryl chlorides. *J. Organomet. Chem.*, **270**, 115–20.
75. Spencer, A. (1984) Selective preparation of non-symmetrically substituted divinylbenzenes by palladium-catalysed arylations of alkenes with bromobenzoic acid derivatives. *J. Organomet. Chem.*, **265**, 323–32.
76. Spencer, A. (1983) Synthesis of styrene and stilbene derivatives by the palladium-catalysed arylation of ethylene with aroyl chlorides. *J. Organomet. Chem.*, **247**, 117–22.
77. Spencer, A. (1983) A highly efficient version of the palladium-catalysed arylation of alkenes with aryl bromides. *J. Organomet. Chem.*, **258**, 101–8.
78. Yao, Q.W., Kinney, E.P. and Yang, Z. (2003) Ligand-free Heck reaction: Pd(OAc)$_2$ as an active catalyst revisited. *J. Org. Chem.*, **68**, 7528–31.
79. Gruber, A.S., Pozebon, D., Monteiro, A.L. and Dupont, J. (2001) On the use of phosphine-free PdCl$_2$(SEt$_2$)$_2$ complex as catalyst precursor for the Heck reaction. *Tetrahedron Lett.*, **42**, 7345–8.
80. Herrmann, W.A., Brossmer, C., Reisinger, C.P. *et al.* (1997) Palladacycles: efficient new catalysts for the Heck vinylation of aryl halides. *Chem. Eur. J.*, **3**, 1357–64.
81. Gruber, A.S., Zim, D., Ebeling, G. *et al.* (2000) Sulfur-containing palladacycles as catalyst precursors for the Heck reaction. *Org. Lett.*, **2**, 1287–90.
82. Schmidt, A.F., Al-Halaiqa, A. and Smirnov, V.V. (2006) Effect of macrokinetic factors on the ligand-free Heck reaction with non-activated bromoarenes. *J. Mol. Cat. A: Chem.*, **250**, 131–7.
83. Beletskaya, I.P. and Cheprakov, A.V. (1999) Palladium-catalyzed reactions: from the art of today to a common tool of tomorrow, in *Transition Metal Catalyzed Reactions* (eds S.-I. Murahashi and S.G. Davies), Blackwell Science, Oxford, pp. 29–54.
84. De Vries, A.H.M., Mulders, J., Mommers, J.H.M. *et al.* (2003) Homeopathic ligand-free palladium as a catalyst in the Heck reaction. A comparison with a palladacycle. *Org. Lett.*, **5**, 3285–8.
85. Blackmond, D.G., Schultz, T., Mathew, J.S. *et al.* (2006) Comprehensive kinetic screening of palladium catalysts for Heck reactions. *Synlett*, 3135–9.
86. Reetz, M.T., Lohmer, G. and Schwickardi, R. (1998) A new catalyst system for the Heck reaction of unreactive aryl halides. *Angew. Chem. Int. Ed.*, **37**, 481–3.
87. Li, J.H., Wang, D.P. and Xie, Y.X. (2005) Pd(OAc)$_2$/DABCO as a highly active catalytic system for the Heck reaction. *Synthesis*, 2193–7.
88. Chen, W., Li, R., Han, B. *et al.* (2006) The design and synthesis of bis(thiourea) ligands and their application in Pd-catalyzed Heck and Suzuki reactions under aerobic conditions. *Eur. J. Org. Chem.*, 1177–84.
89. Kondolff, I., Feuerstein, M., Doucet, H. and Santelli, M. (2007) Synthesis of all-*cis*-3-(2-diphenylphosphinoethyl)-1,2,4-tris(diphenylphosphinomethyl)cyclopentane (Ditricyp) from dicyclopentadiene. *Tetrahedron*, **63**, 9514–21.
90. Doucet, H. and Santelli, M. (2006) *cis,cis,cis*-1,2,3,4-Tetrakis(diphenylphosphinomethyl)cyclopentane: Tedicyp, an efficient ligand in palladium-catalysed reactions. *Synlett*, 2001–15.
91. Feuerstein, M., Doucet, H. and Santelli, M. (2002) A new efficient tetraphosphine/palladium catalyst for the Heck reaction of aryl halides with styrene or vinylether derivatives. *Tetrahedron Lett.*, **43**, 2191–4.

92. Hierso, J.C., Fihri, A., Amardeil, R. *et al.* (2005) Use of a bulky phosphine of weak σ-donicity with palladium as a versatile and highly-active catalytic system: allylation and arylation coupling reactions at 10^{-1}–10^{-4} mol% catalyst loadings of ferrocenyl bis(difurylphosphine)/Pd. *Tetrahedron*, **61**, 9759–66.

93. Fall, Y., Berthiol, F., Doucet, H. and Santelli, M. (2007) Palladium–tetraphosphine catalysed Heck reaction with simple alkenes: influence of reaction conditions on the migration of the double bond. *Synthesis*, 1683–96.

94. Lemhadri, M., Battace, A., Zair, T. *et al.* (2007) Heck arylations of pent-4-enoates or allyl-malonate using a palladium/tetraphosphine catalyst. *J. Organomet. Chem.*, **692**, 2270–81.

95. Battace, A., Zair, T., Doucet, H. and Santelli, M. (2006) Heck vinylations using vinyl sulfide, vinyl sulfoxide, vinyl sulfone, or vinyl sulfonate derivatives and aryl bromides catalyzed by a palladium complex derived from a tetraphosphine. *Synthesis*, 3495–505.

96. Berthiol, F., Doucet, H. and Santelli, M. (2006) Heck reactions of aryl halides with alk-1-en-3-ol derivatives catalysed by a tetraphosphine-palladium complex. *Appl. Organomet. Chem.*, **20**, 855–68.

97. Berthiol, F., Doucet, H. and Santelli, M. (2006) Synthesis of β-aryl ketones by tetraphosphine/palladium catalysed Heck reactions of 2- or 3-substituted allylic alcohols with aryl bromides. *Tetrahedron*, **62**, 4372–83.

98. Kondolff, I., Doucet, H. and Santelli, M. (2004) Direct synthesis of protected arylacetaldehydes by palladium–tetraphosphine-catalyzed arylation of ethyleneglycol vinylether. *Synlett*, 1561–4.

99. Battace, A., Feuerstein, M., Lemhadri, M. *et al.* (2007) Heck reactions of α- or β-substituted enol ethers with aryl bromides catalysed by a tetraphosphane/palladium complex – direct access to acetophenone or 1-arylpropanone derivatives. *Eur. J. Org. Chem.*, 3122–32.

100. Kondolff, I., Doucet, H. and Santelli, M. (2006) Direct synthesis of protected arylacetaldehydes by tetrakis(phosphane)palladium-catalyzed arylation of ethyleneglycol vinyl ether. *Eur. J. Org. Chem.*, 765–74.

101. Battace, A., Zair, T., Doucet, H. and Santelli, M. (2006) Heck reactions of 2-substituted enol ethers with aryl bromides catalysed by a tetraphosphine/palladium complex. *Tetrahedron Lett.*, **47**, 459–62.

102. Lemhadri, M., Doucet, H. and Santelli, M. (2006) Synthesis of (*E*)-1-aryl alk-1-en-3-ones by tetraphosphine/palladium-catalysed Heck reactions of alk-1-en-3-ones with aryl bromides. *Synlett*, 2935–40.

103. Fall, Y., Doucet, H. and Santelli, M. (2007) Heck reaction with an alkenylidenecyclopropane: the formation of arylallylidenecyclopropanes. *Tetrahedron Lett.*, **48**, 3579–81.

104. Feuerstein, M., Doucet, H. and Santelli, M. (2001) Efficient Heck vinylation of aryl halides catalyzed by a new air-stable palladium–tetraphosphine complex. *J. Org. Chem.*, **66**, 5923–5.

105. Molnár, A. and Papp, A. (2006) Efficient heterogeneous palladium–montmorillonite catalysts for Heck coupling of aryl bromides and chlorides. *Synlett*, 3130–4.

106. Chanthateyanonth, R. and Alper, H. (2003) The first synthesis of stable palladium(II) PCP-type catalysts supported on silica – application to the Heck reaction. *J. Mol. Cat. A: Chem.*, **201**, 23–31.

107. Herrmann, W.A., Brossmer, C., Öfele, K. *et al.* (1995) Palladacycles as structurally defined catalysts for the Heck olefination of chloro- and bromoarenes. *Angew. Chem. Int. Ed. Engl.*, **34**, 1844–8.

108. Iyer, S. and Ramesh, C. (2000) Aryl-Pd covalently bonded palladacycles, novel amino and oxime catalysts {di-μ-chlorobis(benzaldehydeoxime-6-C,N)dipalladium(II), di-μ-chlorobis(dimethylbenzylamine-6-C,N)dipalladium(II)} for the Heck reaction. *Tetrahedron Lett.*, **41**, 8981–4.

109. Iyer, S. and Jayanthi, A. (2001) Acetylferrocenyloxime palladacycle-catalyzed Heck reactions. *Tetrahedron Lett.*, **42**, 7877–8.

110. McGuinness, D.S. and Cavell, K.J. (2000) Donor-functionalized heterocyclic carbene complexes of palladium(II): efficient catalysts for C–C coupling reactions. *Organometallics*, **19**, 741–8.

111. Calo, V., Nacci, A., Lopez, L. and Mannarini, N. (2000) Heck reaction in ionic liquids catalyzed by a Pd–benzothiazole carbene complex. *Tetrahedron Lett.*, **41**, 8973–6.

112. Peris, E., Loch, J.A., Mata, J. and Crabtree, R.H. (2001) A Pd complex of a tridentate pincer CNC bis-carbene ligand as a robust homogenous Heck catalyst. *Chem. Commun.*, 201–2.

113. Calo, V., Nacci, A., Monopoli, A. *et al.* (2001) Heck reaction of β-substituted acrylates in ionic liquids catalyzed by a Pd–benzothiazole carbene complex. *Tetrahedron*, **57**, 6071–7.

114. Grundemann, S., Albrecht, M., Loch, J.A. *et al.* (2001) Tridentate carbene CCC and CNC pincer palladium(II) complexes: structure, fluxionality, and catalytic activity. *Organometallics*, **20**, 5485–8.

115. Loch, J.A., Albrecht, M., Peris, E. *et al.* (2002) Palladium complexes with tridentate pincer bis-carbene ligands as efficient catalysts for C–C coupling. *Organometallics*, **21**, 700–6.

116. Beller, M. and Zapf, A. (1998) Phosphites as ligands for efficient catalysis of Heck reactions. *Synlett*, 792–3.

117. Li, G.Y., Zheng, G. and Noonan, A.F. (2001) Highly active, air-stable versatile palladium catalysts for the C–C, C–N, and C–S bond formations via cross-coupling reactions of aryl chlorides. *J. Org. Chem.*, **66**, 8677–81.

118. Zapf, A. and Beller, M. (2001) Palladium catalyst systems for cross-coupling reactions of aryl chlorides and olefins. *Chem. Eur. J.*, **7**, 2908–15.

119. Davidson, J.B., Simon, N.M. and Sojka, S.A. (1984) Palladium catalyzed olefin arylation: extending the scope of the Heck reaction to aryl chlorides. *J. Mol. Cat.*, **22**, 349–52.

120. Buchmeiser, M.R. and Wurst, K. (1999) Access to well-defined heterogeneous catalytic systems via ring-opening metathesis polymerization (ROMP): applications in palladium(II)-mediated coupling reactions. *J. Am. Chem. Soc.*, **121**, 11101–7.

121. Morales-Morales, D., Redón, R., Yung, C. and Jensen, C.M. (2000) High yield olefination of a wide scope of aryl chlorides catalyzed by the phosphinito PCP pincer complex: [PdCl{C$_6$H$_3$(OPPr^i$_2$)$_2$-2,6}]. *Chem. Commun.*, 1619–20.

122. Littke, A.F. and Fu, G.C. (1999) Heck reactions in the presence of P(*t*-Bu)$_3$: expanded scope and milder reaction conditions for the coupling of aryl chlorides. *J. Org. Chem.*, **64**, 10–11.

123. Shaughnessy, K.H., Kim, P. and Hartwig, J.F. (1999) A fluorescence-based assay for high-throughput screening of coupling reactions. Application to Heck chemistry. *J. Am. Chem. Soc.*, **121**, 2123–32.

124. Yang, C.L., Lee, H.M. and Nolan, S.P. (2001) Highly efficient Heck reactions of aryl bromides with *n*-butyl acrylate mediated by a palladium/phosphine–imidazolium salt system. *Org. Lett.*, **3**, 1511–4.

125. Stambuli, J.P., Buhl, M. and Hartwig, J.F. (2002) Synthesis, characterization, and reactivity of monomeric, arylpalladium halide complexes with a hindered phosphine as the only dative ligand. *J. Am. Chem. Soc.*, **124**, 9346–7.

126. Ben-David, Y., Portnoy, M., Gozin, M. and Milstein, D. (1992) Palladium-catalyzed vinylation of aryl chlorides. Chelate effect in catalysis. *Organometallics*, **11**, 1995–6.

127. Littke, A.F., and Fu, G.C. (2002) Palladium-catalyzed coupling reactions of aryl chlorides. *Angew. Chem. Int. Ed.*, **41**, 4176–211.

128. Bedford, R.B., Cazin, C.S.J. and Holder, D. (2004) The development of palladium catalysts for C–C and C–heteroatom bond forming reactions of aryl chloride substrates. *Coord. Chem. Rev.*, **248**, 2283–321.

129. Whitcombe, N.J., Hii, K.K. and Gibson, S.E. (2001) Advances in the Heck chemistry of aryl bromides and chlorides. *Tetrahedron*, **57**, 7449–76.

130. Littke, A.F. and Fu, G.C. (2001) A versatile catalyst for Heck reactions of aryl chlorides and aryl bromides under mild conditions. *J. Am. Chem. Soc.*, **123**, 6989–7000.

131. Netherton, M.R. and Fu, G.C. (2001) Air-stable trialkylphosphonium salts: simple, practical, and versatile replacements for air-sensitive trialkylphosphines. Applications in stoichiometric and catalytic processes. *Org. Lett.*, **3**, 4295–8.

132. Schnyder, A., Indolese, A.F., Studer, M. and Blaser, H.U. (2002) A new generation of air stable, highly active Pd complexes for C–C and C–N coupling reactions with aryl chlorides. *Angew. Chem. Int. Ed.*, **41**, 3668–71.

133. Schnyder, A., Aemmer, T., Indolese, A.E. *et al.* (2002) First application of secondary phosphines as supporting ligands for the palladium-catalyzed Heck reaction: efficient activation of aryl chlorides. *Adv. Synth. Catal.*, **344**, 495–8.

134. Yi, C.Y. and Hua, R.M. (2006) An efficient palladium-catalyzed Heck coupling of aryl chlorides with alkenes. *Tetrahedron Lett.*, **47**, 2573–6.

135. Lo, M.Y., Zhen, C.G., Lauters, M. *et al.* (2007) Organic–inorganic hybrids based on pyrene functionalized octavinylsilsesquioxane cores for application in OLEDs. *J. Am. Chem. Soc.*, **129**, 5808–9.

136. Lo, M.Y. and Sellinger, A. (2006) Highly fluorescent blue-emitting materials from the Heck reaction of triphenylvinylsilane with conjugated dibromoaromatics. *Synlett*, 3009–12.

137. Southard, G.E., Van Houten, K.A. and Murray, G.M. (2006) Heck cross-coupling for synthesizing metal-complexing monomers. *Synthesis*, 2475–7.

138. Southard, G.E. and Murray, G.M. (2005) Synthesis of vinyl-substituted β-diketones for polymerizable metal complexes. *J. Org. Chem.*, **70**, 9036–9.

139. Muraoka, N., Mineno, M., Itami, K. and Yoshida, J.-i. (2005) Rapid synthesis of CDP840 with 2-pyrimidyl vinyl sulfide as a platform. *J. Org. Chem.*, **70**, 6933–6.

140. Leclerc, J.P., Andre, M. and Fagnou, K. (2006) Heck, direct arylation, and hydrogenation: two or three sequential reactions from a single catalyst. *J. Org. Chem.*, **71**, 1711–4.

141. Itami, K., Tonogaki, K., Ohashi, Y. and Yoshida, J.-i. (2004) Rapid construction of multi-substituted olefin structures using vinylboronate ester platform leading to highly fluorescent materials. *Org. Lett.*, **6**, 4093–6.

142. Tonogaki, K., Soga, K., Itami, K. and Yoshida, J.-i. (2005) Versatile synthesis of 1,1-diaryl-1-alkenes using vinylboronate ester as a platform. *Synlett*, 1802–4.

143. Iwasawa, T., Komano, T., Tajima, A. *et al.* (2006) Phosphines having a 2,3,4,5-tetraphenylphenyl moiety: effective ligands in palladium-catalyzed transformations of aryl chlorides. *Organometallics*, **25**, 4665–9.

144. Demchuk, O.M., Yoruk, B., Blackburn, T. and Snieckus, V. (2006) A mixed naphthyl–phenyl phosphine ligand motif for Suzuki, Heck, and hydrodehalogenation reactions. *Synlett*, 2908–13.

145. Gümgüm, B., Biricik, N., Durap, F. *et al.* (2007) Application of *N*,*N*-bis(diphenylphosphino) aniline palladium(II) complexes as pre-catalysts in Heck coupling reactions. *Appl. Organomet. Chem.*, **21**, 711–5.

146. Datta, A., Ebert, K. and Plenio, H. (2003) Nanofiltration for homogeneous catalysis separation: soluble polymer-supported palladium catalysts for Heck, Sonogashira, and Suzuki coupling of aryl halides. *Organometallics*, **22**, 4685–91.

147. Cabri, W., Candiani, I., DeBernardinis, S. *et al.* (1991) Heck reaction on anthraquinone derivatives: ligand, solvent, and salt effects. *J. Org. Chem.*, **56**, 5796–800.

148. Ozawa, F., Kubo, A. and Hayashi, T. (1991) Catalytic asymmetric arylation of 2,3-dihydrofuran with aryl triflates. *J. Am. Chem. Soc.*, **113**, 1417–9.

149. Shaw, B.L. and Perera, S.D. (1998) Chelating diphosphine-palladium(II) dihalides; outstandingly good catalysts for Heck reactions of aryl halides. *Chem. Commun.*, 1863–4.

150. Fu, X.Y., Zhang, S.Y., Yin, J.G. *et al.* (2002) First examples of a tosylate in the palladium-catalyzed Heck cross coupling reaction. *Tetrahedron Lett.*, **43**, 573–6.
151. Trzeciak, A.M. and Ziolkowski, J.J. (2007) Monomolecular, nanosized and heterogenized palladium catalysts for the Heck reaction. *Coord. Chem. Rev.*, **251**, 1281–93.
152. Astruc, D. (2007) Palladium nanoparticles as efficient green homogeneous and heterogeneous carbon–carbon coupling precatalysts: a unifying view. *Inorg. Chem.*, **46**, 1884–94.
153. Biffis, A., Zecca, M. and Basato, M. (2001) Palladium metal catalysts in Heck C–C coupling reactions. *J. Mol. Cat. A: Chem.*, **173**, 249–74.
154. Walter, J., Heiermann, J., Dyker, G. *et al.* (2000) Hexagonal or quasi two-dimensional palladium nanoparticles – tested at the Heck reaction. *J. Catal.*, **189**, 449–55.
155. Moiseev, I.I. and Vargaftik, M.N. (1998) Pd cluster catalysis: a review of reactions under anaerobic conditions. *New J. Chem.*, **22**, 1217–27.
156. Schmid, G., Harms, M., Malm, J.O. *et al.* (1993) Ligand-stabilized giant palladium clusters - promising candidates in heterogeneous catalysis. *J. Am. Chem. Soc.*, **115**, 2046–8.
157. Schmid, G. (1992) Large clusters and colloids – metals in the embryonic state. *Chem. Rev.*, **92**, 1709–27.
158. Benfield, R.E. (1992) Mean coordination numbers and the nonmetal metal transition in clusters. *J. Chem. Soc., Faraday Trans.*, **88**, 1107–10.
159. Li, Y., Boone, E. and El-Sayed, M.A. (2002) Size effects of PVP–Pd nanoparticles on the catalytic Suzuki reactions in aqueous solution. *Langmuir*, **18**, 4921–5.
160. Le Bars, J., Specht, U., Bradley, J.S. and Blackmond, D.G. (1999) A catalytic probe of the surface of colloidal palladium particles using Heck coupling reactions. *Langmuir*, **15**, 7621–5.
161. Davis, J.J., Bagshaw, C.B., Busuttil, K.L. *et al.* (2006) Spatially controlled Suzuki and Heck catalytic molecular coupling. *J. Am. Chem. Soc.*, **128**, 14135–41.
162. Ananikov, V.P., Orlov, N.V., Beletskaya, I.P. *et al.* (2007) New approach for size- and shape-controlled preparation of Pd nanoparticles with organic ligands. Synthesis and application in catalysis. *J. Am. Chem. Soc.*, **129**, 7252–3.
163. Narayanan, R. and El-Sayed, M.A. (2005) Effect of catalysis on the stability of metallic nanoparticles: Suzuki reaction catalyzed by PVP–palladium nanoparticles. *J. Am. Chem. Soc.*, **125**, 8340–7.
164. Heidenreich, R.G., Krauter, E.G.E., Pietsch, J. and Köhler, K. (2002) Control of Pd leaching in Heck reactions of bromoarenes catalyzed by Pd supported on activated carbon. *J. Mol. Cat. A: Chem.*, **182**, 499–509.
165. Köhler, K., Heidenreich, R.G., Krauter, J.G.E. and Pietsch, M. (2002) Highly active palladium/activated carbon catalysts for Heck reactions: correlation of activity, catalyst properties, and Pd leaching. *Chem. Eur. J.*, **8**, 622–31.
166. Albers, P.W., Krauter, J.G.E., Ross, D.K. *et al.* (2004) Identification of surface states on finely divided supported palladium catalysts by means of inelastic incoherent neutron scattering. *Langmuir*, **20**, 8254–60.
167. Klingelhofer, S., Heitz, W., Greiner, A. *et al.* (1997) Preparation of palladium colloids in block copolymer micelles and their use for the catalysis of the Heck reaction. *J. Am. Chem. Soc.*, **119**, 10116–20.
168. Beletskaya, I.P., Khokhlov, A.R., Tarasenko, E.A. and Tyurin, V.S. (2007) Palladium supported on poly(*N*-vinylimidazole) or poly(*N*-vinylimidazole-co-*N*-vinylcaprolactam) as a new recyclable catalyst for the Mizoroki–Heck reaction. *J. Organomet. Chem.*, **692**, 4402–6.
169. Reetz, M.T. and Lohmer, G. (1996) Propylene carbonate stabilized nanostructured palladium clusters as catalysts in Heck reactions. *J. Chem. Soc., Chem. Commun.*, **16**, 1921–2.
170. Saffarzadeh-Matin, S., Kerton, F.M., Lyman, J.M. and Rayner, C.M. (2006) Formation and catalytic activity of Pd nanoparticles on silica in supercritical CO_2. *Green Chem.*, **8**, 965–71.

171. Beller, M., Fischer, H., Kuhlein, K. *et al.* (1996) First palladium-catalyzed Heck reactions with efficient colloidal catalyst systems. *J. Organomet. Chem.*, **520**, 257–9.

172. Shmidt, A.F. and Mametova, L.V. (1996) Main features of catalysis in the styrene phenylation reaction. *Kinet. Catal.*, **37**, 406–8.

173. Caporusso, A.M., Innocenti, P., Aronica, L.A. *et al.* (2005) Functional resins in palladium catalysis: promising materials for Heck reaction in aprotic polar solvents. *J. Catal.*, **234**, 1–13.

174. Köhler, K., Prockl, S.S. and Kleist, W. (2006) Supported palladium catalysts in Heck coupling reactions – problems, potential and recent advances. *Curr. Org. Chem.*, **10**, 1585–601.

175. Polshettiwar, V. and Molnár, A. (2007) Silica-supported Pd catalysts for Heck coupling reactions. *Tetrahedron*, **63**, 6949–76.

176. Sommer, W.J. and Weck, M. (2007) Supported *N*-heterocyclic carbene complexes in catalysis. *Coord. Chem. Rev.*, **251**, 860–73.

177. Yin, L.X. and Liebscher, J. (2007) Carbon–carbon coupling reactions catalyzed by heterogeneous palladium catalysts. *Chem. Rev.*, **107**, 133–73.

178. Alonso, F., Beletskaya, I.P. and Yus, M. (2005) Non-conventional methodologies for transition-metal catalysed carbon–carbon coupling: a critical overview. Part 1: the Heck reaction. *Tetrahedron*, **61**, 11771–835.

179. Herrmann, W.A. (2002) *N*-Heterocyclic carbenes: a new concept in organometallic catalysis. *Angew. Chem. Int. Ed.*, **41**, 1290–309.

180. Mata, J.A., Poyatos, M. and Peris, E. (2007) Structural and catalytic properties of chelating bis- and tris-*N*-heterocyclic carbenes. *Coord. Chem. Rev.*, **251**, 841–59.

181. Kantchev, E.A.B., O'Brien, C.J. and Organ, M.G. (2007) Palladium complexes of *N*-heterocyclic carbenes as catalysts for cross-coupling reactions – a synthetic chemist's perspective. *Angew. Chem. Int. Ed.*, **46**, 2768–813.

182. Kremzow, D., Seidel, G., Lehmann, C.W. and Fürstner, A. (2005) Diaminocarbene- and Fischer-carbene complexes of palladium and nickel by oxidative insertion: preparation, structure, and catalytic activity. *Chem. Eur. J.*, **11**, 1833–53.

183. Dhudshia, B. and Thadani, A.N. (2006) Acyclic diaminocarbenes: simple, versatile ligands for cross-coupling reactions. *Chem. Commun.*, 668–70.

184. Lee, M.T., Lee, H.M. and Hu, C.H. (2007) A theoretical study of the Heck reaction: *N*-heterocyclic carbene versus phosphine ligands. *Organometallics*, **26**, 1317–24.

185. Roland, S., Mangeney, P. and Jutand, A. (2006) Reactivity of Pd(0)(NHC)₂ (NHC: *N*-heterocyclic carbene) in oxidative addition with aryl halides in Heck reactions. *Synlett*, 3088–94.

186. Pytkowicz, J., Roland, S., Mangeney, P. *et al.* (2003) Chiral diaminocarbene palladium(II) complexes: synthesis, reduction to Pd(0) and activity in the Mizoroki–Heck reaction as recyclable catalysts. *J. Organomet. Chem.*, **678**, 166–79.

187. Selvakumar, K., Zapf, A. and Beller, M. (2002) New palladium carbene catalysts for the Heck reaction of aryl chlorides in ionic liquids. *Org. Lett.*, **4**, 3031–3.

188. Wang, R.H., Zeng, Z., Twamley, B. *et al.* (2007) Synthesis and characterization of pyrazolyl-functionalized imidazolium-based ionic liquids and hemilabile (carbene)palladium(II) complex catalyzed Heck reaction. *Eur. J. Org. Chem.*, 655–61.

189. Zhou, L. and Wang, L. (2006) Functionalized ionic liquid as an efficient and recyclable reaction medium for phosphine-free palladium-catalyzed Heck reaction. *Synthesis*, 2653–8.

190. Herrmann, W.A., Reisinger, C.P. and Spiegler, M. (1998) Chelating *N*-heterocyclic carbene ligands in palladium-catalyzed Heck-type reactions. *J. Organomet. Chem.*, **557**, 93–6.

191. Tubaro, C., Biffis, A., Gonzato, C. *et al.* (2006) Reactivity of chelating dicarbene metal complex catalysts, I: an investigation on the Heck reaction. *J. Mol. Cat. A: Chem.*, **248**, 93–8.

192. Scholl, M., Trnka, T.M., Morgan, J.P. and Grubbs, R.H. (1999) Increased ring closing metathesis activity of ruthenium-based olefin metathesis catalysts coordinated with imidazolin-2-ylidene ligands. *Tetrahedron Lett.*, **40**, 2247–50.
193. Scholl, M., Ding, S., Lee, C.W. and Grubbs, R.H. (1999) Synthesis and activity of a new generation of ruthenium-based olefin metathesis catalysts coordinated with 1,3-dimesityl-4,5-dihydroimidazol-2-ylidene ligands. *Org. Lett.*, **1**, 953–6.
194. Huang, J., Stevens, E.D., Nolan, S.P. and Petersen, J.L. (1999) Olefin metathesis-active ruthenium complexes bearing a nucleophilic carbene ligand. *J. Am. Chem. Soc.*, **121**, 2674–8.
195. Yang, C.L. and Nolan, S.P. (2001) A highly efficient palladium/imidazolium salt system for catalytic Heck reactions. *Synlett*, 1539–42.
196. McGuinness, D.S., Green, M.J., Cavell, K.J. *et al.* (1998) Synthesis and reaction chemistry of mixed ligand methylpalladium–carbene complexes. *J. Organomet. Chem.*, **565**, 165–78.
197. McGuinness, D.S., Cavell, K.J., Skelton, B.W. and White, A.H. (1999) Zerovalent palladium and nickel complexes of heterocyclic carbenes: Oxidative addition of organic halides, carbon–carbon coupling processes, and the Heck reaction. *Organometallics*, **18**, 1596–605.
198. Tubaro, C., Biffis, A., Basato, M. *et al.* (2005) A simple route to novel palladium(II) catalysts with oxazolin-2-ylidene ligands. *Organometallics*, **24**, 4153–8.
199. Calo, V., Del Sole, R., Nacci, A. *et al.* (2000) Synthesis and crystal structure of bis(2,3-dihydro-3-methylbenzothiazole-2-ylidene)palladium(II) diiodide: the first palladium complex with benzothiazole carbene ligands suitable for homogeneous catalysis. *Eur. J. Org. Chem.*, 869–71.
200. Yen, S.K., Koh, L.L., Hahn, F.E. *et al.* (2006) Convenient entry to mono- and dinuclear palladium(II) benzothiazolin-2-ylidene complexes and their activities toward Heck coupling. *Organometallics*, **25**, 5105–12.
201. Huynh, H.V., Neo, T.C. and Tan, G.K. (2006) Mixed dicarboxylato-bis(carbene) complexes of palladium(II): synthesis, structures, *trans–cis* isomerism, and catalytic activity. *Organometallics*, **25**, 1298–302.
202. Fiddy, S.G., Evans, J., Neisius, T. *et al.* (2007) Comparative experimental and EXAFS studies in the Mizoroki–Heck reaction with heteroatom-functionalised *N*-heterocyclic carbene palladium catalysts. *Chem. Eur. J.*, **13**, 3652–9.
203. Özdemir, I., Yigit, M., Cetinkaya, E. and Cetinkaya, B. (2006) Synthesis of novel palladium *N*-heterocyclic-carbene complexes as catalysts for Heck and Suzuki cross-coupling reactions. *Appl. Organomet. Chem.*, **20**, 187–92.
204. Herrmann, W.A., Böhm, V.P.W., Gstöttmayr, C.W.K. *et al.* (2001) Synthesis, structure and catalytic application of palladium(II) complexes bearing *N*-heterocyclic carbenes and phosphines. *J. Organomet. Chem.*, **617**, 616–28.
205. Haneda, S., Okui, A., Ueba, C. and Hayashi, M. (2007) An efficient synthesis of 2-arylimidazoles by oxidation of 2-arylimidazolines using activated carbon–O_2 system and its application to palladium-catalyzed Mizoroki–Heck reaction. *Tetrahedron*, **63**, 2414–7.
206. Haneda, S., Ueba, C., Eda, K. and Hayashi, M. (2007) Imidazole and imidazoline derivatives as *N*-donor ligands for palladium-catalyzed Mizoroki–Heck reaction. *Adv. Synth. Catal.*, **349**, 833–5.
207. Taige, M.A., Zeller, A., Ahrens, S. *et al.* (2007) New Pd–NHC-complexes for the Mizoroki–Heck reaction. *J. Organomet. Chem.*, **692**, 1519–29.
208. Demir, S., Özdemir, I. and Cetinkaya, B. (2006) Use of bis(benzimidazolium)–palladium system as a convenient catalyst for Heck and Suzuki coupling reactions of aryl bromides and chlorides. *Appl. Organomet. Chem.*, **20**, 254–9.
209. Wang, R.H., Jin, C.M., Twamley, B. and Shreeve, J.M. (2006) Syntheses and characterization of unsymmetric dicationic salts incorporating imidazolium and triazolium functionalities. *Inorg. Chem.*, **45**, 6396–403.

210. Ahrens, S., Zeller, A., Taige, M. and Strassner, T. (2006) Extension of the alkane bridge in bisNHC–palladium–chloride complexes. Synthesis, structure, and catalytic activity. *Organometallics*, **25**, 5409–15.

211. Baker, M.V., Brown, D.H., Simpson, P.V. *et al.* (2006) Palladium, rhodium and platinum complexes of *ortho*-xylyl-linked bis-*N*-heterocyclic carbenes: synthesis, structure and catalytic activity. *J. Organomet. Chem.*, **691**, 5845–55.

212. Nielsen, D.J., Cavell, K.J., Skelton, B.W. and White, A.H. (2006) Silver(I) and palladium(II) complexes of an ether-functionalized quasi-pincer bis-carbene ligand and its alkyl analogue. *Organometallics*, **25**, 4850–6.

213. Chen, T., Gao, J. and Shi, M. (2006) A novel tridentate NHC–Pd(II) complex and its application in the Suzuki and Heck-type cross-coupling reactions. *Tetrahedron*, **62**, 6289–94.

214. Yagyu, T., Oya, S., Maeda, M. and Jitsukawa, K. (2006) Syntheses and characterization of palladium(II) complexes with tridentate *N*-heterocyclic carbene ligands containing aryloxy groups and their application to Heck reaction. *Chem. Lett.*, **35**, 154–5.

215. Hahn, F.E., Jahnke, M.C. and Pape, T. (2007) Synthesis of pincer-type bis(benzimidazolin-2-ylidene) palladium complexes and their application in C–C coupling reactions. *Organometallics*, **26**, 150–4.

216. Nielsen, D.J., Cavell, K.J., Skelton, B.W. and White, A.H. (2006) Methyl-palladium(II) complexes of pyridine-bridged bis(nucleophilic heterocyclic carbene) ligands: substituent effects on structure, stability, and catalytic performance. *Inorg. Chim. Acta*, **359**, 1855–69.

217. Scherg, T., Schneider, S.K., Frey, G.D. *et al.* (2006) Bridged imidazolium salts used as precursors for chelating carbene complexes of palladium in the Mizoroki–Heck reaction. *Synlett*, 2894–907.

218. Herrmann, W.A., Öfele, K., Schneider, S.K. *et al.* (2006) A carbocyclic carbene as an efficient catalyst ligand for C–C coupling reactions. *Angew. Chem. Int. Ed.*, **45**, 3859–62.

219. Wass, D.F., Haddow, M.F., Hey, T.W. *et al.* (2007) Cyclopropenylidene carbene ligands in palladium C–C coupling catalysis. *Chem. Commun.*, 2704–6.

220. Yao, Q., Zabawa, M., Woo, J. and Zheng, C. (2007) Carbocyclic carbene ligands derived from aromatic nitrones: formation and catalytic activity of their Pd(II) complexes. *J. Am. Chem. Soc.*, **129**, 3088–9.

221. Schneider, S.K., Roembke, P., Julius, G.R. *et al.* (2006) Pyridin-, quinolin- and acridinylidene palladium carbene complexes as highly efficient C–C coupling catalysts. *Adv. Synth. Catal.*, **348**, 1862–73.

222. Shaw, B.L., Perera, S.D. and Staley, E.A. (1998) Highly active, stable, catalysts for the Heck reaction; further suggestions on the mechanism. *Chem. Commun.*, 1361–2.

223. Shaw, B.L. (1998) Speculations on new mechanisms for Heck reactions. *New J. Chem.*, **22**, 77–9.

224. Beletskaya, I.P. and Cheprakov, A.V. (2004) Palladacycles in catalysis – a critical survey. *J. Organomet. Chem.*, **689**, 4055–82.

225. Bedford, R.B. (2003) Palladacyclic catalysts in C–C and C–heteroatom bond-forming reactions. *Chem. Commun.*, 1787–96.

226. Beller, M., Fischer, H., Herrmann, W.A. *et al.* (1995) Palladacycles as efficient catalysts for aryl coupling reactions. *Angew. Chem. Int. Ed. Engl.*, **34**, 1848–9.

227. Herrmann, W.A. and Böhm, V.P.W. (1999) Heck reaction catalyzed by phospha-palladacycles in non-aqueous ionic liquids. *J. Organomet. Chem.*, **572**, 141–5.

228. Beletskaya, I.P., Kashin, A.N., Karlstedt, N.B. *et al.* (2001) NC-palladacycles as highly effective cheap precursors for the phosphine-free Heck reactions. *J. Organomet. Chem.*, **622**, 89–96.

229. Tietze, L.F., Brasche, G., Grube, A. *et al.* (2007) Synthesis of novel spinosyn A analogues by Pd-mediated transformations. *Chem. Eur. J.*, **13**, 8543–63.

230. Albisson, D.A., Bedford, R.B. and Scully, P.N. (1998) Orthopalladated triarylphosphite complexes as highly efficient catalysts in the Heck reaction. *Tetrahedron Lett.*, **39**, 9793–6.
231. Chen, M.T., Huang, C.A. and Chen, C.T. (2006) Synthesis, characterization, and catalytic applications of palladacyclic complexes bearing C,N,S-donor ligands. *Eur. J. Inorg. Chem.*, 4642–8.
232. Blackmond, D.G., Rosner, T. and Pfaltz, A. (1999) Comprehensive kinetic screening of catalysts using reaction calorimetry. *Org. Process Res. Dev.*, **3**, 275–80.
233. Ohff, M., Ohff, A. and Milstein, D. (1999) Highly active Pd(II) cyclometallated imine catalysts for the Heck reaction. *Chem. Commun.*, 357–8.
234. Wu, Y.J., Hou, J.J., Yun, H.Y. *et al.* (2001) Cyclopalladated ferrocenylimines: highly active catalysts for Heck reactions. *J. Organomet. Chem.*, **637**, 793–5.
235. Rocaboy, C. and Gladysz, J.A. (2002) Highly active thermomorphic fluorous palladacycle catalyst precursors for the Heck reaction; evidence for a nanoparticle pathway. *Org. Lett*, **4**, 1993–6.
236. Rocaboy, C. and Gladysz, J.A. (2003) Thermomorphic fluorous imine and thioether palladacycles as precursors for highly active Heck and Suzuki catalysts; evidence for palladium nanoparticle pathways. *New J. Chem.*, **27**, 39–49.
237. Alonso, D.A., Najera, C. and Pacheco, M.C. (2000) Oxime palladacycles: stable and efficient catalysts for carbon–carbon coupling reactions. *Org. Lett.*, **2**, 1823–6.
238. Alonso, D.A., Najera, C. and Pacheco, M.C. (2002) Oxime-derived palladium complexes as very efficient catalysts for the Heck–Mizoroki reaction. *Adv. Synth. Catal.*, **344**, 172–83.
239. Yang, F., Zhang, Y.M., Zheng, R. *et al.* (2002) Cyclopalladated complexes of tertiary arylamines as highly efficient catalysts using in the Heck reactions. *J. Organomet. Chem.*, **651**, 146–8.
240. Gai, X.J., Grigg, R., Ramzan, M.I. *et al.* (2000) Pyrazole and benzothiazole palladacycles: stable and efficient catalysts for carbon–carbon bond formation. *Chem. Commun.*, 2053–4.
241. Wang, H.X., Wu, H.F., Yang, X.L. *et al.* (2007) Highly active ferrocenylamine-derived palladacycles for carbon–carbon cross-coupling reactions. *Polyhedron*, **26**, 3857–64.
242. Kumar, N.S.C.R., Raj, I.V.P. and Sudalai, A. (2007) Sulfonamide- and hydrazine-based palladium catalysts: stable and efficient catalysts for C–C coupling reactions in aqueous medium. *J. Mol. Cat. A: Chem.*, **269**, 218–24.
243. Poverenov, E., Gandelman, M., Shimon, L.J.W. *et al.* (2005) Pincer 'hemilabile' effect. PCN platinum(II) complexes with different amine 'arm length'. *Organometallics*, **24**, 1082–90.
244. Poverenov, E., Gandelman, M., Shimon, L.J.W. *et al.* (2004) Nucleophilic de-coordination and electrophilic regeneration of 'hemilabile' pincer-type complexes: formation of anionic dialkyl, diaryl, and dihydride Pt(II) complexes bearing no stabilizing π-acceptors. *Chem. Eur. J.*, **10**, 4673–84.
245. Ohff, M., Ohff, A., van der Boom, M.E. and Milstein, D. (1997) Highly active Pd(II) PCP-type catalysis for the Heck reaction. *J. Am. Chem. Soc.*, **119**, 11687–8.
246. Sjövall, S., Wendt, O.F. and Andersson, C. (2002) Synthesis, characterisation and catalytic investigation of a new type of PC(sp^3)P pincer Pd(II) complex. *J. Chem. Soc., Dalton Trans.*, 1396–400.
247. Miyazaki, F., Yamaguchi, K. and Shibasaki, M. (1999) The synthesis of a new palladacycle catalyst. Development of a high performance catalyst for Heck reactions. *Tetrahedron Lett.*, **40**, 7379–83.
248. Morales-Morales, D., Grause, C., Kasaoka, K. *et al.* (2000) Highly efficient and regioselective production of trisubstituted alkenes through Heck couplings catalyzed by a palladium phosphinito PCP pincer complex. *Inorg. Chim. Acta*, **300**, 958–63.

249. Arvela, R.K., Leadbeater, N.E., Sangi, M.S. *et al.* (2005) A reassessment of the transition-metal free Suzuki-type coupling methodology. *J. Org. Chem.*, **70**, 161–8.
250. Bergbreiter, D.E., Osburn, P.L., Wilson, A. and Sink, E.M. (2000) Palladium-catalyzed C–C coupling under thermomorphic conditions. *J. Am. Chem. Soc.*, **122**, 9058–64.
251. Jung, I.G., Son, S.U., Park, K.H. *et al.* (2003) Synthesis of novel Pd–NCN pincer complexes having additional nitrogen coordination sites and their application as catalysts for the Heck reaction. *Organometallics*, **22**, 4715–20.
252. Diez-Barra, E., Guerra, J., Hornillos, V. *et al.* (2003) 1,2,4-Triazole-based palladium pincer complexes. A new type of catalyst for the Heck reaction. *Organometallics*, **22**, 4610–2.
253. Bergbreiter, D.E., Osburn, P.L. and Frels, J.D. (2005) Mechanistic studies of SCS–Pd complexes used in Heck catalysis. *Adv. Synth. Catal.*, **347**, 172–84.
254. Bergbreiter, D.E., Osburn, P.L. and Liu, Y.S. (1999) Tridentate SCS palladium(II) complexes: new, highly stable, recyclable catalysts for the Heck reaction. *J. Am. Chem. Soc.*, **121**, 9531–8.
255. Begum, R.A., Powell, D. and Bowman-James, K. (2006) Thioamide pincer ligands with charge versatility. *Inorg. Chem.*, **45**, 964–6.
256. Churruca, F., SanMartin, R., Tellitu, I. and Dominguez, E. (2005) *N*-Heterocyclic NCN-pincer palladium complexes: a source for general, highly efficient catalysts in Heck, Suzuki, and Sonogashira coupling reactions. *Synlett*, 3116–20.
257. Xu, C., Gong, J.F., Yue, S.F. *et al.* (2006) Tricyclohexylphosphine-cyclopalladated ferrocenylimine complexes: synthesis, crystal structures and application in Suzuki and Heck reactions. *Dalton Trans.*, 4730–9.
258. Mino, T., Shirae, Y., Sasai, Y. *et al.* (2006) Phosphine-free palladium catalyzed Mizoroki–Heck reaction using hydrazone as a ligand. *J. Org. Chem.*, **71**, 6834–9.
259. Kawashita, Y., Ueba, C. and Hayashi, M. (2006) A simple synthesis of 2-arylbenzothiazoles and its application to palladium-catalyzed Mizoroki–Heck reaction. *Tetrahedron Lett.*, **47**, 4231–3.
260. Herrera-Álvarez, C., Gómez-Benítez, V., Redón, R. *et al.* (2004) [1,1′-Bis(diphenylphosphino)ferrocene]palladium(II) complexes with fluorinated benzenethiolate ligands: examination of the electronic effects in the solid state, solution and in the Pd-catalyzed Heck reaction with the catalytic system [Pd(dppf)(SR$_F$)$_2$]. *J. Organomet. Chem.*, **689**, 2464–72.
261. Kovala-Demertzi, D., Yadav, P.N., Dernertzis, M.A. *et al.* (2004) First use of a palladium complex with a thiosemicarbazone ligand as catalyst precursor for the Heck reaction. *Tetrahedron Lett.*, **45**, 2923–6.
262. Kawano, T., Shinomaru, T. and Ueda, I. (2002) Highly active Pd(II) catalysts with *trans*-bidentate pyridine ligands for the Heck reaction. *Org. Lett.*, **4**, 2545–7.
263. Nájera, C., Gil-Moltó, J., Karlström, S. and Falvello, L.R. (2003) Di-2-pyridylmethylamine-based palladium complexes as new catalysts for Heck, Suzuki, and Sonogashira reactions in organic and aqueous solvents. *Org. Lett.*, **5**, 1451–4.
264. Durand, J., Gladiali, S., Erre, G. *et al.* (2007) Palladium chemistry of 2-ferrocenyl-1,10-phenanthroline ligand. *Organometallics*, **26**, 810–8.
265. Kwok, T.J. and Virgilio, J.A. (2005) A preparative route to methyl 3-(heteroaryl)acrylates using Heck methodology. *Org. Process Res. Dev.*, **9**, 694–6.
266. Pryjomska, I., Bartosz-Bechowski, H., Ciunik, Z. *et al.* (2006) Chemistry of palladium phosphinite (PPh$_2$(OR)) and phosphonite (P(OPh)$_2$(OH)) complexes: catalytic activity in methoxycarbonylation and Heck coupling reactions. *Dalton Trans.*, 213–20.
267. Pathak, D.D., Maheswaran, H., Prasanth, K.L. and Kantam, M.L. (2007) Design and development of a new chelating bis(phosphinite)-based palladium(II) catalyst and its application to the Heck reaction. *Synlett*, 757–60.

268. Catsoulacos, D.P., Steele, B.R., Heropoulos, G.A. *et al.* (2003) An iminophosphine dendrimeric ligand and its evaluation in the Heck reaction. *Tetrahedron Lett.*, **44**, 4575–8.
269. Del Zotto, A., Zangrando, E., Baratta, W. *et al.* (2005) [MCl(ligand)]$^+$ complexes (M $=$ Ni, Pd, Pt) with a P,N,N terdentate ligand – solid state and solution structures and catalytic activity of the Pd^{II} derivative in the Heck reaction. *Eur. J. Inorg. Chem.*, 4707–14.
270. Schultz, T. and Pfaltz, A. (2005) Palladium(II) complexes with chelating P,O-ligands as catalysts for the Heck reaction. *Synthesis*, 1005–11.

3

Focus on Regioselectivity and Product Outcome in Organic Synthesis

Peter Nilsson, Kristofer Olofsson and Mats Larhed

Department of Medicinal Chemistry, Organic Pharmaceutical Chemistry, Uppsala University, BMC, Uppsala, Sweden

3.1   Introduction

The ability to create new C–C bonds is of principal importance in synthetic organic chemistry, and today there are a number of transition metal-catalysed methods available for that purpose [1]. Palladium(0)-catalysed organic reactions (e.g. Mizoroki–Heck, Sonogashira, Suzuki, Stille, Negishi and carbonylative couplings) belong to the most versatile and popular of these methods, especially for functionalizing aromatic coupling partners [2]. One explanation for the unique appreciation of organopalladium chemistry is the wide array of functionalities that is compatible with the reaction conditions and the fact that the palladium metal, which is very expensive, can be used in catalytic amounts.

In the late 1960s, palladium-mediated arylation and alkenylation of alkenes were independently discovered by Japanese (Moritani–Fujiwara) [3] and American (Heck) [4–9] groups. The palladium(II)-mediated version of the reaction was further refined into a palladium(0)-catalysed transformation employing aryl and vinyl halides, mainly by Heck and Nolley [10] and Mizoroki *et al.* [11]. In the last 35 years, this selective vinylic substitution process, known as the Heck or as the Mizoroki–Heck arylation or alkenylation/vinylation reaction, has been extensively explored and used in several diverse areas, such as the preparation of biologically active compounds and pharmaceuticals, agrochemicals, complex natural products, heterocycles, dyes and novel materials [12–15]. This chapter will focus primarily on factors affecting the regioselectivity [16], the structural outcome

The Mizoroki–Heck Reaction   Edited by Martin Oestreich
© 2009 John Wiley & Sons, Ltd

and the synthetic usefulness of the intermolecular Mizoroki–Heck arylation and vinylation reaction using homogeneous palladium(0) catalysis with electron-rich, allylic and
electron-deficient alkenes.

3.2 Mechanistic Aspects

In modern small-scale implementations, the Mizoroki–Heck reaction is usually performed
in polar solvents (including water) using palladium concentrations of 0.10–10 mol% along
with a stoichiometric amount of an organic or inorganic base [17]. Although a plethora of
heterogeneous catalysts have been successfully used with mainly electron-poor alkenes,
most examples in organic synthesis employ homogeneous catalysts [17–19]. Frequently,
phosphine or nitrogen ligands, together with halide or acetate additives, are included in
the reaction cocktail to control reactivity and selectivity, as well as to prevent precipitation
of elemental palladium(0) in the form of 'palladium black' [15, 16]. The catalytic route
of the palladium(0)-catalysed Mizoroki–Heck reaction has been the subject of extensive
research and a large number of both experimental and theoretical mechanistic studies have
been performed [18–22]. The basic catalytic cycle consists of four fundamental steps:

1. *Oxidative addition.* The organic substrate RX reacts with Pd(0)L$_2$ generating
   RPd(II)XL$_2$ (L = monodentate ligand or solvent, X = leaving group).
2. *π-Complex formation.* Either L or X is displaced and palladium(II) coordinates to the
   double bond of the alkene.
3. *syn-Insertion.* R and Pd(II)XL$_2$ are inserted into the double bond, giving a
   σ-alkylpalladium(II) complex.
4. *syn-β-H-Elimination.* The C–Pd(II) and one of the neighbouring C–H rotate into the *syn*
   conformation. HPd(II)XL$_2$ is eliminated and the Mizoroki–Heck product is produced.

The base is required to regenerate the active Pd(0)L$_2$, which enters a new catalytic cycle.
A more detailed presentation of the Mizoroki–Heck cycle, including two of the possible
mechanistic pathways using bidentate ligands [16], is presented in Figure 3.1.

3.2.1 Oxidative Addition

In the oxidative addition process, palladium(0) is formally oxidized to palladium(II) via
an insertion of an organic R group, most commonly an aryl or vinyl group, and the leaving
group X, which usually is a halide or a pseudohalide (Figure 3.1, step 1) [19]. Examples
of pseudohalides that have been used in palladium(0)-catalysed Mizoroki–Heck reactions
include aryldiazonium salts [23] and iodonium salts [24], triflates [25, 26], nonaflates [27],
tosylates [28, 29], mesylates [29], sulfonyl chlorides [30], acid chlorides [31], phosphates
[32] and sulfoxides [33]. Moreover, aromatic acid anhydrides [34], ester functionalities
[35] or even a free carboxylic acid [36] might serve as oxidative addition precursors
in Mizoroki–Heck protocols. The rate of the oxidative addition is quite sensitive to the
choice of halide or pseudohalide, with the reactivity falling in the series $RN_2^+ > I >$
OTf $\approx$ Br $\gg$ Cl $>$ OTs $\approx$ OPO(OR)$_2$. Iodides are usually reactive at room temperature
even without the addition of phosphine ligands. Aryl or vinyl triflates, nonaflates and
bromides typically need aryl phosphine ligands, unless the reactions are run under phase-
transfer conditions [15, 37]. As expected from considering the dissociation energy of the

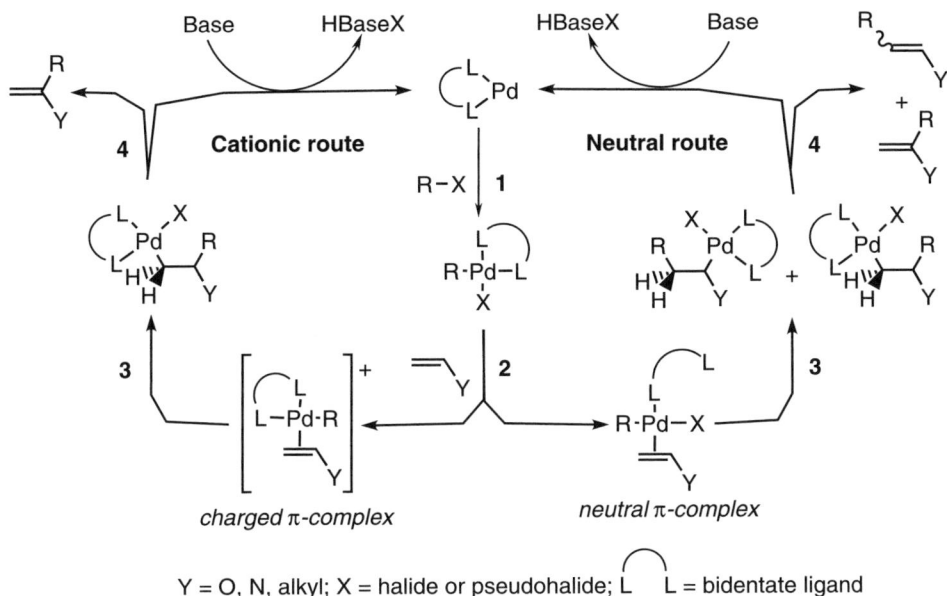

Y = O, N, alkyl; X = halide or pseudohalide; L ⌢ L = bidentate ligand

Figure 3.1   *The cationic and the neutral Mizoroki–Heck routes.*

C—Cl bond compared with the C—I and C—Br bonds, the chlorides are more reluctant to undergo Mizoroki–Heck reactions [38]. There are, however, a number of reports of Mizoroki–Heck couplings where aryl chlorides, as well as even more sluggish tosylates and phosphates, have been utilized [19, 32]. Aryl fluorides are inert under Mizoroki–Heck reaction conditions.

The choice of organic moiety to be coupled with the alkene is limited, as it should not have a *syn*-hydrogen on an sp³-hybridized carbon situated next to the palladium-binding carbon. The desired reaction may be ineffective if this prerequisite is not fulfilled, since Pd–H elimination of the *syn*-hydrogen occurs more readily than the intended insertion of the organopalladium moiety into the double bond. Suitable R groups can, for example be aryl, alkenyl, benzyl, allyl, alkynyl or alkoxycarbonyl-methyl groups [12, 13].

Although a 14-electron Pd(0)L₂ complex is commonly referred to as the reactive species in the oxidative addition of monodentate phosphine-ligated palladium complexes, both a 12-electron species with sterically bulky *tert*-butyl phosphine ligands and an anionic 16-electron complex, Pd(0)(PPh₃)₂(X)⁻ (X = halide or acetate), have been described [19, 20, 39, 40].

3.2.2   π-Complex Formation and Migratory Insertion

It is vital to control the π-complex formation and insertion steps in order to direct the regioselectivity of the Mizoroki–Heck reaction, in which the organic R group will be either added to the internal carbon of the monosubstituted alkene, yielding an α-product, or the terminal, providing *trans*- or *cis*-β-products (Figure 3.1, steps 2–4 and Figure 3.2) [16, 41].

The regioisomeric outcome of the Mizoroki–Heck reaction is determined by two major factors, namely electronic and steric effects. Under classical Mizoroki–Heck conditions,

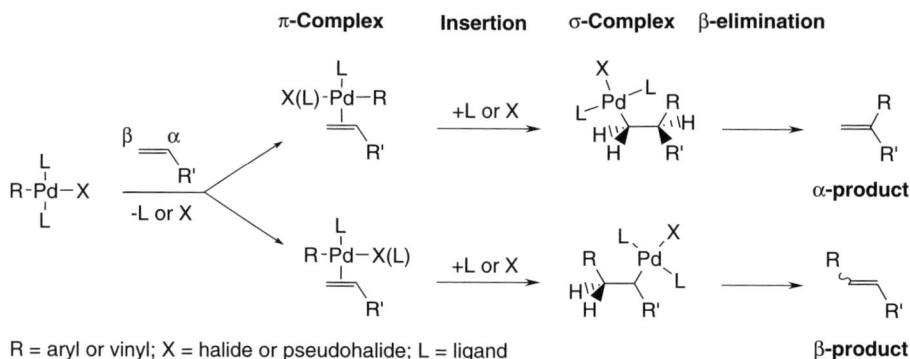

Figure 3.2 *Regiochemical outcome of the intermolecular Mizoroki–Heck reaction with mono-substituted alkenes.*

electron-poor alkenes generally react smoothly, delivering mainly the *trans*-isomer of the β-substituted product. On the other hand, vinyls with an electron-donating substituent, such as a heteroatom or an alkyl (e.g. allylic alkenes), generally form a mixture of the α-substituted product and *cis*- and *trans*-isomers of the β-product in the presence of monodentate ligands (or without ligands) [16, 21].

The regioselectivity for electron-rich alkenes is determined by the nature of the π-complex. In a positively charged π-complex, the organopalladium moiety is generally stabilized by an uncharged bidentate phosphine or nitrogen ligand (Figure 3.1, cationic route) [16]. In a neutral π-complex, palladium coordinates one ligand and one anionic counter ion, usually a halide (Figure 3.1, neutral route). The positively charged π-complex will lead to the α-product, while the neutral π-complex will give rise to a mixture of the α- and β-substituted products (Figures 3.1 and 3.2). This mechanistic hypothesis is not only supported by the preparative isomeric distribution, but also by computational investigations [42].

To obtain a reaction that proceeds via the cationic pathway with electron-rich enol ethers or enamides, strongly coordinating bidentate ligands, such as DPPP (1,3-bis(diphenylphosphino)propane) or dmphen (2,9-dimethyl-[1,10]-phenanthroline), are often used [15, 43, 44]. However, the properties of the counter ion are also very important. When weakly coordinating triflate or tosylate is used as the leaving group, the reaction proceeds via the cationic route. In contrast, with aryl halides as aryl palladium precursors the neutral intermediate predominates. To achieve a high α-selectivity with aryl bromides and iodides, an additive, often a silver or a thallium salt, is required to scavenge the halide [16]. The main drawbacks of these delicate methods to perform 'cationic' Mizoroki–Heck arylations are that aryl triflates are rarely commercially available and that the metal additives are either expensive or very toxic. However, it is promising that new methods have recently been reported that afford the same high α-regioselectivity with both organic bromides and aryl iodides by simply increasing the polarity of the reaction system (ionic liquids [45] or water [46, 47]). Further, aryl chlorides have also been found to be a useful precursor for internal arylation of electron-rich alkenes using DPPP and hydrogen-bond-donating $HNEt_3^+BF_4^-$ [48]. With allylic substrates it is often possible to obtain

		Route				Route	
		Neutral	**Cationic**			**Neutral**	**Cationic**
		Arylation ratio Internal:Terminal				Arylation ratio Internal:Terminal	
1)	On-Bu	mixture of isomers	100:0	4)	OH	0:100	100:0
2)	(pyrrolidinone alkene)	40:60	100:0	5)	(pentene)	20:80	80:20
3)	OAc	mixture of products	95:5	6)	CO_2Me	0:100	0:100

Figure 3.3 *General regiochemical outcome in Mizoroki–Heck arylation proceeding via the neutral or cationic route.*

either high internal or terminal selectivity by careful choice of starting substrates and reaction conditions (cationic or neutral route). Some previously reported regioselectivities for the arylation of electron-rich, neutral and electron-deficient substrates are summarized in Figures 3.3 and 3.4 [16, 44]. Interestingly, the intermolecular arylation of nonfunctionalized alkenes has never been performed without generation of regioisomeric products (Figure 3.3, entry 5).

3.2.3   *β*-H-Elimination and Palladium(0) Recycling

β-Hydride elimination is the step of the Mizoroki–Heck reaction yielding the product (Figure 3.1, step 4). For this process to occur, the insertion complex must be able to rotate to a position where a *β*-hydrogen is aligned *syn* to the palladium(II) centre. The elimination will then result in formation of a reconstituted alkene and a palladium hydride species. The *β*-H-elimination is reversible (see Figures 3.5 and 3.6) and the preferred formation of the thermodynamically more stable *trans*-products is thus explained [12]. There is today no precise knowledge of how palladium(II) is reduced back to catalytically

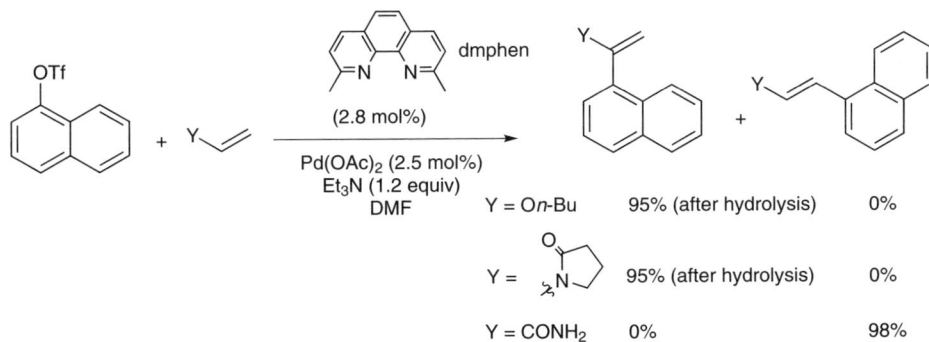

Y = On-Bu	95% (after hydrolysis)	0%	
Y = (pyrrolidinone)	95% (after hydrolysis)	0%	
Y = $CONH_2$	0%	98%	

Figure 3.4 *Demonstration of the regiochemical control achieved using cationic conditions with different alkenes and 1-naphthyl triflate*

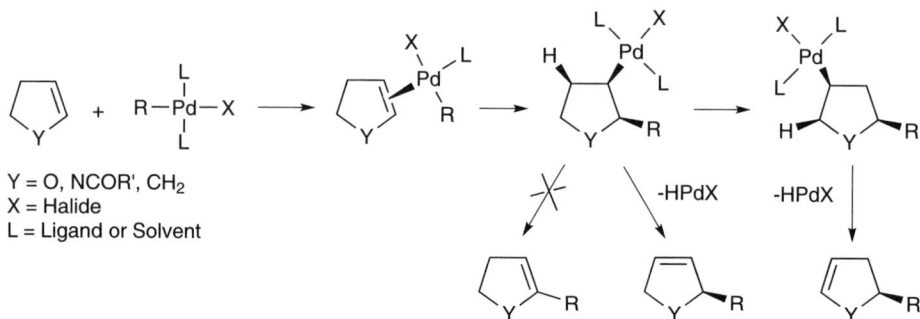

Figure 3.5   *Mizoroki–Heck reaction and subsequent double-bond isomerization of an endo-cyclic alkene.*

active palladium(0). In a report from 1998, a mechanism based on a base-promoted (E2-type) elimination of palladium(II) yielding palladium(0) was suggested to be more energetically favoured than a mechanism where a direct β-H-elimination yields a free HPdL species [49].

3.2.4   Cyclic Alkenes and Double-Bond Migration

The investigation of the intermolecular version of the Mizoroki–Heck reaction with endocyclic alkenes began in the late 1970s [50–52]. It was directly evident that the problem with reversible and repeating β-hydride eliminations/migratory insertions complicated the use of cyclic alkenes in a manner similar to long-chain acyclic 1-alkenes [53]. Mixtures of isomers were obtained. An example of a C—C bond formation–isomerization [54] process is shown in Figure 3.5.

The organopalladium complex inserts in a 1,2-*syn* mode and a σ-complex is produced. After β-elimination, the allylic compound might be liberated. Alternatively, readdition of the hydridopalladium species in the reverse direction eventually leads to the cyclic system with the double bond in conjugation with the heteroatom (or next to the CH$_2$ group) [54, 55]. The isomer with the double bond conjugated to the aromatic or vinylic substituent is not observed, since formation of the aforementioned isomer requires either *anti*-elimination or readdition of the hydridopalladium species at the opposite face of the ring system. As earlier mentioned, a correctly positioned hydrogen accessible for *syn*-elimination is a prerequisite

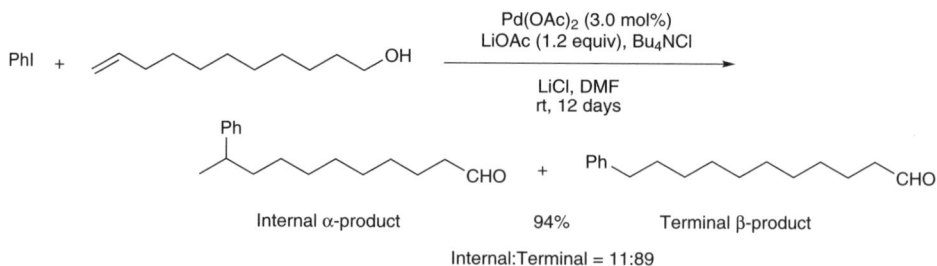

Figure 3.6   *Mizoroki–Heck arylation–double-bond isomerization of 10-undecenyl alcohol.*

for the β-elimination to occur. The elimination of a hydrogen atom in an *anti* fashion to the palladium is very uncommon, and the metal is postulated to be continuously coordinated to the same side of the cyclic alkene until irreversible elimination of HPdX occurs [41].

In a fascinating linear case, 10-undecenyl alcohol reacts with phenyl iodide to provide thermodynamically favoured 11-phenylundecanal as the major product (Figure 3.6) [53]. The highly efficient repeated elimination of HPdX and subsequent readdition, regardless of the regioselectivity in the original PhPdX insertion, is a fine example of 'living' HPdX species.

3.2.5 Aryl Scrambling

Frequently, a considerable amount of side products, derived from facile aryl–aryl exchange in the oxidative addition complex, is formed in a Mizoroki–Heck arylation reaction executed in the presence of triarylphosphine ligands. This process is particularly significant at higher reaction temperatures and with electron-rich aryl halides [56, 57]. Thus, a reaction of 4-bromoanisole with butyl acrylate employing $Pd(OAc)_2/PPh_3$ as catalytic system and sodium acetate as base furnished butyl (E)-cinnamate in addition to the expected coupling product (Figure 3.7a) [58].

An equilibrium reaction involving aryl group exchange, as depicted in Figure 3.7b, is responsible for the product pattern. Interestingly, in particular, electron donor-substituted aryl halides have been found to undergo facile aryl–aryl exchange. The aryl scrambling does not only produce side products, but also degrades the arylphosphine catalyst [58]. Thus, recently developed catalysts, such as palladacycles, nitrogen and carbenene-based ligands, provide catalytic tools to avoid the aryl scrambling process [59].

3.2.6 Mizoroki–Heck Reactions with Vinyl–X Substrates

Vinyl halides and triflates serve as good Mizoroki–Heck vinylating agents, yielding synthetically important diene products [12, 13]. Vinylation occurs in the terminal β-position of electron-deficient alkenes, as in the case of the arylation reactions, and the same

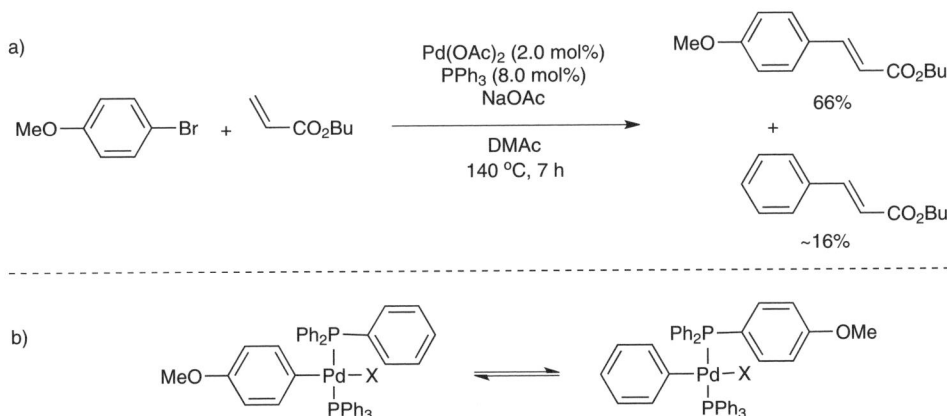

Figure 3.7 (a) Aryl scrambling in the triphenylphosphine-modulated Mizoroki–Heck vinylation of 4-bromoanisole. (b) Aryl migration in the oxidative addition complex.

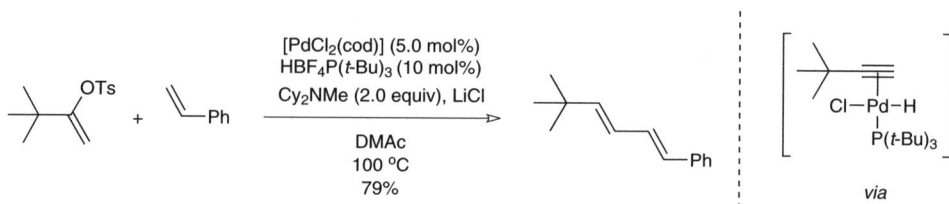

Figure 3.8 *Mizoroki–Heck coupling with 1-tert-butyl vinyl tosylate involving a 1,2-migration of the alkenyl palladium(II) intermediate.*

reaction profiles including double-bond migrations are also encountered [15]. Recent work by Skrydstrup's group has expanded the substrate scope to include vinyl tosylates and phosphates also [32, 60]. Unexpectedly, the normal coupling product was not formed using bulky 1-*tert*-butyl vinyl tosylate and styrene. Rather, the product from an apparent 1,2-migration of the intermediate alkenyl palladium(II) species was obtained (Figure 3.8) [32]. This isomerization was found to be general using different alkenes and was also expanded to include 1-*tert*-butylvinyl *O,O*-diphenyl phosphate. Furthermore, a reasonable reaction path for this rearrangement was proposed using density functional theory calculations.

3.2.7 Reactions Using Low Palladium Catalyst Concentration

Microwave-heated Mizoroki–Heck reactions in water using ultralow palladium catalyst concentrations have been performed by Arvela and Leadbeater. Different catalyst concentrations were investigated using a commercially available 1000 ppm palladium solution as the catalyst source (Figure 3.9) [61]. Impressively, useful Mizoroki–Heck arylations were performed with palladium concentrations as low as 500 ppb.

3.2.8 Carbonylative, Decarbonylative and Desulfitative Mizoroki–Heck Reactions

In contrast to intramolecular carbonylative Mizoroki–Heck cyclizations, the intermolecular carbonylative reaction of aryl halides with alkenes has been much less explored. Figure 3.10 depicts one of these rare examples using a carbon monoxide pressure of 5 atm [62]. Small

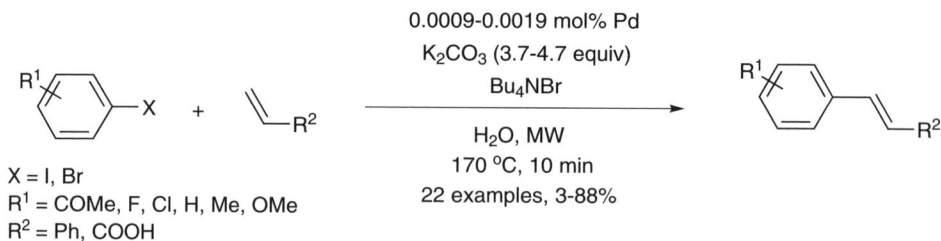

X = I, Br
R^1 = COMe, F, Cl, H, Me, OMe
R^2 = Ph, COOH

Figure 3.9 *Mizoroki–Heck arylation in water using extremely low palladium concentrations.*

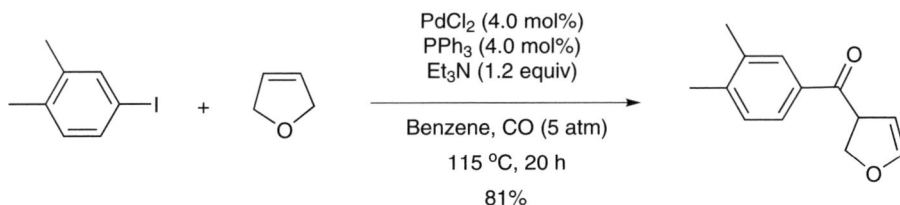

Figure 3.10   *Carbonylative Mizoroki–Heck reaction with 2,5-dihydrofurane.*

amounts of the saturated 3-aroylfuran product were also formed under these reductive conditions.

Aroyl chlorides and aryl sulfonyl chlorides can also be employed as arylating agents under decarbonylative and desulfitative conditions respectively. Improved yields were reported by Dubbaka and Vogel [64] using Herrmann's palladacycle [63], arenesulfonyl chlorides and bulky trioctylmethylammonium chloride under reflux conditions (Figure 3.11).

3.3 Electron-Rich Alkenes

Heteroatom-substituted alkenes, such as vinyl ethers and enamides, constitute easily accessible and versatile building blocks. For example, these vinylic substrates can be – after regiocontrolled arylation/vinylation and optional hydrolysis and/or hydrogenation – easily transformed into various compound classes (Figure 3.12).

3.3.1 Internal Arylations

An entrance to the cationic route is the usage of aryl halides with addition of a halide sequestering agent (e.g. silver or thallium salts) in stoichiometric amount. By employing this methodology, the halide ion is removed from the palladium(II) centre via precipitation, as first demonstrated by Grigg *et al.* [65] and Cabri *et al.* [66], providing essential cationic reaction intermediates (Figure 3.13, lower part). However, for obvious reasons, the usage of these expensive and toxic heavy-metals impelled the development of less waste-intensive protocols.

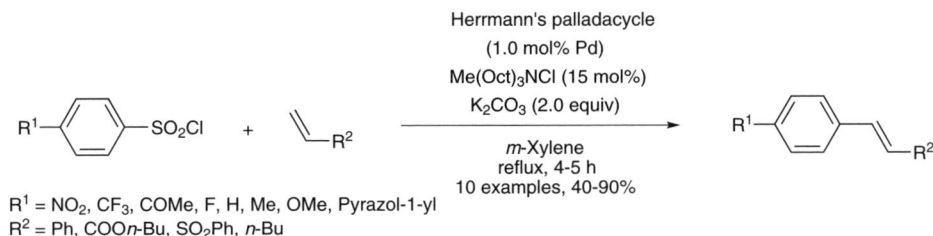

R^1 = NO$_2$, CF$_3$, COMe, F, H, Me, OMe, Pyrazol-1-yl
R^2 = Ph, COO*n*-Bu, SO$_2$Ph, *n*-Bu

Figure 3.11   *Desulfitative Mizoroki–Heck couplings using arylsulfonyl chlorides.*

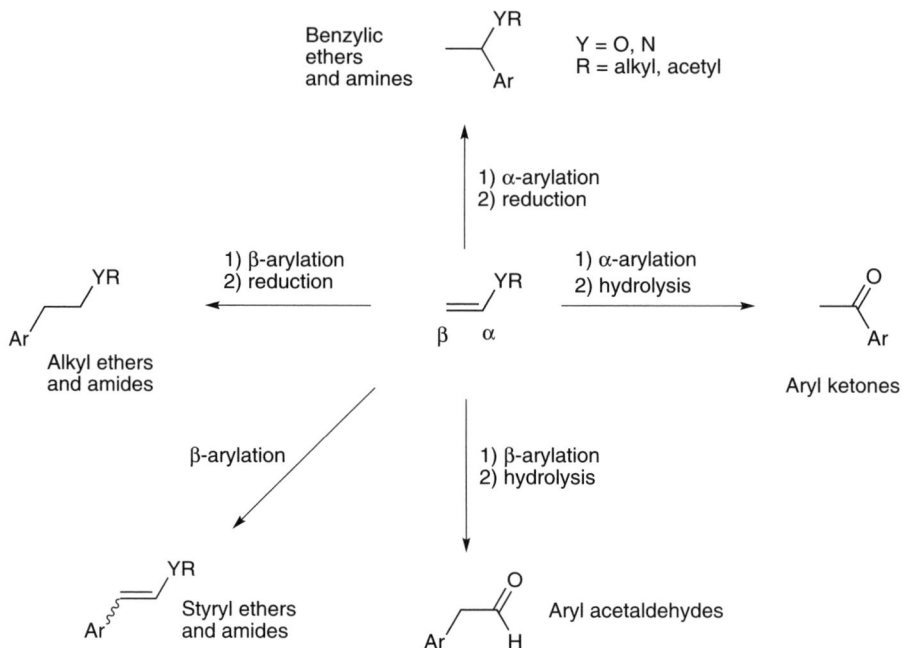

Figure 3.12 *Examples of versatile transformations involving a regioselective Mizoroki–Heck arylation sequence with heteroatom substituted alkenes.*

A very recent report from Mo and Xiao [48] disclosed the utilization of ammonium salts as efficient halide ion scavengers, providing impressive yields of branched products in short reaction times (1.5–16 h). It is believed that the quaternary ammonium salt forms a hydrogen bond to the bromide or iodide, thus preventing possible adverse coordination to the Ar–Pd–alkene complex in the crucial insertion step (Figure 3.14). The protocol

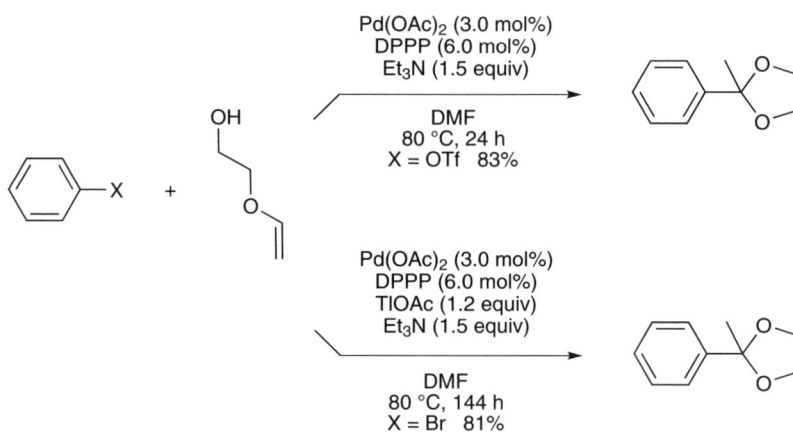

Figure 3.13 *One-pot, two-step arylation–ketalization process using complementary α-selective Mizoroki–Heck protocols.*

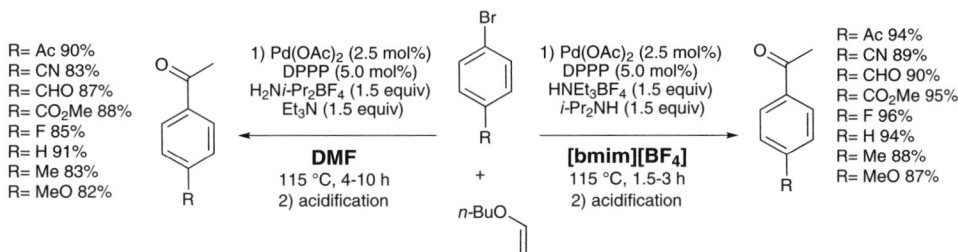

Figure 3.14 Polar solvents in combination with ammonium salts promote internal arylations.

appears to be robust and high yielding, encompassing both enamides and vinyl ethers. Even electron-deficient aryl chlorides proved to be functional, providing high yields, although demanding higher temperature and longer reaction times.

The reactive double bond, recreated after a chemo- and regioselective α-arylation of enamides, can be further manipulated via a carbonylative annulation, furnishing attractive 1-(3-indanonyl)pyrrolidin-2-one structures as exemplified by Wu et al. [67]. In its simplicity, this sequence constitutes an illustrative example of the synthetic versatility of arylated enamides (Figure 3.15).

Decoration and fine-tuning of a molecular framework, demonstrated to be active on a chosen biological target, is an application where palladium transformations are highly competitive. Furthermore, aryl groups are ubiquitous in the majority of approved pharmaceutical compounds. Hence, suitably equipped Ar–X precursors can often be smoothly converted to a variety of functional groups (e.g. via Mizoroki–Heck couplings) in order to optimize the biological activity and to study the structure–activity relationship.

Hultén et al. [68] made use of an enol ether and an iodo-substituted precursor to optimize HIV-1 protease inhibitors in a high-yielding manner using an α-selective Mizoroki–Heck arylation and subsequent hydrolysis (Figure 3.16). This class of reactions using ethylene glycol vinyl ether can now be performed in neat water without the use of toxic TlOAc [47].

An illustrative example of the great applicability of Mizoroki–Heck chemistry is presented in Figure 3.17, where the required indanone moiety is constructed in a one-pot manner starting with an internal arylation. The aminoindanone was obtained in optically

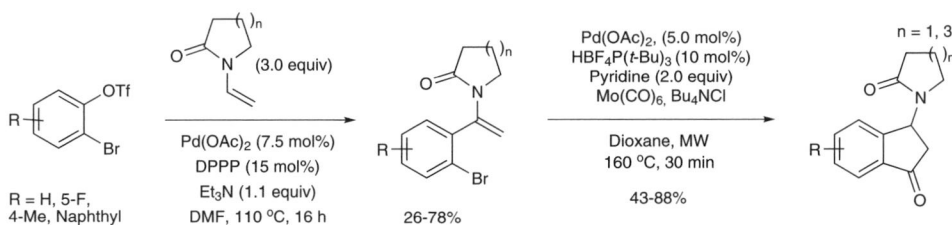

Figure 3.15 Synthesis of branched aryl enamides and their use in carbonylative annulations.

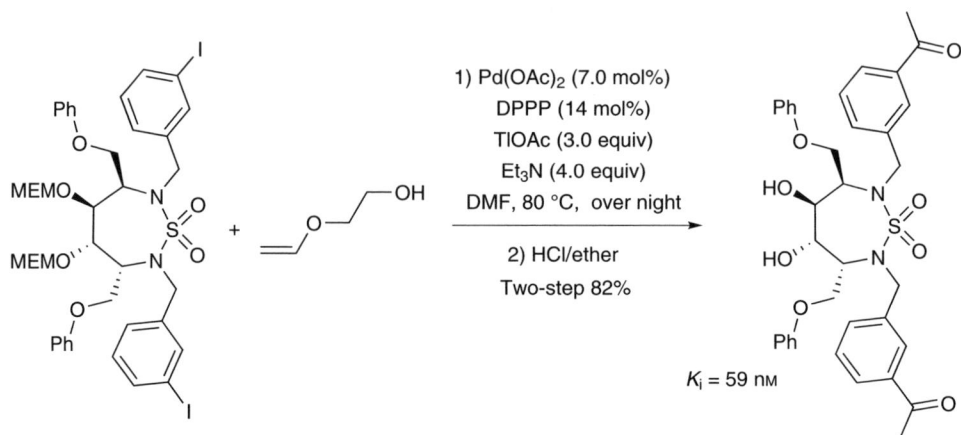

Figure 3.16   *Preparation of a cyclic HIV-1 protease inhibitor employing a regioselective internal arylation with consecutive hydrolysis.*

pure form using enantiopure *tert*-butyl sulfinyl imines as starting materials. After isolation of ring-closed diastereomers, standard silica flash chromatography and subsequent deprotection provided single stereoisomers which were incorporated into the HIV-1 inhibitor structure [69].

Chiral benzylic amines are attractive structural motifs present in approved pharmaceuticals such as Sertralin (Zoloft) and Rivastigmin (Exelon). An α-regioselective Mizoroki–Heck coupling using *N*-acyl enamides allows flexibility in the choice of leaving groups (e.g. triflates, halides) and, thus, nicely complements the alternative enantioselective imine reduction approach. Researchers at Dowpharma have disclosed a high yielding, regioselective (Mizoroki–Heck reaction) and enantioselective (hydrogenation) sequence to manufacture an intermediate required in the synthesis of a pharmaceutically active compound targeting epilepsy (Figure 3.18) [70].

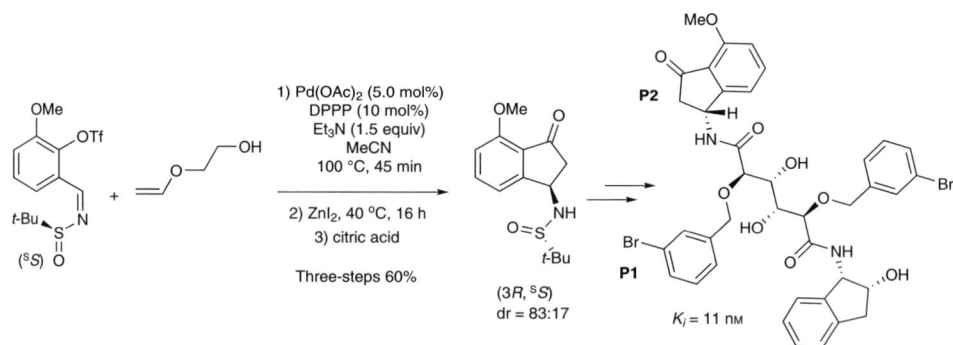

Figure 3.17   *Synthesis of a stereopure linear HIV-1 protease inhibitor P2-building block via arylation of ethylene glycol vinyl ether and a subsequent stereoselective annulation.*

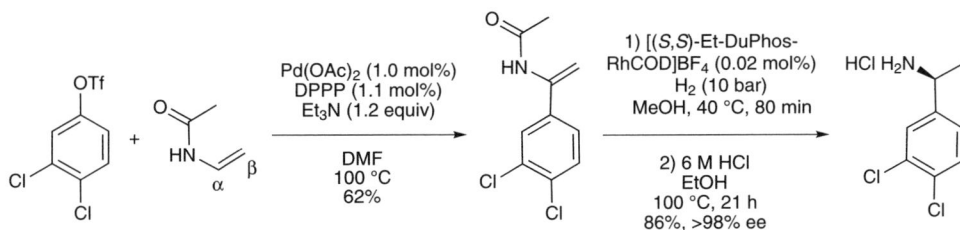

Figure 3.18 *Preparation of an intermediate useful as building block for total synthesis of a GABA-B antagonist, using internal arylation and enantioselective hydrogenation.*

3.3.2 Terminal Arylations

A study on the β-directing influence of chloride ions, promoting a neutral pathway, was conducted by Andersson and Hallberg [26] in 1988. Regrettably, no general method for exploiting aryl chlorides was available at the time; thus, chloride ions had to be introduced separately as additives or, alternatively, aroyl chlorides were used under decarbonylative conditions [71]. Today, the commercially available and air-stable ligand salt [(*tert*-Bu)$_3$PH]BF$_4$ ('Fu salt') offers the possibility to activate aryl chlorides in palladium(0)-catalysed couplings in a general and reliable manner [72]. In one of his papers, Fu showed that the (*tert*-Bu)$_3$P ligand provided a regiochemical outcome of $\beta/\alpha = 91 : 9$ in the reaction between butyl vinyl ether and *p*-chloroacetophenone (87% yield of $\alpha + \beta$) [73]. Inspired by this achievement, the Larhed group decided to investigate this ligand further for terminal arylation of vinyl ethers with aryl chlorides. Thus, a β-selective high-speed microwave protocol was developed and applied in the total synthesis of (*S*)-Betaxolol·HCl, rendering a procedure involving only four separate synthetic steps (17% overall yield, Figure 3.19) [40].

The Mizoroki–Heck reaction is not only applicable for small-scale medicinal chemistry; large-scale production has also been reported involving this methodology [74]. In fact, the starting compound in the thromboxane antagonist synthesis depicted in Figure 3.20 was made by Mizoroki–Heck coupling with ethyl acrylate followed by a regioselective terminal large-scale *N*-vinyl phthalimid arylation ultimately furnishing the target compound in kilogram scale. Although formally a nitrogen-substituted alkene, the carbonyl groups

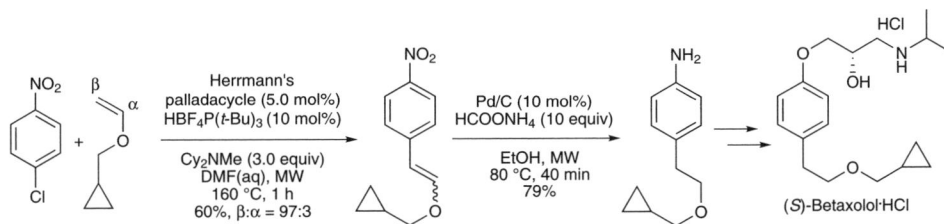

Figure 3.19 *Key-step Mizoroki–Heck arylation of an enol ether using an aryl chloride in the production of the β_1 selective adrenergic blocker (S)-Betaxolol.*

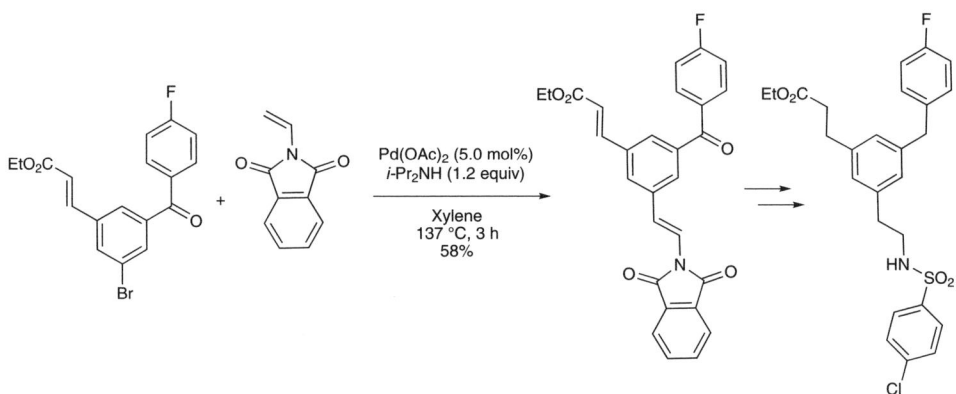

Figure 3.20 *Multi-kilogram scale synthesis of a thromboxane receptor antagonist utilizing β-selective arylation of N-vinylphthalimide under phosphine-free Mizoroki–Heck conditions.*

reduce the electron density of the π-system of the alkene, allowing synthesis of the linear product with full selectivity.

3.3.3   Vinylations

Mizoroki–Heck vinylation protocols seem to be inherently more α-selective than the corresponding arylations using electron-rich alkenes. However, vinylations are more sensitive, and the conversion and regioselectivity under non-optimized reaction conditions can easily deteriorate. Using ligandless conditions, Andersson and Hallberg [75] performed a study on fundamental vinylating agents, substantiating the strong electronic interactions that govern the regiochemical outcome of electron-rich alkenes (Figure 3.21). As expected, the more electron-deficient vinyl bromide preferentially inserted to the more negatively charged terminal β-carbon.

Homocoupling, reduction or aromatization of the vinyl triflate are often competing side-reactions that need to be suppressed. Recently, Larhed and coworkers addressed these issues using microwave heating and efficient fluorous techniques to enable rapid synthesis

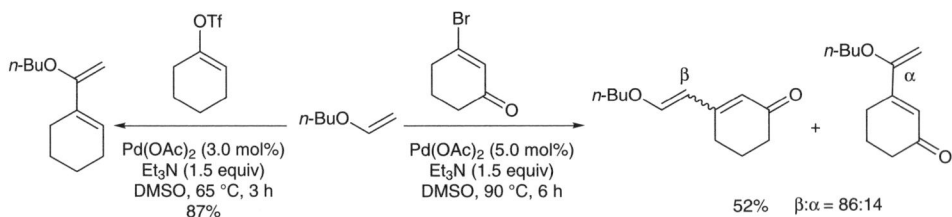

Figure 3.21 *Mizoroki–Heck vinylation providing evidence for the influence of the electron density on the vinylating agent on the regiochemical outcome.*

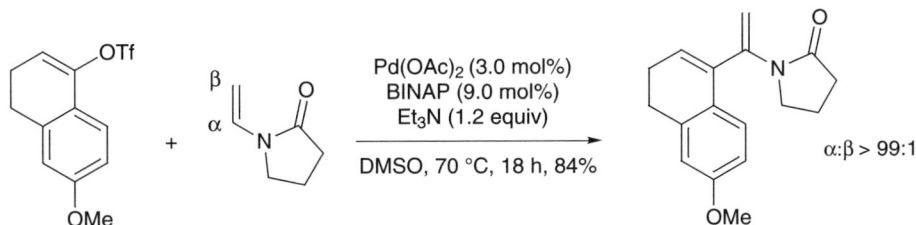

Figure 3.22 *Internal Mizoroki–Heck vinylation using vinyl triflates.*

and purification of the target compounds. Accordingly, enamides were successfully α-vinylated, furnishing 2-acylamino-1,3-diene products (Figure 3.22) [76].

Vinyl tosylates and mesylates are useful substitutes for vinyl triflates [29]. Despite being electron poor, almost complete α-selectivity was obtained using various vinylating substrates. Enol ethers and enamides worked equally well, although cyclic moieties (e.g. *N*-vinyl pyrrolidin-2-one) were found to be unproductive (Figure 3.23).

3.4 Allylic Alkenes

Allylic alkenes have been found to be similar to vinylic alkenes in many ways, giving mainly internal insertion when cationic reaction conditions prevail and terminal under neutral conditions [16, 43]. Unfortunately, the balance between the linear and branched products is often fragile, making any optimization demanding. Coordination and neighbouring-group effects have been demonstrated to be of importance with allylic reactants, as well as with the heteroatom-substituted electron-rich alkenes discussed previously [77, 78].

3.4.1 Terminal Couplings

By far most the popular class of the allylic substrates in the Mizoroki–Heck reaction has been allylic alcohols, partly due to their ubiquitousness, but also due to the possibility of testing and developing catalytic systems for regiocontrol, as this coupling often results in mixtures of isomers. Reactions under Jeffery conditions are known to result in terminal insertions (Figure 3.24) [79], and similar reactions with vinylic triflates instead of halides are also well known (Figure 3.25) [80].

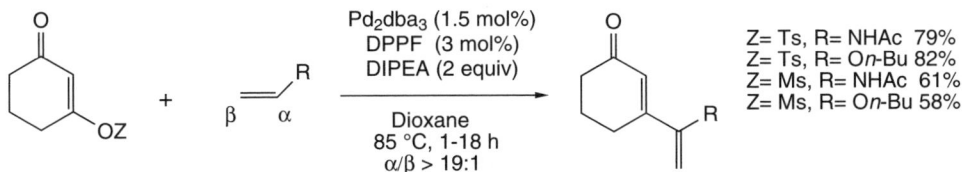

Figure 3.23 *α-Selective Mizoroki–Heck vinylation using vinyl tosylate and mesylate derivatives as coupling partners with enamides and butyl vinyl ether.*

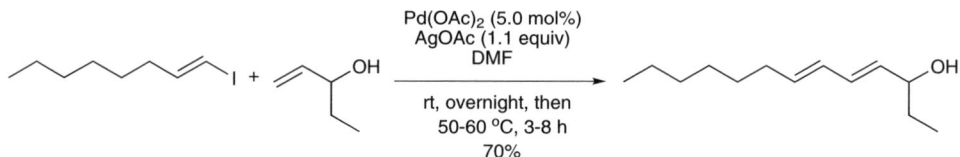

Figure 3.24   *Terminal Mizoroki–Heck reaction between a vinylic iodide and an allylic alcohol.*

Doucet and coworkers have, in a number of papers, described Mizoroki–Heck reactions with allyl alcohol derivatives, many times with very good terminal regioselectivity using aryl bromides as coupling partners and Tedicyp as ligand (Figure 3.26). It was noted that the turnover number (TON) for aryl bromides was actually higher than for the more expensive aryl iodides [81]. Protected allyl alcohols (to circumvent loss of yield due to the thermal instability of 3-arylpropionaldehydes – a possible isomerization product of the coupling) were investigated in a separate paper. The reaction with the trimethylsilane derivative was not successful due to decomposition, but the corresponding coupling with the *tert*-butyldimethylsilane-protected alkene could be used in a series of reactions. Allyl alcohol blocked with the tetrahydropyran (THP) protective group was also investigated, but this formed more complex mixtures of isomers [82]. Further variations on couplings using the Pd/Tedicyp catalytic system have also been published [83, 84].

An alternative choice of reagents can lead to the introduction of an allyl group after elimination of a β-acetoxy group (Figure 3.27). Palladium on carbon and tetrabutylammonium salts together with aryl iodides were found to be effective in this respect. Electron-withdrawing groups on the aryl were associated with shorter reaction times, but were in some cases troubled by lower regioselectivity caused by isomerization of the alkene to the thermodynamically more stable styrene [85].

A relatively recent report on Mizoroki–Heck couplings in tetrabutylammonium bromide (TBAB) used 3-hydroxy-methylenealkanoates to give β-arylketones after a presumed fast decarbomethoxylation (Figure 3.28). Electron-rich and electron-poor aryl bromides were both good substrates. TBAB was found to function better as ionic liquid than butylpyridinium tosylate (which generates a less nucleophilic anion), most likely due to TBAB's superior capacity to stabilize the catalyst [86]. The scope of this reaction was further elucidated in later publications [87, 88].

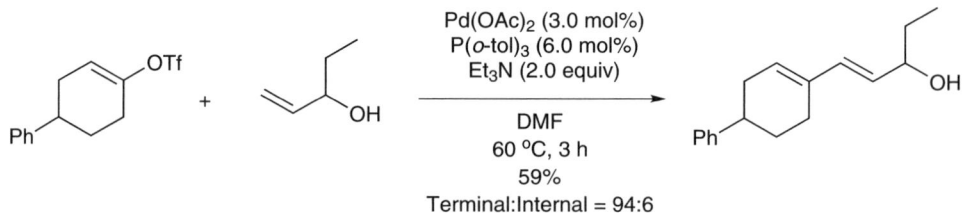

Figure 3.25   *Terminal selectivity in the Mizoroki–Heck reaction with a vinylic triflate and an allylic alcohol.*

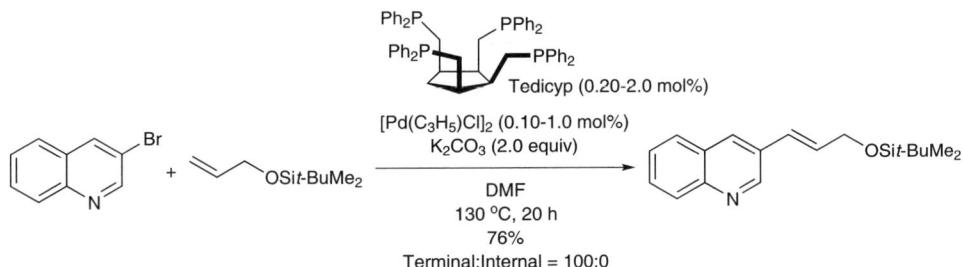

Figure 3.26 *Terminal Mizoroki–Heck arylation with allyloxy-tert-butyldimethylsilane.*

Full terminal regioselectivity was reported by Kang and coworkers using a phosphine-free method with hypervalent iodonium salts. Efforts to produce phenyl-substituted allylic alcohols with iodobenzenes under previously reported reaction conditions did not meet with success, as these produced mixtures of substituted ketones and allylic alcohols. The combination of hypervalent iodonium salts and phosphine-free conditions using Pd(OAc)$_2$ resulted in substituted allylic alcohols as the sole products under mild conditions and with high catalyst efficiency. This suggests that the isomerization that can follow the coupling was suppressed, thus offering an alternative to the previously mentioned method (Figure 3.29) [89].

As a contrast, the Mizoroki–Heck reaction in molten salts using ligandless reaction conditions produced the isomerized β-arylated ketones. Aryl iodides were first studied, but it was soon shown that aryl bromides were reactive at higher reaction temperatures as well. Lower amounts of catalyst resulted in longer reaction times. The example in Figure 3.30 illustrates a single-step synthesis of the nonsteroidal anti-inflammatory drug Nabumethone [90].

An *N*-phenyl-allylamine has been arylated terminally using phosphine-free conditions and a heteroaryl iodide in the synthesis of an H$^+$/K$^+$-ATPase inhibitor (Figure 3.31) [91]. This class of inhibitors has attracted the interest of many medicinal chemistry groups due to the large sales of the blockbuster Losec/Prilosec (Omeprazole). In the cited case, the allylic amine moiety was added at a late stage of the synthesis and the primary amine was protected by Boc to prevent formation of π-allyl palladium complexes. The deprotection was then performed with trifluoroacetic acid.

An interesting application of a large-scale Mizoroki–Heck reaction with allylamines has been reported by Brown Ripin *et al.* Several different amine protecting groups and

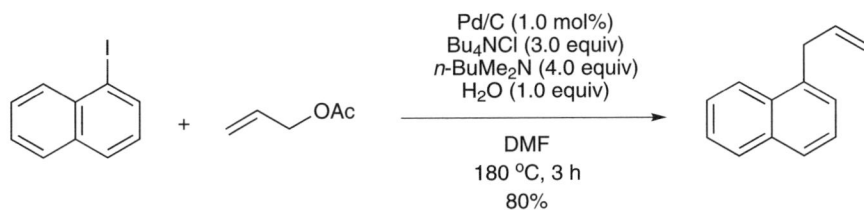

Figure 3.27 *Terminal allylation of 1-iodonaphthalene with allylic acetate using Pd/C as catalyst.*

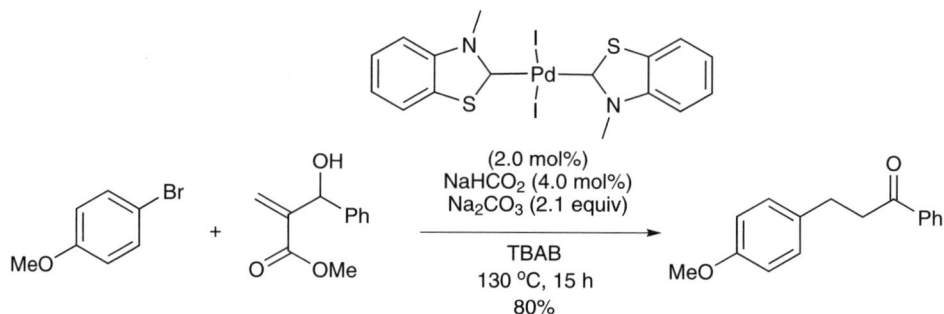

Figure 3.28 *Arylation and subsequent decarbomethoxylation of a hydroxymethylene alkanoate in TBAB.*

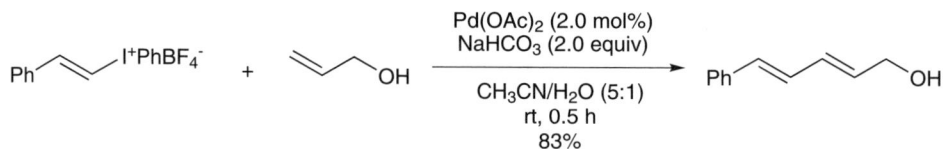

Figure 3.29 *Completely regioselective arylation of allylic alcohol with hypervalent iodonium salts.*

Figure 3.30 *Terminal Mizoroki–Heck arylation of an allylic alcohol in TBAB. Synthesis of the nonsteroidal anti-inflammatory drug Nabumethone.*

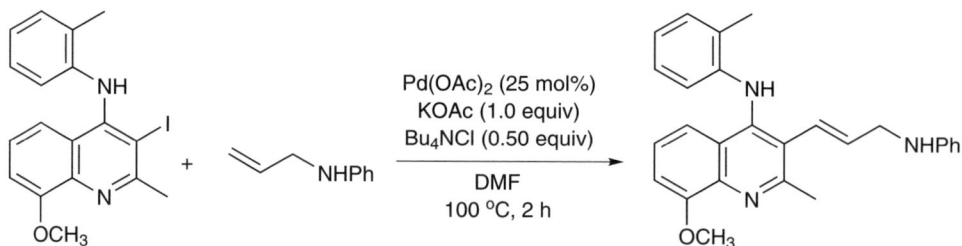

Figure 3.31 *Use of the Mizoroki–Heck reaction in the synthesis of gastric H^+/K^+-ATPase inhibitors.*

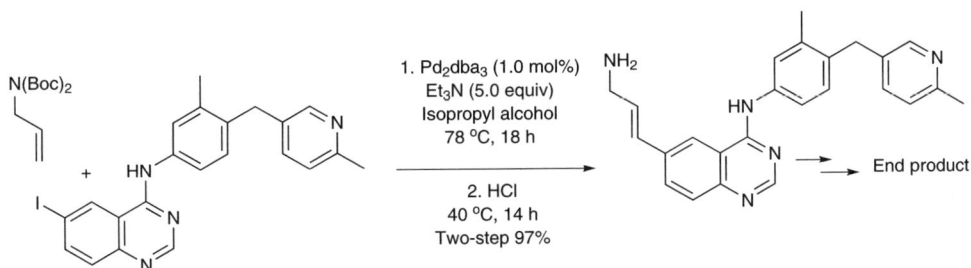

Figure 3.32 *Large-scale Mizoroki–Heck coupling in the synthesis of the oncology drug CP-724714.*

Mizoroki–Heck coupling strategies were evaluated in the kilogram synthesis of the anti-cancer drug CP-724714 (Figure 3.32) [92]. Sonogashira and Suzuki reactions were also discussed in this paper, together with a detailed discussion on the development of the large-scale reaction conditions.

3.4.2 Internal Couplings

High regioselectivity for the internal arylation is induced if the reaction conditions are changed to favour the cationic pathway. Interestingly, the ethyl ether of allyl alcohol was shown to have a lower regioselectivity than allyl alcohol using 1,1′-bis(diphenylphosphino)-ferrocene (DPPF) as the ligand. In fact, the regiochemical outcome of the reaction with allyl ethyl ether was more similar to that of the nonheteroatom-containing alkenes (Figure 3.33) [93].

Coupling of allyl alcohol and a heteroaryl chloride using Pd(OAc)$_2$ and DPPF has been reported to be low yielding (Figure 3.34) [94]. Apparently, the oxidative addition was not the rate-limiting step, as the reaction proceeded smoothly using ethyl acrylate as alkene. Initially, the original idea had been to couple the heteroaryl chloride with allyl alcohol under Jeffery conditions; but, in this case, the reactivity of the chloride was not high enough for this reaction to occur.

Full internal selectivity can be achieved in ionic liquids using bidentate ligands. The cationic pathway has been suggested to be favoured by the polar ionic liquid solvent, and no isomerization of the products to the corresponding carbonyl compounds was observed in Figure 3.35. This method adds nicely to the previously reported routes, as full internal

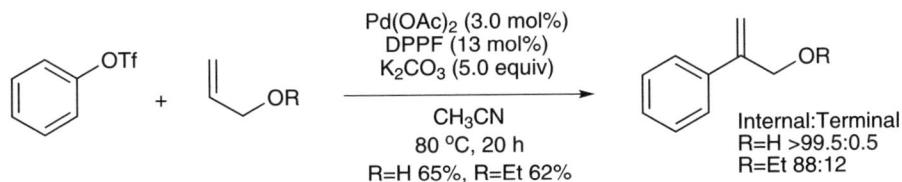

Figure 3.33 *Internal arylation of allyl alcohol and allyl ethyl ether.*

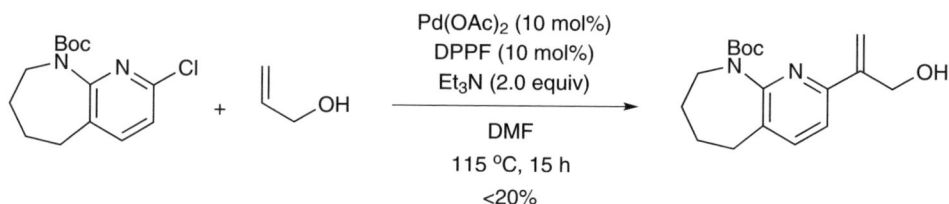

Figure 3.34 *Regioselective Mizoroki–Heck reaction of allylic alcohol and a heteroaryl chloride.*

selectivity can be achieved with relatively cheap and easily available aryl bromides (Figure 3.35). It is to be noted that homoallyl alcohol, and other alkenes with even longer chains between the double bond and the oxygen, produced regioselectivities similar to ordinary alkenes [95]. Similar reactions could also be done with heteroaryl substrates [96].

Other applications of similar couplings include the synthesis of tricyclic indole-2-carboxylic acids as *N*-methyl-D-aspartate (NMDA)–glycine antagonists [97] and 10-deoxyartemisinins against malaria [98].

High internal regioselectivity in the coupling of allyltrimethylsilane and aryl triflates has been reported, where the β-stabilizing effect of silicon in allyltrimethylsilane was suggested to improve the regioselectivity compared with ordinary alkenes (Figure 3.36) [99]. Although the reaction exemplified was performed with oil-bath heating, the reaction could also be conducted under microwave heating.

An important alternative to this protocol is the ionic liquid-mediated reaction presented in Figure 3.37 that allows the use of aryl bromides instead of the often less easily available aryl triflates [45]. The same paper also describes a number of other couplings with different alkenes giving products in good yields and impressive regioselectivities.

Internal arylation of both *N,N*-dialkylallyl amines and protected primary allylamine equivalents can be performed with high regioselectivities [77, 93, 100]. *N*-Boc allylamines could be coupled and deprotected in high yields using aryl triflates as arylpalladium precursors and DPPF as bidentate ligand (Figure 3.38). In many cases the phthalate protective group gave rise to products with lower regioselectivities under the reaction conditions cited [77]. Internally arylated dibenzyl-protected allylamines could also be prepared from aryl triflates using very similar reaction conditions [100].

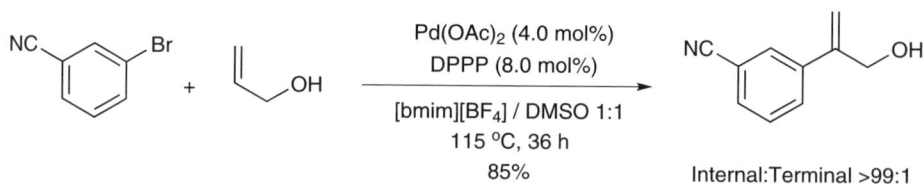

Figure 3.35 *Highly selective internal arylation of allyl alcohol in 1-butyl-3-methylimidazolium tetrafluoroborate ([bmim][BF₄]).*

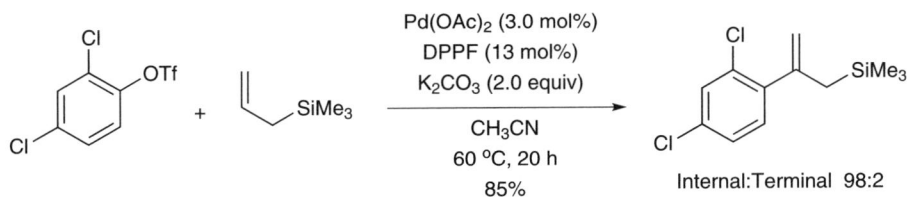

Figure 3.36   *Internal arylation of allyltrimethylsilane with aryl triflates.*

3.5   Electron-Poor Alkenes

3.5.1   Arylations and Double Arylations

Electron-poor alkenes lead almost exclusively to terminal Mizoroki–Heck arylation and vinylation, favoured by both steric and electronic (mesomeric as well as inductive) properties. One of only a few available examples using an alkene with a purely inductive electron-withdrawing group is illustrated in Figure 3.39 [101].

While full terminal selectivity is experienced with mesomerically electron-deficient acrylic acid, acrylic esters, acrylic amides and acrylonitrile, the product ratio arising from the use of styrenes often includes small amounts of branched products [16]. Most commonly, the *trans*-configuration of arylated products dominates.

Trisubstituted, arylated internal alkenes display broad applications in pharmaceutical chemistry. In this respect, the Mizoroki–Heck reaction is a powerful tool, as it does not interfere with a broad range of functional groups. Generally, the low regioselectivity in the formation of the double bond, due to different alternatives for β-hydride elimination with disubstituted alkenes, is a major weakness of the Mizoroki–Heck approach. The intriguing problem of unselective β-elimination was studied in detail by Beller and Riermeier [102], who anticipated that the nature of the base would influence the product distribution upon arylation of α-methyl styrene (Figure 3.40). The best results were, in fact, obtained using *N,N*-diisopropylethylamine as base, affording 65% total yield and an internal/terminal selectivity of 95 : 5.

In general, the Mizoroki–Heck coupling of aryl halides with both electron-rich and electron-poor terminal alkenes affords monoarylated products. With electron-deficient alkenes under selected reaction conditions, such as with excess of the aryl halide, with special catalysts at high temperatures or under high pressure, a twofold terminal arylation to give 1,1-diarylalkene derivatives may occur (Figure 3.41) [103]. Triple arylations

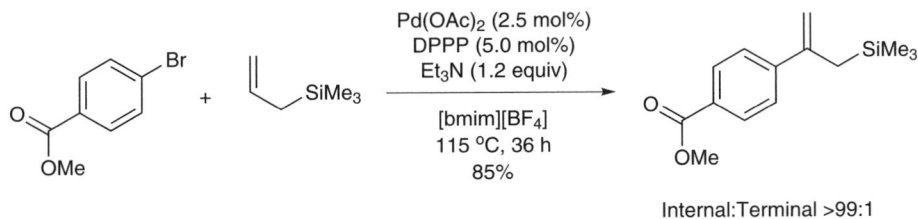

Figure 3.37   *Internal arylation of allyltrimethylsilane with aryl bromides in 1-butyl-3-methylimidazolium tetrafluoroborate ([bmim][BF₄]).*

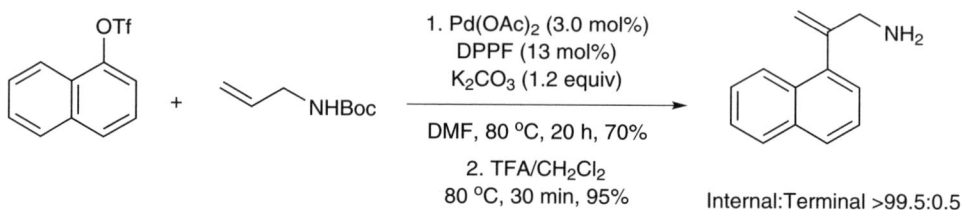

Figure 3.38 *Internal arylation of a primary allyl amine equivalent.*

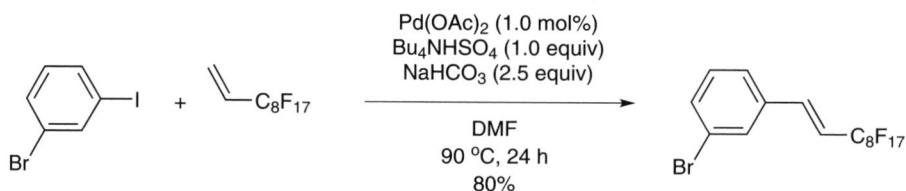

Figure 3.39 *Chemoselective arylation of 1,1,2-trihydrogenperfluorooct-1-ene.*

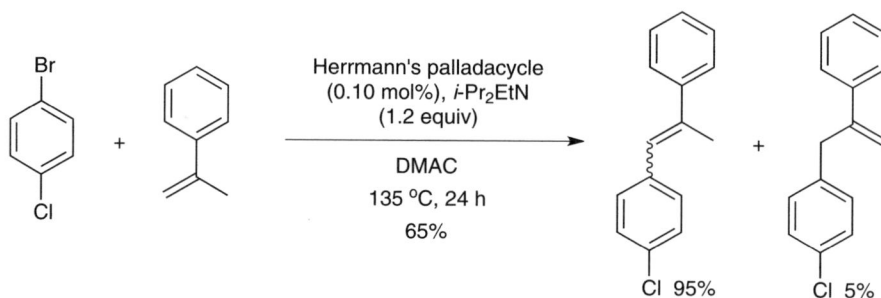

Figure 3.40 *Regioselective arylation of α-methyl styrene with Herrmann's palladacycle providing mainly the internal alkene.*

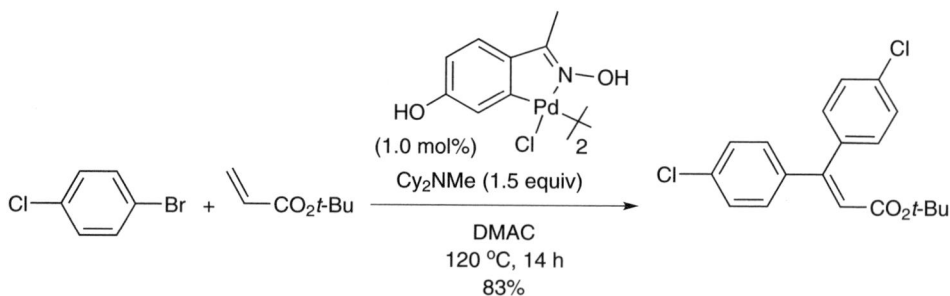

Figure 3.41 *Diarylation of tert-butyl acrylate in a sealed vessel using an oxime-derived palladacycle.*

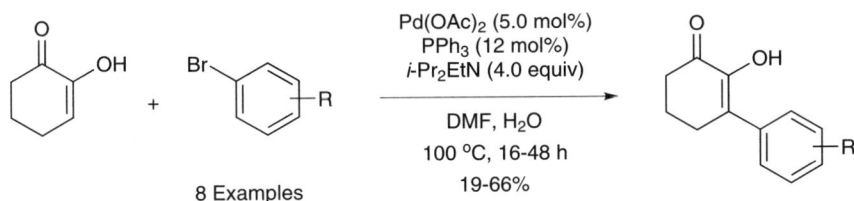

Figure 3.42 *Mizoroki–Heck arylation of 1,2-cyclohexanedione.*

of mono-substituted alkenes have, to the best of our knowledge, only been reported in chelation-accelerated Mizoroki–Heck reactions [104].

The first example of a Mizoroki–Heck arylation of a free enol is depicted in Figure 3.42. Despite the dual electronic characteristics of this unsaturated Mizoroki–Heck substrate, we chose to categorize this reaction as an example of an arylation of an electron-poor alkene due to neutral reaction conditions and the terminal selectivity. Although direct α-arylation to carbonyl groups via enolate anion intermediates is well established [105], the use of a weak base supports the Mizoroki–Heck insertion–elimination mechanism. Two different routes, A and B, were proposed for liberation of the arylated ketol product from the intermediate σ-alkylpalladium(II) complex. According to route A, PdH is eliminated by a direct β-H-elimination, generating a dicarbonyl structure (Figure 3.43) [106]. Alternatively, a palladium enolate is created followed by a *syn*-β-elimination of the benzyl hydrogen, giving the free product (route B).

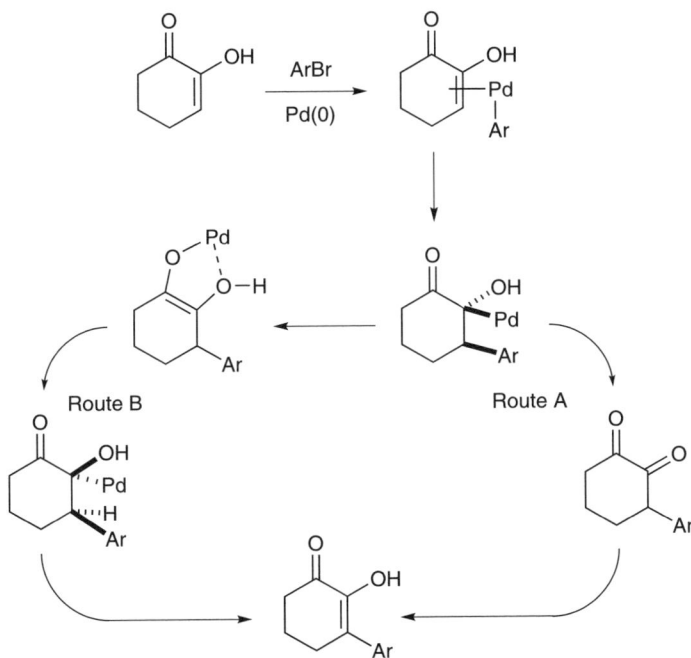

Figure 3.43 *Proposed reaction route with two possible mechanisms for β-elimination.*

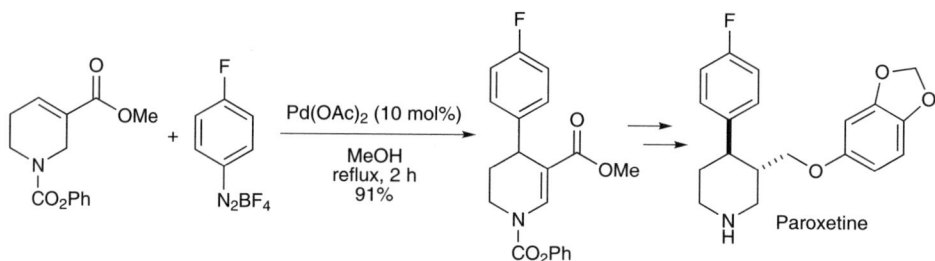

Figure 3.44   *Mizoroki–Heck arylation of a cyclic acrylate with an arenediazonium tetrafluoroborate.*

3.5.2   Examples from Pharmaceutical and Medicinal Chemistry

In 2006, the group of Correia reported a new synthesis of the antidepressant drug (±)-Paroxetine [107]. In this seven-step process, a Mizoroki–Heck arylation using an arene-diazonium tetrafluoroborate as arylating agent was used as the key C—C bond-forming reaction (Figure 3.44). After migratory insertion of the Ar–Pd complex, the need for a *syn*-β-hydrogen directs the double bond into conjugation with the ring nitrogen. Despite using a 1.2 equiv excess of the arenediazonium salt, no double arylated product was detected.

Malaria is caused by protozoal parasites of the genus *Plasmodium*. Of the four species causing human malaria, *Plasmodium falciparum* is the most lethal and it is responsible for almost 3 million deaths annually. Among the new targets for drug intervention are the haemoglobin-degrading aspartic proteases plasmepsin I and II (Plm I and II). Ersmark *et al.* investigated a class of inhibitors incorporating a 1,2-dihydroxyethylene isostere as potential Plm I and II inhibitors. Since previous investigations had suggested a flexible flap region allowing the S1′ pocket to accommodate large residues, one of the strategies to gain activity in both classes of compounds was to extend and vary the P1 and P1′ side chains. Mizoroki–Heck, Suzuki and Sonogashira couplings were used to produce several highly active and novel lead structures (Figure 3.45) [108].

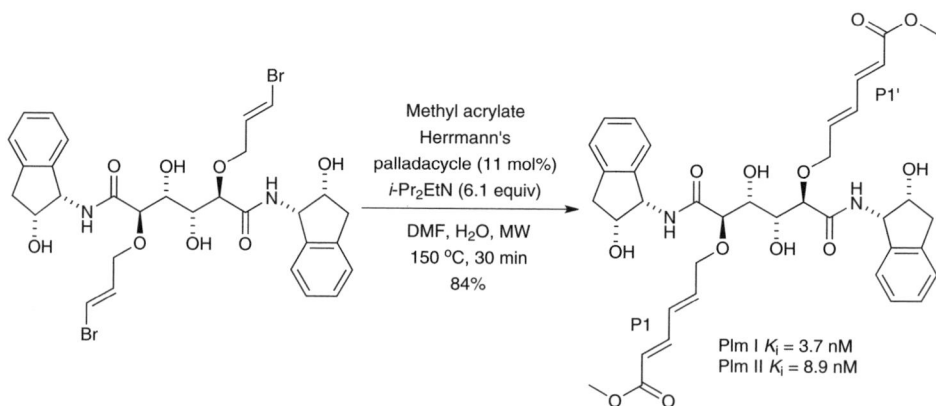

Figure 3.45   *Preparation of a highly potent symmetrical plasmepsin inhibitor by double terminal Mizoroki–Heck vinylation.*

3.6   Summary

Four decades after its discovery, the popularity and usefulness of the Mizoroki–Heck re-action is still increasing, primarily attributed to the high generality of the reaction, the use of only catalytic amounts of palladium, the exceptional tolerance of various func-tional groups and the simplicity of the preparative procedures. In this chapter, we have presented examples of intermolecular Mizoroki–Heck reactions with special emphasis on the regiochemical outcome. Despite not presenting a comprehensive review, we hope that this chapter, together with all other contributions in this volume, will provide some insight into the current status of the Mizoroki–Heck reaction. As progress continues in this field, new chemical and environmental challenges will arise. It is likely that they will be met by innovative approaches that will lead to an ever increased usefulness of this fabulous reaction.

References

1. De Meijere, A. and Diederich, F. (eds) (2004) *Metal-Catalyzed Cross-Coupling Reactions, Second Completely Revised and Enlarged Edition*, Vol 2, Wiley–VCH Verlag GmbH, Weinheim.
2. Negishi, E.-i. (ed.) (2002) *Handbook of Organopalladium Chemistry for Organic Synthesis*, Vol 1, John Wiley & Sons, Inc., New York.
3. Moritani, I. and Fujiwara, Y. (1967) Aromatic substitution of styrene–palladium chloride com-plex. *Tetrahedron Lett.*, **8**, 1119–22.
4. Heck, R.F. (1968) Acylation, methylation, and carboxyalkylation of olefins by Group VIII metal derivatives. *J. Am. Chem. Soc.*, **90**, 5518–26.
5. Heck, R.F. (1968) The palladium-catalyzed arylation of enol esters, ethers, and halides. A new synthesis of 2-aryl aldehydes and ketones. *J. Am. Chem. Soc.*, **90**, 5535–8.
6. Heck, R.F. (1968) The arylation of allylic alcohols with organopalladium compounds. A new synthesis of 3-aryl aldehydes and ketones. *J. Am. Chem. Soc.*, **90**, 5526–31.
7. Heck, R.F. (1968) The addition of alkyl- and arylpalladium chlorides to conjugated dienes. *J. Am. Chem. Soc.*, **90**, 5542–46.
8. Heck, R.F. (1968) Aromatic haloethylation with palladium and copper halides. *J. Am. Chem. Soc.*, **90**, 5538–42.
9. Heck, R.F. (1968) Allylation of aromatic compounds with organopalladium salts. *J. Am. Chem. Soc.*, **90**, 5531–4.
10. Heck, R.F. and Nolley, J.P. Jr. (1972) Palladium-catalyzed vinylic hydrogen substitution reac-tions with aryl, benzyl, and styryl halides. *J. Org. Chem.*, **37**, 2320–2.
11. Mizoroki, T., Mori, K. and Ozaki, A. (1971) Arylation of olefin with aryl iodide catalyzed by palladium. *Bull. Chem. Soc. Jpn.*, **44**, 581.
12. De Meijere, A. and Meyer, F.E. (1994) Fine feathers make fine birds – the Heck reaction in modern garb. *Angew. Chem., Int. Ed. Engl.*, **33**, 2379–411.
13. Beletskaya, I.P. and Cheprakov, A.V. (2000) The Heck reaction as a sharpening stone of palladium catalysis. *Chem. Rev.*, **100**, 3009–66.
14. Whitcombe, N.J., Hii, K.K. and Gibson, S.E. (2001) Advances in the Heck chemistry of aryl bromides and chlorides. *Tetrahedron*, **57**, 7449–76.
15. Larhed, M. and Hallberg, A. (2002) Scope, mechanism, and other fundamental aspects of the intermolecular Heck reaction, in *Handbook of Organopalladium Chemistry for Organic Synthesis*, Vol 1 (ed. E.-i. Negishi), John Wiley & Sons, Inc., New York, pp. 1133–78.

16. Cabri, W. and Candiani, I. (1995) Recent developments and new perspectives in the Heck reaction. *Acc. Chem. Res.*, **28**, 2–7.
17. Alonso, F., Beletskaya, I.P. and Yus, M. (2005) Non-conventional methodologies for transition-metal catalyzed carbon—carbon coupling: a critical overview. Part 1: the Heck reaction. *Tetrahedron*, **61**, 11771–835.
18. Schmidt, A.F., Al Halaiqa, A. and Smirnov, V.V. (2006) Interplays between reactions within and without the catalytic cycle of the Heck reaction as a clue to the optimization of the synthetic protocol. *Synlett*, 2861–78.
19. Knowles, J.P. and Whiting, A. (2007) The Heck–Mizoroki cross-coupling reaction: a mechanistic perspective. *Org. Biomol. Chem.*, **5**, 31–44.
20. Amatore, C. and Jutand, A. (2000) Anionic Pd(0) and Pd(II) intermediates in palladium-catalyzed Heck and cross-coupling reactions. *Acc. Chem. Res.*, **33**, 314–21.
21. Von Schenck, H., Åkermark, B. and Svensson, M. (2003) Electronic control of the regiochemistry in the Heck reaction. *J. Am. Chem. Soc.*, **125**, 3503–8.
22. Deeth, R.J., Smith, A. and Brown, J.M. (2004) Electronic control of the regiochemistry in palladium–phosphine catalyzed intermolecular Heck reactions. *J. Am. Chem. Soc.*, **126**, 7144–51.
23. Kikukawa, K. and Matsuda, T. (1977) Reaction of diazonium salts with transition metals. I. Arylation of olefins with arenediazonium salts catalyzed by zero valent palladium. *Chem. Lett.*, 159–62.
24. Moriarty, R.M., Epa, W.R. and Awasthi, A.K. (1991) Palladium-catalyzed coupling of alkenyl iodonium salts with olefins: a mild and stereoselective Heck-type reaction using hypervalent iodine. *J. Am. Chem. Soc.*, **113**, 6315–7.
25. Chen, Q. and Yang, Z. (1986) Palladium-catalyzed reaction of phenyl fluoroalkanesulfonates with alkynes and alkenes. *Tetrahedron Lett.*, **27**, 1171–4.
26. Andersson, C.M. and Hallberg, A. (1988) Regioselective palladium-catalyzed arylation of vinyl ethers with 4-nitrophenyl triflate. Control by addition of halide ions. *J. Org. Chem.*, **53**, 2112–4.
27. Webel, M. and Reissig, H.U. (1997) Heck reactions starting from silyl enol ethers. A simple one-pot nonaflation-coupling procedure for the synthesis of 1,3-dienes. *Synlett*, 1141–2.
28. Fu, X., Zhang, S., Yin, J. *et al.* (2002) First examples of a tosylate in the palladium-catalyzed Heck cross coupling reaction. *Tetrahedron Lett.*, **43**, 573–6.
29. Hansen, A.L. and Skrydstrup, T. (2005) Regioselective Heck couplings of α,β-unsaturated tosylates and mesylates with electron-rich olefins. *Org. Lett.*, **7**, 5585–7.
30. Miura, M., Hashimoto, H., Itoh, K. and Nomura, M. (1989) Palladium-catalyzed desulfonylative coupling of arylsulfonyl chlorides with acrylate esters under solid–liquid phase-transfer conditions. *Tetrahedron Lett.*, **30**, 975–6.
31. Blaser, H.-U. and Spencer, A. (1982) The palladium-catalyzed arylation of activated alkenes with aroyl chlorides. *J. Organomet. Chem.*, **233**, 267–74.
32. Hansen, A.L., Ebran, J.-P., Ahlquist, M. *et al.* (2006) Heck coupling with nonactivated alkenyl tosylates and phosphates: examples of effective 1,2-migrations of the alkenyl palladium(II) intermediates. *Angew. Chem. Int. Ed.*, **45**, 3349–53.
33. Ruano, J.L.G., Aleman, J. and Paredes, C.G. (2006) Oxidative addition of Pd(0) to Ar–SO$_2$R bonds: Heck-type reactions of sulfones. *Org. Lett.*, **8**, 2683–6.
34. Stephan, M.S., Teunissen, A.J.J.M., Verzijl, G.K.M. and De Vries, J.G. (1998) Heck reactions without salt formation: aromatic carboxylic anhydrides as arylating agents. *Angew. Chem. Int. Ed.*, **37**, 662–4.
35. Gooβen, L.J. and Paetzold, J. (2002) Pd-catalyzed decarbonylative olefination of aryl esters: towards a waste-free Heck reaction. *Angew. Chem. Int. Ed.*, **41**, 1237–41.

36. Myers, A.G., Tanaka, D. and Mannion, M.R. (2002) Development of a decarboxylative palladation reaction and its use in a Heck-type olefination of arene carboxylates. *J. Am. Chem. Soc.*, **124**, 11250–1.
37. Jeffery, T. (1996) On the efficiency of tetraalkylammonium salts in Heck type reactions. *Tetrahedron*, **52**, 10113–30.
38. Littke, A.F. and Fu, G.C. (2002) Palladium-catalyzed coupling reactions of aryl chlorides. *Angew. Chem. Int. Ed*, **41**, 4176–211.
39. Stambuli, J.P., Incarvito, C.D., Buehl, M. and Hartwig, J.F. (2004) Synthesis, structure, theoretical studies, and ligand exchange reactions of monomeric, T-shaped arylpalladium(II) halide complexes with an additional, weak agostic interaction. *J. Am. Chem. Soc.*, **126**, 1184–94.
40. Datta, G.K., von Schenck, H., Hallberg, A. and Larhed, M. (2006) Selective terminal Heck arylation of vinyl ethers with aryl chlorides: a combined experimental—computational approach including synthesis of betaxolol. *J. Org. Chem.*, **71**, 3896–903.
41. Daves, G.D. Jr and Hallberg, A. (1989) 1,2-Additions to heteroatom-substituted olefins by organopalladium reagents. *Chem. Rev.*, **89**, 1433–45.
42. Andappan, M.M.S, Nilsson, P., von Schenck, H. and Larhed, M. (2004) Dioxygen-promoted regioselective oxidative Heck arylations of electron-rich olefins with arylboronic acids. *J. Org. Chem.*, **69**, 5212–8.
43. Cabri, W., Candiani, I., Bedeschi, A. and Santi, R. (1992) Palladium-catalyzed arylation of unsymmetrical olefins. Bidentate phoshine ligand controlled regioselectivity. *J. Org. Chem.*, **57**, 3558–63.
44. Cabri, W., Candiani, I., Bedeschi, A. and Santi, R. (1993) 1,10-Phenanthroline derivatives: a new ligand class in the Heck reaction. Mechanistic aspects. *J. Org. Chem.*, **58**, 7421–6.
45. Mo, J., Xu, L. and Xiao, J. (2005) Ionic liquid-promoted, highly regioselective Heck arylation of electron-rich olefins by aryl halides. *J. Am. Chem. Soc.*, **127**, 751–60.
46. Vallin, K.S.A., Larhed, M. and Hallberg, A. (2001) Aqueous DMF–potassium carbonate as a substitute for thallium and silver additives in the palladium-catalyzed conversion of aryl bromides to acetyl arenes. *J. Org. Chem.*, **66**, 4340–3.
47. Arvela, R.K., Pasquini, S. and Larhed, M. (2007) Highly regioselective internal Heck arylation of hydroxyalkyl vinyl ethers by aryl halides in water. *J. Org. Chem.*, **72**, 6390–6.
48. Mo, J. and Xiao, J.L. (2006) The Heck reaction of electron-rich olefins with regiocontrol by hydrogen-bond donors. *Angew. Chem. Int. Ed.*, **45**, 4152–7.
49. Deeth, R.J., Smith, A., Hii, K.K. and Brown, J.M. (1998) The Heck olefination reaction; a DFT study of the elimination pathway. *Tetrahedron Lett.*, **39**, 3229–32.
50. Heck, R.F. (1971) Electronic and steric effects in the olefin arylation and carboalkoxylation reactions with organopalladium compounds. *J. Am. Chem. Soc.*, **93**, 6896–901.
51. Larock, R.C. and Gong, W.H. (1989) Palladium-catalyzed intermolecular vinylation of cyclic alkenes. *J. Org. Chem.*, **54**, 2047–50.
52. Arai, I. and Daves, G.D. Jr (1979) Palladium-catalyzed phenylation of enol ethers and acetates. *J. Org. Chem.*, **44**, 21–3.
53. Larock, R.C., Leung, W.Y. and Stolz-Dunn, S. (1989) Synthesis of aryl-substituted aldehydes and ketones via palladium-catalyzed coupling of aryl halides and non-allylic unsaturated alcohols. *Tetrahedron Lett.*, **30**, 6629–32.
54. Hii, K.K., Claridge, T.D.W. and Brown, J.M. (1997) Intermediates in the intermolecular, asymmetric Heck arylation of dihydrofurans. *Angew. Chem., Int. Ed. Engl.*, **36**, 984–7.
55. Ozawa, F., Kubo, A. and Hayashi, T. (1991) Catalytic asymmetric arylation of 2,3-dihydrofuran with aryl triflates. *J. Am. Chem. Soc.*, **113**, 1417–9.
56. Kong, K.C. and Cheng, C.H. (1991) Facile aryl–aryl exchange between the palladium center and phosphine ligands in palladium(II) complexes. *J. Am. Chem. Soc.*, **113**, 6313–5.

57. Morita, D.K., Stille, J.K. and Norton, J.R. (1995) Methyl/phenyl exchange between palladium and a phosphine ligand. Consequences for catalytic coupling reactions. *J. Am. Chem. Soc.*, **117**, 8576–81.

58. Herrmann, W.A. (1996) Catalytic carbon–carbon coupling by palladium complexes: Heck reactions, in *Applied Homogeneous Catalysis with Organometallic Compounds* (eds B. Cornils and W.A. Herrmann), Wiley–VCH Verlag GmbH, Weinheim, pp. 712–32.

59. Farina, V. (2004) High-turnover palladium catalysts in cross-coupling and Heck chemistry: a critical overview. *Adv. Synth. Catal.*, **346**, 1553–82.

60. Ebran, J.-P., Hansen, A.L., Gogsig, T.M. and Skrydstrup, T. (2007) Studies on the Heck reaction with alkenyl phosphates: can the 1,2-migration be controlled? Scope and limitations. *J. Am. Chem. Soc.*, **129**, 6931–42.

61. Arvela, R.K. and Leadbeater, N.E. (2005) Microwave-promoted Heck coupling using ultralow metal catalyst concentrations. *J. Org. Chem.*, **70**, 1786–90.

62. Satoh, T., Itaya, T., Okuro, K. *et al.* (1995) Palladium-catalyzed cross-carbonylation of aryl iodides with five-membered cyclic olefins. *J. Org. Chem.*, **60**, 7267–71.

63. Herrmann, W.A., Brossmer, C., Reisinger, C.P. *et al.* (1997) Palladacycles: efficient new catalysts for the Heck vinylation of aryl halides. *Chem. Eur. J.*, **3**, 1357–64.

64. Dubbaka, S.R. and Vogel, P. (2005) Palladium-catalyzed desulfitative Mizoroki–Heck couplings of sulfonyl chlorides with mono- and disubstituted olefins: rhodium-catalyzed desulfitative Heck-type reactions under phosphine- and base-free conditions. *Chem. Eur. J.*, **11**, 2633–41.

65. Grigg, R., Loganathan, V., Santhakumar, V. *et al.* (1991) Suppression of alkene isomerization in products from intramolecular Heck reactions by addition of thallium(I) salts. *Tetrahedron Lett.*, **32**, 687–90.

66. Cabri, W., Candiani, I., Bedeschi, A. *et al.* (1992) Alpha-regioselectivity in palladium-catalyzed arylation of acyclic enol ethers. *J. Org. Chem.*, **57**, 1481–6.

67. Wu, X., Nilsson, P. and Larhed, M. (2005) Microwave-enhanced carbonylative generation of indanones and 3-acylaminoindanones. *J. Org. Chem.*, **70**, 346–9.

68. Hultén, J., Andersson, H.O., Schaal, W. *et al.* (1999) Inhibitors of the C(2)-symmetric HIV-1 protease: nonsymmetric binding of a symmetric cyclic sulfamide with ketoxime groups in the P2/P2′ side chains. *J. Med. Chem.*, **42**, 4054–61.

69. Arefalk, A., Wannberg, J., Larhed, M. and Hallberg, A. (2006) Stereoselective synthesis of 3-aminoindan-1-ones and subsequent incorporation into HIV-1 protease inhibitors. *J. Org. Chem.*, **71**, 1265–8.

70. Harrison, P. and Meek, G. (2004) A complementary method to obtain *N*-acyl enamides using the Heck reaction: extending the substrate scope for asymmetric hydrogenation. *Tetrahedron Lett.*, **45**, 9277–80.

71. Andersson, C.M. and Hallberg, A. (1988) Synthesis of beta-arylvinyl ethers by palladium-catalyzed reaction of aroyl chlorides with vinyl ethers. *J. Org. Chem.*, **53**, 235–9.

72. Netherton, M.R. and Fu, G.C. (2001) Air-stable trialkylphosphonium salts: simple, practical, and versatile replacements for air-sensitive trialkylphosphines. Applications in stoichiometric and catalytic processes. *Org. Lett.*, **3**, 4295–8.

73. Littke, A.F. and Fu, G.C. (2001) A versatile catalyst for Heck reactions of aryl chlorides and aryl bromides under mild conditions. *J. Am. Chem. Soc.*, **123**, 6989–7000.

74. Waite, D.C. and Mason, C.P. (1998) A Scalable synthesis of the thromboxane receptor antagonist 3-[3-[2-(4-chlorobenzenesulfonamido)ethyl]-5-(4-fluorobenzyl)phenyl]propionic acid via a regioselective Heck cross-coupling strategy. *Org. Process Res. Dev.*, **2**, 116–20.

75. Andersson, C.M. and Hallberg, A. (1989) Palladium-catalyzed vinylation of alkyl vinyl ethers with enol triflates. A convenient synthesis of 2-alkoxy 1,3-dienes. *J. Org. Chem.*, **54**, 1502–5.

76. Vallin, K.S.A, Zhang, Q.S., Larhed, M. *et al.* (2003) A new regioselective Heck vinylation with enamides. Synthesis and investigation of fluorous-tagged bidentate ligands for fast separation. *J. Org. Chem.*, **68**, 6639–45.

77. Olofsson, K., Sahlin, H., Larhed, M. and Hallberg, A. (2001) Regioselective palladium-catalyzed synthesis of beta-arylated primary allylamine equivalents by an efficient Pd–N coordination. *J. Org. Chem.*, **66**, 544–9.

78. Oestreich, M. (2005) Neighbouring-group effects in Heck reactions. *Eur. J. Org. Chem.*, 783–92.

79. Jeffery, T. (1991) Palladium-catalysed reaction of vinylic halides with allylic alcohols: a highly chemo-, regio- and stereo-controlled synthesis of conjugated dienols. *J. Chem. Soc., Chem. Commun.*, 324–5.

80. Bernocchi, E., Cacchi, S., Ciattini, P.G. *et al.* (1992) Palladium-catalysed vinylation of allylic alcohols with enol triflates. A convenient synthesis of conjugated dienols. *Tetrahedron Lett.*, **33**, 3073–6.

81. Berthiol, F., Doucet, H. and Santelli, M. (2004) Heck reactions of aryl bromides with alk-1-en-3-ol derivatives catalysed by a tetraphosphine/palladium complex. *Tetrahedron Lett.*, **45**, 5633–6.

82. Berthiol, F., Doucet, H. and Santelli, M. (2005) Heck reaction of protected allyl alcohols with aryl bromides catalyzed by a tetraphosphanepalladium complex. *Eur. J. Org. Chem.*, 1367–77.

83. Berthiol, F., Doucet, H. and Santelli, M. (2005) Heck reaction of aryl bromides with pent-4-en-2-ol, 2-phenylpent-4-en-2-ol, or hept-6-en-3-ol catalysed by a palladium–tetraphosphine complex. *Synthesis*, 3589–602.

84. Berthiol, F., Doucet, H. and Santelli, M. (2006) Synthesis of β-aryl ketones by tetraphosphine/palladium catalysed Heck reactions of 2- or 3-substituted allylic alcohols with aryl bromides. *Tetrahedron*, **62**, 4372–83.

85. Mariampillai, B., Herse, C. and Lautens, M. (2005) Intermolecular Heck-type coupling of aryl iodides and allylic acetates. *Org. Lett.*, **7**, 4745–7.

86. Calò, V., Nacci, A., Lopez, L. and Napola, A. (2001) Arylation of alpha-substituted acrylates in ionic liquids catalyzed by a Pd–benzothiazole carbene complex. *Tetrahedron Lett.*, **42**, 4701–3.

87. Calò, V., Nacci, A., Monopoli, A. and Spinelli, M. (2003) Arylation of allylic alcohols in ionic liquids catalysed by a Pd–benzothiazole carbene complex. *Eur. J. Org. Chem.*, 1382–5.

88. Calò, V., Nacci, A. and Monopoli, A. (2004) Regio- and stereo-selective carbon—carbon bond formation in ionic liquids. *J. Mol. Catal. A*, **214**, 45–56.

89. Kang, S.-K., Lee, H.-W., Jang, S.-B. *et al.* (1996) Complete regioselection in palladium-catalyzed arylation and alkenylation of allylic alcohols with hypervalent iodonium salts. *J. Org. Chem.*, **61**, 2604–5.

90. Bouquillon, S., Ganchegui, B., Estrine, B. *et al.* (2001) Heck arylation of allylic alcohols in molten salts. *J. Organomet. Chem.*, **634**, 153–6.

91. Yum, E.K., Yang, O.-K., Kang, S.K. *et al.* (2004) Synthesis of 4-phenylamino-3-vinylquinoline derivatives as gastric H^+/K^+-ATPase inhibitors. *Bull. Korean Chem. Soc.*, **25**, 1091–4.

92. Brown Ripin, D.H., Bourassa, D.E., Brandt, T. *et al.* (2005) Evaluation of kilogram-scale Sonagashira, Suzuki, and Heck coupling routes to oncology candidate CP-724,714. *Org. Process Res. Dev.*, **9**, 440–50.

93. Olofsson, K., Larhed, M. and Hallberg, A. (2000) Highly regioselective palladium-catalyzed beta-arylation of *N,N*-dialkylallylamines. *J. Org. Chem.*, **65**, 7235–9.

94. Cowden, C.J., Hammond, D.C., Bishop, B.C. *et al.* (2004) An efficient and general synthesis of 3-substituted propionaldehydes using the Suzuki–Miyaura coupling. *Tetrahedron Lett.*, **45**, 6125–8.

95. Mo, J., Xu, L.J., Ruan, J.W. *et al.* (2006) Regioselective Heck arylation of unsaturated alcohols by palladium catalysis in ionic liquid. *Chem. Commun.*, 3591–3.

96. Pei, W., Mo, J. and Xiao, J.L. (2005) Highly regioselective Heck reactions of heteroaryl halides with electron-rich olefins in ionic liquid. *J. Organomet. Chem.*, **690**, 3546–51.

97. Katayama, S., Ae, N. and Nagata, R. (2001) Synthesis of tricyclic indole-2-carboxylic acids as potent NMDA–glycine antagonists. *J. Org. Chem.*, **66**, 3474–83.

98. Khac, V.T., Van, V.N. and Van, T.N. (2005) A new route to novel 10-deoxoartemisinins. *Tetrahedron Lett.*, **46**, 4243–5.

99. Olofsson, K., Larhed, M. and Hallberg, A. (1998) Highly regioselective palladium-catalyzed internal arylation of allyltrimethylsilane with aryl triflates. *J. Org. Chem.*, **63**, 5076–9.

100. Wu, J., Marcoux, J.-F., Davies, I.W. and Reider, P.J. (2001) Beta-regioselective intermolecular Heck arylation of *N,N*-disubstituted allylamines. *Tetrahedron Lett.*, **42**, 159–62.

101. Darses, S., Pucheault, M. and Genet, J.-P. (2001) Efficient access to perfluoroalkylated aryl compounds by Heck reaction. *Eur. J. Org. Chem.*, 1121–8.

102. Beller, M. and Riermeier, T.H. (1998) Palladium-catalyzed reactions for fine chemical synthesis, 4. Phosphapalladacycle-catalyzed Heck reactions for efficient synthesis of trisubstituted olefins: evidence for palladium(0) intermediates. *Eur. J. Inorg. Chem.*, 29–35.

103. Botella, L. and Najera, C. (2005) Mono- and β,β-double-Heck reactions of α,β-unsaturated carbonyl compounds in aqueous media. *J. Org. Chem.*, **70**, 4360–9.

104. Nilsson, P., Larhed, M. and Hallberg, A. (2001) Highly regioselective, sequential, and multiple palladium-catalyzed arylations of vinyl ethers carrying a coordinating auxiliary: an example of a Heck triarylation process. *J. Am. Chem. Soc.*, **123**, 8217–25.

105. Culkin, D.A. and Hartwig, J.F. (2003) Palladium-catalyzed α-arylation of carbonyl compounds and nitriles. *Acc. Chem. Res.*, **36**, 234–45.

106. Garg, N., Larhed, M. and Hallberg, A. (1998) Heck arylation of 1,2-cyclohexanedione and 2-ethoxy-2-cyclohexen-1-one. *J. Org. Chem.*, **63**, 4158–62.

107. Pastre, J.C. and Correia, C.R.D. (2006) Efficient Heck arylations of cyclic and acyclic acrylate derivatives using arenediazonium tetrafluoroborates. A new synthesis of the antidepressant drug ($\pm$)-Paroxetine. *Org. Lett.*, **8**, 1657–60.

108. Ersmark, K., Feierberg, I., Bjelic, S. *et al.* (2004) Potent inhibitors of the *Plasmodium falciparum* enzymes Plasmepsin I and II devoid of Cathepsin D inhibitory activity. *J. Med. Chem.*, **47**, 110–22.

4

Waste-Minimized Mizoroki–Heck Reactions

Lukas Gooßen

Fachbereich Chemie – Organische Chemie, Technische Universität Kaiserslautern, Erwin-Schrödinger-Straße (Gebäude 54), Kaiserslautern, Germany

Käthe Gooßen

Bayer H HealthCare AG, Strategic Planning Pharma, Berlin, Germany

4.1 Introduction

The Mizoroki–Heck reaction is a tremendously powerful tool for the selective vinylation of aryl halides and related compounds, as expounded in many facets in the present book. Its attractiveness lies in the regiospecificity of C—C bond formation as defined by the position of the leaving group X in Ar—X, and the mild conditions that allow its use in the construction even of complex, functionalized molecules. In comparison with traditional arylation methods, (e.g. Friedel—Crafts-type reactions) and with cross-couplings of organometallic reagents with vinyl halides, it can be considered to be an environmentally benign and technologically safe process with comparably small waste volumes. However, the very use of halide leaving groups on the arene, which brings about the great benefit of regiospecificity, inevitably results in the stoichiometric release of strong acids as by-products. These are usually trapped by excess base, leading to the formation of salt waste. Moreover, the best yields are often obtained using polar aprotic solvents, such as dimethylformamide and *N*-methylpyrrolidone (NMP), giving rise to further environmental issues during work-up: both salt and solvent waste are extracted into the aqueous phase, which must be disposed of as waste water, since a recovery would be difficult and not economically viable. While

The Mizoroki–Heck Reaction   Edited by Martin Oestreich
© 2009 John Wiley & Sons, Ltd

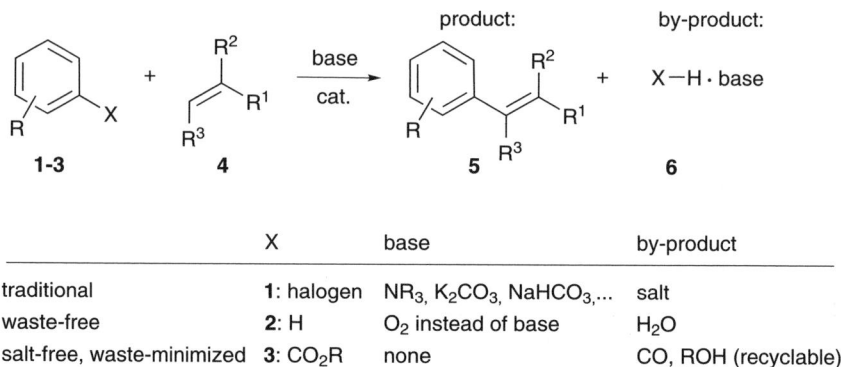

	X	base	by-product
traditional	1: halogen	NR$_3$, K$_2$CO$_3$, NaHCO$_3$,...	salt
waste-free	2: H	O$_2$ instead of base	H$_2$O
salt-free, waste-minimized	3: CO$_2$R	none	CO, ROH (recyclable)

Figure 4.1 *General reaction scheme for traditional and waste-minimized Mizoroki–Heck reactions.*

for most laboratory and low-volume industrial applications the above-mentioned advantages of the Mizoroki–Heck reaction by far outweigh the waste problem, minimizing waste streams becomes an important issue in the manufacture of commodities [1]. Because salt formation is an immediate result of the regiospecificity of the reaction, the development of a waste-free version appears to be futile, leaving the researcher little choice but either to sacrifice the regiospecificity by using arene substrates **2**[2] or to accept the formation of some type of by-product **6** derived from whatever leaving group X is chosen to define the position of C–C bond formation (Figure 4.1).

It should be noted that, for the Mizoroki–Heck reaction, like for most other transformations, deviations from standard reaction protocols but using standard substrates can bring about significant advances towards greener processes, and could thus also be considered as waste-minimized Mizoroki–Heck reactions [3]. Examples include reducing the amounts of reagents employed, better separation of the products, recycling the catalysts, by-products and solvents, minimizing the catalyst loading, and using more environmentally benign solvent systems, such as water [4] or supercritical carbon dioxide [5]. In this chapter, however, we only discuss attempts to address the seemingly intrinsic limitations of the Mizoroki–Heck reaction by exploiting alternative leaving groups X.

4.2 Oxidative Coupling of Arenes

4.2.1 Palladium Catalysts

The first waste-free vinylation of arenes under C–H activation is as old as the Mizoroki–Heck reaction itself: already in 1967, Moritani and Fujiwara [6] revealed a stoichiometric reaction of styrene–palladium(II) chloride dimers with benzene in the presence of acetic acid to give stilbenes in a modest 24% yield. During this process, the palladium(II) precursor is reduced to palladium(0), so that the key to closing the catalytic cycle was to add an efficient reoxidation step to regenerate an active palladium(II) species. One year later, the same group presented a first approach, substoichiometric in palladium,

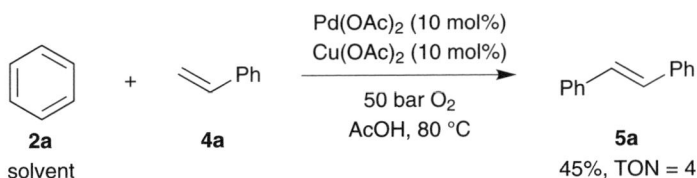

Figure 4.2 *First catalytic oxidative Mizoroki–Heck vinylation by Fujiwara.*

using molecular oxygen together with silver(I) or copper(II) acetate as mediators [7]. For example, styrene (**4a**) was arylated with excess benzene (**2a**) in the presence of acetic acid and 10 mol% of both palladium and copper acetate under 50 atm oxygen to give a modest 45% yield of *trans*-stilbene (**5a**) (Figure 4.2). The results obtained in the vinylation of substituted arenes already reveal the main drawback of this approach: Similar to standard electrophilic aromatic substitutions, the products are usually obtained as mixtures of regioisomers. For example, in the vinylation of toluene with styrene, 58% *para*- and 3% *ortho*-product were isolated.

In an improved version of this prototype reaction, Shue [8] used 20 atm oxygen in the absence of mediators and achieved a turnover number (TON) = 11. With other substrates, the turnover also remained modest, a record of TON = 20 having been achieved with furan derivatives [9]. Again, substituted arenes gave mixtures of regioisomers, with a slight preference for *para*-substitution.

Based on experimental observations, a mechanism has been proposed (Figure 4.3) which is helpful in understanding the main features of such transformations. The initial step is believed to be an electrophilic substitution of the aromatic C—H bond of **2** with the cationic [Pd(OAc)]⁺-species ***b*** – generated by protonolysis from ***a*** – under formation of the σ-arylpalladium complex ***c***. The intermediacy of cationic palladium such as ***b*** is supported by the observation that strongly coordinating halides retard the reaction, and that protic media (i.e. the presence of carboxylic acids) are essential for the catalyst productivity. The function of the metal salt or Brønsted acid additives may lie in enhancing the electrophilic nature of the active palladium species [7, 10]. It is easily conceivable that, due to the size and charge of compound ***b***, the regioselectivity of this step is influenced by the sum of electronic and steric effects, which may, however, be overridden by the directing effect of coordinating substituents. Still, the unusually high ratio of *meta*- to *ortho/para*-substitution for substrates bearing *ortho/para*-directing groups has fuelled doubts over a purely electrophilic mechanism.

Evidence for the intermediacy of σ-arylpalladium acetate complexes ***c*** was provided by the isolation of their trinuclear dialkyl sulfide adducts [11]. The two following steps, insertion of the alkene **4** and β-hydride elimination, correspond to the classical Mizoroki–Heck reaction pathway. The resulting palladium(0) species, which is likely to be stabilized in the form of a hydridopalladium carboxylate ***e***, is then reoxidized by molecular oxygen to the initial palladium(II) acetate (***a***) under liberation of water. The precise mechanism of this reoxidation is not yet fully understood, but it seems that, at elevated oxygen pressures, it is not rate-determining even in the absence of promoters. Mechanistic studies by Jacobs and coworkers [10] indicate that the beneficial effect of adding transition metal salts, originally intended to facilitate this oxidation step, in fact arises from an acceleration

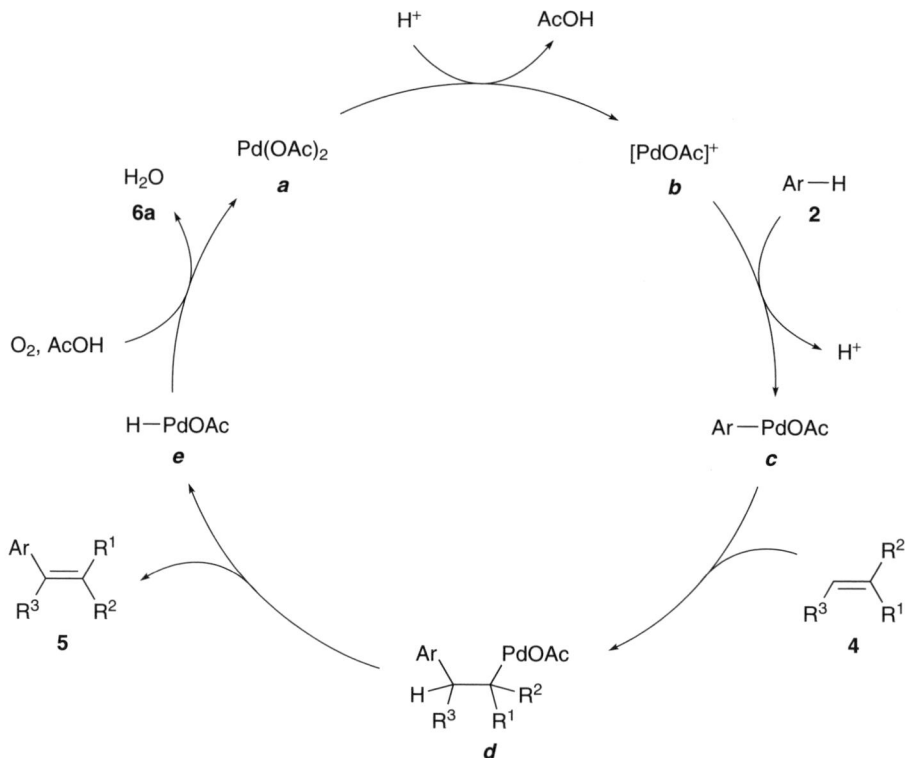

Figure 4.3   *Mechanism of the oxidative vinylation of arenes according to Fujiwara.*

of the induction step by creating a more electrophilic catalyst, thus promoting the rate-determining insertion into the C–H bond (see above). They found that benzoic acid in catalytic amounts was an even more effective additive than manganese acetate or other metal salts, reaching the record TON of 762 (Figure 4.4).

Recently, a major advance in the development of the reoxidation step was achieved through the use of heteropolyacids. Ishii and coworkers [12] found that the oxidative coupling of benzenes with electron-deficient alkenes can be performed at ambient pressure under oxygen and even under an air atmosphere when using molybdovanadophosphoric

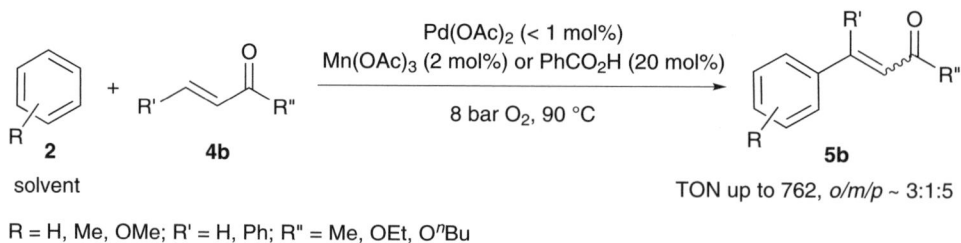

R = H, Me, OMe; R' = H, Ph; R" = Me, OEt, O^{n}Bu

Figure 4.4   *Efficient protocol for the oxidative arylation under pressurized oxygen.*

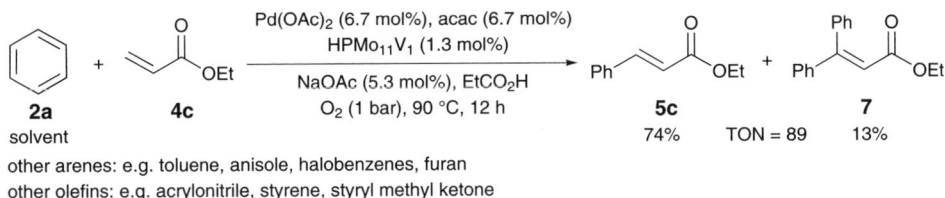

Figure 4.5   *Oxidative vinylation at ambient oxygen pressure using molybdovanadophosphoric acid.*

acid (Figure 4.5). Again, the acidity and composition of the reaction medium had a decisive effect on the reaction yields, with the best results being obtained in propionic acid in the presence of sodium acetate and acetylacetone. Remarkably, the *meta-* and *para*-products were formed in almost the same amounts ($o/m/p = 14:42:44$) when toluene was used as the substrate, which again indicates that more than one effect must be responsible for the regioselectivity. Unfortunately, undesired side products still arise from double arylation and oxidative acetoxylation, especially under conditions giving higher TONs.

As can be seen from the general procedure for the coupling of benzene with ethyl acrylate, this transformation has clearly left the stage of a laboratory curiosity and has grown into a synthetically applicable process [12b]: a solution of Pd(OAc)$_2$ (0.1 mmol), H$_4$PMo$_{11}$V$_1$O$_{40}$·nH$_2$O (HPMo$_{11}$V$_1$, $n = 30$) (46.7 mg, ~0.02 mmol), NaOAc (0.08 mmol), acetylacetone (0.1 mmol), benzene (30 mmol) and ethyl acrylate (1.5 mmol) in propionic acid (5 mL) was placed in a round-bottom flask (30 mL) equipped with a balloon filled with O$_2$, and the mixture was allowed to react under stirring at 90 °C for 2.5 h; the reaction gave ethyl cinnamate (**5c**), ethyl β-phenylcinnamate (**7**) and 3-propionylacrylate in 74%, 13% and 5% yields respectively.

In parallel developments, the aspect of a waste-free transformation was temporarily neglected in favor of improvements in the selectivity and efficiency of the C–H function-alization. Using *tert*-butyl perbenzoate as the oxidant, which is converted to benzoic acid and *tert*-butanol in the process, the catalytic turnover was substantially enhanced up to TON = 67 with furan substrates in comparison with the original protocol [13].

The addition of benzoquinone led to the development of more active reoxidation systems. A substoichiometric amount is sufficient when *tert*-butyl hydroperoxide is present as a co-oxidant, leading to TONs of up to 280 in the reaction of benzene with ethyl cinnamate (Figure 4.6) [14]. Under these conditions, the reaction was found to be applicable to various arenes in combination with several electron-deficient alkenes. Hydrogen peroxide was also found to be a suitable co-oxidant, although the resulting environmentally benign protocol, in which water is formed as the only by-product, gave lower turnover (TON = 35).

Kinetic investigations showed that the reactivity of the arenes follows the order furans > indoles > naphthalene > anisole > toluene/benzene, supporting the proposed electrophilic character of the reaction, while the reactivity of the alkenes is in the order PhCH=CHCO$_2$Et, PhCH=CHCOMe > CH$_2$=CHCO$_2$Et > CH$_3$CH=CHCO$_2$Et > CH$_2$=CHCN, indicating the sensitivity of the catalyst with regard to coordinating functional groups such as cyano. The regioselectivity was higher than in previous protocols: the reaction led to product mixtures with a preference for the *para*-substituted products, a maximum having been

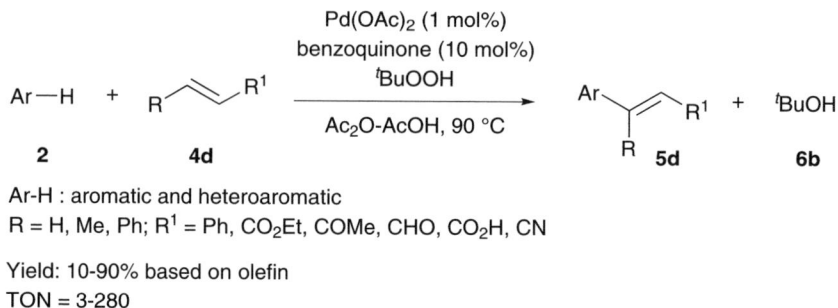

Figure 4.6 *Oxidative arene vinylation using* tert-butyl hydroperoxide *according to Fujiwara.*

achieved for anisole ($o/m/p = 1 : 1 : 6$). Common side-reactions include double vinylation for particularly reactive arenes, and the formation of double bond isomers for substrates that allow two different pathways in the β-hydride elimination step.

Clearly, regioselectivity is less of a problem in intramolecular reactions of substrates bearing an alkene substituent adjacent to an unsubstituted position on the aromatic ring, allowing synthetically valuable applications in ring-closure reactions. Thus, Stoltz and coworkers developed an interesting synthesis of benzofurans [15] and indoles [16] via an intramolecular oxidative arylation of alkenes, in the presence of palladium acetate using electron-poor pyridines as the ligands. In the case of the indoles (e.g., **8**), these reactions proceeded in good yields with TON > 7 and 1 atm oxygen as the sole oxidant, while most examples with benzofurans were reported using stoichiometric amounts of benzoquinone (Figure 4.7).

Another approach to control the regioselectivity of the vinylation was reported by de Vries and coworkers [17], who made use of amide substituents to direct the oxidative coupling exclusively into the *ortho*-position (Figure 4.8). Although this represents a great advance, their method does not allow for discrimination between two nonidentical C–H bonds in *ortho*-position of the amide. The protocol still calls for benzoquinone as the stoichiometric reoxidant, since a waste-free version using hydrogen peroxide gave substantially lower yields. Electron-rich arenes **2b** were found to be most reactive, but the *ortho*-directed oxidative arene vinylation has also been applied to halo-substituted acetanilides [18].

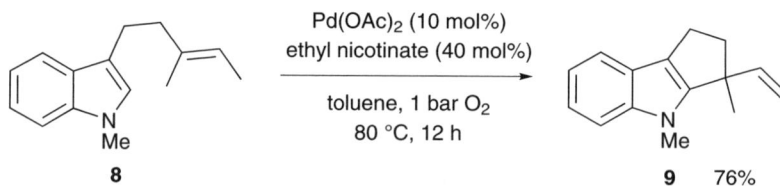

Figure 4.7 *Intramolecular oxidative vinylation of indoles.*

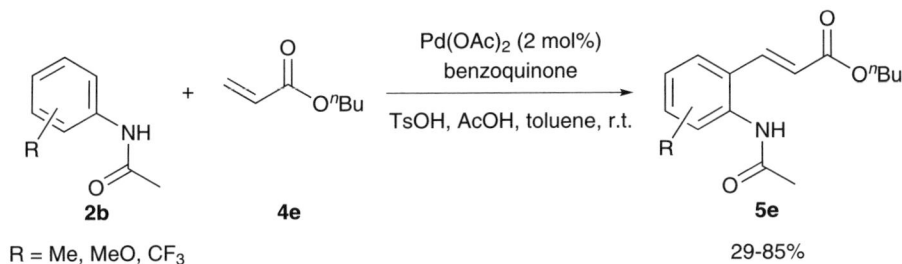

Figure 4.8 ortho-*Directed oxidative arene vinylation using benzoquinone as the oxidant.*

4.2.2 Rhodium and Ruthenium Catalysts

Palladium is, of course, the traditional metal used in Mizoroki–Heck chemistry, but due to the similarity of the overall process, a small selection of related transformations based on other metals merit coverage in this context. The rhodium-catalyzed developments were triggered by a side-reaction observed in the carbonylation of ethylene using $Rh_4(CO)_{12}$ or $Rh_6(CO)_{16}$ as the catalyst under carbon monoxide (20–25 bar) and ethylene pressure (30 bar) at 200–250 °C: when arenes **2** were used as the solvent, vinylation products **5f** were obtained along with the main product, pentan-3-one (**10**, Figure 4.9) [19]. Here, the alkene **4f** also functioned as the dehydrogenating agent, so that the addition of another oxidant was not required.

A rhodium catalyst, $RhCl(CO)PMe_3$, was also used in the photochemical coupling of benzene (**2a**) with methyl acrylate (**4g**). In this case, the product alkene **5g** served as a hydrogen scavenger giving rise to the formation of the saturated product (**11**, Figure 4.10) [20].

The first arylation of ethylene (**4f**) to produce styrene (**5h**), in which rhodium is reoxidized with molecular oxygen, was reported by Matsumoto and Yoshida [21] (Figure 4.11): using

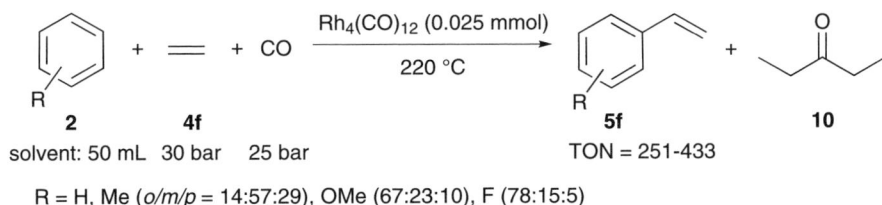

Figure 4.9 *Early example of a rhodium-catalyzed oxidative vinylation.*

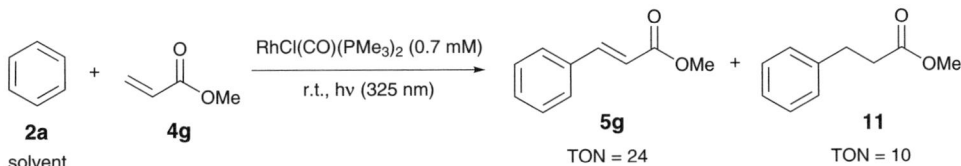

Figure 4.10 *Example of a photochemical rhodium-catalyzed arene vinylation.*

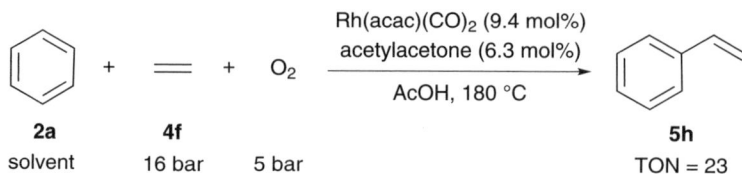

Figure 4.11   *Atom-efficient oxidative vinylation with O_2 as the reoxidant.*

Rh(acac)(CO)$_2$ as the catalyst in acetic acid/acetylacetone under an overall pressure of 21 bar, styrene was obtained in high selectivity, albeit in low yield (TON = 23, turnover frequency TOF = $1.88 \times 10^{-2}\,s^{-1}$). The presence of both acetic acid and acetylacetone is crucial; the nature of the rhodium(I) precursor is less critical, almost the same results having been obtained with Wilkinson's catalyst RhCl(PPh$_3$)$_3$.

The mechanism is proposed to be analogous to that of the Fujiwara process, in that the initial step consists of an electrophilic attack of the aromatic C—H bond by rhodium(III) under formation of a σ-arylrhodium intermediate followed by insertion of ethylene into the C—Rh bond, β-hydride elimination to release the product and, finally, proton release/reoxidation of rhodium(I) to rhodium(III).

Ruthenium has a rich chemistry of hydroarylation reactions [22], but it has also been used successfully by Milstein and coworkers [23] as a catalyst for oxidative couplings of the Fujiwara–Moritani type (Figure 4.12). Under an atmosphere of carbon monoxide (6 bar), various ruthenium precursors effectively promoted the reaction of acrylates (e.g., **4g**) with benzene (**2a**) to give a 1 : 1 ratio of the (*E*)-cinnamate **5i** and methyl propionate **12**, rather than the expected hydroarylation product methyl 3-phenylpropionate. Added oxygen (2 bar) could partly take over the role of the reoxidant from the alkene, resulting in an increase in the incorporation of the alkene into the cinnamate product, giving a ratio of up to 3 : 1 of the arylated to the reduced acrylate.

Hydroquinone was beneficial to the reaction outcome, its effect being attributed to the suppression of radical side-reactions. The proposed mechanism consists of initial electrophilic attack of the metal onto the C—H bond of the arene to give an arylruthenium species with concomitant proton release, followed by alkene insertion into the Ru—C bond, β-hydride elimination to liberate the cinnamate and a ruthenium hydride, and finally regeneration of the active catalyst either by insertion of another alkene into the Ru—H bond, protonation of the alkylruthenium complex and elimination of methyl propionate (under an

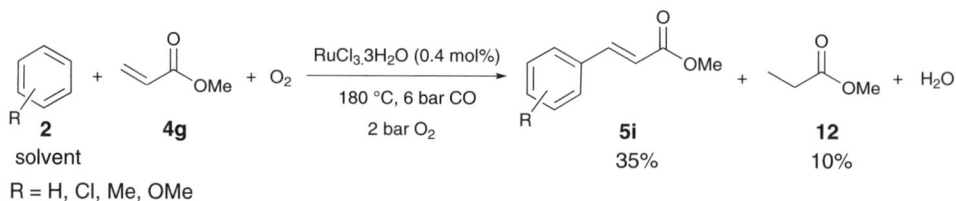

Figure 4.12   *Ruthenium-catalyzed vinylation with partly oxygen and partly the substrate as the reoxidants.*

inert atmosphere), or by oxidation with molecular oxygen when present. Besides acrylates, ethylene was also used as the alkene, and various substituted arenes were employed. The formation of almost statistical mixtures of regioisomers was again observed.

4.3   Mizoroki–Heck Reactions of Carboxylic Acid Derivatives

Even if all selectivity and productivity issues of the oxidative Mizoroki–Heck-type reactions of arenes were to be solved, an inherent drawback common to most C—H functionalization processes will remain: the absence of a leaving group, defining the position of bond formation. If this fundamental advantage of the original Mizoroki–Heck reaction, namely its regiospecificity, is to be kept, then some sort of positional marker is required, and the challenge becomes to chose one which will lead to the ecologically and economically most advantageous process. Atom economy is one important factor in this context, but it should not be the only one considered, as the physical properties and the toxicity as well as the recyclability of the by-product have to be taken into account. From a process standpoint, all by-products should ideally be water or volatile molecules.

4.3.1   Acid Chlorides

In the standard Mizoroki–Heck reaction, the waste salt arises from trapping the acid released by an added base. Thus, a straightforward approach towards reducing the waste volume is to distil off this acid instead of neutralizing it. While this has not yet been achieved when starting from aryl halides, Miura and coworkers demonstrated this concept for acid chlorides **13**. As an example, benzoyl chloride and styrene (**4a**) were coupled in 83% yield in the presence of only 0.05 mol% of [RhCl(C$_2$H$_4$)$_2$]$_2$ in refluxing *o*-xylene, while the hydrochloric acid and carbon monoxide by-products were removed in a stream of nitrogen gas (Figure 4.13) [24]. This protocol was also applied to other arenes and alkenes with high TONs and reasonable chemoselectivities. Recently, a palladium-catalyzed version of this reaction was reported by the same group [25].

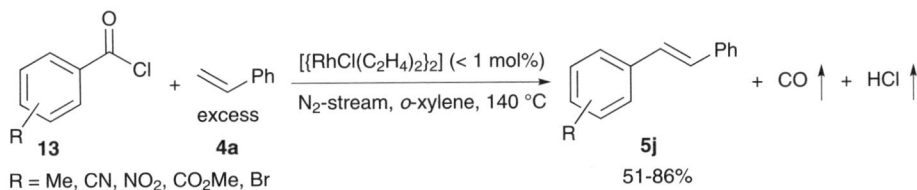

Figure 4.13   *Rhodium-catalysed base-free decarbonylative Mizoroki–Heck reaction.*

   Although this is an interesting concept and may find specific applications, the overall process does not yet offer a great advantage from an environmental standpoint, as the production of acyl chlorides is particularly waste-intensive.

4.3.2   Carboxylic Anhydrides

The discovery by de Vries and coworkers that acyl complexes formed in the oxidative addition of aromatic carboxylic anhydrides to palladium catalysts decarbonylate in the

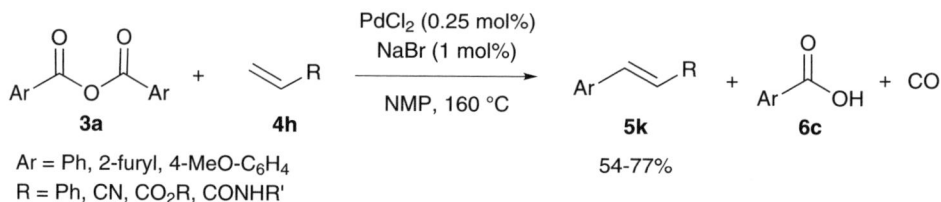

Figure 4.14 *Decarbonylative Mizoroki–Heck reaction of carboxylic anhydrides.*

presence of halides at elevated temperatures became the basis of the DSM approach to waste-minimized Mizoroki–Heck reactions: benzoic anhydride was reacted with *n*-butyl acrylate under loss of carbon monoxide in the presence of a ligand-free palladium catalyst stabilized solely by excess halide, to give (*E*)-*n*-butyl cinnamate along with benzoic acid (Figure 4.14) [26]. In a consecutive step, they planned to recycle the by-product benzoic acid back to the anhydride by thermal dehydration, so that, overall, only the combustible gas carbon monoxide and water would be released. However, while the decarbonylative coupling step was successfully applied to a range of anhydrides and alkenes, the supposedly straightforward thermal dehydration step proved to be troublesome even for benzoic acid itself and could not be extended to other carboxylic acids. Thus, a productive, waste-minimized overall process starting from easily available carboxylic acids following this elegant route could not be demonstrated to function.

This was unfortunate, as aromatic carboxylic acids are widely available and can, in principle, be prepared in an environmentally benign fashion by (air) oxidation of the toluene derivatives and, therefore, possess an intrinsic ecological advantage over aryl halides. As the activation of carboxylic acids via a thermal generation of anhydrides proved to be so difficult, we decided to assist it by adding a dehydrating agent in order to be able to at least extend the scope of the Mizoroki–Heck reaction to this attractive substrate class. This was achieved in a one-pot process, in which aromatic carboxylic acids **3b** are converted *in situ* into the mixed *tert*-butyloxycarbonyl anhydrides in the presence of di-*tert*-butyldicarbonate (**14**), and subsequently alkenylated under release of carbon monoxide, carbon dioxide and *tert*-butanol (**6b**) as the only by-products (Figure 4.15) [27]. Selected examples can be found in Table 4.1.

This approach represents a convenient method for utilizing carboxylic acid substrates in small-scale laboratory applications. However, while it avoids the formation of stoichiometric salt waste that needs to be separated from the products, the reagent

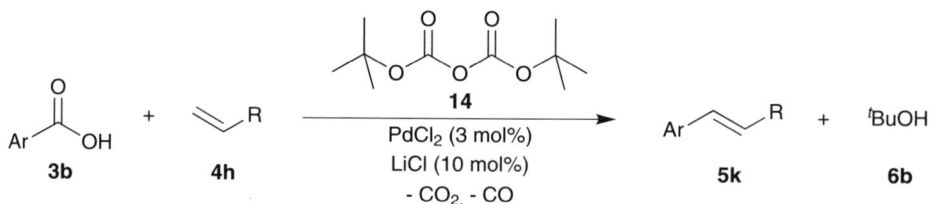

Figure 4.15 *Salt-free decarbonylative Mizoroki–Heck reaction of carboxylic acids in the presence of BOC₂O.*

Table 4.1 *Overview of selected results for three versions of the decarbonylative vinylation reaction*

	Yield (%)[a] [Selectivity][b]		
Product	R = H, Method A[c]	R = 4-nitrophenyl, Method B[d]	R = isopropenyl, Method C[e]
	87 [13:1]	80 [10:1]	85 [9:1]
	48 [10:1]	54 [9:1]	92 [8:1]
	72 [20:1]	88 [20:1]	96 [10:1]
	88 [28:1]	79 [13:1]	99 [15:1]
		85 [13:1]	98 [9:1]
	78 [14:1]	47 [10:1]	75 [10:1]
		90 [13:1]	
		83 [13:1]	63 [9:1]
	R[1] = H, R[3]=[t]Bu 51 [23:1]	R[1] = H, R[3]=[n]Bu 95 [20:1]	R[1] = 4-CN, R[3]=[n]Bu 98 [>50:1]

[a] Isolated yields.
[b] Ratio of 1,2- to 1,1-substituted alkenes as determined by gas chromatography.
[c] 1.00 mmol carboxylic acid, 1.20 mmol alkene, 3.00 mmol BOC$_2$O, 0.03 mmol Pd(OAc)$_2$, 0.10 mmol LiCl, 0.10 mmol isoquinoline, 5 ml NMP, 120 °C, 16 h.
[d] 1.00 mmol 4-nitrophenyl ester, 1.20 mmol alkene, 0.03 mmol PdCl$_2$, 0.09 mmol LiCl, 0.30 mmol isoquinoline, NMP, 160 °C.
[e] 1.00 mmol isopropenyl ester, 2.00 mmol alkene, 0.03 mmol PdBr$_2$, 0.03 mmol HOCH$_2$CH$_2$N(^{n}Bu)$_3$$^+Br^-$, 4 ml NMP, 160 °C, 16 h.

di-*tert*-butyldicarbonate is costly and the total weight of volatile by-products is high, so that the remaining degree of waste minimization in the overall process is rather low.

4.3.3   Carboxylic Esters

It became clear from the problems that arose in these earlier approaches that, in order to allow an efficient recycling step, the Mizoroki–Heck reaction had to start from less reactive carboxylate derivatives, such as esters, that can more easily be regenerated from carboxylic acids. A promising strategy towards achieving a truly waste-minimized overall process thus lies in extending the decarbonylative vinylation to esters that are directly accessible from carboxylic acids and alcohols via azeotropic esterification. This general concept, which is a focus of our research, is illustrated in Figure 4.16.

The most difficult step in the catalytic cycle is the first: an oxidative insertion of the palladium catalyst *a* into the C—O bond of the carboxylic ester **3c**, leading to the acyl complex *b*. This was unprecedented even for activated esters, and is increasingly endothermic with increasing basicity of the alkoxide leaving group. In analogy to the de Vries process, an

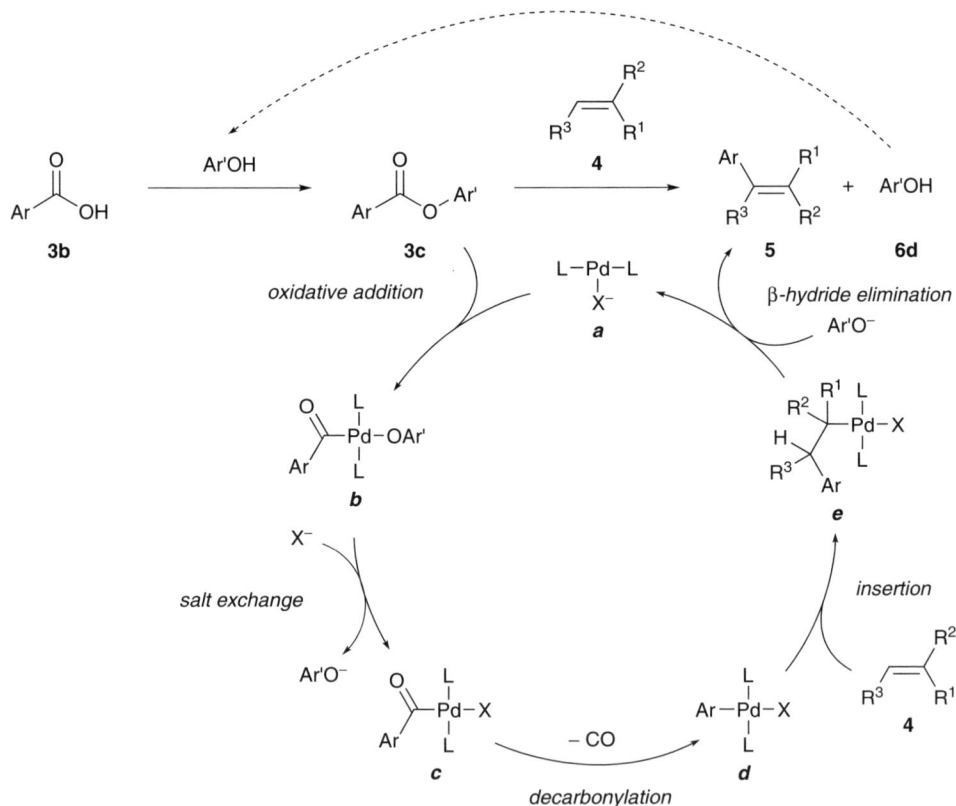

Figure 4.16   *Mechanism for a waste-minimized Mizoroki–Heck reaction starting from carboxylic acids.*

exchange of the alkoxide with halide ligands under formation of complex **c** is necessary to promote the decarbonylation step, leading to the same arylpalladium species **d** as found in traditional Mizoroki–Heck reactions. The remaining steps, namely the insertion of the alkene **4** to give intermediate **e** and β-hydride elimination leading back to **a**, correspond to the standard mechanism. The alkoxide released earlier acts as the base in the final step, scavenging the proton to regenerate the original palladium species **a**. The recycling of the alcohol **6d** is aided by opposite factors to this vinylation process: the more basic the alcohol, the easier it is to convert it back to the starting ester with another molecule of the carboxylic acid **3b**. In an ideal process, this esterification would take place directly under the reaction conditions, so that only a catalytic amount of alcohol would be required to mediate a single-step net process, in which a carboxylic acid reacts with an alkene to give the vinyl arene along with carbon monoxide and water.

This concept was proven to be viable using the example of 4-nitrophenyl esters: in the presence of a palladium(II) chloride/lithium chloride/isoquinoline catalyst system, the 4-nitrophenyl esters **3c** of various aromatic, heteroaromatic and vinylic carboxylic acids were converted to the corresponding vinyl arenes **5k** along with 4-nitrophenol (Figure 4.17). The latter was demonstrated to react with benzoic acid at the same temperature as required for the vinylation step (160 °C) to regenerate the corresponding ester [28], thus demonstrating that at least a two-step waste-minimized Mizoroki–Heck reaction is feasible.

The reaction was successfully applied to both electron-rich and electron-poor 4-nitrophenyl carboxylates; among them, the conversion of the electron-deficient esters was found to be faster and more efficient. Many functional groups are tolerated on both the side of the carboxylic ester (halo, keto, formyl, ester, cyano, nitro and protected amino groups, heterocyclic and α,β-unsaturated carboxylic esters) and of the alkene (electron-rich alkyl-substituted alkenes, electron-poor acrylate derivatives, trimethylvinylsilane as an ethylene surrogate). The cinnamate derivatives could become particularly useful substrates, since the availability of the synthetically equivalent vinyl halides is rather limited. In analogy to conventional Mizoroki–Heck chemistry, linear (*E*)-substituted alkenes are predominantly but not exclusively obtained. Selected examples are shown in Table 4.1.

Beside 4-nitrophenyl esters, other activated esters and amides (e.g. of pentafluorophenol, imidazole and meanwhile even 3-chlorophenol) have been shown to be viable substrates. However, substantial additional progress in catalyst development is required to extend the scope of the reaction further to carboxylic acid alkyl esters, which can be regenerated by *in situ* esterification. This is strictly required, as a two-step process is not likely to be able to compete with the standard Mizoroki–Heck reaction from a purely economical standpoint.

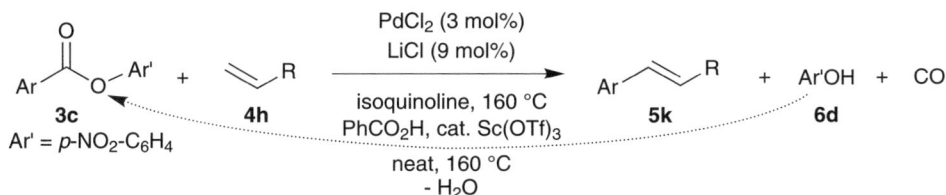

Figure 4.17 *Proof of concept for a Mizoroki–Heck-type process releasing carbon monoxide and water as the only waste.*

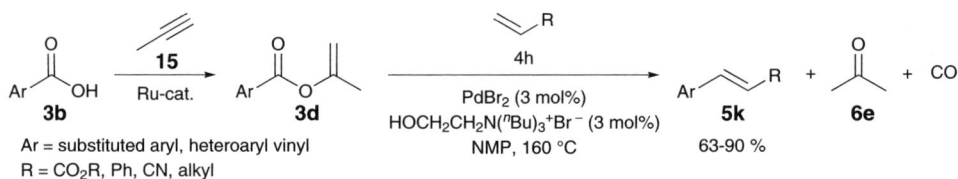

Figure 4.18 *Two-step vinylation process utilizing alkynes for carboxylic acid activation.*

The discovery that even enol esters **3d** can be alkenylated in the presence of catalyst systems consisting of palladium(II) bromide/hydroxy-functionalized tetra-*n*-alkylammonium bromides became the basis of yet another approach towards salt-free Mizoroki–Heck reactions (Figure 4.18): the carboxylic acids **3b** are activated by their waste-free addition to propyne (**15**), a side component of the C$_3$-fraction in steam-cracking. This can be seen as a green alternative [29] to the waste-intensive conversion of carboxylic acids into their acid chlorides using thionyl chloride. The resulting isopropenyl esters **3d** are then vinylated to give the desired vinyl arenes **5k** along only with low molecular weight volatile by-products, namely carbon monoxide and acetone (**6e**) [30]. These can be incinerated, covering part of the energy requirements of the process. Overall, this can be considered to be environmentally benign in comparison with an alternative process that produces salt waste and still uses fossil fuels as its energy source, especially as, over the whole sequence, no stoichiometric inorganic reagents have to be solubilized, so that the amount of solvent can drastically be reduced.

The substrate scope was found to be very similar to that described above for the 4-nitrophenyl esters, although the selectivity for the linear alkene products in the Mizoroki–Heck reaction was somewhat lower. The reaction tolerates many functionalities, including esters, ethers, nitro, keto, trifluoromethyl and even formyl groups (Table 4.1).

It is worth noting that the palladium catalyst could successfully be recycled: palladium in reduced form was allowed to precipitate on added celite over the course of the reaction, the solid was later removed by centrifugation, and the palladium reoxidized with the corresponding amount of bromine. This version of the Mizoroki–Heck reaction may not qualify as an attractive alternative to standard protocols for everyday laboratory use, but it can be advantageous especially for companies with an in-house supply of carboxylic acids and propyne gas.

4.4   Summary and Outlook

While all these approaches point to various possible and often elegant solutions to the waste problem of the Mizoroki–Heck reaction, each still has its drawbacks. Thus, in the short term, customized solutions to minimize the waste effluent will be required for specific synthetic applications, until, in the long run, the ideal, generally applicable method is found. This could on the one hand ensue from the development of shape-selective C—H activation catalysts that by themselves are able to define the position of functionalization, to be used in the oxidative arylation of alkenes with either molecular oxygen or hydrogen peroxide as the reoxidants. On the other hand, waste-minimized versions of the decarbonylative Mizoroki–Heck reaction of carboxylic acids could one day result from the development of more active catalysts that allow an insertion into the C—O bond of simple esters or even carboxylic acids.

References

1. Tucker, C.E. and de Vries, J.G. (2002) Homogeneous catalysis for the production of fine chemicals. Palladium- and nickel-catalyzed aromatic carbon-carbon bond formation. *Top. Catal.*, **19**, 111–8.
2. For a review on C—H activation chemistry, see: Ritlend, V., Sirlin, C. and Pfeffer, M. (2002) Ru-, Rh-, and Pd-catalyzed C—C bond formation involving C—H activation and addition on unsaturated substrates: reactions and mechanistic aspects. *Chem. Rev.*, **102**, 1731–69.
3. (a) Beletskaya, I.P. and Cheprakov, A.V. (2000) The Heck reaction as a sharpening stone of palladium catalysis. *Chem. Rev.*, **100**, 3009–66; (b) Alonso, F., Beletskaya, I.P. and Yus, M. (2005) Non-conventional methodologies for transition-metal catalyzed carbon–carbon coupling: A critical overview. Part 1: the Heck reaction. *Tetrahedron*, **61**, 11771–835.
4. (a) Jeffery, T. (1984) Palladium-catalysed vinylation of organic halides under solid–liquid phase transfer conditions. *J. Chem. Soc., Chem. Commun.*, 1287–8; (b) Botella, L. and Najera, C. (2004) Controlled mono and double Heck reactions in water catalyzed by an oxime-derived palladacycle. *Tetrahedron Lett.*, **45**, 1833–6.
5. (a) Carroll, M.A. and Holmes, A.B. (1998) Palladium-catalysed carbon—carbon bond formation in supercritical carbon dioxide. *J. Chem. Soc., Chem. Commun.*, 1395–6; (b) Morita, D.K., Pesiri, D.R., David, S.A. *et al.* (1998) Palladium-catalyzed cross-coupling reactions in supercritical carbon dioxide. *J. Chem. Soc., Chem. Commun.*, 1397–8.
6. Moritani, I. and Fujiwara, Y. (1967) Aromatic substitution of styrene-palladium chloride complex. *Tetrahedron Lett.*, 1119–22.
7. Fujiwara, Y., Moritani, I., Danno, S. *et al.* (1969) Aromatic substitution of olefins. VI. Arylation of olefins with palladium(II) acetate. *J. Am. Chem. Soc.*, **91**, 7166–9.
8. Shue, R.S. (1971) Catalytic coupling of aromatics and olefins by homogeneous palladium(II) compounds under oxygen. *J. Chem. Soc., Chem. Commun.*, 1510–1.
9. Maruyama, O., Yoshidomi, M., Fujiwara, Y. and Taniguchi, H. (1979) Pd(II)—Cu(II)-catalyzed synthesis of mono- and dialkenyl-substituted five-membered aromatic heterocycles. *Chem. Lett.*, 1229–30.
10. Dams, M., de Vos, D.E., Celen, S. and Jacobs, P.A. (2003) Toward waste-free production of Heck products with a catalytic palladium system under oxygen. *Angew. Chem. Int. Ed.*, **42**, 3512–5.
11. Fuchita, Y., Hiraki, K., Kamogawa, Y. *et al.* (1989) Activation of aromatic carbon–hydrogen bonds by palladium(II) acetate–dialkyl sulfide systems. Formation and characterization of novel diphenyltripalladium(II) complexes. *Bull. Chem. Soc. Jpn.*, **62**, 1081–5.
12. (a) Yokota, T., Tani, M., Sakaguchi, S. and Ishii, Y. (2003) Direct coupling of benzene with olefin catalyzed by Pd(OAc)$_2$ combined with heteropolyoxometalate under dioxygen. *J. Am. Chem. Soc.*, **125**, 1476–7; (b) Tani, M., Sakaguchi, S. and Ishii, Y. (2004) Pd(OAc)$_2$-catalyzed oxidative coupling reaction of benzenes with olefins in the presence of molybdovanadophosphoric acid under atmospheric dioxygen and air. *J. Org. Chem.*, **69**, 1221–6.
13. Tsuji, J. and Nagashima, H. (1984) Palladium-catalyzed oxidative coupling of aromatic compounds with olefins using *tert*-butyl perbenzoate as a hydrogen acceptor. *Tetrahedron*, **40**, 2699–702.

14. Jia, C., Lu, W., Kitamura, T. and Fujiwara, Y. (1999) Highly efficient Pd-catalyzed coupling of arenes with olefins in the presence of *tert*-butyl hydroperoxide as oxidant. *Org. Lett.*, **1**, 2097–100.

15. Zhang, H., Ferreira, E.M. and Stoltz, B.M. (2004) Direct oxidative Heck cyclizations: intramolecular Fujiwara–Moritani arylations for the synthesis of functionalized benzofurans and dihydrobenzofurans. *Angew. Chem. Int. Ed.*, **43**, 6144–8.

16. Ferreira, E.M. and Stoltz, B.M. (2003) Catalytic C—H bond functionalization with palladium(II): aerobic oxidative annulations of indoles. *J. Am. Chem. Soc.*, **125**, 9578–9.

17. Boele, M.D.K., van Strijdonck, G.P.F., de Vries, A.H.M. *et al.* (2002) Selective Pd-catalyzed oxidative coupling of anilides with olefins through C—H bond activation at room temperature. *J. Am. Chem. Soc.*, **124**, 1586–7.

18. Lee, G.T., Jiang, X., Prasad, K. *et al.* (2005) Pd-catalyzed *ortho*-selective oxidative coupling of halogenated acetanilides with acrylates. *Adv. Synth. Catal.*, **347**, 1921–4.

19. (a) Hong, P. and Yamazaki, H. (1979) Reactions of ethylene and benzenes catalyzed by rhodium carbonyls under carbon monoxide. The formation of styrenes and 3-pentanone. *Chem. Lett.*, 1335–6; (b) Hong, P. and Yamazaki, H. (1984) Rhodium carbonyl-catalyzed activation of carbon–hydrogen bonds for application in organic synthesis: V. Phenylation of olefins with benzenes. *J. Mol. Catal.*, **26**, 297–311.

20. Sasaki, K., Sakakura, T., Tokunaga, Y. *et al.* (1988) C≡C double bond insertion in catalytic C—H activation. Dehydrogenative cross coupling of arenes with olefins. *Chem. Lett.*, 685–8.

21. Matsumoto, T. and Yoshida, H. (2000) Oxidative arylation of ethylene with benzene to produce styrene. *Chem. Lett.*, **29**, 1064–5.

22. The hydroarylation of alkynes, which has emerged as a powerful alternative to the oxidative arylation of alkene, is not discussed in this chapter as it is not a true Heck alkenylation. Jia, C., Piao, D., Oyamada, J. *et al.* (2000) Efficient activation of aromatic C—H bonds for addition to C—C multiple bonds. *Science*, **287**, 1992–5.

23. Weissman, H., Song, X. and Milstein, D. (2001) Ru-catalyzed oxidative coupling of arenes with olefins using O$_2$. *J. Am. Chem. Soc.*, **123**, 337–8.

24. Sugihara, T., Satoh, T., Miura, M. and Nomura, M. (2004) Rhodium-catalyzed coupling reaction of aroyl chlorides with alkenes. *Adv. Synth. Catal.*, **346**, 1765–72.

25. Sugihara, T., Satoh, T. and Miura, M. (2005) Mizoroki–Heck type arylation of alkenes using aroyl chlorides under base-free conditions. *Tetrahedron Lett.*, **46**, 8269–71.

26. Stephan, M.S., Teunissen, A.J.J.M., Verzijl, G.K.M. and de Vries, J.G. (1998) Heck reactions without salt formation: aromatic carboxylic anhydrides as arylating agents. *Angew. Chem. Int. Ed.*, **37**, 662–4.

27. Goossen, L.J., Paetzold, J. and Winkel, L. (2002) Pd-catalyzed decarbonylative Heck olefination of aromatic carboxylic acids activated *in situ* with di-*tert*-butyl dicarbonate. *Synlett*, 1721–3.

28. Goossen, L.J. and Paetzold, J. (2002) Pd-catalyzed decarbonylative olefination of aryl esters: towards a waste-free Heck reaction. *Angew. Chem. Int. Ed.*, **41**, 1237–41.

29. Bruneau, C., Neveux-Duflos, M. and Dixneuf, P.H. (1999) Utilization of an industrial feedstock without separation. Ruthenium-catalysed hydrocarboxylation of propadiene and propyne. *Green Chem.*, **1**, 183–5.

30. Goossen, L.J. and Paetzold, J. (2004) Decarbonylative Heck olefination of enol esters: salt-free and environmentally friendly access to vinyl arenes. *Angew. Chem. Int. Ed.*, **43**, 1095–8.

5

Formation of Carbocycles

Axel B. Machotta and Martin Oestreich

*Organisch-Chemisches Institut, Westfälische Wilhelms-Universität Münster,
Corrensstraße 40, D-48149 Münster, Germany*

5.1   Introduction

In a narrower sense, this review covers intramolecular Mizoroki–Heck [1] reactions forming carbocycles [2]; that is, the palladium-catalyzed intramolecular coupling of vinyl/aryl (pseudo-)halides with an alkene tethered by a hydrocarbon chain. Ring closures furnishing heterocycles are covered in Chapter 6; also beyond the scope of this chapter are the domino/cascade or tandem (Chapter 8) and asymmetric processes (Chapters 12 and 16) dealing with formation of a carbocycle.

Within the past three decades [3], the intramolecular Mizoroki–Heck reaction has emerged as a particularly versatile and reliable C—C bond-forming process, which allows for construction of the whole spectrum of ring sizes: small ($n = 3$ or 4), normal ($n = 5, 6$ or 7), medium ($n = 8$–14) and large ($n > 14$). In general, the ring-closing event can occur in an *exo* mode or in a competing *endo* mode, depending on the favoured ring size (Scheme 5.1) [4]. A pivotal feature of the intramolecular (in contrast to the intermolecular) scenario is that tri- and even tetra-substituted C—C double bonds do react.

Cyclopropanes and cyclobutanes were only accessible via *exo*-type cyclizations, the former only within domino/cascade reactions (see Chapter 8) and the latter being realized by Bräse [5a] in desymmetrizing Mizoroki–Heck cyclizations (see Chapter 13) and by Mulzer *et al.* [5b]; five-membered and larger rings are formed in both modes.

For the common ring sizes ($n = 5$–7), the selected representative examples are subdivided into sections according to the specific (formal) cyclization mode; those leading to larger cycles are summarized in a single paragraph.

The Mizoroki–Heck Reaction   Edited by Martin Oestreich
© 2009 John Wiley & Sons, Ltd

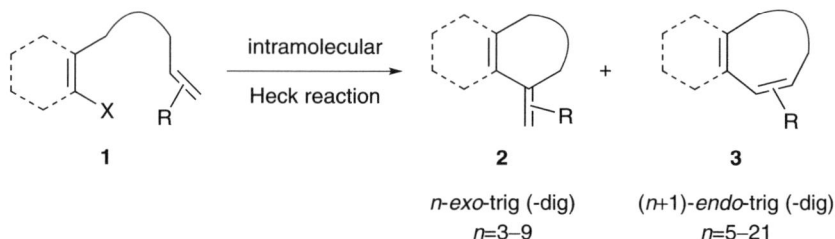

Scheme 5.1

5.2 Formation of Carbocycles by 5-*exo*-trig Cyclization

In the late 1970s, the first intramolecular Mizoroki–Heck reactions were explored primarily to identify new routes to indoles or other heterocycles (see Chapter 6) [3, 6, 7]. Several years later, Grigg *et al.* [8, 9] began to study the formation of carbocycles systematically. The first generation of substrates, open-chain 2-bromo-1,6-heptadienes **4a–d**, were cyclized in good yields, providing a facile access to 1,2-dimethylenecyclopentanes **5a–d**, which are suitable for subsequent Diels–Alder reactions and were also employed in domino Mizoroki–Heck–Diels–Alder processes (see Chapter 8); regioselectivity was notoriously difficult to control, though, yielding 5-*exo*-trig along with 6-*endo*-trig products (**4a–d** → **5a–d**, Scheme 5.2). **4a** gave a 5 : 4 ratio of **5a** and **6a** in 67% yield, using Pd(OAc)$_2$ (2 mol%) and Ph$_3$P (4 mol%) with K$_2$CO$_3$ at 80 °C in MeCN. Regioselectivity was enhanced to 5 : 1 in 63% yield with (Ph$_3$P)$_3$RhCl (5 mol%) (Chapter 10) and, unexpectedly, was reversed with (Ph$_3$P)$_4$Pd (3 mol%) as the catalyst to 1 : 10 in 74% yield. Later, de Meijere and coworkers [10] overcame this problem using Ag$_2$CO$_3$ as the base, enabling exclusive formation of **5a** in 95% yield (**4a** → **5a**, Scheme 5.2). On the other hand, Genêt and coworkers [11] succeeded in performing an immaculate 6-*endo*-trig ring closure (**4a** → **6a**) in moderate yield (30%) using PdCl$_2$ (10 mol%) and a water-soluble triphenylphosphine derivative in an MeCN:H$_2$O mixture (not shown).

The related substrates **4b–d** gave **5b–d** with low regiocontrol (2.5 : 1, 2.8 : 1 and 1.3 : 1) in 73%, 74% and 94% yield respectively. Two fluorine-derived substrates (not shown) were tested by Grigg *et al.* [8, 9] cyclizing regioselectively in a 5-*exo* mode with up to 90% yield.

Substitution at the alkene moiety enabled regioselective cyclization: cyclopentanes **5e** and **5f** were isolated in 88% and 50% yields respectively, albeit as a mixture of double-bond isomers (**4e–f** → **5e–f**) [9]. A substrate with two terminal methyl groups (not shown) also led to exclusive 5-*exo*-trig cyclization in the presence of Pd(OAc)$_2$ and Ph$_3$P with Ag$_2$CO$_3$ as the base in 83% yield; but, in this case, β-hydride elimination was steered away from the ring [12]. With an internal methyl substitution, the formal 6-*endo*-trig ring closure occurred quantitatively (**7** → **8**) [9]. However, Negishi and coworkers [13] showed with related substrates that an apparent 6-*endo*-trig cyclization is likely to stem from a 5-*exo*-mode cyclization/cyclopropanation cascade and subsequent rearrangement (see Section 5.5, Schemes 5.30 and 5.31).

Grigg *et al.* [14, 15] introduced a different approach to achieve exclusive 5-*exo*-trig cyclization: dibromide **9a** entered an uncommon reaction pathway via β-bromide

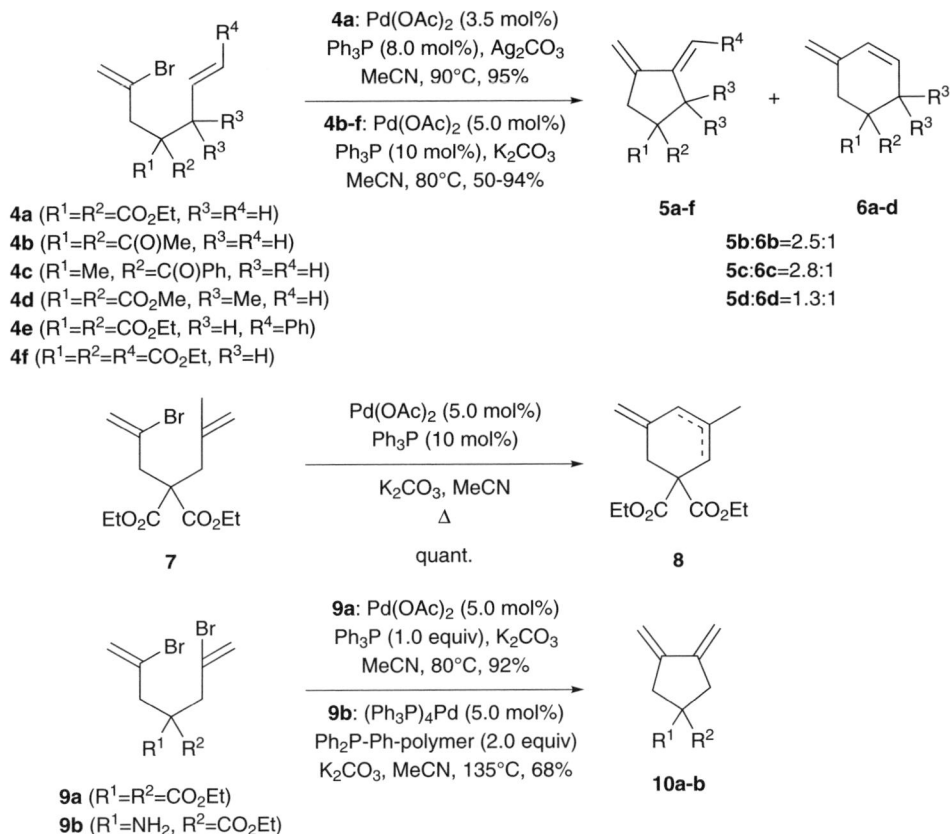

4a (R¹=R²=CO₂Et, R³=R⁴=H)
4b (R¹=R²=C(O)Me, R³=R⁴=H)
4c (R¹=Me, R²=C(O)Ph, R³=R⁴=H)
4d (R¹=R²=CO₂Me, R³=Me, R⁴=H)
4e (R¹=R²=CO₂Et, R³=H, R⁴=Ph)
4f (R¹=R²=R⁴=CO₂Et, R³=H)

9a (R¹=R²=CO₂Et)
9b (R¹=NH₂, R²=CO₂Et)

Scheme 5.2

elimination (**9a** → **10a**, Scheme 5.2). Although still catalytic in Pd(OAc)₂ (5 mol%), a stoichiometric amount of Ph₃P was required to provide **10a** in excellent yield (92%) and without any double-bond migration. Later, Pleixats and coworkers [16] applied this methodology in the reaction of **9b**, affording glycine derivative **10b** in 68% yield (**9b** → **10b**). Cyclization of a corresponding acetyl acetate derivative (not shown) resulted in 84% yield under these conditions. Later, Møller and Undheim [17] applied this reaction type to bislactim ethers (not shown), merely obtaining moderated yields (48%). Sinou and coworkers [18] studied Mizoroki–Heck reactions with β-alkoxide elimination; an example is given in Chapter 6.

Another special case in intramolecular Mizoroki–Heck chemistry was disclosed by Negishi and coworkers [19], using benzyl chlorides as substrates: with (Ph₃P)₄Pd as the catalyst, **11** gave **12** in good yield (82%) (**11** → **12**, Scheme 5.3). Instead of chloride, bromide and mesylate are also suitable leaving groups, but the corresponding iodide provided **12** as a mixture of double-bond isomers. The *ortho*-allyl group in **11** could be replaced by an *ortho*-2-cyclohexenyl moiety (not shown), which gave an annulated tricyclic product in respectable 77% isolated yield. Under identical conditions, **13** furnished spirocyclic **14**

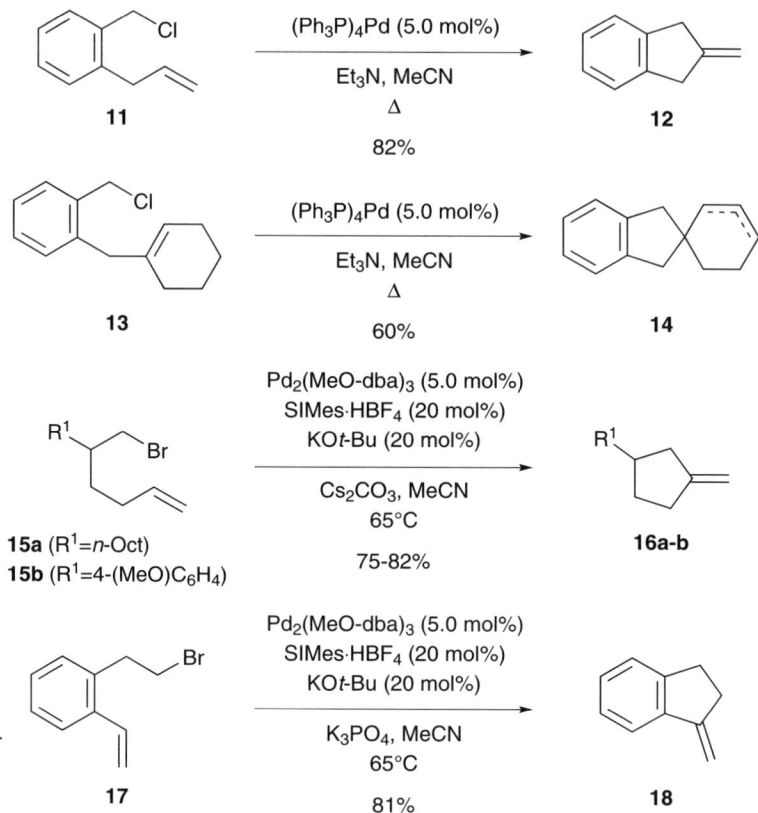

Scheme 5.3

in good yield (60%), yet as a 70 : 30 mixture of double-bond isomers (**13** → **14**). Very recently, Firmansjah and Fu [20] revealed these so-called 'alkyl-Mizoroki–Heck' reactions for a novel class of alkyl bromides and chlorides (not shown), which are usually suscepti-ble to β-hydride elimination within the reaction. However, under the conditions described above, this side-reaction was completely prevented and **15a** and **15b** were cyclized in 75% and 82% yields respectively (**15a–b** → **16a–b**); even **17** provided **18** selectively in good yield (**17** → **18**).

Of course, syntheses of annulated aromatic compounds via classic Mizoroki–Heck reaction of aryl (pseudo)halides are prevalent: a selection of substrates with open-chain allyl moieties is depicted in Scheme 5.4. Simple aryl iodide **19a** was cyclized in high yield (90%), but in a 60 : 40 mixture of exocyclic and endocyclic double-bond isomers (**19a** → **20a**) [21]. A structural modification, conjugation with an ester group, a recurrent motif in Mizoroki–Heck chemistry, allowed for preparation of isomerically pure **20b** in 74% yield (**19b** → **20b**) [22]. In the second example, reaction of pyridine **21** leads to an intermediate of the synthesis of the alkaloid (±)-oxerine (**23**) in useful yield (80%) (**21** → **22**) [23]. No double-bond isomerization was observed in this case, using Pd(OAc)$_2$ and Ph$_3$P in MeCN at 70 °C. Excellent yields (96%) were achieved in the cyclization of naphthalene

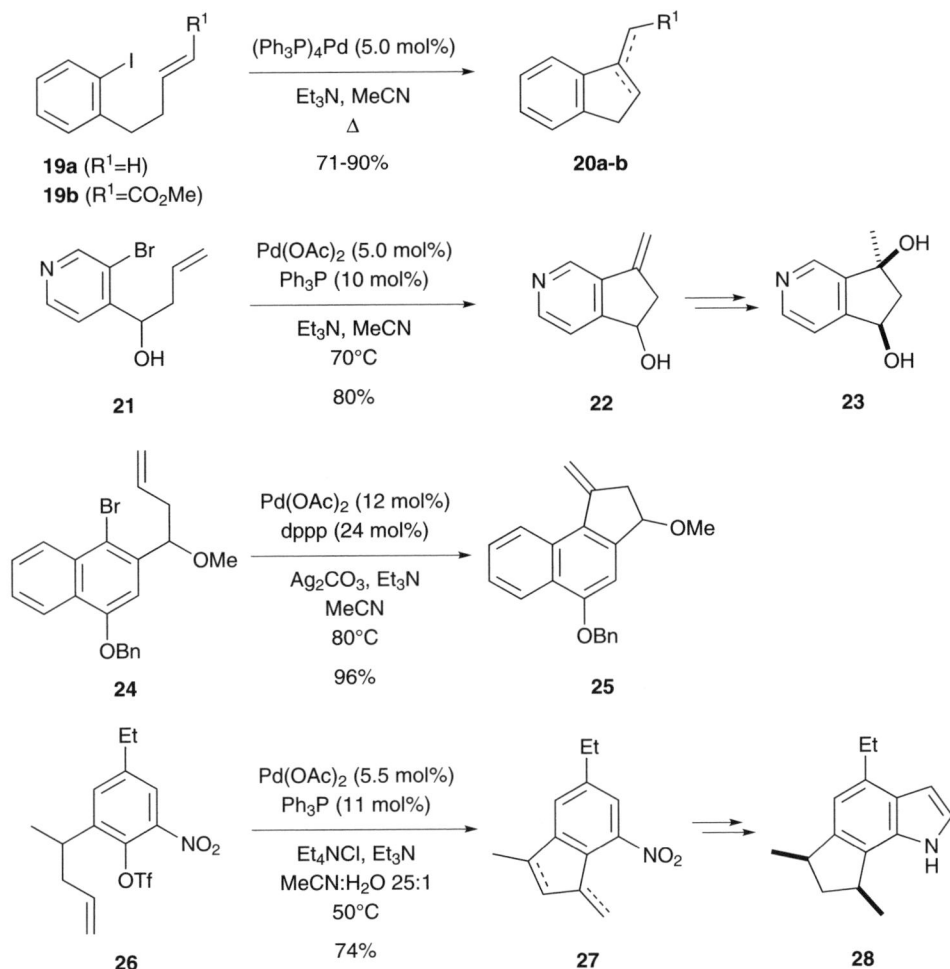

Scheme 5.4

derivative **24** by Boger and Turnbull [24], using a combination of Ag$_2$CO$_3$ and Et$_3$N (**24 → 25**). Cyclization of **26** is the key step in Blechert's (±)-*cis*-trikentrin A (**28**) synthesis and underlines the functional group tolerance of Mizoroki–Heck processes. The product was obtained in 74% yield accompanied by substantial alkene migration; **27** was formed as a 1.3 : 1 mixture of exocyclic to endocyclic double-bond isomers (**26 → 27**) [25].

Furthermore, Kündig *et al.* [26] investigated intramolecular Mizoroki–Heck reactions at planar chiral arene tricarbonyl chromium complexes **29a–c**, giving indanes substituted in the benzylic position in good yield (80% for R^1 = Me, 78% for R^1 = OH and 85% for R^1 = OMe) (**29a–c → 30a–c**, Scheme 5.5).

Examples of Mizoroki–Heck cyclizations forming five-membered carbocycles are scarce for cyclic vinyl halides or triflates, presumably due to the strained reaction products. Grigg *et al.* [27] tested **31**, a cyclic derivative of **4a** (Scheme 5.2), giving regioselectively the

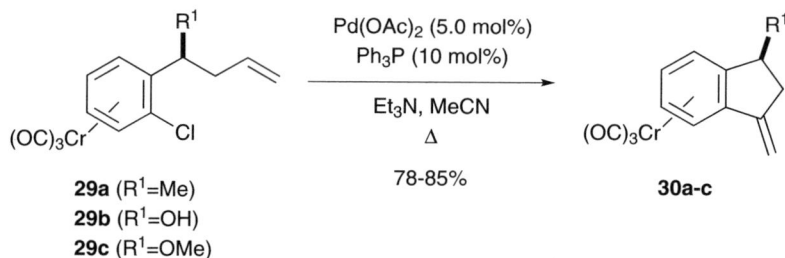

29a (R^1=Me)
29b (R^1=OH)
29c (R^1=OMe)

Scheme 5.5

5-*exo*-trig product **32** in moderate yield (53%) and as a 1.3 : 1 mixture of exo- and endo-cyclic double-bond isomers (**31** → **32**, Scheme 5.6). More complex substrates **33a** and **33b** were utilized by Liang and Paquette [28] to investigate the regiochemical outcome of the β-hydride elimination. **34a** was obtained in an isomerically pure form, with a terminal nonconjugated double-bond in 91% yield (**33a** → **34a**). In the cyclization of the corresponding α,β-unsaturated ester **33b**, the thermodynamically more stable conjugated diene (82%) was favoured over the nonconjugated one (16%) (**33b** → **34b**).

In contrast to the substrate-type presented in Scheme 5.6, intramolecular Mizoroki–Heck reactions with cyclic alkene moieties are quite common. Negishi and coworkers [21, 29] screened numerous substrates with different substitution patterns, out of which four are shown in Scheme 5.7. Cyclization of aryl iodide **35** proceeded well and furnished tricyclic **36** in good yield, including 10% of a double-bond isomer (not shown) (**35** → **36**). Mizoroki–Heck reactions of cyclohexenones **37** and **39** provided 68% and 82% yields respectively and, probably, due to conjugation with the carbonyl group in isomerically pure form (**37**, **39** → **38**, **40**). The two analogous cyclohexenone derivatives of aryl iodide **35** (not shown) cyclized under identical conditions in 50% and 71% yields respectively. Substrate **41a** even allowed for formation of spirocyclic **42a** in good yield, yet with poor

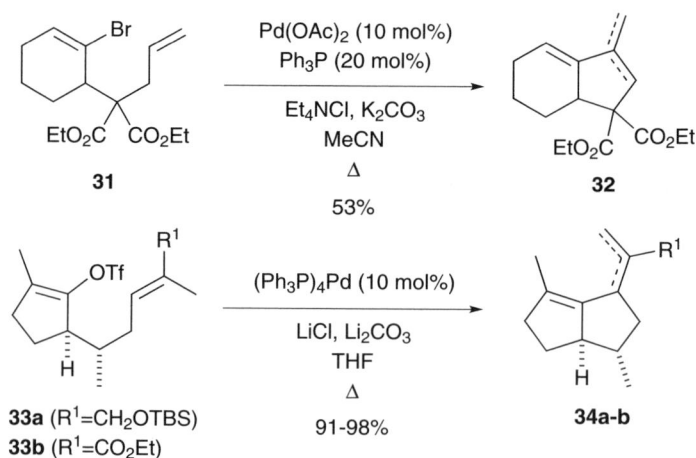

33a (R^1=CH$_2$OTBS)
33b (R^1=CO$_2$Et)

Scheme 5.6

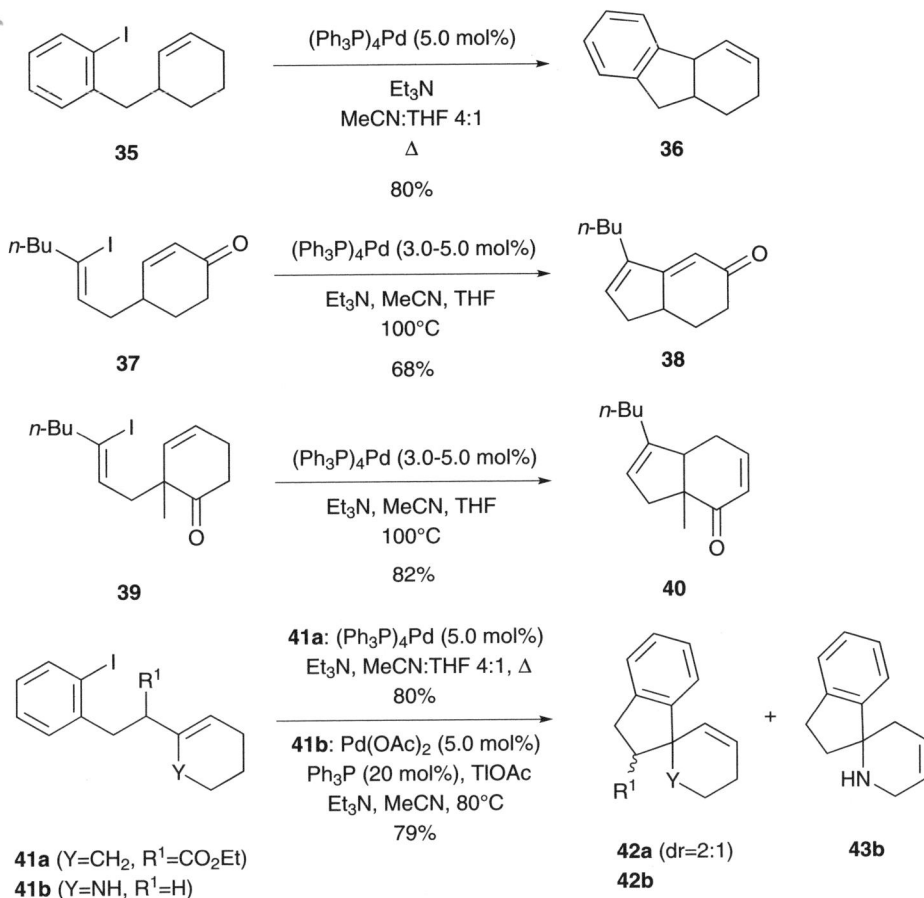

Scheme 5.7

diastereocontrol and significant post-Mizoroki–Heck double-bond migration (1 : 1 mixture, other isomer not shown) (**41a → 42a**). Instead, Ripa and Hallberg [30] attempted to control the alkene migration in the cyclization of enamine **41b** by finely tuning the reaction conditions. Under the reported conditions, post-Mizoroki–Heck double-bond migration was almost completely thwarted and **42b** was formed in good yield (**41b → 42b**). In the absence of any phosphine ligand and TlOAc, with K_2CO_3 as base at $100\,°C$ in MeCN, the reaction became sluggish, but the isomeric enamine of **42b** (not shown) is formed in 54% yield and with fair selectivity. With $Pd(OAc)_2$ (10 mol%), 2,2′-bis(diphenylphosphino)-1,1-binaphthyl (BINAP; 20 mol%) as the ligand and cyclohexene (20 mol%) as an additive, the third isomer **43b** is predominantly formed in moderate yield (54%). Related reactions were accomplished for a substrate with an elongated hydrocarbon tether (not shown), providing the corresponding six-membered cycles in only moderate yields (44–52%).

In general, reactions forming annulated five-membered rings are highly *cis*-selective, due to the strong ring strain in case of *trans*-annulation. This structural constraint also

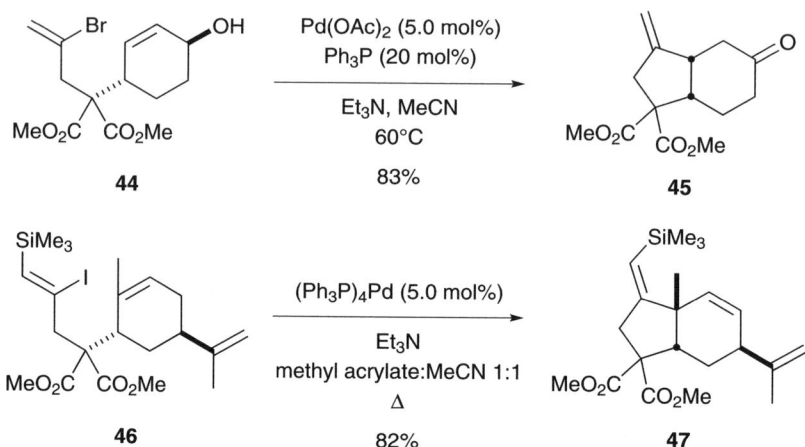

Scheme 5.8

applies to diastereoselective Mizoroki–Heck cyclizations (Scheme 5.8). Intramolecular reaction of allyl alcohol **44** proceeded well, giving diastereomerically pure ketone **45** in good yield (**44** → **45**) [31]. In this particular case, an enol is generated upon β-hydride elimination, which eventually tautomerizes. Mizoroki–Heck cyclization of an open-chain 2-buten-4-oyl-substituted malonate (not shown) furnished at 80 °C the corresponding aldehyde in respectable 60% yield. Reaction of terpene-derived malonate **46** provided the bicyclo[4.3.0]nonane skeleton in good yield under formation of a quaternary stereocenter (**46** → **47**) [32]. Notably, the analogous radical cyclization gave **47** in 75% yield, but as a mixture of *E*- and *Z*-configured vinyl silanes.

Under more drastic reaction conditions, cyclization of **48a** allowed for the formation of a bridged bicyclic framework in 66% yield with K_2CO_3 in 8.6 : 1 : 1.4 ratio of conjugated **49a**, **50a** and an undesired 6-*endo*-trig product with a benzannulated bicyclo[2.2.2]octane core (not shown) (**48a** → **49a**, **50a**, Scheme 5.9) [33]. The 6-*endo*-trig cyclization was thwarted by the additional methyl group in **48b**, but still post-Mizoroki–Heck double-bond migration occurred, furnishing a 7.4 : 5.5 : 1 ratio of **49b**, **50b** and **51b** (**48b** → **49b**, **50b**, **51b**). The double-bond migration was restricted by changing the base from KOAc to

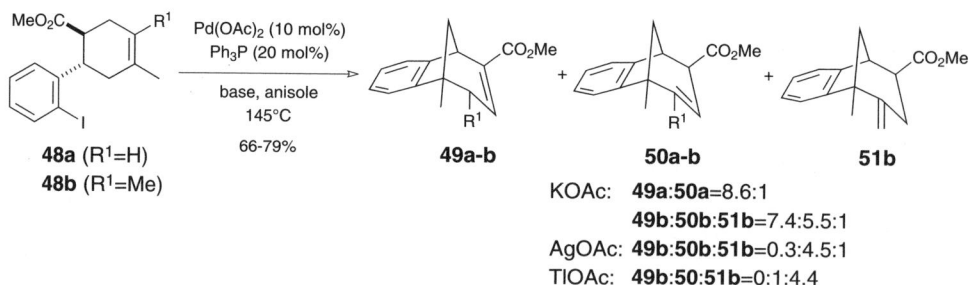

Scheme 5.9

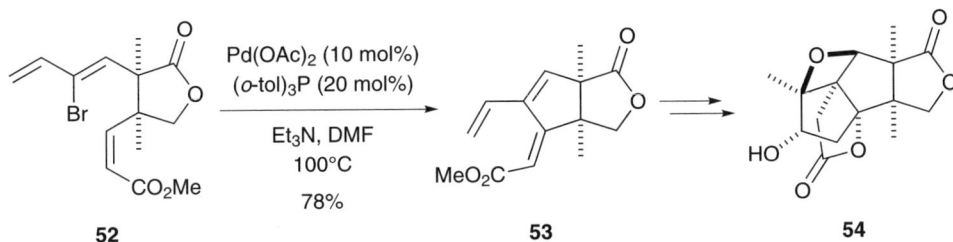

Scheme 5.10

AgOAc, providing a 0.3 : 4.5 : 1 mixture in 79% yield, and was completely prevented with TlOAc, furnishing only **50b** and **51b** in 74% yield and a 1 : 4.4 ratio of isomers.

As already mentioned above, the Mizoroki–Heck reaction tolerates almost any functional group and was, therefore, elegantly applied in numerous complex molecule syntheses (Chapter 16). Fukuyama and coworkers [34] recently disclosed a short synthesis of the AB ring motif towards the synthesis of merrilactone A (**54**), employing the intramolecular Mizoroki–Heck cyclization of **52** as the key step (**52** → **53**, Scheme 5.10). The B ring, possessing a structurally interesting array of conjugated C—C double bonds, is formed in 78% yield and under complete stereocontrol.

During the studies en route to the synthesis of monocerin (**60**), Marsden and coworkers [35] found an intriguing ligand effect on the regiochemical outcome of the Mizoroki–Heck reaction of **55** (**55** → **56**, Scheme 5.11). Under somewhat standard conditions, catalytic in Pd(OAc)$_2$ but stoichiometric in Ph$_3$P in refluxing MeCN, the normally disfavoured (formal) 6-*endo*-trig product **57** was produced in 67% yield exclusively (**55** → **57**). Conversely, the use of 1,1'-bis(diphenylphosphino)ferrocene as ligand provided an access to the desired five-membered ring in good yield (**55** → **56**). The authors assumed that both steric hindrance and, more importantly, the electron-rich nature of **55** account for the unexpected six-membered ring formation, as cyclization of the electron-neutral, less hindered aryl bromide **58** gave **59** as the sole product under identical reaction conditions (**58** → **59**). A mechanism involving 5-*exo*-trig migratory insertion for both substrates is suggested; for **55**, subsequent β-hydride elimination is sterically hampered, while the intermediate

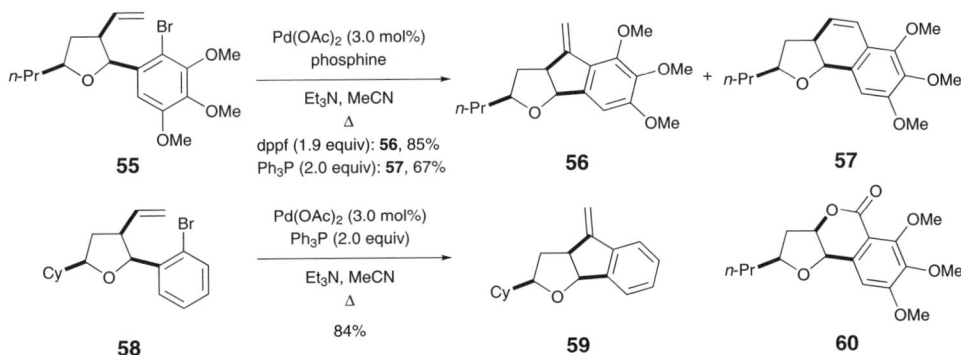

Scheme 5.11

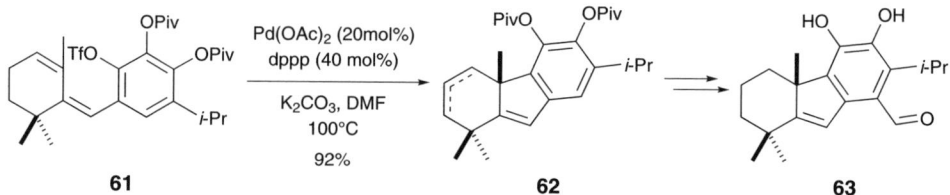

Scheme 5.12

σ-alkylpalladium(II) might attack at the electron-rich arene to form a phenonium ion, which, in turn, rearranges to give the 6-*endo* product.

Triflate **61** is the key intermediate in the total synthesis of (±)-dichroanal B (**63**), recently reported by Node and coworkers [36] (Scheme 5.12). It was cyclized in particularly high yield (92%), forming a quaternary carbon, but as a mixture of inseparable double-bond isomers (**61** → **62**). These were selectively hydrogenated in the presence of the tertiary double bond afterwards. A cognate test substrate with an unsubstituted phenyl triflate moiety (not shown) also provided its tricyclic product in 92% yield.

The following example stems from an expeditious formal (±)-aphidicolin (**66**) synthesis by Fukumoto and coworkers [37, 38]. The requisite reaction conditions were defined after a screening of the analogous dehydroxylated precursor (not shown), cyclizing in 86% yield. The bridged bicyclo[3.2.1]octane core is efficiently constructed by Mizoroki–Heck reaction of **64** in high yield and without any post-Mizoroki–Heck double-bond migration (**64** → **65**, Scheme 5.13).

5.3   Formation of Carbocycles by 5-*endo*-trig Cyclization

According to Baldwin's rules [4], 5-*endo*-trig cyclizations are normally disfavoured; nevertheless, this reaction mode can be realized by Mizoroki–Heck reactions. Although they were outshone by the many 5- and 6-*exo*-type transformations, quite a number of examples following this cyclization mode were reported. This was mostly for the construction of five-membered heteroaromatic compounds (Chapter 6), but in these cases an alternative mechanistic scenario sometimes seems likely [39]. However, a few syntheses of carbocycles via (formal) 5-*endo*-trig/dig Mizoroki–Heck cyclization were disclosed as well.

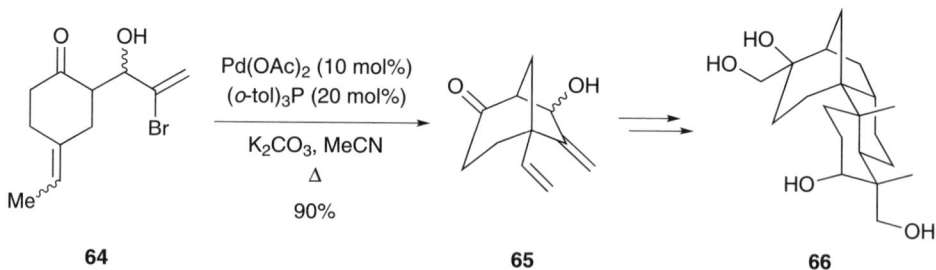

Scheme 5.13

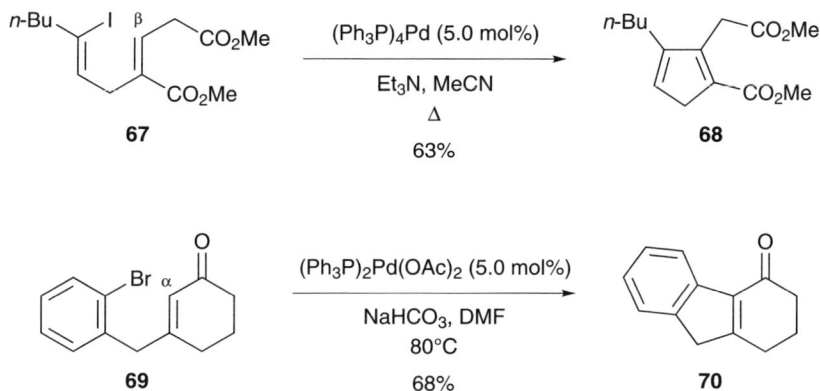

Scheme 5.14

Cyclopentadiene **68** was formed in 63% yield by the Mizoroki–Heck cyclization of open-chain vinyl iodide **67** under classic reaction conditions (**67** → **68**, Scheme 5.14) [22]. The analogous unsubstituted aryl iodide (not shown) provided a comparable yield (65%). Reaction in the β-position of the α,β-unsaturated carbonyl or carboxyl compound is not mandatory, as intramolecular Mizoroki–Heck reaction of **69** also proceeded well, forming tricyclic ketone **70** in 68% yield (**69** → **70**).

It is interesting to note that the substitution pattern of nonactivated alkene functions also has only minor influence on the 5-*endo*-trig Mizoroki–Heck cyclization (Scheme 5.15). The Mizoroki–Heck reaction of alkene **71** without terminal substitution proceeded well and provided indanone **72** in 65% yield (**71** → **72**) [40]. Furthermore, terminal substitution is tolerated; *E*-configured **73** was cyclized in moderate yield (52%) (**73** → **74**) [31], while the *Z*-isomer of **73** (not shown) furnished **74** in 48% yield. The products of this pair **72** and **74** possess keto groups, which originate from tautomeric enol intermediates generated by β-hydride elimination.

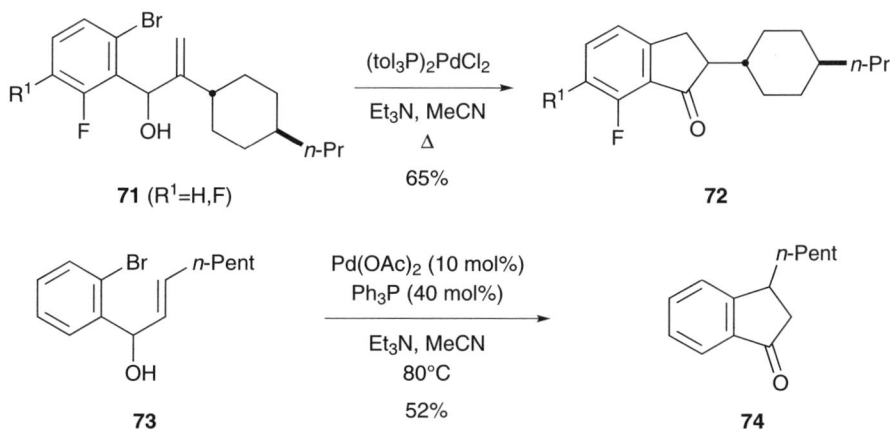

Scheme 5.15

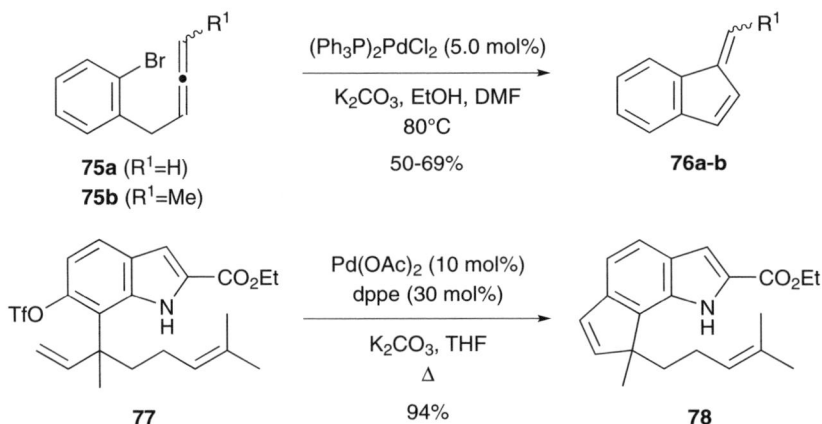

Scheme 5.16

Scheme 5.16 presents two conceptually interesting examples of Mizoroki–Heck reactions. In the first one, Ma and Negishi [41] showed that allenes are also suitable for intramolecular Mizoroki–Heck reactions: **75a** was cyclized in a 5-*endo*-dig mode, providing benzofulvene (**76a**) in 69% yield (**75a** → **76a**, Scheme 5.16). Its cognate, **75b**, gave 50% yield. The second was recently reported by Tanner and coworkers [42], showing for the first time, that a 5-*endo*-trig pathway was preferred over a 6-*exo*-trig-pathway in a cyclization precursor containing two different alkene units (**77** → **78**). Furthermore, cyclization of **77** occurred in remarkably high yield (94%).

Ichikawa *et al.* [43] reported a reaction involving a β-fluoride elimination, which is briefly mentioned in Chapter 6.

5.4   Formation of Carbocycles by 6-*exo*-trig Cyclization

The 6-*exo*-trig mode cyclization is – next to the 5-*exo*-trig mode – most commonly seen in intramolecular Mizoroki–Heck chemistry and has been studied in all imaginable facets. It has, therefore, had major impact on complex molecule synthesis.

When investigating the first syntheses of five-membered carbocycles (Section 5.2), Grigg *et al.* [8, 9] at the same time also revealed Mizoroki–Heck cyclizations forming the structurally related six-membered ring systems. Cyclization of 2-bromo-1,7-octadiene **79** gave **80** with a 4 : 1 ratio of exocyclic to endocyclic double-bond isomers in 86% yield, and without any 7-*endo*-trig product (**79** → **80**, Scheme 5.17). Reaction of **81** was less successful, only providing 66% yield and a 2.5 : 1 ratio of double-bond isomers (**81** → **82**). Both products **80** and **82** were later successfully employed in Diels–Alder reactions [9]. In analogy to the synthesis of isomerically pure cyclopentanes (Section 5.2, Scheme 5.2), Grigg *et al.* [15] tested the 2,7-dibromo derivative of **79** (not shown), but in this case Mizoroki–Heck reaction was sluggish and **80** was obtained in only 28% yield. In the cyclization of **83**, the fully conjugated product **84** was formed exclusively in excellent yield (91%) (**83** → **84**) [22]. Replacement of the vinyl iodide by an aryl iodide

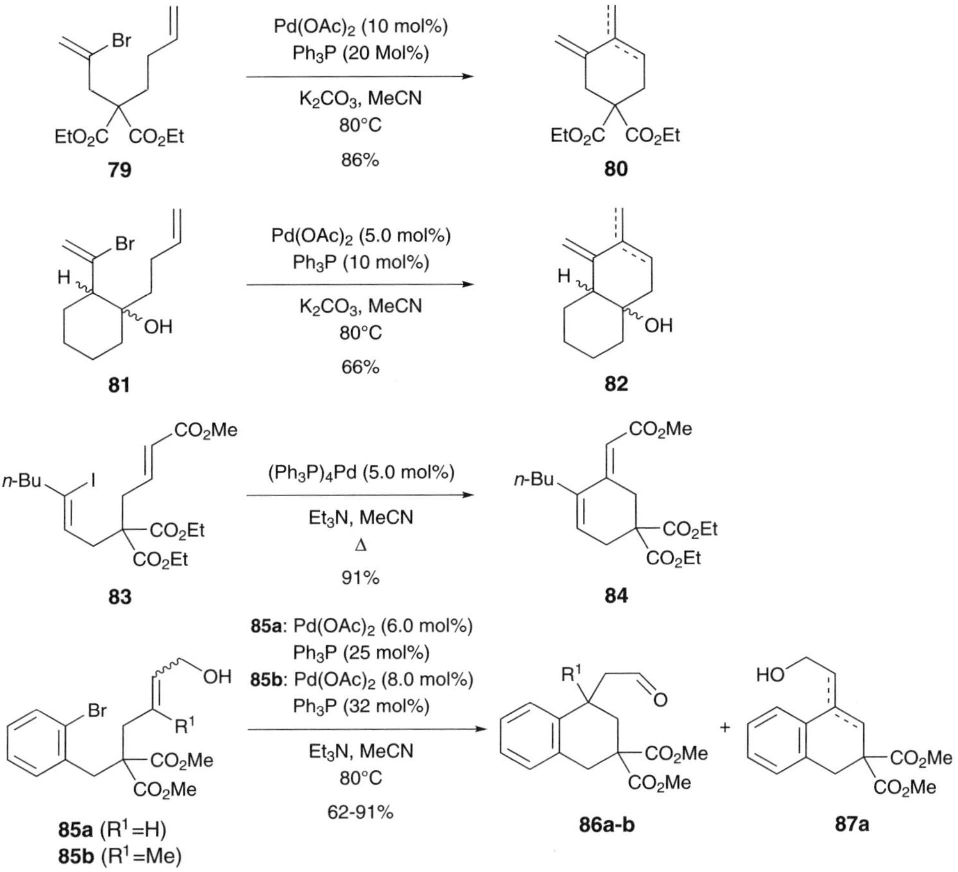

Scheme 5.17

moiety (not shown) resulted in a small decrease of the yield; only 82% was obtained. Intramolecular Mizoroki–Heck reaction of **85a** was unselective and resulted in a 1 : 1 : 1 mixture of aldehyde **86a** and two further double-bond isomers **87a**, but in high overall yield (91%) (**85a → 86a, 87a**) [31]. However, in the cyclization of cognate **85b**, a quaternary carbon is formed, which inhibits double-bond migration (**85b → 86b**). Consequently, **86b** is the sole product formed, in 62% yield.

Tietze and Modi [44] undertook an interesting approach to control the position of the double bond, envisioning a trimethylsilyl group to direct β-hydride elimination. Triflate **88a** with a simple alkene terminus reacted unselectively, giving a 1.4 : 1 regioisomeric mixture of **89a** and **90a** in 80% yield (**88a → 89a, 90a**, Scheme 5.18). In contrast, cyclization of **88b** provided vinyl silane **89b** exclusively in 90% yield, yet without control of the alkene geometry ($E : Z = 60 : 40$) (**88b → 89b**). The $E : Z$ selectivity was enhanced to 96 : 4 by replacing triflate in **88b** by iodide (not shown), but then **89b** was only formed in moderate yield (56%). Tietze and Modi also investigated the formation of an analogous five- (93%, $E : Z = 90 : 10$) and seven-membered ring (61%, $E : Z = 53 : 47$), by changing

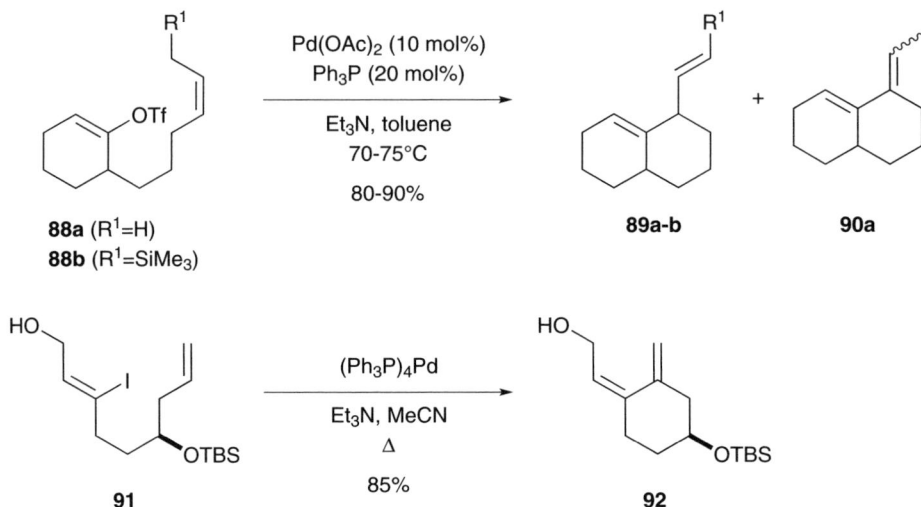

Scheme 5.18

the length of the tether in **88b** (not shown). Mizoroki–Heck cyclization of precursor **91** was explored by Granja and coworkers [45] and provided exclusively diene **92** in high yield (**91 → 92**).

Intramolecular Mizoroki–Heck reaction of **93** was used by Shimizu and coworkers [46] for the synthesis of the A ring synthon **94** of $1\alpha,25$-dihydroxyvitamin D_3 (**95**), the hormonally active form of vitamin D_3 (**93 → 94**, Scheme 5.19). The six-membered ring was formed in excellent yield (86%) and with the required configuration of the acrylate fragment. Later, Chen and Crich [47] investigated the *E*-isomer of **93** (not shown), which was cyclized in 92% yield in the presence of $(Ph_3P)_4Pd$ (5 mol%). Alibés and coworkers [48] envisioned Mizoroki–Heck reaction of **96**, which was realized in 88% yield (**96 → 97**), as a key step in the synthesis of tetracyclic (−)-allonorsecurinine (**98**). A cognate piperidine derivative **99** was cyclized in 78% yield, providing an access to securinine (**101**).

En route to the synthesis of mensacarcin (**104**), a polyfunctionalized hexahydroanthracene, showing cytostatic and cytotoxic activity, Tietze *et al.* [49] devised a Mizoroki–Heck cyclization for the formation of the tricyclic core. Several substrates with different protecting groups and substitution patterns were tested, out of which **102** turned out to be the best, affording **103** in 94% yield under optimized reaction conditions (**102 → 103**, Scheme 5.20). A similar strategy was pursued by Banerjee and coworkers [50] for the synthesis of tetrahydroanthracenes leading to umbrosone (**107**). Variation of the substituents in **105** had a minor effect, the reactions proceeded smoothly in 84–86% yield (**105 → 106**), and subsequent elimination of H_2O provided the tetrahydroanthracene core. When diene **108** was employed, the aromatic system in **109** was directly installed by double-bond migration (**108 → 109**).

In Mizoroki–Heck cyclizations, aromatic annulation is usually achieved starting from 1,2-substituted arenes. In addition, it is also possible to realize ring closures having other geometrical relationships (Scheme 5.21). Cyclization of aryl bromide **110** provided tricyclic

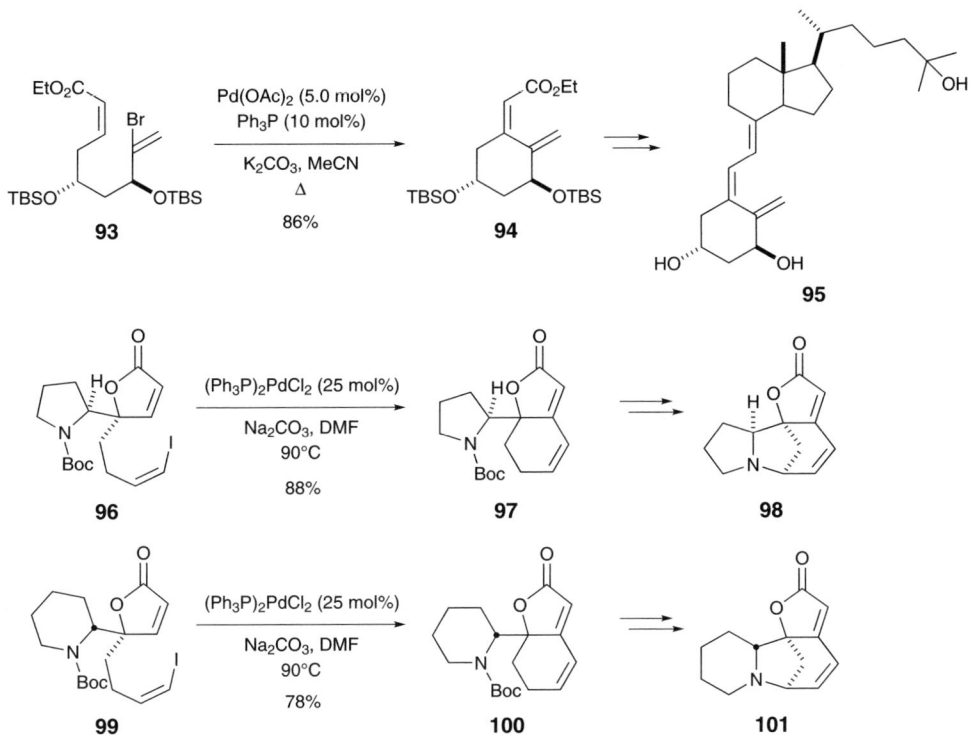

Scheme 5.19

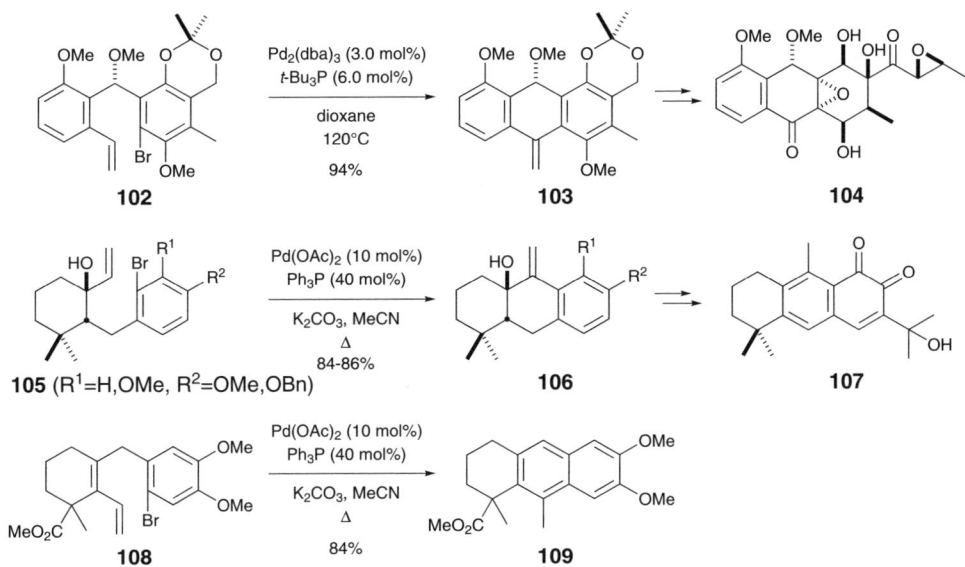

Scheme 5.20

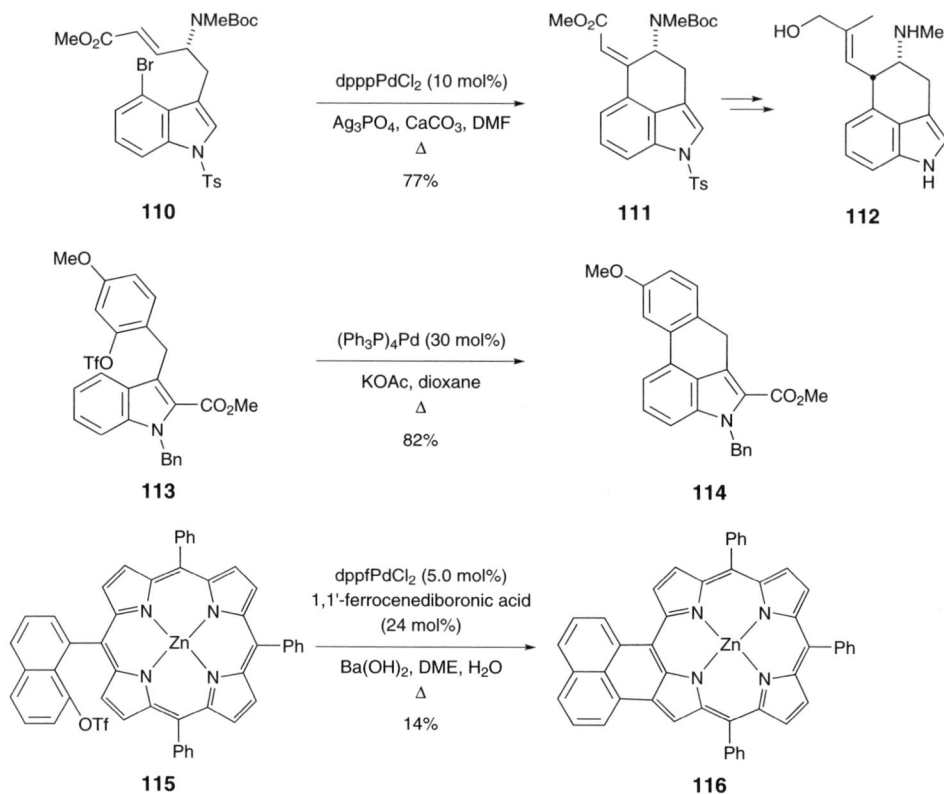

Scheme 5.21

111, an intermediate en route to chanolavine-I (**112**), in 77% yield (**110** → **111**) [51]. It is noteworthy that this structural motif can be constructed by an inverse approach, in which the aromatic ring functions as the alkene coupling partner (NB: intramolecular C–H bond activation might also be a reaction pathway [52]). Triflate **113** was cyclized in a remarkable 82% yield (**113** → **114**); its analogous aryl bromide (not shown) gave **114** in the presence of (Ph₃P)₄Pd (15 mol%) in 74% yield [53]. Formal Mizoroki–Heck cyclization of **115** seems to follow an unusual, modified reaction mechanism, as 1,1′-ferrocenediboronic acid is an essential additive to obtain any porphyrin derivative **116** (**115** → **116**) [54].

Naturally, 6-*exo*-trig Mizoroki–Heck cyclization was widely used in a diastereoselective sense within many complex molecule syntheses; for example, in the total synthesis of (−)-guanacastepene E (**119a**) and (−)-heptemerone B (**119b**) recently reported by Trauner and coworkers [55] (Scheme 5.22). Intramolecular Mizoroki–Heck reaction of enantiomerically enriched **117** enabled under Jeffery's conditions [56] the formation of the six-membered cycle in 75% yield and installed the quaternary stereocenter of **118** with good diastereoselectivity (dr = 5.1 : 1). Mizoroki–Heck reaction of allyl alcohol **120** was employed by Trost and Corte [57] to construct the 2,4-*cis*-disubstituted 3,3-dimethylcyclohexane ring in **121**, the central part of highly cytotoxic saponaceolides (**120** → **121**): aldehyde **121** was obtained in good yield (75%), 2.4 : 1 *cis* : *trans* ratio and was applied in the total synthesis

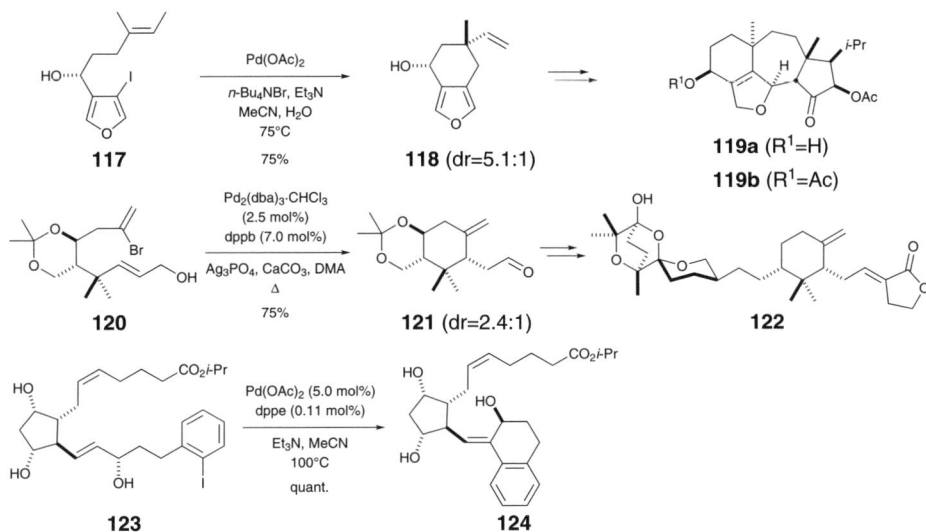

Scheme 5.22

of (+)-saponaceolide B (**122**). Hacksell and coworkers [58] utilized diastereoselective Mizoroki–Heck reaction of **123** in the preparation of substituted prostaglandin $F_{2\alpha}$ analogues. Allyl alcohol **124** was formed in diastereomerically pure form and excellent yield along with 20% of the isomeric homoallyl alcohol (not shown) (**123** → **124**).

The first example in Scheme 5.23 stems from the (±)-duocarmycin SA (**127**) synthesis by Natsume and coworkers [59]: cyclization of **125** provided the expected product **126** in 82% yield along with 11% the isomeric enamine (not shown) (**125** → **126**, Scheme 5.23). In the Mizoroki–Heck reaction of **128**, an eight-membered ring was exclusively *trans*-annulated in 64% yield, resembling a steroid-like polycyclic structure (**128** → **129**) [45, 60].

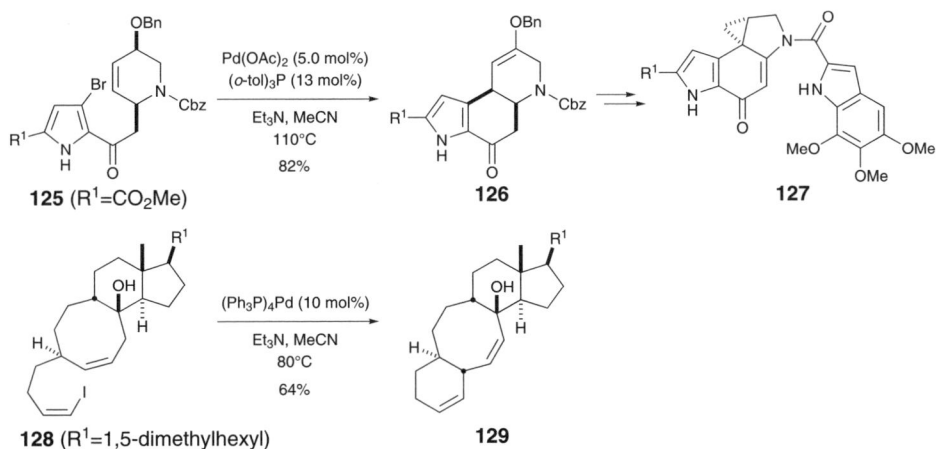

Scheme 5.23

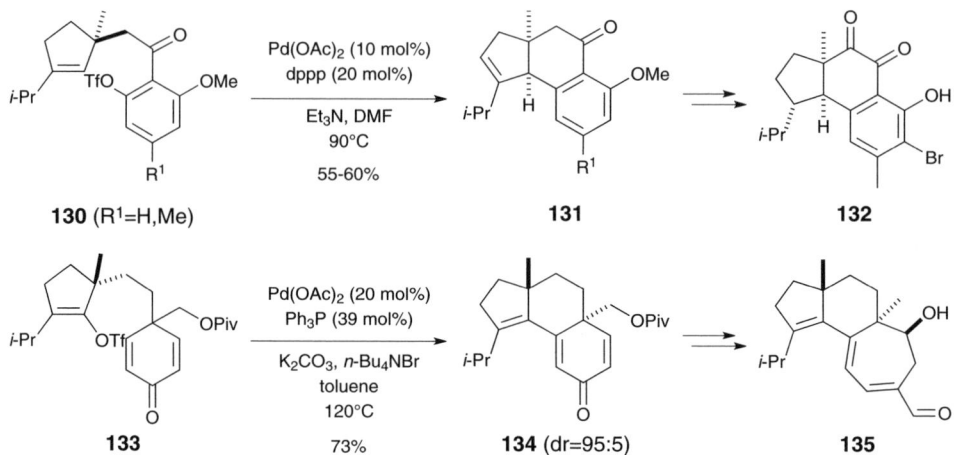

Scheme 5.24

Within the total synthesis of 6-*epi*-(−)-hamigeran B (**132**), Mehta and Shinda [61] investigated the Mizoroki–Heck cyclization of aryl triflate **130** (**130** → **131**, Scheme 5.24). Ring formation occurred completely *cis* selective, providing **131** in good yield along with 30% of deoxygenated starting material. A different approach to a structurally very similar tricycle was realized recently by Desmaële and coworkers [62, 63] during synthetic studies on cyathin terpenoids; for example, allocyathin B$_2$ (**135**). Cyclohexadienone **134** was formed in 73% yield and with high diastereoselection (**133** → **134**).

Intramolecular Mizoroki–Heck reactions forming a *cis*-fused decalin system have been thoroughly explored by Overman and coworkers [64] in connection with studies towards the total synthesis of complex cardenolides such as (−)-ouabagenin (**140**). In a model reaction, vinyl triflate **136** was cyclized in excellent yield and entirely *cis*-selective, forming a quaternary setereocenter (**136** → **137**, Scheme 5.25) [65]. Product **137** was obtained in a 3 : 1 mixture of double-bond isomers. Highly functionalized precursor **138a** provided, under optimized conditions, the steroid **139a** in 90% yield with complete diastereoselection (**138a** → **139a**) [66]. The related thioether **138b** furnished pentacyclic **139b**, possessing the delicate functionalization at C-11, still in 70% yield (**138b** → **139b**) [67].

Mizoroki–Heck reaction of **141** was used by Overman and coworkers [68] in the synthesis of a phenolic aphidicolin analogue (**143**), affording the spirocyclic product **142** in good yield (67%), but with mediocre diastereoselectivity (**141** → **142**, Scheme 5.26). Further spirocyclic products (not shown) were synthesized by Ripa and Hallberg [30] in moderate yields (44–52%) (discussed in Section 5.2).

Aside from the aforementioned demanding examples, the 6-*exo*-trig Mizoroki–Heck cyclization also allows for the formation of bridged polycyclic compounds (Scheme 5.27). Vinyl iodide **144** was cyclized in good yield and produced **145** as a 9 : 1 mixture of double-bond isomers (minor isomer not shown) (**144** → **145**) [21]. The analogous aryl iodide (not shown) furnished in the presence of 5 mol% (Ph$_3$P)$_4$Pd the product in 81% yield, but in a 7 : 3 mixture of double-bond isomers. Coe *et al.* [69] used the Mizoroki–Heck reaction

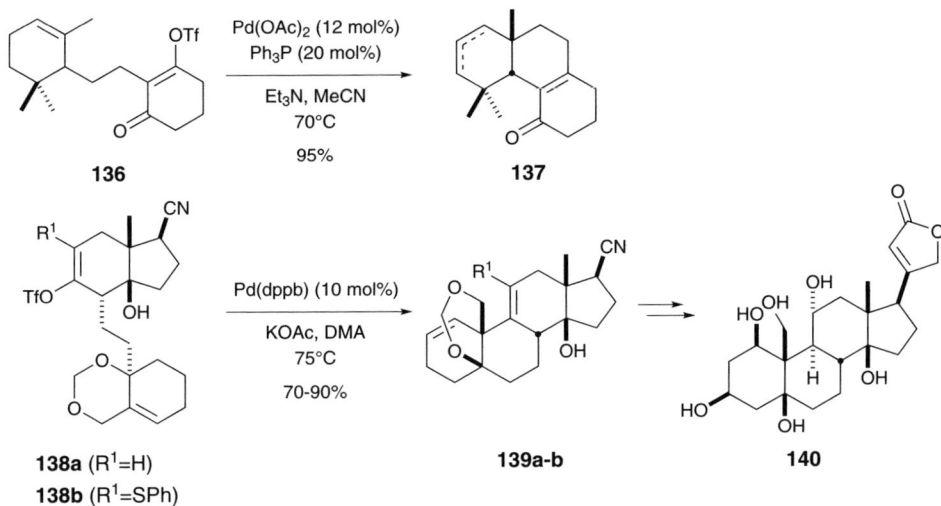

136 → **137**

Pd(OAc)$_2$ (12 mol%)
Ph$_3$P (20 mol%)

Et$_3$N, MeCN
70°C

95%

138a (R^1=H)
138b (R^1=SPh)

Pd(dppb) (10 mol%)

KOAc, DMA
75°C

70-90%

139a-b

140

Scheme 5.25

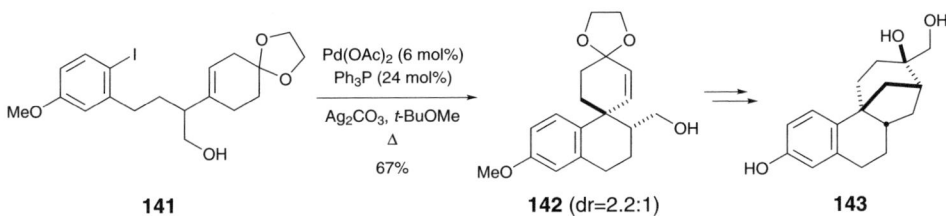

141

Pd(OAc)$_2$ (6 mol%)
Ph$_3$P (24 mol%)

Ag$_2$CO$_3$, *t*-BuOMe
Δ

67%

142 (dr=2.2:1)

143

Scheme 5.26

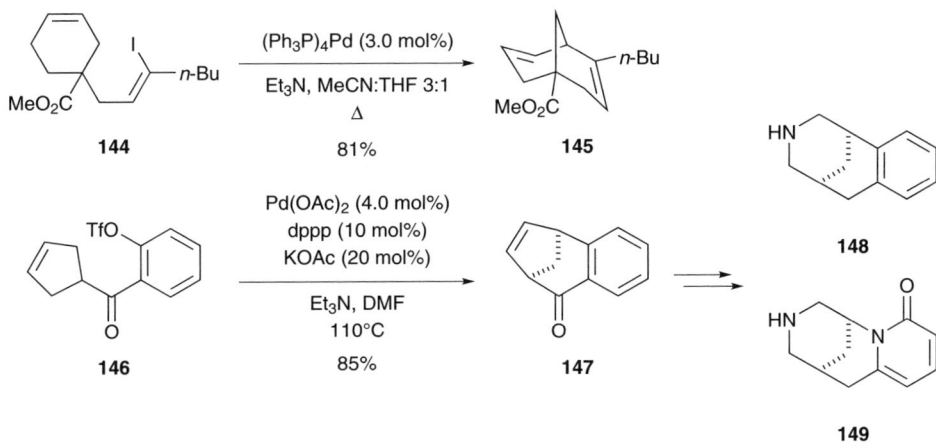

144

(Ph$_3$P)$_4$Pd (3.0 mol%)

Et$_3$N, MeCN:THF 3:1
Δ

81%

145

148

146

Pd(OAc)$_2$ (4.0 mol%)
dppp (10 mol%)
KOAc (20 mol%)

Et$_3$N, DMF
110°C

85%

147

149

Scheme 5.27

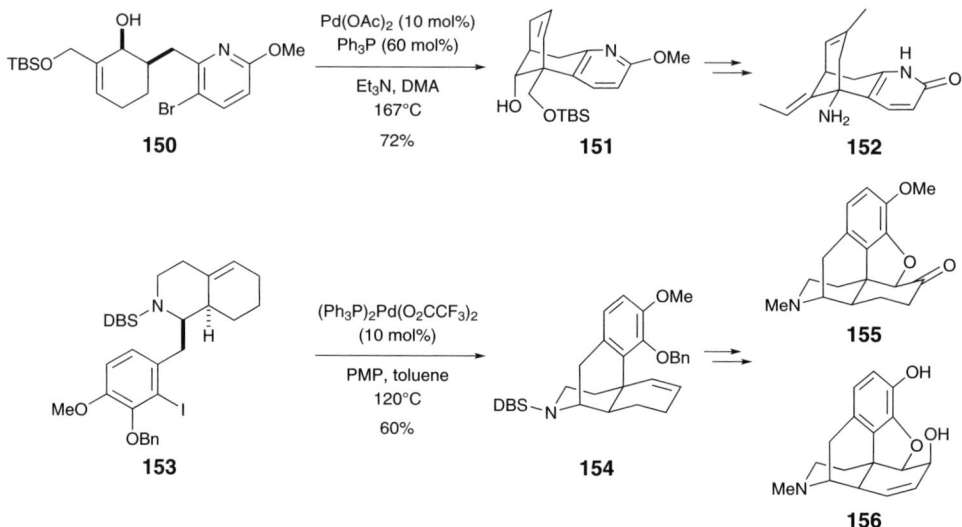

Scheme 5.28

of **146** for the synthesis of a carbon analogue **148** of (−)-cytisine (**149**) (**146** → **147**); the tricyclic ketone **147** was obtained in 85% yield.

For the formal total synthesis of huperzine A (**152**), an acetylcholinesterase inhibitor and potential drug for Alzheimer's disease, several substrates with different substitution patterns were tested by Mann and coworkers [70, 71]. The best results were obtained with **150**, as its cyclization selectively follows the planned 6-*exo*-trig and not a competing 7-*endo*-trig pathway (**150** → **151**, Scheme 5.28). The product **151** was produced in good yield (72%) and without any alkene isomerization [71]. Reaction of the epimeric secondary alcohol (not shown) was equally selective, but gave only 57% yield. Overman and coworkers [72] utilized the Mizoroki–Heck cyclization of **153** as key step in the total synthesis of (−)-dihydrocodeinone (**155**) and (−)-morphine (**156**) (**153** → **154**). The bridged tetracyclic product **154** was elegantly constructed in 60% yield under formation of a quaternary stereocenter.

5.5   Formation of Carbocycles by 6-*endo*-trig (dig) Cyclization

Intramolecular Mizoroki–Heck reactions following a (formal) 6-*endo*-trig pathway are, as with the 5-*endo*-trig cyclizations (see above), not very common; this circumstance is usually rationalized by Baldwin's rules [4]. Apart from occasional side-reactions of 5-*exo*-trig-type cyclizations (see Section 5.2, Scheme 5.2), there were also several selective transformations reported.

For instance, Mizoroki–Heck reaction of vinyl iodide **157** furnished exclusively **158** in 70% yield in a 7:3 mixture of double-bond isomers (**157** → **158**, Scheme 5.29) [21]. The substitution pattern of **159** is quite similar to **7** (Scheme 5.2) and, likewise, the six-membered ring was formed exclusively (**159** → **160**) [31]. **160** was obtained in excellent

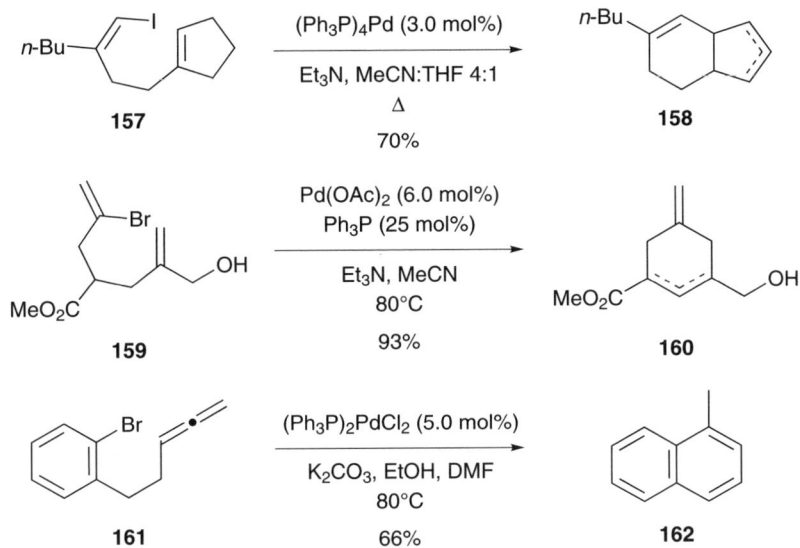

Scheme 5.29

yield with a 1 : 1 ratio of double-bond isomers. As allenes performed well in 5-*endo*-dig cyclizations (see Scheme 5.16), **161** was also a suitable substrate for 6-*endo*-dig cyclization; in this case, α-methylnaphthalene (**162**) was formed in 66% yield after double-bond migration (**161** → **162**) [41].

However, a straightforward 6-*endo*-trig cyclization is at least sometimes in question or might even be ruled out. As demonstrated by Negishi and coworkers [13] in the reactions of **163** and **165** (Scheme 5.30), Mizoroki–Heck cyclization of **163** proceeded well, forming

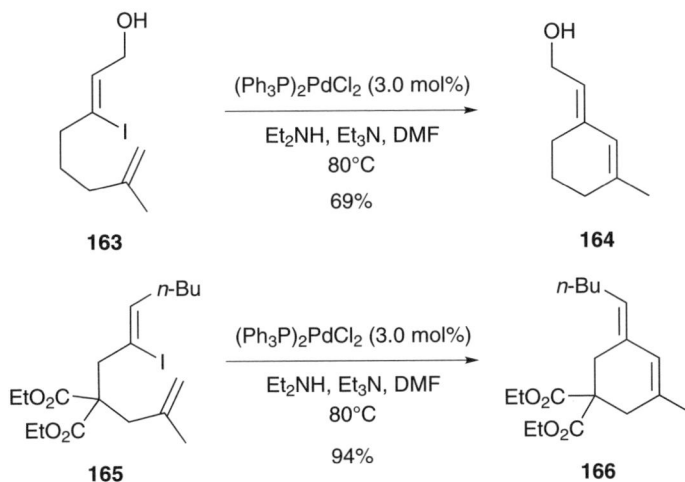

Scheme 5.30

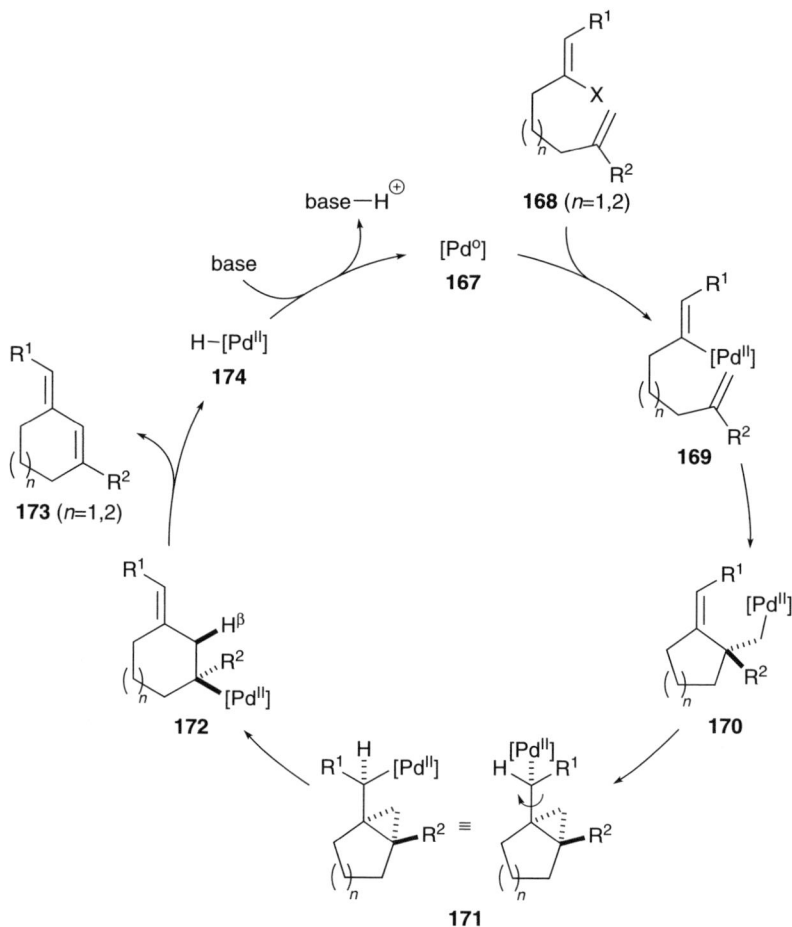

Scheme 5.31

164 in 69% yield, but with inverted (!) geometry at the former vinyl iodide (**163** → **164**), which cannot be reconciled with the classic mechanism of Mizoroki–Heck reactions (see Chapter 1). In analogy to **163**, reaction of **165** provided **166** in high yield (94%) and changed double bond geometry (**165** → **166**).

In agreement with related observations [13], these authors proposed an alternative cascade mechanism for the 6- and 7-*endo*-trig cyclizations outlined in Scheme 5.31: vinyl palladium species **169**, generated from **168** by oxidative addition of catalyst **167**, undergoes a 5-*exo*-trig cyclization to give **170** (**169** → **170**). If β-hydride elimination is not possible or disfavoured, then another migratory insertion will form a cyclopropane (**170** → **171**), which then, after internal rotation, will undergo a cyclopropylcarbinyl-to-homoallayl rearrangement (**171** → **172**). Subsequent β-hydride elimination releases the product **173** and reductive elimination closes the catalytic cycle (**172** → **174** → **167**).

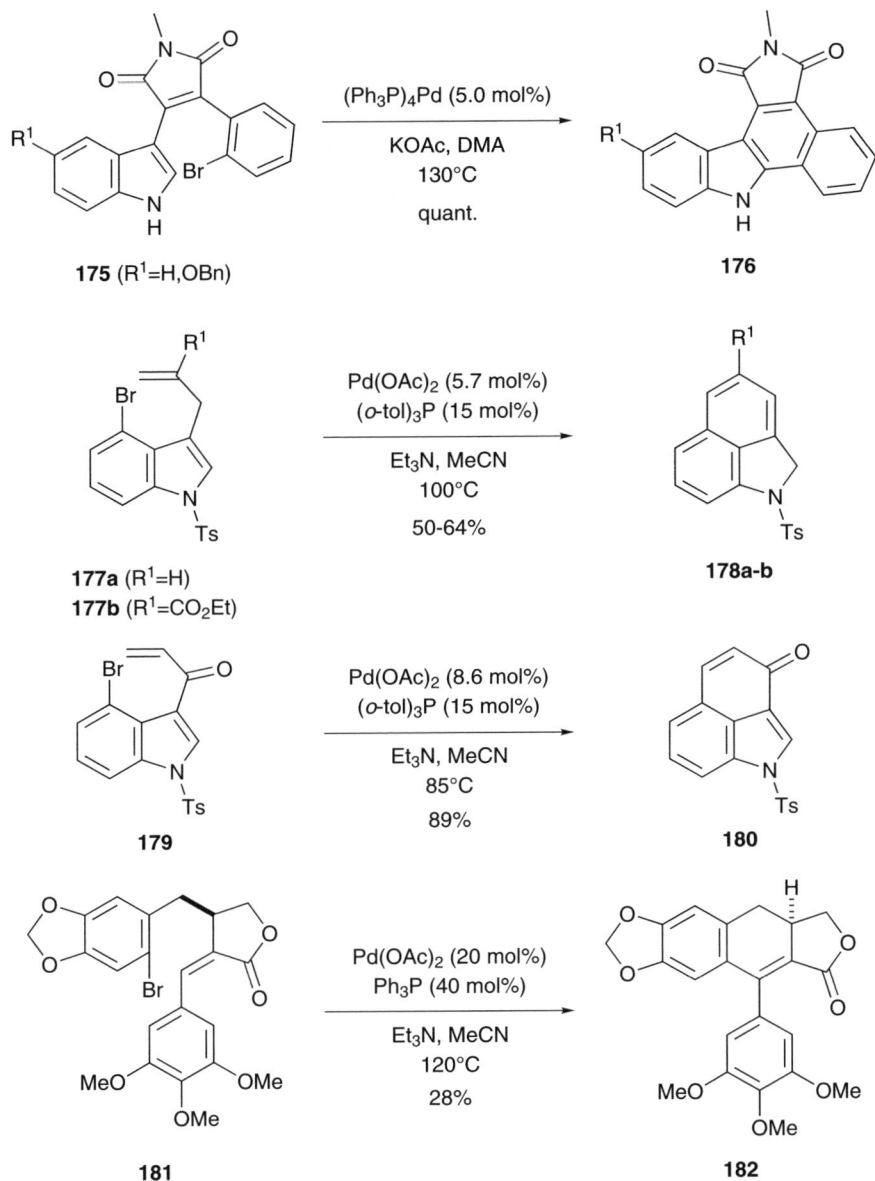

175 (R^1=H,OBn)

176

177a (R^1=H)
177b (R^1=CO$_2$Et)

178a-b

179

180

181

182

Scheme 5.32

The 6-*endo*-trig Mizoroki–Heck cyclization was also used in a few syntheses of more complex systems, especially for the formation of phenylcarbazols of type **176**, which are potential anticancer agents (Scheme 5.32). In the cyclization of aryl bromide **175**, a new aromatic ring is formed which might allow for high yields (**175** → **176**) [73]. In addition, 2-naphthyl triflates [74], as well as naphthyl bromides, 5-methoxyphenyl bromide and thiophenyl bromides [75] with varied substitution patterns of the indole moiety, were also successfully used as substrates (not shown). Hegedus *et al.* [76] envisioned Mizoroki–Heck cyclization of **177** and **179** for the synthesis of 3,4-substituted indoles. Reaction of **177a** provided **178a** in 50% yield, and cyclization of cognate acrylate **177b** furnished 64% yield (**177a–b** → **178a–b**). In both cases, isomerization to the corresponding naphthalene occurred, whereas reaction of ketone **179** selectively provided indole **180** in high yield (89%) (**179** → **180**). Ishibashi *et al.* [77] used the intramolecular Mizoroki–Heck reaction of **181** for the synthesis of podophyllotoxin derivatives **182**. Though **182** was the sole product, the isolated yield was only low (28%) (**181** → **182**).

5.6   Formation of Carbocycles by 7-*exo*-trig (dig) Cyclization

Aside from the common formation of five- and six-membered cycles, the intramolecular Mizoroki–Heck reaction also proved to be a reliable tool for the formation of medium-sized rings.

For instance, Negishi *et al.* [78] studied the Mizoroki–Heck reaction of **183** towards the synthesis of (±)-7-*epi*-β-bulnesene (**210**) (see Scheme 5.36): the product **184** was obtained without double bond isomerization in 80% yield (**183** → **184**, Scheme 5.33).

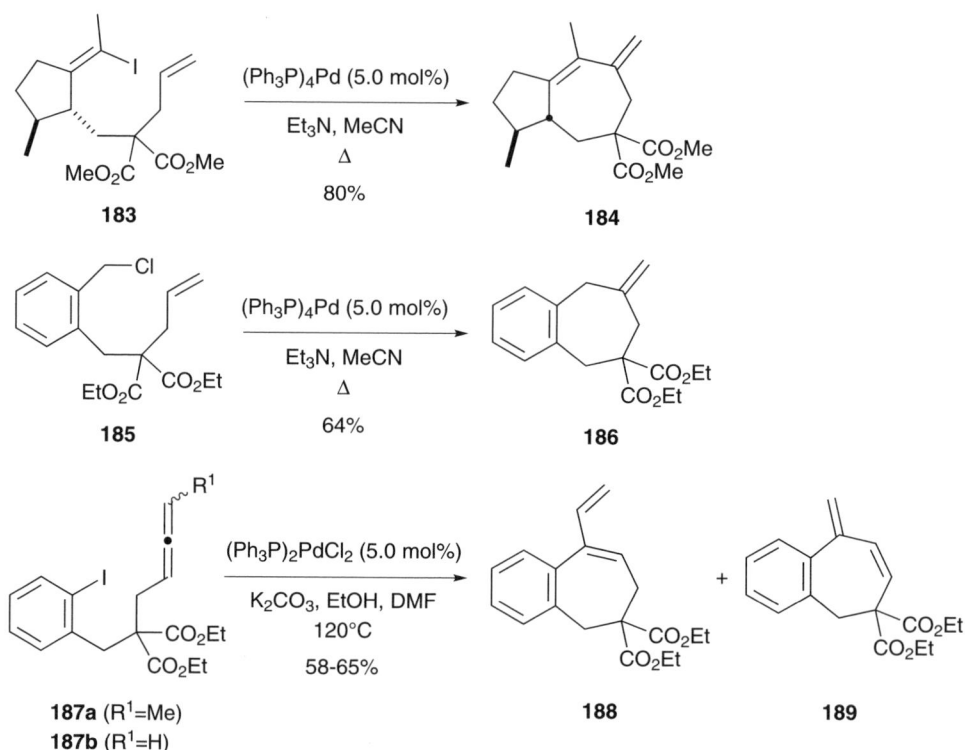

Scheme 5.33

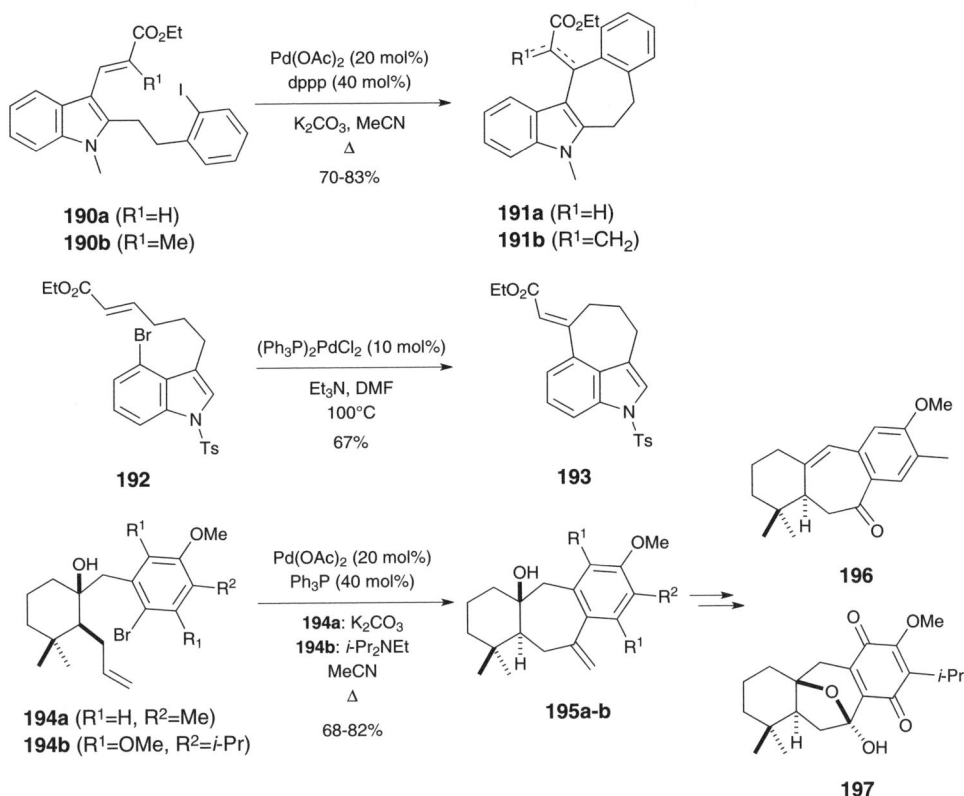

Scheme 5.34

As for the smaller ring sizes, substrate scope was extended to benzyl chlorides, with **185** being cyclized in 64% yield (**185** → **186**) [19], and to allenes, with reaction of **187a** giving product **188** in 58% yield via 7-*exo*-dig cyclization (**187a** → **188**) [41]. Without the terminal methyl group, selective 7-*endo*-dig cyclization occurred, giving **189** in 65% yield (**187b** → **189**). Analogous vinyl iodides of **187b** (not shown) were also tested, cyclizing in comparable yields.

Mizoroki–Heck reaction of **190a** provided annulated indole **191a** under optimized conditions in 83% yield and was not accompanied by partial 8-*endo*-trig cyclization (**190a** → **191a**, Scheme 5.34) [79, 80]. Under identical conditions, **190b** cyclized in high yield (70%); β-hydride elimination or subsequent alkene migration was selectively steered towards the less substituted double bond (**190b** → **191b**). Synthesis of tricyclic 3,4-disubstituted indole derivative **193** was accomplished by Mizoroki–Heck reaction of aryl bromide **192** in 67% yield (**192** → **193**) [81]; a corresponding radical cyclization failed for this substrate. Banerjee and coworkers [82] utilized intramolecular Mizoroki–Heck reaction of **194a** within the synthesis of (±)-faveline methyl ether (**196**). Styrene **195a** was obtained as the sole product in good yield (82%) (**194a** → **195a**). The highly crowded substrate **194b** required more forcing reaction conditions, but tricyclic **195b**, an intermediate in the

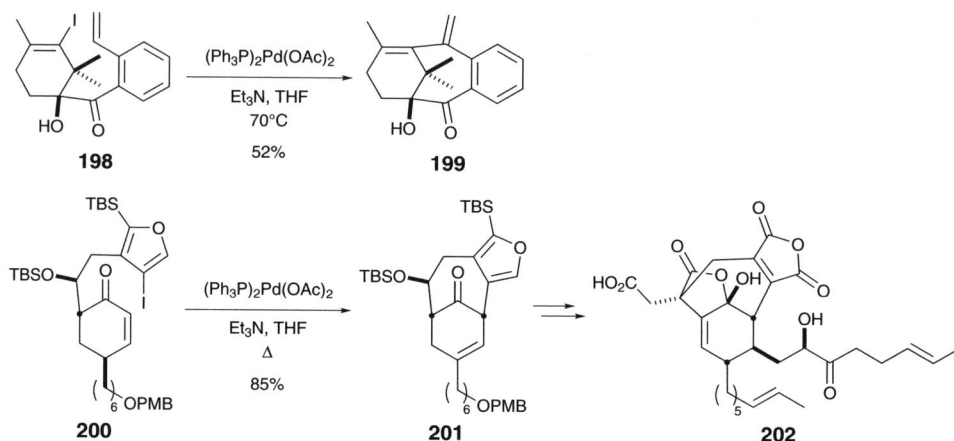

Scheme 5.35

(±)-komaroviquinone (**197**) synthesis, was still produced in 68% yield (**194b** → **195b**). A synthesis of estradiol derivatives by 7-*exo*-trig cyclization by Tietze *et al.* [83] is discussed in context with its homologue (see Section 5.8, Scheme 5.37).

Mizoroki–Heck cyclization of **198** was part of the studies towards the total synthesis of taxol (**224**) (see Scheme 5.38) by Danishefsky and coworkers [84] (**198** → **199**, Scheme 5.35). The strained seven-membered ring is formed in the presence of the hydroxyketone moiety in 52% yield and unconsumed **198** was re-isolated quantitatively. Danishefsky and coworkers [85] employed the high-yielding Mizoroki–Heck reaction of **200** to assemble the bridged carbon skeleton in **201**, an intermediate en route to the synthesis of CP-225,917 (**202**) and related CP-263,114 (not shown), which are squalene synthetase and farnesyl transferase inhibitors (**200** → **201**).

5.7   Formation of Carbocycles by 7-*endo*-trig Cyclization

Although intramolecular Mizoroki–Heck cyclizations via a 7-*endo*-trig reaction pathway are extremely rare, this reaction type was recently applied very elegantly as the pivotal step in the total synthesis of (+)-guanacastepene N (**207**) (Scheme 5.36) [86]. Within the studies towards complex cardenolides such as ouabagenin (**140**) by Overman and coworkers [66, 67], a 6-*exo*/3-*exo* cascade cyclization sequence [13] was envisioned in the reaction of **203**, but the cyclization occurred selectively in a 7-*endo*-mode, forming **204** in 70% yield (**203** → **204**). The obvious similarity of the ABC ring system of **204** with the guanacastepene skeleton led to the design of **205**, which was cyclized in remarkably high yield (75%) (**205** → **206**). The corresponding methyl ester (not shown) provided even 83% yield [86]. Negishi *et al.* [78] used cyclization of **208** in the synthesis of (±)-7-*epi*-β-bulnesene (**210**); the seven-membered ring was formed in 53% yield (**208** → **209**). In analogy to formal 6-*endo*-trig Mizoroki–Heck cyclization of **163** and

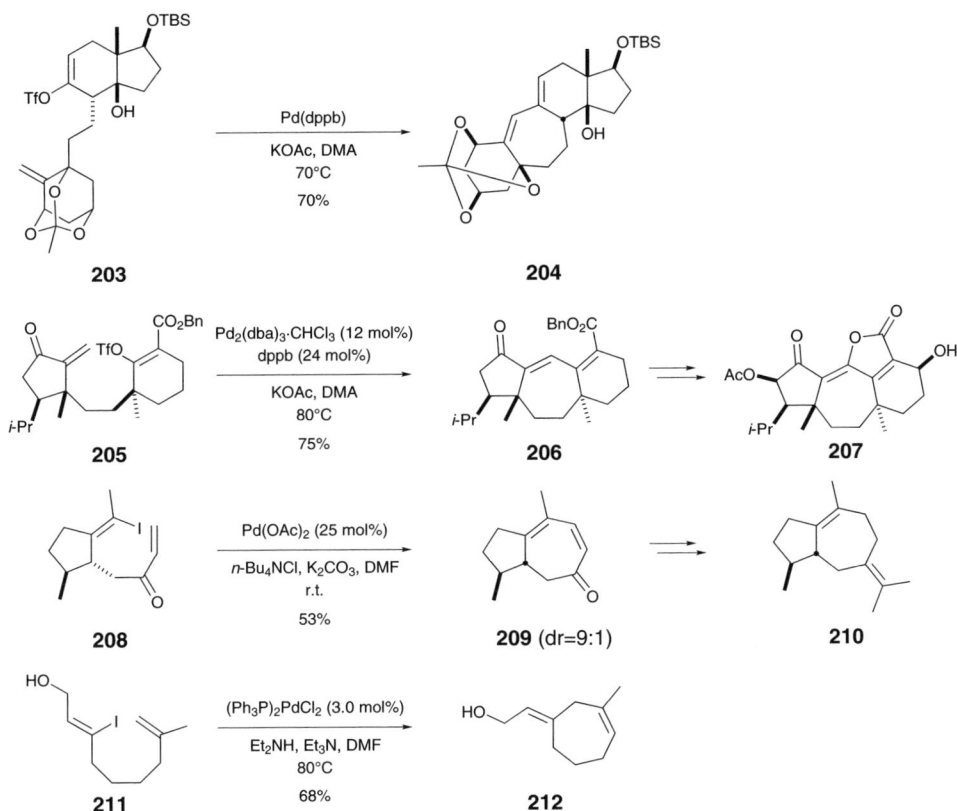

Scheme 5.36

165, the cascade reaction mechanism was proposed for the cyclization of **211** as well (see Section 5.5, Schemes 5.30 and 5.31), which occurred in 68% yield under complete inversion of double-bond geometry (**211** → **212**) [13]. Despite 7-*endo*-trig cyclization, there is a single study containing 7-*endo*-dig cyclizations, which were already discussed in Scheme 5.33 (Section 5.6) [41].

5.8 Formation of Eight-Membered and Larger Carbocycles

The intramolecular Mizoroki–Heck reaction also allows for the synthesis of medium-sized and even large rings; the latter were usually synthesized under high dilution conditions or on a solid support (Chapter 14).

Tietze *et al.* [83] were interested in the synthesis of medium-ring-sized estradiol derivatives such as **214** (Scheme 5.37). Mizoroki–Heck reaction of **213a** in the presence of the Herrmann–Beller catalyst proceeded well and provided **214a** without 8-*endo*-trig side-product in high yield (80%) and a 4 : 1 ratio of exocyclic to endocyclic (not shown)

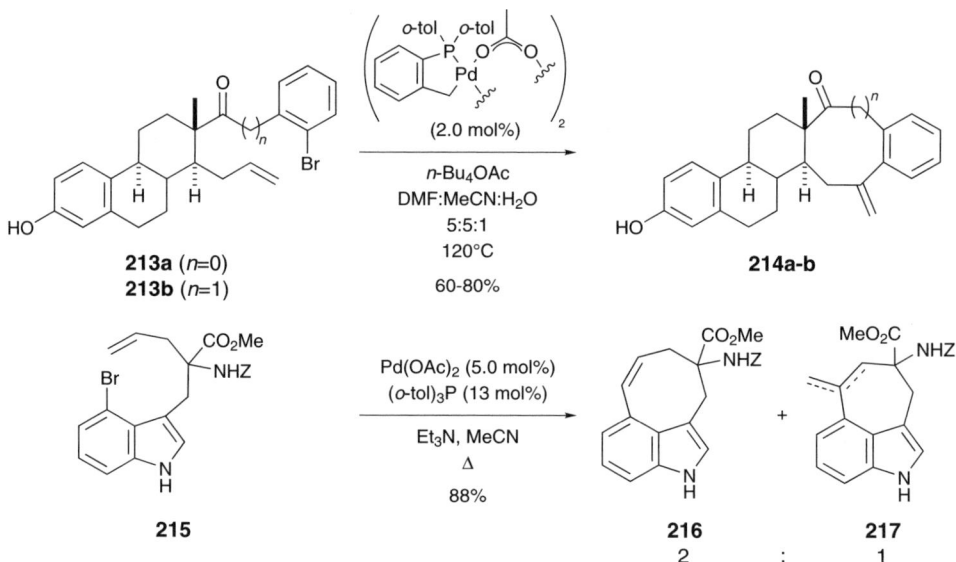

Scheme 5.37

double-bond isomers (**213a** → **214a**). Homologous **214b** was obtained in 60% yield along with 14% of the 9-*endo*-trig side-product (not shown) (**213b** → **214b**). Cyclization precursor **215** was utilized by Roberts and coworkers [87] for the synthesis of conformationally constrained tryptophanes (**215** → **216**). The regioselectivity of the reaction was low, due to the competing 7-*exo*-trig cyclization; thus, desired **216** was furnished in a 2 : 1 mixture with **217**, but still in excellent overall yield.

In the early 1990s, taxol (**224**) was, owing to its challenging tetracyclic structure and significant cytotoxic properties, the Holy Grail in natural product synthesis. En route to its synthesis, Danishefsky and coworkers [84, 88, 89] tested several substrates with varying substitution patterns for the demanding ring closure of the central eight-membered ring (Scheme 5.38). The promising cyclization of **218** occurred strictly in an 8-*exo* sense in 80% yield, whereupon the alcohol moiety was oxidized (**218** → **219**) [84]. Curiously enough, the analogous *cis*-acetal (not shown) did not undergo Mizoroki–Heck cyclization at all. The structurally related triethylsilylethers **220a** and **220b**, the latter with an oxygenated aryl moiety, gave 70% and 78% yields respectively (**220a–b** → **221a–b**) [89]. Finally, intramolecular Mizoroki–Heck reaction of highly functionalized **222** was applied in the total synthesis of taxol (**224**), forming the tetracyclic core in 49% yield (**222** → **223**) [89, 90]. A related Mizoroki–Heck reaction for the synthesis of a cholesterol–baccatin III hybrid **227** gave 50% yield under similar conditions (**225** → **226**) [91].

Ma and Negishi [41] also extended the scope of *endo*-selective Mizoroki–Heck reactions of allenes to the formation of medium-sized rings and macrocycles. A selection of representative reactions is depicted in Scheme 5.39. Cyclizations of precursors **228a** and **228b** provided the corresponding eight- and nine-membered rings in 56% and 62% yields

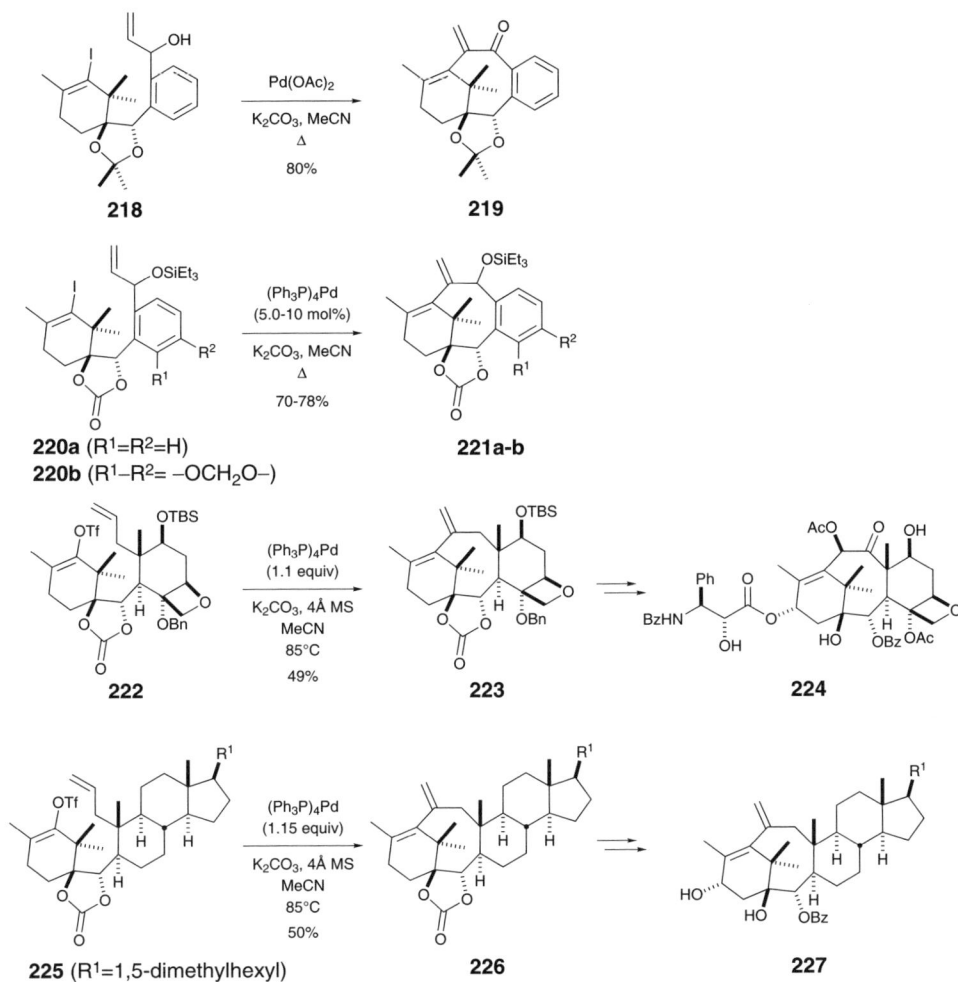

Scheme 5.38

respectively, under high dilution conditions $(2 \times 10^{-3}\,\mathrm{mol\,l^{-1}})$ (**228a–b** → **229a–b**). Substrates **228c** and **228d** reacted in 48% and 40% yields respectively, forming 10- and 11-membered carbocycles (**228c–d** → **229c–d**). Reaction of **230a** formed a 12-membered ring in 50% yield (**230a** → **231a**), and cyclization of **230b** a 20-membered macrocycle in remarkable 86% yield (**230b** → **231b**). Replacing the allenyl moiety by an allyl group in **230b** (not shown) allowed for 21-*endo*-trig cyclization in 66% yield.

Dyker and Grundt [92] used the Mizoroki–Heck reaction of **232** for the synthesis diketone **233**, which provides access to the steroid skeleton via double transannular cyclization (**232** → **233**, Scheme 5.40). The 13-membered macrocycle was formed in 61% yield along with 17% of a dimer (not shown).

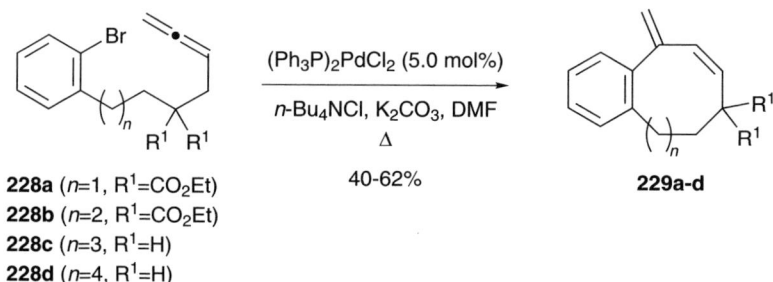

228a (*n*=1, R^1=CO$_2$Et)
228b (*n*=2, R^1=CO$_2$Et)
228c (*n*=3, R^1=H)
228d (*n*=4, R^1=H)

(Ph$_3$P)$_2$PdCl$_2$ (5.0 mol%)

n-Bu$_4$NCl, K$_2$CO$_3$, DMF
Δ

40–62%

229a-d

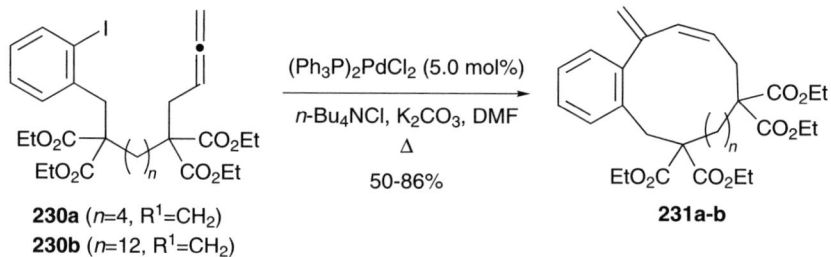

230a (*n*=4, R^1=CH$_2$)
230b (*n*=12, R^1=CH$_2$)

(Ph$_3$P)$_2$PdCl$_2$ (5.0 mol%)

n-Bu$_4$NCl, K$_2$CO$_3$, DMF
Δ

50–86%

231a-b

Scheme 5.39

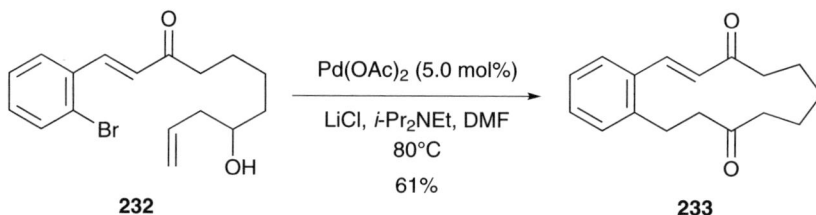

232

Pd(OAc)$_2$ (5.0 mol%)

LiCl, *i*-Pr$_2$NEt, DMF
80°C

61%

233

Scheme 5.40

References

1. Link, J.T. (2002) The intramolecular Heck reaction, in *Organic Reactions*, Vol. **60** (ed. L.E. Overman), John Wiley & Sons Inc., New York, pp. 157–534.
2. Bräse, S. and de Meijere, A. (2002) Synthesis of carbocycles, in *Handbook of Organopalladium Chemistry for Organic Synthesis*, Vol. **1** (eds E.-i. Negishi and A. de Meijere), John Wiley & Sons Inc., New York, pp. 1223–54.
3. Mori, M., Chiba, K. and Ban, Y. (1977) The reactions and syntheses with organometallic compounds. V. A new synthesis of indoles and isoquinolines by intramolecular palladium-catalyzed reactions of aryl halides with olefinic bonds. *Tetrahedron Lett.*, 1037–40.
4. Baldwin, J.E. (1976) Rules for ring closure. *J. Chem. Soc., Chem. Commun.*, 734–6.
5. (a) Bräse, S. (1999) Synthesis of bis(enolnonaflates) and their 4-*exo*-trig-cyclizations by intramolecular Heck reactions. *Synlett*, 1654–6; (b) Innitzer, A., Brecker, L. and Mulzer, J. (2007) Functionalized cyclobutanes via Heck cyclization. *Org. Lett.*, **9**, 4431–4.
6. Terpko, M.O. and Heck, R.F. (1979) Rearrangement in the palladium-catalyzed cyclization of α-substituted *N*-acryloyl-*o*-bromoanilines. *J. Am. Chem. Soc.*, **101**, 5281–3.

7. Odle, R., Blevins, B., Ratcliff, M. and Hegedus, L.S. (1980) Conversion of 2-halo-*N*-allylanilines to indoles via palladium(0) oxidative addition – insertion reactions. *J. Org. Chem.*, **45**, 2709–10.
8. Grigg, R., Stevenson, P. and Worakun, T. (1984) Rhodium- and palladium-catalysed formation of conjugated mono- and bis-exocyclic dienes. 5-*exo*-trig versus 6-*endo*-trig cyclisation. *J. Chem. Soc., Chem. Commun.*, 1073–5.
9. Grigg, R., Stevenson, P. and Worakun, T. (1988) The regioselectivity of rhodium- and palladium-catalysed cyclisations of 2-bromo-1,6- and 1,7-dienes. *Tetrahedron*, **44**, 2033–48.
10. Meyer, F.E., How Ang, K., Steinig, A.G. and de Meijere, A. (1994) Sequential Heck and Diels–Alder reactions: facile construction of bicyclic systems in a single synthetic operation. *Synlett*, 191–3.
11. Lemaire-Audoire, S., Savignac, M., Dupuis, C. and Genêt, J.-P. (1996) Intramolecular Heck-type reactions in aqueous medium. Dramatic change in regioselectivity. *Tetrahedron Lett.*, **37**, 2003–6.
12. How Ang, K., Bräse, S., Steinig, A.G., *et al.* (1996) Versatile synthesis of bicyclo[4.3.0]nonenes and bicyclo[4.4.0]decenes by a domino Heck–Diels–Alder reaction. *Tetrahedron*, **52**, 11503–28.
13. Owczarczyk, Z., Lamaty, F., Vawter, E.J. and Negishi, E.-i. (1992) Apparent *endo*-mode cyclic carbopalladation with inversion of alkene configuration via *exo*-mode cyclization–cyclopropanation–rearrangement. *J. Am. Chem. Soc.*, **114**, 10091–2.
14. Grigg, R., Stevenson, P. and Worakun, T. (1985) Palladium catalysed intra- and intermolecular coupling of vinyl halides. Regiospecific formation of 1,3-dienes. *J. Chem. Soc., Chem. Commun.*, 971–2.
15. Grigg, R., Stevenson, P. and Worakun, T. (1988) Regiospecific formation of 1,3-dienes by the palladium catalysed intra- and intermolecular coupling of vinyl halides. *Tetrahedron*, **44**, 2049–54.
16. Moreno-Mañas, M., Pleixats, R. and Roglans, A. (1995) Ethyl *N*-(diphenylmethylene)glycinate as anionic glycine equivalent transition metal mediated preparation of bicyclic and tricyclic α,α-disubstituted α-amino acids and derivatives. *Liebigs Ann.*, 1807–14.
17. Møller, B. and Undheim, K. (1998) Cyclic α-amino acids by Pd-mediated cycloisomerization and coupling reactions. *Tetrahedron*, **54**, 5789–804.
18. Nguefack, J.-F., Bolitt, V. and Sinou, D. (1997) Palladium-mediated cyclization on carbohydrate templates. 1. Synthesis of enantiopure bicyclic compounds. *J. Org. Chem.*, **62**, 1341–7.
19. Wu, G., Lamaty, F. and Negishi, E.-i. (1989) Palladium-catalyzed cyclization of benzyl halides and related electrophiles containing alkenes and alkynes as a novel route to carbocycles. *J. Org. Chem.*, **54**, 2507–8.
20. Firmansjah, L. and Fu, G.C. (2007) Intramolecular Heck reactions of unactivated alkyl halides. *J. Am. Chem. Soc.*, **129**, 11340–1.
21. Negishi, E.-i., Zhang, Y. and O'Connor, B. (1988) Efficient synthesis of carbocyclic and carbopolycyclic compounds via intramolecular carbopalladation catalyzed by palladium–phosphine complexes. *Tetrahedron Lett.*, **29**, 2915–8.
22. O'Connor, B., Zhang, Y., Negishi, E.-i. *et al.* (1988) Palladium-catalyzed cyclization of alkenyl and aryl halides containing α,β-unsaturated carbonyl groups via intramolecular carbopalladation. *Tetrahedron Lett.*, **29**, 3903–6.
23. Zhao, J., Yang, X., Jia, X. *et al.* (2003) Novel total syntheses of (±)-oxerine by intramolecular Heck reaction. *Tetrahedron*, **59**, 9379–82.
24. Boger, D.L. and Turnbull, P. (1998) Synthesis and evaluation of a carbocyclic analogue of the CC-1065 and doucarmycin alkylation subunits: role of the vinylogous amide and implications on DNA alkylation catalysis. *J. Org. Chem.*, **63**, 8004–11.
25. Wiedenau, P., Monse, B. and Blechert, S. (1995) Total synthesis of (±)-*cis*-trikentrin A. *Tetrahedron*, **51**, 1167–76.

26. Kündig, E.P., Ratni, H., Crousse, B. and Bernardinelli, G. (2001) Intramolecular Pd-catalyzed carbocyclization, Heck reactions, and aryl-radical cyclizations with planar chiral arene tricarbonyl chromium complexes. *J. Org. Chem.*, **66**, 1852–60.

27. Grigg, R., Sridharan, V., Stevenson, P. *et al.* (1990) The synthesis of fused ring nitrogen heterocycles via regiospecific intramolecular Heck reactions. *Tetrahedron*, **46**, 4003–18.

28. Liang, S. and Paquette, L.A. (1992) Palladium-catalyzed intramolecular cyclization of vinyl and aryl triflates. Associated regioselectivity of the β-hydride elimination step. *Acta Chem. Scand.*, **46**, 597–605.

29. Zhang, Y., O'Connor, B. and Negishi, E.-i. (1988) Palladium-catalyzed procedures for [3 + 2] annulation via intramolecular alkenylpalladation and arylpalladation. *J. Org. Chem.*, **53**, 5588–90.

30. Ripa, L. and Hallberg, A. (1996) Controlled double-bond migration in palladium-catalyzed intramolecular arylation of enamidines. *J. Org. Chem.*, **61**, 7147–55.

31. Gaudin, J.-M. (1991) Intramolecular Heck reaction with substrates possessing an allylic alcohol moiety. *Tetrahedron Lett.*, **32**, 6113–6.

32. Apte, S., Radetich, B., Shin, S. *et al.* (2004) Silylstannylation of highly functionalized acetylenes. Synthesis of precursors for annulations via radical or Heck reactions. *Org. Lett.*, **6**, 4053–6.

33. Grigg, R., Santhakumar, V., Sridharan, V. *et al.* (1991) The synthesis of bridged-ring carbo- and heterocycles via palladium catalysed regiospecific cyclisation reactions. *Tetrahedron*, **47**, 9703–20.

34. Harada, K., Kato, H. and Fukuyama, Y. (2005) Synthetic studies toward merrilactone A: a short synthesis of AB ring motif. *Tetrahedron Lett.*, **46**, 7407–10.

35. Cassidy, J.H., Farthing, C.N., Marsden, S.P. *et al.* (2006) A concise, convergent total synthesis of monocerin. *Org. Biomol. Chem.*, **4**, 4118–26.

36. Planas, L., Mogi, M., Takita, H. *et al.* (2006) Efficient route to 4a-methyltetrahydrofluorenes: a total synthesis of (±)-dichroanal B via intramolecular Heck reaction. *J. Org. Chem.*, **71**, 2896–8.

37. Toyota, M., Nishikawa, Y. and Fukumoto, K. (1994) An expeditious and efficient formal synthesis of (±)-aphidicolin. *Tetrahedron Lett.*, **35**, 6495–8.

38. Toyota, M., Nishikawa, Y. and Fukumoto, K. (1994) Aphidicolin synthesis (II) – an expeditious and efficient formal synthesis of (±)-aphidicolin. *Tetrahedron*, **50**, 11153–66.

39. Grigg, R. and Savic, V. (2000) Palladium catalysed synthesis of pyrroles from enamines. *Chem. Commun.*, 873–4.

40. Bremer, M. and Lietzau, L. (2005) 1,1,6,7-Tetrafluoroindanes: improved liquid crystals for LCD-TV application. *New J. Chem.*, **29**, 72–4.

41. Ma, S. and Negishi, E.-i. (1995) Palladium-catalyzed cyclization of ω-haloallenes. A new general route to common, medium, and large ring compounds via cyclic carbopalladation. *J. Am. Chem. Soc.*, **117**, 6345–57.

42. Vital, P., Norrby, P.-O. and Tanner, D. (2006) An intramolecular Heck reaction that prefers a 5-*endo*- to a 6-*exo*-trig cyclization pathway. *Synlett*, 3140–4.

43. Ichikawa, J., Sakoda, K., Mihara, J. and Ito, N. (2006) Heck-type 5-*endo*-trig cyclization promoted by vinylic fluorines: ring-fluorinated indene and 3*H*-pyrrole syntheses from 1,1-difluoro-1-alkenes. *J. Fluorine Chem.*, **127**, 489–504.

44. Tietze, L.F. and Modi, A. (2000) Regioselective silane-terminated intramolecular Heck reaction with alkenyl triflates and alkenyl iodides. *Eur. J. Org. Chem.*, 1959–64.

45. García-Fandiño, R., José Aldegunde, M., Codesido, E.M. *et al.* (2005) RCM for the construction of novel steroid-like polycyclic systems. 1. Studies on the synthesis of a PreD$_3$–D$_3$ transition state analogue. *J. Org. Chem.*, **70**, 8281–90.

46. Nagasawa, K., Zako, Y., Ishihara, H. and Shimizu, I. (1991) Stereoselective synthesis of 1α-hydroxyvitamin D$_3$ A-ring synthons by palladium-catalyzed cyclization. *Tetrahedron Lett.*, **32**, 4937–40.

47. Chen, C. and Crich, D. (1993) The cyclization route to the calcitriol A-ring: a formal synthesis of (+)-1α,25-dihydroxyvitamin D₃. *Tetrahedron*, **49**, 7943–54.

48. Alibés, R., Ballbé, M., Busqué, F. *et al.* (2004) A new general access to either type of *Securinega* alkaloids: synthesis of securinine and (−)-allonorsecurinine. *Org. Lett.*, **6**, 1813–6.

49. Tietze, L.F., Stewart, S.G. and Polomska, M.E. (2005) Intramolecular Heck reactions for the synthesis of novel antibiotic mensacarcin: investigation of catalytic, electronic and conjugative effects in the preparation of the hexahydroanthracene core. *Eur. J. Org. Chem.*, 1752–9.

50. Sengupta, S., Mukhopadhyay, R., Achari, B. and Banerjee, A.K. (2005) Intramolecular Heck reaction strategy for the synthesis of functionalised tetrahydroanthracenes: a facile formal total synthesis of the linear abietane diterpene, umbrosone. *Tetrahedron Lett.*, **46**, 1515–9.

51. Yokoyama, Y., Kondo, K., Mitsuhashi, M. and Murakami, Y. (1996) Total synthesis of optically active chanoclavine-I. *Tetrahedron Lett.*, **37**, 9309–12.

52. Fishwick, C.W.G., Grigg, R., Sridharan, V. and Virica, J. (2003) Sequential azomethine imine cycloaddition–palladium catalysed cyclisation processes. *Tetrahedron*, **59**, 4451–68.

53. Miki, Y., Shirokoshi, H., Asai, M. *et al.* (2003) Synthesis of naphth[3,2,1-*cd*]indole by Heck cyclization of 2-methoxycarbonyl-3-benzoylindoles. *Heterocycles*, **60**, 2095–101.

54. Cammidge, A.N., Scaife, P.J., Berber, G. and Hughes, D.L. (2005) Cofacial prophyrin–ferrocene dyads and a new class of conjugated porphyrin. *Org. Lett.*, **7**, 3413–6.

55. Miller, A.K., Hughes, C.C., Kennedy-Smith, J.J. *et al.* (2006) Total synthesis of (−)-heptemerone B and (−)-guanacastepene E. *J. Am. Chem. Soc.*, **128**, 17057–62.

56. Jeffery, T. (1996) On the efficiency of tetraalkylammonium salts in Heck type reactions. *Tetrahedron*, **52**, 10113–30.

57. Trost, B.M. and Corte, J.R. (1999) Total synthesis of (+)-saponaceolide B. *Angew. Chem. Int. Ed.*, **38**, 3664–6.

58. Liljebris, C., Resul, B. and Hacksell, U. (1995) Ligand-controlled palladium-catalyzed intramolecular reactions of phenyl-substituted prostaglandin F₂α analogues. *Tetrahedron*, **51**, 9139–54.

59. Muratake, H., Abe, I. and Natsume, M. (1994) Total synthesis of an antitumor antibiotic, (±)-duocarmycin SA. *Tetrahedron Lett.*, **35**, 2573–6.

60. Codesido, E.M., Rodríguez, J.R., Castedo, L. and Granja, J.R. (2002) Toward an analogue of the transition state of PreD₃–D₃ isomerization: stereoselective synthesis of linearly fused 6–8–6 carbocyclic systems. *Org. Lett.*, **4**, 1651–4.

61. Mehta, G. and Shinde, H.M. (2003) Enantiospecific total synthesis of 6-*epi*-(−)-hamigeran B. Intramolecular Heck reaction in a sterically constrained environment. *Tetrahedron Lett.*, **44**, 7049–53.

62. Drège, E., Morgant, G. and Desmaële, D. (2005) Asymmetric synthesis of the tricyclic core of cyathin diterpenoids via intramolecular Heck reaction. *Tetrahedron Lett.*, **46**, 7263–6.

63. Drège, E., Tominiaux, C., Morgant, G. and Desmaële, D. (2006) Synthetic studies on cyathin terpenoids: enantioselective synthesis of the tricyclic core of cyathin through intramolecular Heck cyclisation. *Eur. J. Org. Chem.*, 4825–40.

64. Deng, W., Jensen, M.S., Overman, L.E. *et al.* (1996) A strategy for total synthesis of complex cardenolides. *J. Org. Chem.*, **61**, 6760–1.

65. Laschat, S., Narjes, F. and Overman, L.E. (1994) Application of intramolecular Heck reactions to the preparation of steroid and terpene intermediates having *cis* A–B ring fusions. Model studies for the total synthesis of complex cardenolides. *Tetrahedron*, **50**, 347–58.

66. Overman, L.E. and Rucker, P.V. (1998) Enantioselective synthesis of cardenolide precursors using an intramolecular Heck reaction. *Tetrahedron Lett.*, **39**, 4643–46.

67. Hynes, J. Jr., Overman, L.E., Nasser, T. and Rucker, P.V. (1998) Intramolecular Heck cyclization of α-sulfenyl enol triflates. Asymmetric synthesis of a pentacyclic cardenolide precursor having functionality at C-11. *Tetrahedron Lett.*, **39**, 4647–50.

68. Abelman, M.M., Kado, N., Overman, L.E. and Sarkar, A.K. (1997) Synthesis of a phenolic analog of aphidicolin. Diminished stereoselection in intramolecular 6-*exo* Heck reactions of substrates having a hydrocarbon tether. *Synlett*, 1469–71.

69. Coe, J.W., Vetelino, M.G., Bashore, C.G. *et al.* (2005) In pursuit of $\alpha 4 \beta 2$ nicotinic receptor partial agonists for smoking cessation: carbon analogs of (−)-cytisine. *Bioorg. Med. Chem. Lett.*, **15**, 2974–9.

70. Kelly, S.A., Foricher, Y., Mann, J. and Bentley, J.M. (2003) A convergent approach to huperzine A and analogues. *Org. Biomol. Chem.*, **1**, 2865–76.

71. Lucey, C., Kelly, S.A. and Mann, J. (2007) A concise and convergent (formal) total synthesis of huperzine A. *Org. Biomol. Chem.*, **5**, 301–6.

72. Hong, C.Y., Kado, N. and Overman, L.E. (1993) Asymmetric synthesis of either enantiomer of opium alkaloids and morphinans. Total synthesis of (−)- and (+)-dihydrocodeinone and (−)- and (+)-morphine. *J. Am. Chem. Soc.*, **115**, 11028–9.

73. Routier, S., Mérour, J.-Y., Dias, N. *et al.* (2006) Synthesis and biological evaluation of novel phenylcarbazoles as potential anticancer agents. *J. Med. Chem.*, **49**, 789–99.

74. Routier, S., Coudert, G. and Mérour, J.-Y. (2001) Synthesis of naphthopyrrolo[3,4-*c*]carbazoles. *Tetrahedron Lett.*, **42**, 7025–8.

75. Sanchez-Martinez, C., Faul, M.M., Shih, C. *et al.* (2003) Synthesis of aryl- and heteroaryl[*a*]pyrrolo[3,4-*c*]carbazoles. *J. Org. Chem.*, **68**, 8008–14.

76. Hegedus, L.S., Sestrick, M.R., Michaelson, E.T. and Harrington, P.J. (1989) Palladium-catalyzed reactions in the synthesis of 3- and 4-substituted indoles. *J. Org. Chem.*, **54**, 4141–6.

77. Ishibashi, H., Ito, K., Tabuchi, M. and Ikeda, M. (1991) Synthetic studies on podophyllum lignans: tributyltin hydride-induced radical cyclization and intramolecular Heck reaction of α-benzylidene-β-(*o*-bromobenzyl)-γ-lactones. *Heterocycles*, **32**, 1279–82.

78. Negishi, E.-i., Ma, S., Sugihara, T. and Noda, Y. (1997) Synthesis of hydrazulenes via Zr-promoted bicyclization of enynes and transition metal-catalyzed or radical cyclization of alkenyl iodides. Efficient synthesis of (±)-7-*epi*-β-bulnesene. *J. Org. Chem.*, **62**, 1922–3.

79. Cornec, O., Joseph, B. and Mérour, J.-Y. (1995) Intramolecular Heck reaction: synthesis of potential antiinflammatory agents. *Tetrahedron Lett.*, **36**, 8587–90.

80. Joseph, B., Cornec, O. and Mérour, J.-Y. (1998) Intramolecular Heck reaction: synthesis of benzo[4,5]cyclohepta[*b*]indole derivatives. *Tetrahedron*, **54**, 7765–76.

81. Yokoyama, Y., Matsushima, H., Takashima, M. *et al.* (1997) A new route for the tricyclic indole system: a useful intermediate for ergot alkaloids. *Heterocycles*, **46**, 133–6.

82. Sengupta, S., Drew, M.G.B., Mukhopadhyay, R. *et al.* (2005) Stereoselective syntheses of (±)-komaroviquinone and (±)-faveline methyl ether through intramolecular Heck reaction. *J. Org. Chem.*, **70**, 7694–700.

83. Tietze, L.F., Sommer, K.M., Schneider, G. *et al.* (2003) Novel medium ring sized estradiol derivatives by intramolecular Heck reactions. *Synlett*, 1494–6.

84. Masters, J.J., Jung, D.K., Bornmann, W.G. and Danishefsky, S.J. (1993) A concise synthesis of a highly functionalized C-aryl taxol analog by an intramolecular Heck olefination reaction. *Tetrahedron Lett.*, **34**, 7253–6.

85. Kwon, O., Su, D.-S., Meng, D. *et al.* (1998) Total syntheses of CP-225,917 and CP-263,114: creation of a matrix structure by sequential aldol condensation and intramolecular Heck ring closure. *Angew. Chem. Int. Ed.*, **37**, 1877–80.

86. Iimura, S., Overman, L.E., Paulini, R. and Zakarian, A. (2006) Enantioselective total synthesis of guanacastepene N using an uncommon 7-*endo* Heck cyclization as a pivotal step. *J. Am. Chem. Soc.*, **128**, 13095–101.

87. Horwell, D.C., Nichols, P.D., Ratcliffe, G.S. and Roberts, E. (1994) Synthesis of conformationally constrained tryptophan derivatives. *J. Org. Chem.*, **59**, 4418–23.

88. Danishefsky, S.J., Masters, J.J., Young, W.B. *et al.* (1996) Total synthesis of baccatin III and taxol. *J. Am. Chem. Soc.*, **118**, 2843–59.
89. Young, W.B., Masters, J.J. and Danishefsky, S. (1995) Sterocontrolled syntheses of C-aryl taxanes by intramolecular Heck olefination. Novel instances of diastereofacial guidance by proximal coordination. *J. Am. Chem. Soc.*, **117**, 5528–34.
90. Masters, J.J., Link, J.T., Snyder, L.B. *et al.* (1995) A total synthesis of taxol. *Angew. Chem., Int. Ed. Engl.*, **34**, 1723–6.
91. Masters, J.J., Jung, D.K., Danishefsky, S.J. *et al.* (1995) A novel intramolecular Heck reaction: synthesis of a cholesterol–baccatin III hybrid. *Angew. Chem., Int. Ed. Engl.*, **34**, 452–5.
92. Dyker, G. and Grundt, P. (1999) Construction of the steroid framework via a functionalized macrocyclic compound. *Eur. J. Org. Chem.*, 323–7.

6

Formation of Heterocycles

Thierry Muller and Stefan Bräse

*Institut für Organische Chemie, Universität Karlsruhe (TH), Fritz-Haber-Weg 6, D-76131,
Karlsruhe, Germany*

6.1 Introduction

Heterocycles are of outstanding importance in organic synthesis because they are part of innumerable natural products having interesting properties, notably in the pharmaceutical and crop science sectors. As a consequence, their preparation has evoked large synthetic interest. Among the transition-metal-catalysed annulations, the Mizoroki–Heck reaction, the most renowned palladium-based synthetic tool [1–5], occupies a front seat. Besides several multicomponent syntheses of heterocycles via the Mizoroki–Heck reaction [6], the intramolecular version is still the most widespread method, even though it had a relatively stumbling start.

The intramolecular Mizoroki–Heck reaction, although it had been known since 1977, when it was first used in the synthesis of heterocycles [7], started to be explored properly only around the mid 1980s. Its large use for the preparation of carbocycles began only in the late 1980s [8]. Since then, however, there has been a tremendous increase of interest in this intramolecular cyclization, culminating in its widespread utilization in total synthesis [9]. This was mostly due to the development of asymmetric versions by two major contributors, namely Overman for the construction of heterocycles [9] (Chapter 12) and Shibasaki focusing on carbocyclic systems [10] (Chapter 16).

The formation of heterocycles by intramolecular Mizoroki–Heck reactions has been reviewed several times as a specific topic [11–16]. In this chapter, a classification by ring size has been established. The examples in the different subsections are organized according to specific substructures of the starting materials and according to types of heteroatom.

The Mizoroki–Heck Reaction   Edited by Martin Oestreich
© 2009 John Wiley & Sons, Ltd

The proper asymmetric synthesis of heterocycles via the Mizoroki–Heck reaction will be dealt with in Chapters 12 and 16.

6.2 'Cyclic' Peculiarities and Constraints

6.2.1 Mechanistic Considerations and the Role of Additives

The intramolecular version of the Mizoroki–Heck reaction often proceeds with high regio- and stereoselectivity, mainly depending on ring size of the heterocycles formed and the reaction conditions [13]. In order to understand or predict the outcome of the reaction, a brief overview of the most important factors governing or influencing stereo- and regio- nselectivity is provided below. Stereoselectivity in the intramolecular Mizoroki–Heck re- action is generally not a concern in the sense that, for obvious geometric constraints in the *endo*-mode ring closure, the newly formed double bond has *cis*-stereochemistry in nearly all cases. If the intramolecular Mizoroki–Heck reaction follows an *exo*-type cyclization, however, the stereochemistry of the *exo*-cyclic double bond is difficult to control.

In order to understand fully the extent of the regioselectivity issues, some mechanistic considerations are unavoidable. It is not the purpose of this chapter to provide detailed mechanistic investigations (Chapter 1); however, in order to give a concise overview on the formation of heterocycles by means of the Mizoroki–Heck reaction, some mechanis- tic aspects with specific relevance to heterocyclic formation have to be addressed. The generally accepted mechanism of the classic Mizoroki–Heck reaction is believed to pass through five major steps: (i) oxidative addition of an aryl or alkenyl (pseudo)halide to a palladium(0) species to form a σ-aryl- or σ-alkenylpalladium(II) complex; (ii) formation of a π-complex of the palladium(II) species with an alkene; (iii) *syn*-insertion of the σ-aryl- or σ-alkenylpalladium(II) bond into the C=C double bond to give a σ-(β-aryl)- or σ-(β- alkenyl)alkylpalladium(II) complex; (iv) with a β-hydrogen available, *syn*-β-elimination of hydridopalladium(II) (pseudo)halide (HPdX) after internal C—C bond rotation; (v) re- ductive elimination of HX from the HPdX, assisted by a base, to regenerate palladium(0) and to start another catalytic cycle [17].

In this section, only examples of Mizoroki–Heck reactions where a proper addition of the σ-aryl- or σ-alkenylpalladium(II) complex to a double bond of an alkene or alkyne occurs are considered. As a consequence, an often-met deviation from the classic Mizoroki–Heck mechanism, the so-called cyclopalladation, will not be treated in further detail [12, 18]. However, as it is of some importance, especially in heterocycle formation and mainly be- cause it will be encountered later during polycyclization cases, it shall be mentioned briefly below. Palladacycles are assumed to be intermediates in intramolecular Mizoroki–Heck re- actions when β-elimination of the formed intermediate cannot occur. These are frequently postulated as intermediates during intramolecular aryl–aryl Mizoroki–Heck reactions un- der dehydrohalogenation (Scheme 6.1). The reactivity of these palladacycles is strongly correlated to their size. Six-membered and larger palladacycles quickly undergo reductive elimination, whereas the five-membered species can, for example, lead to Mizoroki–Heck- type domino or cascade processes [18, 19].

Returning to the classical Mizoroki–Heck reaction and regarding the regioselectiv- ity of its intramolecular version, one particularly appreciated feature is the commonly

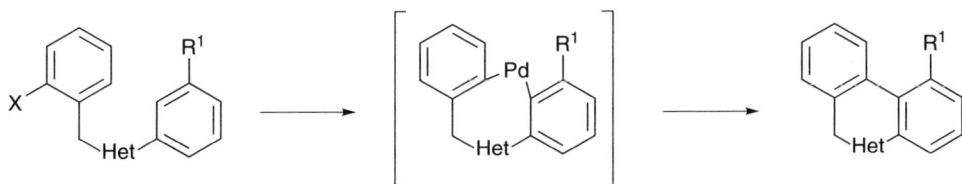

Scheme 6.1 *Example of an intramolecular aryl–aryl Mizoroki–Heck reaction proceeding via a palladacycle.*

experienced exclusive formation of a sole regiochemical isomer, in most cases the *exo*-cyclization product. The regiocontrol seems to be largely governed by the size of the cycle to be formed, with 5-*exo* and 6-*exo* being particularly favoured. In the Cabri–Hayashi model, the coordination–carbopalladation event is considered to be irreversible and the complex should ideally be in an eclipsed conformation; for example, the alkene should be coplanar with the C(sp^2)–Pd(II) bond of the aryl or vinyl species (Figure 6.1) [8, 20].

In the intermolecular Mizoroki–Heck reaction, due to steric and electronic effects, an eclipsed form is generally favoured. In the case of the intramolecular version, one has to add conformational constraints to the already existing ones, which are often limiting the eclipsed form to a single possibility and, thus, leading to a single regioisomer, generally the *exo*-cyclization product. It is only for the formation of larger ring systems that the conformational constraints become less important and that the generally experienced directing rules in the intermolecular Mizoroki–Heck reaction become dominant again [13, 21].

Following the simplified mechanistic situation detailed above (steps (i) to (v)), the outcome of the Mizoroki–Heck reaction is consistent with a *syn*-addition of an σ-aryl- or σ-alkenyl-palladium species to an alkene, followed by a *syn*-elimination of hydridopalladium halide to re-establish a double bond. Concerning the regiocontrol with respect to the position of the C=C double bond forming during the intramolecular Mizoroki–Heck reaction, however, the situation is somewhat more complex. This is mainly due to the frequent lack of conformationally available β-hydrogen atoms in the cyclic intermediate that could undergo a *syn*-elimination with the palladium and due to the ease of double-bond isomerization reactions in cyclic unsaturated compounds [11]. Another concern is the competition between β- and β'-hydride elimination. This can be avoided to a certain extent by careful selection of the reaction conditions; but, in most cases, this problem is circumvented by employing substrates having no duly positioned β'-hydrogen atoms [10]. Even if the desired isomer is obtained through *syn*-elimination, a further difficulty lies in

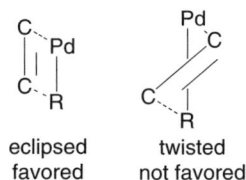

eclipsed
favored

twisted
not favored

Figure 6.1 *Coordination of the palladium species to the double bond.*

the reversibility of the latter, which often results in a reinsertion of the double bond into the Pd–H bond and can lead to a mixture of isomers (post-Mizoroki–Heck double-bond migration). Isomerization can, however, be suppressed or avoided by addition of silver [10] and thallium salts [22] or by using nitrogen- and phosphorous-based ligands [21].

A number of intramolecular Mizoroki–Heck reactions yield the product consistent with a formal *anti*-elimination of the HPdX [11]. These experimental findings are in opposition to the generally accepted mechanism of a *syn*-elimination; however, a reasonable explanation is at hand in most cases. There are two main types of alkenyl derivatives which, if added to an σ-aryl- or σ-alkenylpalladium(II) complex, deliver the formal *anti*-elimination product. The first case is intramolecular Mizoroki–Heck reactions with α,β-unsaturated carbonyl systems which result in the product of a formal 1,4-addition. The initially formed σ-(β-aryl)- or σ-(β-alkenyl)alkylpalladium complex should be long-lived enough to epimerize through a palladium(II) enolate intermediate and, thus, deliver the formal *anti*-elimination product through conventional *syn*-elimination (Scheme 6.2).

Intramolecular Mizoroki–Heck reactions with styrene-type alkenes constitute the second frequently met case. Finding suitable rationales in theses cases seems to be more intricate. There is some evidence for a mechanism involving a radical intermediate [11]. Most explanations, however, cite a facile epimerization at the benzylic position and, thus, furnishing the required *cis*-stereochemistry or a base-assisted reductive elimination of the palladium species (Scheme 6.3) [11]. In some cases, suitable substrates or conditions can lead to the *anti*-elimination products via an E1cb-type mechanism [23, 24].

As seen before, regioselectivity in the intramolecular Mizoroki–Heck reaction can be controlled or directed by cautious choice of substrates and reaction conditions. In reality, the mechanism of the Mizoroki–Heck reaction is much more complex than the oversimplified

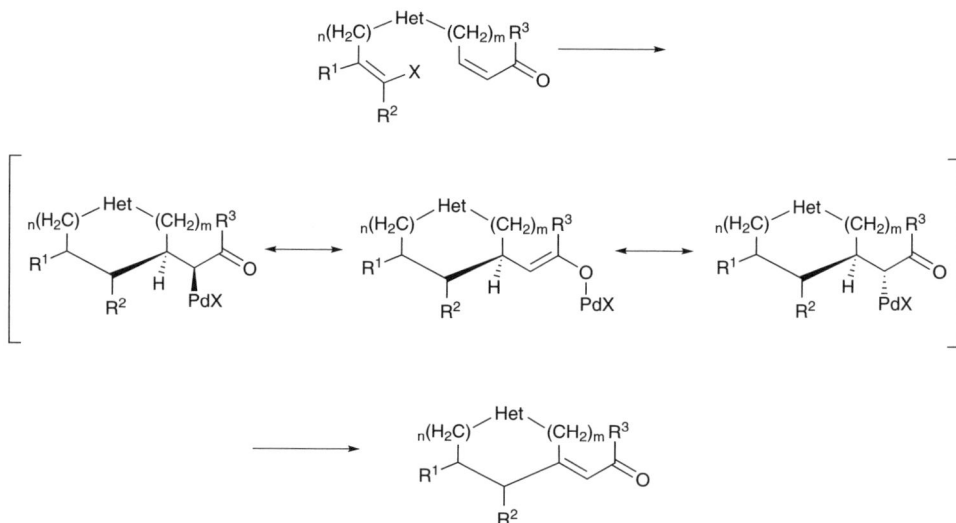

Scheme 6.2   *Mizoroki–Heck reaction proceeding via syn-elimination through a palladium(II) enolate intermediate.*

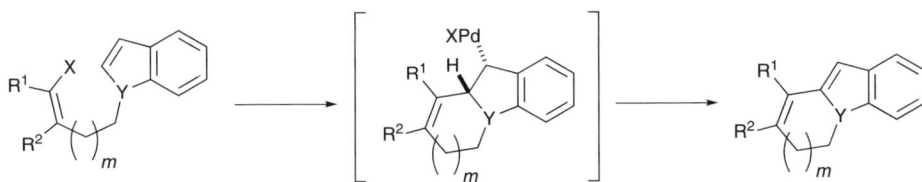

Scheme 6.3   *Suggested base-assisted β-elimination during a Mizoroki–Heck reaction.*

general catalytic cycle usually cited. There are probably at least three different mechanistic pathways [8, 20], termed neutral, cationic and anionic, which were introduced by Cabri and by Hayashi and further applied to the asymmetric intramolecular Mizoroki–Heck reaction by Overman's group [9]; for a review, see Zeni and Larock [14]. Depending on the pathway, the outcome of the reaction is often different.

Besides the different mechanistic pathways mainly depending on the substrate types, the Mizoroki–Heck reaction can also be run under several conditions with diverse additives which affect the outcome. A major achievement in this field was the discovery of the so-called Jeffery conditions [25–28]. Mizoroki–Heck reactions are indeed greatly accelerated by the use of inorganic bases in combination with a phase-transfer agent which allows lowered reaction temperatures. In some cases, Jeffery conditions can even lead to the *endo*-cyclized product, whereas standard conditions give the *exo*-cyclized compound [8].

In the same way, Mizoroki–Heck reactions followed by subsequent reduction, also referred to as reductive Mizoroki–Heck reactions, have become a valuable tool for the construction of heterocycles. This process generally takes place in the presence of formic acid in combination with a base and yields the reduced heterocycle; that is, without the reformation of a double bond. These conditions have been pioneered by Grigg and coworkers [29], who introduced the terms of anion capture or anion trapping. The scope of this method is not restricted to the sole Mizoroki–Heck products obtained by hydride capture; the σ-(β-aryl)- or σ-(β-alkenyl)alkylpalladium(II) complex can also be trapped by various other neutral or anionic nucleophiles, such as organometallics [30]. Furthermore, these conditions can also deliver the 'normal' Mizoroki–Heck product by the use of alkynes. They constitute, in fact, an elegant method to avoid the frequently encountered double-bond shifts instead of adding silver or thallium salts. Intramolecular additions to triple bonds can indeed be conducted under reducing conditions to yield the desired product as a single stereoisomer [30].

It is the simplicity of adapting the reaction conditions and the ease of switching from one pathway to another by the appropriate catalyst cocktail leading to a whole range of different products, which makes the intramolecular Mizoroki–Heck reaction the method of choice for the construction of complex heterocycles, including spirocyclic compounds, bridged ring systems and even congested quaternary centres [15].

6.2.2   Structural Prerequisites and Substrate Classification

At the beginning of the 1980s, heterocycles via intramolecular Mizoroki–Heck reaction were prepared almost exclusively starting from haloarenes. Nowadays, a wide range of

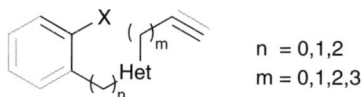

Figure 6.2   *The precursors encountered most often during intramolecular Mizoroki–Heck reactions [12].*

vinylic or aryl halides bearing heteroatoms, as well as neighbouring C=C double bonds, are used to obtain the desired heterocycles [14]. Suitable substrates for heterocyclic intramolecular Mizoroki–Heck reactions have to fulfil some simple but nonetheless essential requirements: they have to be composed of a halide function capable of oxidative addition to a palladium(0) catalyst and a heteroatom-containing hydrocarbon chain tethering either an alkene or an alkyne. A lot of compounds fulfil these criteria; the substructures most frequently encountered lead to five-, six- or seven-membered rings and are summarized in Figure 6.2 [12]. Nitrogen-containing heterocycles represent by far the largest product class and 5-*exo*-cyclizations are undoubtedly the most commonly found ring-closing reactions, followed by 6-*exo*-cyclizations. It is only for the formation of larger heterocycles that *endo*-reactions become the dominating cyclization mode.

6.3   Formation of Five-Membered Cycles

Figure 6.3 recalls the most frequently encountered substrate substructures for five-membered ring closures.

This is not a comprehensive listing, and other substructures leading to five-membered cycles will be presented at the end of this section.

6.3.1   5-*exo*-Cyclization

There are more examples of 5-*exo*-cyclizatzion via intramolecular Mizoroki–Heck reaction than of any other compound class, among which substructure **A** (Figure 6.3) with an allyl side chain constitutes the most widespread substrate.

Indoles or heterocycles containing an indole skeleton were the first products to be prepared by intramolecular heterocyclic Mizoroki–Heck reaction. In 1977, the first intramolecular Mizoroki–Heck reaction was reported by Mori *et al.* [7], who prepared indole **3** by cyclization of (*E*)-methyl 4-(*N*-(2-bromophenyl)acetamido)but-2-enoate (**1**) in the presence of palladium acetate, triphenylphosphine and *N*,*N*,*N'*,*N'*-tetramethylethylenediamine (Scheme 6.4).

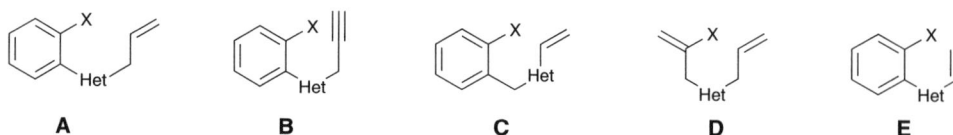

| A | B | C | D | E |

Figure 6.3   *Most frequently met substrate structures for five-membered ring formations.*

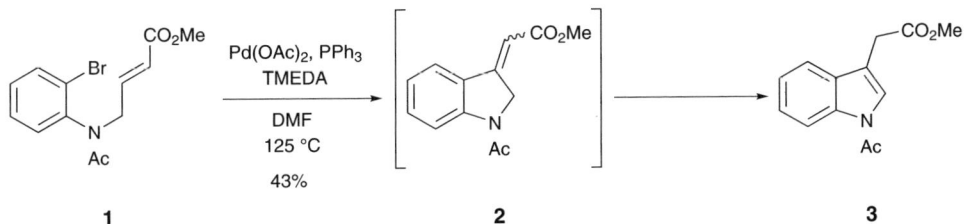

Scheme 6.4 *First synthesis of heterocycles via intramolecular Mizoroki–Heck reaction.*

There are two things worth mentioning about this first example of a heterocyclic in-tramolecular Mizoroki–Heck reaction. First, different halogen derivatives of **1** were tested; as expected, the chloride derivative did not give any cyclized product, but, interestingly, the aryl iodide gave only 24% of the desired indole along with 47% of deallylated starting material. Second, and most important, according to the mechanism of the Mizoroki–Heck reaction, the two *exo*-cyclic double-bond isomers of **2** completely isomerized to the more stable *endo*-cyclic position by a reinsertion of the hydridopalladium halide species into the double bond followed by a second *syn*-elimination. This is a commonly observed phenomenon for 5-*exo*-cyclizations, which can, however, be suppressed by adapting the reaction conditions. Addition of silver carbonate during the reaction of aryl iodide **4**, for example, delivered exclusively the 3-methyleneindole **5** derivate in good yield (Scheme 6.5) [31].

Inoue *et al.* [32] even managed to produce diastereoselectively both *exo*-cyclic double-bond isomers. The 3-alkylideneoxindole substructures **7** and **9** were obtained under iden-tical reaction conditions with complete inversion of the double-bond geometry, although exclusive formation of the (*E*)-isomer **9** proceeded with only 12% yield (79% recovery of (*Z*)-**8**) (Scheme 6.6). The authors claimed that the stereoselectivity is mainly due to the distinct functions of the two nitrogen protecting groups. The NBoc group at carbon atom C_A stabilizes an s-*cis*-conformation which organizes the reacting centres into spatial proximity, whereas the steric bulk of the C_B–N protecting group prevents isomerization by suppressing readdition and elimination of the hydridopalladium halide species.

Another way to prevent isomerization of the double bond to the thermodynamically more stable *endo*-cyclic position is the use of the Jeffery conditions. Tietze and Grote [33] showed that, depending on the reaction conditions, either *endo*- or *exo*-cyclic double-bond formation can be favoured (Scheme 6.7). Under standard Mizoroki–Heck conditions using

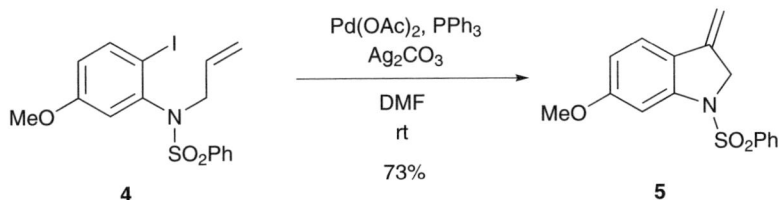

Scheme 6.5 *Exclusive exo-cyclic double-bond formation by addition of silver salt.*

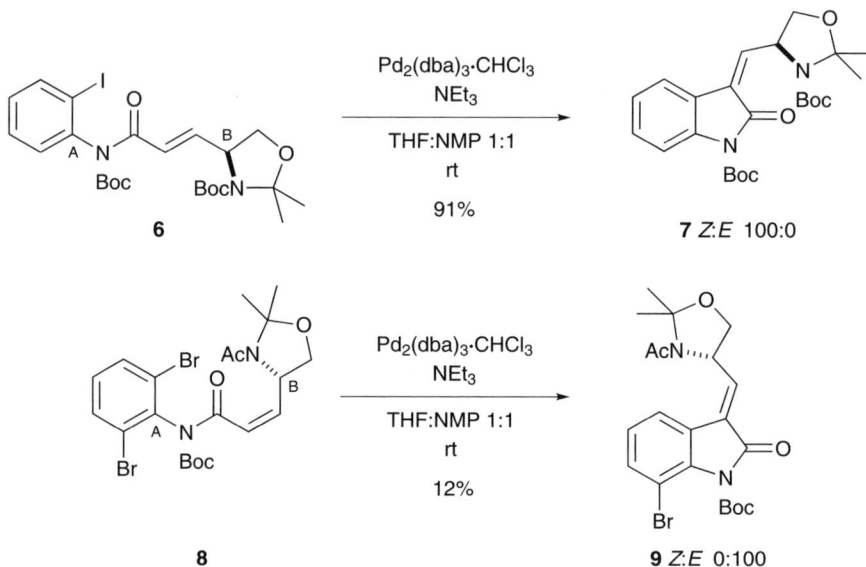

Scheme 6.6 *Exclusive stereoselective synthesis of both* exo-*cyclic double-bond isomers of a 3-alkylideneoxindole structure.*

tetrakis(triphenylphosphine)palladium(0) and triethylamine in acetonitrile, a ratio of 1 : 7 in favour of the more stable compound **12** is obtained. Under Jeffery conditions, however, the major compound is retaining an *exo*-cyclic double bond. Against all expectations, the ratio of both isomers was not time dependent; all attempts to convert **11** into the thermodynamically more stable compound by heating at 100 °C for several days under Jeffrey conditions were unproductive.

A recent procedure describes a one-pot, two-step mutlicomponent reaction of bromoanilines **13**, aldehydes **14**, acids **15** and isocyanides **16** yielding polysubstituted indoles **18** [34]. It is based on a Ugi four-component reaction yielding the precursor **17** for the *in situ* performed classical intramolecular Mizoroki–Heck reaction, thereby providing a facile access to highly substituted dihydro-indoles, 1*H*-indoles and 1*H*-indole-2-carboxylic

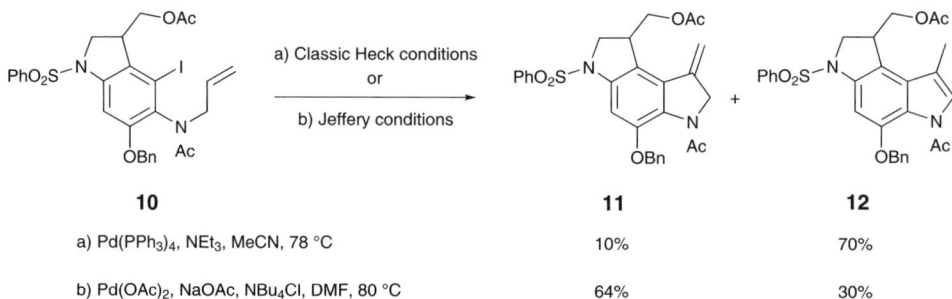

a) Pd(PPh₃)₄, NEt₃, MeCN, 78 °C

b) Pd(OAc)₂, NaOAc, NBu₄Cl, DMF, 80 °C

Scheme 6.7 *Standard Mizoroki–Heck versus Jeffery conditions.*

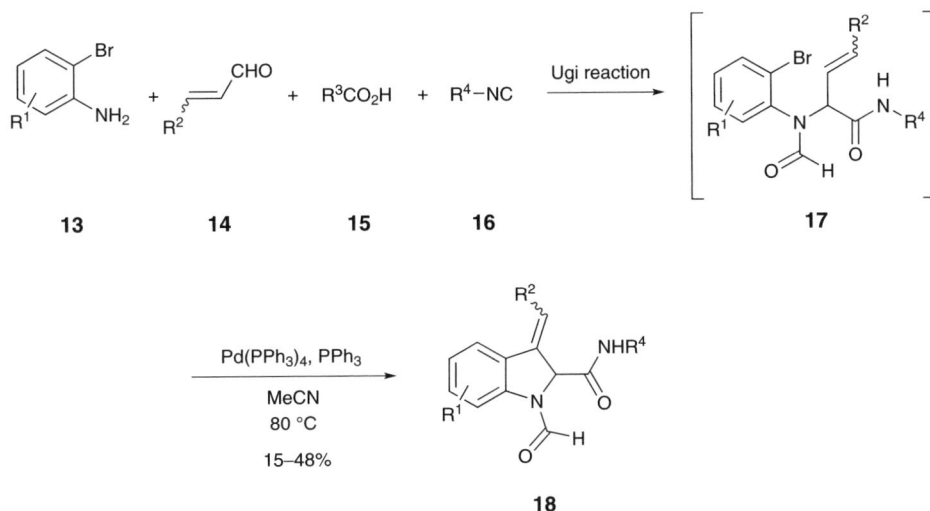

Scheme 6.8 *One-pot Ugi–Mizoroki–Heck synthesis of highly substituted indole derivatives.*

acid building blocks from cheap starting material in yields ranging from 15 to 48% (Scheme 6.8).

A huge number of similar examples preparing indoles or related derivatives via intramolecular Mizoroki–Heck reactions can be found in the literature [35–39]; enumerating them all would be impossible.

All substructures of type **A** (Figure 6.3) seen so far had either terminal or 1,2-disubstituted double bonds. If we consider precursors incorporating a 1,1-disubstituted double bond, the intermediately formed σ-(β-aryl)alkylpalladium complex lacks a β-hydrogen to eliminate the hydridopalladium halide species. As a consequence, the complex formed will readily undergo palladium-catalysed polycyclization or polycyclization–anion capture reactions [40–42]. These reactions have been intensively investigated by Grigg *et al.* [43], who were among the first to study polycyclizations leading to spiro systems (Scheme 6.9) [44].

As illustrated in Scheme 6.9, formation of the second ring involves attack of σ-alkylpalladium(II) species **20** at the hetarene. An interesting fact is that most examples of an addition of the corresponding palladium species to a double bond involve 'electron-poor' or 'neutral' alkenes. The indole and pyrrole moieties presented in this example are, however, even when bearing a carboxaldehyde substituent, 'electron-rich' species. Considering this, the excellent yield of 91% for the spirocyclizations via two successive 5-*exo*-trig cyclizations seems even more impressive.

Grigg and coworkers also pioneered the work on cyclization–anion capture reactions using hydride-, [45] carbonyl- [46] or cyanide- [47] capture (Scheme 6.10). The terminating anion capture is, however, not limited to these three examples; in fact, the organopalladium(II) intermediate can be captured by a wide variety of anionic, neutral and organometallic agents [47].

As a lot of natural products contain fused *N*-heterocycles, intramolecular Mizoroki–Heck reactions have also been extensively used in total synthesis. Magallanesine (**28**) [48], an

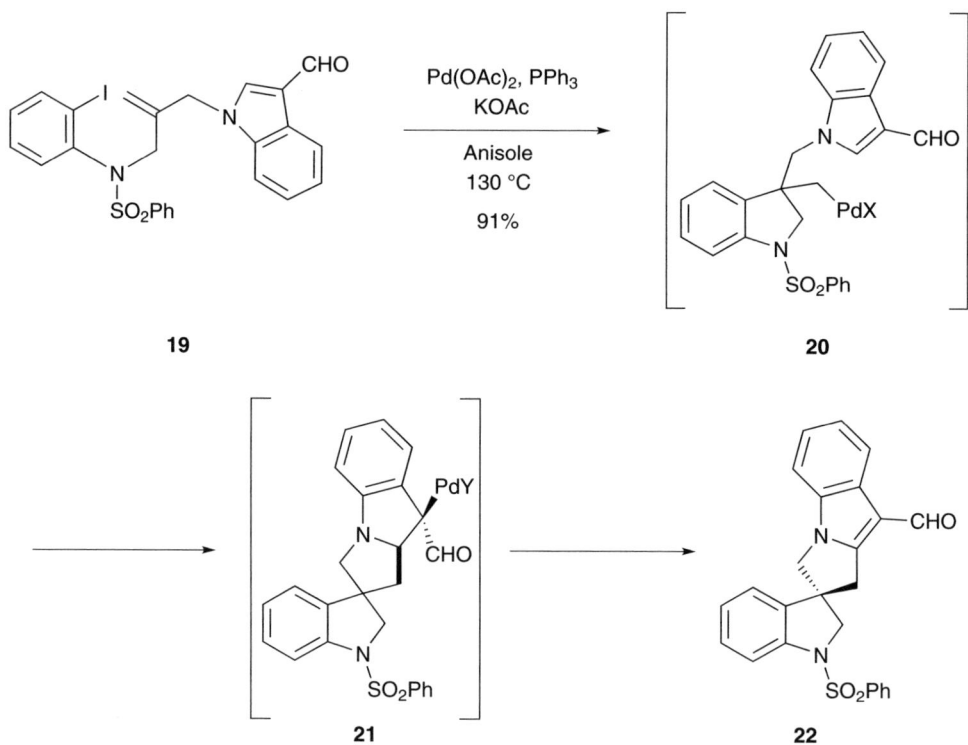

Scheme 6.9   *Polycyclization via Mizoroki–Heck reaction building spirocycle* **22**.

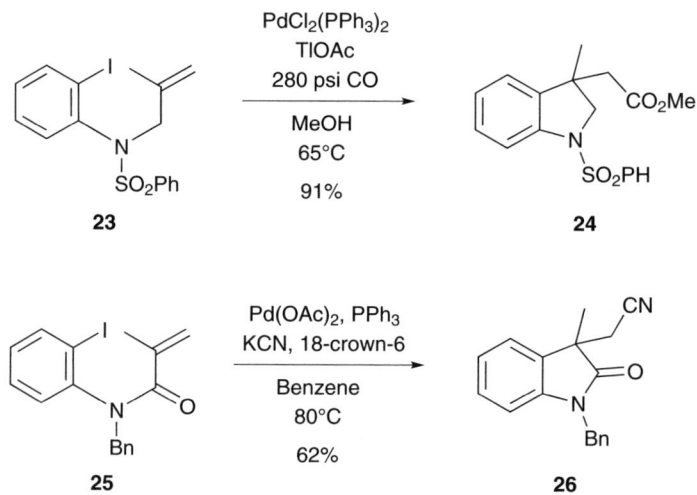

Scheme 6.10   *Examples of cascade cyclization—carbonylation or —cyanide capture.*

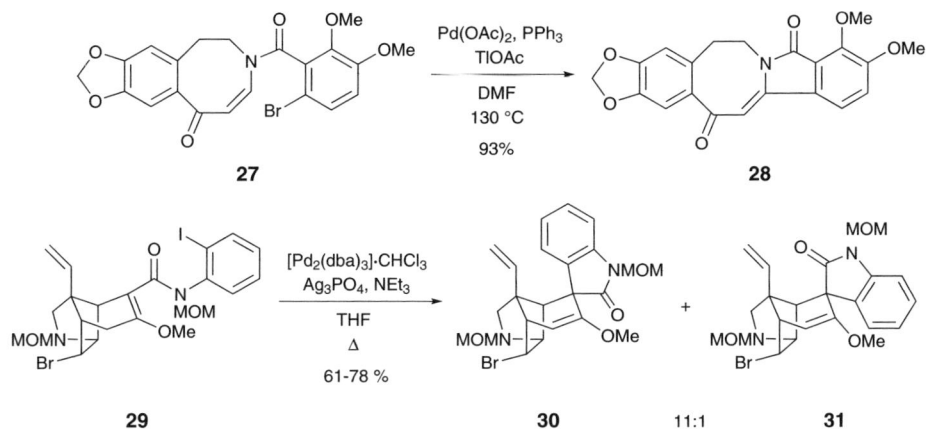

Scheme 6.11   *The Mizoroki–Heck reaction in the total synthesis of complex natural products.*

alkaloid extracted from the plant family *Berberidaceae*, as well as key precursors in the syntheses of the alkaloids gelsemine (**30**) [49, 50] and gelsedine [51, 52], three particularly challenging structures from a synthetic point of view, have been prepared in this way (Scheme 6.11).

Although most examples of 5-*exo* cyclizations of substructure **A** (Figure 6.3) lead to *N*-heterocycles, a number of examples also treat the preparation of oxygen-containing cycles. Several groups studied the formation of benzofurans and derivatives thereof via intramolecular Mizoroki–Heck reaction [53–57]. As already seen for the preparation of indoles and their derivatives, the benzofuran skeleton with its *endo*-cyclic double bond is assumed to be formed under Jeffery conditions by readdition of the hydridopalladium halide species to the *exo*-cyclic product followed by a second elimination [53].

Scheme 6.12 illustrates some typical 5-*exo* cyclizations conducted in ionic liquid [55]. In general, aryl substituents ($R^1 \neq H$) result in lower yields, whereas substituents on the allyl chain ($R^2 \neq H$) do not affect this generally high-yielding cyclization [53, 55].

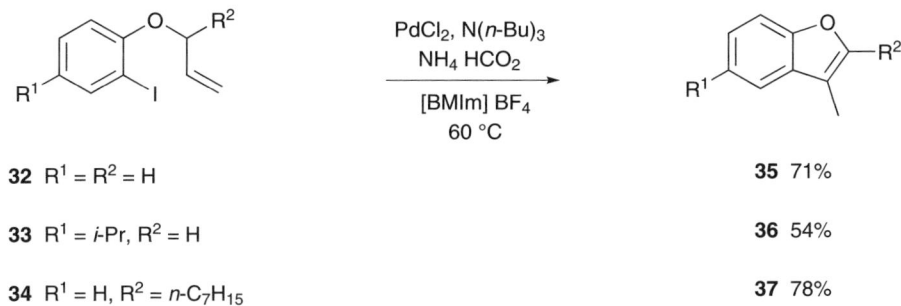

32 $R^1 = R^2 = H$

33 $R^1 = i\text{-Pr}, R^2 = H$

34 $R^1 = H, R^2 = n\text{-C}_7\text{H}_{15}$

35 71%

36 54%

37 78%

[BMlm] BF_4 = 1-*n*-butyl-3-methylimidazolium tetraborate
$NH_4\ HCO_2$ = ammonium formate

Scheme 6.12   *Formation of benzofurans in ionic liquid by intramolecular Mizoroki–Heck reaction.*

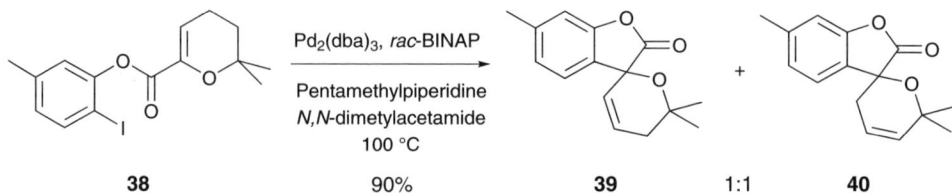

38   90%   **39**   1:1   **40**

Scheme 6.13   *The 1 : 1 mixture of double bond isomers via intramolecular Mizoroki–Heck reaction.*

Vickers and Keay [58] generated a stereogenic quaternary centre through intramolecular Mizoroki–Heck reaction in excellent yield. However, they obtained a 1 : 1 mixture of the two possible regioisomers (Scheme 6.13).

As for its nitrogen analogs, cascade Mizoroki–Heck–anion capture procedures have also been developed for oxygen-containing heterocycles starting from 1,1-disubstituted allyl ether derivatives [59, 60]. Grigg and coworkers [60] synthesized spiro compound **42** through a double intramolecular Mizoroki–Heck reaction with formation of two heterocycles (Scheme 6.14). The second Mizoroki–Heck reaction is a 6-*exo* cyclization involving an alkyne 'captured' by tributyltinhydride, leading to an *N*-heterocycle.

Lautens and Fang [24] have reported an unusual intramolecular Mizoroki–Heck reaction with dihydronaphthalene substrate **43** involving a base-assisted *anti*-hydride elimination that gives tetracycle **45** in moderate yield (Scheme 6.15).

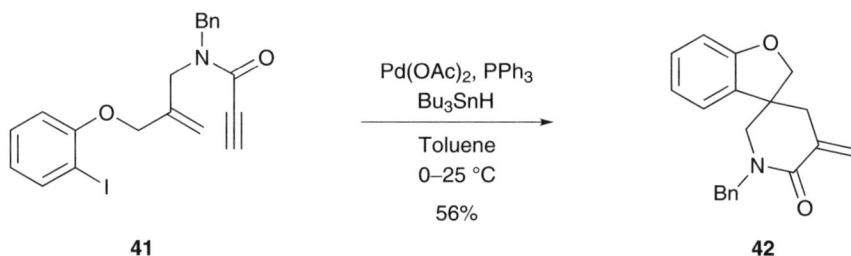

41   56%   **42**

Scheme 6.14   *Cascade of two intramolecular heterocyclic-forming Mizoroki–Heck reactions with anion capture.*

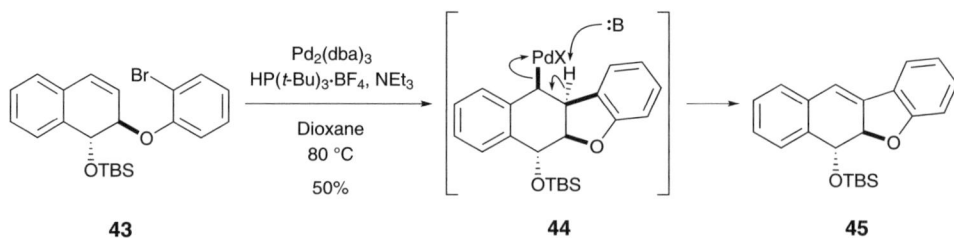

43   50%   **44**   **45**

Scheme 6.15   *Intramolecular Mizoroki–Heck reaction involving a base-assisted anti-hydride elimination.*

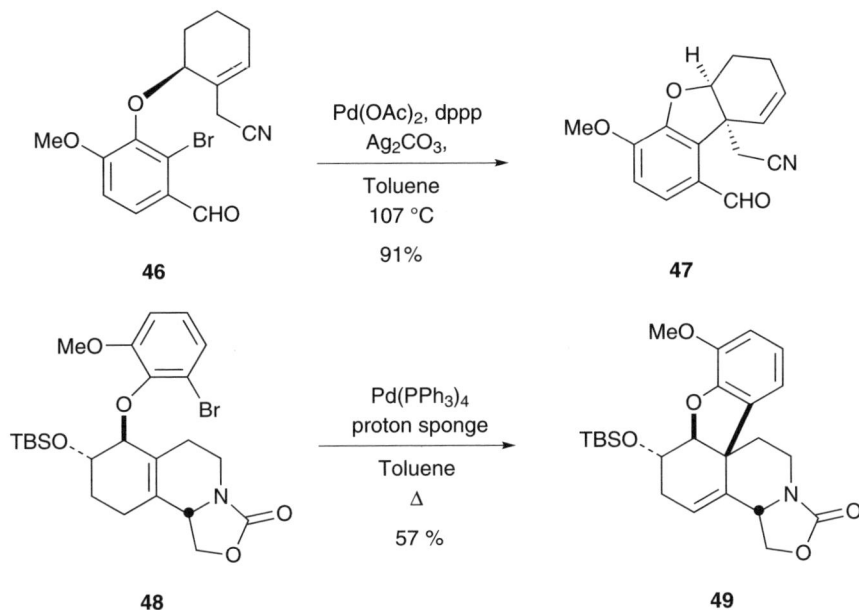

Scheme 6.16 *Key steps in the syntheses of (−)-galanthamine and noroxymorphine.*

Substructure type **A** (Figure 6.3) has also been employed in total synthesis of some *O*-heterocyclic-containing natural products. Key intermediates **47** and **49** in the syntheses of (−)-galanthamine [61, 62] and morphine or noroxymorphone [62, 63] have been prepared by means of an intramolecular Mizoroki–Heck reaction (Scheme 6.16).

One of the rare examples of an intramolecular Mizoroki–Heck cyclization forming thiocycles also involves substructure **A** (Figure 6.3) by reacting allyl *o*-iodophenyl sulfides **50** and **51** under classic Mizoroki–Heck conditions to produce 3-substituted benzo[*b*]thiophenes **52** and **53** respectively (Scheme 6.17) [64].

Substrates corresponding to the general structure of propargylic derivative **B** (Figure 6.3) are, of course, closely related to allyl substructure **A** (Figure 6.3), although there are not that many examples found in the literature. These substrates are frequently used in

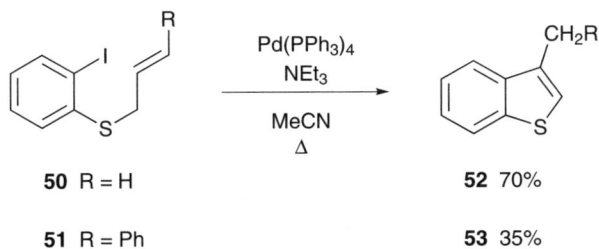

Scheme 6.17 *Synthesis of benzo[b]thiophenes via intramolecular Mizoroki–Heck reaction.*

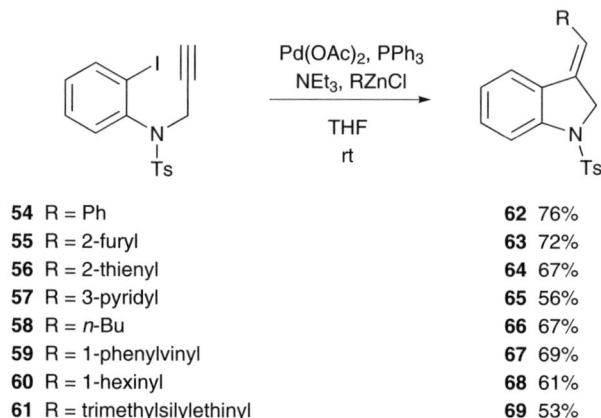

54 R = Ph	62 76%
55 R = 2-furyl	63 72%
56 R = 2-thienyl	64 67%
57 R = 3-pyridyl	65 56%
58 R = *n*-Bu	66 67%
59 R = 1-phenylvinyl	67 69%
60 R = 1-hexinyl	68 61%
61 R = trimethylsilylethinyl	69 53%

Scheme 6.18 *Synthesis of (Z)-3-methylene-2,3-dihydroindole derivatives via reductive Mizoroki–Heck reaction.*

reductive Mizoroki–Heck reactions and represent an elegant method to avoid double-bond isomerization.

Luo and Wang [65] showed that the intermediate palladium(II) species can be trapped by a variety of organozinc chlorides to give exclusively the corresponding (Z)-3-methylene-2,3-dihydroindole derivatives **62–69** in fair to good yields (Scheme 6.18).

Grigg and coworkers also reported the exclusive formation of the (Z)-isomers in their palladium domino cyclization–anion capture processes using substrates of general substructure **B** (Figure 6.3), independently of whether the intermediates were trapped with organozinc, -boron [66] or -tin reagents [67]. Substrates of propargylic substructure **B** can also be used in domino cyclizations as demonstrated by Grigg *et al.* [68] during the synthesis of hippadine using a Pd(0)/(Bu₃Sn)₂ catalyst system (Scheme 6.19). In this special case, however, no traces of the *exo*-cyclic alkene **71** could be detected and the authors concluded that it had been completely isomerized to the more stable *endo*-cyclic alkene **72** by the Pd(0)/(Bu₃Sn)₂ catalyst system [68].

Among the limited examples of thioheterocycles, unstable 3-methylene-2,3-dihydrobenzo[*b*]thiophene (**74**) was prepared in modest yield starting from propargyl *o*-iodophenyl sulfide **73** under classic hydride capture conditions with formic acid (Scheme 6.20) [64].

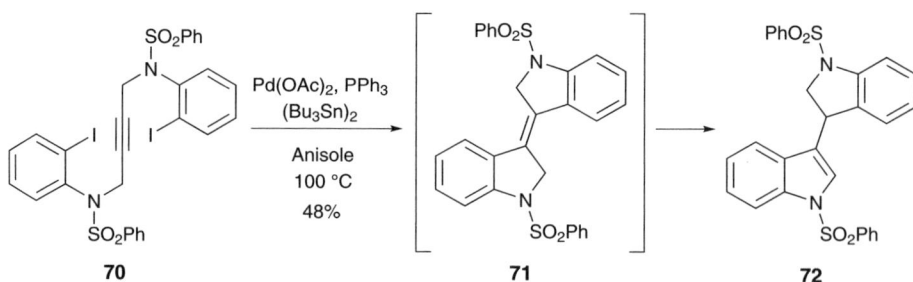

Scheme 6.19 *Polycyclization of **70** followed by isomerization of **71**.*

73   40%   **74**

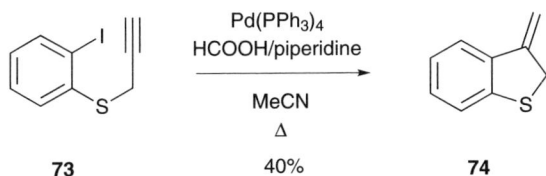

Scheme 6.20 *Synthesis of 3-methylene-2,3-dihydrobenzo[b]thiophene (74) via reductive Mizoroki–Heck reaction.*

Vinyl-substituted substructures **C** (Figure 6.3) are also typical for a variety of substrates leading to five-membered heterocycles.

Grigg *et al.* [69] used enamides **75** and **77** to prepare heterocycles **76** and **78** respectively (Scheme 6.21). Besides the asymmetric induction, these two examples show once more that a judicious choice of the conditions and substrates will control the outcome of the reaction. The first example is a spiro-cyclization by β'-hydride elimination. In the second example, the less stable isomer **78** was obtained, which did not isomerize to the thermodynamically more stable product because of the use of silver salt.

Comins *et al.* [70] showed that *N*-acetyl-2,3-dihydro-4-pyridone reacts in a regio- and stereoselective manner to give the 5-*exo*-cyclization products. Product **80** is obtained under Jeffrey conditions, whereas hydride capture delivers the formal 1,4-addition product **81** (Scheme 6.22).

Cyclizations of substructure type **C** (Figure 6.3) are also used in total synthesis of natural products bearing five-membered heterocycles, as shown by the example of (*S*)-camptothecin (**83**), a natural anticancer agent (Scheme 6.23) [71]. This pentacyclic alkaloid was synthesized in 10 steps, the last one being a 5-*exo*-intramolecular Mizoroki–Heck reaction.

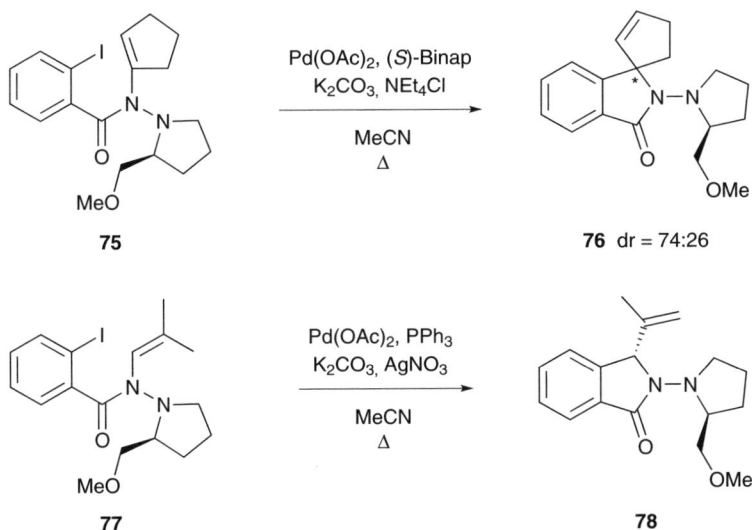

75   **76**   dr = 74:26

77   **78**

Scheme 6.21 *Two examples of enamide substrates for intramolecular Mizoroki–Heck reactions.*

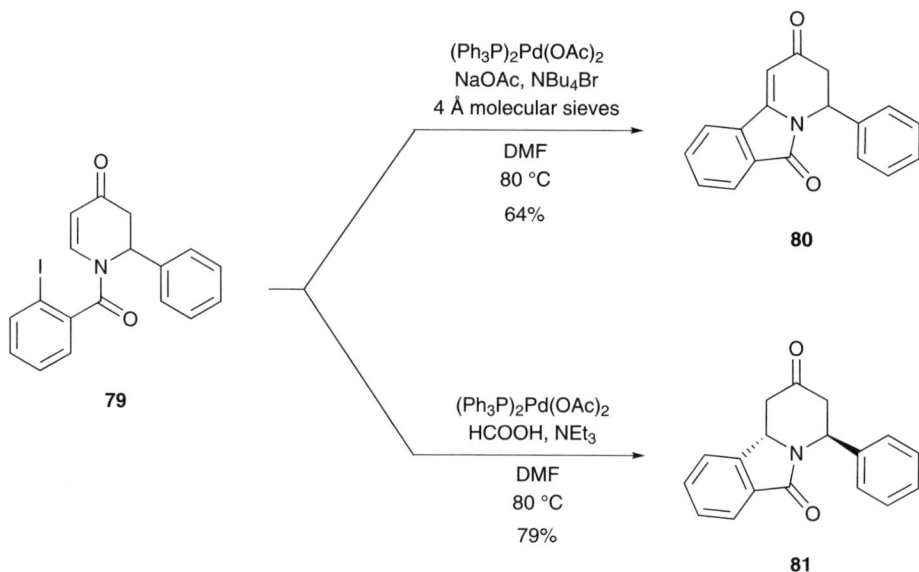

Scheme 6.22   *The 5-exo-cyclization of **79** under different conditions.*

Substrates of substructure **C** (Figure 6.3) have been used to prepare *O*-heterocycles as well. A recent survey studies the microwave-accelerated generation of conformationally restricted spiro[cyclohexane-1,1′-isobenzofuran] derivatives via 5-*exo*-cyclization of the corresponding cyclohexenyl *o*-iodobenzyl ethers (Scheme 6.24) [21, 72]. The double-bond position in the hexacycles could be controlled by thoughtful choice of starting material and reactions conditions. This study also constitutes one of the rare examples of electron-rich alkenes used in Mizoroki–Heck reactions [21, 72].

Another often encountered substructure for intramolecular 5-*exo*-Mizoroki–Heck reactions is allyl side chain containing vinyl halide **D** (Figure 6.3) [73, 74]. Based on this substrate structure, a series of 3-substituted pyrrolo[2,3-*b*]quinoxalines have been prepared via intramolecular Mizoroki–Heck reaction under Jeffery's 'ligand-free' conditions in moderate to excellent yields (Scheme 6.25) [75].

Substructure **D** (Figure 6.3), furthermore, is the basis for the cyclization of carbohydrate templates generating bicyclic [76] or tricyclic [77] compounds (Scheme 6.26). These

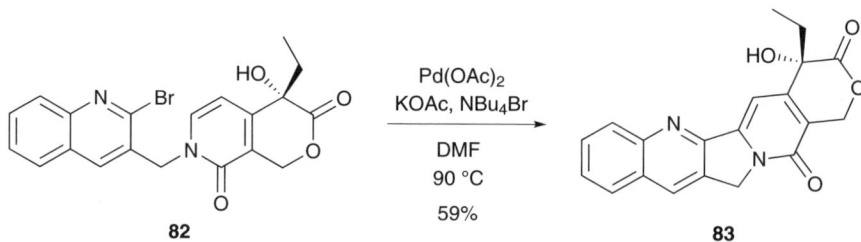

Scheme 6.23   *Last step in the total synthesis of (S)-camptothecin.*

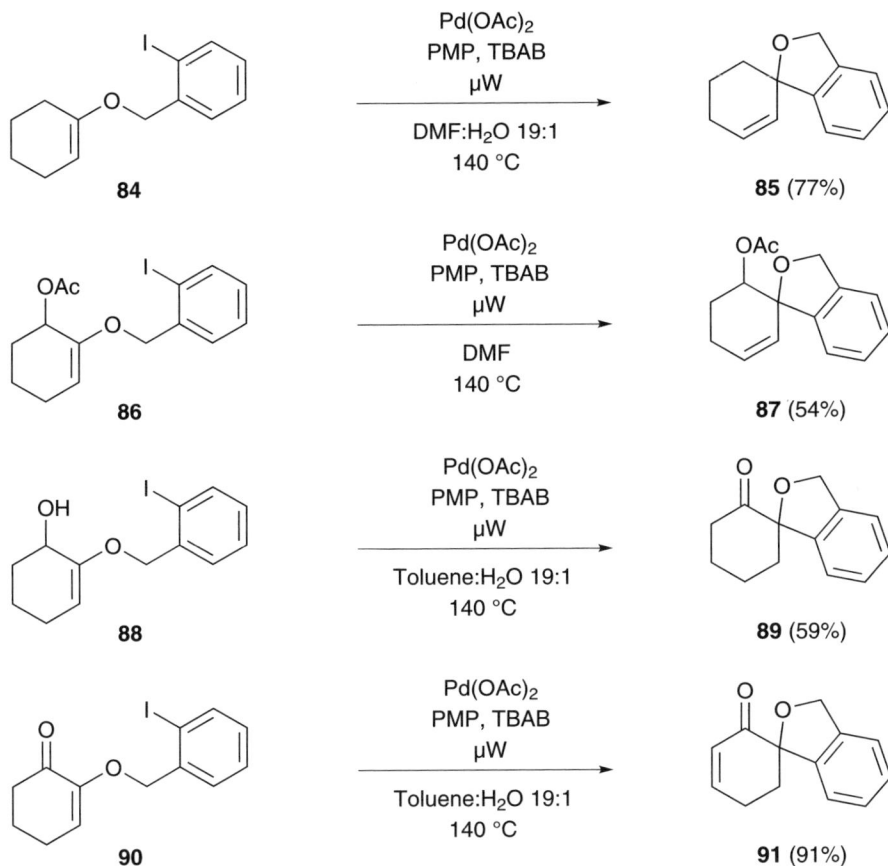

Scheme 6.24 *Microwave-accelerated spiro-cyclization of o-iodobenzyl cyclohexyl ethers.*

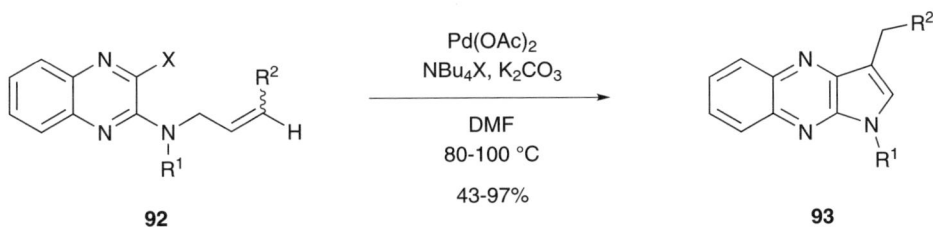

X = Br, Cl; R^1 = H, allyl, benzyl; R^2 = H, alkyl, phenyl, benzyloxymethyl

Scheme 6.25 *Synthesis of 3-substituted pyrrolo[2,3-b]quinoxalines via intramolecular Mizoroki–Heck reaction.*

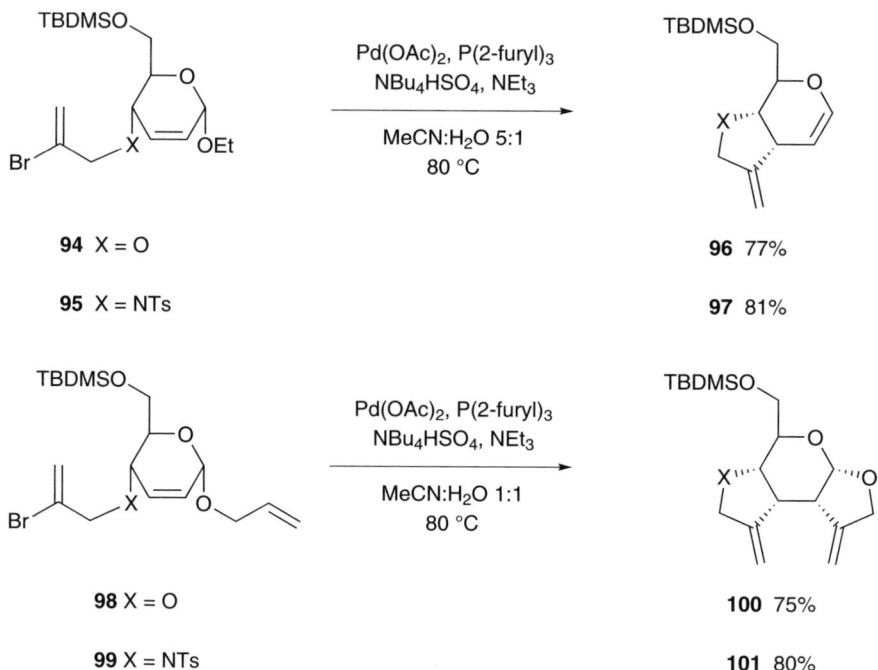

Scheme 6.26   *Mizoroki–Heck-mediated hetero-cyclizations of pyranoside templates.*

palladium-mediated cyclizations in the field of sugar chemistry have mainly been explored by Sinou and coworkers [76, 77]. The bi- or tri-cyclic systems are generally obtained under mild conditions and in good yields starting from the corresponding unsaturated carbohydrates.

Aside from these most frequently met substrate types leading to 5-*exo*-cyclizations discussed above, there are several examples of further substructures that also give rise to intramolecular Mizoroki–Heck reactions via 5-*exo*-cyclizations.

An example of such structures is an α-haloamide skeleton bearing a double bond. In this context, Mori *et al.* [78] reported the synthesis of pyrrolizidine derivatives **103**, **104** and **105** from *N*-iodoacetyl-2-vinylpyrrolidine (**102**) (Scheme 6.27).

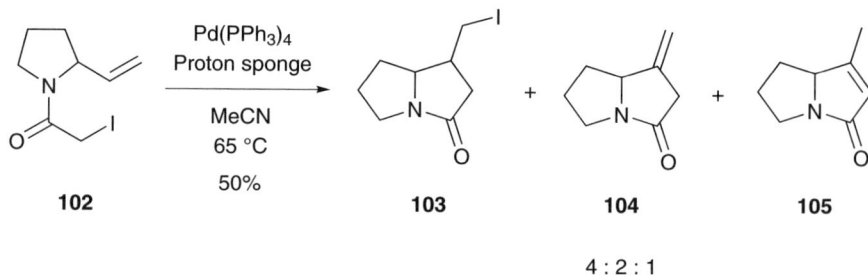

4 : 2 : 1

Scheme 6.27   *Synthesis of pyrrolizidine derivatives.*

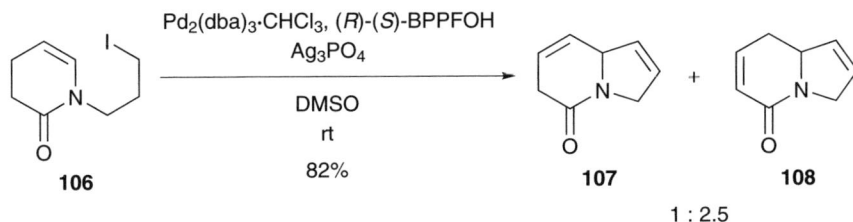

Scheme 6.28   *Synthesis of indolizidine derivatives.*

In another example, a vinyl iodide bearing a cyclic alkene on the heteroatom-containing side chain was employed (Scheme 6.28) [79]. It describes the asymmetric synthesis of indolizidine derivatives with optimized conditions to get asymmetric **108** with good ee of 81%. As compound **107** can be readily isomerized to **108** using Pd/C in methanol, this method gives ready access to a whole range of indolizidines.

6.3.2   5-*endo*-Cyclization

Intramolecular Mizoroki–Heck reactions via 5-*endo*-cyclization, disfavoured according to the Baldwin rules, are by far not as common as the *exo*-cyclic variant. There are, nevertheless, some cases where this cyclization type can occur, notably when substrates of substructure **E** (Figure 6.3) are used in the presence of Mizoroki–Heck conditions.

To our knowledge, there are only examples of *N*-heterocycles being formed by a 5-*endo*-cyclization, most of which start with enaminone derivatives [80–85]. Sørensen and Pombo-Villar [86] reported a microwave-accelerated procedure where 3,4-dihydro-4-methylcyclopentyl[*b*]indol-1(2*H*)-one (**110**) was obtained in quantitative yield in 30 min compared to 16 h under classic conditions (Scheme 6.29).

Barluenga *et al.* [87] developed a one-pot synthesis for indoles from *o*-bromoanilines and alkenyl bromides via a cascade alkenyl amination/Mizoroki–Heck reaction. In order to improve the yields of their procedure, they studied the cyclization of preformed imine **113** to indole **114** in the presence of DavePhos and X-Phos as supporting ligands (Scheme 6.30).

Another novel one-pot synthesis of 2-aryl/vinylindoles based on a ruthenium-catalysed hydroamination and a palladium-catalysed Mizoroki–Heck reaction involves terminal or internal alkynes and 2-chloroaniline (Scheme 6.31) [88]. This method is particularly interesting because it uses readily available and relatively cheap chloro-derivatives. The

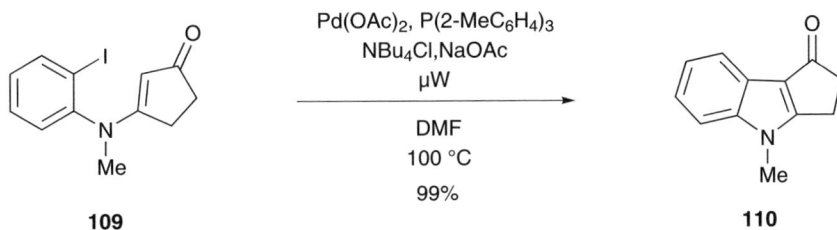

Scheme 6.29   *Microwave-accelerated synthesis of 3,4-dihydro-4-methylcyclopental[b]indol-1(2H)-one **110**.*

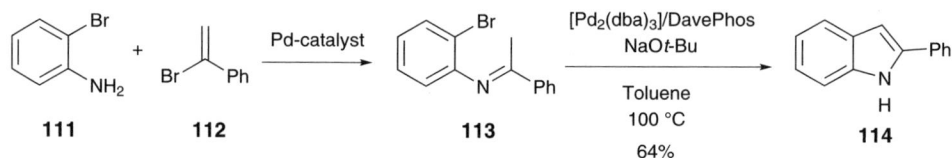

Scheme 6.30   *Synthesis of indole 114.*

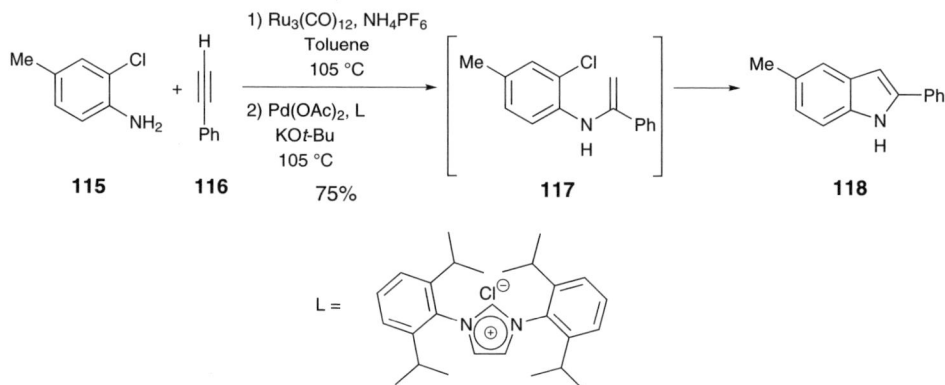

Scheme 6.31   *One-pot synthesis of indole 118.*

intramolecular Mizoroki–Heck reaction of the *in situ*-generated 2-chloroanilino enamine **117** was accomplished using an *N*-heterocyclic carbene-derived ligand L.

Another Mizoroki–Heck-type 5-*endo*-cyclization involves 1,1-difluoro- and 2,2,2-trifluoroalkenes during which a β-fluorine is eliminated after insertion of the double bond. Ichikawa *et al.* [89] pioneered this process and used it to synthesize 4-difluoromethylene-1-pyrroline **120** and 5-fluoro-3*H*-pyrrole [91] **122** (Scheme 6.32).

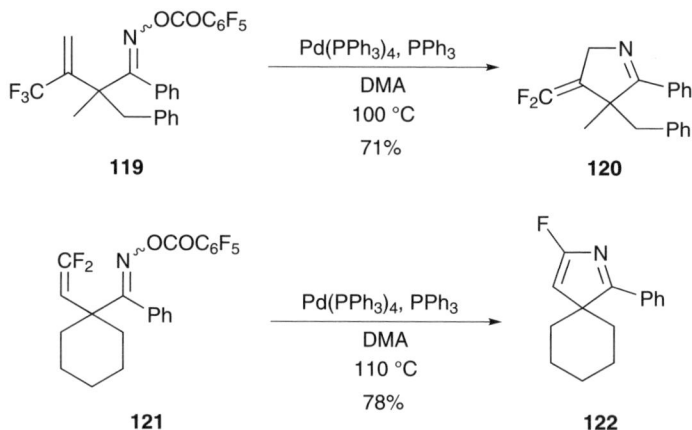

Scheme 6.32   *Synthesis of 4-difluoromethylene-1-pyrroline 120 and 5-fluoro-3H-pyrrole 122.*

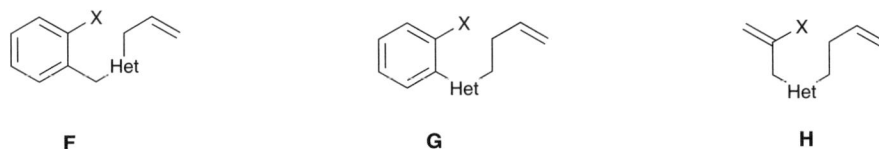

F   G   H

Figure 6.4 *Most frequently encountered precursors for the formation of six-membered rings.*

6.4 Formation of Six-Membered Cycles

The three most widespread substrate substructures for six-membered ring closure are represented in Figure 6.4. This inventory is non exhaustive and other substructures giving six-membered cycles will be cited at the end of this section.

6.4.1 6-*exo*-Cyclizations

Formation of six-membered cycles is the second most common case of heterocyclic Mizoroki–Heck reactions, with, as for the formation of five-membered rings, *exo*-cyclization reactions being largely favoured compared to *endo*-cyclization reactions. As already noted for the five-membered cycles, there are more examples of *N*-heterocyclic formation than of oxygen-containing heterocycles.

Three substrate substructures are particularly often mentioned for giving 6-*exo*-cyclizations, among which allyl side chain containing skeleton **F** is most prevalent (Figure 6.4) [91–95]. Some representative examples of substructure **F** 6-*exo*-cyclizations are treated in detail below.

In order to optimize the conditions for a one-pot synthesis of 4-spiroannulated tetrahydroisoquinolines, Broggini and coworkers [96] had a deeper look at the Mizoroki–Heck conditions to convert precursor **123** into tetrahydroisoquinoline derivative **124** (Scheme 6.33). No isomerized product with the thermodynamically more stable *endo*-cyclic double bond could be detected.

Monteiro and coworkers [97] reported the synthesis of pyrrolo[2,1,*c*]pyrrolidine derivate **126**, which was again the sole isolated product under optimized conditions; no traces of the *endo*-cyclic double-bond isomer could be detected (Scheme 6.34).

Herradón and coworkers [98] showed that the stereochemistry of the newly formed *exo*-cyclic double bond is dependent on substrate and reaction conditions in selected cases. Cyclization of **127** in the presence of a silver salt gave pyrido[1,2-*b*]-isoquinoline derivative **128** quantitatively as a single (*Z*)-isomer at the *exo*-cyclic double bond; no conversion was seen at all under Jeffery conditions (Scheme 6.35). Conversely, the deoxo substrate **129** was

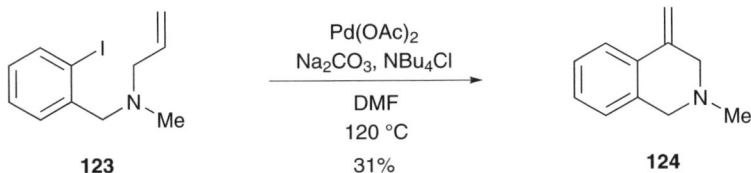

123   31%   124

Scheme 6.33 *Synthesis of tetrahydroisoquinoline derivative **124**.*

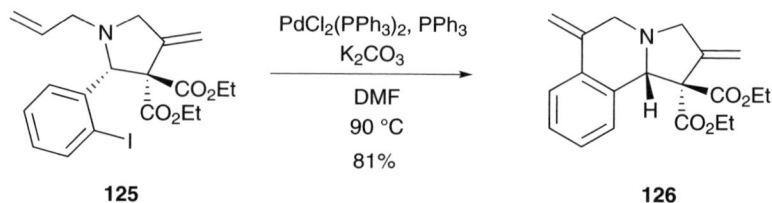

Scheme 6.34   *Synthesis of pyrrolo[2,1,c]pyrrolidine derivative* **126**.

unreactive under the previously successful experimental setup, but it delivered the expected tricyclic product as a 9 : 1 mixture of the (Z)-**130** and (E)-**131** under Jeffery conditions.

Substructure **F** (Figure 6.4) has found application in total synthesis as well, notably during the synthesis of lycoricidine, a member of the Amaryllidaceae alkaloids. Several groups worked quasi-simultaneously on the synthesis of this polycyclic alkaloid [99, 100]. The best results concerning the 6-*exo*-cyclization of key intermediate **132** were obtained by Ogawa and coworkers [101] and delivered the 2-epimer of (+)-lycoricidine **133** in 68% yield (Scheme 6.36). The use of thallium(I) acetate was crucial for the cyclization to proceed properly.

O-Heterocycles have also been prepared from substrate substructures **F** (Figure 6.4). Sinou and coworkers [102] reported the synthesis of enantiopure *cis*-fused pyrano[2,3,*c*]pyrans from the corresponding glycals via an intramolecular Mizoroki–Heck cyclization either followed by classic β-hydride or by β-alkoxy elimination (Scheme 6.37).

Bankston *et al.* [103] studied the synergistic effect of rhodium(I) addition during the intramolecular Mizoroki–Heck reaction of some crotyl ethers (Scheme 6.38).

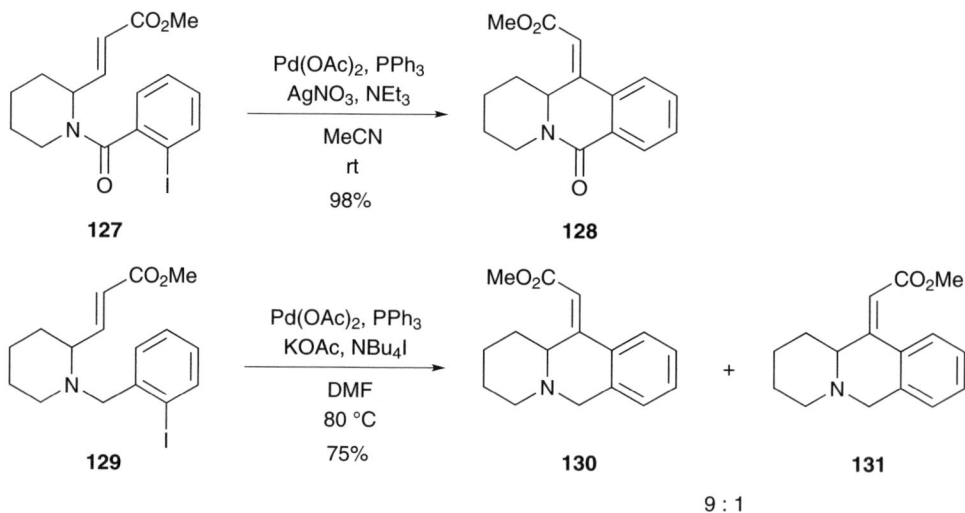

Scheme 6.35   *Synthesis of pyrido[1,2-b]-isoquinoline derivative* **128** *and its deoxo analogues* **130** *and* **131**.

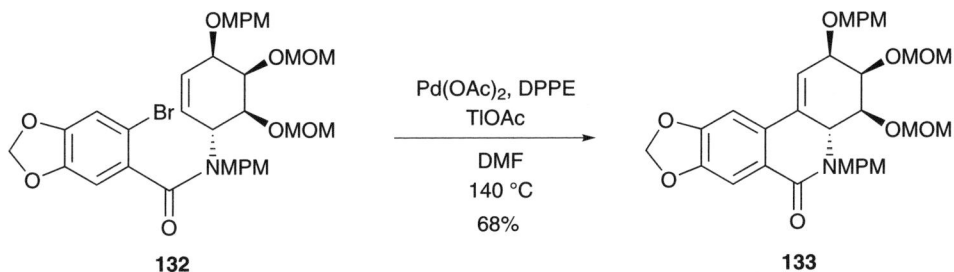

Scheme 6.36 Intramolecular Mizoroki–Heck reaction as one key step in the total synthesis of the 2-epimer of (+)-lycoricidine.

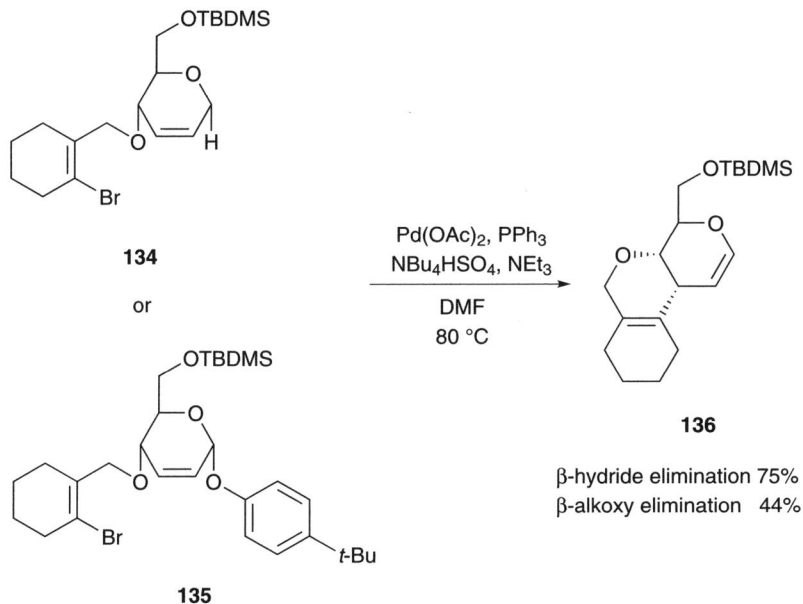

Scheme 6.37 Synthesis of pyrano[2,3,c]pyrans through classic β-hydrogen or via β-alkoxy elimination.

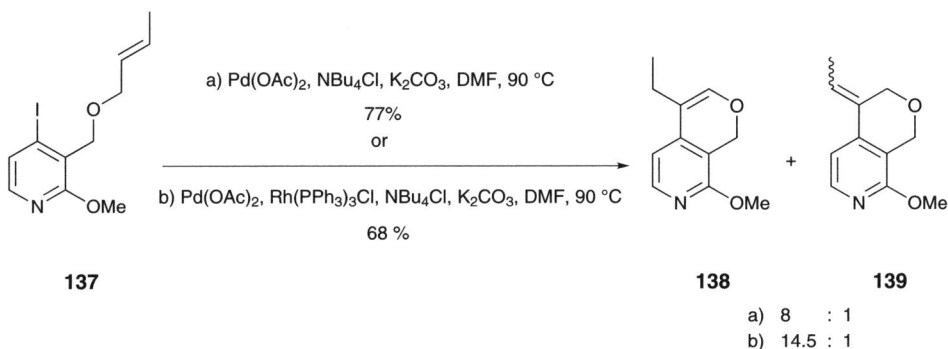

a) Pd(OAc)$_2$, NBu$_4$Cl, K$_2$CO$_3$, DMF, 90 °C

77%

or

b) Pd(OAc)$_2$, Rh(PPh$_3$)$_3$Cl, NBu$_4$Cl, K$_2$CO$_3$, DMF, 90 °C

68 %

a) 8   : 1
b) 14.5 : 1

Scheme 6.38 Synergistic effect of Rh(I) addition during intramolecular Mizoroki–Heck reaction of crotyl ether **137**.

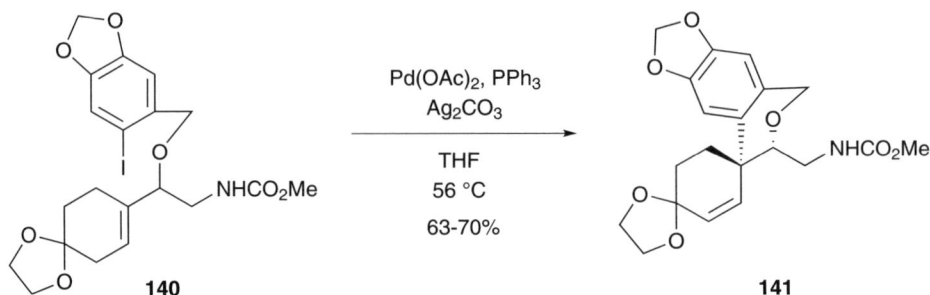

Scheme 6.39   *Synthesis of key intermediate **141** in the total synthesis of (±)-tazettine.*

A greater selectivity at faster rate along with greater reproducibility compared to classical Mizoroki–Heck conditions were observed in favour of the *endo*-cyclic product **138**.

Other *O*-heterocyclic ring formations from substructure **F** (Figure 6.4) concern the preparation of isochromanes [104] or the study of the effect of ether versus ester tethering during the cyclization [105].

There are also examples citing substructure **F** in the total synthesis of natural products containing *O*-heterocycles, namely in a synthetic approach to (20*S*)-camptothecin [106] or in the synthesis of the Amaryllidaceae alkaloids (±)-tazettine and (±)-6a-epipretazettine (Scheme 6.39) [107].

Another structure mentioned several times in the literature as a suitable substrate for 6-*exo*-cyclizations is substructure **G** (Figure 6.4) [108, 109]. Larock and Babu [110] prepared 4-methylisoquinoline (**143**) from *N*-(but-3-enyl)-2-iodoaniline (**142**) under very mild conditions in nearly quantitative yield (Scheme 6.40). The cyclized compound completely isomerizes to the aromatic isoquinoline; no *exo*-cyclic product could be detected. It is worth mentioning that the reaction proceeds cleanly without a protecting group on the secondary amine, a group which has been known to cause problems during intramolecular cyclizations if left unprotected [8].

An intramolecular microwave-assisted Mizoroki–Heck reaction has been described by Smalley and Mills [111] for the preparation of quinol-2-one derivatives. When amide **144** was subjected to standard Mizoroki–Heck conditions under microwave irradiation,

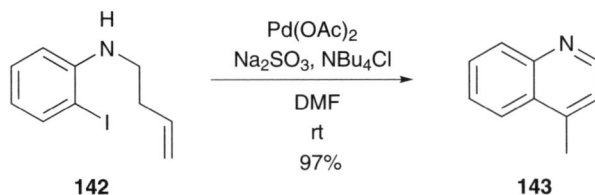

Scheme 6.40   *Synthesis of 4-methylisoquinoline (**143**) by intramolecular Mizoroki–Heck reaction.*

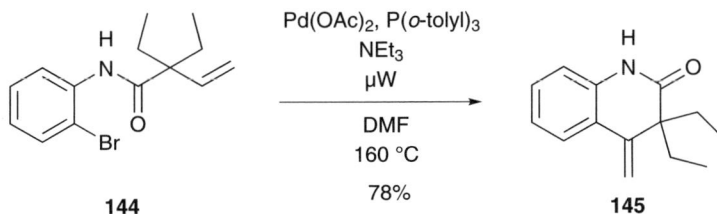

Scheme 6.41 *Microwave-assisted synthesis of 3,3-diethyl-4-(methylene)-1-quinol-2-one (145).*

3,3-diethyl-4-(methylene)-1-quinol-2-one (**145**) was obtain as sole product in 78% after only 15 min, without traces of a 7-*endo*-cyclization product (Scheme 6.41).

Guillou *et al.* [112] used the same type of 6-*exo*-cyclization in the total synthesis of (±)-galanthamine, a natural product that we have already come across in the section on 5-*exo*-cyclizations.

In the substructure **G**-series (Figure 6.4), *O*-heterocycles have also been prepared. Söderberg *et al.* [113] reacted compound **146** under standard Mizoroki–Heck conditions to give an 18:4:1 mixture of the three isomeric 5-nitro-1-benzopyrans **147**:**148**:**149** (Scheme 6.42).

An intramolecular Mizoroki–Heck reaction of substructure **G** (Figure 6.4) is also the key step in the formal total synthesis of (±)-lycoramine from the Amaryllidaceae family [114]. Cyclization of **150** was best accomplished in nonclassical conditions using a bidentate ligand to give α,β-unsaturated ketone **151** in 50% yield (Scheme 6.43).

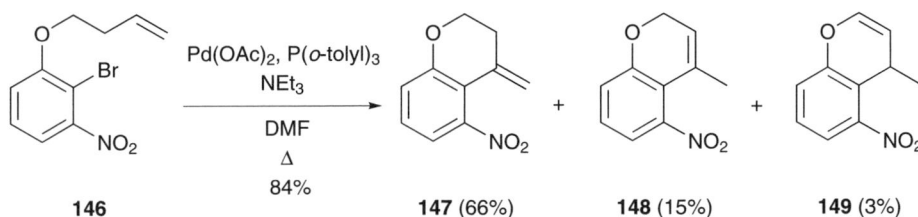

Scheme 6.42 *Preparation of 5-nitro-1-benzopyrans.*

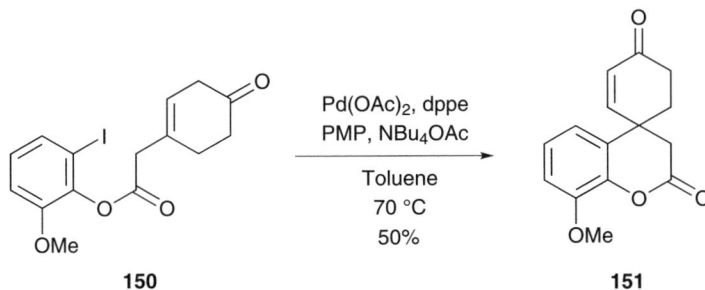

Scheme 6.43 *Formation of spiro compound 151 via intramolecular Mizoroki–Heck reaction.*

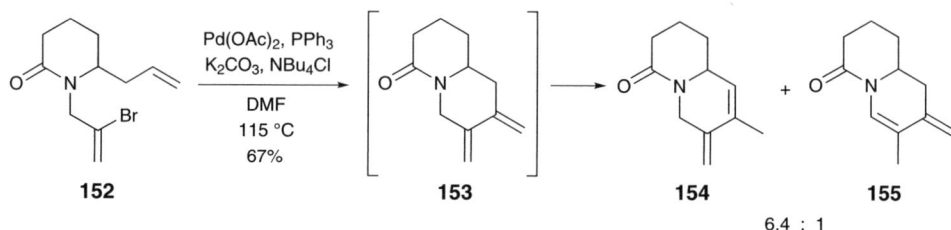

Scheme 6.44 *Synthesis of quinolizidinones 154 and 155.*

The last, relatively often-cited, substructure leading to 6-*exo*-cyclizations is vinyl halide **H** (Figure 6.4). Santos and Pilli [110] reported the synthesis of quinolizidinones **154** and **155** from lactam **152** (Scheme 6.44). The reaction probably proceeds via an exclusive 6-*exo*-cyclization, giving *exo*-bismethylene **153**, followed by either thermally promoted proton shift or through a sequence of readdition and elimination of the hydridopalladium species.

The ultimate step in the total synthesis of (±)-dehydrotubifoline (**157**), a *Strychnos* alkaloid, is an intramolecular Mizoroki–Heck reaction of **H**-like (Figure 6.4) substrate **156** (Scheme 6.45) [115]. This ring closure preserved the integrity of the double-bond stereochemistry and delivered, after the excepted enamine–imine tautomerization, the desired product **157** in excellent yield.

In addition, substructure **H** (Figure 6.4) has also been employed in other total syntheses, namely in a novel route to the geissoschizine skeleton [116] and in the enantioselective total synthesis of (−)-strychnine [117].

As was the case for 5-*exo*-cyclizations, aside from the more frequently encountered substructures that undergo 6-*exo*-cyclization, many miscellaneous examples can be found throughout the literature [118–120]. One study concerns the effect of the stereochemistry of vinyl bromo substrates **158** and **160** on the ratio of possible double-bond isomers in the product (Scheme 6.46) [121]. Starting from the *cis*-compound **158**, bisunsaturated indolizidinone **159** was obtained along with only traces of the bis-*exo*-cyclic diene. The *trans*-compound **160** reacted twice as fast and gave a 4 : 1 mixture of **161** : **162**. Product **161** is probably the first to be formed, and its more stable isomer **162** is most likely formed by a readdition and elimination of HPdBr.

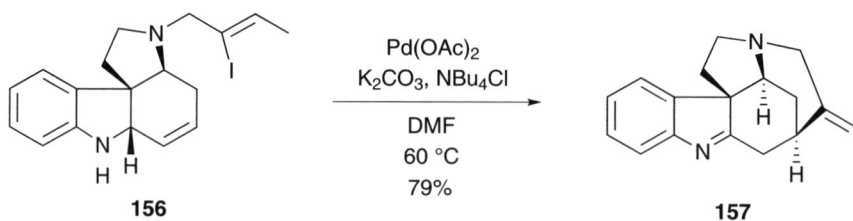

Scheme 6.45 *Synthesis of (±)-dehydrotubifoline 157.*

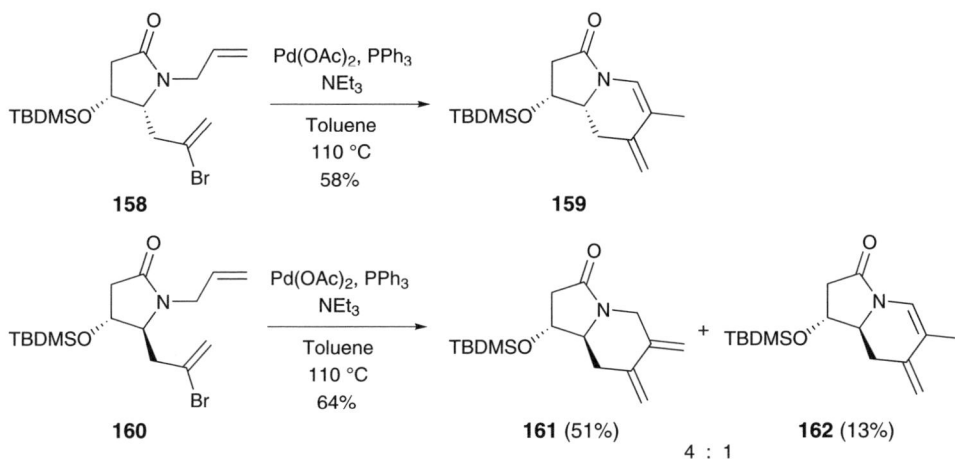

Scheme 6.46   *Synthesis of indolizidinone derivatives.*

Another example is the generation of a tricyclic structure starting from alkyne **163** (Scheme 6.47) [122]. The reaction proceeds firstly through an 6-*exo*-cyclization to generate alkyl-palladium species **164**. Subsequent elimination yields the conjugated triene intermediate **165** whose isomerization is suppressed by silver salt which enables an electrocyclic rearrangement to give tricycle **166**.

Heterocycles containing two covalently linked heteroatoms can also be prepared via intramolecular Mizoroki–Heck reaction, proving once more the huge application domain of palladium-catalysed processes. Danishefsky and coworkers [123], when submitting (*Z*)-enol ether **167** to palladium on charcoal in hot acetonitrile, observed a conversion to aldehyde **168** along with an unprecedented stereochemical inversion of the N–O bond oxygen relative to the dioxolane ring (Scheme 6.48). When alkene **169** was cyclized to **170** under similar conditions, no such inversion was observed. The authors state that the formation of the aldehyde from the enol ether proceeds via an intermediate that undergoes an Arbuzov-like rearrangement leading directly to an aldehyde. The stereochemistry of the

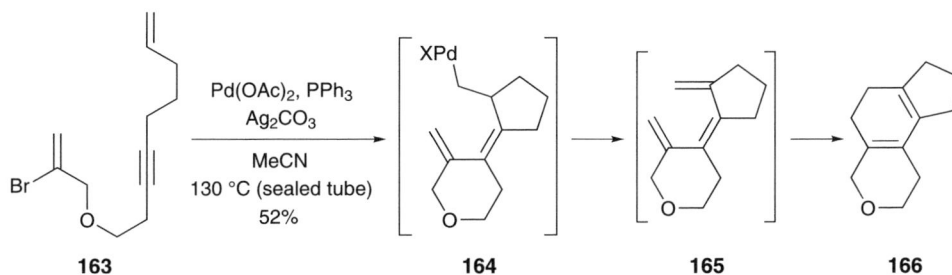

Scheme 6.47   *Synthesis of tricycle **166**.*

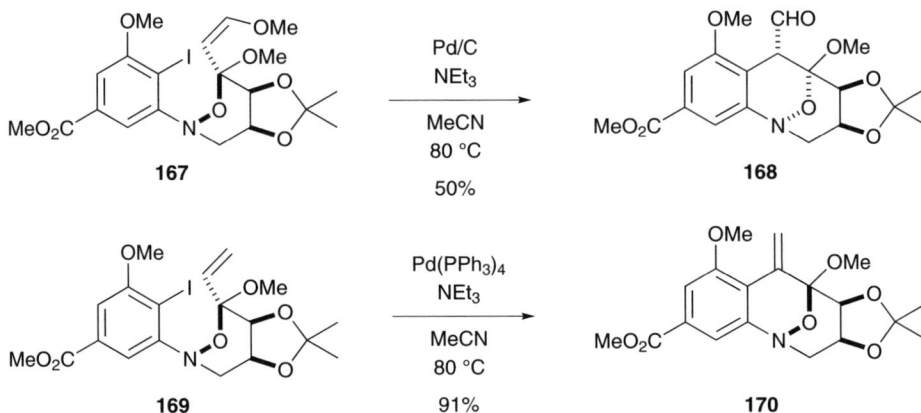

Scheme 6.48 *Unprecedented stereochemical inversion during intramolecular Heck–Mizoroki reaction.*

aldehyde arises from a face-specific attack of the aryl group *anti* to the methoxy function at C-8.

Cyclic sulfonamides can also be prepared via intramolecular Mizoroki–Heck reaction starting from pyrrolidine derivatives [124]. Evans [125] showed that alcohol **171**, under standard Mizoroki–Heck conditions, gave a 50 : 50 mixture of regioisomers **172** and **173** resulting from an unselective 6-*exo*-cyclization at both the 3- and 4-positions (Scheme 6.49). If protected alcohol **174** is used, more 4-isomer **175** was obtained, probably due to the sterically demanding pivaloyl group.

6.4.2 6-*endo*-Cyclization

As for 5-*endo*- versus 5-*exo*-cyclization, the 6-*exo*-cyclization is largely favoured compared to the 6-*endo* mode. There are, nevertheless, some examples of specific 6-*endo*-cyclizations using, as expected, substructures previously mentioned to undergo 5-*exo*-cyclizations [126–128]. They remain an exception, however, and are generally accompanied by other ring sizes.

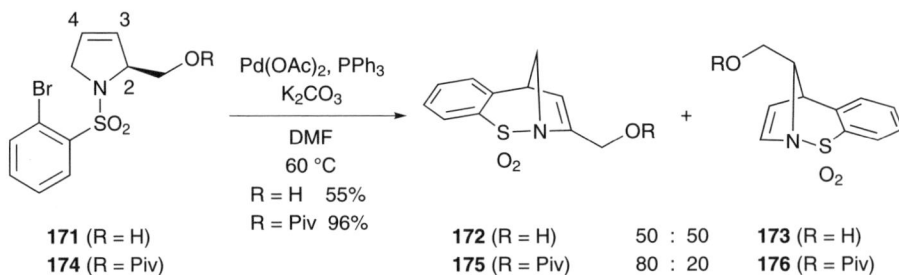

Scheme 6.49 *Synthesis of cyclic sulfonamides.*

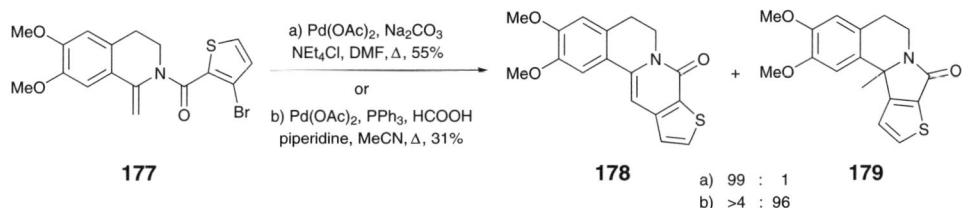

Scheme 6.50 *The 6-endo- versus 5-exo-cylization.*

An interesting case of 6-*endo*- versus 5-*exo*-cylization has been studied by Bombrun and Sageot [129], who investigated intramolecular Mizoroki–Heck cyclization of enamides in a thiophene series (Scheme 6.50). If the reaction was carried out under Jeffery conditions, then it proceeded nearly exclusively via the 6-*endo*-cyclization mode; however, when run under reductive Mizoroki–Heck conditions (e.g. in the presence of a hydride source), a 5-*exo*-cyclization is mostly achieved and the lack of a β-hydrogen in this case leads to product **179**.

Another example giving exclusive 6-*endo*-cyclization in excellent yield is the cyclization of methylenephtalimidine precursor **180** to tetracyclic amide **181** (Scheme 6.51) [130].

6.5 Formation of Seven-Membered Cycles

The preparation of seven-membered heterocycles via intramolecular Mizoroki–Heck reaction is by far not as frequent as the syntheses of five- and six-membered rings. There are, however, some examples, although these reactions started to be properly explored only in the late 1990s. The substrate substructures which one often comes across are summarized in Figure 6.5.

As already mentioned, the increase in the ring size goes along with an enhanced preference for the *endo*-cyclization mode. In the case of seven-membered rings, one seems to have reached the summit, as there are approximately equal amounts of *exo*- and *endo*-cyclization cases reported in the literature.

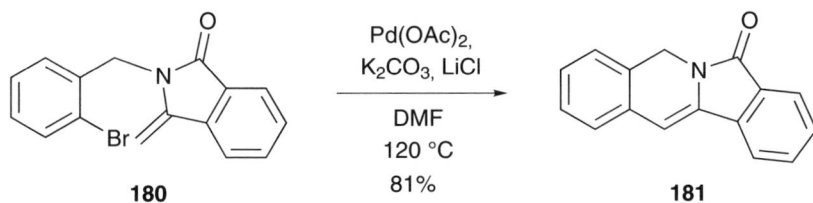

Scheme 6.51 *High-yielding 6-endo-cylization.*

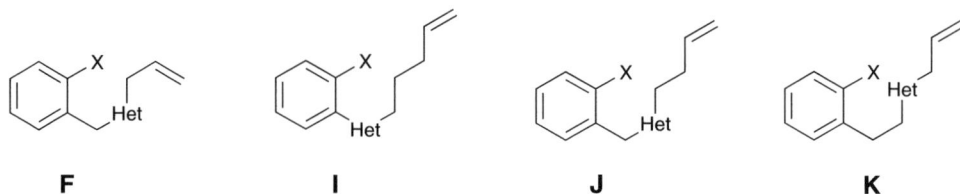

Figure 6.5 *Widespread precursors for the formation of seven-membered rings.*

6.5.1 7-*exo*-Cyclizations

Substrates for 7-*exo*-cyclizations can be classified into three major substructures: **I**, **J** and **K** (Figure 6.5).

Skrydstrup and coworkers [131] used substrate **I**-type **182** in their formal total synthesis of PKC inhibitor balanol to prepare lactone **183**; this was, however, accompanied by a certain amount of 8-*endo*-cyclization product **184** (Scheme 6.52).

Boger and Turnbull [132] studied the effect of two electron-rich and one electron-deficient acceptor alkenes on the cyclization. In all three cases, careful removal of the oxygen from the reaction mixture provided reproducible results and avoided palladium-hydride-induced isomerization (Scheme 6.53). Substrates bearing an activated electron-deficient acceptor alkene (e.g. **187** and **189**) delivered the desired products in high yield, whereas electron-rich acceptor alkene **185** failed to react.

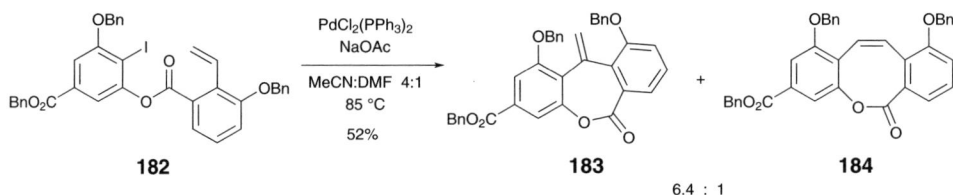

Scheme 6.52 *Preparation of lactone 183 via 7-exo-cyclization.*

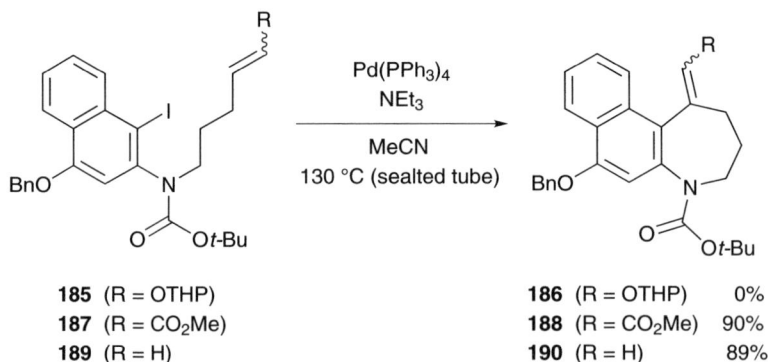

185 (R = OTHP)	**186** (R = OTHP) 0%
187 (R = CO₂Me)	**188** (R = CO₂Me) 90%
189 (R = H)	**190** (R = H) 89%

Scheme 6.53 *Seven-membered ring closure.*

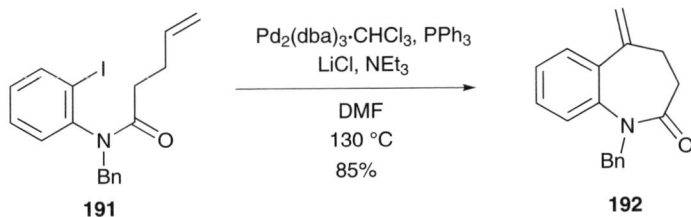

Scheme 6.54   *Synthesis of benzolactam **192**.*

Recently, Hii and coworkers [133] showed that the cyclization towards medium-sized rings was not only dependent on the catalytic conditions and nature of the substrate, but was also to a large extent reliant on the nature of the ligand. The latter was found to affect profoundly the chemo- and regio-selectivity and competitive double-bond migration. By employing a metal-to-ligand ratio of 1 : 4, the attempted cyclization to benzolactam **192** proceeded exclusively in an *exo* mode in good yield and no migration or isomerization occurred during the course of reaction (Scheme 6.54) [133].

There are also several examples of substructures **J** (Figure 6.5) giving 7-*exo*-cyclic compounds in good to excellent yields [134], three of which are presented below. Guillou and coworkers [135] reported the synthesis of the spirocyclic seven-membered compound **194** via an intramolecular Mizoroki–Heck reaction in good yield (Scheme 6.55).

Seven-membered lactam **196** was prepared in excellent yield by a microwave-assisted specific 7-*exo*-Mizoroki–Heck-cyclization reaction in only 25 min (Scheme 6.56) [136].

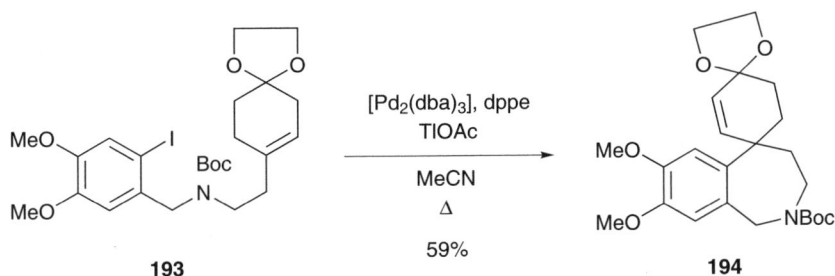

Scheme 6.55   *Synthesis of spiro-compound **194**.*

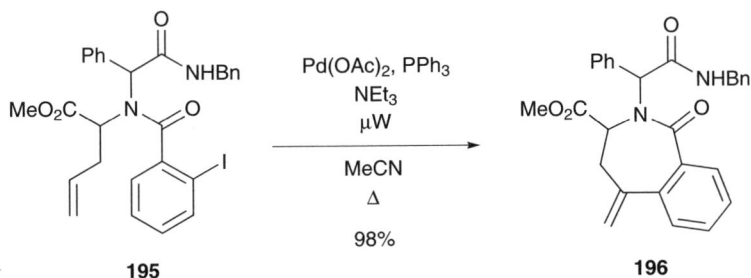

Scheme 6.56   *Microwave-assisted specific 7-exo-Mizoroki–Heck-cyclization.*

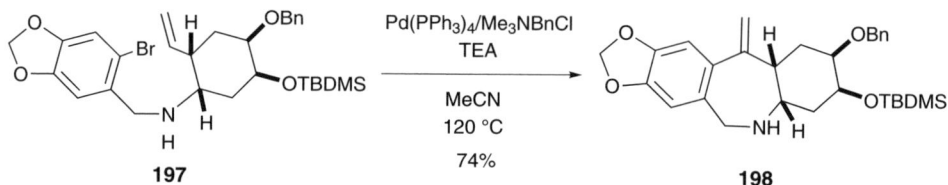

Scheme 6.57   *Specific 7-exo-Mizoroki–Heck-cyclization of bromo olefin **198**.*

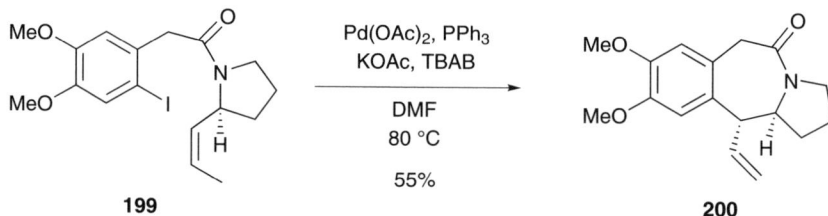

Scheme 6.58   *Diastereo- and regioselective 7-exo-Mizoroki–Heck-cyclization.*

Jin and Weinreb [137, 138] reported enantioselective total syntheses of the 5,11-methanomorphanthridine Amaryllidaceae alkaloids (−)-pancracine and (−)-coccinine. They used a specific Mizoroki–Heck-cyclization to convert bromo olefin **197** into seven-membered exocyclic alkene **198** in good yield (Scheme 6.57).

The last substrate substructure encountered relatively often is of type **K** (Figure 6.5). Several groups used this substructure to prepare benzazepine derivatives [139–143]. Tietze *et al.* [144], could avoid isomerization of the double bond through optimization of the reaction conditions and lactam **200** could be obtained in a diastereo- and regioselective Mizoroki–Heck reaction in 55% yield (Scheme 6.58).

6.5.2   7-*endo*-Cyclizations

Seven-membered ring closures via *endo*-cyclization are by far more widespread than the analogous *endo*-modes for five- or six-membered cycles. There are some examples using substructure **G** (Figure 6.4) as substrate. Joseph and coworkers [145] reported the synthesis of fused heterocycles bearing a benzazepinone moiety. Tetracycle **202** was obtained in excellent yield starting from indole derivative **201** in the presence of silver salt (Scheme 6.59).

The substrate of choice for a 7-*endo*-cyclization, however, seems to be substructure **F** (Figure 6.5), as corroborated by the number of recent publications in this domain [146–148].

Joseph and coworkers [147] cyclized compound **203** under identical conditions as depicted above to give bent tetracycle **204** in nearly quantitative yield (Scheme 6.60).

Lamaty and coworkers [148] investigated microwave-assisted Mizoroki–Heck reactions in poly(ethylene glycol) for the synthesis of benzazepines and developed a rapid, generally high-yielding preparation method for differently substituted benzazepines **206** (Scheme 6.61).

Finally, Ohishi and coworkers [149] reported the synthesis of novel furo[2,3,4-*jk*][2]benzazepin-4(3*H*)-one derivatives via 7-*endo*-cyclization in modest to good yields.

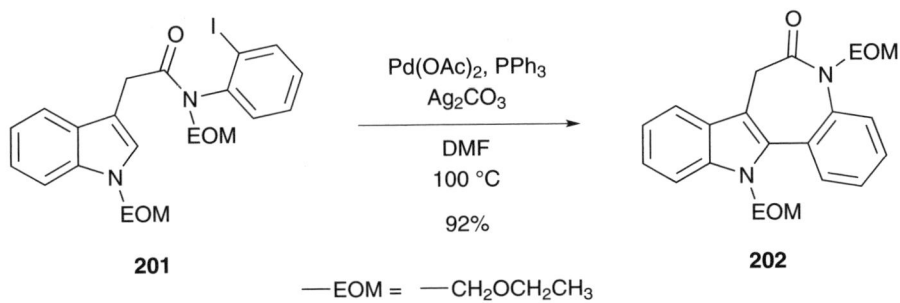

Scheme 6.59 *Synthesis of tetracycle* **202**.

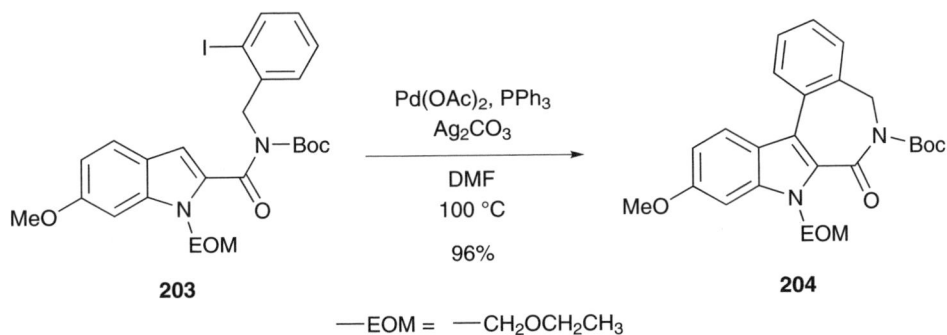

Scheme 6.60 *Synthesis of tetracycle* **204**.

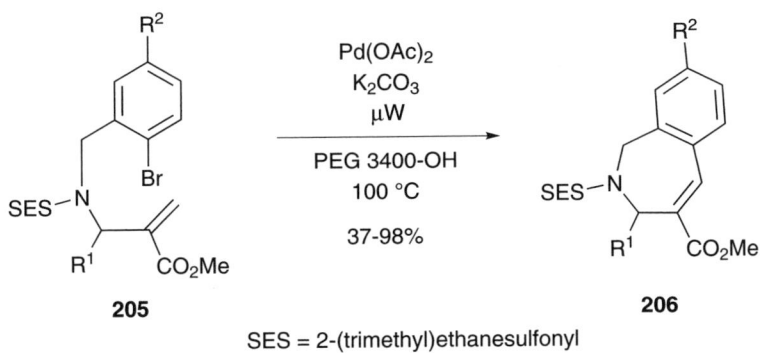

SES = 2-(trimethyl)ethanesulfonyl

Scheme 6.61 *Microwave-assisted high-yielding synthesis of substituted benzazepines.*

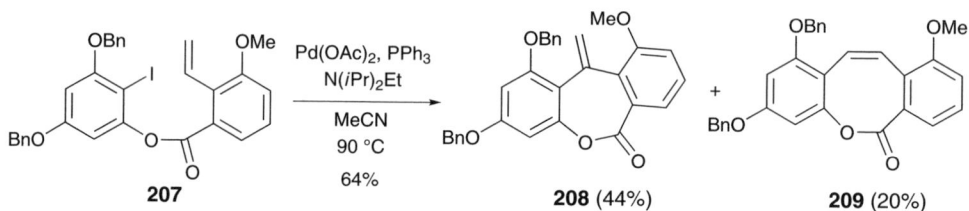

Scheme 6.62   *The 8-endo- versus 7-exo-cyclization.*

6.6   Formation of Eight- and Nine-Membered Cycles

The concise formation of medium-size eight- and nine-membered cycles is relatively intricate; however, there are a few recent literature examples [150–154].

In general, 8-*endo*-cyclizations give mixtures due to competing 7-*exo*-ring closure. Skrydstrup and coworkers [155] showed that the 8-*endo*-cyclization product **209** could only be isolated in 20% yield, the major product being the 7-*exo*-cyclization product (Scheme 6.62).

Hii and coworkers [133] developed an exclusive 8-*exo*-cyclization which allows one to obtain benzazocinone **211** as the sole isomer in excellent yield starting from amide **210** (Scheme 6.63).

One of the rare reports on nine-ring closure describes a mixture of compounds being obtained in modest yield from competing 8-*exo*- and 9-*endo*-cyclization (Scheme 6.64) [156].

Gibson *et al.* [152], in their bromoarene-based approach to phenylalanine analogs, describe a 9-*endo*-cyclization occurring with an impressive yield of 69% for a typical preparative reaction scale (Scheme 6.65).

6.7   Formation of Macrocycles

Large rings are often encountered in natural product synthesis and there are, to date, relatively few methods to prepare them in adequate yields. One tool regularly employed is the

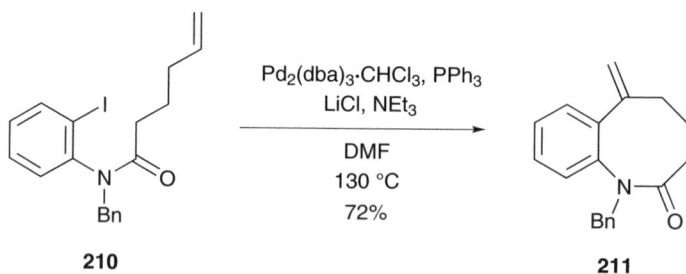

Scheme 6.63   *Exclusive high-yielding 8-exo-cyclization.*

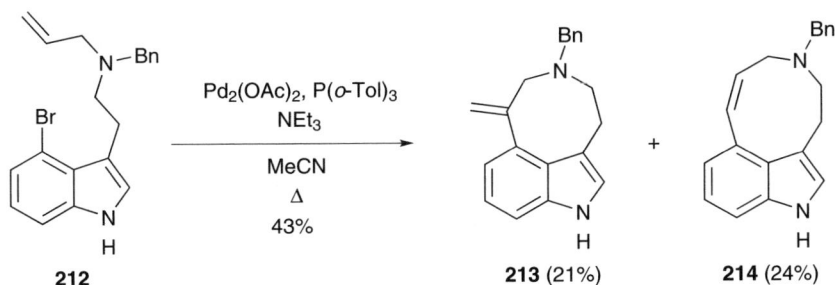

Scheme 6.64   *The 9-endo- versus 8-exo-cyclization.*

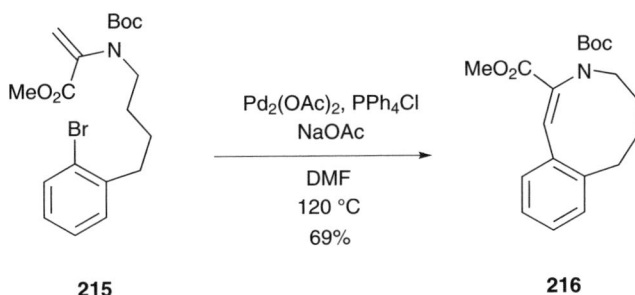

Scheme 6.65   *High yielding 9-endo-cyclization.*

macrocyclic Mizoroki–Heck reaction, which generally delivers the products in acceptable yields by exclusive *endo*-cyclization [157–160].

The first example of a Mizoroki–Heck-mediated macroheterocyclization has been reported by Ziegler *et al.* [161], who prepared conjugated dienone **218** from terminal vinylic iodide **217** (Scheme 6.66).

Regioselective Mizoroki–Heck macrocyclization was also a key step in the synthesis of macrocyclic taxoids. Ojima and coworkers [162] prepared macrocyclic taxoid **220** under very mild standard Mizoroki–Heck conditions in an excellent yield of 87% over two steps (Scheme 6.67).

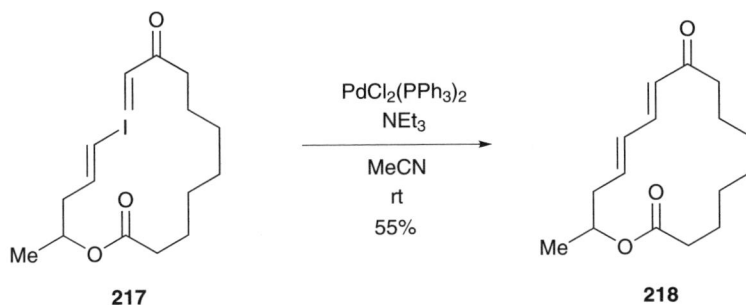

Scheme 6.66   *Synthesis of dienone 218.*

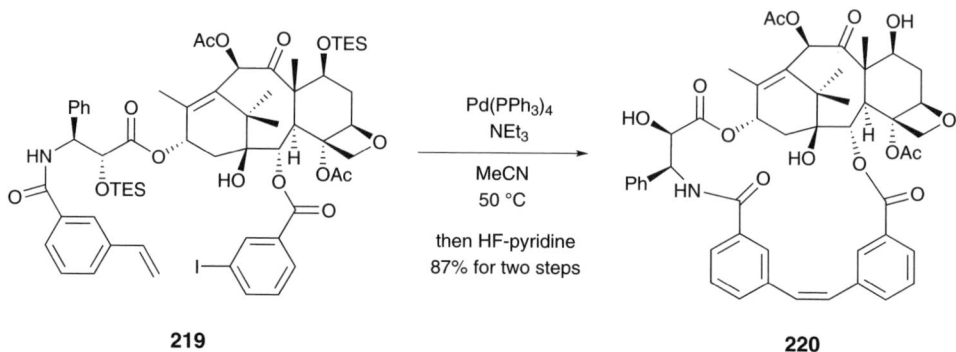

219   **220**

Scheme 6.67   *High-yielding synthesis of macrocyclic taxoid* **220**.

This example of a macrocyclization proves once again the high flexibility and tolerance towards a wide range of functional groups of the Mizoroki–Heck reaction.

6.8 Conclusion

The synthesis of heterocycles via intramolecular Mizoroki–Heck reaction is still an expanding field of current research due to the significant importance of heterocycles, thereby engendering the need for reliable, specific and high-yielding preparation methods. The Mizoroki–Heck reaction is one of the chemist's synthetic tools matching best these criteria and has only begun to reveal its full synthetic potential in this specific domain.

References

1. Bräse, S. and de Meijere, A. (2004) Cross coupling of organic halides with alkenes: the Heck reaction, in *Metal-Catalyzed Cross-Coupling Reactions*, Vol. **1** (eds A. de Meijere and F. Diederich), Wiley–VCH Verlag GmbH, Weinheim, pp. 217–315.
2. Bräse, S. and de Meijere, A. (2002) Background, in *Handbook of Organo-Palladium Chemistry* (ed. E.-i. Negishi), John Wiley & Sons Inc., New York, pp. 1123–32.
3. Bräse, S. and de Meijere, A. (2002) Double and multiple Heck reaction, in *Handbook of Organo-Palladium Chemistry* (ed. E.-i. Negishi), John Wiley & Sons Inc., New York, pp. 1179–208.
4. Bräse, S. and de Meijere, A. (2002) Palladium-catalyzed tandem and cascade carbopalladation – termination with alkenes, arenes and related π-compounds, in *Handbook of Organo-Palladium Chemistry* (ed. E.-i. Negishi), John Wiley & Sons Inc., New York, pp. 1369–405.
5. Bräse, S., Köbberling, J. and Griebenow, N. (2002) Organopalladium reactions in combinatorial chemistry, in *Handbook of Organo-Palladium Chemistry* (ed. E.-i. Negishi), John Wiley & Sons Inc., New York, pp. 3031–128.
6. Balme, G., Bossharth, E. and Monteiro, N. (2003) Pd-assisted multicomponent synthesis of heterocycles. *Eur. J. Org. Chem.*, 4101–11.

7. Mori, M., Chiba, K. and Ban, Y. (1977) New synthesis of indoles and isoquinolines by intramolecular palladium-catalyzed reactions of aryl halides with olefinic bonds. *Tetrahedron Lett.*, 1037–40.

8. Gibson, S.E. and Middleton, R.J. (1996) The intramolecular Heck reaction. *Contemp. Org. Synth.*, **3**, 447–71.

9. Dounay, A.B. and Overman, L.E. (2003) The asymmetric intramolecular Heck reaction in natural product total synthesis. *Chem. Rev.*, **103**, 2945–63.

10. Shibasaki, M., Boden, C.D.J. and Kojima, A. (1997) The asymmetric Heck reaction. *Tetrahedron*, **53**, 7371–95.

11. Ikeda, M., El Bialy, S.A.A. and Yakura, T. (1999) Synthesis of heterocycles using the intramolecular Heck reaction involving a 'formal' anti-elimination process. *Heterocycles*, **51**, 1957–70.

12. Dyker, G. (2002) Synthesis of heterocycles, in *Handbook of Organopalladium Chemistry for Organic Synthesis* (ed. E.-i. Negishi), John Wiley & Sons Inc., New York, pp. 1255–82.

13. Arnold, L.A., Luo, W. and Guy, R.K. (2004) Synthesis of medium ring heterocycles using an intramolecular Heck reaction. *Org. Lett.*, **6**, 3005–7.

14. Zeni, G. and Larock, R.C. (2006) Synthesis of heterocycles via palladium-catalyzed oxidative addition. *Chem. Rev.*, **106**, 4644–80.

15. Guiry, P.J. and Kiely, D. (2004) The development of the intramolecular asymmetric Heck reaction. *Curr. Org. Chem.*, **8**, 781–94.

16. Grigg, R., Santhakumar, V., Sridharan, V. *et al.* (1991) The synthesis of bridged-ring carbo- and hetero-cycles via palladium catalysed regiospecific cyclisation reactions. *Tetrahedron*, **47**, 9703–20.

17. Beletskaya, I.P. and Cheprakov, A.V. (2000) The Heck reaction as sharpening stone of palladium catalysis. *Chem. Rev.*, **100**, 3009–66.

18. Dyker, G. (1997) Palladacycles as reactive intermediates. *Chem. Ber./Recueil*, **130**, 1567–78.

19. Bräse, S., Wertal, H., Frank, D. *et al.* (2005) Intramolecular Heck couplings and cycloisomerizations of bromodienes and enynes with 1′,1′-disubstituted methylenecyclopropane terminators: efficient syntheses of [3]dendralenes. *Eur. J. Org. Chem.*, 4167–78.

20. Cabri, W. and Candiani, I. (1995) Recent developments and new perspectives in the Heck reaction. *Acc. Chem. Res.*, **28**, 2–7.

21. Svennebring, A. (2006) *Fast Microwave-enhanced Intra, Pseudo-intra- and Intermolecular Heck Reactions, Digital Comprehensive Summaries of the Uppsala Dissertations from the Faculty of Pharmacy 45*, Uppsala University.

22. Grigg, R., Loganathan, V., Santhakumar, V. *et al.* (1991) Suppression of alkene isomerisation in products from intramolecular Heck reactions by addition of Tl(I) salts. *Tetrahedron Lett.*, **32**, 687–90.

23. Maeda, K., Farrington, E.J., Galardon, E. *et al.* (2002) Competing regiochemical pathways in the Heck arylation of 1,2-dihydronaphthalene. *Adv. Synth. Catal.*, **344**, 104–9.

24. Lautens, M. and Fang, Y.-Q. (2003) Synthesis of novel tetracycles via an intramolecular Heck reaction with anti-hydride elimination. *Org. Lett.*, **5**, 3679–82.

25. Jeffery, T. (1984) Palladium-catalysed vinylation of vinylic halides under solid–liquid phase transfer conditions. *J. Chem. Soc., Chem. Commun.*, 1287–9.

26. Jeffery, T. (1985) Highly stereospecific palladium-catalysed vinylation of vinylic halides under solid–liquid phase transfer conditions. *Tetrahedron Lett.*, **26**, 2667–70.

27. Jeffery, T. (1996) On the efficiency of tetraalkylammonium salts in Heck type reactions. *Tetrahedron*, **52**, 10113–30.

28. Jeffery, T. (2000) Tetraalkylammonium salt-based catalyst systems for directing Heck-type reactions. Arylation of allyltrimethylsilane. *Tetrahedron Lett.*, **41**, 8445–9.

29. Burns, B., Grigg, R., Sridharan, V. and Worakun, T. (1988) Palladium catalysed tandem cyclisation-anion capture processes. Hydride ion capture by vinylpalladium species. *Tetrahedron Lett.*, **29**, 4325–8.

30. de Meijere, A. and Meyer, F.E. (1994) Fine feathers make fine birds: the Heck reaction in modern garb. *Angew. Chem., Int. Ed. Engl.*, **33**, 2379–411.

31. Sakamoto, T., Kondo, Y., Uchiyama, M. and Yamanaka, H. (1993) Concise synthesis of CC-1065/duocarmycin pharmacophore using the intramolecular Heck reaction. *J. Chem. Soc., Perkin Trans. 1*, 1941–2.

32. Inoue, M., Takahashi, T., Furuyama, H. and Hirama, M. (2006) Structural factors governing stereoselective Heck reaction for the construction of the oxindole portion of TMC-95A. *Synlett*, 3037–40.

33. Tietze, L.F. and Grote, T. (1994) Synthesis of (±)-N^2-(Benzenesulfonyl)-CPI, the protected A-unit of the antitumor antibiotic CC-1065, by two metal-initiated cyclizations. *J. Org. Chem.*, **59**, 192–6.

34. Kalinski, C., Umkehrer, M., Schmidt, J. *et al.* (2006) A novel one-pot synthesis of highly diverse indole scaffolds by the Ugi/Heck reaction. *Tetrahedron Lett.*, **47**, 4683–6.

35. Macor, J.E., Ogilvie, R.J. and Wythes, M.J. (1996) An improved synthesis of the unique anti-migraine agent CP-122,288: a bromine atom passenger in an intramolecular Heck reaction. *Tetrahedron Lett.*, **37**, 4289–92.

36. Wensbo, D., Annby, U. and Gronowitz, S. (1995) Indole-3-acetic acids and hetero analogues by one pot synthesis including Heck cyclisation. *Tetrahedron*, **51**, 10323–42.

37. Arcadi, A., Cacchi, S., Marinelli, F. and Pace, P. (1993) 5-Alkyl-5-[2-(o-iodophenylcarbamoyl)vinyl] derivatives of Meldrum's acid as substrates for the intramolecular Heck reactions: application to the synthesis of carbazoles. *Synlett*, 743–4.

38. Ohno, H., Aso, A., Kadoh, Y., Fujii, N. *et al.* (2007) Heck-type cyclization of oxime ethers: stereoselective carbon–carbon bond formation with aryl halides to produce heterocyclic oximes. *Angew. Chem. Int. Ed.*, **46**, 6325–8.

39. Knepper, K., Ziegert, R.E. and Bräse, S. (2003) Solid-phase indole synthesis. *PharmaChem*, **6**, 4–7.

40. Ang, K.H., Bräse, S., Steinig, A.G. *et al.* (1996) Versatile synthesis of bicyclo[4.3.0]nonenes and bicyclo[4.4.0]decenes by a domino Heck–Diels–Alder reaction. *Tetrahedron*, **52**, 11503–28.

41. De Meijere, A. and Bräse, S. (1999) Palladium in action: domino coupling and allylic substitution reactions for the efficient construction of complex organic molecules. *J. Organomet. Chem.*, **576**, 88–110.

42. De Meijere, A., von Zezschwitz, P. and Bräse, S. (2005) The virtue of palladium-catalyzed domino reactions – diverse oligocyclizations of acyclic 2-bromoenynes and 2-bromoenediynes. *Acc. Chem. Res.*, **38**, 413–22.

43. Grigg, R., Sridharan, V., Stevenson, P. *et al.* (1990) The synthesis of fused ring nitrogen heterocycles via regiospecific intramolecular Heck reactions. *Tetrahedron*, **46**, 4003–18.

44. Grigg, R., Fretwell, P., Meerholtz, C. and Sridharan, V. (1994) Palladium catalysed synthesis of spiroindolines. *Tetrahedron*, **50**, 359–70.

45. Burns, B., Grigg, R., Santhakumar, V. *et al.* (1992) Palladium catalysed tandem cyclisation–anion capture processes. Part 1. Background and hydride ion capture by alkyl- and π-allyl-palladium species. *Tetrahedron*, **48**, 7297–320.

46. Grigg, R., Kennewell, P. and Teasdale, A.J. (1992) Palladium catalysed cascade cyclisation–carbonylation processes. Rate enhancement by Tl(I) salts. *Tetrahedron Lett.*, **33**, 7789–92.

47. Grigg, R., Santhakumar, V. and Sridharan, V. (1993) Palladium catalysed cascade cyclisation – cyanide ion capture. *Tetrahedron Lett.*, **34**, 3163–4.

48. Yoneda, R., Sakamoto, Y., Oketo, Y. *et al.* (1994) A total synthesis of magallanesine via [1,2]-Meisenheimer rearrangement. *Tetrahedron Lett.*, **35**, 3749–52.
49. Madin, A. and Overman, L.E. (1992) Controlling stereoselection in intramolecular Heck reactions by tailoring the palladium catalyst. *Tetrahedron Lett.*, **33**, 4859–62.
50. Madin, A., O'Donnell, C.J., Oh, T. *et al.* (1999) Total synthesis of (±)-gelsemine. *Angew. Chem. Int. Ed.*, **38**, 2934–6.
51. Beyersbergen van Henegouwen, W.G., Fieseler, R.M., Rutjes, F.P.J.T. and Hiemstra, H. (1999) Total synthesis of (+)-gelsedine. *Angew. Chem. Int. Ed.*, **38**, 2214–7.
52. Beyersbergen van Henegouwen, W.G., Fieseler, R.M., Rutjes, F.P.J.T. and Hiemstra, H. (2000) First total synthesis of *ent*-gelsedine via a novel iodide-promoted allene *N*-acyliminium ion cyclization. *J. Org. Chem.*, **65**, 8317–25.
53. Larock, R.C. and Stinn, D.E. (1988) Synthesis of benzofurans via palladium-promoted cyclization of *ortho*-substituted aryl allyl ethers. *Tetrahedron Lett.*, **29**, 4687–90.
54. Cho, S.Y., Kim, S.S., Park, K.-H. *et al.* (1996) Synthesis of 3-alkylfluoropyridines via palladium-catalyzed cyclization of iodopyridinyl allyl ethers. *Heterocycles*, **43**, 1641–52.
55. Xie, X., Chen, B., Lu, J. *et al.* (2004) Synthesis of benzofurans in ionic liquid by PdCl$_2$-catalyzed intramolecular Heck reaction. *Tetrahedron Lett.*, **45**, 6235–37.
56. Morice, C., Domostoj, M., Briner, K. *et al.* (2001) Synthesis of constrained arylpiperidines using intramolecular Heck or radical reactions. *Tetrahedron Lett.*, **42**, 6499–502.
57. Fukuyama, Y., Yuasa, H., Tonoi, Y. *et al.* (2001) First syntheses of 1,13- and 1,15-dihydroxyherbertenes, and herbertenolide by applying intramolecular Heck reaction for the construction of adjacent quaternary centers. *Tetrahedron*, **57**, 9299–307.
58. Vickers, T.D. and Keay, B.A. (2003) Synthesis of (+/−)-curcumene ether. *Synlett*, 1349–51.
59. Schmidt, B. and Hoffmann, H.M.R. (1991) On the way to aflatoxins and related structure types. Regio-controlled annulations by application of homogenous palladium catalysis, urethane tether and *ortho,ortho′*-diiodide effect. *Tetrahedron*, **47**, 9357–68.
60. Casaschi, A., Grigg, R., Sansano, J.M. *et al.* (1996) Palladium catalysed cascade hydrostannylation-bis-cyclisation–intramolecular anion capture. Routes to bridged- and spirocyclic small and macrocyclic heterocycles. *Tetrahedron Lett.*, **37**, 4413–6.
61. Trost, B.M. and Tang, W. (2002) An efficient enantioselective synthesis of (−)-galanthamine. *Angew. Chem. Int. Ed.*, **41**, 2795–7.
62. Trost, B.M., Tang, W. and Toste, F.D. (2005) Divergent enantioselective synthesis of (−)-galanthamine and (−)-morphine. *J. Am. Chem. Soc.*, **127**, 14785–803.
63. Frey, D.A., Duan, C. and Hudlicky, T. (1999) Model study for a general approach to morphine and noroxymorphone via a rare Heck cyclization. *Org. Lett.*, **1**, 2085–7.
64. Arnau, N., Moreno-Mañas, M. and Pleixats, R. (1993) Preparation of benzo[*b*]thiophenes by Pd(0)-catalyzed intramolecular cyclization of allyl (and propargyl) *o*-iodophenyl sulfides. *Tetrahedron*, **49**, 11019–28.
65. Luo, F.-T. and Wang, R.-T. (1991) Conversion of 2-iodoaniline into (*Z*)-3-methylene-2,3-dihydroindole derivatives. *Heterocycles*, **32**, 2365–72.
66. Burns, B., Grigg, R., Sridharan, V. *et al.* (1989) Regiospecific palladium catalysed tandem cyclisation–anion capture processes. Stereospecific group transfer from organozinc and organoboron reagents. *Tetrahedron Lett.*, **30**, 1135–8.
67. Burns, B., Grigg, R., Ratananukul, P. *et al.* (1988) Regiospecific palladium catalysed tandem cyclisation–anion capture processes. Stereospecific group transfer from organotin reagents. *Tetrahedron Lett.*, **29**, 5565–8.
68. Grigg, R., Teasdale, A. and Sridharan, V. (1991) Palladium catalysed intramolecular coupling of aryl and benzylic halides and related tandem cyclisations. A simple synthesis of hippadine. *Tetrahedron Lett.*, **32**, 3859–62.

69. Grigg, R., Dorrity, M.J.R., Malone, J.F. *et al.* (1990) Asymmetric induction in the creation of tri- and tetra-substituted carbon stereocentres by the intramolecular Heck reaction. *Tetrahedron Lett.*, **31**, 3075–76.

70. Comins, D.L., Joseph, S.P. and Zhang, Y.-m. (1996) Regio- and stereoselective intramolecular Heck reaction of *N*-acyl-2,3-dihydro-4-pyridones. *Tetrahedron Lett.*, **37**, 793–6.

71. Comins, D.L., Baevsky, M.F. and Hong, H. (1992) A 10-step, asymmetric synthesis of (*S*)-camptothecin. *J. Am. Chem. Soc.*, **114**, 10971–2.

72. Svennebring, A., Nilsson, P. and Larhed, M. (2007) Microwave-accelerated spiro-cyclizations of *o*-halobenzyl cyclohexenyl ethers by palladium(0) catalysis. *J. Org. Chem.*, **72**, 5851–4.

73. Huwe, C.M. and Blechert, S. (1994) Synthesis of nitrogen heterocycles from vinyl glycine derivatives via palladium catalysis. *Tetrahedron Lett.*, **35**, 9537–40.

74. Lemaire-Audoire, S., Savignac, M., Dupuis, C. and Genêt, J.-P. (1996) Intramolecular Heck-type reactions in aqueous medium. Dramatic change on regioselectivity. *Tetrahedron. Lett.*, **37**, 2003–6.

75. Li, J.J. (1999) Synthesis of novel 3-substituted pyrrolo[2,3-*b*]quinoxalines via an intramolecular Heck reaction on an aminoquinoxaline scaffold. *J. Org. Chem.*, **64**, 8425–7.

76. Nguefack, J.-F., Bolitt, V. and Sinou, D. (1997) Palladium-mediated cyclization on carbohydrate templates. 1. Synthesis of enantiopure bicyclic compounds. *J. Org. Chem.*, **62**, 1341–7.

77. Nguefack, J.-F., Bolitt, V. and Sinou, D. (1997) Palladium-mediated cyclization on carbohydrate templates. 2. Synthesis of enantiopure tricyclic compounds. *J. Org. Chem.*, **62**, 6827–32.

78. Mori, M., Kanda, N., Oda, I. and Ban, Y. (1985) New synthesis of heterocycles by use of palladium catalyzed cyclization of α-haloamide with internal double bond. *Tetrahedron*, **41**, 5465–74.

79. Nukui, S., Sodeoka, M. and Shibasaki, M. (1993) Catalytic asymmetric synthesis of a functionalized indolizidine derivative. A useful intermediate suitable for the synthesis of various glycosidase inhibitors. *Tetrahedron Lett.*, **34**, 4965–8.

80. Iida, H., Yuasa, Y. and Kibayashi, C. (1980) Intramolecular cyclization of enaminones involving arylpalladium complexes. Synthesis of carbazoles. *J. Org. Chem.*, **45**, 2938–42.

81. Michael, J.P., Chang, S.-F. and Wilson, C. (1993) Synthesis of pyrrolo[1,2-*a*]indoles by intramolecular Heck reaction of *N*-(2-bromoaryl) enaminones. *Tetrahedron Lett.*, **34**, 8365–8.

82. Edmondson, S.D., Mastracchio, A. and Parmee, E.R. (2000) Palladium-catalyzed coupling of vinylogous amides with aryl halides: applications to the synthesis of heterocycles. *Org. Lett.*, **2**, 1109–12.

83. Harris, J.M. and Padwa, A. (2003) A new β-carbolinone synthesis using a Rh(II)-promoted [3 + 2]-cycloaddition and Pd(0) cross-coupling/Heck cyclization chemistry. *Org. Lett.*, **5**, 4195–7.

84. Micheal, J.P., de Koning, C.B., Peterson, R.L. and Stanbury, T.V. (2001) Asymmetric synthesis of a tetracyclic model for the aziridinomitosenes. *Tetrahedron Lett.*, **42**, 7513–6.

85. Mmutlane, E.M., Harris, J.M. and Padwa, A. (2005) 1,3-Dipolar cycloaddition chemistry for the preparation of novel indolizinone-based compounds. *J. Org. Chem.*, **70**, 8055–63.

86. Sørensen, U.S. and Pombo-Villar, E. (2004) Synthesis of cyclopenta[*b*]indol-1-ones and carbazol-4-ones from *N*-(2-halophenyl)-substituted enaminones by intramolecular Heck reaction. *Helv. Chim. Acta*, **87**, 82–9.

87. Barluenga, J., Fernández, M.A., Aznar, F. and Valdés, C. (2005) Cascade alkenyl amination/Heck reaction promoted by a bifunctional palladium catalyst: a novel one-pot synthesis of indoles from *o*-haloanilines and alkenyl halides. *Chem. Eur. J.*, **11**, 2276–83.

88. Ackermann, L. and Althammer, A. (2006) One-pot 2-aryl/vinylindole synthesis consisting of a ruthenium-catalyzed hydroamination and a palladium-catalyzed Heck reaction using 2-choroaniline. *Synlett*, 3125–9.

89. Ichikawa, J., Nadano, R. and Ito, N. (2006) 5-*endo* Heck-type cyclization of 2-(trifluoromethyl)allyl ketone oximes: synthesis of 4-difluoromethylene-substituted 1-pyrrolines. *Chem. Commun.*, 4425–7.

90. Ichikawa, J., Sakoda, K., Mihara, J. and Ito, N. (2006) Heck-type 5-*endo*-trig cyclizations promoted by vinylic fluorines: ring-fluorinated indene and 3*H*-pyrrole syntheses from 1,1-difluoro-1-alkenes. *J. Fluorine Chem.*, **127**, 489–504.

91. Hudlicky, T. and Olivo, H.F. (1992) A short synthesis of (+)-lycoricidine. *J. Am. Chem. Soc.*, **114**, 9694–96.

92. Tietze, L.F. and Burkhardt, O. (1995) Stereo- and regioselective intramolecular Heck reaction of α-amino acid derivatives for the synthesis of enantiopure 3,4-dihydroisoquinolinones. *Liebigs Ann.*, 1153–7.

93. Mohanakrishnan, A.K. and Srinivasan, P.C. (1996) Synthesis of 4-substituted-1,2,3,4-tetrahydro-β-carbolines via intramolecular radical cyclisation and Heck reaction. *Tetrahedron Lett.*, **37**, 2659–62.

94. Oikawa, M., Takeda, Y., Naito, S. *et al.* (2007) A three-component approach to isoquinoline derivatives by cycloaddition/Heck reaction sequence. *Tetrahedron Lett.*, **48**, 4255–8.

95. Bräse, S. (1999) Intramolecular transition-metal catalyzed cyclizations of electron rich chloroarenes. *Tetrahedron Lett.*, **40**, 6757–9.

96. Beccalli, E.M., Broggini, G., Martinelli, M. *et al.* (2006) New 4-spiroannulated tetrahydroiso-quinolines by a one-pot sequential procedure. Isolation and characterization of σ-alkylpalladium Heck intermediates. *Org. Lett.*, **8**, 4521–4.

97. Clique, B., Fabritius, C.-H., Couturier, C. *et al.* (2003) Unexpected isolation, and structural characterization, of a β-hydrogen-containing σ-alkylpalladium halide complex in the course of an intermolecular Heck reaction. Synthesis of polycyclic isoquinoline derivatives. *Chem. Commun.*, 272–3.

98. Sánchez-Sancho, F., Mann, E. and Herradón, B. (2001) Efficient synthesis of chiral isoquino-line and pyrido[1,2-*b*]-isoquinoline derivatives via intramolecular Heck reactions. *Adv. Synth. Catal.*, **343**, 360–8.

99. Martin, S.F. and Tso, H.-H. (1993) Synthetic studies on the narciclasine alkaloids. A synthesis of (±)-lycoricidine. *Heterocycles*, **35**, 85–8.

100. McIntosh, M.C. and Weinreb, S.M. (1993) An approach to total synthesis of (+)-lycoricidine. *J. Org. Chem.*, **58**, 4823–32.

101. Chida, N., Ohtsuka, M. and Ogawa, S. (1993) Total synthesis of (+)-lycoricidine and its 2-epimer from D-glucose. *J. Org. Chem.*, **58**, 4441–7.

102. Bedjeguelal, K., Bolitt, V. and Sinou, D. (1999) Intramolecular Heck cyclisation–β-alkoxy elimination in carbohydrate chemistry. A simple route to enantiopure annelated dioxatricyclic compounds. *Synlett*, 762–4.

103. Bankston, D., Fang, F., Huie, E. and Xie, S. (1999) Palladium(II) acetate–tris (triphenylphosphine) rhodium(I) chloride: a novel catalytic couple for the intramolecular Heck reaction. *J. Org. Chem.*, **64**, 3461–6.

104. Tietze, L.F., Burkhardt, O. and Henrich, M. (1997) Intramolecular Heck reaction for the synthesis of isochromanes under ambient and high pressure. *Liebigs Ann./Recueil*, 887–91.

105. Woodcock, S.R. and Branchaud, B.P. (2005) Effect of ether versus ester tethering on Heck cyclizations. *Tetrahedron Lett.*, **46**, 7213–5.

106. Josien, H., Ko, S.-B., Bom, D. and Curran, D.P. (1998) A general synthetic approach to the (20*S*)-camptothecin family of antitumor agents by a regiocontrolled cascade radical cyclization of aryl isonitriles. *Chem. Eur. J.*, **4**, 67–83.

107. Abelman, M.M., Overman, L.E. and Tran, V.D. (1990) Construction of quaternary carbon cen-ters by palladium-catalyzed intramolecular alkene insertions. Total synthesis of the Amarylli-dacea alkaloids (±)-tazettine and (±)-6a-epipretazettine. *J. Am. Chem. Soc.*, **112**, 6959–64.

108. Di Fabio, R., Alvaro, G., Bertani, B. and Giacobbe, S. (2000) Straightforward synthesis of new tetrahydroquinoline derivatives. *Can. J. Chem.*, **78**, 809–15.
109. Santos, L.S. and Pilli, R.A. (2002) The intramolecular Heck reaction and the synthesis of indolizidinone, quinolizidinone and benzoazepinone derivatives. *Synthesis*, 87–93.
110. Larock, R.C. and Babu, S. (1987) Synthesis of nitrogen heterocycles via palladium-catalyzed intramolecular cyclization. *Tetrahedron Lett.*, **28**, 5291–4.
111. Smalley, T.L. Jr and Mills, W.Y. (2005) Regioselective synthesis of 3,3-diethyl-4-(methylene)-1-quinol-2-ones by an intramolecular microwave assisted Heck reaction. *J. Heterocycl. Chem.*, **42**, 327–31.
112. Guillou, C., Beunard, J.-L., Gras, E. and Thal, C. (2001) An efficient total synthesis of (±)-galanthamine. *Angew. Chem. Int. Ed.*, **40**, 4745–6.
113. Söderberg, B.C.G., Hubbard, J.W., Rector, S.R. and O'Neil, S.N. (2005) Synthesis of fused indoles by sequential palladium-catalyzed Heck reaction and *N*-heteroannulation. *Tetrahedron*, **61**, 3637–49.
114. Gras, E., Guillou, C. and Thal, C. (1999) A formal synthesis of (±)-lycoramine via an intramolecular Heck reaction. *Tetrahedron Lett.*, **40**, 9243–4.
115. Rawal, V.H., Michoud, C. and Monestel, R.F. (1993) General strategy for the stereocontrolled synthesis of *Strychnos* alkaloids: a concise synthesis of (±)-dehydrotubifoline. *J. Am. Chem. Soc.*, **115**, 3030–1.
116. Birman, V.B. and Rawal, V.H. (1998) A novel route to the geissoschizine skeleton: the influence of ligands on the diastereoselectivity of the Heck cyclization. *Tetrahedron Lett.*, **39**, 7219–22.
117. Solé, D., Bonjoch, J., García-Rubio, S. *et al.* (2000) Enantioselective total synthesis of Wieland–Gumlich aldehyde and (−)-strychnine. *Chem. Eur. J.*, **6**, 655–65.
118. Kirschbaum, S. and Waldmann, H. (1997) Construction of the tricyclic benzoquinolizine ring system by combination of a tandem Mannich–Michael reaction with a Heck reaction. *Tetrahedron Lett.*, **38**, 2829–32.
119. Kirschbaum, S. and Waldmann, H. (1998) Three-step access to the tricyclic benzo[*a*]quinolizine ring system. *J. Org. Chem.*, **63**, 4936–46.
120. Beccalli, E.M., Broggini, G., Marchesini, A. and Rossi, E. (2002) Inramolecular Heck reaction of 2- and 3-iodoindole derivatives for the synthesis of β- and γ-carbolinones. *Tetrahedron*, **58**, 6673–8.
121. Lennartz, M. and Steckhan, E. (2000) Synthesis of bicyclic lactams via ring closing olefin metathesis and intramolecular Heck reaction. *Synlett*, 319–22.
122. Meyer, F.E., Parson, P.J. and de Meijere, A. (1991) Palladium-catalyzed polycyclization of dienynes: surprisingly facile formation of tetracyclic systems containing a three-membered ring. *J. Org. Chem.*, **56**, 6487–8.
123. McClure, K.F., Danishefsky, S.J. and Schulte, G.K. (1994) A remarkable stereochemical inversion in some Heck arylation reactions. A mechanistic proposal. *J. Org. Chem.*, **59**, 355–60.
124. Paquette, L.A., Dura, R.D., Fosnaugh, N. and Stepanian, M. (2006) Direct comparison of the response of bicyclic sultam and lactam dienes to photoexcitation. Concerning the propensity of differing bond types to bridgehead nitrogen for homolytic cleavage. *J. Org. Chem.*, **71**, 8438–45.
125. Evans, P. (2007) The double reduction of cyclic sulfonamides for the synthesis of (4*S*-phenylpyrrolidin-2*R*-yl)methanol and 2*S*-methyl-4*S*-phenylpyrrolidine. *J. Org. Chem.*, **72**, 1830–3.
126. Dankwardt, J.W. and Flippin, L.A. (1995) Palladium-mediated 6-*endo*-trig intramolecular cyclization of *N*-acryloyl-7-bromoindolines. A regiochemical variant of the intramolecular Heck reaction. *J. Org. Chem.*, **60**, 2312–3.
127. Mamouni, A., Daïch, A. and Decroix, B. (1998) Palladium(II) catalyzed cyclization: a facile access to new difunctionalized thieno[3,2-*f*]indolizine. *Synth. Commun.*, **28**, 1839–46.

128. Sánchez-Sancho, F., Mann, E. and Herradón, B. (2000) Efficient syntheses of polyannular heterocycles featuring microwave-accelerated Bischler–Napieralski reaction, stereoselective Heck cyclization, and Claisen rearrangement. *Synlett*, 509–13.

129. Bombrun, A. and Sageot, O. (1997) Palladium-catalyzed cyclization of 2-heteroyl-1-methylene-1,2,3,4-tetrahydroisoquinolines. Studies on 6-*endo*- versus 5-*exo*-trig cyclization. *Tetrahedron Lett.*, **38**, 1057–60.

130. Kim, G., Kim, J.H., Kim, W.-j. and Kim, Y.A. (2003) Intramolecular Heck reaction of methylenephthalimidine derivatives: a simple route to lennoxamine and chilenine. *Tetrahedron Lett.*, **44**, 8207–9.

131. Laursen, B., Denieul, M.-P. and Skrydstrup, T. (2002) Formal total synthesis of the PKC inhibitor, balanol: preparation of the fully protected benzophenone fragment. *Tetrahedron*, **58**, 2231–8.

132. Boger, D.L. and Turnbull, P. (1997) Synthesis and evaluation of CC-1065 and duocarmycin analogs incorporating the 1,2,3,4,11,11a-hexahydrocyclopropa[*c*]naphtho[2,1-*b*]azepin-6-one (CNA) alkylation subunit: structural features that govern reactivity and reaction regioselectivity. *J. Org. Chem.*, **62**, 5849–63.

133. Cropper, E.L., White, A.J.P., Ford, A. and Hii, K.K. (2006) Ligand effects in the synthesis of *N*-heterocycles by intramolecular Heck reactions. *J. Org. Chem.*, **71**, 1732–5.

134. Lormann, M.E.P., Nieger, M. and Bräse, S. (2006) Desymmetrisation of a bicyclo[4.4.0]decadiens: A planar-chiral complex proved to be most effective in an asymmetric Heck reaction. *J. Organomet. Chem.*, **691**, 2159–61.

135. Bru, C., Thal, C. and Guillou, C. (2003) Concise total synthesis of (±)-maritidine. *Org. Lett.*, **5**, 1845–6.

136. Gracias, V., Moore, J.D. and Djuric, S.W. (2004) Sequential Ugi/Heck cyclization strategies for the facile construction of highly functionalized *N*-heterocyclic scaffolds. *Tetrahedron Lett.*, **45**, 417–20.

137. Jin, J. and Weinreb, S.M. (1997) Enantioselective total syntheses of the 5,11-methanomorphanthridine class of Amaryllidaceae alkaloids (−)-pancracine and (−)-coccinine. *J. Am. Chem. Soc.*, **119**, 2050–1.

138. Jin, J. and Weinreb, S.M. (1997) Application of a stereospecific intramolecular allenylsilane imino ene reaction to enantioselective total synthesis of the 5,11-methanomorphanthridine class of Amaryllidaceae alkaloids. *J. Am. Chem. Soc.*, **119**, 5773–84.

139. Tietze, L.F. and Schimpf, R. (1993) Efficient synthesis of 2,3,4,5-tetrahydro-1*H*-3-benzazepines by intramolecular Heck reaction. *Synthesis*, 876–80.

140. Donets, P.A. and van der Eycken, E.V. (2007) Efficient synthesis of the 3-benzazepine framework via intramolecular Heck reductive cyclization. *Org. Lett.*, **9**, 3017–20.

141. Tietze, L.F. and Schirok, H. (1997) Highly efficient synthesis of cephalotaxine by two palladium-catalyzed cyclizations. *Angew. Chem., Int. Ed. Engl.*, **36**, 1124–5.

142. Tietze, L.F., Schirok, H. and Wöhrmann, M. (2000) Palladium-catalyzed synthesis of cephalotaxine analogues. *Chem. Eur. J.*, **6**, 510–8.

143. Worden, S.M., Mapitse, R. and Hayes, C.J. (2002) Towards a total synthesis of (−)-cephalotaxine: construction of the BCDE-tetracyclic core. *Tetrahedron Lett.*, **43**, 6011–4.

144. Tietze, L.F., Burkhardt, O. and Henrich, M. (1997) Diastereo- and regioselective intramolecular Heck reaction of α-amino alcohol derivatives for the synthesis of enantiomerically pure isoquinolines and benzazepines at ambient and high pressure. *Liebigs Ann./Recueil*, 1407–13.

145. Joucla, L., Putey, A. and Joseph, B. (2005) Synthesis of fused heterocycles with a benzazepinone moiety via intramolecular Heck coupling. *Tetrahedron Lett.*, **46**, 8177–9.

146. Ribière, P., Declerck, V., Nédellec, Y. *et al.* (2006) Synthesis of novel poly(ethylene glycol) supported benzazepines: the crucial role of PEG on the selectivity of an intramolecular Heck reaction. *Tetrahedron*, **62**, 10456–66.

147. Putey, A., Joucla, L., Picot, L. *et al.* (2007) Synthesis of latonduine derivatives via intramolecular Heck reaction. *Tetrahedron*, **63**, 867–79.
148. Declerck, V., Ribière, P., Nédellec, Y. *et al.* (2007) A microwave-assisted Heck reaction in poly(ethylene glycol) for the synthesis of benzazepines. *Eur. J. Org. Chem.*, 201–8.
149. Ando, K., Akai, Y., Kunitomo, J.-i. *et al.* (2007) Synthesis and biological activities of novel furo[2,3,4-*jk*][2]benzazepin-4(3*H*)-one derivatives. *Org. Biomol. Chem.*, **5**, 655–63.
150. Gibson, S.E. and Middleton, R.J. (1995) Synthesis of 7-,8- and 9-membered rings via *endo* Heck cyclisations of amino acid derived substrates. *J. Chem. Soc., Chem. Commun.*, 1743–4.
151. Gibson, S.E., Guillo, N., Middleton, R.J. *et al.* (1997) Synthesis of conformationally constrained phenylalanine analogues via 7-, 8- and 9-*endo* Heck cyclisations. *J. Chem. Soc., Perkin Trans. 1*, 447–55.
152. Gibson, S.E., Jones, J.O., McCague, R., Tozer, M.J. and Whitcombe, N.J. (1999) A bromoarene based approach to phenylalanine analogues Hic and Nic. *Synlett*, 954–6.
153. Denieul, M.-P. and Skrydstrup, T. (1999) Application of an intramolecular Heck reaction for the construction of the balanol aryl core structure. *Tetrahedron Lett.*, **40**, 4901–4.
154. Enders, D., Lenzen, A., Backes, M. *et al.* (2005) Asymmetric total synthesis of the 1-*epi*-aglycon of the cripowellins A and B. *J. Org. Chem.*, **70**, 10538–51.
155. Denieul, M.-P., Laursen, B., Hazell, R. and Skrydstrup, T. (2000) Synthesis of the benzophenone fragment of balanol via an intramolecular cyclization event. *J. Org. Chem.*, **65**, 6052–60.
156. Kalinin, A.V., Chauder, B.A., Rakhit, S. and Snieckus, V. (2003) *seco*-C/D ring analogues of ergot alkaloids. Synthesis via intramolecular Heck and ring-closing metathesis reactions. *Org. Lett.*, **5**, 3519–21.
157. Stocks, M.J. , Harrison, R.P. and Teague, S.J. (1995) Macrocyclic ring closures employing the intramolecular Heck reaction. *Tetrahedron Lett.*, **36**, 6555–8.
158. Jeong, S., Chen, X. and Harran, P.G. (1998) Macrocyclic triarylethylenes via Heck endocyclization: a system relevant to diazonamide synthesis. *J. Org. Chem.*, **63**, 8640–1.
159. Akaji, K., Teruya, K., Akaji, M. and Aimoto, S. (2001) Synthesis of cyclic RGD derivatives via solid phase macrocyclization using the Heck reaction. *Tetrahedron*, **57**, 2293–303.
160. Reddy, P.R., Balraju, V., Madhavan, G.R. *et al.* (2003) Synthesis of small cyclic peptides via intramolecular Heck reactions. *Tetrahedron Lett.*, **44**, 353–6.
161. Ziegler, F.E., Chakraborty, U.R. and Weisenfeld, R.B. (1981) A palladium-catalyzed carbon–carbon bond formation of conjugated dienones. *Tetrahedron*, **37**, 4035–40.
162. Geng, X., Miller, M.L., Lin, S. and Ojima, I. (2003) Synthesis of novel C2–C3′N-linked macrocyclic taxoids by means of highly regioselective Heck macrocyclization. *Org. Lett.*, **5**, 3733–6.

7

Chelation-Controlled Mizoroki–Heck Reactions

Kenichiro Itami

Department of Chemistry and Research Center for Materials Science, Nagoya University,
Chikusa-ku, Nagoya, Japan

Jun-ichi Yoshida

Department of Synthetic Chemistry and Biological Chemistry, Graduate School of Engineering,
Kyoto University, Nishikyo-ku, Kyoto, Japan

7.1 Introduction

Where there is a lack of reactivity and/or selectivity in a certain metal-catalysed or -mediated process, it is commonplace to tune the stereoelectronics of catalyst or metal-containing reagent by adjusting the ligand field for enhancing the reactivity or selectivity. Alternatively, a catalyst-directing group (a functional group – often a coordinating heteroatom – that is suitably attached on a substrate) could provide a powerful tool for enhancing the efficiency of an otherwise sluggish process and for steering the course of the reaction by exploitation of attractive substrate–catalyst interactions [1]. During the last two decades, such chelation control has emerged as a useful strategy to enhance the utility of the Mizoroki–Heck reaction in organic synthesis [2]. In this chapter, we will summarize the state of the art of so-called chelation-controlled Mizoroki–Heck reactions.

The Mizoroki–Heck Reaction   Edited by Martin Oestreich
© 2009 John Wiley & Sons, Ltd

7.2   Overview of Chelation-Controlled Mizoroki–Heck Reactions

7.2.1   Hallberg's Benchmark Demonstration

Although there are a number of examples that exhibit chelation effects in palladium-mediated transformations, it was not until Hallberg and coworkers' benchmark report in 1990 that synthetic chemists started to utilize chelation control strategically in Mizoroki–Heck chemistry [3]. Hallberg and coworkers demonstrated that the presence of a β-amino substituent exerts a profound influence on the regioselectivity of the arylation of vinyl ethers (Figure 7.1). With the aid of attractive coordination of the NMe$_2$ group to organopalladium intermediates, the otherwise α-selective arylation can be reversed to being β-selective.

7.2.2   Representative Substrates for Chelation-Controlled Mizoroki–Heck Reactions

Since Hallberg and coworkers' report, many examples demonstrating chelation (directing) effects have appeared that substantially raise the potential of the Mizoroki–Heck reaction in organic synthesis. Representative substrates are shown in Figure 7.2. These substrates might be roughly classified into either of two broad categories: (i) substrates that bear directing heteroatoms within the main substrate structures and (ii) substrates that bear directing groups attached through suitable linking atoms. The latter directing groups are somewhat 'strategic' in nature and might be regarded as 'removable' or 'functionalizable' directing groups [4].

Allylic alcohols and amines are typical substrates that bear directing heteroatoms within the main substrate structure. Under suitable conditions, oxygen and nitrogen atoms in allylic alcohols and amines participate in the Mizoroki–Heck reaction through O—Pd or N—Pd coordination, bringing about some chelation effect in the outcome of reactions.

Hallberg's group deliberately introduced β-amino tethers for the β-selective arylation of vinyl ethers. They further extended this system to chiral substrates for diastereoselective arylation processes. Similar to their systems, Badone introduced β-phosphino tethers for the β-selective arylation of vinyl ethers. Tamaru has reported carbamate directing

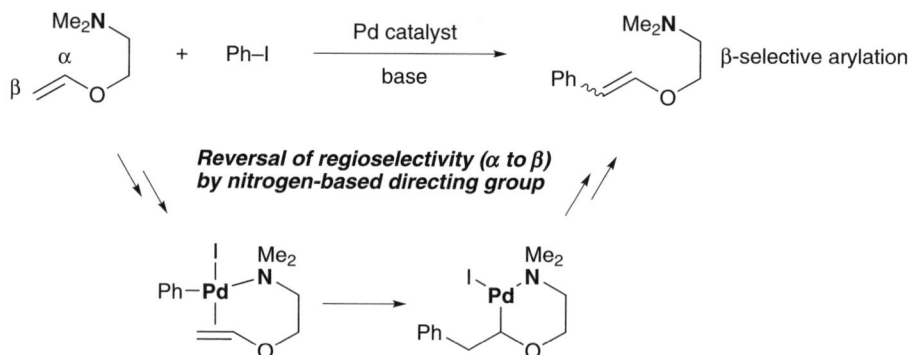

Figure 7.1   *Hallberg's pioneering work on chelation-controlled Mizoroki–Heck reaction.*

Typical Substrates Bearing Directing Heteroatoms

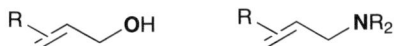

Strategic Substrates with Removable Directing Groups

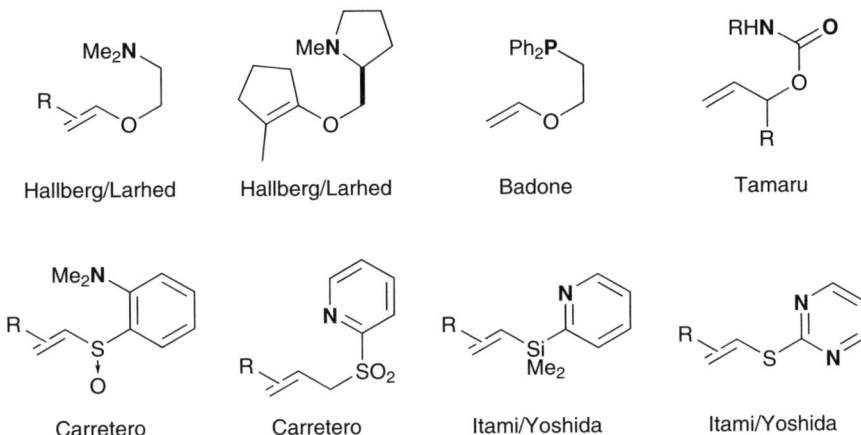

| Hallberg/Larhed | Hallberg/Larhed | Badone | Tamaru |

| Carretero | Carretero | Itami/Yoshida | Itami/Yoshida |

Figure 7.2 *Representative substrates for chelation-controlled Mizoroki–Heck reactions.*

groups for the terminal-selective arylation of allylic alcohols. Carretero has introduced (*o*-amino)phenylsulfinyl groups for the Mizoroki–Heck reaction of vinyl sulfoxides. A number of asymmetric (diastereoselective) processes are made possible by utilizing related sulfur-stereogenic groups. Carretero also reported use of the 2-pyridyl group for a chelation-controlled Mizoroki–Heck reaction of allylic sulfones. Itami and Yoshida have reported 2-pyridyl- and 2-pyrimidyl-directed Mizoroki–Heck arylations of vinyl silanes and vinyl sulfides. Multiple arylations, enabled by introducing these groups, have been extensively utilized in the programmed and diversity-oriented synthesis of multisubstituted olefins.

7.2.3 Chelation Effects in the Mizoroki–Heck Reaction

When an olefinic substrate bearing a directing group is employed, there are several chelation effects possible. A simplified textbook mechanism of Mizoroki–Heck arylation is shown in Figure 7.3: (i) oxidative addition of an aryl halide to palladium(0) complex **A**, generating ArPdX **B**; (ii) π-complexation of C=C, giving C=C-coordinated ArPdX **C**; (iii) insertion (carbopalladation) of the Ar—Pd bond into the coordinated C=C bond, yielding alkylpalladium complex **D**; (iv) β-hydrogen elimination of **D**, liberating arylation product and HPdX **E**; (v) regeneration of palladium(0) complex **A** by base-mediated abstraction of HX from HPdX.

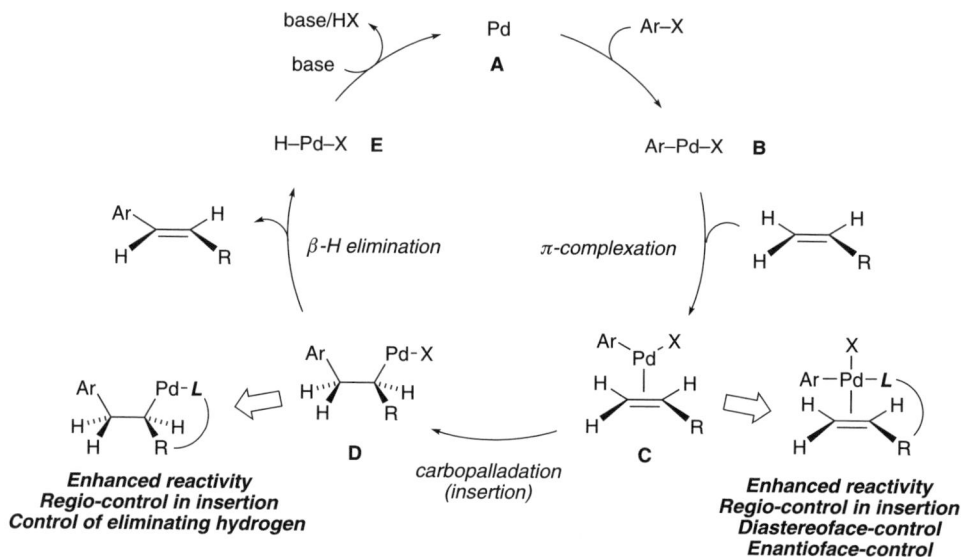

Figure 7.3　*A simplified mechanism and key chelation-controlled intermediates.*

Among these elementary reactions, the C=C π-complexation, carbopalladation (insertion) and β-elimination steps are particularly influenced by the presence of coordinating groups (L) embedded into substrates (Figure 7.3). In general, the Pd—L coordination would render the C=C π-complexation to palladium(II) intramolecular in nature, thereby often facilitating the reaction as a whole. Such a directing strategy is particularly useful in achieving an efficient intermolecular Mizoroki–Heck reaction of, for electronic and/or steric reasons, otherwise unreactive substrates. In many instances, the regioselectivity at the carbopalladation step is controlled through a Pd—L coordination as well. Such chelation control is capable of either enhancing or even reversing the regioselectivity of electronically biased substrates. A Pd—L coordination at an alkylpalladium complex **D** often assists selecting either of the conformationally accessible hydrogen atoms in the subsequent β-elimination step. Chelation control is also useful in the selection of a C=C face or group, among others. There are several reports of diastereoselective and enantioselective Mizoroki–Heck reactions utilizing Pd—L coordination.

7.3　Allylic Alcohols and Related Systems

Since the early reports by Mizoroki, Heck and others, allylic alcohols **1** have been elusive substrates in the Mizoroki–Heck reaction (Figure 7.4) [5]. When an insertion of ArPdX to C=C results in an organopalladium intermediate with palladium on the β-carbon of allylic alcohol, which is often the case, there are hydrogen atoms available at the α- (H_α) and γ-carbon (H_γ) in the subsequent β-elimination step (Figure 7.4). It has been known from the early days that the treatment of allylic alcohol (**1**) with iodobenzene under typical

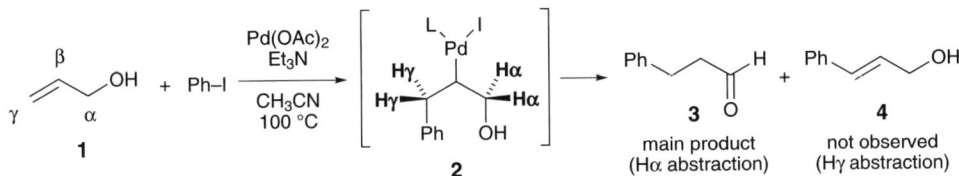

Figure 7.4 *Arylation of allylic alcohol under typical conditions.*

Mizoroki–Heck conditions [Pd(OAc)$_2$, Et$_3$N, CH$_3$CN] affords 3-phenylpropanal (**3**) as the result of abstracting H$_\alpha$ in the β-elimination step (Figure 7.4). The H$_\gamma$-abstracted product, cinnamyl alcohol (**4**), is usually not observed.

Ortar and coworkers [6] reported a reagent-based approach to selectively yield H$_\gamma$-abstracted products. When an alkenyl triflate **5** is used as an alkenylating agent in the Mizoroki–Heck reaction of allylic alcohol, the β-elimination of H$_\gamma$ becomes possible, thereby furnishing a dienyl alcohol **7** as the main product (Figure 7.5). The cationic nature of organopalladium intermediate **6** allows for coordination of the hydroxy group to palladium, which effectively prevents C$_\alpha$–H$_\alpha$ from a *syn*-relationship with the C$_\beta$–Pd(II) required for β-hydrogen elimination (Figure 7.5).

Jeffery has reported an alternative additive-based solution to yield H$_\gamma$-abstracted products. Mizoroki–Heck reaction of allylic alcohols with aryl or alkenyl halides in the presence of silver salts (AgOAc or Ag$_2$CO$_3$) results in selective H$_\gamma$-abstraction [7]. Similar hydroxy-coordination to the cationic organopalladium intermediates are believed to be involved in this system. In this regard, the use of hypervalent iodonium salts is also effective for generating cationic palladium species [8].

Tamaru and coworkers [9] have reported an alternative method of realizing a selective H$_\gamma$-abstraction without using silver salts. Allylic carbamates **8** can be transformed into the corresponding cinnamyl derivatives **10** by the Mizoroki–Heck arylation with iodobenzene (Figure 7.6). A six-membered chelate intermediate **9** has been proposed for the selective H$_\gamma$-abstraction.

Kang *et al.* [10] have observed similar chelation-controlled regioselective β-hydrogen elimination in the Mizoroki–Heck arylation of allylic diols **12** (Figure 7.7). Supporting evidence for the requirement of diol functionality in controlling regioselection during β-hydrogen elimination (**13**) is provided.

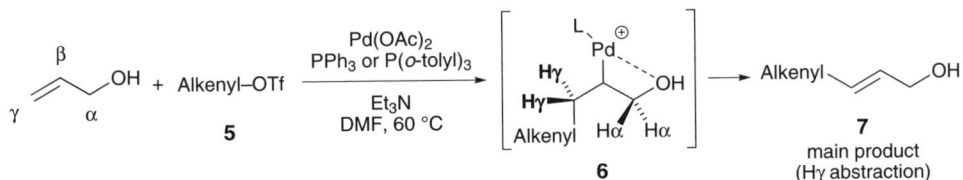

Figure 7.5 *Hydroxy-directed regioselective β-hydrogen elimination.*

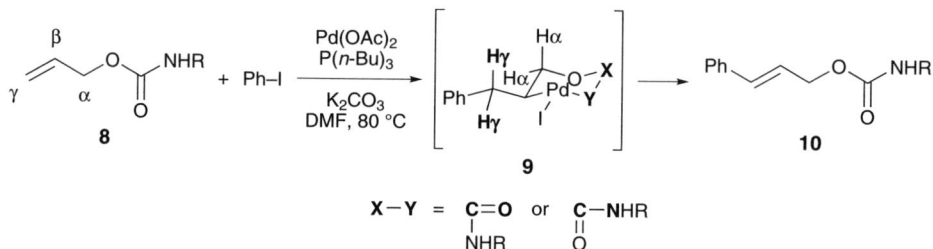

Figure 7.6   *Regioselective β-hydrogen elimination in protected allylic alcohols.*

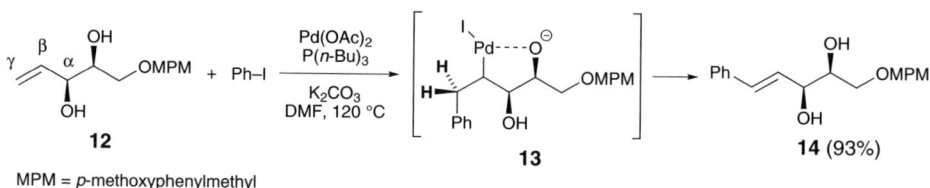

MPM = *p*-methoxyphenylmethyl

Figure 7.7   *Regioselective β-hydrogen elimination in allylic diol system.*

7.4 Allylic Amines

Nilsson and Hallberg [11] have reported the Mizoroki–Heck arylation of allylic amine **15** as a key reaction in the synthesis of preclamol, a dopamine autoreceptor agonist. The arylation of **15** with aryl iodide **16** using Pd(OAc)$_2$/P(*o*-tolyl)$_3$/AgNO$_3$/Et$_3$N occurs exclusively at the β-position (Figure 7.8). The high preference for β- over γ-arylation ($\beta : \gamma > 93 : 7$) might be due to the formation of a stabilized five-membered (bicyclic) organopalladium intermediate **17**. The arylation of corresponding allylic carbamate **19** results in low regioselectivity, suggesting the importance of the electronic nature of the amine nitrogen for the chelation control. Similar directing effects in controlling the regioselectivity are also proposed in the arylation of acyclic allylic amine systems [12].

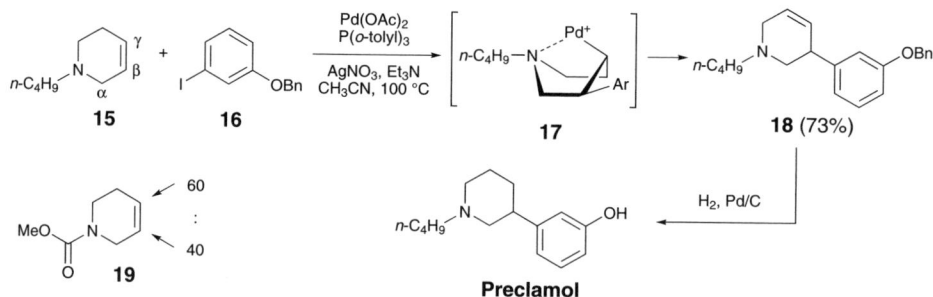

Figure 7.8   *Directed Mizoroki–Heck reaction of allylic amines.*

7.5 Vinyl Ethers

7.5.1 Regioselective Arylation

Vinyl ethers are another type of elusive substrate in controlling the regioselectivity in Mizoroki–Heck arylation. Under typical conditions, preferential arylation at α-positions is observed. Hallberg and coworkers [3, 13] have elegantly disclosed that the α-selectivity of vinyl ethers can be reversed to β-selectivity by using a series of nitrogen-based coordinating groups (tertiary amines and pyridine) attached to the substrate structure (Figures 7.1 and 7.9). As mentioned in previous sections, this is the first clear demonstration of chelation control in Mizoroki–Heck chemistry. Systematic control experiments strongly support that the high level of regiocontrol is attributed to the attractive Pd–N interactions during the reaction; ring-size effects are also observed (Figure 7.9). The only drawback seems to be modest stereoselectivities (E/Z).

Baldone and Guzzi [14] have reported the chelation-controlled Mizoroki–Heck arylation of vinyl ether **23** bearing a phosphino directing group. Under the influence of Pd(OAc)$_2$ and proton sponge (1,8-bis(dimethylamino)naphthalene), **23** undergoes β-selective arylation with a range of aryl triflates (Figure 7.10). Although the reactivity was significantly lower than that of **23**, cognate **25** also undergoes β-selective arylation. As in the case of aminoethyl-tagged vinyl ethers, the stereoselectivities (E/Z) remain modest (1/1–1/2.5).

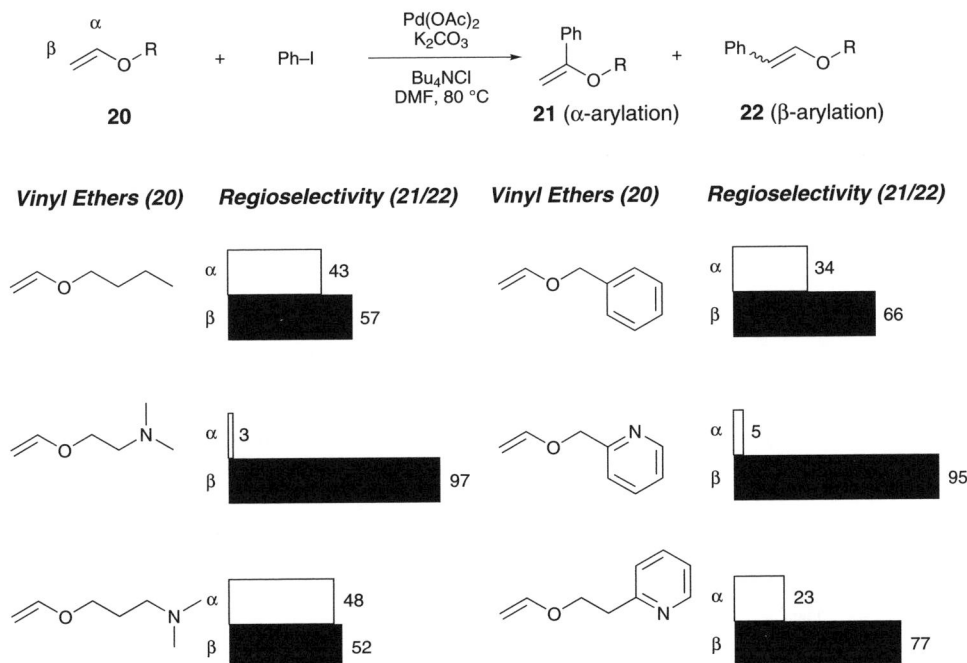

Figure 7.9 *Amino-directed β-selective arylation of vinyl ethers.*

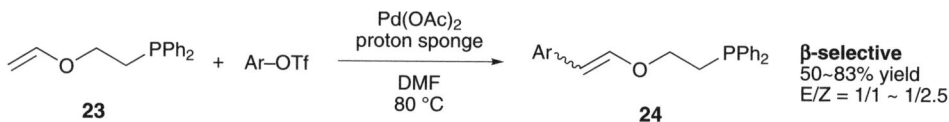

Reactivity comparison: Arylation using 1-naphthyl triflate

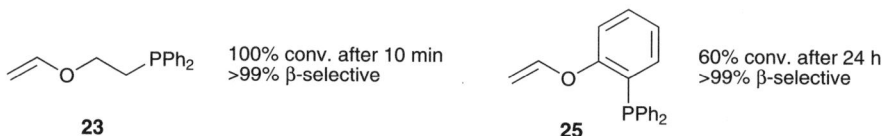

Figure 7.10 *Phosphino-directed β-selective arylation of vinyl ethers.*

7.5.2 Multiple Arylation

The traditional intermolecular Mizoroki–Heck arylation has been essentially limited to the monoarylations of olefins. By taking advantage of attractive substrate–catalyst interactions, multiple arylations have become possible by using an olefin bearing a coordinating auxiliary. In 2001, Hallberg and coworkers [15] and Itami *et al.* [16] independently discovered such processes. The latter processes will be discussed in the sections dealing with vinyl silanes (Sections 7.6) and vinyl sulfides (Section 7.7).

Hallberg and coworkers [15] demonstrated that their aminoethyl-tagged vinyl ether **26** underwent stepwise (two-pot) β,β-diarylation, giving **28** in moderate to good yields and modest diastereocontrol (Figure 7.11). The subsequent hydrolysis of **28** under microwave irradiation affords the corresponding diaryl acetaldehydes **29** in excellent yields. A range of aryl iodides was utilized in this reaction; Herrmann's palladacycle under microwave heating allowed for the use of aryl bromides in β,β-diarylation of **26** [17]. The chelation-assisted mechanism (Figure 7.1) is again strongly supported from several control experiments, as well as by X-ray analyses of model palladium complexes [15].

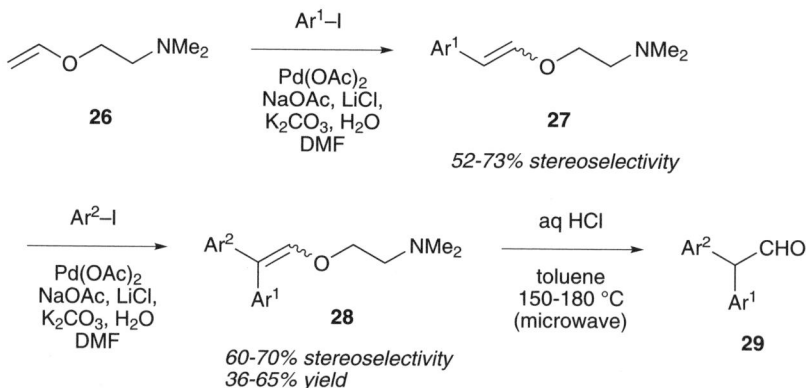

Figure 7.11 *Directed β,β-diarylation of vinyl ethers.*

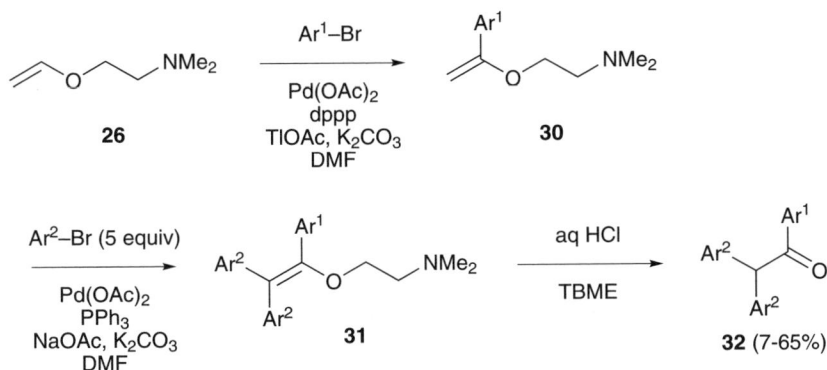

Figure 7.12 *Directed α,β,β-triarylation of vinyl ethers.*

Furthermore, a rare Mizoroki–Heck triarylation process becomes possible with **26** (Figure 7.12) [15]. The process initiates with α-arylation of **26** followed by β,β-diarylation. The α-arylation products **30** are obtained through the reaction of **26** and aryl bromides (Ar^{1}Br) with the agency of Pd(OAc)$_2$/dppp/TlOAc (dppp = 1,3-bis(diphenylphosphino)propane), following the procedure developed by Cabri and Candiani [18]. After aqueous workup, alkenyl ethers **30** are allowed to react with a fivefold excess of aryl bromides (Ar^{2}Br) in the presence of Pd(OAc)$_2$/PPh$_3$ to furnish triarylated vinyl ethers **31**, which are further hydrolysed to the corresponding triaryl ketones **32**.

7.5.3 Asymmetric Arylation

Hallberg and coworkers [19] have reported an asymmetric chelation-controlled Mizoroki–Heck arylation of prolinol-derived vinyl ether **33** (Figure 7.13). In the presence of a catalytic amount of Pd(OAc)$_2$, a regio- and stereo-selective α-arylation of **33** occurs with

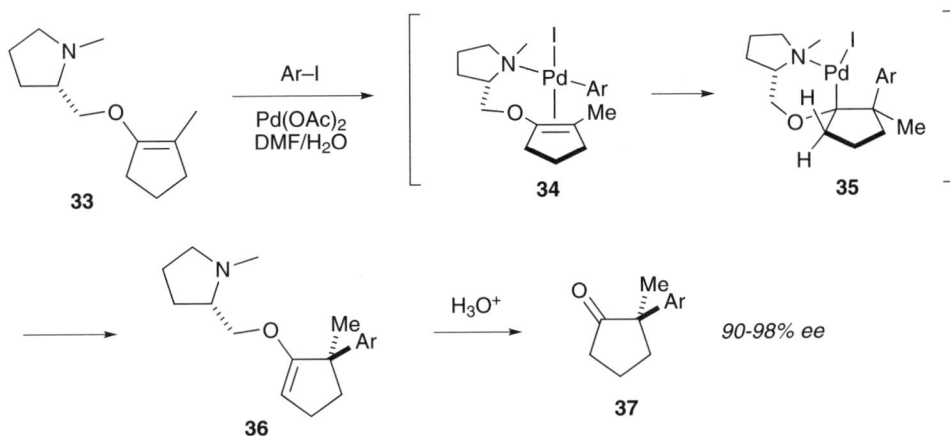

Figure 7.13 *Asymmetric arylation of vinyl ethers.*

a variety of aryl iodides providing vinyl ethers **36** in good yields with excellent diastereose-lectivities. The reactions are likely to proceed through face-selective C=C π-complexation and carbopalladation mediated by N-chelation (**34**→**35**) followed by *syn* β-hydrogen elimination. Notably, acid-mediated hydrolysis of **36** furnishes cyclopentenones **37** having a quaternary chiral centre at the α-position in high enantiomeric excesses (90–98% ee).

7.6 Vinyl Silanes

7.6.1 Reactivity Enhancement by the 2-Pyridyl Group on Silicon

In 2000, Itami and coworkers [20, 21] reported that the efficient Mizoroki–Heck arylation of vinyl silanes could be accomplished by appending a catalyst-directing 2-pyridyl group on silicon (Figure 7.14). In the presence of a palladium catalyst and a base, the Mizoroki–Heck reaction of alkenyl(2-pyridyl)silanes **38** with aryl halides takes place smoothly, giving the β-arylated products **39** in high yields with virtually complete stereoselectivity (>99% *E*). A suitably positioned nitrogen atom of the pyridyl group accelerates the C=C π-complexation and successive carbopalladation events (**40**→**41**→**42**). NMR and X-ray crystal structure analyses of several key palladium complexes provide conclusive evidence for the existence of pyridyl-to-palladium coordination during the reaction [16, 22]. Other related vinyl silanes, such as α-substituted vinyl(2-pyridyl)silane [16] and vinyl(2-pyrimidyl)silane [23], also efficiently undergo the Mizoroki–Heck reaction with high stereoselectivity.

7.6.2 Multiple Arylation

A 'hard-to-achieve' double Mizoroki–Heck reaction also proceeds with vinyl(2-pyridyl)silane and vinyl(2-pyrimidyl)silane **43** (Figure 7.15) [16, 21, 23]. The incoming aryl group always occupies a position *trans* to the silyl group, which is in agreement with *syn* carbopalladation and *syn* β-hydrogen elimination (Figure 7.14). This double Mizoroki–Heck reaction enables the installing of two different aryl groups at the two β-C—H bonds in

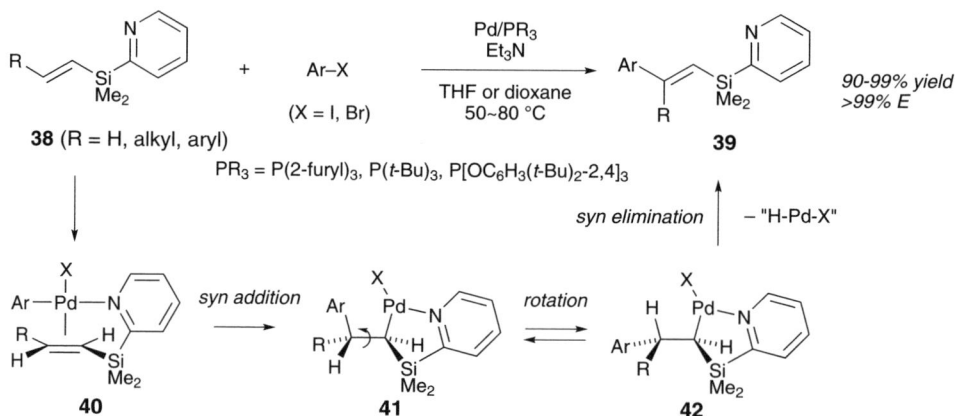

Figure 7.14   *Chelation-assisted Mizoroki–Heck reaction of alkenyl(2-pyridyl)silanes.*

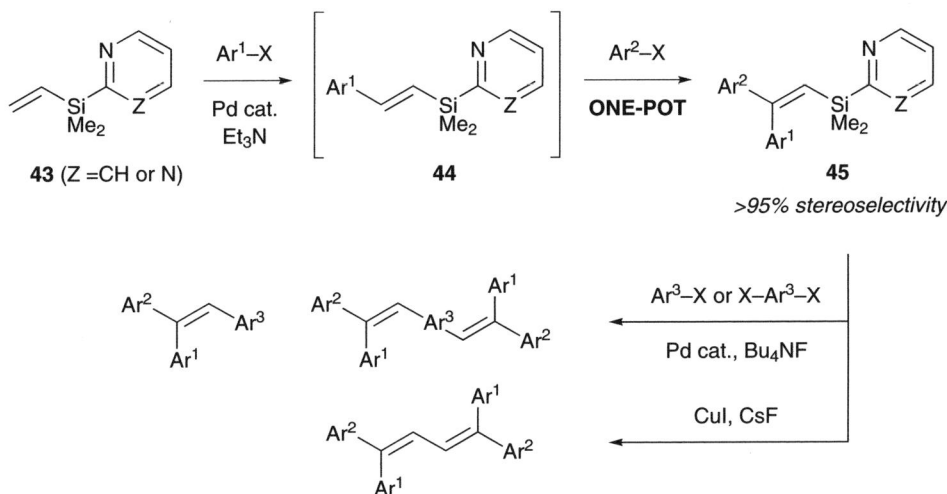

Figure 7.15 *One-pot double Mizoroki–Heck reaction of vinylsilanes.*

one pot. The one-pot double Mizoroki–Heck reaction also proceeds smoothly with α-substituted vinyl(2-pyridyl)silanes [16]. By merging these one-pot double Mizoroki–Heck arylations with the subsequent palladium-catalysed silicon-based cross-coupling or copper-mediated homocoupling, a range of functional multisubstituted olefins is accessible in a programmed and diversity-oriented fashion (Figure 7.15) [24].

7.7   Vinyl Sulfides

In 2004, Itami *et al.* [25] reported that the Mizoroki–Heck arylation of vinyl sulfides can be achieved by appending a catalyst-directing 2-pyrimidyl group on sulfur (Figure 7.16).

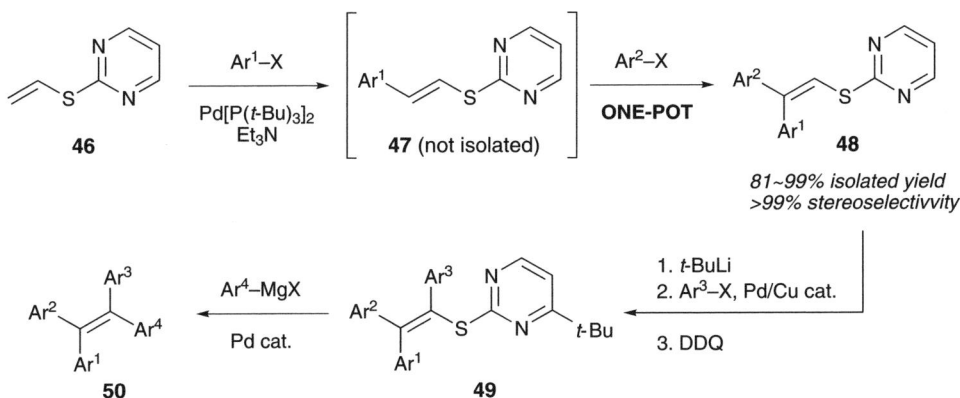

Figure 7.16 *One-pot double arylation of vinyl 2-pyrimidyl sulfide.*

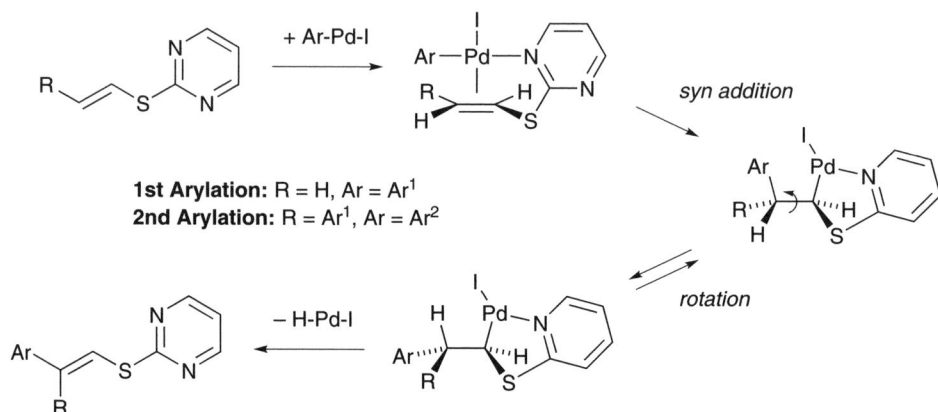

Figure 7.17 *Stereochemical rationale for double Mizoroki–Heck arylation of vinyl 2-pyrimidyl sulfide.*

Under the catalytic influence of Pd[P(*t*-Bu)$_3$]$_2$ and Et$_3$N, efficient Mizoroki–Heck arylation of vinyl 2-pyrimidyl sulfide (**46**) takes place with a range of aryl and heteroaryl iodides, furnishing β-arylated products **47** in excellent yield. Again, owing to the presence of a catalyst-directing 2-pyrimidyl group on sulfur, the demanding one-pot double Mizoroki–Heck arylation can be accomplished giving β,β-diarylated products **48** with virtually complete regio- and stereo-selectivities.

Similar to the case of vinyl(2-pyridyl)silane and vinyl(2-pyrimidyl)silane, the incoming aryl group always occupies a position *trans* to the sulfur group, which clearly proves the diastereospecificity of the carbopalladation/β-elimination sequence in the present reaction (Figure 7.17).

By merging this one-pot double Mizoroki–Heck arylation with the subsequent α-lithiation/cross-coupling and sulfur-based cross-coupling, a programmed and diversity-oriented synthesis of tetrasubstituted olefins **50** can be realized (Figure 7.16) [25]. Moreover, by using this one-pot double Mizoroki–Heck arylation as a key step, CDP840 (a selective phosphodiesterase-4 inhibitor) can be synthesized very rapidly [26].

7.8   Vinyl Sulfoxides

7.8.1   Reactivity Enhancement by the *o*-Aminophenyl Group on Sulfur

Carretero and coworkers [27] have introduced an *o*-aminophenyl group as a versatile directing group for chelation-controlled Mizoroki–Heck reactions, including an asymmetric reaction, which will be discussed below. The Mizoroki–Heck arylation of *o*-aminophenyl-equipped alkenyl sulfoxide **51** with iodobenzene under Pd(OAc)$_2$/dppf/Ag$_2$CO$_3$ (dppf = 1,1'-bis(diphenylphosphino)ferrocene) occurs exclusively at the β-position, yielding **52** with complete control of the double-bond geometry (Figure 7.18) [28]. Replacement of the *o*-aminophenyl group with a simple *p*-tolyl group is highly detrimental to this process (<40% conversion after prolonged reaction time). The enhanced rate is likely due to

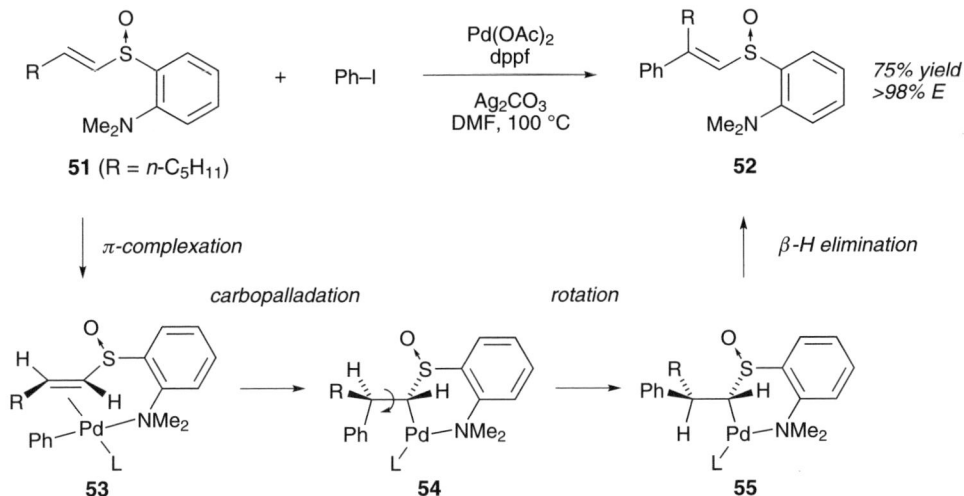

Figure 7.18 *Mizoroki–Heck reaction of alkenyl sulfoxides promoted by o-aminophenyl group.*

the coordination of the Me$_2$N group to palladium(II) intermediates, which facilitates both olefin π-complexation and carbopalladation steps (**53**→**54**→**55**).

7.8.2 Asymmetric Arylation

Carretero and coworkers [27] reported that an intermolecular diastereoselective Mizoroki–Heck reaction of alkenyl sulfoxides could be achieved with the aid of the *o*-aminophenyl group as a chelating chiral auxiliary (Figure 7.19). The arylation of *o*-aminophenyl-equipped cyclic alkenyl sulfoxide **56** with iodobenzene takes place efficiently under Pd(OAc)$_2$/dppp/Ag$_2$CO$_3$ to afford **58** with 94% diastereoselectivity. Interestingly, the reaction using the corresponding alkenyl phenyl sulfoxide leads to the arylation product favouring the opposite diastereomer in 77% selectivity. As in the aforementioned Hallberg system, chelation-controlled face-selective C=C π-complexation (**57**) for the reaction of **56** is again a likely scenario, while the reaction of phenyl analogue occurs under conventional

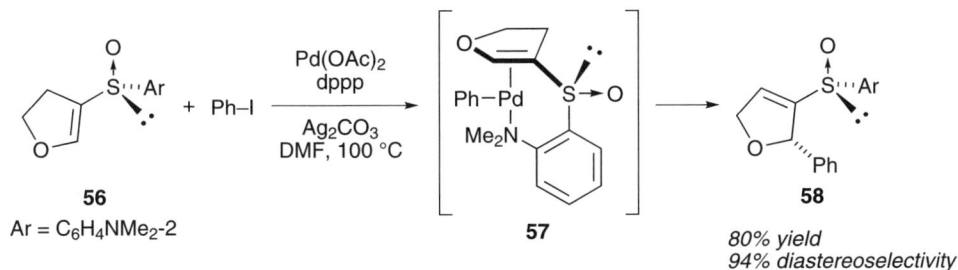

Figure 7.19 *Diasteroselective arylation directed by o-aminophenyl sulfoxide moiety.*

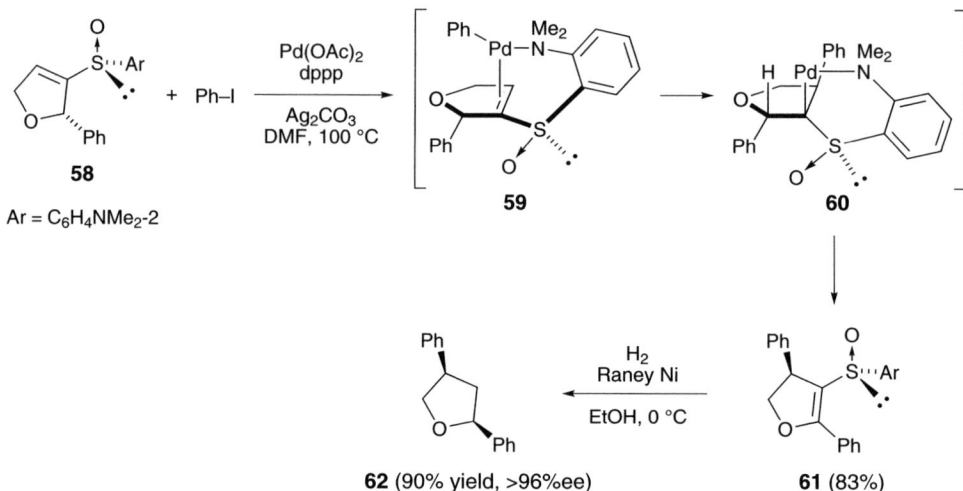

Figure 7.20 Second arylation and removal of sulfinyl group.

steric control. Related chelation-controlled diastereoselective arylation of cyclopentyl sulfoxide was also reported [29].

Furthermore, the second Mizoroki–Heck reaction of the thus-obtained **58** with iodobenzene yields **61**, which is then transformed into enantiopure diphenyltetrahydrofuran **62** by hydrogenation and C—S cleavage with Raney-Ni (Figure 7.20). The complete asymmetric induction might be rationalized by assuming a chelate model (**59→60**) similar to that proposed for **56**.

The *o*-aminophenyl-directed asymmetric Mizoroki–Heck reaction can also be applied to intramolecular reactions [30]. For example, when alkenyl iodide **63** is subjected to palladium catalyst, Mizoroki–Heck cyclization furnishes **64** with almost complete diastereoselectivity (Figure 7.21). An asymmetric reaction utilizing an enantiomerically pure sulfoxide unit is also possible.

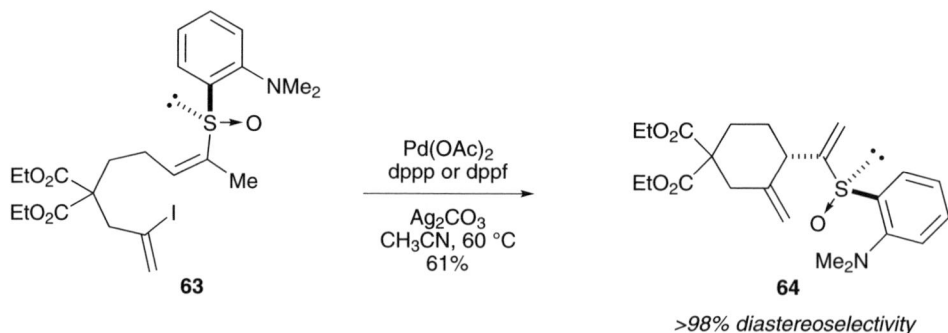

Figure 7.21 Diastereoselective cyclization directed by o-aminophenylsulfinyl group.

7.9 Vinyl and Allylic Sulfones

During the course of studies on *o*-aminophenyl-directed reactions, Carretero and coworkers [31] reported that the *o*-aminophenyl group also facilitated the Mizoroki–Heck arylation of alkenyl sulfones. In a quite similar manner to the corresponding sulfoxides, alkenyl sulfones equipped with the *o*-aminophenyl group on sulfur (**65**) react with iodobenzene under a Pd(OAc)$_2$/Ag$_2$CO$_3$ system giving **66** in high yields with high stereoselectivities (Figure 7.22).

As before, Pd–N coordination during a catalytic cycle (**67**→**68**→**69**) is indicated by the fact that the absence of coordinating *o*-amino group results in a completely different outcome giving substituted dihydrophenanthrene **71** (Figure 7.23). The formation of **71**

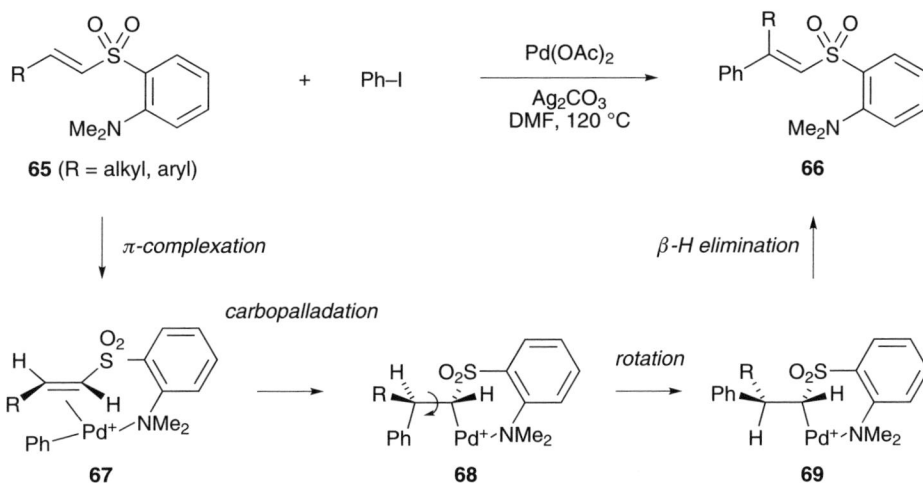

Figure 7.22 *Mizoroki–Heck reaction of alkenyl sulfones promoted by o-aminophenyl group.*

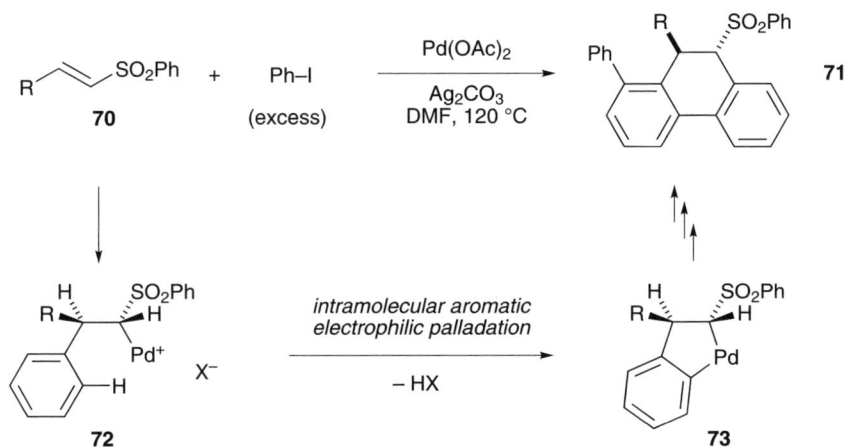

Figure 7.23 *Cascade arylation with simple alkenyl sulfones.*

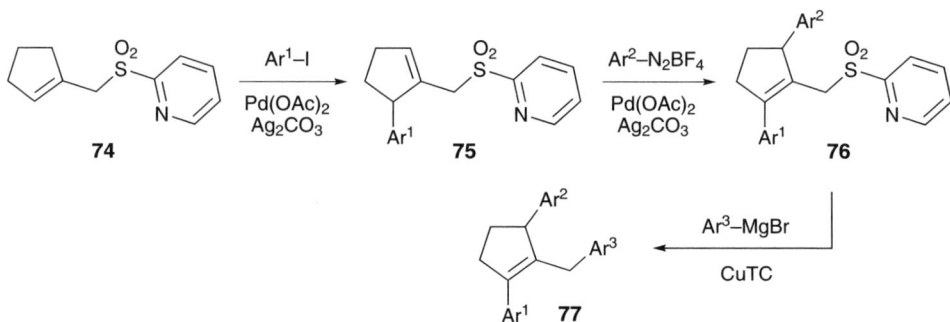

Figure 7.24 *Mizoroki–Heck arylation of allylic sulfones directed by 2-pyridyl group on sulfur.*

is proposed to proceed through intramolecular aromatic electrophilic palladation after the phenylpalladation step (**72→73**). It is assumed that coordination of the *o*-amino group suppresses this intramolecular aromatic electrophilic palladation (Figure 7.22).

Introduction of the coordinating 2-pyridyl group on sulfur greatly accelerates the Mizoroki–Heck reaction of cyclic allylic sulfones [32]. For example, the cyclopentene derivative **74** reacts with aryl iodides (Ar^{1}I) in the presence of palladium catalyst to afford **75** in excellent yields (Figure 7.24). The resultant alkenes **75** are further arylated with aryldiazonium salts under the influence of Pd(OAc)$_2$/Ag$_2$CO$_3$ to furnish **76** in excellent yields. Both of these processes are likely to be assisted by the effective coordination of the 2-pyridyl group to palladium intermediates in the Mizoroki–Heck process. Finally, the allylic sulfones **76** thus obtained are subjected to copper-catalysed allylic substitutions with Grignard reagents to produce densely substituted cyclopentenes **77** in good yields (Figure 7.24).

7.10 Other Chelation-Controlled Systems

As is obvious from the numerous examples in previous sections, the presence of coordinating functionalities in substrate structures greatly influences the outcome of Mizoroki–Heck reactions. In this last section, we will briefly mention other representative chelation-controlled systems.

Hallberg and coworkers [33] have reported an efficient Mizoroki–Heck reaction of 1,2-cyclohexanedione and 2-ethoxy-2-cyclohexanone (**78**) with aryl bromide as a new synthetic route to synthetically useful arylated cyclohexenone derivatives **79** (Figure 7.25). Since the primarily formed organopalladium intermediate **80** cannot directly suffer β-hydrogen elimination for conformational reasons, the reaction most likely proceeds through enolization (**80→81→82**), as depicted in Figure 7.25. The formation of palladium enolate **81** might be facilitated by additional stabilization through the hydroxy group and the ethoxy group respectively. Indeed, under otherwise identical conditions, the arylation does not occur with simple cyclohexenone.

During the course of synthetic study of diazonamide, Harran and coworkers [34] reported an interesting example of Mizoroki–Heck macrocyclization of **83** driven by a phenolic hydroxy group (Figure 7.26). Use of the corresponding methyl ether results in a marked

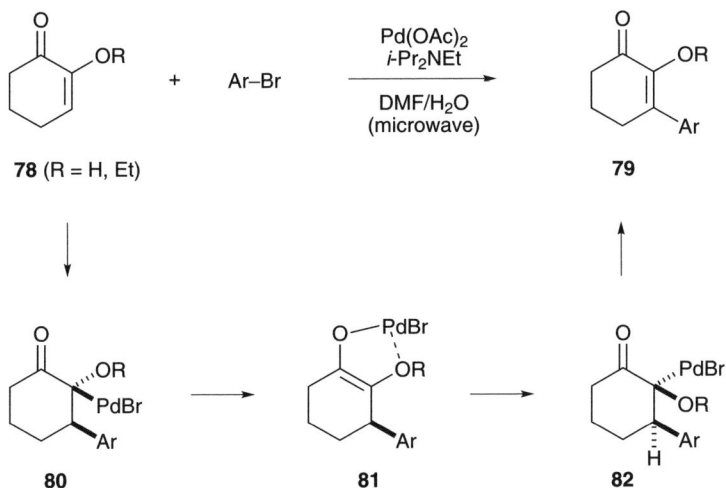

Figure 7.25 *Directed Mizoroki–Heck reaction of cyclohexenone derivatives.*

decrease in cyclization. The formation of palladium phenoxide (**85**) from a free hydroxy group appears to be pivotal for preorganization prior to macrocyclization.

During the study of desymmetrizing Mizoroki–Heck cyclizations, Oestreich *et al.* [35] found an interesting role of catalyst-directing hydroxy group in achieving a highly enantios-elective process. For example, when bis-homoallylic alcohol tethered with phenyl triflate

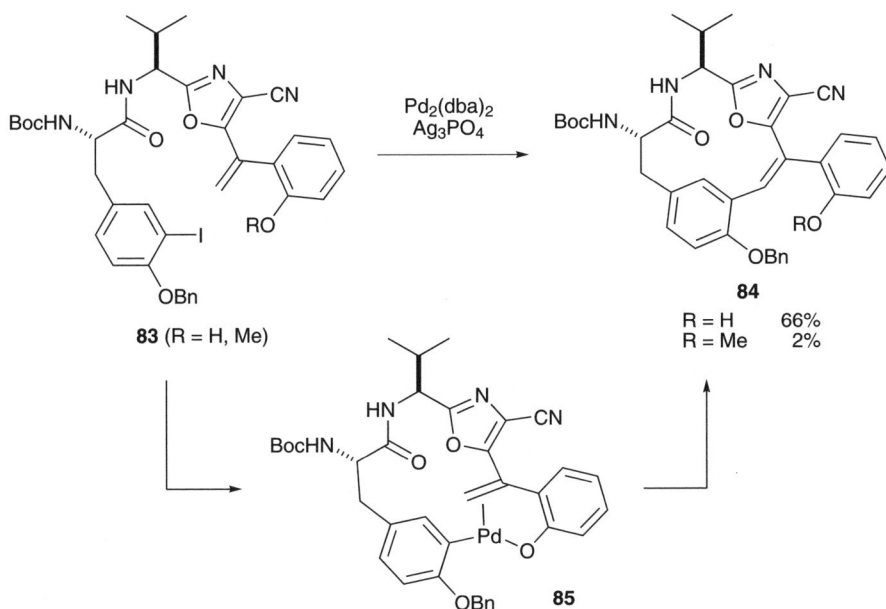

Figure 7.26 *Mizoroki–Heck macrocyclization driven by phenolic hydroxy group.*

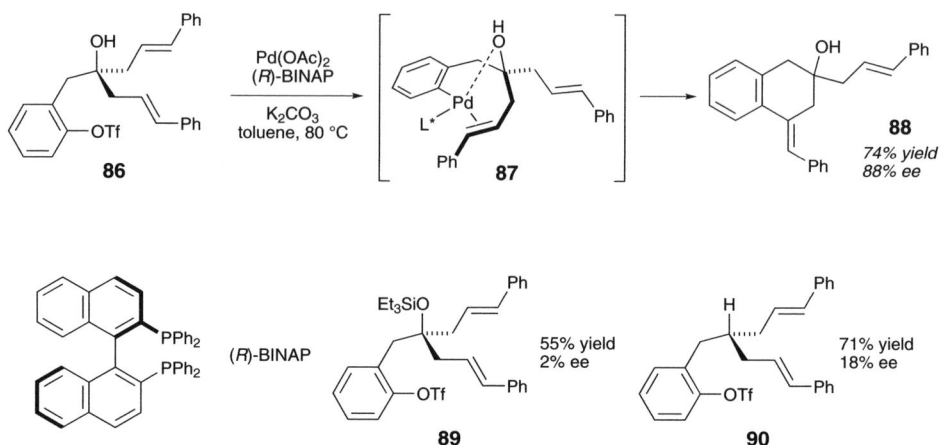

Figure 7.27   *Enantioselective cyclization directed by hydroxy group.*

(**86**) is subjected to 2,2′-bis(diphenylphosphino)-1,1′-binaphthyl (BINAP)-modified palladium catalyst, an intramolecular Mizoroki–Heck reaction takes place to give **88** in good yield with high enantioselectivity (88% ee) (Figure 7.27). Interestingly, replacement of a hydroxy group in the substrate structure to triethylsiloxy (**89**) and hydrogen atom (**90**) substantially erodes the enantioselectivities (2% ee and 18% ee respectively). The pronounced effect of the hydroxy group indicates that it might act as a catalyst-directing group in the stereochemistry-determining step (**87**) of the ring closure of **86**. In addition to the enhanced control in enantioselectivity, the presence of the hydroxy group greatly accelerates the reaction.

7.11   Summary

The chelation-controlled Mizoroki–Heck reaction offers various opportunities in organic synthesis. In many cases, the presence of additional coordinating auxiliaries results in facile olefin–π-complexation and carbopalladation, thereby enhancing the overall reactivity of alkenes in the Mizoroki–Heck reaction. A suitably positioned coordinating group can be used to control regioselectivity during the carbopalladation event through attractive heteroatom–palladium interactions. Other than heteroatom-based σ-donors, carbon-based π-donors, such as alkenyl groups, can also be used as chelation elements in Mizoroki–Heck reactions [36]. By taking advantage of the chelation-stabilized conformation, position-selective abstraction of the hydrogen atom from an organopalladium intermediate becomes possible in the β-elimination event. Moreover, chelation control can be utilized in diastereoselective and enantioselective Mizoroki–Heck reactions. Another important point in chelation-controlled Mizoroki–Heck reaction is that the chelating group can be used for other purposes, such as a phase tag for separation of products and recovery of catalysts [37]. The examples demonstrated here bode well for the potential of chelation-controlled Mizoroki–Heck reactions for further investigation.

References

1. Reviews on directed reactions: (a) Hoveyda, A.H., Evans, D.A. and Fu, G.C. (1993) Substrate-directable chemical reactions. *Chem. Rev.*, **93**, 1307–70; (b) Beak, P. and Meyers, A.I. (1986) Stereo- and regiocontrol by complex induced proximity effects: reactions of organolithium compounds. *Acc. Chem. Res.*, **19**, 356–63; (c) Snieckus, V. (1990) Directed *ortho* metalation. Tertiary amide and *O*-carbamate directors in synthetic strategies for polysubstituted aromatics. *Chem. Rev.*, **90**, 879–933; (d) Beak, P., Basu, A., Gallagher, D.J. *et al.* (1996) Regioselective, diastereoselective, and enantioselective lithiation-substitution sequences: reaction pathways and synthetic applications. *Acc. Chem. Res.*, **29**, 552–60; (e) Whisler, M.C., MacNeil, S., Snieckus, V. and Beak, P. (2004) Beyond thermodynamic acidity: a perspective on the complex-induced proximity effect (CIPE) in deprotonation reactions. *Angew. Chem. Int. Ed.*, **43**, 2206–25.

2. (a) Oestreich, M. (2005) Neighbouring-group effects in Heck reactions. *Eur. J. Org. Chem.*, 783–92; (b) Oestreich, M. (2008) Directed Mizoroki–Heck reactions, in *Topics in Organometallic Chemistry*, Vol. 24 (ed. N. Chatani), Springer-Verlag, Heidelberg, pp. 169–92.

3. Andersson, C.-M., Larsson, J. and Hallberg, A. (1990) Chelation-controlled, palladium-catalyzed vinylic substitution reactions of vinyl ethers. 2-Arylethanal equivalents from aryl halides. *J. Org. Chem.*, **55**, 5757–61.

4. The representative examples using removable directing groups for metal-catalysed and -mediated reactions: (a) Breit, B. (2000) Controlling stereoselectivity with the aid of a reagent-directing group: hydroformylation, cuprate addition, and domino reaction sequences. *Chem. Eur. J.*, **6**, 1519–24; (b) Jun, C.-H., Moon, C.W. and Lee, D.-Y. (2002) Chelation-assisted carbon–hydrogen and carbon–carbon bond activation by transition metal catalysts. *Chem. Eur. J.*, **8**, 2422–8; (c) Itami, K., Mitsudo, K., Fujita, K. *et al.* (2004) Catalytic intermolecular Pauson–Khand-type reaction: strong directing effect of pyridylsilyl and pyrimidylsilyl groups and isolation of Ru complexes relevant to catalytic reaction. *J. Am. Chem. Soc.*, **126**, 11058–66; (d) Adrio, J. and Carretero, J.C. (1999) The *tert*-butylsulfinyl group as a highly efficient chiral auxiliary in asymmetric Pauson–Khand reactions. *J. Am. Chem. Soc.*, **121**, 7411–2; (e) Chatani, N., Tatamidani, H., Ie, Y. *et al.* (2001) The ruthenium-catalyzed reductive decarboxylation of ethers: catalytic reactions involving the cleavage of acyl–oxygen bonds of esters. *J. Am. Chem. Soc.*, **123**, 4849–50; (f) Krauss, I.J., Wang, C.C.Y. and Leighton, J.L. (2001) Highly regioselective and diastereoselective directed hydroformylation of allylic ethers: a new approach to propionate aldol synthesis. *J. Am. Chem. Soc.*, **123**, 11514–5; (g) Ko, S., Na, Y. and Chang, S. (2002) A novel chelation-assisted hydroesterification of alkenes via ruthenium catalysis. *J. Am. Chem. Soc.*, **124**, 750–1; (h) Pastine, S.J., Gribkov, D.V. and Sames, D. (2006) sp³ C—H bond arylation directed by amidine protecting group: α-arylation of pyrrolidines and piperidines. *J. Am. Chem. Soc.*, **128**, 14220–1.

5. (a) Melpolder, J.B. and Heck, R.F. (1976) A palladium-catalyzed arylation of allylic alcohols with aryl halides. *J. Org. Chem.*, **41**, 265–72; (b) Chalk, A.J. and Magennis, S.A. (1976) Palladium-catalyzed vinyl substitution reactions. I. A new synthesis of 2- and 3-phenyl substituted allylic alcohols, aldehydes, and ketones from allylic alcohols. *J. Org. Chem.*, **41**, 273–8.

6. (a) Cacchi, S., Morera, E. and Ortar, G. (1984) Palladium-catalyzed vinylation of enol triflates. *Tetrahedron Lett.*, **25**, 2271–4; (b) Bernocchi, E., Cacchi, S., Ciattini, P.G. *et al.* (1992) Palladium-catalyzed vinylation of allylic alcohols with enol triflates. A convenient synthesis of conjugated dienols. *Tetrahedron Lett.*, **33**, 3073–6.

7. (a) Jeffery, T. (1991) Palladium-catalyzed arylation of allylic alcohols: highly selective synthesis of β-aromatic carbonyl compounds or β-aromatic α,β-unsaturated alcohols. *Tetrahedron Lett.*, **32**, 2121–24; (b) Jeffery, T. (1991) Palladium-catalyzed reaction of vinylic halides with allylic alcohols: a highly chemo-, regio- and stereo-controlled synthesis of conjugated dienols. *J. Chem.*

Soc., Chem. Commun, 324–5; (c) Jeffery, T. (1993) Palladium-catalyzed direct synthesis of optically active dienols. *Tetrahedron Lett.*, **34**, 1133–6.

8. Kang, S.-K., Lee, H.-W., Jang, S.-B. *et al.* (1996) Complete regioselection in palladium-catalyzed arylation and alkenylation of allylic alcohols with hypervalent iodonium salts. *J. Org. Chem.*, **61**, 2604–5.

9. Ono, K., Fugami, K., Tanaka, S. and Tamaru, Y. (1994) Palladium catalyzed arylation of *N*-alkyl *O*-allylic carbamates: synthesis of cinnamyl alcohols via Heck arylation. *Tetrahedron Lett.*, **35**, 4133–6.

10. Kang, S.-K., Jung, K.-Y., Park, C.-H. *et al.* (1995) Palladium-catalyzed arylation of allylic diols: highly selective synthesis of phenyl-substituted allylic diols. *Tetrahedron Lett.*, **35**, 6287–90.

11. Nilsson, K. and Hallberg, A. (1992) Synthesis of 1-propyl-3-(3-hydroxyphenyl)piperidine by regiocontrolled palladium-catalyzed arylation. *J. Org. Chem.*, **57**, 4015–7.

12. (a) Olofsson, K., Larhed, M. and Hallberg, A. (2000) Highly regioselective palladium-catalyzed β-arylation of *N,N*-dialkylallylamines. *J. Org. Chem.*, **65**, 7235–9; (b) Olofsson, K., Sahlin, H., Larhed, M. and Hallberg, A. (2001) Regioselective palladium-catalyzed synthesis of β-arylated primary allylamine equivalents by an efficient Pd—N coordination. *J. Org. Chem.*, **66**, 544–9.

13. (a) Larhed, M., Andersson, C.-M., and Hallberg, A. (1993) Chelation-controlled, palladium-catalyzed arylation of vinyl ethers. *Acta Chem. Scand.*, **47**, 212–7; (b) Larhed, M., Andersson, C.-M., and Hallberg, A. (1994) Chelation-controlled, palladium-catalyzed arylation of enol ethers with aryl triflates. Ligand control of selection for α- or β-arylation of [2-(dimethylamino)ethoxy]ethene. *Tetrahedron*, **50**, 285–304; (c) Stadler, A., von Schenck, H., Vallin, K.S.A. *et al.* (2004) Terminal Heck vinylations of chelating vinyl ethers. *Adv. Synth. Catal.*, **346**, 1773–81.

14. Badone, D. and Guzzi, U. (1993) Palladium-catalyzed β-arylation of modified vinyl ethers with aryl triflates. *Tetrahedron Lett.*, **34**, 3603–6.

15. Nilsson, P., Larhed, M. and Hallberg, A. (2001) Highly regioselective, sequential, and multiple palladium-catalyzed arylations of vinyl ethers carrying a coordinating auxiliary: an example of a Heck triarylation process. *J. Am. Chem. Soc.*, **123**, 8217–25.

16. Itami, K., Nokami, T., Ishimura, Y. *et al.* (2001) Diversity-oriented synthesis of multisubstituted olefins through the sequential integration of palladium-catalyzed cross-coupling reactions. 2-Pyridyldimethyl(vinyl)silane as a versatile platform for olefin synthesis. *J. Am. Chem. Soc.*, **123**, 11577–85.

17. Svennebring, A., Nilsson, P. and Larhed, M. (2004) Microwave-promoted and chelation-controlled double arylations of terminal olefinic carbon of vinyl ethers. *J. Org. Chem.*, **69**, 3345–9.

18. Cabri, W. and Candiani, I. (1995) Recent developments and new perspectives in the Heck reaction. *Acc. Chem. Res.*, **28**, 2–7.

19. Nilsson, P., Larhed, M. and Hallberg, A. (2003) A new highly asymmetric chelation-controlled Heck arylation. *J. Am. Chem. Soc.*, **125**, 3430–1.

20. Itami, K., Mitsudo, K., Kamei, T. *et al.* (2000) Highly efficient carbopalladation across vinylsilane: dual role of the 2-PyMe₂Si group as a directing group and as a phase tag. *J. Am. Chem. Soc.*, **122**, 12013–4.

21. (a) Itami, K., Nokami, T. and Yoshida, J. (2001) Palladium-catalyzed cross-coupling reaction of alkenyldimethyl(2-pyridyl)silanes with organic halides: complete switch from the carbometalation pathway to the transmetalation pathway. *J. Am. Chem. Soc.*, **123**, 5600–1; (b) Nokami, T., Itami, K. and Yoshida, J. (2004) Aqueous photo-dimerization using 2-pyridylsilyl group as a removable hydrophilic group. *Chem. Lett.*, **33**, 596–7; (c) Itami, K., Ushiogi, Y., Nokami, T. *et al.* (2004) Stereoselective synthesis of multisubstituted butadienes through directed Mizoroki–Heck reaction and homocoupling reaction of vinyl(2-pyridyl)silane. *Org. Lett.*, **6**, 3695–8.

22. Itami, K., Kamei, T. and Yoshida, J. (2001) Unusually accelerated silylmethyl transfer from tin in Stille coupling: implication of coordination-driven transmetalation. *J. Am. Chem. Soc.*, **123**, 8773–9.

23. Itami, K., Ohashi, Y. and Yoshida, J. (2005) Triarylethene-based extended π-systems: programmable synthesis and photophysical properties. *J. Org. Chem.*, **70**, 2778–92.

24. (a) Itami, K. and Yoshida, J. (2006) Multisubstituted olefins: platform synthesis and applications to materials science and pharmaceutical chemistry. *Bull. Chem. Soc. Jpn.*, **79**, 811–24; (b) Itami, K. and Yoshida, J. (2006) Platform synthesis: a useful strategy for rapid and systematic generation of molecular diversity. *Chem. Eur. J.*, **12**, 3966–74.

25. Itami, K., Mineno, M., Muraoka, N. and Yoshida, J. (2004) Sequential assembly strategy for tetrasubstituted olefin synthesis using vinyl 2-pyrimidyl sulfide as a platform. *J. Am. Chem. Soc.*, **126**, 11778–9.

26. Muraoka, N., Mineno, M., Itami, K. and Yoshida, J. (2005) Rapid synthesis of CDP840 with 2-pyrimidyl vinyl sulfide as a platform. *J. Org. Chem.*, **70**, 6933–6.

27. Buezo, N.D., Alonso, I. and Carretero, J.C. (1998) Sulfinyl group as a novel chiral auxiliary in asymmetric Heck reactions. *J. Am. Chem. Soc.*, **120**, 7129–30.

28. Alonso, I. and Carretero, J.C. (2001) Highly stereoselective synthesis of trisubstituted α,β-unsaturated sulfoxides by Heck reaction. *J. Org. Chem.*, **66**, 4453–6.

29. Buezo, N.D., de la Rosa, J.C., Priego, J. *et al.* (2001) Sulfoxides as stereochemical controllers in intermolecular Heck reactions. *Chem. Eur. J.*, **7**, 3890–900.

30. Buezo, N.D., Mancheño, O.G., and Carretero, J.C. (2000) The 2-(*N,N*-dimethylamino) phenylsulfinyl group as an efficient chiral auxiliary in intramolecular Heck reactions. *Org. Lett.*, **2**, 1451–4.

31. Mauleón, P., Núñez, A.A., Alonso, I. and Carretero, J.C. (2003) Palladium-catalyzed cascade reaction of α,β-unsaturated sulfones with aryl iodides. *Chem. Eur. J.*, **9**, 1511–20.

32. Llamas, T., Gómez Arrayás, R. and Carretero, J.C. (2004) Chelation-induced catalytic multiple arylation of allylic 2-pyridyl sulfones. *Adv. Synth. Catal.*, **346**, 1651–4.

33. Garg, N., Larhed, M. and Hallberg, A. (1998) Heck arylation of 1,2-cyclohexanedione and 2-ethoxy-2-cyclohexenone. *J. Org. Chem.*, **63**, 4158–62.

34. Jeong, S., Chen, X. and Harran, P.G. (1998) Macrocyclic triarylethylenes via Heck endocyclization: a system relevant to diazonamide synthesis. *J. Org. Chem.*, **63**, 8640–1.

35. Oestreich, M., Sempere-Culler, F. and Machotta, A.B. (2005) Catalytic desymmetrizing intramolecular Heck reaction: evidence for an unusual hydroxy-directed migratory insertion. *Angew. Chem. Int. Ed.*, **44**, 149–52.

36. (a) Madin, A. and Overman, L.E. (1992) Controlling stereoselection in intramolecular Heck reactions by tailoring the palladium catalyst. *Tetrahedron Lett.*, **33**, 4859–62; (b) Madin, A., O'Donnell, C.J., Oh, T. *et al.* (2005) Use of the intramolecular Heck reaction for forming congested quaternary carbon stereocenters. Stereocontrolled total synthesis of ($\pm$)-gelsemine. *J. Am. Chem. Soc.*, **127**, 18054–65.

37. Yoshida, J. and Itami, K. (2002) Tag strategy for separation and recovery. *Chem. Rev.*, **102**, 3693–716.

8

The Mizoroki–Heck Reaction in Domino Processes

Lutz F. Tietze and Laura M. Levy

Institut für Organische und Biomolekulare Chemie, Georg-August-Universität, Göttingen,
Tammannstraße 2, Göttingen, Germany

8.1 Introduction

Domino reactions represent a new concept in organic synthesis, allowing the highly efficient preparation of complex structures in a few steps starting from simple substrates [1]. They are defined as *processes of two or more bond-forming reactions under identical reaction conditions, in which the latter transformations take place at the functionalities obtained in the former bond-forming reactions.* The greater the number of individual steps and the higher the complexity of the product is, the more valuable is such a domino process. Moreover, domino reactions can be performed as one-, two- and multi-component reactions, of which the latter allow the synthesis of highly diversified substance libraries. They are time-resolved transformations, which can be nicely illustrated by domino tiles, where one tile tips over the next, which tips over the next, and so on. Domino reactions can be classified according to the mechanism of the different steps, which would lead, for example, for a threefold domino process taking eight different types of transformations into account to 512 new types of domino process (Table 8.1).

   Clearly, domino reactions are not only elegant, but they are also beneficial to our environment, since they preserve our natural resources and can reduce the amount of waste produced in a synthesis. That is the reason why they are also highly appropriate for industry; moreover, they save time, allowing one to produce more compounds per period of time compared with stepwise procedures.

The Mizoroki–Heck Reaction   Edited by Martin Oestreich
© 2009 John Wiley & Sons, Ltd

Table 8.1   *Classification of domino reactions.*

I. Transformation	II. Transformation	III. Transformation
1. Cationic	1. Cationic	1. Cationic
2. Anionic	2. Anionic	2. Anionic
3. Radical	3. Radical	3. Radical
4. Pericyclic	4. Pericyclic	4. Pericyclic
5. Photochemical	5. Photochemical	5. Photochemical
6. Transition-metal-catalysed	6. Transition-metal-catalysed	6. Transition-metal-catalysed
7. Oxidative or reductive	7. Oxidative or reductive	7. Oxidative or reductive
8. Enzymatic	8. Enzymatic	8. Enzymatic

Catalytic domino reactions are very attractive, especially those which use transition-metal-catalysed transformations. Thus, an increasing number of domino processes starting with such a reaction have been published in recent years. In particular, a wide range of palladium-catalysed domino processes have been used.

In general, palladium-catalysed transformations are of major importance in synthetic organic chemistry, since they allow the formation of C—C, C—O, C—N and other bonds under mild reaction conditions and, moreover, they tolerate many functional groups.

One distinguishes palladium(0)- and palladium(II)-catalysed reactions. The most common palladium(0) transformations are the Mizoroki–Heck[1] and the cross-coupling transformations such as the Suzuki–Miyaura,[2] the Stille and the Sonogashira reactions, which allow the arylation or alkenylation of C=C double bonds, boronic acid derivates, stannanes and alkynes respectively [2]. Another important palladium(0) transformation is the nucleophilic substitution of usually allylic acetates or carbonates known as the Tsuji–Trost reaction [3]. The most versatile palladium(II)-catalysed transformation is the Wacker oxidation, which is industrially used for the synthesis of acetaldehyde from ethylene [4]. It should be noted that many of these palladium-catalysed transformations can also be performed in an enantioselective way [5].

8.1.1 General Aspects

In most of the palladium-catalysed domino processes known so far, the Mizoroki–Heck reaction – the palladium(0)-catalysed reaction of aryl halides or triflates as well as of alkenyl halides or triflates with alkenes or alkynes – has been applied as the starting transformation accordingly to our classification (Table 8.1). It has been combined with another Mizoroki–Heck reaction [6] or a cross-coupling reaction [7], such as Suzuki, Stille or Sonogashira reactions. In other examples, a Tsuji–Trost reaction [8], a carbonylation, a pericyclic or an aldol reaction has been employed as the second step. On the other hand, cross-coupling reactions have also been used as the first step followed by, for example, a Mizoroki–Heck reaction; or Tsuji–Trost reactions, palladation of alkynes or allenes [9], carbonylations [10], aminations [11] or palladium(II)-catalysed Wacker-type reactions [12] were employed as the first step. A novel illustrative example of the latter procedure is the efficient enantioselective synthesis of vitamin E [13].

---

[1] In the following, only one name is used to avoid confusion in the domino nomenclature.
[2] As footnote 1.

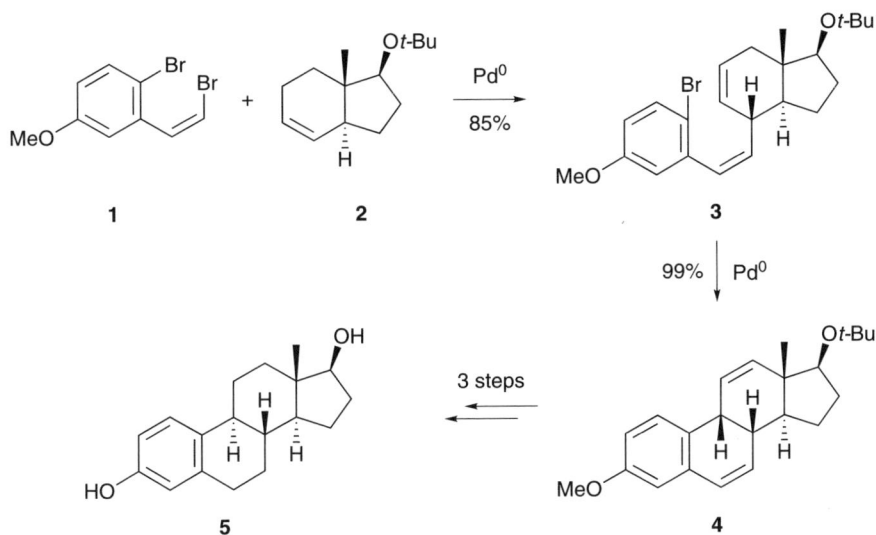

Scheme 8.1   *Synthesis of enantiopure estradiol (5).*

In transition-metal-catalysed domino reactions, often more than one catalyst is employed. In Tietze's definition and classification of domino reactions, no differentiation is made between domino reactions, whether it is just one or more than one transition-metal catalyst used for the different steps, as long as the steps take place.

As already mentioned, palladium has the advantage of being compatible with many functional groups. Therefore, it is an ideal catalyst for domino reactions. However, as in all processes of this type, an adjustment of the reactivity of the functionalities involved in the different steps is necessary. This can be done by taking advantage of, for example, the different reactivities of aryl iodides compared with aryl bromides or of vinyl bromides compared with aryl bromides. This concept has been used by Tietze and coworkers in their synthesis of estradiol (**5**) (Scheme 8.1). They employed two Mizoroki–Heck reactions starting with the substituted anisol **1**, which undergoes a Mizoroki–Heck reaction with the indene derivative **2** to give **3** [14]. The transformation takes place exclusively at the vinyl bromide moiety in **1**. In a second Mizoroki–Heck reaction, **3** was transformed into **4**, which led to the enantiopure estradiol (**5**) in three more steps (Scheme 8.1).

Another possibility to achieve a differentiation between two functionalities which can react with palladium is the formation of different ring sizes in the different steps. Thus, an intramolecular cyclization to give a medium-sized ring is disfavoured compared with an intermolecular reaction followed by an intramolecular reaction to afford a macrocycle.

Similarly, in the reaction of **6**, a four- or a six-membered ring could be obtained in the first step with clear preference towards the formation of the six-membered ring. Thus, one of the early examples of two successive Mizoroki–Heck reactions is the synthesis of the condensed bicyclic compound **7** from the acyclic precursor **6** by Overman and coworkers [15] (Scheme 8.2). Later, it has been shown by Negishi that a multitude of rings as in **9** can be formed according to this scheme, starting from **8** [16] (Scheme 8.2). The cycloisomerization of enynes and allenes catalysed by palladium has also proven to be an interesting process for the synthesis of tricycle **13**, as shown by Trost and coworkers [17].

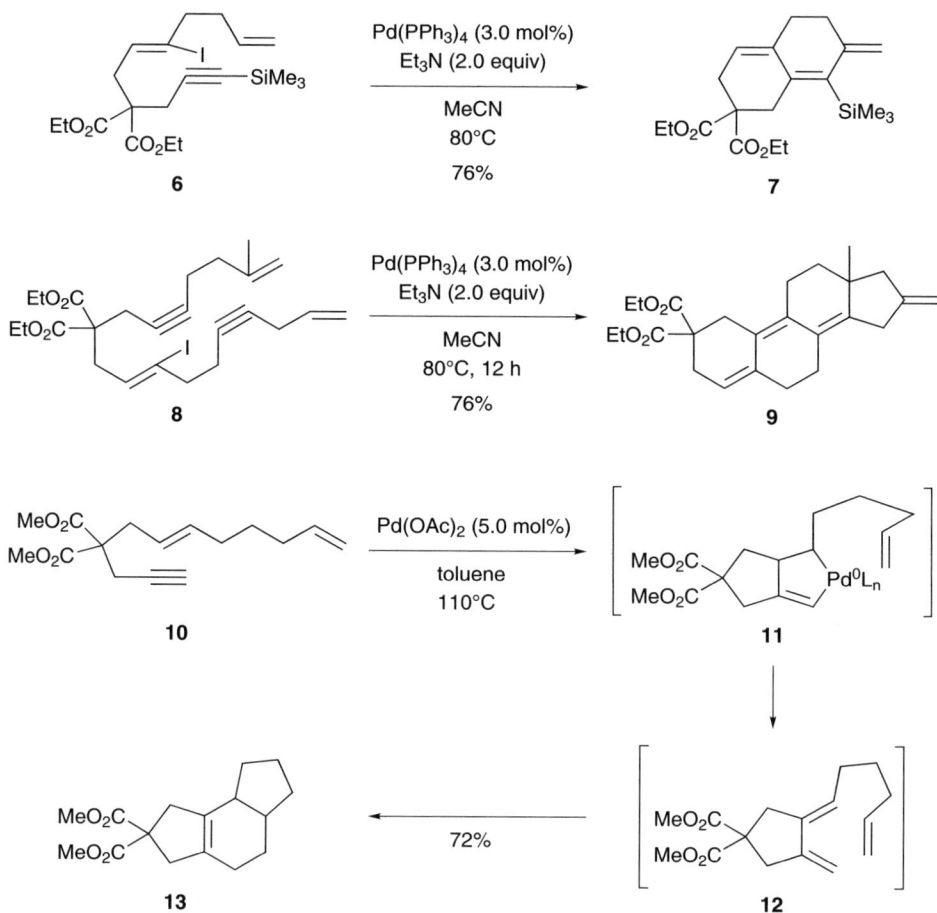

Scheme 8.2   *Formation of polycyclic compounds.*

Thus, transformation of the dienyne **10** using Pd(OAc)$_2$ led to **13** in 72% yield, in which the last step is a Diels–Alder reaction of the intermediate **12** (Scheme 8.2). The variety of products can be increased by an anion capturing process of the intermediate palladium compounds being intensively explored by Grigg and coworkers [18].

8.2   Domino Mizoroki–Heck Reactions

A field of great interest at the moment is the synthesis of overcrowded tetrasubstituted alkenes which can act as switches and molecular motors [19]. They are also an interesting class of compounds for the development of optical data storage. Tietze and coworkers [1c] have prepared a wide variety of these types of compounds, such as **16**, by domino Mizoroki–Heck processes in high yield and complete control of the configuration of the double bond formed using aryl bromides **14a–f** as substrates, which contain a triple bond and an allylsilane moiety. The best results were obtained using the Herrmann–Beller catalyst (**15**) [20]. It can be assumed that the palladium species **18a–c** are intermediates,

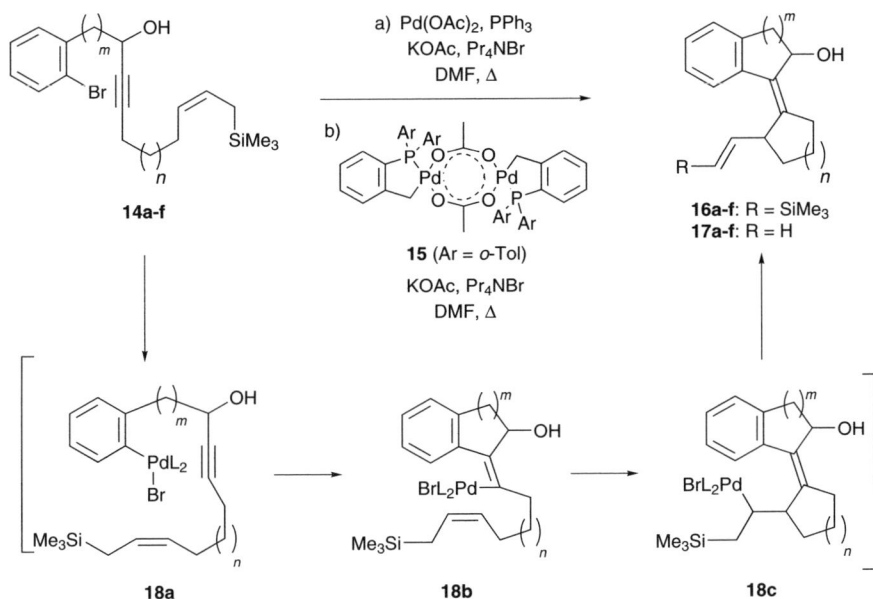

Scheme 8.3   *Allylsilane-terminated domino Mizoroki–Heck reactions of* **14a–f**.

Entry	Substrate	m/n	Conditions	Temp. [°C]	Time [h]	dr of **16**	Yield [%] **16** (**17**)
1	**14a**	1/1	a	80	4.5	1.2:1	71 (3)
2	**14b**	2/1	a	95	22	1.3:1	66 (5)
3	**14c**	3/1	b	130	21	2.6:1	43 (3)
4	**14d**	1/2	a	80	15	4.5:1	62 (8)
5	**14e**	2/2	a	100	17	9.4:1	61 (18)
6	**14f**	3/2	b	130	21	20:1	43 (7)

whereby the allylsilane moiety in **18c** is responsible for the selective elimination step with formation of the tertiary stereogenic centre. Thus, by employing allylsilanes, the inherent disadvantage of Mizoroki–Heck reactions using alkenes with α- and α'-hydrogen atoms to give mixtures of double-bond isomers can be avoided [21]. In such cases, the XL_2–Pd–H elimination of the intermediate palladium complex is highly regioselective. Under irradiation with a high-pressure mercury lamp, (*E*)-**16** gave a 1 : 1-mixture of the (*Z*)- and (*E*)-isomers of the inner double bond (Scheme 8.3).

Very recently, the same group [22] performed the synthesis of acenaphthylenes **20** by a combination of a Mizoroki–Heck reaction and an unexpected C–H activation by treatment of a series of alkynes **19** with palladium(0). Acenaphthylenes **20** were obtained as single products in high yields in those cases where R is an aryl moiety. When aliphatic alkynes are used, the yield drops considerably. It can be assumed that, after oxidative addition, a *cis*-carbopalladation of the triple bond takes place to give a vinyl–palladium intermediate which undergoes the C–H insertion into the adjacent naphthalene and not into the aryl ether moiety in **19a–e** (Scheme 8.4).

Also a multitude of novel heterocyclic compounds, such as **25** and **26**, can be prepared by a double Mizoroki–Heck approach [23]. Thus, the palladium-catalysed reaction of **21** and

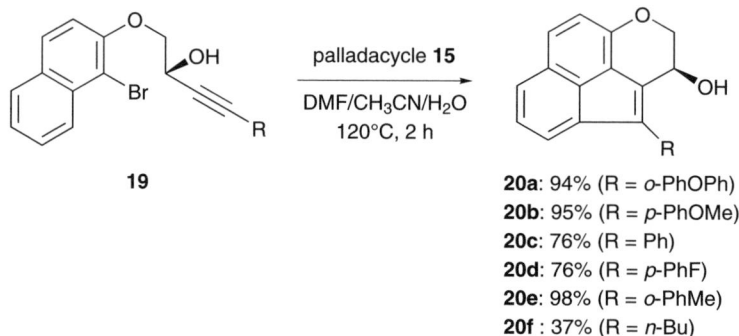

Scheme 8.4   *Synthesis of acenaphthylenes.*

the cyclic enamide **22** gave a 1.2 : 1-mixture of **23** and **24**, which in a second Mizoroki–Heck reaction using the palladacycle **15** led to **25** and **26** in an overall yield of 44–49%. The synthesis can also be performed as a domino process using a mixture of Pd(OAc)$_2$ and the palladacycle **15**.

In the same context, Kaufmann and coworkers [24] have applied a three-component domino Mizoroki–Heck reductive transformation for the preparation of epibatidine analogues **29**. In the presence of triphenylarsine as ligand, Pd(OAc)$_2$, iodobenzene and trimethylsilylacetylene, **27** led to **28** diastereoselectively (Schemes 8.5 and 8.6).

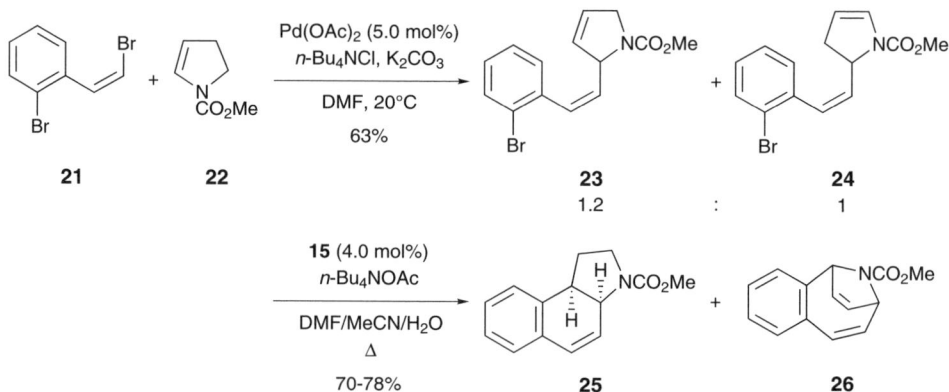

Scheme 8.5   *Double Mizoroki–Heck reaction for the synthesis of heterocycles.*

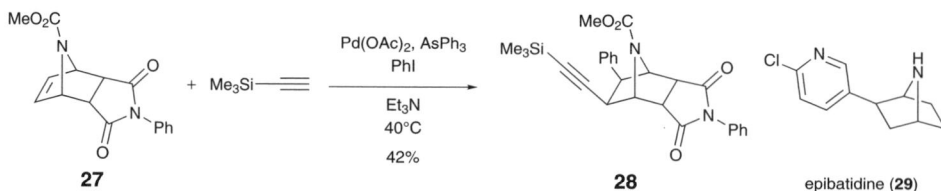

Scheme 8.6   *Synthesis of epibatidine analogues.*

In 1999, Overman and coworkers [25] described a neat synthesis of (+)- and (−)-scopadulcic acid A (**30**) using a palladium-catalysed domino process of the methylene cycloheptene iodide **36**. By this, they generated **37** containing the B, C and D rings of the scopadulan skeleton as a single stereoisomer in 90% yield according to the retrosynthetic analysis of **30** to give **31**. The synthesis of **30** started with **32**, which was enantioselectively reduced to the corresponding secondary alcohol in 94% yield and 88% ee using (*R*)-*B*-isopinocampheyl-9-borabicyclo[3.3.1]nonane ((*R*)-Alpine-Borane) following the procedure by Brown and Pai [26a] and Midland and Graham [26b]. Then, the RCH$_2$I group was transformed into the corresponding organo-lithium derivate followed by coupling with the amide **33** to afford **34**. Cope rearrangement of the corresponding enoxysilane of **34** then led to **35** and further to **36**. In the following domino Mizoroki–Heck reaction of **36**, Ag$_2$CO$_3$ was used as an additive to suppress migration of the double bond in the initially formed intermediate. In a seven-step procedure, **37** was then transformed into **38**, which in eight further steps afforded the desired product **30** (Scheme 8.7).

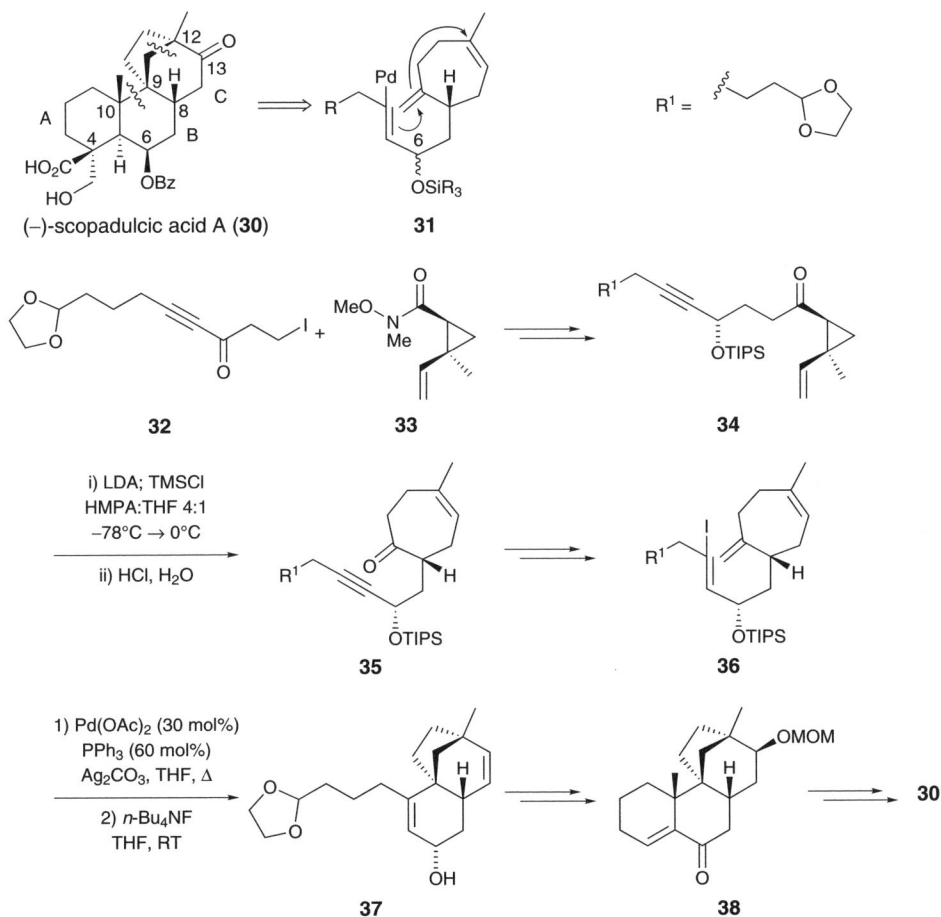

Scheme 8.7 *Synthesis of (−)-scopadulcic acid A (30).*

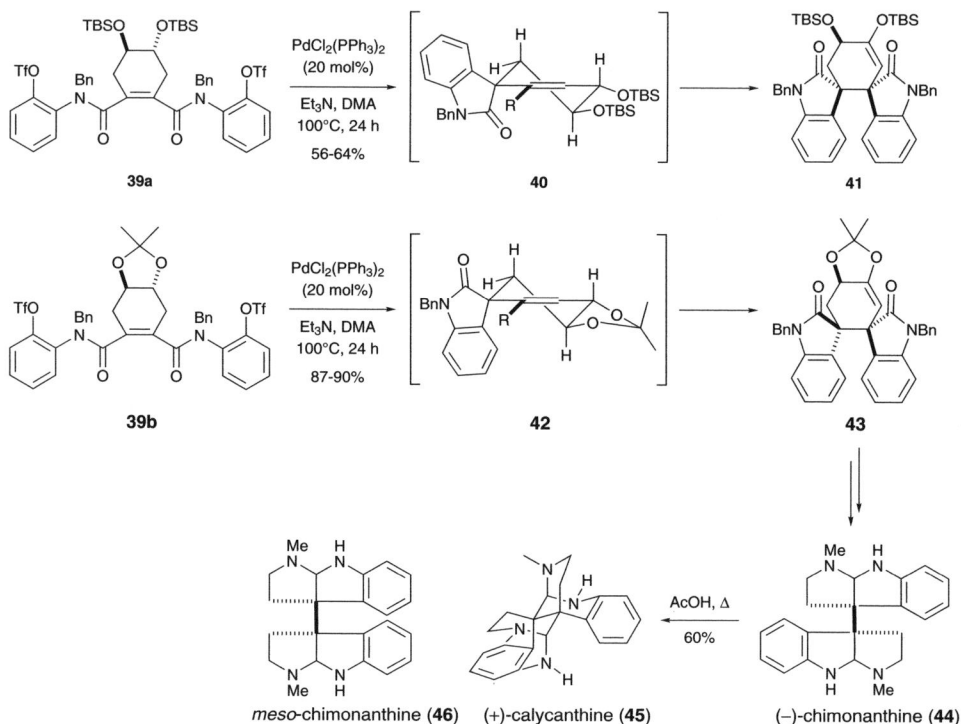

Scheme 8.8   *Synthesis of bis-spiroindolines alkaloids **44–46**.*

Again making use of an intramolecular domino Mizoroki–Heck process, the same group accomplished the total synthesis of three natural bis-spiroisoindoline alkaloids **44**, **45** and **46** in a stereo- and enantio-controlled manner [27]. During the double cyclization steps, two contiguous quaternary centres are formed with complete diastereoselectivity and in excellent yield. Interestingly, the protecting group of the cyclohexenediol moiety in the substrate **39** is of paramount importance for the stereocontrol and outcome of this process. Employing **39a** containing two *tert*-butyldimethylsilyl groups, the *cis*-compound **41** was formed; however, when the silyl ethers were replaced by an acetonide, not only did the yield improve from 70 to 90%, but also the *trans*-compound **43** was obtained with excellent selectivity. A later study revealed that, in the intermediates **40** and **42**, it is mainly steric hindrance after the first insertion that guides the formation of the preferred conformers, which accounts for the high diastereoselectivity [28, 29] (Scheme 8.8).

An enantioselective total synthesis of (+)-xestoquinone (**49**) was developed by the Keay group [30] using an asymmetric double Mizoroki–Heck reaction with 2,2′-bis(diphenylphosphino)-1,1′-binaphthyl (BINAP) as chiral ligand. Reaction of **47a** with *in situ* prepared Pd[(S)-BINAP]₂ gave **48** in 82% yield and 68% ee, which was transformed into xestoquinone (**49**). In the second Mizoroki–Heck reaction, an *endo*-trig cyclization takes place to give a six-membered ring instead of the seemingly more favoured *exo*-trig cyclization, which would form a five-membered ring. Density functional theory calculations

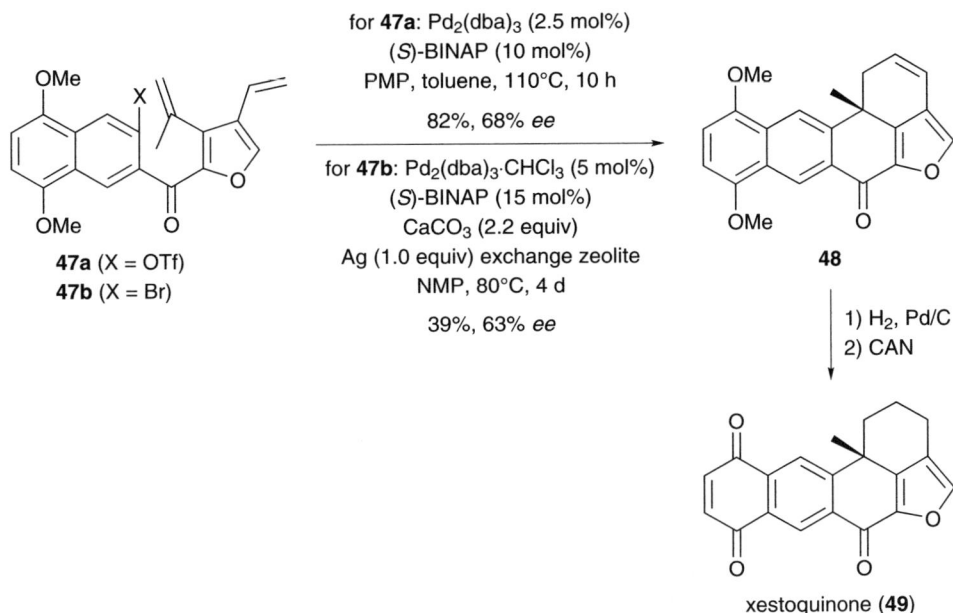

Scheme 8.9 *Synthesis of xestoquinone (49).*

by Ziegler and coworkers [31] have revealed that this selectivity, at both thermodynamic and kinetic levels, is controlled by the relative stability of the cyclic system after the migratory insertion of the palladium into the exocyclic double bond. Later, Shibasaki and coworkers [32] showed that the cyclization step can also be applied to the aryl bromide **47b** when silver salts are used as additives (Scheme 8.9).

The intramolecular Mizoroki–Heck reaction is a convenient method for the synthesis of all types of cyclic compounds, such as normal, medium and large rings [33]. Using ω-haloalkenes, 20-membered carbocycles can be obtained in over 85% yield; however, when applying the same reaction conditions to the synthesis of 13-membered rings starting from the corresponding shorter ω-haloalkenes, these were only obtained in 18% yield. In agreement with these observations, the palladium(0)-catalysed transformation of substrate **50** does not undergo an intramolecular cyclization; rather, there is first an intermolecular C—C bond formation to afford **51** which then cyclizes to yield the 26-membered carbocycle **52** in 54% yield [34] (Scheme 8.10).

The remarkable effect that minor changes in reaction conditions and substrate composition can have on the outcome of a palladium-catalysed transformation is very well illustrated in the following examples. The Heck group [35] described the palladium-catalysed reaction of iodobenzene and twofold excess of tolane **53** in the presence of triphenylphosphane and triethylamine to give the substituted naphthalene **54**. Dyker and Kellner [36] obtained the phenanthrene **55** when using a 2 : 1-mixture of iodobenzene and **53** in the presence of *n*-Bu$_4$NBr and K$_2$CO$_3$. Furthermore, Cacchi's group [37] employed almost identical

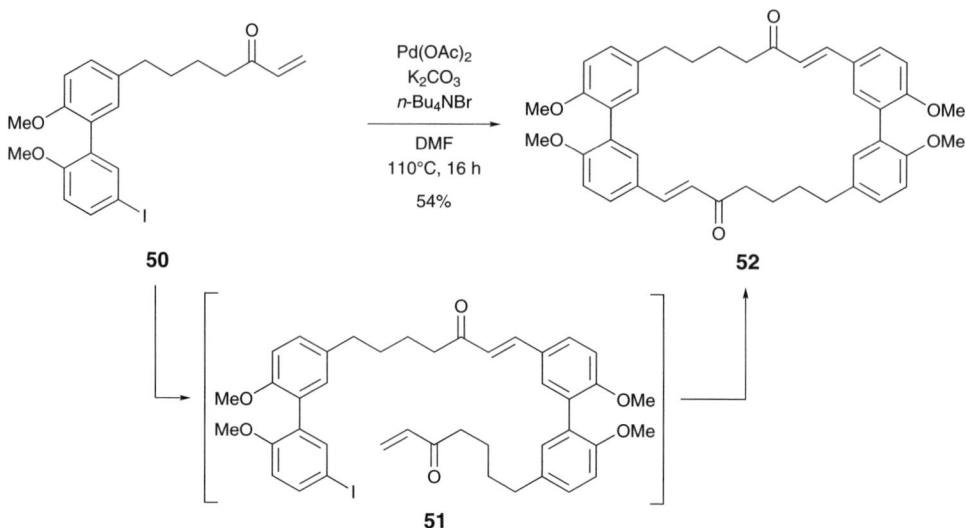

Scheme 8.10   *Synthesis of macrocyclic compounds.*

conditions but added formic acid, which led to the formation of triphenylethene (**56**) (Scheme 8.11).

Moreover, Larock and Tian [38] have recently reported an efficient synthesis of 9-alkylidene- and 9-benzylidene-9*H*-fluorenes **57**, again using iodobenzene and **53** as substrates. The best results were obtained with sodium acetate and *n*-Bu₄NCl, which allowed one to obtain **57** in 62% yield. A proposed mechanism is given in Scheme 8.12, suggesting a migration of palladium from a vinylic onto the aromatic ring [39].

Finally, the same group carried out a mono and a double annulation of arynes to synthesize fluoren-2-ones **60** and polycyclic aromatic systems **62** respectively [40]. These methods involve the *in situ* formation of highly reactive arynes by reaction of 2-(trimethylsilyl)aryl triflates **58** and CsF. In the presence of Pd(dba)₂ (dba = dibenzylideneacetone), a cross-coupling and cyclization with a 2-halobenzaldehyde occur in the former case and C—H activation and cyclization occur in the latter case (Scheme 8.13).

Grigg *et al.* [41] first described a cyclization–cyclopropanation process which was later on developed further by de Meijere's group. It is a nice example of a domino process with four C—C bonds being formed in a single transformation [42]. Thus, reaction of **64** with Herrmann–Beller catalyst (**15**) furnished **66** as the only product. It can be assumed that the palladium compound **65** is an intermediate (Scheme 8.14).

Another cyclization process was observed in the palladium(0)-catalysed transformation of substrates of type **67**, which gave annulated naphthalines **68** as described by Grigg *et al.* [43]. The reaction can also be performed as a two-component transformation involving a combination of an intra- and intermolecular process (Scheme 8.15).

The same group [44] also developed a double Mizoroki–Heck reaction which was then terminated by a formal Friedel–Crafts alkylation to give **70** from **69** involving an

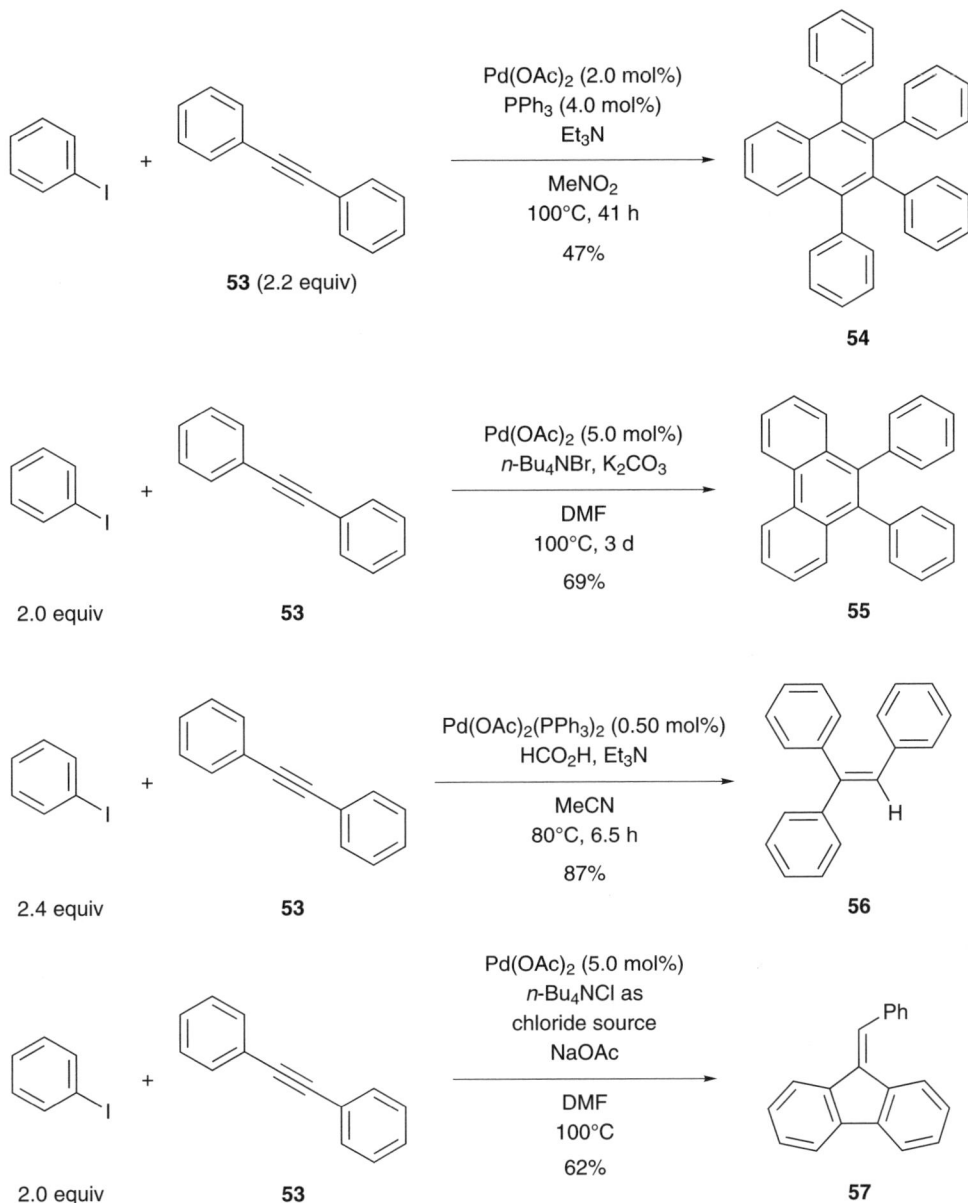

Scheme 8.11   *Reaction of iodobenzene and **53** under different conditions.*

attack of an alkylpalladium(II) intermediate on an aryl or heteroaryl moiety. Notewor-thy is the finding that the Friedel–Crafts alkylation occurs on both electron-rich and electron-poor heteroaromatic rings as well as on substituted phenyl rings. Furthermore, single Mizoroki–Heck/Friedel–Crafts alkylation combinations have also been performed (Scheme 8.16).

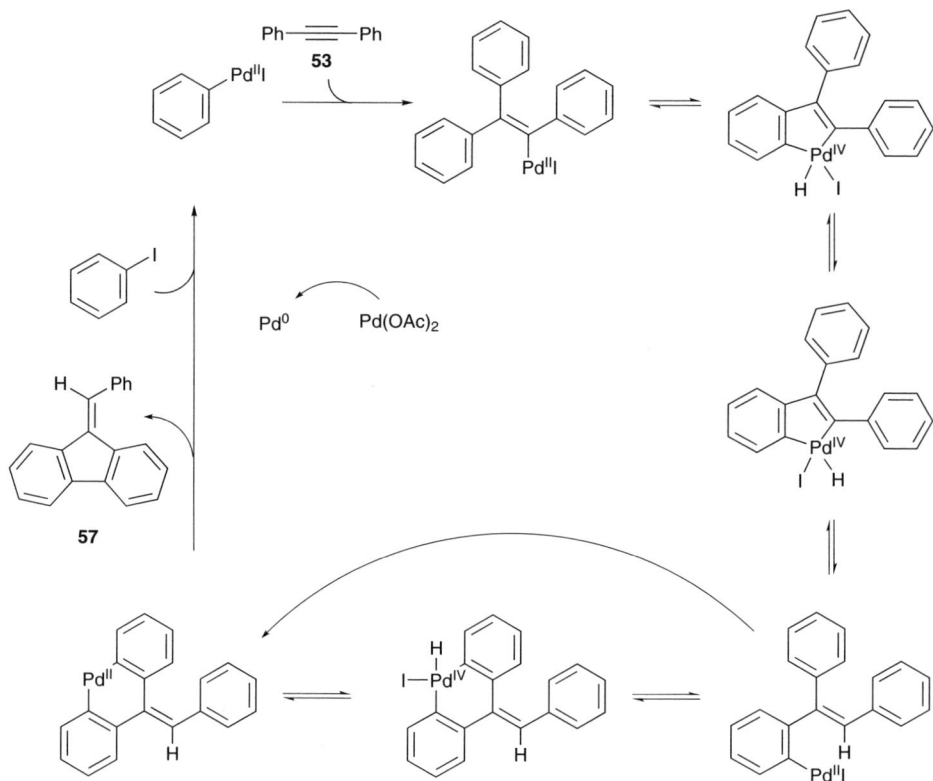

Scheme 8.12   *Proposed mechanism for the formation of **57** from iodobenzene and **53**.*

In another interesting example by Grigg and Sridharan [45], the reaction of the alkynyl aryl iodide **71** with norbornene in the presence of Pd(OAc)$_2$, triphenylphosphine and triethylamine led to the cyclopropanated norbornene derivate **75** as a single diastereomer in 40% yield. It can be assumed that, first, the alkenyl palladium species **72** is formed stereoselectively, which undergoes a Mizoroki–Heck reaction with norbornene to give **73** in a *syn*-fashion followed by formation of the cyclopropanated intermediate **74**, which loses HPdI (Scheme 8.17).

As shown in the preceding examples, the intramolecular palladium-catalysed poly-cyclization is a well-established procedure; however, there are only a few examples of polycyclizations where the first step is an intermolecular process. In this respect, the palladium(0)-catalysed domino reaction of allenes in the presences of iodobenzene reported by Tanaka and coworkers [46] is an intriguing transformation. Palladium-catalysed reaction of **76** in the presence of iodobenzene led to the tetracyclic **77** in 49% yield, allowing the formation of three rings in one sequence (Scheme 8.18).

Domínguez and coworkers [47] used a twofold Mizoroki–Heck reaction for the construction of annulated *N*-heterocycles such as **80a–c** starting from the enamides **78**, which can easily be obtained from the corresponding amines and *o*-iodobenzoic acid chloride (Scheme 8.19).

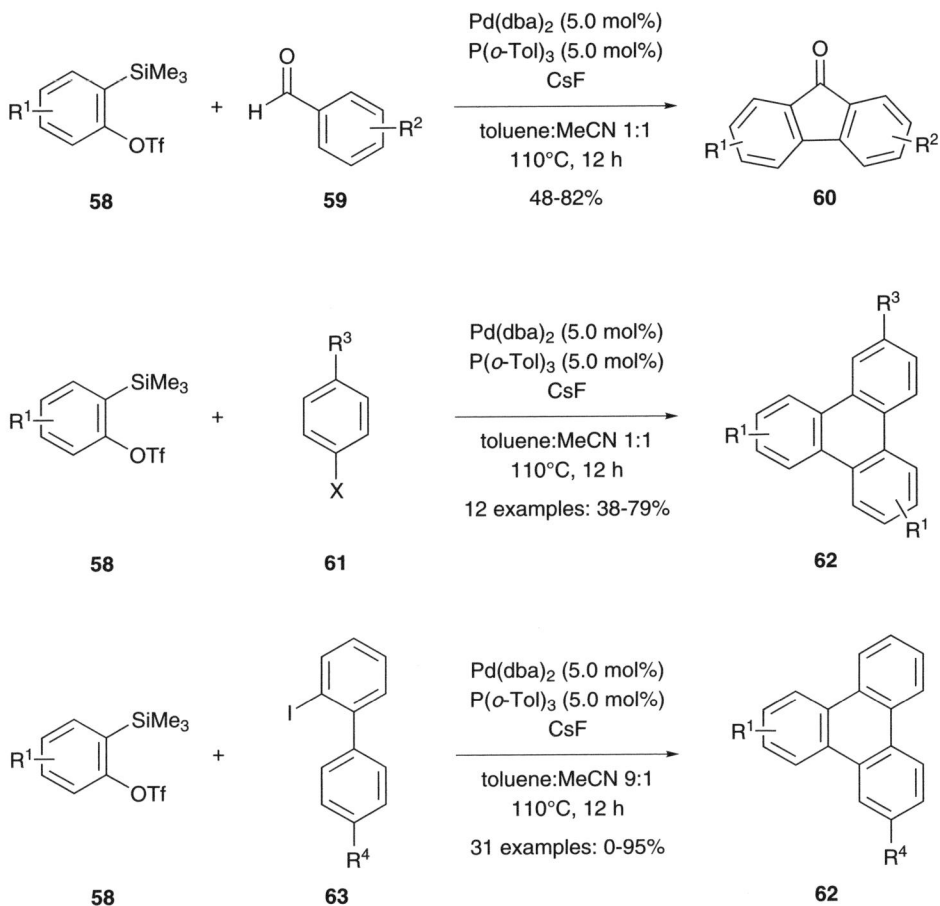

Scheme 8.13   *Annulation of arynes.*

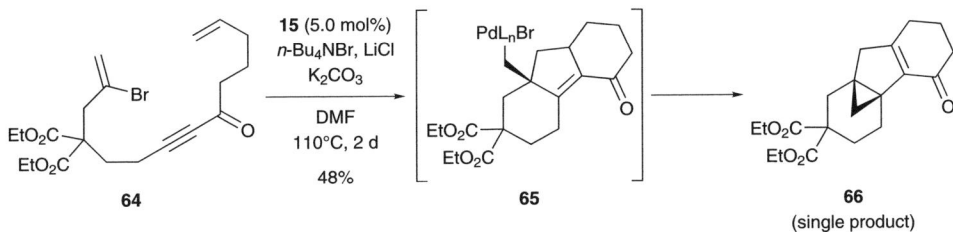

Scheme 8.14   *Synthesis of a tetracyclic cyclopropane derivative.*

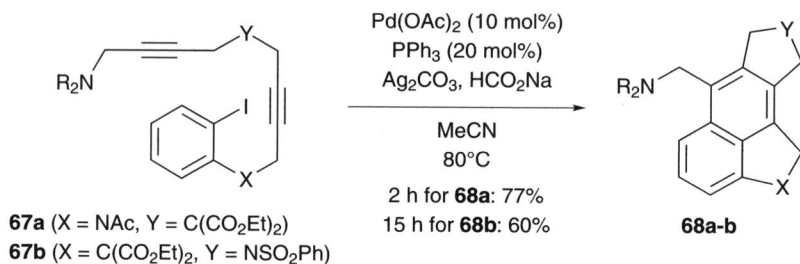

67a (X = NAc, Y = C(CO$_2$Et)$_2$)
67b (X = C(CO$_2$Et)$_2$, Y = NSO$_2$Ph)

Pd(OAc)$_2$ (10 mol%)
PPh$_3$ (20 mol%)
Ag$_2$CO$_3$, HCO$_2$Na

MeCN
80°C

2 h for **68a**: 77%
15 h for **68b**: 60%

68a-b

Scheme 8.15   Cyclotrimerization of **67**.

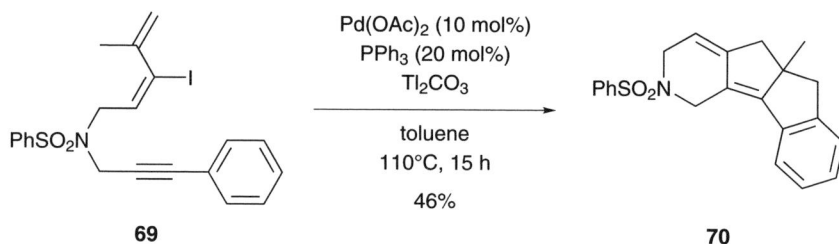

Pd(OAc)$_2$ (10 mol%)
PPh$_3$ (20 mol%)
Tl$_2$CO$_3$

toluene
110°C, 15 h

46%

69

70

Scheme 8.16   Double Mizoroki–Heck/Friedel–Crafts reaction.

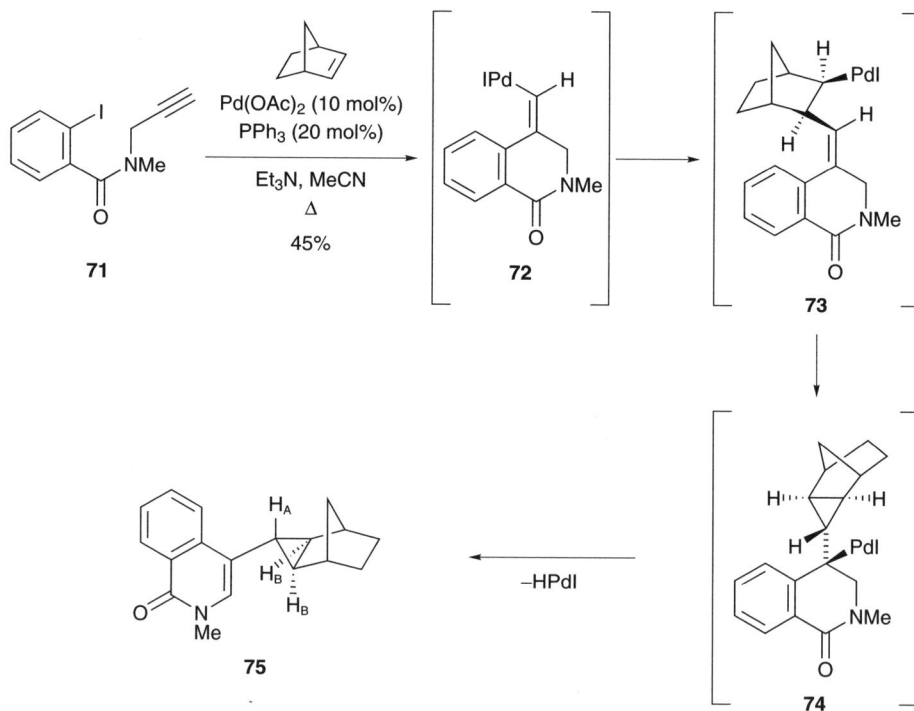

Pd(OAc)$_2$ (10 mol%)
PPh$_3$ (20 mol%)

Et$_3$N, MeCN
Δ

45%

71

72

73

–HPdI

74

75

Scheme 8.17   Triple Mizoroki–Heck reaction.

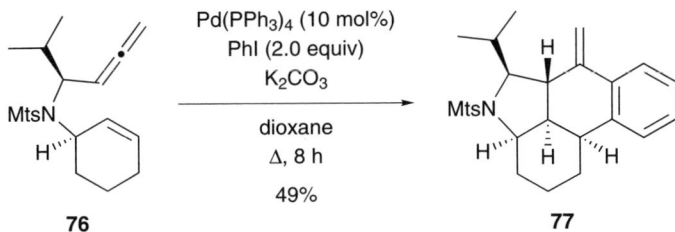

Scheme 8.18   *Intermolecular polycyclization.*

Another double Mizoroki–Heck reaction was reported by Pan and coworkers [48], who prepared substituted pyrrolines **83** using benzylchlorides **81** and the bisallylamine **82** as substrates (Scheme 8.20).

Keay and coworkers [30, 49] observed an unusual remote substituent effect on the enantioselective intramolecular domino Mizoroki–Heck processes to give the tetracyclic products **85a–d** from **84a–d**. The enantiomeric excess varies greatly with the substitution pattern; thus, the transformation of **84a** and **84c** containing one methyl group gave the corresponding products **85a** and **85c** with excellent enantioselectivity up to 96% *ee*, whereas the substrates with no or two methyl groups gave much lower ee-values (Scheme 8.21).

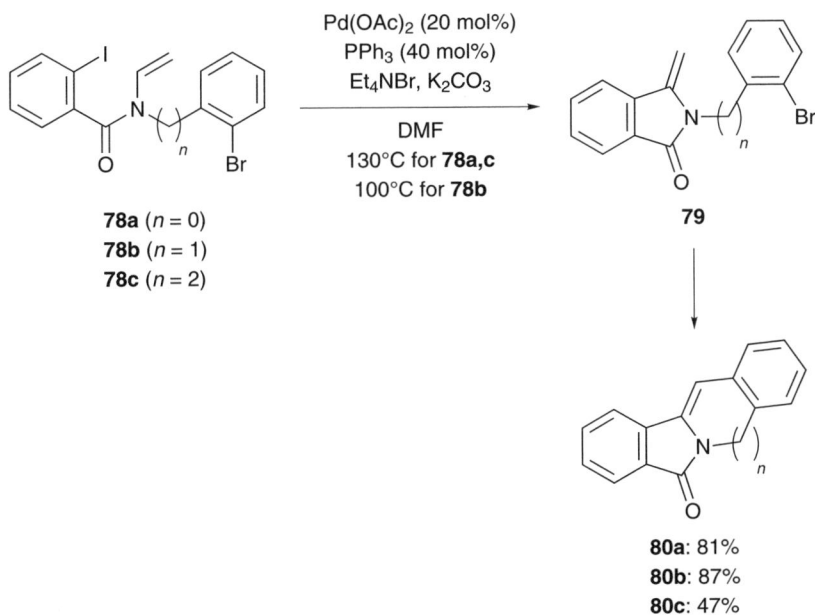

Scheme 8.19   *Synthesis of annulated heterocycles.*

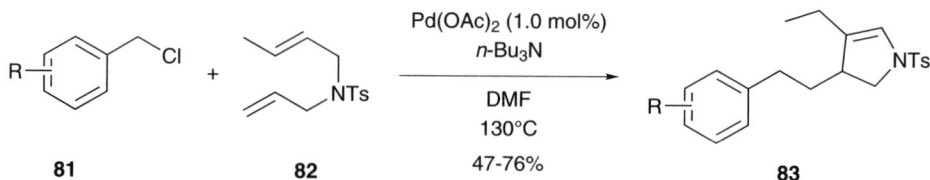

Scheme 8.20   *Synthesis of pyrrolines.*

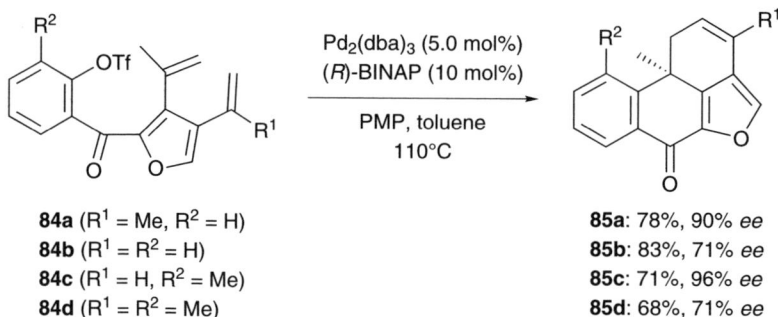

84a (R^1 = Me, R^2 = H)   85a: 78%, 90% *ee*
84b (R^1 = R^2 = H)   85b: 83%, 71% *ee*
84c (R^1 = H, R^2 = Me)   85c: 71%, 96% *ee*
84d (R^1 = R^2 = Me)   85d: 68%, 71% *ee*

Scheme 8.21   *Influence of the substitution pattern on the enantioselectivity in intramolecular double Mizoroki–Heck reaction.*

8.3   Mizoroki–Heck/Cross-Coupling Reactions

Although the combination of a Mizoroki–Heck and a cross coupling reaction has not been yet exploited very much, its application and development have recently been increased. In some examples, the advantage of the comparable high reaction rate of the *cis*-palladation of alkynes over alkenes is used to carry out a Suzuki, Stille or Sonogashira reaction after the oxidative addition and the *cis*-palladation of an alkyne. Moreover, the first step is quite often an intramolecular transformation taking advantage of the entropic effect, but this is not a requirement.

Ternary palladium-catalysed coupling reactions of bicyclic olefins (mainly norbornadiene) with aryl and vinyl halides and various nucleophiles have been investigated intensively over recent years [50]. A new approach in this field is a combination of a Mizoroki–Heck and a Suzuki reaction using a mixture of iodobenzene, phenyl boronic acid and the norbornadiene dicarboxylate. Optimizing the conditions led to 84% of the desired biphenylnorbornene dicarboxylate [51]. Also, substituted iodobenzenes and phenylboronic acids can be used. However, the variation at the norbornadiene moiety is greatly limited.

Cossy and coworkers [52] described a combination of a Mizoroki–Heck and a Suzuki reaction using ynamides and boronic acids to give indole and 7-azaindole derivatives **88**, and its application to the total synthesis of the isoindolobenzazepine alkaloid lennoxamine (**89**) [53]. Thus, reaction of **86** with **87** using Pd(OAc)$_2$ as catalyst led to **88** in moderate to high yield. This strategy was also applied to the synthesis of indoline-based kinase inhibitors by Player and coworkers [54]. Moreover, it has been shown that a carbonylation

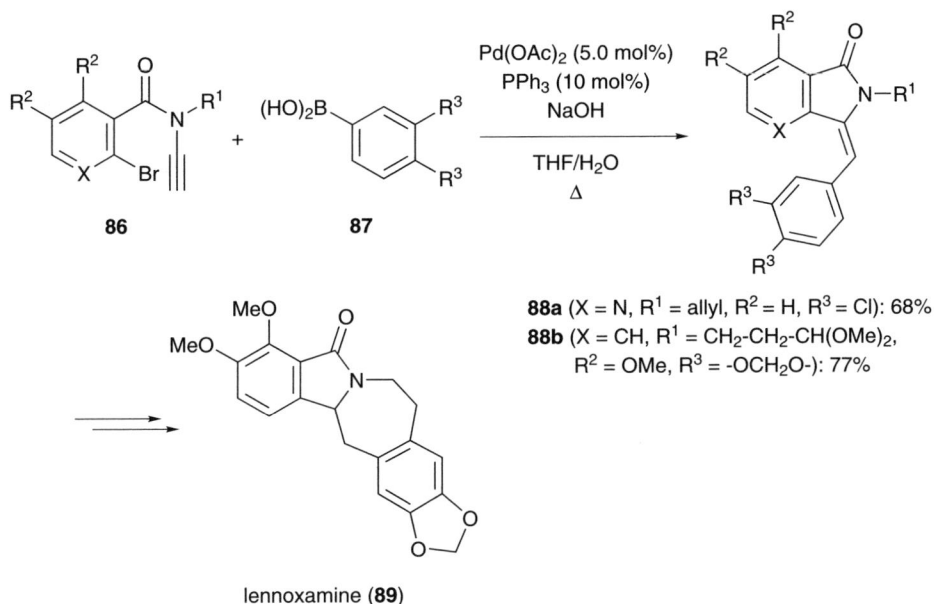

88a (X = N, R^1 = allyl, R^2 = H, R^3 = Cl): 68%
88b (X = CH, R^1 = CH$_2$-CH$_2$-CH(OMe)$_2$,
R^2 = OMe, R^3 = -OCH$_2$O-): 77%

lennoxamine (89)

Scheme 8.22   *Synthesis of azaindoles.*

step can also be performed with these substrates in a Mizoroki–Heck-carbonylation–Suzuki process [55] (Scheme 8.22).

It has been observed that the presence of coordinating groups can prevent the competing β-hydride elimination after the insertion. Ahn and coworkers [56] used a Mizoroki–Heck reaction of the vinyl bromide **90** in the presence of an arylboronic acid for the preparation of 4-alkylidene-3-arylmethylpyrrolidines **91**. The alkylpalladium intermediate is stabilized by coordination with one of the *N*-sulfonyl oxygen atoms, allowing the Suzuki reaction to proceed (Scheme 8.23).

Recently, Zhou and Larock [57] have also used a domino Mizoroki–Heck/Suzuki process in their syntheses of several tamoxifen analogues. They described a three-component coupling reaction of readily available aryl iodides, internal alkynes and arylboronic acids to give the expected tetrasubstituted olefins in good yields. For instance, the treatment of a mixture of iodobenzene, alkyne **92** and phenylboronic acid with catalytic amounts of PdCl$_2$(PhCN)$_2$ gave **93** in 90% yield. In this process, substituted aryliodides and

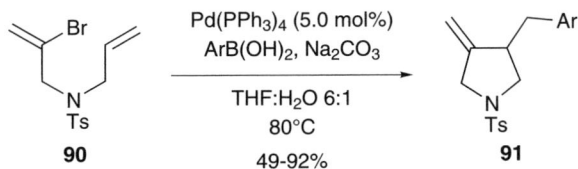

Scheme 8.23   *Synthesis of pyrrolidines.*

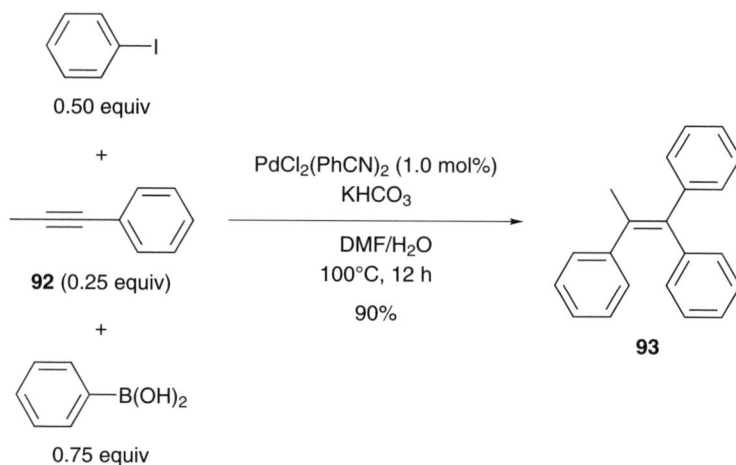

Scheme 8.24 *Synthesis of tamoxifen analogues.*

heteroaromatic boronic acids may also be employed. It can be assumed that, after the palladium(0)-catalysed oxidative addition of the aryl iodide, a *cis*-carbopalladation of the internal alkyne takes place to form a vinylic palladium intermediate which reacts with the ate complex of the arylboronic acid in a transmetalation followed by a reductive elimination (Scheme 8.24).

A combination of an intramolecular palladium(0)-catalysed alkenylation of an alkyne and a Stille reaction followed by an 8π-electrocyclization was developed by Salem and Suffert [58]. In a representative example, treatment of diol **94** with stannyldiene **95** in the presence of Pd(PPh$_3$)$_4$ led to **97** in 16% yield via the intermediate **96**. At higher temperature, **97** can undergo a 4π-conrotatory ring opening to afford a mixture of **99** and **100** via **98** (Scheme 8.25).

Application of microwave irradiation techniques also permitted the use of acyclic γ-bromopropargylic diols and, thus, broadened the scope of this domino transformation [59]. Interestingly, when cyclic γ-bromopropargylic diol **102** bearing an aromatic moiety is employed as substrate, a C—H activation after the Mizoroki–Heck reaction takes place followed by a Stille reaction to yield a mixture of atropoisomeres **103** and **104** instead of a direct Stille coupling to give **101** [60]. This type of transformation was only observed with **102**; therefore, the scope of it is quite narrow. On the other hand, different alkenyl, aryl or heteroaryl tributylstannanes can be used with yields up to 70%. Initial deuterium incorporation experiments and a computational study performed by Debieu and coworkers [61] indicated that there is a complete transfer of the H-1 hydrogen to the vinylic position via a 1,5-Pd–vinyl or aryl shift (Scheme 8.26).

In another combination of a Mizoroki–Heck type reaction followed by a Stille coupling, Grigg and coworkers [62] described the synthesis of tricyclic copounds **108** using an ion capturing approach. The vinylstannes **106** are generated *in situ* via regioselective hydrostannylation–cyclization of terminal 1,6-diynes **105**. Then, the domino process takes place upon addition of aryl iodide **107** and tri-furyl phosphine as ligand to furnish heterocycles **108** in good yields (Scheme 8.27).

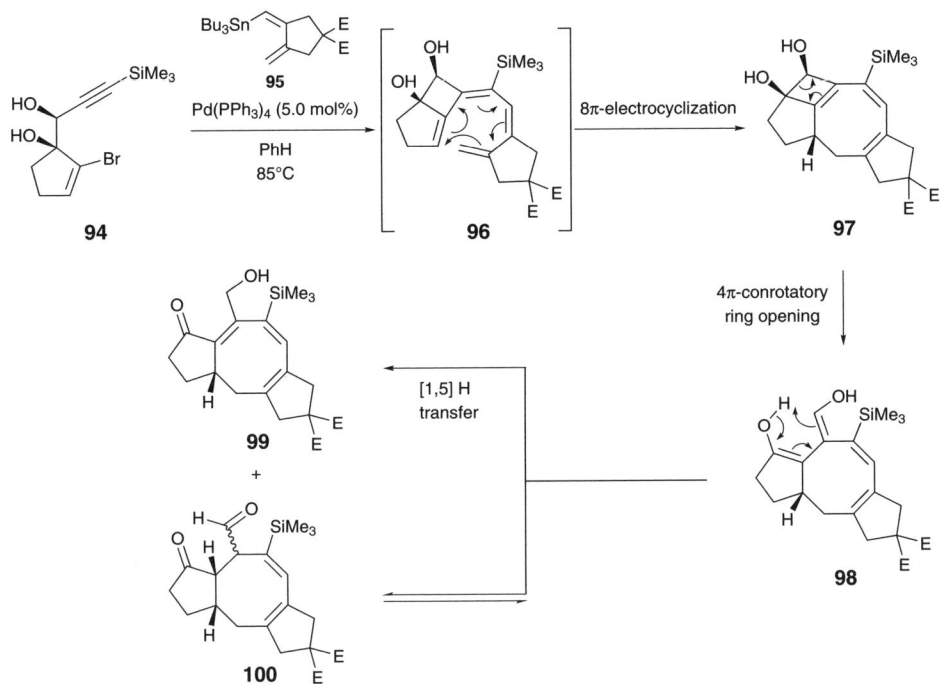

Scheme 8.25   Combination of a domino Mizoroki–Heck/Stille reaction with an electrocyclization.

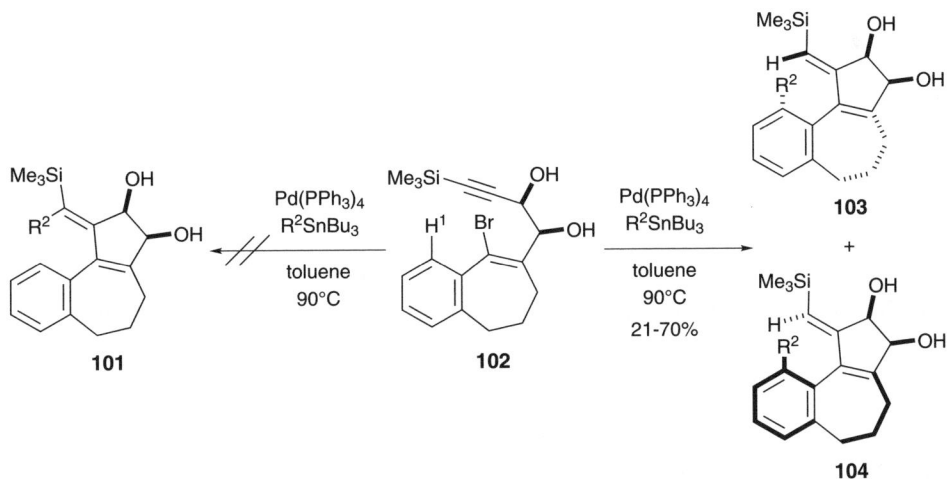

Scheme 8.26   Combination of a domino Mizoroki–Heck/Stille reaction with C–H activation.

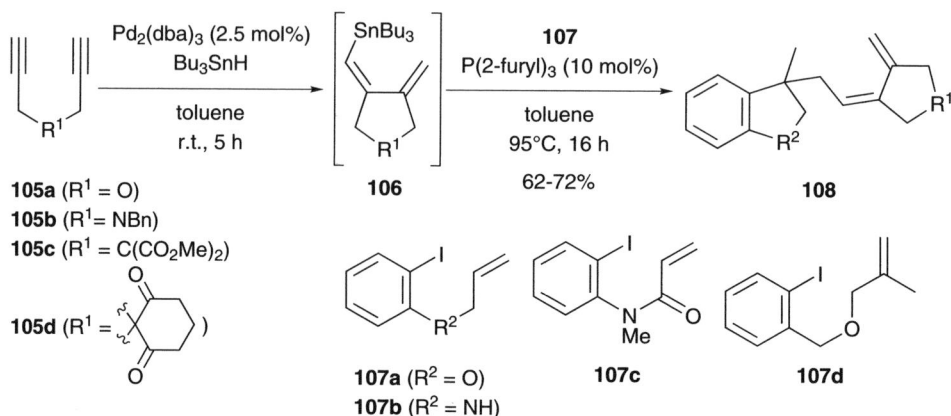

105a (R^1 = O)
105b (R^1= NBn)
105c (R^1 = C(CO$_2$Me)$_2$)

105d (R^1 = [structure])

106

107a (R^2 = O)
107b (R^2 = NH)

107c

107d

108

62-72%

Scheme 8.27   Domino Mizoroki–Heck/Stille process.

Müller and coworkers [63] developed a nice domino process in which a termination of an intramolecular arylation of an alkyne was achieved by a Sonogashira alkynylation. The transformation could even be carried on using an intramolecular cycloaddition. Treatment of a mixture of **109** and **110** in the presence of catalytic amounts of PdCl$_2$(PPh$_3$)$_2$ and CuI in toluene/butyronitrile/triethylamine at reflux led to hitherto unknown spiro compounds **111** in up to 86% yield (Scheme 8.28).

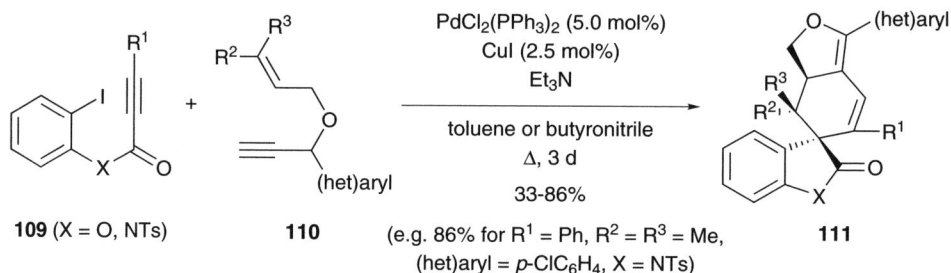

109 (X = O, NTs)

110

(e.g. 86% for R^1 = Ph, R^2 = R^3 = Me,
(het)aryl = *p*-ClC$_6$H$_4$, X = NTs)

33-86%

111

*Scheme 8.28   Synthesis of spiro compounds **111**.*

8.4   Mizoroki–Heck/Nucleophilic Substitution (Tsuji–Trost Reactions)

There are only a few examples so far where a Mizoroki–Heck reaction has been combined with a palladium(0)-catalysed nucleophilic substitution. Most of them make use of an amine as the nucleophile in the substitution of the π-allylpalladium(II) complex formed after the initial Mizoroki–Heck reaction; however, there are also examples for the use of carbon nucleophiles. In an example, Weinreb and coworkers [64] described a three-component process consisting of a vinylbromide, an alkene and malonate ester as the nucleophile forming two new C—C bonds consecutively. Earlier, Larock *et al.* [65] had described a similar transformation.

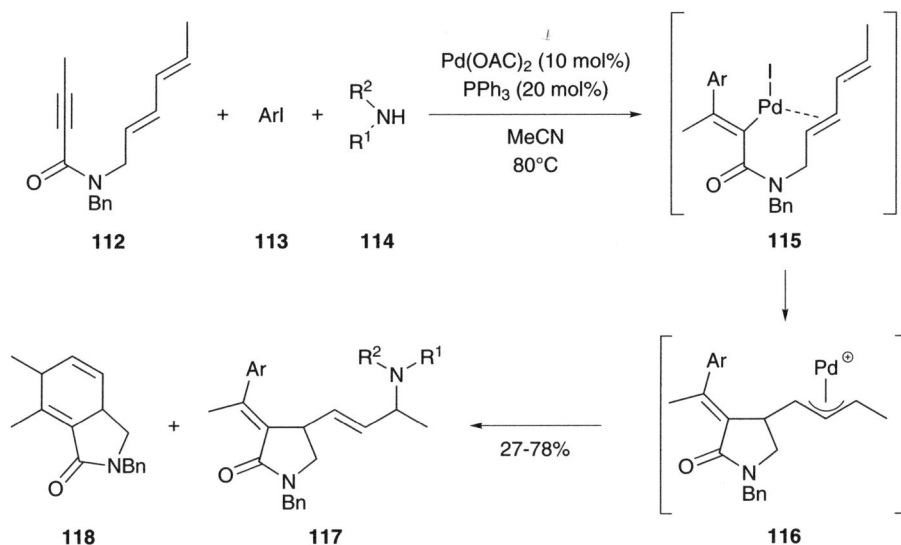

Scheme 8.29 *Synthesis of pyrrolidinones.*

Xie and Lu [66] described the synthesis of α-alkylidene-γ-lactams using an amine as nucleophile. Thus, treating alkyne **112** with aryl halide **113** and amine **114** gave the substituted pyrrolidinone **117** via the proposed intermediates **115** and **116**. As a side product, **118** is formed by a ring closure of **112** (Scheme 8.29).

On the basis of an enantioselective version of a transformation first studied by Larock *et al.* [67], Flubacher and Helmchen [68] employed α,ω-amino-1,3-dienes as substrates. By using palladium complexes with chiral phosphino-oxazolines L* as catalysts, an enantiomeric excess of up to 80% was achieved. In a typical experiment, a suspension of Pd(OAc)$_2$, the chiral ligand **121**, the aminodiene **119** and an aryltriflate was heated in dimethylformamide at 100 °C for 10 days. Proceeding through the chiral palladium complex, the resulting cyclic amine derivative **120** was obtained in 47% yield and 80% *ee* (Scheme 8.30). The reaction time is shorter and the yield higher (61%) using aryliodides,

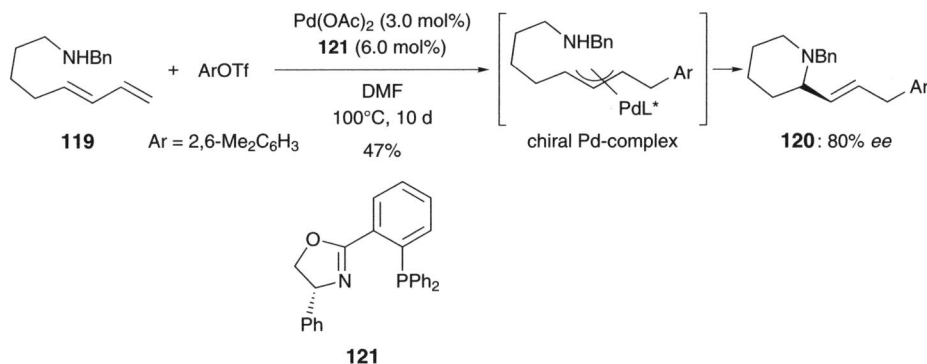

Scheme 8.30 *Enantioselective Mizoroki–Heck allylation process.*

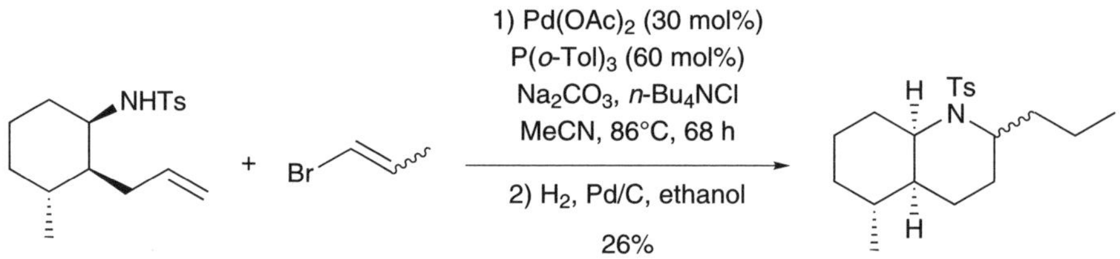

Scheme 8.31   *Towards the synthesis of pumiliotoxin C.*

but the enantiomeric excess is lower (67% *ee*). An even lower *ee*-value of only 12% was obtained with BINAP as a chiral ligand in the palladium(0)-catalysed transformation of **119** with an aryliodide.

As part of the total synthesis of the neurotoxin (−)-pumiliotoxin C [69], Minnaard and coworkers [70] used a domino Mizoroki–Heck/Tsuji–Trost reaction as the key step to prepare the perhydroquinoline **124** in 26% yield from **122** and **123** after hydrogenation (Scheme 8.31). Similarly, acyclic ω-olefinic *N*-tosyl amides with vinyl bromides have been used to give the corresponding lactams in 49–82% yield [71].

8.5   Mizoroki–Heck/CO-Insertions

The insertion of CO into a $C(sp^3)$–Pd or $C(sp^2)$–Pd species is a common procedure and it is applicable to a wide range of reactions, such as hydroformylation, carbonylation, hydroesterification and so on, thus being an ideal terminating transformation within a domino process starting, for example, with a Mizoroki–Heck reaction.

In an approach towards a total synthesis of the marine ascidian metabolite perophoramidine (**125**) and communesins (**126**) [72], Artman and Weinreb [73] developed a domino Mizoroki–Heck/carbonylation process. This allowed the construction of the C,E,F-ring system of **125** together with the C-20 quaternary centre and the introduction of a functionality at C-4. Thus, reaction of **127** in the presence of catalytic amounts of Pd(OAc)$_2$ and P(*o*-Tol)$_3$ under a CO atmosphere in *N,N*-dimethylacetamide/MeOH gave **128** in 77% yield (Scheme 8.32).

Bicyclic lactones such as **131** were synthesized by Copéret and Negishi [74] using a domino Mizoroki–Heck carbopalladation as the key step of vinyl halides such as **129** to give **130**. The product can be further transformed into the desired lactone **131** in a few additional steps (Scheme 8.33).

The synthesis of polycyclic compounds can also be accomplished by a palladium(0)-catalysed transformation of enediynes [75]. In a process described by Negishi and coworkers, enetetraynes of type **132** undergo a tetracyclization followed by a carbonylative esterification to form the fifth ring leading to **133** in 66% yield. In this case, the initial alkenyl–palladium species is obtained by an oxidative addition of palladium(0) to the alkenyliodide moiety in **132** [76] (Scheme 8.34).

perophoramidine (**125**)   cummunesins (**126**)

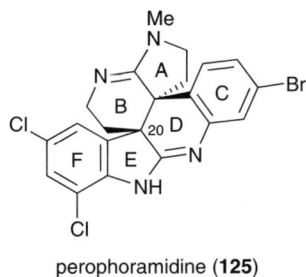

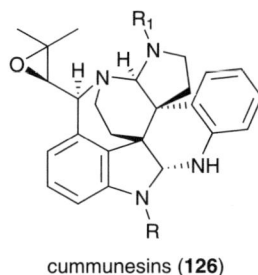

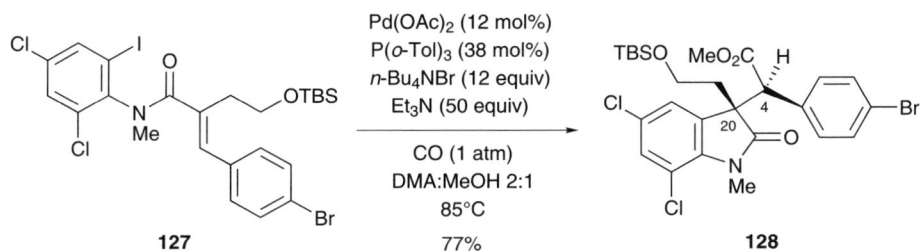

Scheme 8.32   *Synthesis of the C,E,F-ring system of perophoramidine (**125**) and cummunesins (**126**).*

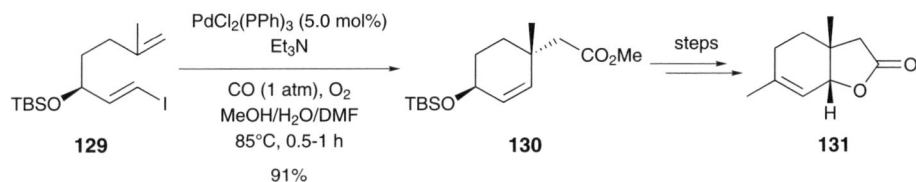

Scheme 8.33   *Synthesis of bicyclic lactones.*

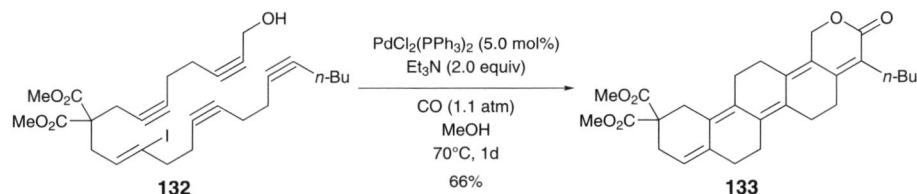

Scheme 8.34   *Polycyclization of enetetraynes.*

8.6   Mizoroki–Heck/C–H Activations

The combination of a Mizoroki–Heck reaction with a C–H activation is a domino process that has been encountered with increasing frequency in the past few years. This type of transformation takes place if the organo-palladium intermediate does not react in the usual way but activates an aryl– or vinyl–H bond in close vicinity, forming a new palladium(II) species that can react further with electrophiles or nucleophiles and giving rise to the formation of another C–C or C–heteroatom bond.

When the initial aryl–palladium species has more than one possible proton in the vicinity, the C—H activation can take place several times. Carretero and coworkers [77] observed three C—H activations after the initial Mizoroki–Heck reaction using α,β-unsaturated sulfones **134** and iodobenzene. Under normal conditions, the expected Mizoroki–Heck product **135** is formed; however, using an excess of iodobenzene, **136** is obtained in high yield. In this transformation, three molecules of iodobenzene are incorporated into the final product. On the other hand, a later study showed that, under the same conditions, similar electron-deficient alkenes as enones do not undergo this domino reaction; instead, only the Mizoroki–Heck product is obtained [78]. A computational analysis of the transformation explained this finding with the difference in the energy of the transition states that ultimately lead to the five-membered ring palladacycle **PdC1** (Scheme 8.35).

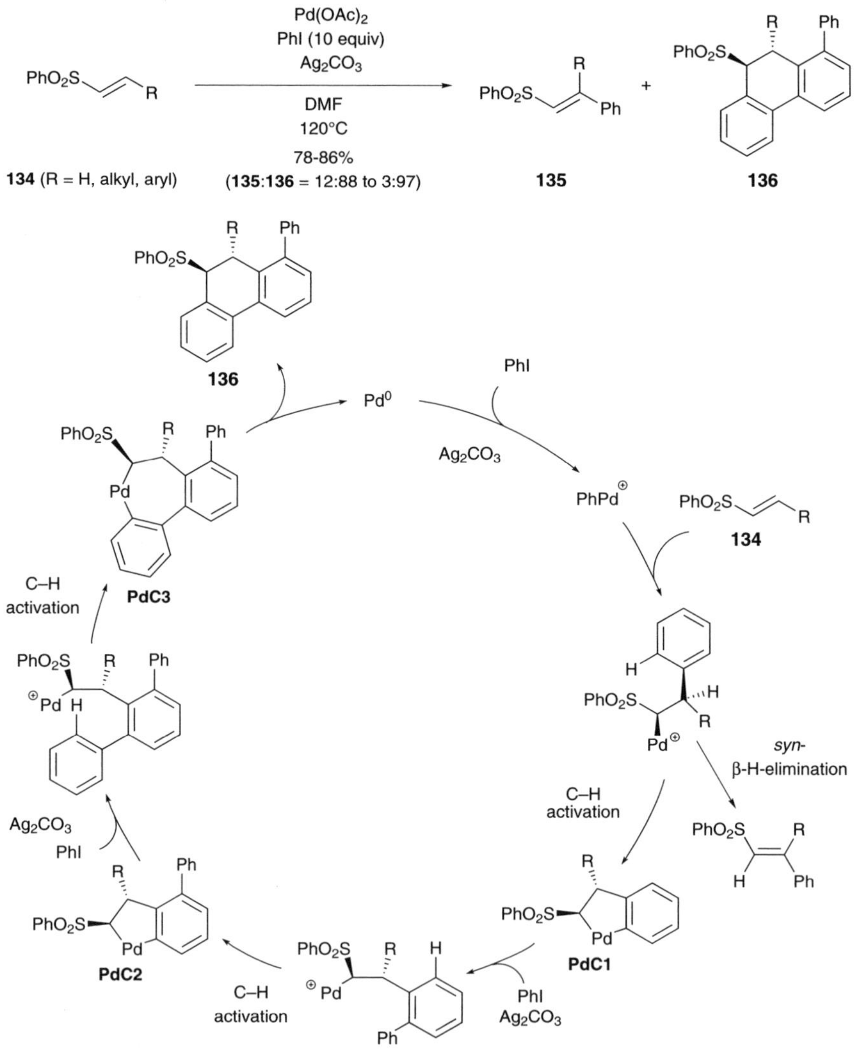

Scheme 8.35 *Domino Mizoroki–Heck reaction/C—H activation of α,β-unsaturated sulfones and PhI.*

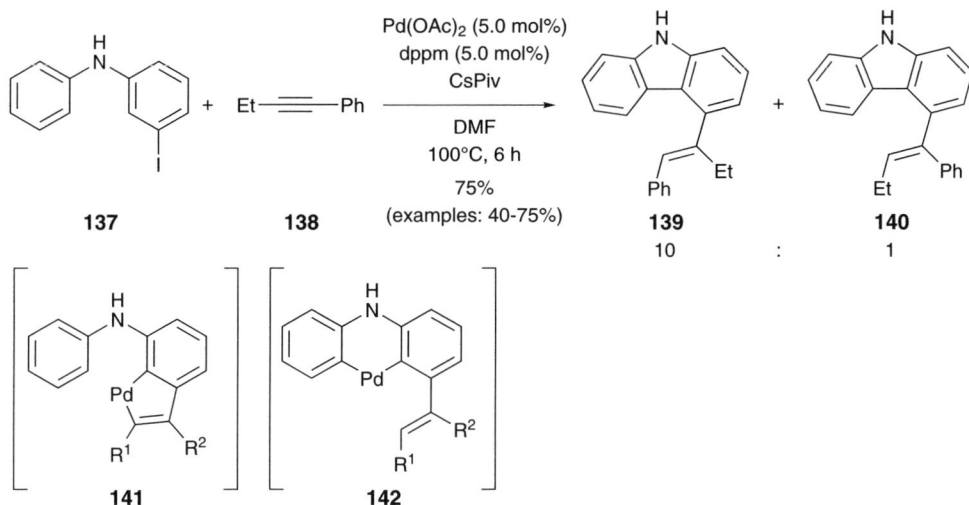

Scheme 8.36 *Synthesis of carbazoles.*

Zhao and Larock [79] found a palladium-catalysed double C—H activation by reaction of *N*-(3-iodophenyl)anilines such as **137** with alkynes such as **138**. Thus, reaction of **137** with **138** in the presence of palladium(0) furnished **139** in a reasonable yield. It can be assumed that the two palladacycles **141** and **142** act as intermediates. The procedure allows the efficient synthesis of substituted carbazoles (Scheme 8.36).

Based on a transformation described by Catellani and coworkers [80], Lautens's group [81] developed a series of syntheses of carbocycles and heterocycles from aryl iodide, alkyl halides and Mizoroki–Heck acceptors. In an early example, the authors described a three-component domino reaction catalysed by palladium for the synthesis of benzo-annulated oxacycles **144** (Scheme 8.37). To do so, they used an *m*-iodoaryl iodoalkyl ether **143**, an alkene substituted with an electron-withdrawing group, such as *t*-butyl acrylate and an iodoalkane such as *n*-BuI in the presence of norbornene. It is proposed that, after the oxidative addition of the aryliodide, a Mizoroki–Heck-type reaction with norbornene and a C—H activation first takes place to form a palladacycle **PdC1**, which is then alkylated with the iodoalkane (Scheme 8.37). A second C—H activation occurs and then, via the formation of the oxacycle **OC1**, norbornene is eliminated. Finally, the aryl–palladium species obtained reacts with the acrylate. The alkylation step of palladacycles of the type **PdC1** and **PdC2** was studied in more detail by Echavarren and coworkers [82] using computational methods. They concluded that, after a C—H activation, the formation of a $C(sp^3)$—$C(sp^2)$ bond between the palladacycle **PdC1** and an iodoalkane presumably proceeds by oxidative addition to form a palladium(IV) species to give **PdC2**. This stays, in contrast with the reaction between a $C(sp^2)$—X electrophile (vinyl or aromatic halide) and **PdC1**, to form a new $C(sp^2)$—$C(sp^2)$ bond which takes place through a transmetallation.

Lately, Catellani and coworkers [83] have described a similar process for the synthesis of 6*H*-dibenzopyran derivatives **147**. They investigated the reaction of *o*-bromophenols **145**, iodoarenes and acceptor-substituted alkenes in the presence of norbornene. The

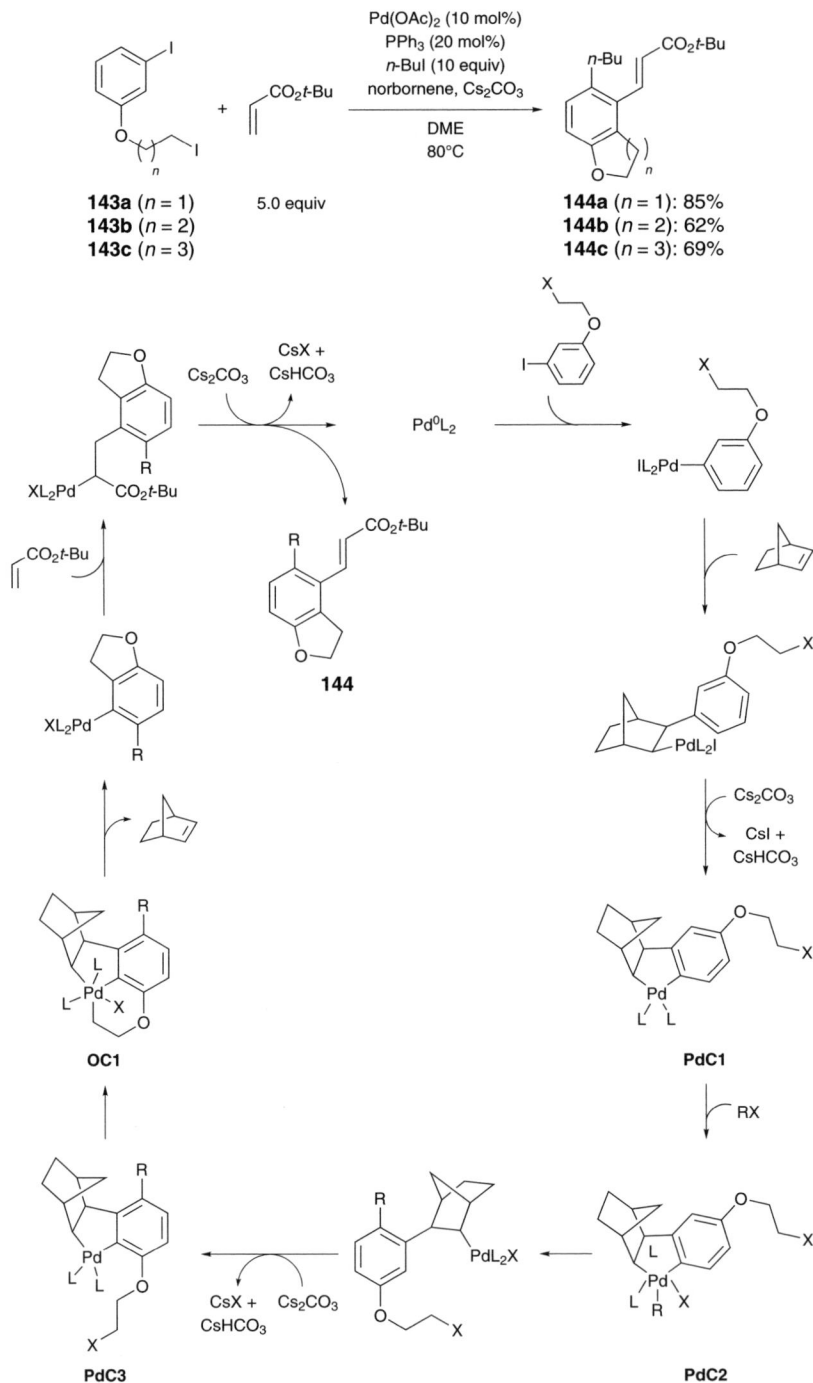

Scheme 8.37   *Postulated mechanism for the reaction of* **143** *to give annulated oxacycles* **144**.

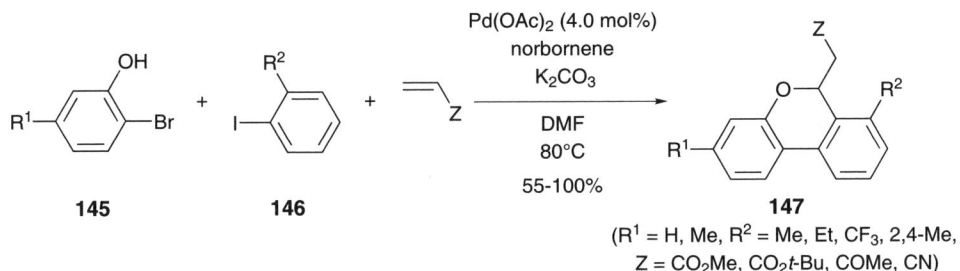

Scheme 8.38 *Synthesis of dibenzopyran derivatives.*

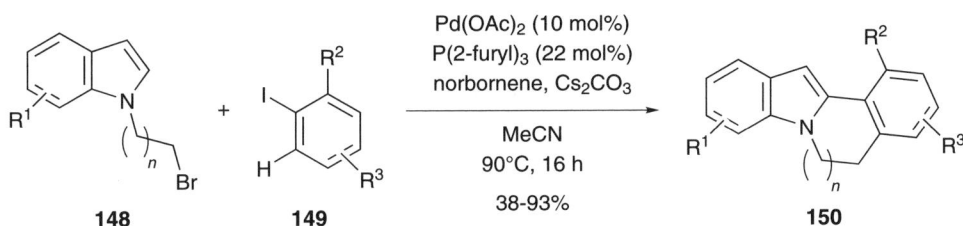

Scheme 8.39 *Synthesis of* **150**.

transformation consists of an aryl coupling followed by alkene insertion and subsequent intramolecular Michael-type ring closure (Scheme 8.38).

A similar transformation was described later by the Lautens group [84]. With a modification of the sequence, annulated indoles **150** were obtained by using *N*-bromoalkyl indoles of type **148** and aryl iodides **149** [85]. 2,3-Disubstituted indoles and thiophenes can also be prepared in the same way [86] (Scheme 8.39).

8.7 Mizoroki–Heck/Pericyclic Transformations

Since Mizoroki–Heck reactions allow the straightforward formation of 1,3-butadienes, the most common combination of a Mizoroki–Heck reaction in a domino process with a nonpalladium-catalysed transformation is that with a pericyclic reaction, especially a Diels–Alder cycloaddition.

The de Meijere group has investigated this domino process extensively [87]. They prepared interesting spiro-compounds containing a cyclopropyl moiety using a combination of a Mizoroki–Heck and a Diels–Alder reaction with bicyclopropylidene **151** as starting material. The transformation can be performed as a three-component process. Thus, reaction of **151**, iodobenzene and acrylate gave **152** in excellent yield. With vinyliodide, the tricyclic compound **153** was obtained (Scheme 8.40).

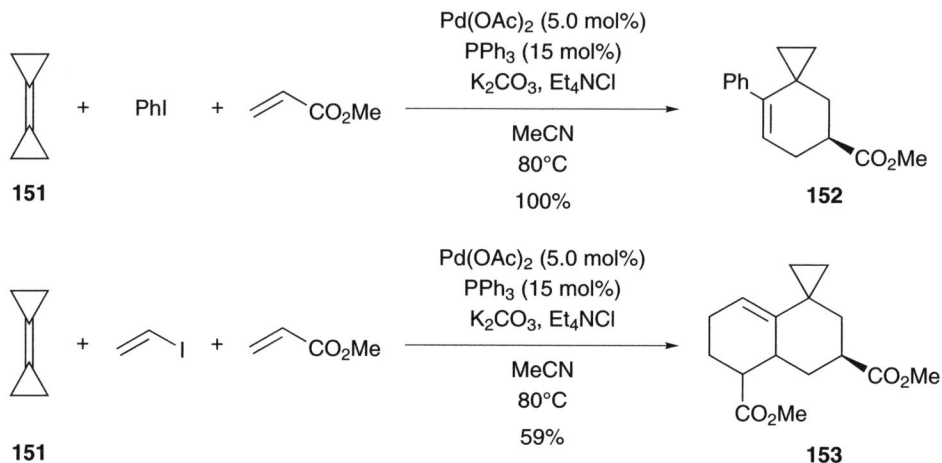

Scheme 8.40 Domino Mizoroki–Heck reaction/cycloaddition.

A domino Mizoroki–Heck/Diels–Alder process described by the same group [88] implies the palladium(0)-catalysed reaction of **154** in the presence of acrylate or methyl vinyl ketone to give the corresponding bicyclic compounds **156** and **157** via the transient **155**. Good yields were obtained only in the presence of potassium carbonate as base (Scheme 8.41).

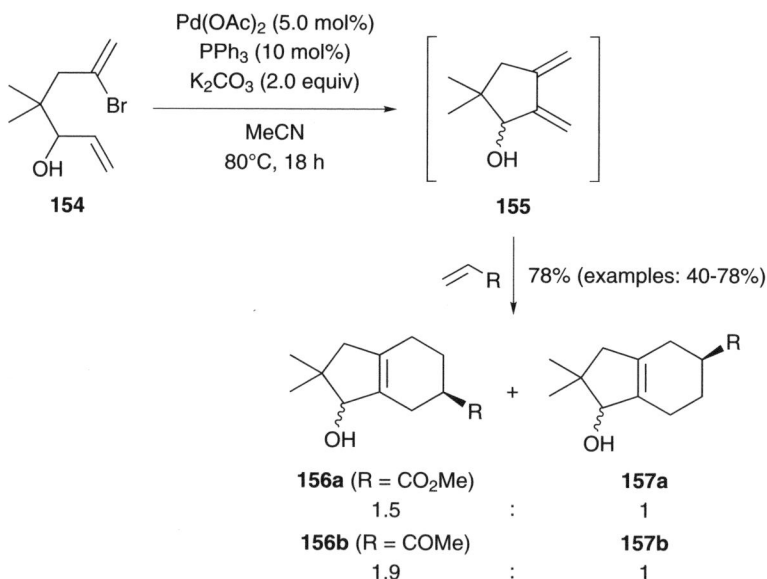

Scheme 8.41 Synthesis of indene derivatives.

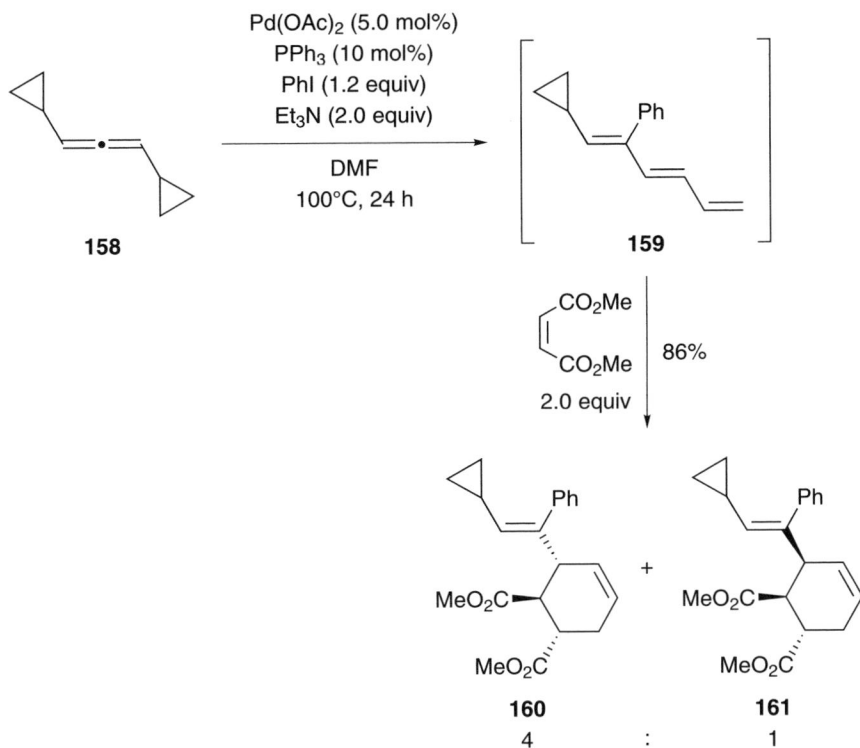

Scheme 8.42   *Synthesis of cyclopropane derivatives.*

Similarly, the palladium-catalysed arylation of 1,3-dicyclopropyl-1,2-propadiene **158** with iodobenzene in the presence of dimethyl maleate led to the diastereomeric cyclopropane derivatives **160** and **161** via **159** in 86% yield as a 4 : 1-mixture [89]. Several other aryl halides and dienophiles have been used in this reaction. A number of other examples of the use of bicyclic propylidenes in a Mizoroki–Heck/pericyclic domino process were also reported later on by the same group [90] (Scheme 8.42).

Also, electrocyclic transformations can be combined with a Mizoroki–Heck reaction, as described by Suffert *et al.* [91] for a carbopalladative ring closure followed by the cyclization of an intermediate 1,3,5-hexatriene. Thus, treatment of a mixture of the *trans*-bis(tributylstannyl)ethylene and the alkyne **165a**, which can be obtained from 2-bromocyclohexenone **162a** and the lithium compound **163**, with catalytic amounts of Pd(PPh$_3$)$_4$ at 90 °C led to **167a** in 62% yield, presumably via the cyclobutane **166**. The ring size of the starting material seems to have a pronounced influence on the reaction, since **164b** obtained from **162b** gave **167b** only in 24% yield (Scheme 8.43).

Thiemann *et al.* [92] prepared steroid analogues containing an areno-annulation as in **169** (Scheme 8.44). They also used a combination of a Mizoroki–Heck reaction and an electrocyclic ring closure of an intermediately formed 1,3,5-hexatriene. Thus, reaction of

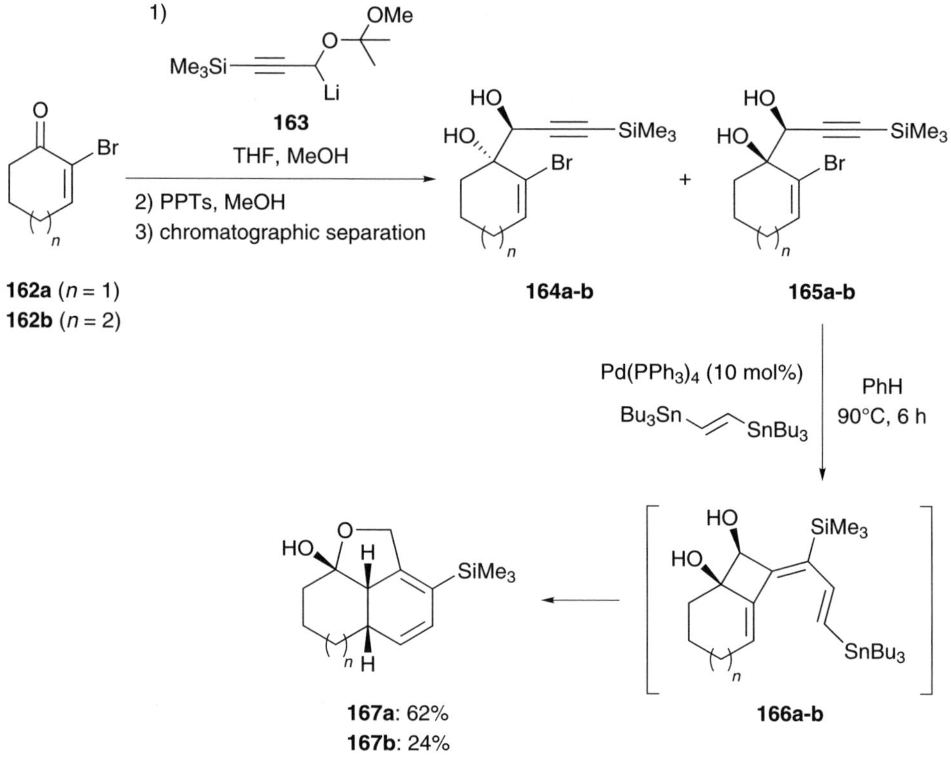

Scheme 8.43   *Synthesis of tricyclic 167.*

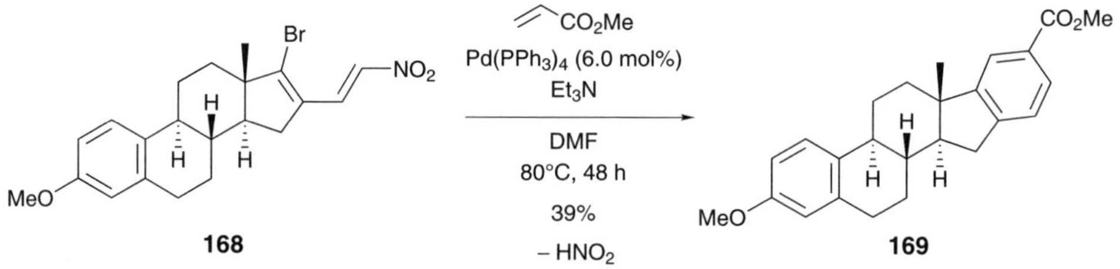

Scheme 8.44   *Synthesis of areno-annulated steroids.*

168 in the presence of palladium(0) led to **169** in 39% yield. The final step is the removal of the nitro group with formation of the aromatic ring system.

8.8   Mizoroki–Heck Reactions/Mixed Transformations

Mizoroki–Heck reactions can also be combined with other processes, such as anion capture processes, aminations, metatheses, aldol and Michael reactions, and isomerizations. The

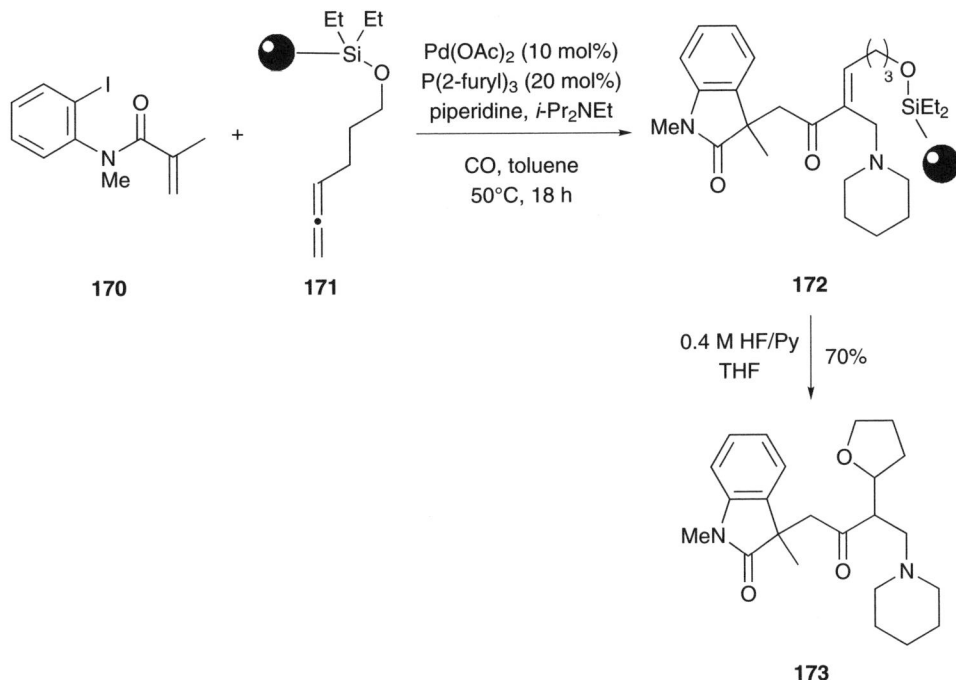

Scheme 8.45 *Domino Mizoroki–Heck/anion-capture reaction.*

anion capture process has also been widely used with other palladium-catalysed transformations. Outstanding examples of many different combinations have been developed by Grigg and coworkers, amongst them the Mizoroki–Heck/Tsuji–Trost reaction/1,3-dipolar cycloaddition [93], the Mizoroki–Heck reaction/metathesis [94], and several other processes involving a Mizoroki–Heck transformation [95, 96]. A first example of an anion capture approach, which was performed on solid phase, is the reaction of **170** and **171** in the presence of CO and piperidine to give **172**, involving an intramolecular Mizoroki–Heck reaction, a carbonylation, allene coupling and a nucleophilic attack by the pyridine on the palladium–allene complex formed (Scheme 8.45). Liberation from solid phase was achieved with HF, yielding **173** [97]. The same group also extended this protocol to a three-component process involving an aryl iodide, allenes and nitrogen nucleophiles bound to a resin [98].

Another interesting example includes a carbonylation, Mizoroki–Heck reaction, metathesis and a 1,3-dipolar cycloaddition. Thus, reaction of a mixture of iodothiophene, allylamine **174** and allene in the presence of 10 mol% Pd(OAc)$_2$, 20 mol% PPh$_3$ and K$_2$CO$_3$ in toluene at 80 °C for 36 h under 1 atm CO gave the enone **175**. Subsequent metathesis reactions using Grubbs' second-generation catalyst **176** led to the Δ^3-pyrroline **177**, which can undergo a 1,3-dipolar cycloaddition [99]. Thus, in the presence of AgOAc and imine **178**, cycloadduct **180** was obtained via the intermediate formation of 1,3-dipole **179** (Scheme 8.46).

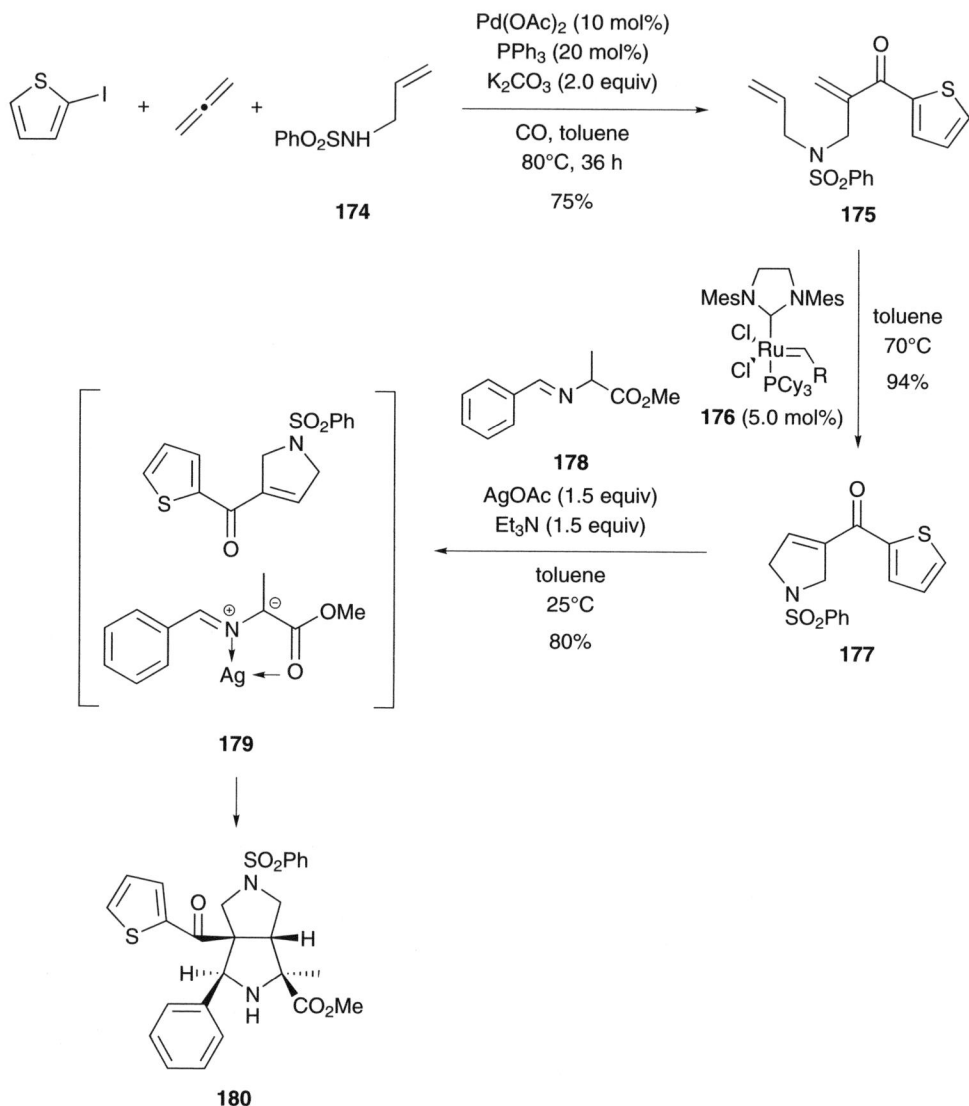

Scheme 8.46   *Synthesis of **180**.*

The Balme group studied the cyclization–coupling reaction of alkene-containing car-bonucleophiles with organic halides using an anion capture approach [100] in the prepa-ration of bi- and tri-cyclic compounds. A short entry to triquinanes, such as **183**, via a transient palladium(II) complex formed by a Mizoroki–Heck reaction of the vinyl bromide **181**, which reacts with the carbanion of a malonate moiety present in the substrate, is an illustrative example (Scheme 8.47). However, depending on the halide used, the products

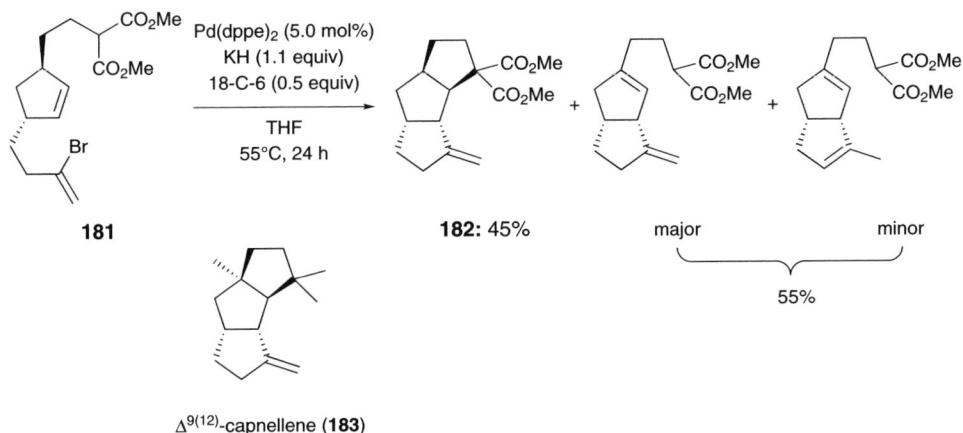

Scheme 8.47 *Synthesis of triquinanes.*

of the normal Mizoroki–Heck reaction are also formed. The methodology was applied to the total synthesis of marine sesquiterpene $(\pm)\Delta^{9(12)}$ capnellene [101].

A combination of a Mizoroki–Heck reaction with an intramolecular aldol condensation is observed on treatment of aromatic aldehydes or ketones such as **184** with allylic alcohols **185** as described by Dyker and coworkers [102]. The palladium-catalysed reaction led to **187** via **186** in 55% yield (Scheme 8.48).

Hallberg and coworkers [103] developed a synthesis of monoprotected 3-hydroxyindan-1-ones **191** in moderate to good yields using salicylaldehyde triflates **188** and

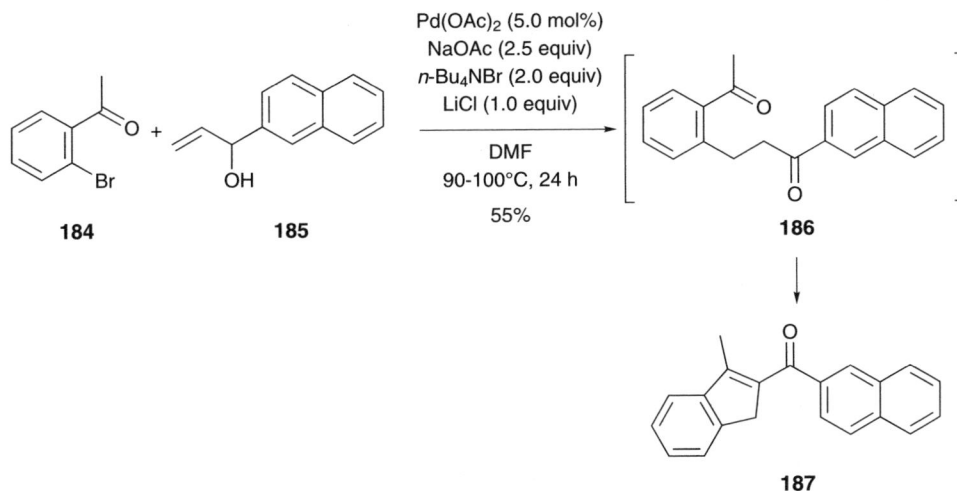

Scheme 8.48 *Domino Mizoroki–Heck/aldol reaction.*

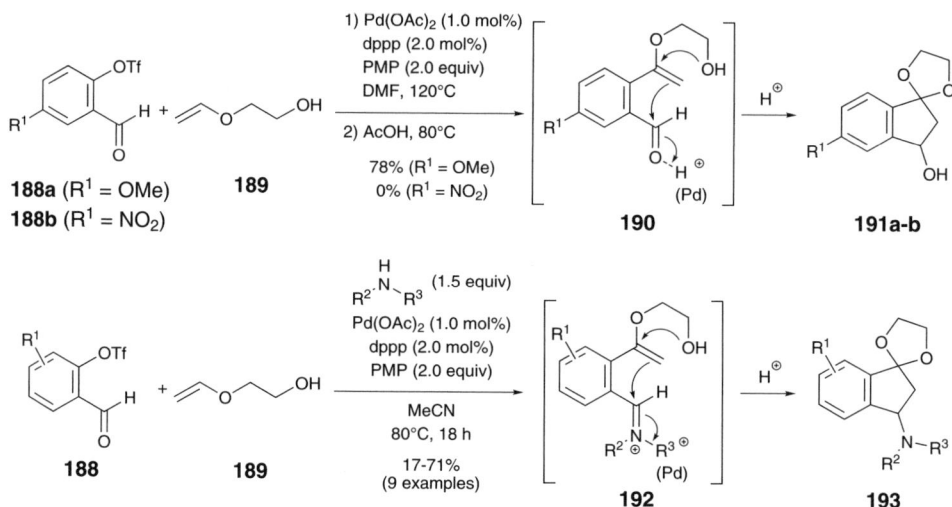

Scheme 8.49 *Synthesis of hydroxy- and amino-indanones.*

2-hydroxyethyl vinyl ether (**189**) presumably via a transient **190** (Scheme 8.49). The best yields were obtained using 1,2,2,6,6-pentamethylpiperidine as base; however, sali-cylaldehydes with electron-withdrawing groups always gave lower yield, and the desired product was not obtained in the reaction of a nitro salicylaldehyde. Later on, the same group extended the method to the preparation of the corresponding *N*-protected 3-aminoindane analogues **193** in the presence of nucleophilic secondary amines [104]. One can propose that the reaction proceeds via an iminium ion **192** similar to the above-mentioned interme-diate **190** followed by a cyclization step.

A combination of a Mizoroki–Heck reaction and a C–N coupling on solid phase was reported by Yamazaki *et al.* [105] in the synthesis of indole carboxylate **197** employing immobilized *N*-acetyldehydroalanine (**194**) and a bifunctionalized arene **195**. In these trans-formations, **196** is the intermediate. The best results were obtained with *o*-dibromoarenes such as **195b** and **195d** using the Pd$_2$(dba)$_3$/P-*t*-Bu$_3$/Cy$_2$NMe catalytic system developed by Littke and Fu [106]. The protocol was also extended to the synthesis of isoquinolines [107] (Scheme 8.50).

The Cacchi group [108] developed a palladium-catalysed domino process of *o*-alkynyltrifluoroacetanilides **198** and aryl or alkenyl halides which led to disubsti-tuted indoles. This methodology was successfully applied by Saulnier *et al.* [109] to the preparation of indolo[2,3-*a*]carbazoles such as **200** using *N*-benzyl-3,4-dibromomaleimide (**199**) as the alkenyl halide (Scheme 8.51). The indolocarbazol unit is found in sev-eral bioactive natural products, such as arcyriaflavin A and the cytotoxic rebecca-mycin.

Dongol and Tay [110] have recently made use of an intramolecular amination in the development of a one-pot synthesis of isoxazolidines starting from *O*-homoallyl hydrox-ylamines **201** and aryl iodides. After a Mizoroki–Heck reaction, a subsequent C–N

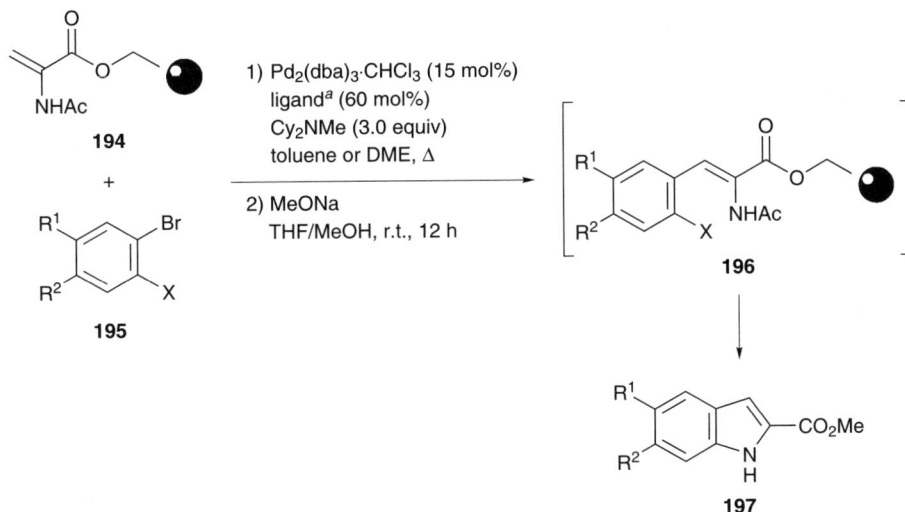

195	R¹	R²	X	197	R¹	R²	Yield [%]^{a,b}
a	H	H	I	a	H	H	46
b	H	H	Br	a	H	H	78
c	H	H	OTf	a	H	H	41
d	Me	Me	Br	b	Me	Me	82
e	OMe	H	Br	c	OMe	H	39
				d	H	OMe	31

^{a} A 0.50 M stock solution of Pt-Bu$_3$ in toluene was used for **195a-b** and t-Bu$_3$PHBF$_4$ was used for **195c-e**.
^{b} Isolated yield of **197** after SiO$_2$ column chromatography based on the loading of **194**.

Scheme 8.50 *Synthesis of indoles.*

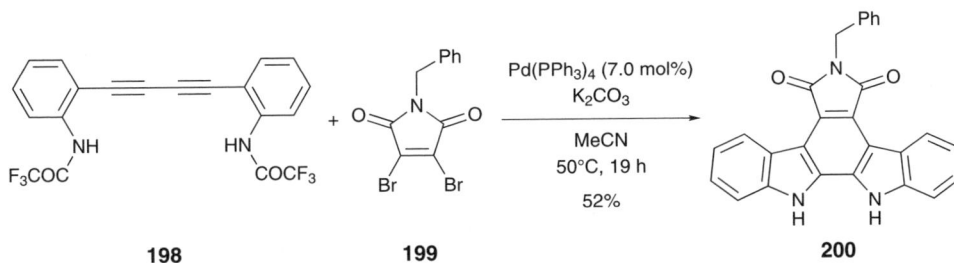

Scheme 8.51 *Synthesis of the indolocarbazole skeleton.*

bond formation takes place to furnish the target compounds with up to 79% GC-yield (Scheme 8.52).

In another example, Cvetovich and coworkers [111] described a Mizoroki–Heck reaction followed by an amidation–cyclization process and its application to the synthesis of p38 MAP kinase inhibitors. The reaction between substituted acrylanilides **205**

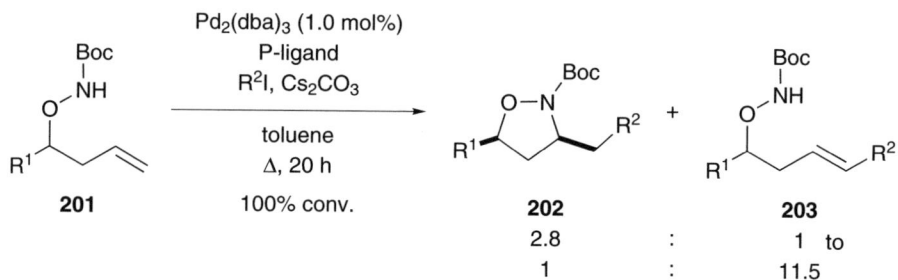

Scheme 8.52   *Synthesis of isoxazolines.*

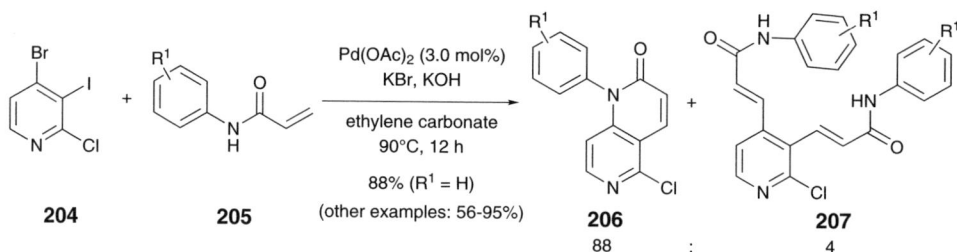

Scheme 8.53   *Synthesis of naphthyridinones.*

and halo-substituted pyridines **204** affords a series of naphthyridinones **206** in good yield (Scheme 8.53). However, in the case of electron-rich acrylanilides, the double Mizoroki–Heck product **207** was primarily obtained in a high ratio with respect to **206**. By studying the effect of different bases, this selectivity could eventually be considerably increased, with KOH giving the best results.

An unusual behaviour in the reaction between acrolein and 8-haloquinoline **210** was observed by Pinel and coworkers [112]. While the reaction of 8-bromoquinoline (**210**) with acrolein diethyl acetal using the Herrmann–Beller catalyst (**15**) led to the formation of the corresponding saturated ester **209** and small amounts of **208**, the reaction with acrolein afforded a tricyclic quinolinone **211** (Scheme 8.54). To explain this outcome, the authors proposed that, after the insertion step, a six-membered palladacycle is formed by coordination to the nitrogen. As a consequence, only the formyl hydrogen is conformationally accessible, thus forming a reactive ketene **214** upon β-hydride elimination. In the next step, a *syn* H—Pd addition to the C—N bond occurs followed by the formation of the intermediate **215** that undergoes reductive elimination to generate the tricyclic product **211**. However, there could also be another explanation for the formation of **211**.

The cyanide ion can serve as a terminating nucleophile in a palladium-catalysed process that starts with a Mizoroki–Heck reaction. This domino Mizoroki–Heck/cyanation transformation was first reported by Torii *et al.* [113] and shortly after by Grigg *et al.*

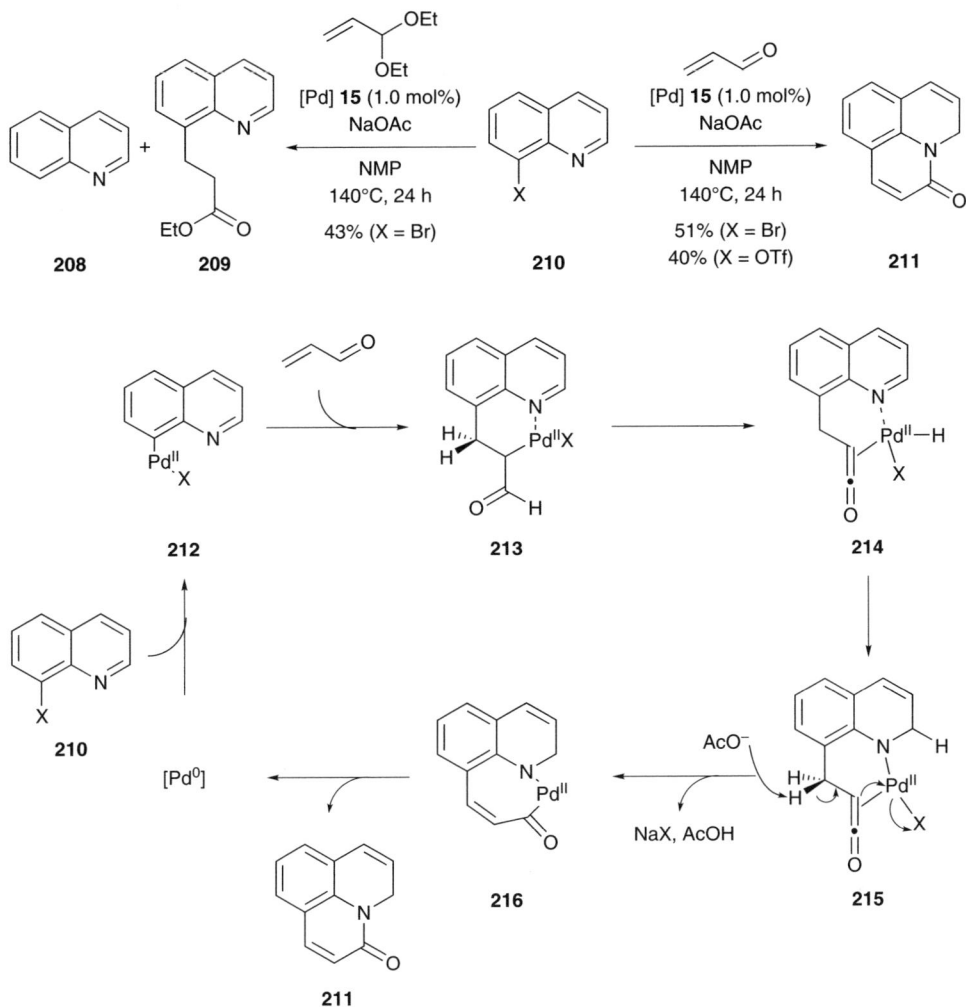

Scheme 8.54 *Synthesis of tricyclic quinolinones.*

[114]. Very recently, an enantioselective version of this transformation and its application towards the synthesis of alkaloids esermethole (**217**) and physostigmine (**218**) has been described by Zhu and coworkers [115]. In the new protocol, KCN is replaced by the less toxic $K_4[Fe(CN)_6]$, $Pd(OAc)_2$ is used as catalyst and (*S*)-DIFLUORPHOS (**221**) as ligand, leading to oxindole **220** in up to 78% yield and 77% *ee* (Scheme 8.55). Lautens and coworkers [116] also made use of this general concept by combining a Mizoroki–Heck/C–H activation/cyanation step for the preparation of polycyclic benzonitriles under microwave irradiation.

esermethole (**217**)   physostigmine (**218**)

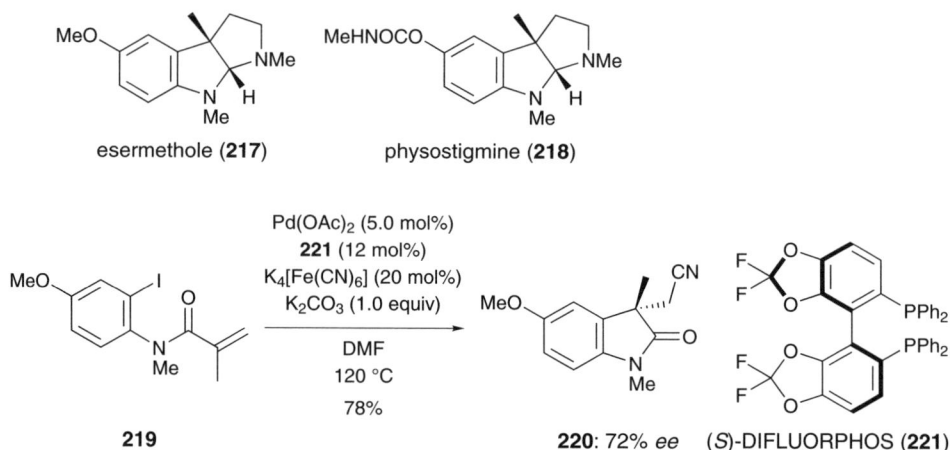

Scheme 8.55 *Mizoroki–Heck/cyanation process of oxindoles.*

8.9 Cross-Coupling/Mizoroki–Heck Reactions

In domino transition-metal-catalysed processes, cross-coupling reactions can also be used as the starting transformation. Most often, Suzuki, Stille and Sonogashira reactions are employed in this context. They can be combined with a Mizoroki–Heck reaction and other palladium-catalysed transformations.

8.9.1 Suzuki/Mizoroki–Heck Reactions

Fortunately, by fine tuning of the reactivity of the different functionalities, not only the combination of a Mizoroki–Heck with a Suzuki reaction but also a Suzuki with a Mizoroki–Heck reaction is possible, even in an enantioselective manner.

Shibasaki and coworkers [117] described the first enantioselective combination of this type in their synthesis of the natural product halenaquinone (**225**) possessing antibiotic, cardiotonic and protein tyrosine kinase inhibitory activities. The key step is an intermolecular Suzuki reaction of **222** and **223** followed by an enantioselective Mizoroki–Heck reaction in the presence of (*S*)-BINAP to construct the third ring and the stereogenic quaternary centre present in **224**. The reaction proceeded with a good *ee*-value of 85% but with a yield of only 20% (Scheme 8.56).

In a different approach by the same group, the low outcome of the aforementioned domino transformation was considerably improved by using triphenylarsine as coligand [118]. In the synthesis of tricyclic **228**, the reaction of **226** and **227** first gave the corresponding boron compound *in situ* by addition of 9-borabicyclo[3.3.1]nonane, which then undergoes cross-coupling with **227** followed by a Mizoroki–Heck reaction to give **228** (Scheme 8.57). Without any additional ligand, lower yields (53%) and several side products were obtained.

Polycyclic aromatic hydrocarbons such as fluoranthene or C_{60}-fullerene are structures of great interest. An easy entrance to analogues and partial structure thereof respectively has now been developed by de Meijere and coworkers [119] using a combination of a Suzuki and a Mizoroki–Heck-type coupling. Thus, reaction of 1,8-dibromophenanthren (**229**) and

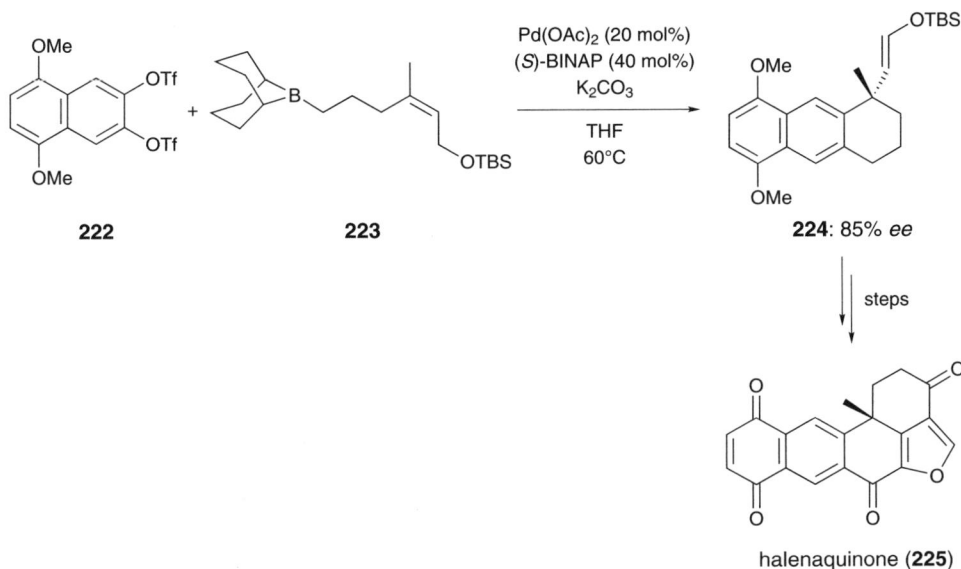

Scheme 8.56 *Synthesis of halenaquinone (225).*

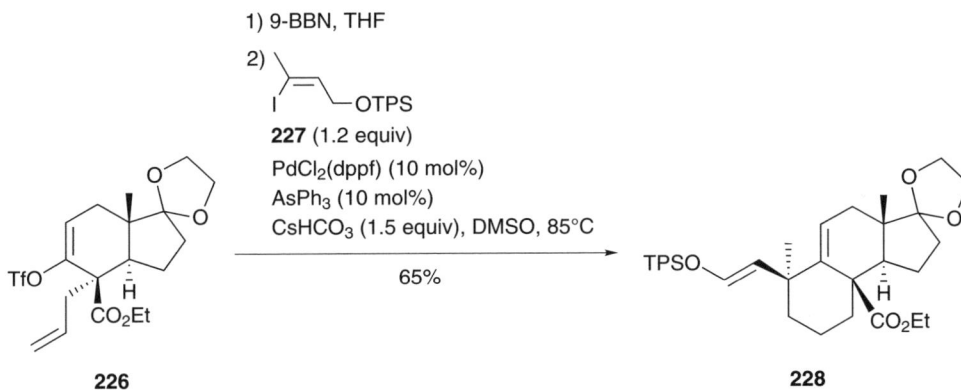

Scheme 8.57 *Synthesis of tricyclic 228.*

o-bromphenylboronic acid (**230**) employing 20 mol% of a palladium(0) catalyst led to **231** and **232** in 54% yield as a 1 : 1-mixture [120] (Scheme 8.58).

Recently, Nobile and coworkers reported a synthesis of polyfluorenylenevinylenes **234**, a potential material in active layers in light-emitting diodes (LEDs) [121]. In this Suzuki/Mizoroki–Heck protocol, potassium vinyltrifluoroborate reacts first with **233** in the presence of Pd(PPh₃)₄ as catalyst and K₂CO₃ as base. Although most of the solvent systems tested did not promote polymerization, up to 77% of the desired polymer were obtained when using dioxane under reflux (Scheme 8.59).

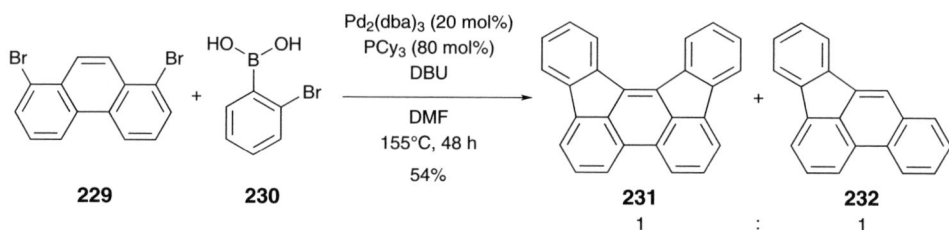

Scheme 8.58   *Synthesis of fluoranthene.*

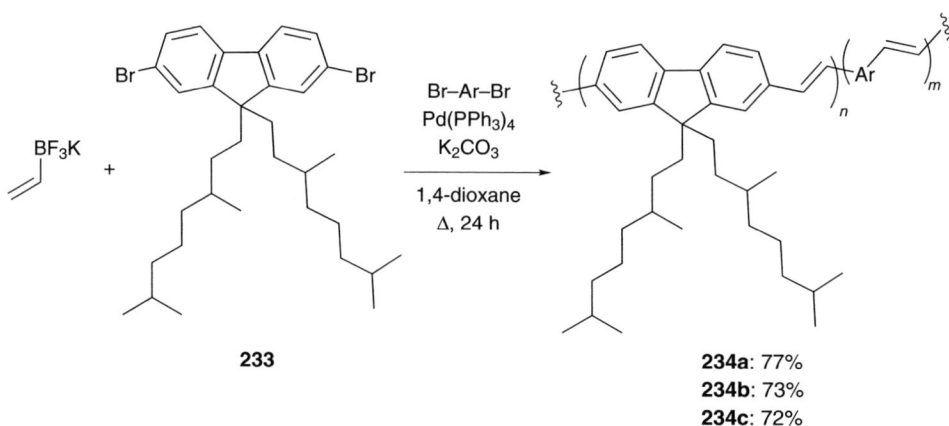

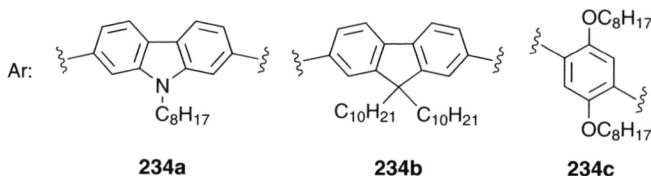

Scheme 8.59   *Synthesis of polymers by a Suzuki/Mizoroki–Heck reaction.*

8.9.2   Sonogashira/Mizoroki–Heck Reactions

In domino processes starting with a Sonogashira reaction, the second is usually an amination or a hydroxylation to give γ-lactones, furans or indoles; however, there are also a few examples where a Mizoroki–Heck reaction is performed as the second step.

In a simple example, Alami and coworkers [122] have shown that benzylhalides such as **235** can react with 1-alkynes such as **236** in the presence of palladium(0) and CuI in a Sonogashira reaction which is followed by a Mizoroki–Heck and yet another Sonogashira reaction to give tetrasubstituted alkenes such as **237** (Scheme 8.60).

Useful sequential one-pot transformations for the synthesis of heterocycles employing two palladium-catalysed steps were developed by Balme and coworkers [123]. A recent

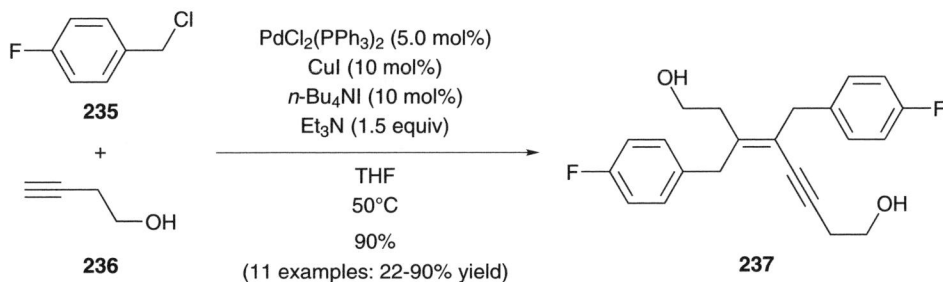

Scheme 8.60 *Synthesis of tetrasubstituted alkenes.*

example is the synthesis of furo[2,3-*b*]pyridones **242** [124, 125] by reaction of iodopyridone **238** and the alkyne **239** in the presence of catalytic amounts of PdCl$_2$(PPh$_3$)$_2$ and CuI in MeCN/Et$_3$N followed by addition of the aryl iodide **241** (Scheme 8.61). A wide variety of other pyridones, alkynes and arylhalides have been used as substrates, with yields ranging from 6 to 90% depending on the electron density of the aromatic halide, with electron-deficient arenes being the most reactive.

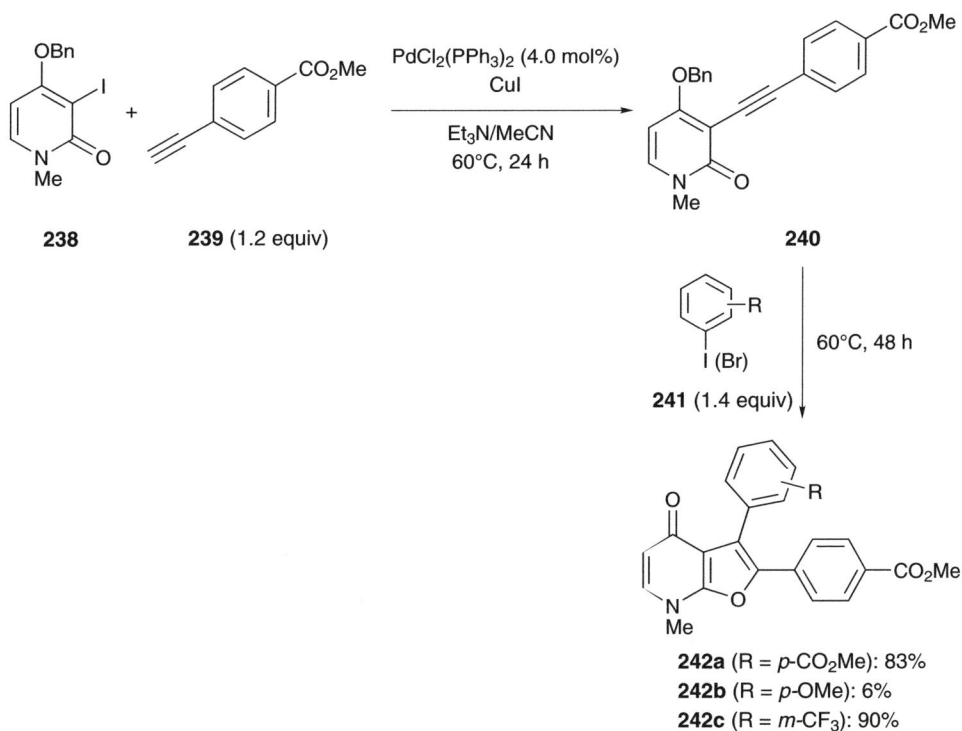

242a (R = *p*-CO$_2$Me): 83%
242b (R = *p*-OMe): 6%
242c (R = *m*-CF$_3$): 90%

Scheme 8.61 *Synthesis of furopyridones.*

8.10   Allylic Substitution (Tsuji–Trost)/Mizoroki–Heck Reaction

The palladium(0)-catalysed nucleophilic substitution of allylic acetates, carbonates or halides, also known as the Tsuji–Trost reaction, is a powerful procedure for the formation of C—C, C—O and C—N bonds. One of the early impressive examples, where this transformation had been combined with a pallada–ene reaction, was developed by Oppolzer and Gaudin [126]. Although, in general, the Tsuji–Trost reaction can be combined with other palladium-catalysed transformations, there are only a few examples where it is combined with a Mizoroki–Heck transformation.

In their enantioselective total synthesis of the alkaloid cephalotaxine (**246**), Tietze and Schirok [127] used a combination of a Tsuji–Trost and a Mizoroki–Heck reaction (Scheme 8.62). It was necessary to adjust the reactivity of the two palladium-catalysed transformations to allow a controlled process. Reaction of **243a** using Pd(PPh₃)₄ as catalyst led to **244**, which furnished **245** in a second palladium-catalysed reaction. In this process, the nucleophilic substitution of the allylic acetate is faster than the oxidative addition of the arylbromide moiety in **243a**; however, if one uses the iodide **243b**, then the yield drops dramatically due to an increased rate of the oxidative addition.

In another example from the same group the two transformations could be accomplished using only one catalytic system [128]. The tricyclic systems **248** were synthesized in a single process from **247**. The electronic density of the arylhalide had a significant influence on the reaction rate of the oxidative addition in comparison with the nucleophilic substitution. Thus, substrate **247d** gave the best result due to the electron-donating group at the aromatic ring. Recently, this process has been extended to even more complex substrates in the synthesis of tetracyclic compounds [129] (Scheme 8.63).

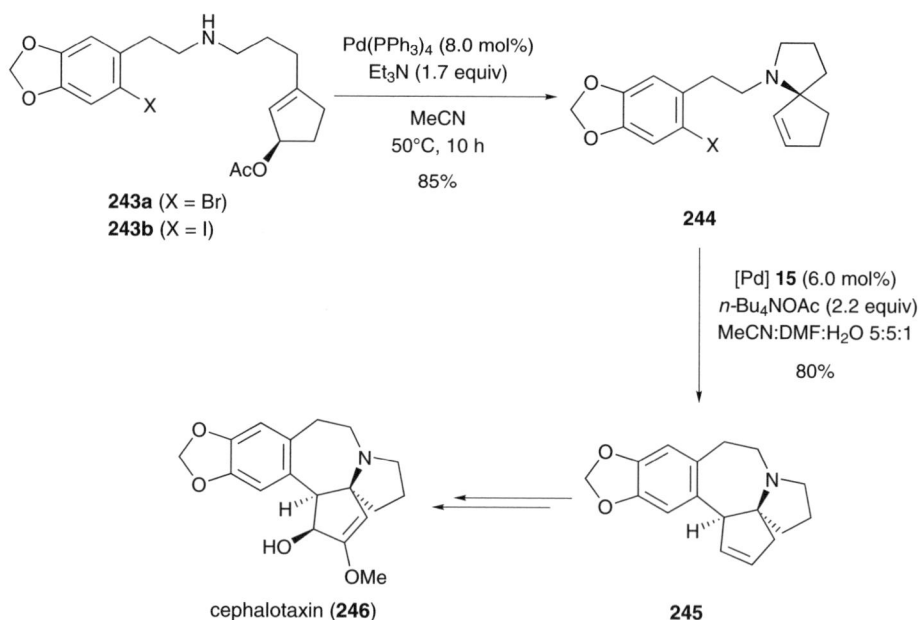

Scheme 8.62   *Enantioselective synthesis of cephalotaxine (246).*

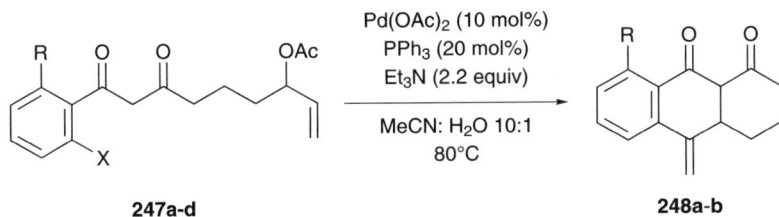

247	R	X	248	Yield [%]
a	H	Br	a	49
b	H	I	a	23
c	OMe	Br	b	77
d	OMe	I	b	89

Scheme 8.63 *Domino Tsuji-Trost/Mizoroki–Heck reaction.*

A combination of a Tsuji–Trost and a Mizoroki–Heck reaction was also used by Poli and coworkers [130] for the synthesis of the aza analogues **254** of the antimicotic podophyllotoxin (**249a**) and the cytotoxic etoposide (**249b**) (Scheme 8.64) [131]. For the synthesis of **254**, compound **250**, obtained from piperonal and the amide **251** containing an allyl acetate moiety, was transformed into **252**, which served as the starting material for the

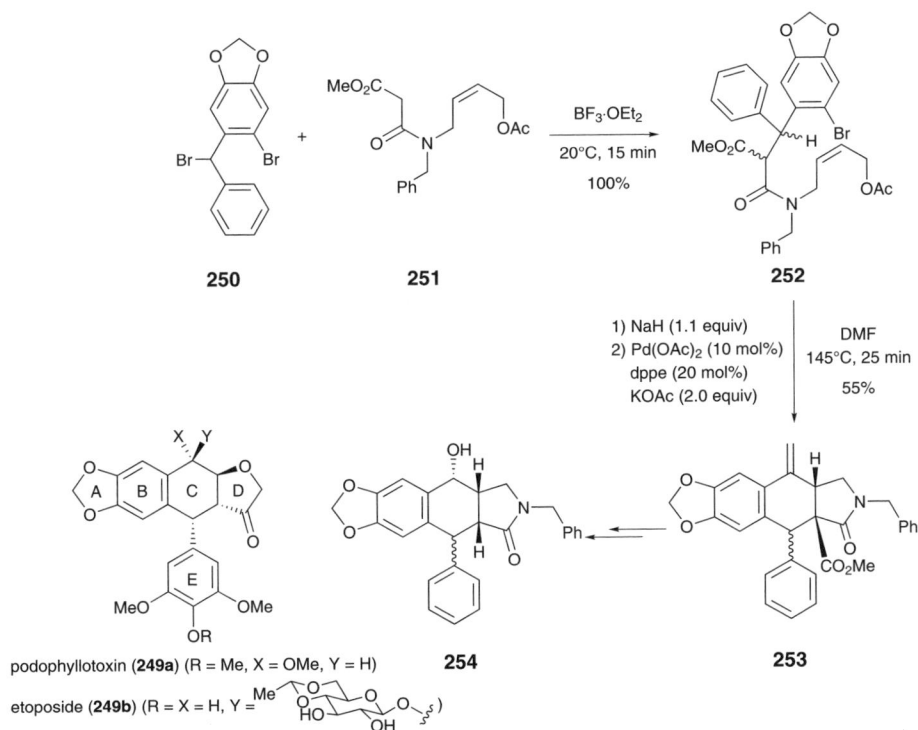

Scheme 8.64 *Synthesis of aza analogues of podophyllotoxin (**249a**) and etoposide (**249b**).*

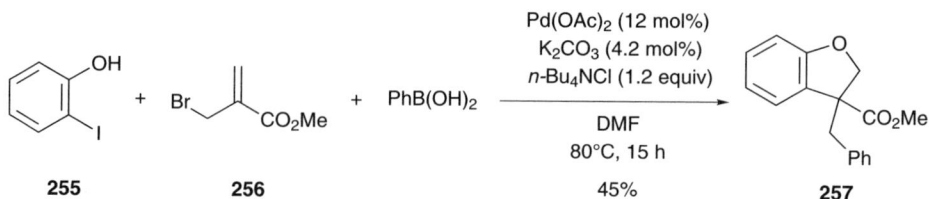

Scheme 8.65 *Synthesis of benzofurans.*

domino reaction. Treatment of **252** with a catalytic amounts of Pd(OAc)$_2$ in the presence of 1,2-bis(diphenylphosphino)ethane at 145 °C led to the tetracyclic structure **253** in 55% yield as a 75 : 25 mixture of two (out of four possible) diastereomers. It has been shown that the nucleophilic substitution of the allyl acetate moiety takes place already at 85 °C, whereas the following Mizoroki–Heck reaction needs a higher reaction temperature.

Lamaty and coworkers [132] described a nice combination of three palladium-catalysed transformations: first, an intermolecular nucleophilic substitution of an allylic bromide to form an aryl ether; second, an intramolecular Mizoroki–Heck type transformation in which the intermediate palladium(II) species is intercepted by a phenylboronic acid. Thus, reaction of a mixture of 2-iodophenol (**255**), methyl 2-bromomethylacrylate (**256**) and phenylboronic acid in the presence of catalytic amounts of Pd(OAc)$_2$ led to 3,3-disubstituted 2,3-dihydrobenzofuran **257**. Besides phenylboronic acid, several substituted boronic acids have also been used in this process (Scheme 8.65).

8.11   Other Palladium(0)-Catalysed Processes Involving a Mizoroki–Heck-Type Reaction

Since the insertion of CO in palladium-catalysed domino reactions can also be part of the first step after an oxidative addition, the combination with a Mizoroki–Heck reaction as the second step has been also investigated. In this context, Okuro and Alper [133] and Grigg *et al.* [134] described a palladium-catalysed transformation of *o*-iodophenols and *o*-iodoanilines with allenes in the presence of CO (Scheme 8.66). Reaction of **258** or **259** with **260** in the presence of palladium(0) under a CO atmosphere (1 atm) led to the

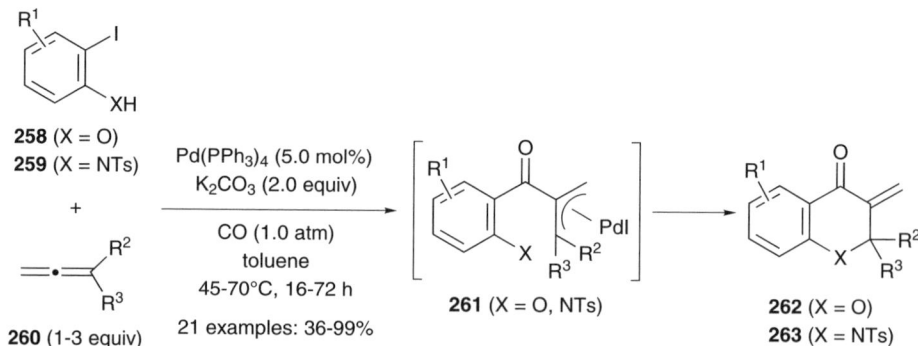

Scheme 8.66 *Synthesis of chromanones and quinolones.*

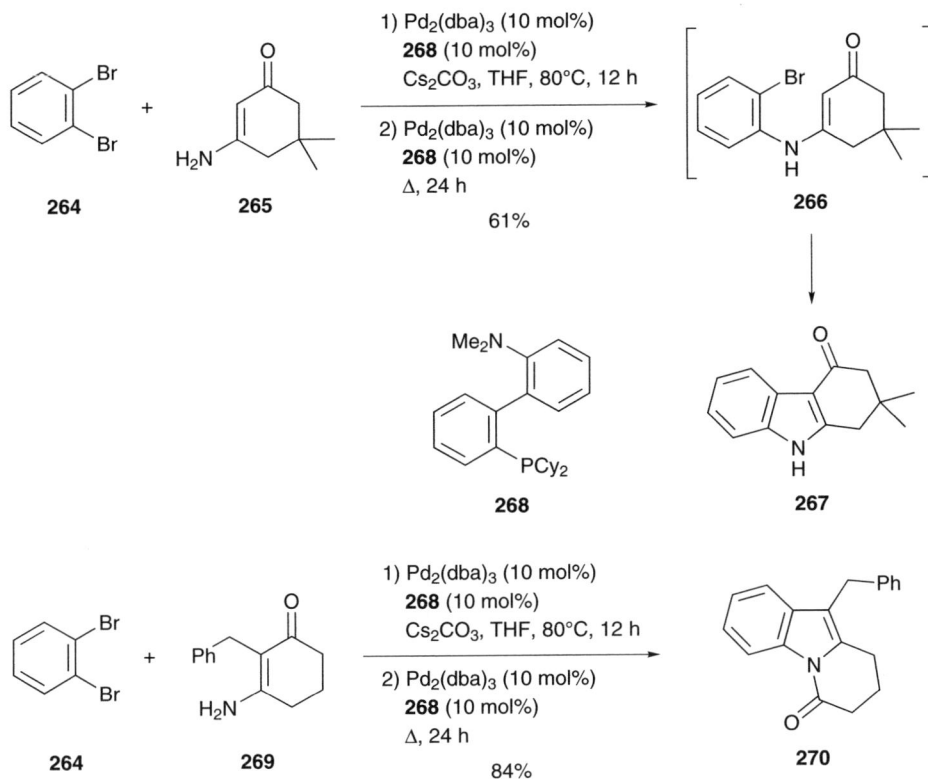

Scheme 8.67 *Synthesis of indoles.*

chromanones **262** and quinolones **263** respectively via the π-allyl-η^3-palladium complex **261**. The enones obtained can be transformed by a Michael addition with amines followed by reduction to give γ-aminoalcohols. Quinolones and chromanones are of interest due to their pronounced biological activity as antibacterials [135], antifungals [136] and neurotrophic factors [137].

Edmondson *et al.* [138] developed a synthesis of 2,3-disubstituted indoles using an amination as the first step (Scheme 8.67). The reaction of **264** and **265** with catalytic amounts of Pd$_2$(dba)$_3$ and the ligand **268** gave compound **267** in high yield. To get to the final indole, a second charge of palladium had to be added after 12 h, otherwise only the amination product was isolated. It can be assumed that in the first step the enaminone **266** is formed, which then cyclizes in a Mizoroki–Heck-type reaction to give **267**. In a similar way, reaction of **264** and **269** led to **270** by an acyl migration in the Mizoroki–Heck cyclization product.

Similarly, Barluenga and coworkers [139] made use of the chemoselectivity observed in the amination of alkenyl bromides in the presence of aryl bromides to carry out the synthesis of 2- and 3-substituted indoles **273** and **275** using *o*-haloanilines **271** and alkenyl halides **272** as well as **274** (Scheme 8.68). After an optimization process, the change from DavePhos (**276**) to X-Phos (**277**) allowed even the conversion of the less reactive *o*-chloroanilines in moderate yields and higher selectivity with respect to the competitive imine formation.

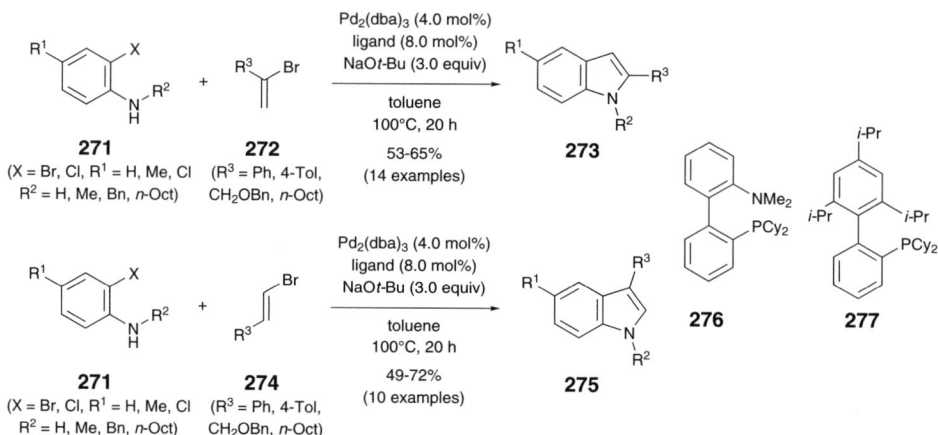

Scheme 8.68   *Synthesis of indoles.*

Ortho-gem-Dihalovinylanilines **278** were also used in another example of a Buchwald–Hartwig-type/Mizoroki–Heck reaction for the synthesis of 2-vinylic indoles **279** (Scheme 8.69). Lautens and coworkers [140] recently illustrated a domino coupling under Jeffery's condition where the aniline nitrogen undergoes an amination step followed by a Mizoroki–Heck coupling with various alkenes. In this process, electronic factors and steric hindrance of the different substituents had only a small effect on the yield; however, in the formation of 3-substituted indoles using this method only very poor yields were obtained. The procedure can also be performed in an intramolecular mode leading to tricyclic compounds such as pyridino and azepino indoles **281** and **282** (Scheme 8.69).

Curran and coworkers [141] developed a palladium(0)-catalysed domino process for the synthesis of the very potent anticancer natural product (*S*)-camptothecin (**283**) [142] and its analogues (Scheme 8.70). Camptothecin (**283**) contains an 11*H*-indolizino[1,2-*b*]quinolin-9-one skeleton, which is also found in mappicine [143] and the promising new analogue DB-67 (**287**) [144]. A domino–radical reaction has been used for its construction in 40–60% yield [145]. However, the product is also accessible from the isonitrile **284**

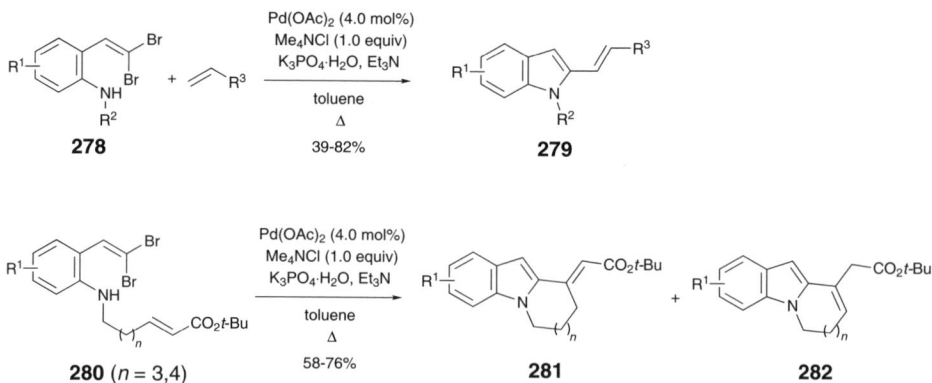

Scheme 8.69   *Synthesis of vinylic indoles.*

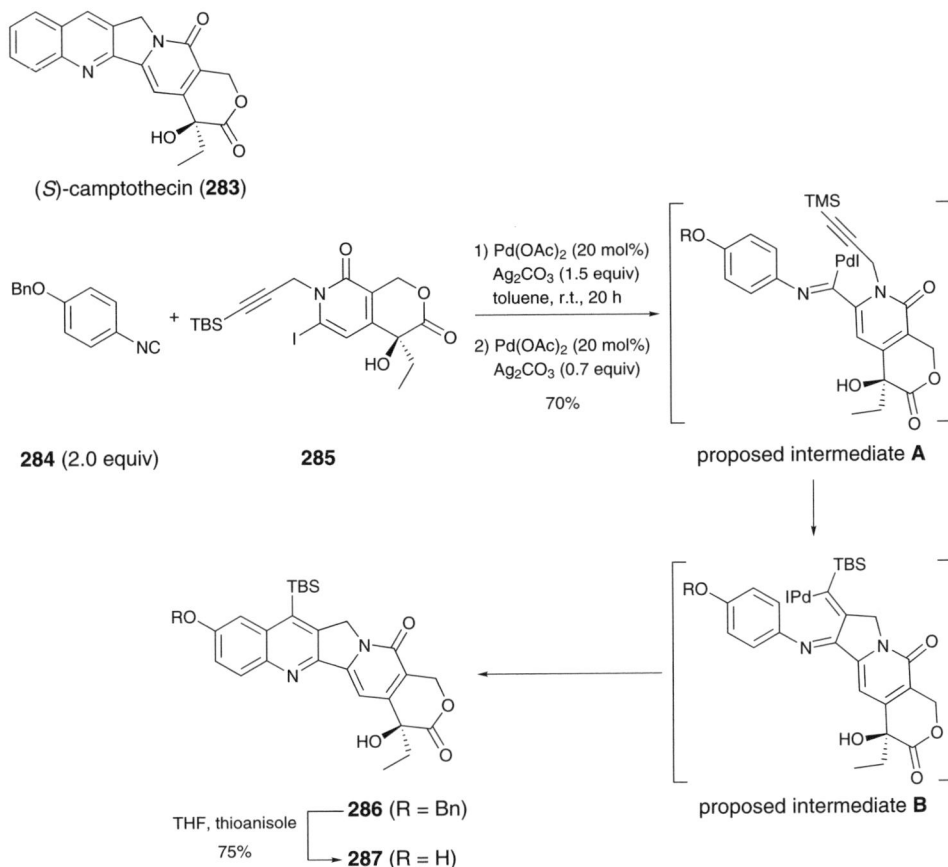

Scheme 8.70   *Synthesis of the camptothecin analogue DB-67 (287).*

and the iodopyridone **285** in a palladium-catalysed domino process. Although initially the formation of **286** did not go to completion using 1.5 equiv Ag$_2$CO$_3$ and 20 mol% Pd(OAc)$_2$ in toluene, the problem could be solved by filtration of the reaction mixture as well as evaporation of the solvent and starting again using a new charge of Pd(OAc)$_2$ and Ag$_2$CO$_3$. Using this procedure, the anticancer agent DB-67 (**287**) was obtained in 53% yield from **284** and **285** after deprotection.

8.12   Palladium(II)-Catalysed Transformations Involving a Mizoroki–Heck-Type Reaction

The Wacker oxidation [146], amongst other nucleophilic additions to alkenes, is the most important reaction based on a palladium(II) catalysis. It is also used industrially for the synthesis of acetaldehyde from ethene and water. This oxidative process has been combined with a Mizoroki–Heck reaction by Tietze and coworkers [13] for an enantioselective total synthesis of vitamin E (**293**) [147] using BOXAX ligand **291** [148]. In this way the chromane ring and parts of the side chain of vitamin E (**293**) can be introduced in one

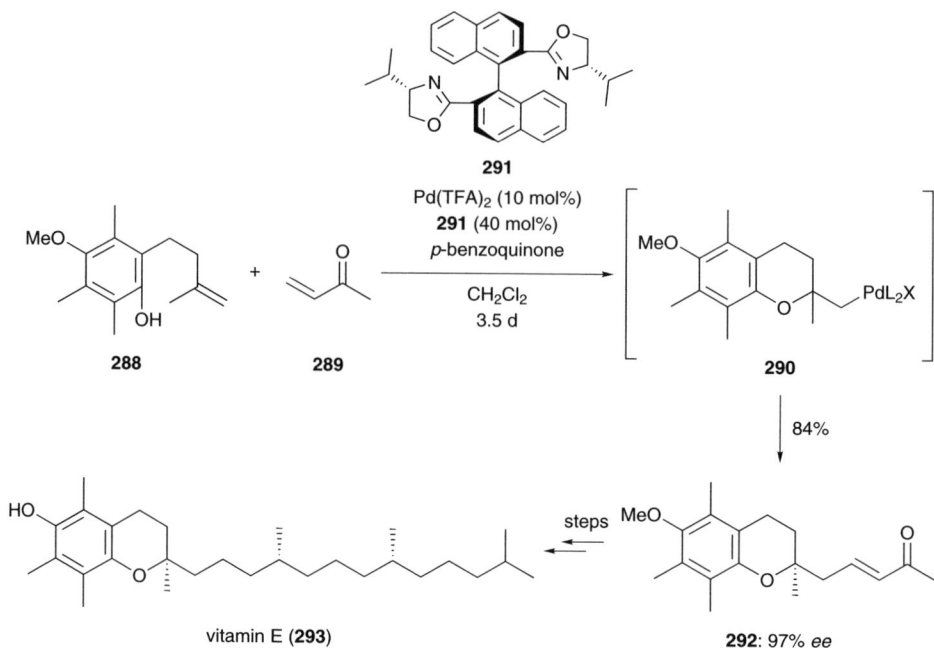

Scheme 8.71 *Enantioselective synthesis of vitamin E (293).*

process. Thus, reaction of **288** with methyl vinyl ketone (**289**) in the presence of **291**, benzoquinone and Pd(TFA)$_2$ (TFA = trifluoroacetate) gave **292** in 84% yield and 97% *ee*. It can be assumed that **290** is an intermediate, which reacts with the α,β-unsaturated ketone (Scheme 8.71). **292** was then transformed into vitamin E (**293**) in a few steps [149]. The same group also applied this domino process to the preparation of various 2,3-dihydrobenzo[1,4]-dioxins and -oxazins [150].

Tri- and tetra-substituted dihydropyranones and furanones have also been synthesized by a Wacker/Mizoroki–Heck domino process (Scheme 8.72). Gouverneur and coworkers [151] very recently prepared these kinds of compounds using different palladium sources, copper acetate as redox mediator, oxygen as oxidant and lithium bromide as additive. The coupling of two electron-poor substrates, ethyl acrylate and β-hydroxy alkynones **294**, led to the desired compounds **295a–e** in moderate yields.

Yang and coworkers [152] described another example of an oxidative cyclization followed by Mizoroki–Heck coupling under a palladium(II) catalysis using nitrogen-based nucleophiles. In this new aza–Wacker/Mizoroki–Heck domino process the intramolecular cyclization of substituted acrylanilide **296** takes place enantioselectively in the presence of (−)-sparteine as chiral ligand and molecular oxygen as the sole oxidant (Scheme 8.73). Enantioselectivities of the reactions range from 75 to 91% *ee* and and the yields of the reactions range from 60 to 70%.

In the synthesis of the steroid (+)-equilenin (**302**) Nemoto *et al.* [153] also used a palladium(II)-catalysed domino process which includes a ring expansion followed by a Mizoroki–Heck-type reaction (Scheme 8.74). Interestingly, almost complete reversion of

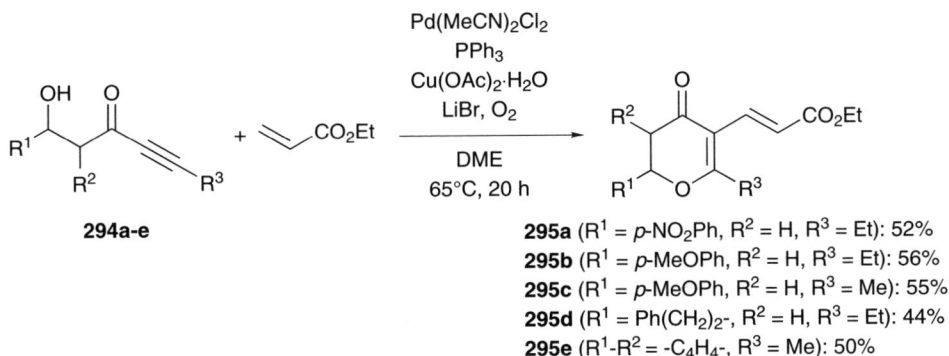

Scheme 8.72 *Synthesis of dihydropyranones.*

295a (R^1 = *p*-NO$_2$Ph, R^2 = H, R^3 = Et): 52%
295b (R^1 = *p*-MeOPh, R^2 = H, R^3 = Et): 56%
295c (R^1 = *p*-MeOPh, R^2 = H, R^3 = Me): 55%
295d (R^1 = Ph(CH$_2$)$_2$-, R^2 = H, R^3 = Et): 44%
295e (R^1-R^2 = -C$_4$H$_4$-, R^3 = Me): 50%

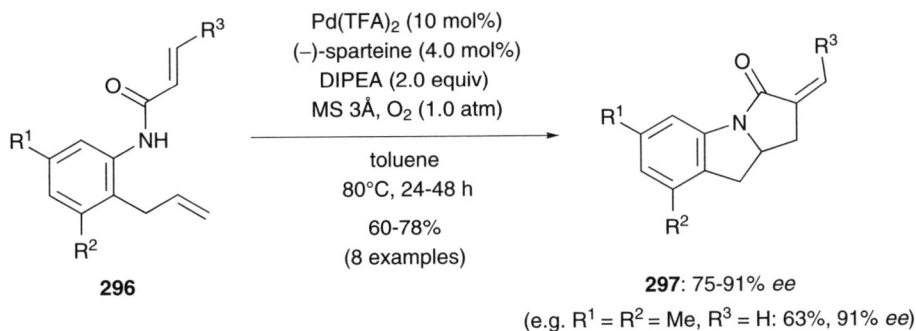

297: 75-91% *ee*
(e.g. R^1 = R^2 = Me, R^3 = H: 63%, 91% *ee*)

Scheme 8.73 *Enantioselective synthesis of indolines.*

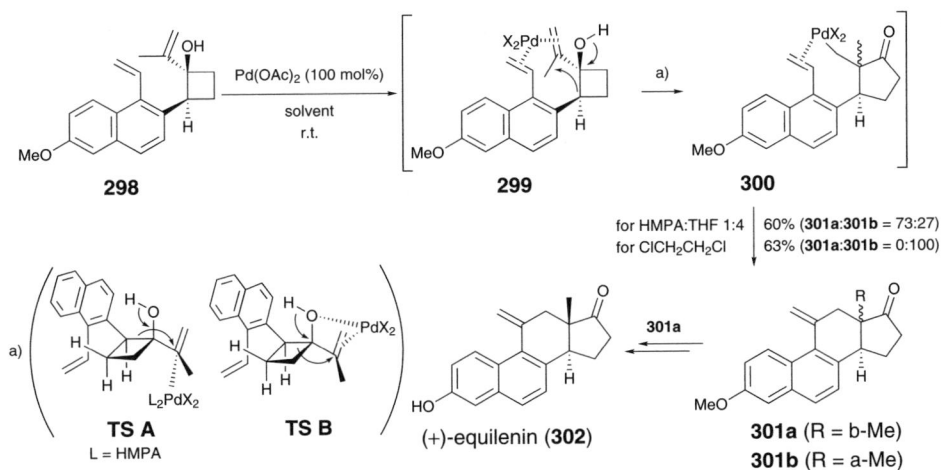

for HMPA:THF 1:4 | 60% (**301a:301b** = 73:27)
for ClCH$_2$CH$_2$Cl | 63% (**301a:301b** = 0:100)

TS A
L = HMPA

TS B

(+)-equilenin (**302**)

301a (R = b-Me)
301b (R = a-Me)

Scheme 8.74 *Synthesis of (+)-equilenin (**302**).*

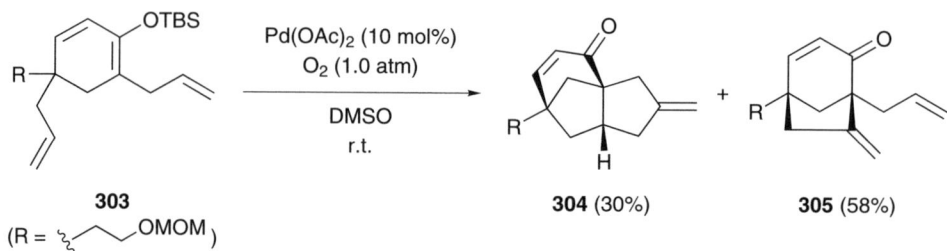

Scheme 8.75 *Palladium-catalysed domino cycloalkenylation.*

the facial selectivity using different solvents was observed. Thus, reaction of compound **298** with Pd(OAc)$_2$ at room temperature in hexamethylphosphoric triamide–tetrahydrofuran (1 : 4) gave the two diastereomers **301a** and **301b** in a 73 : 27 ratio in 60% yield, whereas in only **301b** was formed dichloroethane in 63% yield exclusively. As intermediates, the palladium complexes **299** and **300** can be assumed and the two transition-state structures **TS A** and **TS B** explain the different facial selectivities.

Toyota *et al.* [154] used a combination of a Wacker and a Mizoroki–Heck-type transformation for the construction of the cedrane skeleton. Thus, reaction of **303** using 10 mol% Pd(OAc)$_2$ under an oxygen atmosphere led to the domino product **304** in 30% yield (Scheme 8.75). In addition, 58% of the monocyclized compound **305** was obtained.

Many domino Mizoroki–Heck reactions start with the formation of vinyl palladium species, which are generally formed by an oxidative addition of vinylic halides or triflates to palladium(0). Such an intermediate can also be obtained from an addition of a nucleophile to a divalent palladium-coordinated alkyne or allene. In most of these cases some oxidant must be added to regenerate palladium(II) from palladium(0) in order to achieve a catalytic cycle. However, Liu and Lu [155] have successfully applied a protonolysis reaction of the C–Pd bond formed in the presence of excess halide ions to quench the C–Pd bond with regeneration of a palladium(II) species. In this way, reaction of **306** and acrolein in the presence of Pd(OAc)$_2$ and LiBr gave predominantly **307** (Scheme 8.76). Depending on the substitution pattern and reaction conditions, **308** was formed as a side product.

The same group [156] applied this protonolysis approach to an inter- and intramolecular enyne coupling that starts with an acetoxypalladation of alkynes followed by cyclization to yield a wide variety of carbo- and heterocyclic compounds. Moreover, the

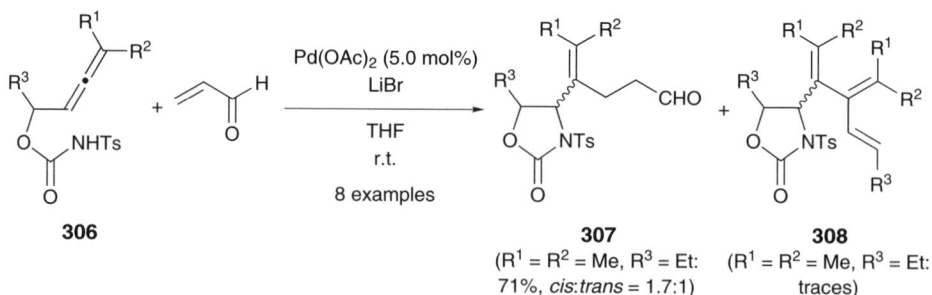

Scheme 8.76 *Synthesis of urethanes.*

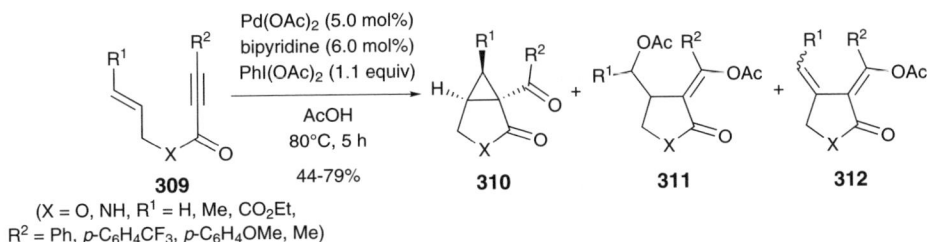

Scheme 8.77 *Synthesis of cyclopropylketones.*

organo-palladium(II) intermediate can be oxidized even further to a palladium(IV) species and by doing so promote a nucleophilic attack. Very recently, Sandford and coworkers [157] reported a novel palladium-catalysed oxidative reaction for the conversion of enynes into cyclopropyl ketones **310** in moderate to good yields (Scheme 8.77). They utilized PhI(OAc)$_2$ to promote the oxidation step and bipyridine as ligand. The authors proposed that, after the formation of a palladium(IV) intermediate, the cyclopropanation step proceeds by an intramolecular nucleophilic attack of the electron-rich olefin onto the C–Pd^{IV} bond with inversion of the configuration at Cα. The side product **311** would derive from the same intermediate by an intermolecular attack of an acetate molecule. Finally, the formation of compound **312** is a consequence of a β-H-elimination of the palladium(II) intermediate formed first.

Alcaide *et al.* [158] have used a similar approach for the synthesis of the tricyclic β-lactam **314** starting from **313** (Scheme 8.78). In this domino process, a π-allyl–palladium complex is formed by a nucleophilic attack of bromide at a divalent palladium-coordinated allene. Thereafter, the nitrogen nucleophile of the urethane moiety reacts with the primarily

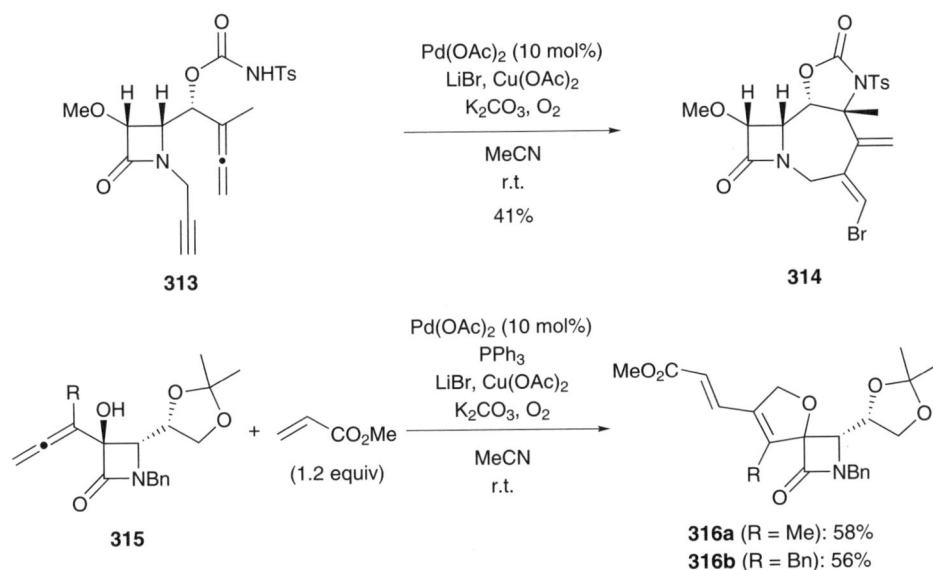

Scheme 8.78 *Synthesis of tricyclic β-lactams.*

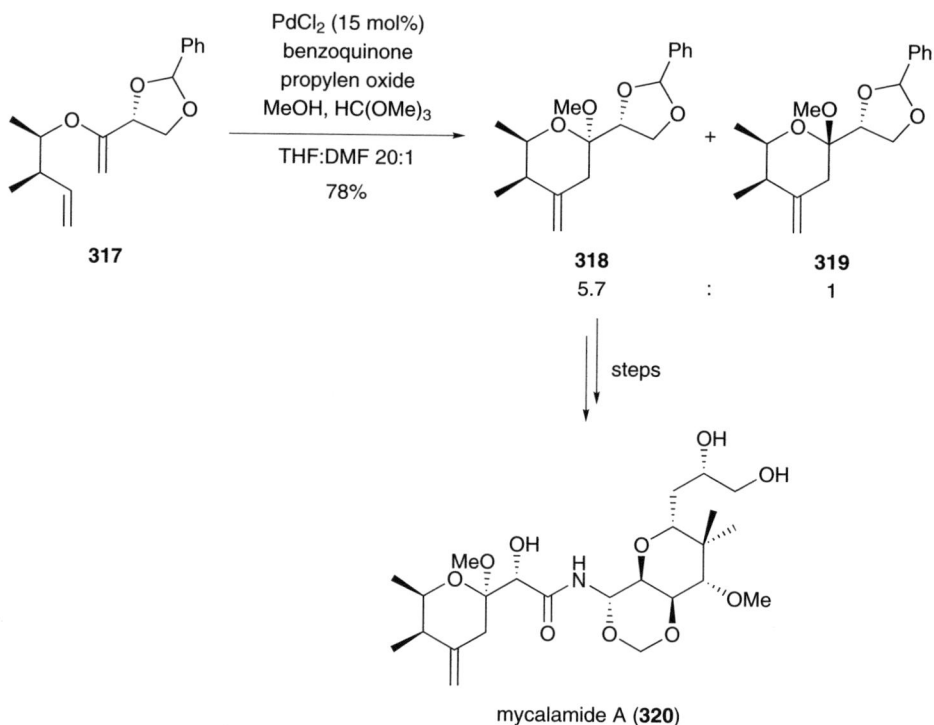

Scheme 8.79   *Synthesis of mycalamide A (**320**).*

obtained π-allyl–palladium complex to form a C–N bond and a vinyl halide. The final step is an intramolecular Mizoroki–Heck-type reaction of the vinyl halide with the alkyne moiety with replacement of palladium by bromide to yield tricycle **314**. In this transformation, $O_2/Cu(OAc)_2$ was used to reoxidize the palladium(0) to palladium(II). In an analogous procedure, but using a hydroxyl moiety as the nucleophile and methyl acrylate instead of the alkyne group in the presence of triphenyl phosphine spirolactams, **316** was prepared from α-allenols **315** in moderate yields [159].

Another example of the palladium(II)-catalysed Wacker/Mizoroki–Heck methodology has recently been published by Rawal and coworkers. In a total synthesis of mycalamide A (**320**), an intermolecular Wacker oxidation with methanol acting as nucleophile followed by ring closure via Mizoroki–Heck reaction furnished the tetrahydropyrans **318** and **319** in a 5.7 : 1 diastereomeric mixture, of which **318** was further transformed into **320** [160] (Scheme 8.79).

References

1. (a) Tietze, L.F., Brasche, G. and Gericke, K. (2006) *Domino Reactions in Organic Synthesis*, Wiley–VCH Verlag GmbH, Weinheim; (b) Tietze, L.F. (1996) Domino reactions in organic synthesis. *Chem. Rev.*, **96**, 115–36; (c) Tietze, L.F. and Beifuss, U. (1993) Sequential transformations in organic chemistry: a synthesis strategy with a future. *Angew. Chem., Int. Ed. Engl.*, **32**, 131–63.

2. (a) Tietze, L.F., Ila, H. and Bell, H.P. (2004) Enantioselective palladium-catalyzed transformations. *Chem. Rev.*, **104**, 3453–516; (b) Bräse, S. and de Meijere, A. (1998) *Metal-Catalyzed Cross-Coupling Reactions* (eds F. Diederich and P.J. Stang), Wiley–VCH Verlag GmbH, Weinheim, pp. 99–166; (c) Poli, G., Giambastiani, G. and Heumann, A. (2000) Palladium in organic synthesis: fundamental transformations and domino processes. *Tetrahedron*, **56**, 5959–89; (d) Beller, M., Riermeir, T.H. and Stark, G. (1998) *Transition Metals for Organic Synthesis* (eds M. Beller and C. Bolm), Wiley–VCH Verlag GmbH, Weinheim, pp. 208–40; (e) Tsuji, J. (2002) Palladium-catalyzed nucleophilic substitution involving allylpalladium, propargylpalladium, and related derivatives, in *Handbook of Organopalladium Chemistry for Organic Synthesis* (eds E.-i. Negishi and A. de Meijere), John Wiley & Sons, Inc., New York, pp. 1669–87.

3. (a) Tsuji, J. (1995) *Palladium Reagents and Catalysts: Innovations in Organic Synthesis*, John Wiley & Sons, Inc., New York; (b) Heumann, A. and Réglier, M. (1995) The stereochemistry of palladium-catalysed cyclisation reactions, part B: addition to π-allyl intermediates. *Tetrahedron*, **51**, 975–1015; (c) Frost, C.G., Howarth, J. and Williams, J.M.J. (1992) Selectivity in palladium catalysed allylic substitution. *Tetrahedron: Asymmetry*, **3**, 1089–122; (d) Acemoglu, L. and Williams, J.M.J. (2002) Synthetic scope of the Tsuji–Trost reaction with allylic halides, carboxylates, ethers, and related oxygen nucleophiles as starting compounds, in *Handbook of Organopalladium Chemistry for Organic Synthesis* (ed. E.-i. Negishi), John Wiley & Sons, Inc., New York, pp. 1689–1705; (e) Godleski, S.A. (1991) Nucleophiles with allyl–metal complexes, in *Comprehensive Organic Synthesis* (ed. B.M. Trost), Pergamon Press, Oxford, pp. 585–661.

4. (a) Zeni, G. and Larock, R.C. (2004) Synthesis of heterocycles via palladium π-olefin and π-alkyne chemistry. *Chem. Rev.*, **104**, 2285–309; (b) Takacs, J.M. and Jiang, X-t. (2003) The Wacker reaction and related alkene oxidation reactions. *Curr. Org. Chem.*, **7**, 369–96; (c) Stahl, S. (2004) Palladium oxidase catalysis: selective oxidation of organic chemicals by direct dioxygen-coupled turnover. *Angew. Chem. Int. Ed.*, **43**, 3400–20, and references cited therein.

5. (a) Tietze, L.F. and Lotz, F. (2006) Asymmetric Heck and other palladium-catalyzed reactions, in *Asymmetric Synthesis – The Essentials* (eds M. Christmann and S. Bräse), Wiley–VCH Verlag GmbH, Weinheim, pp. 147–52; (b) Trost, B.M. (2002) Pd asymmetric allylic alkylation (AAA). A powerful synthetic tool. *Chem. Pharm. Bull.*, **50**, 1–14.

6. (a) Dounay, A.B. and Overman, L.E. (2003) The asymmetric intramolecular Heck reaction in natural product total synthesis. *Chem. Rev.*, **103**, 2945–63; (b) Beletskaya, I.P. and Cheprakov, A.V. (2000) The Heck reaction as a sharpening stone of palladium catalysis. *Chem. Rev.*, **100**, 3009–66; (c) Bräse, S. and de Meijere, A. (1998) Palladium-catalyzed coupling of organyl halides to alkenes – the Heck reaction, in *Metal-Catalyzed Cross-Coupling Reactions* (eds F. Diederich and P.J. Stang), Wiley–VCH Verlag GmbH, Weinheim, pp. 99–166; (d) Oestreich, M. (2005) Neighbouring-group effects in Heck reactions. *Eur. J. Org. Chem.*, 783–92; (e) Whitcombe, N.J., Hii, K.K. and Gibson, S.E. (2001) Advances in the Heck chemistry of aryl bromides and chlorides. *Tetrahedron*, **57**, 7449–76.

7. (a) Nicolaou, K.C., Bulger, P.G. and Sarlah, D. (2005) Palladium-catalyzed cross-coupling reactions in total synthesis. *Angew. Chem. Int. Ed.*, **44**, 4442–89; (b) Schröter, S., Stock, C. and Bach, T. (2005) Regioselective cross-coupling reactions of multiple halogenated nitrogen-, oxygen-, and sulfur-containing heterocycles. *Tetrahedron*, **61**, 2245–67; (c) Christmann, U. and Vilar, R. (2005) Monoligated palladium species as catalysts in cross-coupling reactions. *Angew. Chem. Int. Ed.*, **44**, 366–74; (d) Espinet, P. and Echavarren, A.M. (2004) The mechanisms of the Stille reaction. *Angew. Chem. Int. Ed.*, **43**, 4704–34; (e) Tykwinski, R.R. (2003) Evolution in the palladium-catalyzed cross-coupling of sp- and sp^2-hybridized carbon atoms. *Angew. Chem. Int. Ed.*, **42**, 1566–8.

8. Negishi, E.-i., Copéret, C., Ma, S. *et al.* (1996) Cyclic carbopalladation. A versatile synthetic methodology for the construction of cyclic organic compounds. *Chem. Rev.*, **96**, 365–93.

9. Ma, S. (2004) Pd-catalyzed coupling reactions involving propargylic/allenylic species. *Eur. J. Org. Chem.*, 1175–83.

10. Morimoto, T. and Kakiuchi, K. (2004) Evolution of carbonylation catalysis: no need for carbon monoxide. *Angew. Chem. Int. Ed.*, **43**, 5580–8.

11. Charles, M.D., Schultz, P. and Buchwald, S.L. (2005) Efficient Pd-catalyzed amination of heteroaryl halides. *Org. Lett.*, **7**, 3965–8.

12. (a) Punniyamurthy, T., Velusamy, S. and Iqbal, J. (2005) Recent advances in transition metal catalyzed oxidation of organic substrates with molecular oxygen. *Chem. Rev.*, **105**, 2329–63; (b) de Bruin, B., Budzelaar, P.H.M. and Gal, A.W. (2004) Functional models for rhodium-mediated olefin-oxygenation catalysis. *Angew. Chem. Int. Ed.*, **43**, 4142–57.

13. (a) Tietze, L.F., Sommer, K.M., Zinngrebe, J. and Stecker, F. (2005) Palladium-catalyzed enantioselective domino reaction for the efficient synthesis of vitamin E. *Angew. Chem. Int. Ed.*, **44**, 257–9; (b) Tietze, L.F., Stecker, F., Zinngrebe, J. and Sommer, K.M. (2006) Enantioselective palladium-catalyzed total synthesis of vitamin E by employing a domino Wacker–Heck reaction. *Chem. Eur. J.*, **12**, 8770–6.

14. (a) Tietze, L.F., Nöbel, T. and Spescha, M. (1996) Stereoselective synthesis of steroids with the Heck reaction. *Angew. Chem., Int. Ed. Engl.*, **35**, 2259–61; (b) Tietze, L.F., Nöbel, T. and Spescha, M. (1998) Synthesis of enantiopure estrone via a double Heck reaction. *J. Am. Chem. Soc.*, **120**, 8971–7.

15. (a) Carpenter, N.E., Kucera, D.J. and Overman, L.E. (1989) Palladium-catalyzed polyene cyclizations of trienyl triflates. *J. Org. Chem.*, **54**, 5846–8; (b) Abelman, M.M. and Overman, L.E. (1988) Palladium-catalyzed polyene cyclizations of dienyl aryl iodides. *J. Am. Chem. Soc.*, **110**, 2328–9; (c) Zhang, Y. and Negishi, E.-i. (1989) Palladium-catalyzed cascade carbometalation of alkynes and alkenes as an efficient route to cyclic and polycyclic structures. *J. Am. Chem. Soc.*, **111**, 3454–6; (d) Wu, G.-z., Lamaty, F. and Negishi, E.-i. (1989) Palladium-catalyzed cyclization of benzyl halides and related electrophiles containing alkenes and alkynes as a novel route to carbocycles. *J. Org. Chem.*, **54**, 2507–8.

16. Zhang, Y., Wu, G.-z., Agnel, G. and Negishi, E.-i. (1990) One-step construction of fused tricyclic and tetracyclic structures from acyclic precursors via cyclic carbopalladation. *J. Am. Chem. Soc.*, **112**, 8590–2.

17. (a) Trost, B.M. (1990) Palladium-catalyzed cycloisomerizations of enynes and related reactions. *Acc. Chem. Res.*, **23**, 34–42; (b) Trost, B.M. and Shi, Y. (1992) Cycloisomerization for atom economy. Polycycle construction via tandem transition metal catalyzed electrocyclic processes. *J. Am. Chem. Soc.*, **114**, 791–2; (c) Trost, B.M. and Shi, Y. (1991) A palladium-catalyzed zipper reaction. *J. Am. Chem. Soc.*, **113**, 701–3; (d) Trost, B.M., Lautens, M., Chan, C. *et al.* (1991) Annulation via alkylation–Alder ene cyclizations. Pd-catalyzed cycloisomerization of 1,6-enynes. *J. Am. Chem. Soc.*, **113**, 636–44.

18. (a) Burns, B., Grigg, R., Sridharan, V. and Worakun, T. (1988) Palladium catalysed tandem cyclisation–anion capture processes. Hydride ion capture by vinylpalladium species. *Tetrahedron Lett.*, **29**, 4325–8; (b) Burns, B., Grigg, R., Ratananukul, P. *et al.* (1988) Palladium catalysed tandem cyclisation–anion capture processes. Hydride ion capture by alkyl- and π-allyl–palladium species. *Tetrahedron Lett.*, **29**, 4329–32.

19. (a) Feringa, B.L., Jager, W.F. and de Lange, B. (1993) Organic materials for reversible optical data storage. *Tetrahedron*, **49**, 8267–310; (b) Browne, W.R. and Feringa, B.L. (2006) Making molecular machines work. *Nature Nanotechnol.*, **1**, 25–35.

20. Herrmann, W.A., Brossmer, C., Öfele, K. *et al.* (1995) Palladacycles as structurally defined catalysts for the Heck olefination of chloro- and bromoarenes. *Angew. Chem., Int. Ed. Engl.*, **34**, 1844–8.

21. (a) Tietze, L.F., Heitmann, K. and Raschke, T. (1997) Allylsilane terminated domino Heck reaction. *Synlett*, 35–7; (b) Tietze, L.F. and Raschke, T. (1996) Enantioselective total synthesis

and absolute configuration of the natural norsesquiterpene 7-demethyl-2-methoxycalamenene by a silane-terminated intramolecular Heck reaction. *Liebigs Ann.*, 1981–7; (c) Tietze, L.F. and Raschke, T. (1995) Enantioselective total synthesis of a natural norsesquiterpene of the calamenene group by a silane-terminated intramolecular Heck reaction. *Synlett*, 597–8; (d) Tietze, L.F. and Schimpf, R. (1994) Regio- and enantioselective silane-terminated intramolecular Heck reactions. *Angew. Chem., Int. Ed. Engl.*, **33**, 1089–91.

22. Tietze, L.F. and Lotz, F. (2006) Pd-catalysed domino arylation/C—H activation for the synthesis of acenaphthylenes. *Eur. J. Org. Chem.*, 4676–84.

23. Tietze, L.F. and Ferraccioli, R. (1998) Efficient synthesis of aza-heterocycles by a domino process using an inter- and an intramolecular Heck reaction. *Synlett*, 145–6.

24. Yolacan, Ç., Bagdatli, E., Öcal, N. and Kaufmann, D.E. (2006) Epibatidine alkaloid chemistry: 5. Domino–Heck reactions of azabicyclic and tricyclic systems. *Molecules*, **11**, 603–14.

25. Fox, M.E., Li, C., Marino, J.P. Jr. and Overman, L.E. (1999) Enantiodivergent total syntheses of (+)- and (−)-scopadulcic acid A. *J. Am. Chem. Soc.*, **121**, 5467–80.

26. (a) Brown, H.C. and Pai, G.G. (1985) Selective reductions. 37. Asymmetric reduction of prochiral ketones with *B*-(3-pinanyl)-9-borabicyclo[3.3.1]nonane. *J. Org. Chem.*, **50**, 1384–94; (b) Midland, M.M. and Graham, R.S. (1990) Asymmetric reduction of α,β-acetylenic ketones with *B*-3-pinanyl-9-borabicyclo[3.3.1]nonane: (*R*)-(+)-1-octyn-3-ol. *Org. Synth.*, **63**, 402–6.

27. Overman, L.E., Paone, D.V. and Stearns, B.A. (1999) Direct stereo- and enantiocontrolled synthesis of vicinal stereogenic quaternary carbon centers. Total syntheses of *meso*- and (−)-chimonanthine and (+)-calycanthine. *J. Am. Chem. Soc.*, **121**, 7702–3.

28. Overman, L.E. and Watson, D.A. (2006) Diastereoselection in the formation of spirocyclic oxindoles by the intramolecular Heck reaction. *J. Org. Chem.*, **71**, 2587–99.

29. Overman, L.E. and Watson, D.A. (2006) Diastereoselection in the formation of contiguous quaternary carbon stereocenters by the intramolecular Heck reaction. *J. Org. Chem.*, **71**, 2600–8.

30. Maddaford, S.P., Andersen, N.G., Cristofoli, W.A. and Keay, B.A. (1996) Total synthesis of (+)-xestoquinone using an asymmetric palladium-catalyzed polyene cyclization. *J. Am. Chem. Soc.*, **118**, 10766–73.

31. Balcells, D., Maseras, F., Keay, B.A. and Ziegler, T. (2004) Polyene cyclization by a double intramolecular Heck reaction. A DFT study. *Organometallics*, **23**, 2784–96.

32. Miyazaki, F., Uotsu, K. and Shibasaki, M. (1998) Silver salt effects on an asymmetric Heck reaction. Catalytic asymmetric total synthesis of (+)-xestoquinone. *Tetrahedron*, **54**, 13073–8.

33. (a) Stocks, M.J., Harrison, R.P. and Teague, S.J. (1995) Macrocyclic ring closures employing the intramolecular Heck reaction. *Tetrahedron Lett.*, **36**, 6555–8; (b) Tietze, L.F., Sommer, K.M., Schneider, G. *et al.* (2003) Novel medium ring sized estradiol derivatives by intramolecular Heck reactions. *Synlett*, 1494–6; (c) Ma, S. and Negishi, E.-i. (1995) Palladium-catalyzed cyclization of ω-haloallenes. A new general route to common, medium, and large ring compounds via cyclic carbopalladation. *J. Am. Chem. Soc.*, **117**, 6345–57.

34. Harrowven, D.C., Woodcock, T. and Howes, P.D. (2002) A tandem Heck reaction leading to a 26-membered carbocycle. *Tetrahedron Lett.*, **43**, 9327–9.

35. Wu, G., Rheingold, A.L., Geib, S.J. and Heck, R.F. (1987) Palladium-catalyzed annelation of aryl iodides with diphenylacetylene. *Organometallics*, **6**, 1941–6.

36. Dyker, G. and Kellner, A. (1994) A palladium catalyzed domino coupling process to substituted phenanthrenes. *Tetrahedron Lett.*, **35**, 7633–6.

37. Cacchi, S., Felici, M. and Pietroni, B. (1984) The palladium-catalyzed reaction of aryl iodides with mono and disubstituted acetylenes: a new synthesis of trisubstituted alkenes. *Tetrahedron Lett.*, **25**, 3137–40.

38. Larock, R.C. and Tian, Q. (2001) Synthesis of 9-alkylidene-9*H*-fluorenes by a novel, palladium-catalyzed cascade reaction of aryl halides and 1-aryl-1-alkynes. *J. Org. Chem.*, **66**, 7372–9.

39. (a) Dyker, G., Nerenz, F., Siemsen, P. *et al.* (1996) A palladium-catalyzed domino coupling process leading to annelated pentafulvenes. *Chem. Ber.*, **129**, 1265–9; (b) Dyker, G., Siemsen, P., Sostman, S. *et al.* (1997) Synthesis of polycyclic hydrocarbons by palladium-catalyzed cross-coupling reactions of vinylic bromides with diphenylacetylene. *Chem. Ber.*, **130**, 261–5.

40. (a) Liu, Z. and Larock, R.C. (2007) Highly efficient route to fused polycyclic aromatics via palladium-catalyzed aryne annulation by aryl halides. *J. Org. Chem.*, **72**, 223–32; (b) Zhang, X. and Larock, R.C. (2005) Palladium-catalyzed annulation of arynes by 2-halobenzaldehydes: synthesis of fluoren-9-ones. *Org. Lett.*, **7**, 3973–6.

41. (a) Grigg, R., Sridharan, V. and Sukirthalingam, S. (1991) Alkylpalladium(II) species. Reactive intermediates in a bis-cyclisation route to strained polyfused ring systems. *Tetrahedron Lett.*, **32**, 3855–8; (b) Brown, A., Grigg, R., Ravishankar, T. and Thornton-Pett, M. (1994) Palladium catalysed cyclisation–cyclopropanation and cyclisation–anion capture processes of vinyl triflates. *Tetrahedron Lett.*, **35**, 2753–6; (c) Grigg, R., Sakee, U., Sridharan, V. *et al.* (2006) Palladium catalysed bis- and tris-cyclisations furnishing fused cyclopropyl carbo/heterocycles. *Tetrahedron*, **62**, 9523–32.

42. Schweizer, S., Song, Z.-Z., Meyer, F.E. *et al.* (1999) Two new modes of Pd-catalyzed domino-tetracyclization of bromodienynes – 5-*exo*-trig cyclization wins over β-hydride elimination. *Angew. Chem. Int. Ed.*, **38**, 1452–4.

43. Grigg, R., Loganathan, V. and Sridharan, V. (1996) Palladium catalysed cascade alkyne–arene vinylation/alkylation approach to polyfused heterocycles. *Tetrahedron Lett.*, **37**, 3399–402.

44. Brown, D., Grigg, R., Sridharan, V. and Tambyrajah, V. (1995) A palladium catalysed cascade cyclisation–Friedel–Crafts alkylation approach to angularly fused ring systems. *Tetrahedron Lett.*, **36**, 8137–40.

45. Grigg, R. and Sridharan, V. (1992) Palladium catalysed intermolecular cascade cyclisation–cyclopropanation reactions. *Tetrahedron Lett.*, **33**, 7965–8.

46. Ohno, H., Miyamura, K., Takeoka, Y. and Tanaka, T. (2003) Palladium(0)-catalyzed tandem cyclization of allenenes. *Angew. Chem. Int. Ed.*, **42**, 2647–50.

47. García, A., Rodríguez, D., Castedo, L. *et al.* (2001) Synthesis of fused rings at a pivotal nitrogen: tandem Heck reactions of *N*-vinyl-2-iodobenzamides. *Tetrahedron Lett.*, **42**, 1903–5.

48. Hu, Y.-m., Zhou, J., Long, X.-t. *et al.* (2003) Palladium-catalyzed cascade reactions of benzyl halides with *N*-allyl-*N*-(2-butenyl)-*p*-toluenesulfonamide. *Tetrahedron Lett.*, **44**, 5009–10.

49. Lau, S.Y.W. and Keay, B.A. (1999) Remote substituent effects on the enantiomeric excess of intramolecular asymmetric palladium-catalyzed polyene cyclizations. *Synlett*, 605–7.

50. (a) Catellani, M. and Chiusoli, G.P. (1983) The termination step in palladium-catalyzed insertion reactions. *J. Organomet. Chem.*, **250**, 509–15, and references cited therein; (b) Catellani, M., Chiusoli, G.P. and Mari, A. (1984) Palladium- or nickel-catalyzed sequential reaction of organic bromides, bicyclo[2.2.1]hept-2-ene or bicyclo[2.2.1]hepta-2,5-diene and alkynes. *J. Organomet. Chem.*, **275**, 129–38; (c) Catellani, M., Chiusoli, G.P. and Concari, S. (1989) A new palladium catalyzed synthesis of *cis,exo*-2,3-diarylsubstituted bicyclo[2.2.1]heptanes or bicyclo[2.2.1]hept-2-enes. *Tetrahedron*, **45**, 5263–8; (d) Larock, R.C., Hershberger, S.S., Takagi, K. and Mitchell, M.A. (1986) Synthesis and chemistry of alkylpalladium compounds prepared via vinylpalladation of bicyclic alkenes. *J. Org. Chem.*, **51**, 2450–7, and references cited therein; (e) Larock, R.C. and Johnson, P.L. (1989) Palladium-catalysed intermolecular arylation and alkenylation of bicyclic alkenes. *J. Chem. Soc., Chem. Commun.*, 1368–70; (f) Torii, S., Okumoto, H., Ozaki, H. *et al.* (1990) Palladium-catalyzed tandem assembly of norbornene, vinylic halides, and cyanide nucleophile leading to *cis-exo*-2,3-disubstituted norbornanes. *Tetrahedron Lett.*, **31**, 5319–22; (g) Kang, S.-K., Kim, J.-S., Choi, S.-C. and Lim, K.-H. (1998) Three-component coupling reaction: palladium-catalyzed

coupling of norbornadiene and iodonium salts or diazonium salts with organostannanes, alkynes, and sodium tetraphenylborate. *Synthesis*, 1249–51; (h) Kosugi, M., Tamura, H., Sano, H. and Migita, T. (1987) Palladium catalyzed reactions of organic halides with organotin compounds involving insertion of norbornene. Synthesis of 2,3-disubstituted norbornane. *Chem. Lett.*, 193–4; (i) Kosugi, M., Tamura, H., Sano, H. and Migita, T. (1989) Palladium-catalyzed reaction of organic halides with organotin compounds involving olefin insertion: synthesis of 2,3-disubstituted norbornanes. *Tetrahedron*, **45**, 961–7; (j) Kosugi, M., Kimura, T., Oda, H. and Migita, T. (1993) Synthesis of *cis*-olefins via palladium-catalyzed coupling of organic halides, norbornadiene, and organotin compounds. *Bull. Chem. Soc. Jpn.*, **66**, 3522–4; (k) Oda, H., Ito, K., Kosugi, M. and Migita, T. (1994) Palladium-catalyzed ternary coupling reaction of aryl halide, phenyltributyltin, and norbornadiene. Catalyst tailoring. *Chem. Lett.*, 1443–4.

51. (a) Goodson, F.E. and Novak, B.M. (1997) Palladium-mediated soluble precursor route into poly(arylethynylenes) and poly(arylethylenes). *Macromolecules*, **30**, 6047–55; (b) Shaulis, K.M., Hoskins, B.L., Townsend, J.R. and Goodson, F.E. (2002) Tandem Suzuki coupling–norbornadiene insertion reactions. A convenient route to 5,6-diarylnorbornene compounds. *J. Org. Chem.*, **67**, 5860–3.

52. Couty, S., Liégault, B., Meyer, C. and Cossy, J. (2004) Heck–Suzuki–Miyaura domino reactions involving ynamides. An efficient access to 3-(arylmethylene)isoindolinones. *Org. Lett.*, **6**, 2511–4.

53. (a) Couty, S., Liégault, B., Meyer, C. and Cossy, J. (2006) Synthesis of 3-(arylmethylene)-isoindolin-1-ones from ynamides by Heck–Suzuki–Miyaura domino reactions. Application to the synthesis of lennoxamine. *Tetrahedron*, **62**, 3882–95; (b) Couty, S., Liégault, B., Meyer, C. and Cossy, J. (2006) A short synthesis of lennoxamine via ynamides. *Tetrahedron Lett.*, **47**, 767–9.

54. Cheung, W.S., Patch, R.J. and Player, M.R. (2005) A tandem Heck-carbocyclization/Suzuki-coupling approach to the stereoselective syntheses of asymmetric 3,3-(diarylmethylene)indolinones. *J. Org. Chem.*, **70**, 3741–4.

55. Yanada, R., Obika, S., Inokuma, T. *et al.* (2005) Stereoselective synthesis of 3-alkylideneoxindoles via palladium-catalyzed domino reactions. *J. Org. Chem.*, **70**, 6972–5.

56. Lee, C.-W., Oh, K.S., Kim, K.S. and Ahn, K.H. (2000) Suppressed β-hydride elimination in palladium-catalyzed cascade cyclization-coupling reactions: an efficient synthesis of 3-arylmethylpyrrolidines. *Org. Lett.*, **2**, 1213–6.

57. Zhou, C. and Larock, R.C. (2005) Regio- and stereoselective route to tetrasubstituted olefins by the palladium-catalyzed three-component coupling of aryl iodides, internal alkynes, and arylboronic acids. *J. Org. Chem.*, **70**, 3765–77.

58. Salem, B. and Suffert, J. (2004) A 4-*exo*-dig cyclocarbopalladation/8π electrocyclization cascade: Expeditions access to the Tricyclic core structures of the ophiobolins and aleurodiscal. *Angew. Chem. Int. Ed.*, **43**, 2826–30.

59. Bour, C. and Suffert, J. (2006) 4-*exo-dig* cyclocarbopalladation: a straightforward synthesis of cyclobutanediols from acyclic γ-bromopropargylic diols under microwave irradiation conditions. *Eur. J. Org. Chem.*, 1390–5.

60. Bour, C. and Suffert, J. (2005) Cyclocarbopalladation: sequential cyclization and C—H activation/Stille cross-coupling in the Pd–5-*exo-dig* reaction. *Org. Lett.*, **7**, 653–6.

61. Mota, A.J., Dedieu, A., Bour, C. and Suffert, J. (2005) Cyclocarbopalladation involving an unusual 1,5-palladium vinyl to aryl shift as termination step: theoretical study of the mechanism. *J. Am. Chem. Soc.*, **127**, 7171–82.

62. Anwar, U., Fielding, M.R., Grigg, R. *et al.* (2006) Palladium catalysed tandem cyclisation–anion capture processes. Part 8 [1]: in situ and preformed organostannanes. Carbamyl chlorides and other starter species. *J. Organomet. Chem.*, **691**, 1476–87.

63. D'Souza, D.M., Rominger, F. and Müller, T.J.J. (2005) A domino sequence consisting of insertion, coupling, isomerization, and Diels–Alder steps yields highly fluorescent spirocycles. *Angew. Chem. Int. Ed.*, **44**, 153–8.

64. Nylund, C.S., Smith, D.T., Klopp, J.M. and Weinreb, S.M. (1995) A palladium-mediated tandem carbon–carbon bond forming method featuring nucleophilic substitution of intermediate π-allylpalladium complexes produced via the Heck reaction. *Tetrahedron*, **51**, 9301–18.

65. Larock, R.C., Lu, Y.-d., Bain, A.C. and Russell, C.E. (1991) Palladium-catalyzed coupling of aryl iodides, nonconjugated dienes and carbon nucleophiles by palladium migration. *J. Org. Chem.*, **56**, 4589–90.

66. Xie, X. and Lu, X. (1999) Palladium(0)-catalyzed tandem cyclization of N-(2′,4′-dienyl)alkynamides to α-alkylidene-γ-lactams. *Tetrahedron Lett.*, **40**, 8415–8.

67. Larock, R.C., Yang, H., Weinreb, S.M. and Herr, R.J. (1994) Synthesis of pyrrolidines and piperidines via palladium-catalyzed coupling of vinylic halides and olefinic sulfonamides. *J. Org. Chem.*, **59**, 4172–8.

68. Flubacher, D. and Helmchen, G. (1999) Enantioselective domino Heck–allylic amination reactions. *Tetrahedron Lett.*, **40**, 3867–8.

69. (a) Daly, J.W. (2003) Amphibian skin: a remarkable source of biologically active arthropod alkaloids. *J. Med. Chem.*, **46**, 445–52; (b) Daly, J.W., Kaneko, T., Wilham, J. *et al.* (2002) Bioactive alkaloids of frog skin: combinatorial bioprospecting reveals that pumiliotoxins have an arthropod source. *Proc. Natl. Acad. Sci. USA*, **99**, 13996–4001.

70. Dijk, E.W., Panella, L., Pinho, P. *et al.* (2004) The asymmetric synthesis of (−)-pumiliotoxin C using tandem catalysis. *Tetrahedron*, **60**, 9687–93.

71. Pinho, P., Minnaard, A.J. and Feringa, B.L. (2003) The tandem Heck–allylic substitution reaction: a novel route to lactams. *Org. Lett.*, **5**, 259–61.

72. (a) Verbitski, S.M., Mayne, C.L., Davis, R.A. *et al.* (2002) Isolation, structure determination, and biological activity of a novel alkaloid, perophoramidine, from the philippine ascidian *Perophora namei*. *J. Org. Chem.*, **67**, 7124–6; (b) Seo, J.H., Artman, G.D. and Weinreb, S.M. (2006) Synthetic studies on perophoramidine and the communesins: construction of the vicinal quaternary stereocenters. *J. Org. Chem.*, **71**, 8891–900.

73. Artman, G.D. III and Weinreb, S.M. (2003) An approach to the total synthesis of the marine ascidian metabolite perophoramidine via a halogen-selective tandem Heck/carbonylation strategy. *Org. Lett.*, **5**, 1523–6.

74. Copéret, C. and Negishi, E.-i. (1999) Palladium-catalyzed highly diastereoselective cyclic carbopalladation–carbonylative esterification tandem reaction of iododienes and iodoarylalkenes. *Org. Lett.*, **1**, 165–7.

75. Trost, B.M. and Shi, Y. (1992) Cycloisomerization for atom economy. Polycycle construction via tandem transition metal catalyzed electrocyclic processes. *J. Am. Chem. Soc.*, **114**, 791–2.

76. Sugihara, T., Copéret, C., Owczarczyk, Z. *et al.* (1994) Deferred carbonylative esterification in the Pd-catalyzed cyclic carbometalation–carbonylation cascade. *J. Am. Chem. Soc.*, **116**, 7923–4.

77. Mauleón, P., Alonso, I. and Carretero, J.C. (2001) Unusual palladium-catalyzed cascade arylation of α,β-unsaturated phenyl sulfones under Heck reaction conditions. *Angew. Chem. Int. Ed.*, **40**, 1291–3.

78. Alonso, I., Alcamí, M., Mauleón, P. and Carretero, J.C. (2006) Understanding sulfone behavior in palladium-catalyzed domino reactions with aryl iodides. *Chem. Eur. J.*, **12**, 4576–83.

79. Zhao, J. and Larock, R.C. (2005) Synthesis of substituted carbazoles by a vinylic to aryl palladium migration involving domino C—H activation processes. *Org. Lett.*, **7**, 701–4.

80. This type of reaction was first reported by Catellani and coworkers: (a) Catellani, M. and Fagnola, M.C. (1994) Palladacycles as intermediates for selective dialkylation of arenes and

subsequent fragmentation. *Angew. Chem., Int. Ed. Engl.*, **33**, 2421–2; (b) Catellani, M., Frignani, F. and Rangoni, A. (1997) A complex catalytic cycle leading to a regioselective synthesis of *o,o'*-disubstituted vinylarenes. *Angew. Chem., Int. Ed. Engl.*, **36**, 119–22; (c) Catellani, M., Mealli, C., Motti, E. *et al.* (2002) Palladium–arene interactions in catalytic intermediates: an experimental and theoretical investigation of the soft rearrangement between η^1 and η^2 coordination modes. *J. Am. Chem. Soc.*, **124**, 4336–46; (d) Catellani, M. (2003) Catalytic multistep reactions via palladacycles. *Synlett*, 298–313.

81. (a) Pache, S. and Lautens, M. (2003) Palladium-catalyzed sequential alkylation-alkenylation reactions: new three-component coupling leading to oxacycles. *Org. Lett.*, **5**, 4827–30; (b) Alberico, D. and Lautens, M. (2006) Palladium-catalyzed alkylation–alkenylation reactions: rapid access to tricyclic mescaline analogues. *Synlett*, 2969–72.

82. (a) Cárdenas, D.J., Martín-Matute, B. and Echavarren, A.M. (2006) Aryl transfer between Pd(II) centers or Pd(IV) intermediates in Pd-catalyzed domino reactions. *J. Am. Chem. Soc.*, **128**, 5033–40; (b) García-Cuadrado, D., Braga, A.A.C., Maseras, F. and Echavarren, A.M. (2006) Proton abstraction mechanism for the palladium-catalyzed intramolecular arylation. *J. Am. Chem. Soc.*, **128**, 1066–7; (c) García-Cuadrado, D., de Mendoza, P., Braga, A.A.C. *et al.* (2007) Proton-abstraction mechanism in the palladium-catalyzed intramolecular arylation: substituent effects. *J. Am. Chem. Soc.*, **129**, 6880–6.

83. Motti, E., Faccini, F., Ferrari, I. *et al.* (2006) Sequential unsymmetrical aryl coupling of *o*-substituted aryl iodides with *o*-bromophenols and reaction with olefins: palladium-catalyzed synthesis of 6*H*-dibenzopyran derivatives. *Org. Lett.*, **8**, 3967–70.

84. (a) Alberico, D., Paquin, J.-F. and Lautens, M. (2005) Palladium-catalyzed sequential alkylation–alkenylation reactions: application towards the synthesis of polyfunctionalized fused aromatic rings. *Tetrahedron*, **61**, 6283–97; (b) Martins, A., Marquardt, U., Kasravi, N. *et al.* (2006) Synthesis of substituted benzoxacycles via a domino ortho-alkylation/Heck coupling sequence. *J. Org. Chem.*, **71**, 4937–42; (c) Alberico, D., Rudolph, A. and Lautens, M. (2007) Synthesis of tricyclic heterocycles via a tandem aryl alkylation/Heck coupling sequence. *J. Org. Chem.*, **72**, 775–81.

85. Bressy, C., Alberico, D. and Lautens, M. (2005) A route to annulated indoles via a palladium-catalyzed tandem alkylation/direct arylation reaction. *J. Am. Chem. Soc.*, **127**, 13148–9.

86. Mitsudo, K., Thansandote, P., Wilhelm, T. *et al.* (2006) Selectively substituted thiophenes and indoles by a tandem palladium-catalyzed multicomponent reaction. *Org. Lett.*, **8**, 3939–42.

87. (a) Nüske, H., Bräse, S., Kozhushkov, S.I. *et al.* (2002) A new highly efficient three-component domino Heck–Diels–Alder reaction with bicyclopropylidene: rapid access to spiro[2.5]oct-4-ene derivatives. *Chem. Eur. J.*, **8**, 2350–69; (b) Yücel, B., Arve, L. and de Meijere, A. (2005) A two-step four-component queuing cascade involving a Heck coupling, π-allylpalladium trapping and Diels–Alder reaction. *Tetrahedron*, **61**, 11355–73; (c) Knoke, M. and de Meijere, A. (2005) Domino Heck–Diels–Alder reactions of differently substituted cyclopropylallenes. *Eur. J. Org. Chem.*, 2259–68; (d) von Zezschwitz, P. and de Meijere, A. (2006) Domino Heck–pericyclic reactions, in *Topics in Organometallic Chemistry*, Vol. **19** (ed. T.J.J. Müller), Springer-Verlag, Heidelberg, pp. 49–89.

88. Körbe, S., de Meijere, A. and Labahn, T. (2002) Construction of bi- and tricyclic skeletons by domino-*Heck–Diels–Alder* reactions. *Helv. Chim. Acta*, **85**, 3161–75.

89. Knoke, M. and de Meijere, A. (2003) A versatile access to 1-cyclopropyl-2-aryl-1,3,5-hexatrienes – domino Heck–Diels–Alder reactions of 1,3-dicyclopropyl-1,2-propadiene. *Synlett*, 195–8.

90. (a) Bräse, S., Wertal, H., Frank, D. *et al.* (2005) Intramolecular Heck couplings and cycloisomerizations of bromodienes and enynes with 1′,1′-disubstituted methylenecyclopropane terminators: efficient syntheses of [3]dendralenes. *Eur. J. Org. Chem.*, 4167–78;

(b) Storsberg, J., Yao, M.-L., Öcal, N. *et al.* (2005) Palladium-catalyzed, stereoselective rearrangement of a tetracyclic allyl cyclopropane under arylation. *Chem. Commun.*, 5665–6; (c) Schelper, M. and de Meijere, A. (2005) Facile construction of spirocyclopropanated bi-, tri- and tetracyclic skeletons by novel cascades involving intra- and intermolecular Heck reactions of 2-bromo-1,6-enynes and bicyclopropylidene. *Eur. J. Org. Chem.*, 582–92.

91. Suffert, J., Salem, B. and Klotz, P. (2001) Cascade cyclization: carbopalladative cyclization followed by electrocyclic closure as a route to complex polycycles. *J. Am. Chem. Soc.*, **123**, 12107–8.

92. Thiemann, T., Watanabe, M. and Mataka, S. (2001) Areno-annelated estra-1,3,5(10),6,8,11,14,16-octaenes. *New J. Chem.*, **25**, 1104–7.

93. Grigg, R., Millington, M.L. and Thornton-Pett, M. (2002) Spiro-oxindoles via bimetallic [Pd(0)/Ag(I)] catalytic intramolecular Heck–1,3-dipolar cycloaddition cascade reactions. *Tetrahedron Lett.*, **43**, 2605–8.

94. (a) Grigg, R., Sridharan, V. and York, M. (1998) Sequential and cascade olefin metathesis–intramolecular Heck reaction. *Tetrahedron Lett.*, **39**, 4139–42; (b) Grigg, R. and York, M. (2000) Bimetallic catalytic cascade ring closing metathesis–intramolecular Heck reactions using a fluorous biphasic solvent system or a polymer-supported palladium catalyst. *Tetrahedron Lett.*, **41**, 7255–8.

95. Dondas, H.A., Balme, G., Clique, B. *et al.* (2001) Synthesis of heterocycles via sequential Pd/Ru-catalysed allene insertion–nucleophile incorporation–olefin metathesis. *Tetrahedron Lett.*, **42**, 8673–5.

96. Grigg, R., Sridharan, V. and Zhang, J. (1999) Sequential one-pot Rh(I)/Pd(0) catalysed cycloaddition–cyclisation–anion capture. Assembly of polyfunctional compounds. *Tetrahedron Lett.*, **40**, 8277–80.

97. Grigg, R., Hodgson, A., Morris, J. and Sridharan, V. (2003) Sequential Pd/Ru-catalysed allenylation/olefin metathesis/1,3-dipolar cycloaddition route to novel heterocycles. *Tetrahedron Lett.*, **44**, 1023–6.

98. Grigg, R. and Cook, A. (2006) 'Catch and release' cascades: a resin-mediated three-component cascade approach to small molecules. *Tetrahedron*, **62**, 12172–81.

99. (a) Grigg, R., MacLachlan, W. and Rasparini, M. (2000) Palladium catalysed tetramolecular queuing cascades of aryl iodides, carbon monoxide, amines and a polymer supported allene. *J. Chem. Soc., Chem. Commun.*, 2241–2; (b) Grigg, R., Martin, W., Morrisa, J. and Sridharan, V. (2005) Four-component Pd-catalysed cascade/ring closing metathesis. Synthesis of heterocyclic enones. *Tetrahedron*, **61**, 11380–92.

100. (a) Vittoz, P., Bouyssi, D., Traversa, C. *et al.* (1994) A general solution to the synthesis of triquinanes by a palladium catalyzed process. *Tetrahedron Lett.*, **35**, 1871–4; (b) Bruyère, D., Monteiro, N., Bouyssi, D. and Balme, G. (2003) New developments in the Pd-catalysed cyclisation–coupling reaction of alkene-containing carbonucleophiles with organic halides (and triflates). The first examples of asymmetric catalysis. *J. Organomet. Chem.*, **687**, 466–72; (c) Balme, G., Bouyssi, D., Lomberget, T. and Monteiro, N. (2003) Cyclisations involving attack of carbo- and heteronucleophiles on carbon–carbon π-bonds activated by organopalladium complexes. *Synthesis*, 2115–34.

101. Balme, G. and Bouyssi, D. (1991) Total synthesis of the triquinane marine sesquiterpene $(\pm)\Delta^{9(12)}$ capnellene using a palladium-catalyzed bis-cyclization step. *Tetrahedron*, **50**, 403–14.

102. (a) Dyker, G. and Grundt, P. (1996) Annulated ring-systems by domino-Heck–aldol–condensation and domino-Heck–Michael–addition processes. *Tetrahedron Lett.*, **37**, 619–22; (b) Dyker, G. and Markwitz, H. (1998) A palladium-catalyzed domino process to 1-benzazepines. *Synthesis*, 1750–4.

103. Bengtson, A., Larhed, M. and Hallberg, A. (2002) Protected indanones by a Heck–aldol annulation reaction. *J. Org. Chem.*, **67**, 5854–6.

104. Arefalk, A., Larhed, M. and Hallberg, A. (2005) Masked 3-aminoindan-1-ones by a palladium-catalyzed three-component annulation reaction. *J. Org. Chem.*, **70**, 938–42.

105. Yamazaki, K., Nakamura, Y. and Kondo, Y. (2003) Solid-phase synthesis of indolecarboxylates using palladium-catalyzed reactions. *J. Org. Chem.*, **68**, 6011–19.

106. Littke, A.F. and Fu, G.C. (2001) A versatile catalyst for Heck reactions of aryl chlorides and aryl bromides under mild conditions. *J. Am. Chem. Soc.*, **123**, 6989–7000.

107. Chattopadhyay, S.K., Maity, S., Pal, B.K. and Panja, S. (2002) A new (3 + 3) annulation route to isoquinoline-3-carboxylates. *Tetrahedron Lett.*, **43**, 5079–81.

108. Battistuzzi, G., Cacchi, S. and Fabrizi, G. (2002) The aminopalladation/reductive elimination domino reaction in the construction of functionalized indole rings. *Eur. J. Org. Chem.*, 2671–81, and references cited therein.

109. Saulnier, M.G., Frennesson, D.B., Deshpande, M.S. and Vyas, D.M. (1995) Synthesis of a rebeccamycin-related indolo[2,3-a]carbazole by palladium(0) catalyzed polyannulation. *Tetrahedron Lett.*, **36**, 7841–4.

110. Dongol, K.G. and Tay, B.Y. (2006) Palladium(0)-catalyzed cascade one-pot synthesis of isoxazolidines. *Tetrahedron Lett.*, **47**, 927–30.

111. (a) Cvetovich, R.J., Reamer, R.A., DiMichele, L. *et al.* (2006) Unique tandem Heck–lactamization naphthyridinone ring formation between acrylanilides and halogenated pyridines. *J. Org. Chem.*, **71**, 8610–13; (b) Chung, J.Y.L., Cvetovich, R.J., McLaughlin, M. *et al.* (2006) Synthesis of a naphthyridone p38 MAP kinase inhibitor. *J. Org. Chem.*, **71**, 8602–9.

112. Noël, S., Pinel, C. and Djakovitch, L. (2006) Direct synthesis of tricyclic 5*H*-pyrido[3,2,1-ij]quinolin-3-one by domino palladium catalyzed reaction. *Org. Biomol. Chem.*, **4**, 3760–2.

113. (a) Torii, S., Okumoto, H., Ozaki, H. *et al.* (1990) Palladium-catalyzed tandem assembly of norbornanes, vinylic halides, cyanide nucleophile leading to *cis-exo*-2,3-disubstituted norbornanes. *Tetrahedron Lett.*, **31**, 5319–22; (b) Torii, S., Okumoto, H., Ozaki, H. *et al.* (1992) Asymmetric multipoint control by diastereo-differentiative assembly of three components with palladium catalyst. *Tetrahedron Lett.*, **33**, 3499–502.

114. Grigg, R., Santhakumar, V. and Sridharan, V. (1993) Palladium catalysed cascade cyclisation–cyanide ion capture. *Tetrahedron Lett.*, **34**, 3163–4.

115. Pinto, A., Jia, Y., Neuville, L. and Zhu, J. (2007) Palladium-catalyzed enantioselective domino Heck–cyanation sequence: development and application to the total synthesis of esermethole and physostigmine. *Chem. Eur. J.*, **13**, 961–7.

116. Mariampillai, B., Alberico, D., Bidau, V. and Lautens, M. (2006) Synthesis of polycyclic benzonitriles via a one-pot aryl alkylation/cyanation reaction. *J. Am. Chem. Soc.*, **128**, 14436–7.

117. Kojima, A., Takemoto, T., Sodeoka, M. and Shibasaki, M. (1996) Catalytic asymmetric synthesis of halenaquinone and halenaquinol. *J. Org. Chem.*, **61**, 4876–7.

118. Kojima, A., Honzawa, S., Boden, C.D.J. and Shibasaki, M. (1997) Tandem Suzuki cross-coupling–Heck reactions. *Tetrahedron Lett.*, **38**, 3455–8.

119. Wegner, H.A., Scott, L.T. and de Meijere, A. (2003) A new Suzuki–Heck-type coupling cascade: indeno[1,2,3]-annelation of polycyclic aromatic hydrocarbons. *J. Org. Chem.*, **68**, 883–7.

120. Wegner, H.A., Reisch, H., Rauch, K. *et al.* (2006) Oligoindenopyrenes: a new class of polycyclic aromatics. *J. Org. Chem.*, **71**, 9080–7.

121. Grisorio, R., Mastrorilli, P., Nobile, C.F. *et al.* (2005) A novel synthetic protocol for poly(fluorenylenevinylene)s: a cascade Suzuki–Heck reaction. *Tetrahedron Lett.*, **46**, 2555–8.

122. Pottier, L.R., Peyrat, J.-F., Alami, M. and Brion, J.-D. (2004) Unexpected tandem Sonogashira–carbopalladation–Sonogashira coupling reaction of benzyl halides with terminal alkynes: a novel four-component domino sequence to highly substituted enynes. *Synlett*, 1503–8.

123. Clique, B., Vassiliou, S., Monteiro, N. and Balme, G. (2002) Integrated transition metal catalysed reactions: synthesis of polysubstituted 4-(phenoxymethyl)-3-pyrrolines and their isomers

by one-pot coupling of propargylamines, vinyl sulfones (or nitroalkenes) and phenols. *Eur. J. Org. Chem.*, 1493–9.

124. Bossharth, E., Desbordes, P., Monteiro, N. and Balme, G. (2003) Palladium-mediated three-component synthesis of furo[2,3-*b*]pyridones by one-pot coupling of 3-iodopyridones, alkynes, and organic halides. *Org. Lett.*, **5**, 2441–4.

125. Chaplin, J.H. and Flynn, B.L. (2001) A multi-component coupling approach to benzo[*b*]furans and indoles. *Chem. Commun.*, 1594–5.

126. Oppolzer, W. and Gaudin, J.-M. (1987) 134. Catalytic intramolecular palladium–ene reactions. Preliminary communication. *Helv. Chim. Acta*, **70**, 1477–81.

127. Tietze, L.F. and Schirok, H. (1997) Highly efficient synthesis of cephalotaxine by two palladium-catalyzed cyclizations. *Angew. Chem., Int. Ed. Engl.*, **36**, 1124–5.

128. Tietze, L.F. and Nordmann, G. (2001) A novel palladium-catalyzed domino Tsuji–Trost–Heck process for the synthesis of tetrahydroanthracenes. *Eur. J. Org. Chem.*, 3247–53.

129. Tietze, L.F., Redert, T., Bell, H.P. *et al.* (2008) Efficient synthesis of the structural core of tetracyclines by a palladium-catalyzed domino Tsuji–Trost–Heck–Mizoroki reaction. *Chem. Eur. J.*, **14**, 2527–35.

130. (a) Poli, G. and Giambastiani, G. (2002) An epiisopicropodophyllin aza analogue via palladium-catalyzed pseudo-domino cyclization. *J. Org. Chem.*, **67**, 9456–9; (b) Lemaire, S., Prestat, G., Giambastiani, G. *et al.* (2003) Palladium-catalyzed pseudo-domino cyclizations. An easy entry toward polycondensed pyrrolidone derivatives. *J. Organomet. Chem.*, **687**, 291–300; (c) Poli, G., Giambastiani, G. and Pacini, B. (2001) Pd(0)-catalysed allylic alkylation/Heck coupling in domino sequence. *Tetrahedron Lett.*, **42**, 5179–82; (d) Maitro, G., Vogel, S., Prestat, G. *et al.* (2006) Aryl sulfoxides via palladium-catalyzed arylation of sulfenate anions. *Org. Lett.*, **8**, 5951–4.

131. (a) Damayanthi, Y. and Lown, J.W. (1998) Podophyllotoxins: current status and recent developments. *Curr. Med. Chem.*, **5**, 205–52; (b) Zhang, Y. and Lee, K.-H. (1994) Recent progress in the development of novel antitumor etoposide analogs. *Chin. Pharm. J.*, **46**, 319–69; (c) Ramos, A.C., Peláez-Lamamiéde Clairac, R. and Medarde, M. (1999) Heterolignans. *Heterocycles*, **51**, 1443–70; (d) Ward, R.S. (1982) The synthesis of lignans and neolignans. *Chem. Soc. Rev.*, 75–125; (e) Ward, R.S. (1992) Synthesis of podophyllotoxin and related compounds. *Synthesis*, 719–30.

132. Szlosek-Pinaud, M., Diaz, P., Martinez, J. and Lamaty, F. (2003) Palladium-catalyzed cascade allylation/carbopalladation/cross coupling: a novel three-component reaction for the synthesis of 3,3-disubstituted-2,3-dihydrobenzofurans. *Tetrahedron Lett.*, **44**, 8657–9.

133. Okuro, K. and Alper, H. (1997) Palladium-catalyzed carbonylation of *o*-iodophenols with allenes. *J. Org. Chem.*, **62**, 1566–7.

134. Grigg, R., Liu, A., Shaw, D. *et al.* (2000) Synthesis of quinol-4-ones and chroman-4-ones via a palladium-catalysed cascade carbonylation–allene insertion. *Tetrahedron Lett.*, **41**, 7125–8.

135. (a) Grohe, K. (1992) Antibiotics – the new generation including nalidixic acid and fluoroquinolones such as ciprofloxacin. *Chem. Brit.*, **28**, 34–6; (b) Wentland, M.P. and Cornett, J.B. (1985) Quinolone antibacterial agents. *Ann. Rep. Med. Chem.*, **20**, 145–54.

136. Ward, F.E., Garling, D.L., Buckler, R.T. *et al.* (1981) Antimicrobial 3-methylene flavanones. *J. Med. Chem.*, **24**, 1073–7.

137. Hayakawa, Y., Yamamoto, H., Tsuge, N. and Seto, H. (1996) Structure of a new microbial metabolite, neuchromenin. *Tetrahedron Lett.*, **37**, 6363–4.

138. Edmondson, S.D., Mastracchio, A. and Parmee, E.R. (2000) Palladium-catalyzed coupling of vinylogous amides with aryl halides: applications to the synthesis of heterocycles. *Org. Lett.*, **2**, 1109–12.

139. Barluenga, J., Fernández, M.A., Aznar, F. and Valdés, C. (2005) Cascade alkenyl amination/Heck reaction promoted by a bifunctional palladium catalyst: a novel one-pot synthesis of indoles from *o*-haloanilines and alkenyl halides. *Chem. Eur. J.*, **11**, 2276–83.

140. Fayol, A., Fang, Y.-Q. and Lautens, M. (2006) Synthesis of 2-vinylic indoles and derivatives via a Pd-catalyzed tandem coupling reaction. *Org. Lett.*, **8**, 4203–6.

141. Curran, D.P. and Du, W. (2002) Palladium-promoted cascade reactions of isonitriles and 6-iodo-*N*-propargylpyridones: synthesis of mappicines, camptothecins, and homocamptothecins. *Org. Lett.*, **4**, 3215–8.

142. Liehr, J.G., Giovanella, B.C. and Verschraegen, C.F. (eds) (2000) The camptothecins: unfolding their anticancer potential. *Ann. N. Y. Acad. Sci.*, **922**.

143. Pirillo, A., Verotta, L., Gariboldi, P. *et al.* (1995) Constituents of *nothapodytes foetida*. *J. Chem. Soc., Perkin Trans.*, **1**, 583–7.

144. Bom, D., Curran, D.P., Kruszewski, S. *et al.* (2000) The novel silatecan 7-*tert*-butyldimethylsilyl-10-hydroxycamptothecin displays high lipophilicity, improved human blood stability, and potent anticancer activity. *J. Med. Chem.*, **43**, 3970–80.

145. Josien, H., Ko, S.-B., Bom, D. and Curran, D.P. (1998) A general synthetic approach to the (20*S*)-camptothecin family of antitumor agents by a regiocontrolled cascade radical cyclization of aryl isonitriles. *Chem. Eur. J.*, **4**, 67–83.

146. (a) Trend, R.M., Ramtohul, Y.R., Ferreira, E.M. and Stoltz, B.M. (2003) Palladium-catalyzed oxidative Wacker cyclizations in nonpolar organic solvents with molecular oxygen: A stepping stone to asymmetric aerobic cyclizations. *Angew. Chem. Int. Ed.*, **42**, 2892–5; (b) Arai, M.A., Kuraishi, M., Arai, T. and Sasai, H. (2001) A new asymmetric Wacker-type cyclization and tandem cyclization promoted by Pd(II)-spiro bis(isoxazoline) catalyst. *J. Am. Chem. Soc.*, **123**, 2907–8.

147. Netscher, T. (1996) Stereoisomers of tocopherols-syntheses and analytics. *Chimia*, **50**, 563–7.

148. Nelson, T.D. and Meyers, A.I. (1994) The asymmetric Ullmann reaction. 2. The synthesis of enantiomerically pure C$_2$-symmetric binaphthyls. *J. Org. Chem.*, **59**, 2655–8.

149. Tietze, L.F. and Sommer, K. (2005) Synthesis of vitamin E from (3-methyl-3-butenyl)hydroquinone derivatives. Ger. Offen. DE 102004011265.

150. Tietze, L.F., Wilckens, K.F., Yilmaz, S. *et al.* (2006) Synthesis of 2,3-dihydrobenzo[1,4]dioxins and -oxazins via a domino Wacker–Heck reaction. *Heterocycles*, **70**, 309–19.

151. Silva, F., Reiter, M., Mills-Webb, R. *et al.* (2006) Pd(II)-catalyzed cascade Wacker–Heck reaction: chemoselective coupling of two electron-deficient reactants. *J. Org. Chem.*, **71**, 8390–4.

152. Yip, K.-T., Yang, M., Law, K.-L. *et al.* (2006) Pd(II)-catalyzed enantioselective oxidative tandem cyclization reactions. Synthesis of indolines through C—N and C—C bond formation. *J. Am. Chem. Soc.*, **128**, 3130–1.

153. Nemeto, H., Yoshida, M., Fukumoto, K. and Ihara, M. (1999) A novel strategy for the enantioselective synthesis of the steroidal framework using cascade ring expansion reactions of small ring systems – asymmetric total synthesis of (+)-equilenin. *Tetrahedron Lett.*, **40**, 907–10.

154. Toyota, M., Rudyanto, M. and Ihara, M. (2002) Some aspects of palladium-catalyzed cycloalkenylation: developments of environmentally benign catalytic conditions and demonstration of tandem cycloalkenylation. *J. Org. Chem.*, **67**, 3374–86.

155. Liu, G. and Lu, X. (2001) Palladium(II)-catalyzed tandem reaction of intramolecular aminopalladation of allenyl *N*-tosylcarbamates and conjugate addition. *Org. Lett.*, **3**, 3879–82.

156. Zhao, L., Lu, X. and Xu, W. (2005) Palladium(II)-catalyzed enyne coupling reaction initiated by acetoxypalladation of alkynes and quenched by protonolysis of the carbon-palladium bond. *J. Org. Chem.*, **70**, 4059–63.

157. Welbes, L.L., Lyons, T.W., Cychosz, K.A. and Sanford, M.S. (2007) Synthesis of cyclopropanes via Pd(II)/(IV)-catalyzed reactions of enynes. *J. Am. Chem. Soc.*, **129**, 5836–7.

158. Alcaide, B., Almendros, P. and Aragoncillo, C. (2002) Additions of allenyl/propargyl organometallic reagents to 4-oxoazetidine-2-carbaldehydes: novel palladium-catalyzed domino reactions in allenynes. *Chem. Eur. J.*, **8**, 1719–29.
159. Alcaide, B., Almendros, P. and Rodríguez-Acebes, R. (2005) Pd—Cu bimetallic catalyzed domino cyclization of α-allenols followed by a coupling reaction: new sequence leading to functionalized spirolactams. *Chem. Eur. J.*, **11**, 5708–12.
160. Sohn, J.-H., Waizumi, N., Zohng, H.M. and Rawal, V.H. (2005) Total synthesis of mycalamide A. *J. Am. Chem. Soc.*, **127**, 7290–1.

9

Oxidative Heck-Type Reactions (Fujiwara–Moritani Reactions)

Eric M. Ferreira, Haiming Zhang and Brian M. Stoltz

Division of Chemistry and Chemical Engineering, M/C 164–30, California Institute of Technology, 1200 East California Boulevard, Pasadena, CA, USA

9.1 Introduction

Over the past three decades, the Mizoroki–Heck reaction has proven to be one of the most fundamental and versatile metal-catalysed processes for C—C bond formation in organic synthesis [1]. While the reaction provides desirable products containing a C=C double bond by the coupling of an aryl or vinyl halide with an alkene and extrusion of a hydrohalic acid, the overall process involves two discrete functionalization events, namely (1) the formation of an aryl or vinyl halide and (2) the palladium(0)-catalysed C–C bond formation of the reaction partners (Figure 9.1a). One can easily envision a potentially more efficient process involving oxidative coupling of an unfunctionalized arene directly with an alkene, thus obviating the necessity for prehalogenation of the substrate (Figure 9.1b). Although high C—H bond strengths (e.g. benzene: $110 \, \text{kcal} \, \text{mol}^{-1}$) significantly limit the reactivity of arenes toward further functionalization, tremendous effort has been devoted to this research area and significant progress has been achieved, which constitutes the well-known palladium(II)-catalysed direct oxidative coupling of arenes with alkenes (the Fujiwara–Moritani or oxidative Heck reaction) [2].

The Mizoroki–Heck Reaction   Edited by Martin Oestreich
© 2009 John Wiley & Sons, Ltd

a) Mizoroki-Heck Reaction:

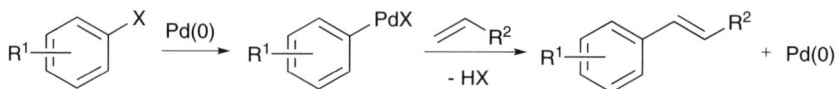

b) Fujiwara-Moritani / Oxidative Heck Reaction

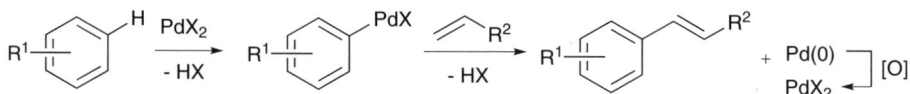

Figure 9.1   *The intermolecular Mizoroki–Heck and the Fujiwara–Moritani reactions.*

9.2 Intermolecular Fujiwara–Moritani Reactions via Electrophilic Palladation

9.2.1 Stoichiometric Fujiwara–Moritani Reactions

The first example of a direct oxidative coupling of arenes with alkenes was described by Fujiwara and Moritani, wherein the double bond of alkenes undergoes substitution reactions with aromatic compounds in the presence of palladium(II) salts [3]. Although stoichiometric and with very low yields, this novel reaction opened a new area of palladium(II)-catalysed reactions between aromatic compounds and alkenes.

Fujiwara and Moritani discovered in 1967 that when styrene–palladium chloride complex **1** was heated in benzene (**2a**), toluene (**2b**) or *p*-xylene (**2c**), in the presence of acetic acid, *trans*-stilbene (**3a**), *trans*-4-methylstilbene (**3b**) or *trans*-2,5-dimethylstilbene (**3c**) were produced respectively, albeit in low yields (Equation (9.1)) [3]. No stilbene derivative was obtained, however, in the case of mesitylene, which may be attributed to the steric hindrance of three methyl groups on the benzene ring [3]. Shortly thereafter, Fujiwara and coworkers [4] found that arylation of styrene (**4a**) occurred much more efficiently in the presence of stoichiometric palladium acetate instead of the styrene–palladium chloride complex (**1**). Thus, equimolar amounts of styrene (**4a**) and palladium acetate were refluxed in benzene (**2a**) in the presence of acetic acid, affording a 90% yield of *trans*-stilbene (**3a**). In the case of toluene (**2b**) and *p*-xylene (**2c**), *trans*-4-methylstilbene (**3b**, 58%) and *trans*-2,5-dimethylstilbene (**3c**, 47%) were obtained (Equation (9.2)).

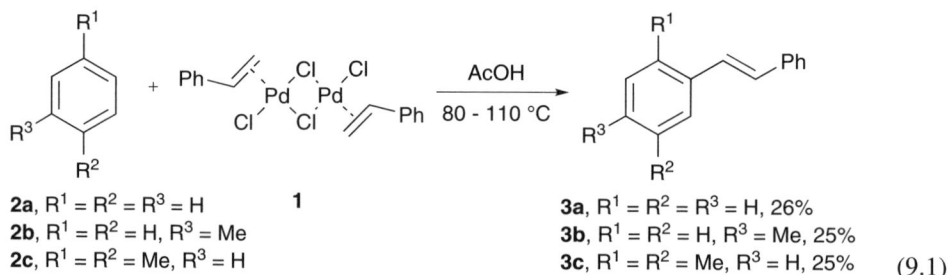

2a, $R^1 = R^2 = R^3 = H$ **1**
2b, $R^1 = R^2 = H$, $R^3 = Me$
2c, $R^1 = R^2 = Me$, $R^3 = H$

3a, $R^1 = R^2 = R^3 = H$, 26%
3b, $R^1 = R^2 = H$, $R^3 = Me$, 25%
3c, $R^1 = R^2 = Me$, $R^3 = H$, 25%   (9.1)

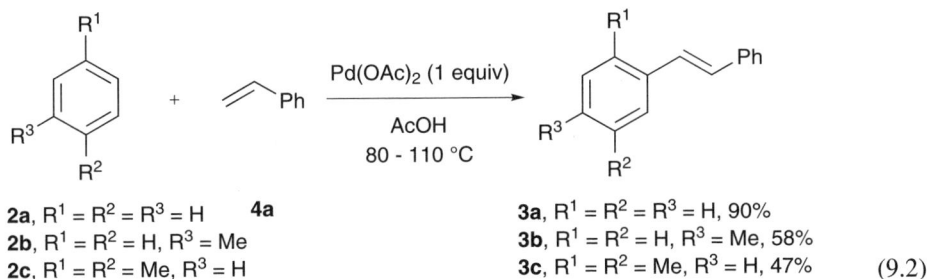

2a, R¹ = R² = R³ = H   **4a**
2b, R¹ = R² = H, R³ = Me
2c, R¹ = R² = Me, R³ = H

3a, R¹ = R² = R³ = H, 90%
3b, R¹ = R² = H, R³ = Me, 58%
3c, R¹ = R² = Me, R³ = H, 47%   (9.2)

The effect of various arene substituents on the oxidative coupling reaction between styrene and monosubstituted benzenes was studied [5]. When a strongly electron-donating group (e.g. a methoxy group) is on the aromatic ring, substitution occurs predominantly at the ortho and para positions. A nitro group, which is electron withdrawing, directs substitution at the meta position. Weakly electron-releasing groups, such as methyl, ethyl or chloride, are moderately ortho and para directing, although considerable amounts of the meta-substituted stilbenes were formed with these arenes. These observed trends are similar to those of electrophilic aromatic substitution reactions. Furthermore, with unsymmetrical alkenes, the aryl group adds predominantly to the less-substituted carbon of the double bond. Increasing substitution on the vinylic carbon atoms decreases the reactivity of the alkene in the palladium(II)-mediated vinylation reaction. For example, the reactions of *trans*-stilbene (**3a**) and triphenylethylene (**5**) with benzene (**2a**) using stoichiometric palladium acetate afforded triphenylethylene (**5**) and tetraphenylethylene (**6**) in only 28% and 13% yields respectively (Equation (9.3)) [4b]. Interestingly, in the case of the reaction of ethylene (**7**) with benzene (**2a**), *trans*-stilbene (**3a**) is formed as a major product, suggesting that styrene derived from the arylation of ethylene undergoes a further arylation reaction with benzene (Equation (9.4)) [4b].

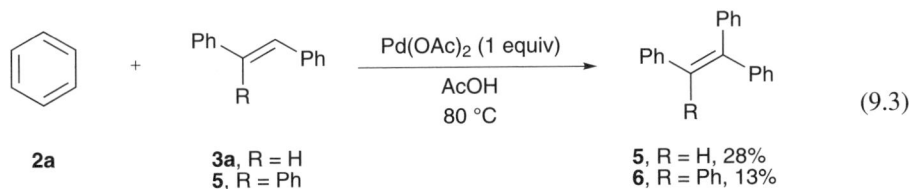

2a

3a, R = H
5, R = Ph

5, R = H, 28%
6, R = Ph, 13%   (9.3)

2a   **7**   **3a**, 14%   **4a**, 6%   (9.4)

The direct alkenylation of aromatics tolerates a variety of alkenes. Besides the alkenes mentioned above, lower alkenes (e.g. 1-butene, propene) [6], alkenes containing polar groups (e.g. acrylonitrile) [4b] and cyclic alkenes such as cyclooctene [7] all participate in the palladium(II)-promoted substitution of arenes. In most cases, however, the yields of these reactions were prohibitively low (less than 20%).

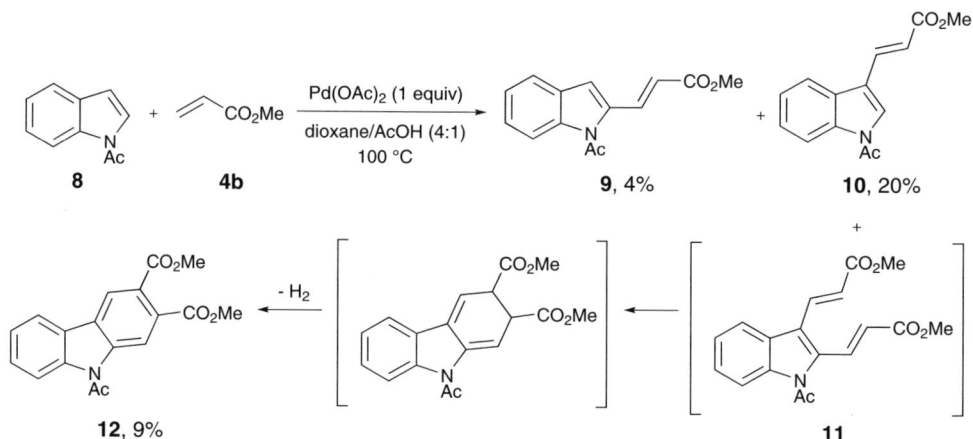

Scheme 9.1

The arene substrates are not limited to simple benzene derivatives. A variety of heteroarenes can also participate in alkene arylations to generate the desired coupling products. Stoichiometric oxidative coupling of aromatic heterocycles such as furan, thiophene, selenophene, *N*-methylpyrrole, benzofuran and benzothiophene with a variety of alkenes, including acrylonitrile, styrene and methyl acrylate, have been extensively studied by Fujiwara and coworkers [8]. Furan, thiophene, selenophene and *N*-methylpyrrole are easily alkenylated with alkenes to give 2-alkenylated and 2,5-dialkenylated heterocycles in relatively low yields (3–46%) [8a], while the reactions of benzofuran and benzothiophene with alkenes produced a mixture of 2- and 3-alkenylated products [8b].

Palladium(II)-mediated oxidative coupling reactions involving the indole nucleus have been studied extensively in the literature. Fujiwara *et al.* [8b] reported that the reaction of *N*-acetylindole (**8**) with methyl acrylate (**4b**) gives (*E*)-methyl 3-(1-acetyl-1*H*-indol-2-yl)acrylate (**9**, 4%) and (*E*)-methyl 3-(1-acetyl-1*H*-indol-3-yl)acrylate (**10**, 20%), along with *N*-acetyl-2,3-bis(methoxycarbonyl)carbazole (**12**, 9%) which was believed to be generated by an electrocyclization and subsequent dehydrogenation of a 2,3-dialkenylated indole intermediate (**11**, Scheme 9.1).

Itahara *et al.* [9] found that *N*-2,6-dichlorobenzoylindole (**13**) was oxidatively coupled with methyl acrylate (**4b**) in the presence of stoichiometric Pd(OAc)$_2$ in acetic acid, affording the 3-alkenylated product **14** in 25% yield. Similarly, *N*-tosylindole (**15**) reacted with ethyl acrylate (**4c**) to generate the 3-alkenylated indole **16** in 48% yield (Scheme 9.2) [10]. Unlike what Fujiwara *et al.* [8b] had observed in the reaction with *N*-acetylindole (**8**) and methyl acrylate (**4b**), both cases did not produce any 2-alkenylated indoles, presumably due to the steric hindrance of the relatively bulky 2,6-dichlorobenzoyl and tosyl groups.

Importantly, one can effect a chemoselective direct alkenylation of aromatic halides without reacting with the carbon–halogen bond [11]. Murakami and coworkers [12] reported a chemoselective C-3 alkenylation of 4-bromo-1-tosylindole (**17**) with methyl 2-*tert*-butoxycarbamoylacrylate (**18**) in the presence of stoichiometric Pd(OAc)$_2$ and 1 equiv chloranil. The reaction smoothly generated 4-bromodehydrotryptophan derivative

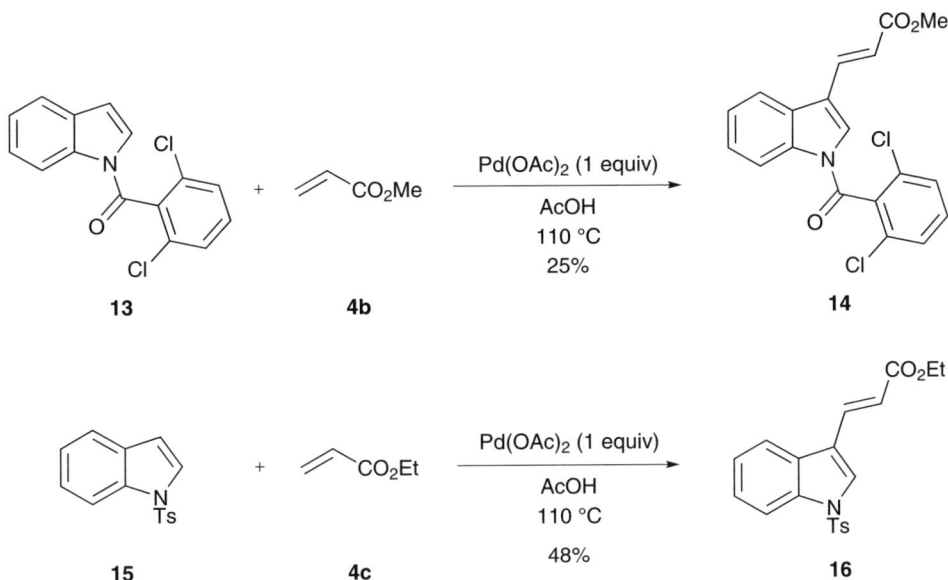

Scheme 9.2

19 in good yield (Equation (9.5)). This method was successfully applied to the total synthesis of optically active ergot alkaloids costaclavine [13], chanoclavine-I [14], DMAT [15] and clavicipitic acid [16].

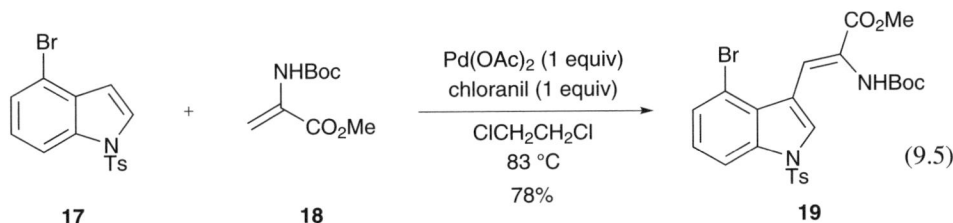

$$(9.5)$$

Interestingly, ferrocene (**20**), a 'nonbenzoid' aromatic system, also underwent substitution reactions with vinyl compounds (e.g. styrene, acrylonitrile and methyl acrylate), albeit in lower yields [17]. For example, ferrocene (**20**) and styrene (**4a**) were oxidatively coupled in the presence of 1 equiv Pd(OAc)$_2$, producing a 20% yield of alkenylated ferrocene derivative **21** (Equation (9.6)).

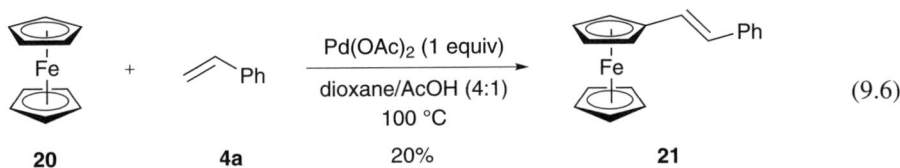

$$(9.6)$$

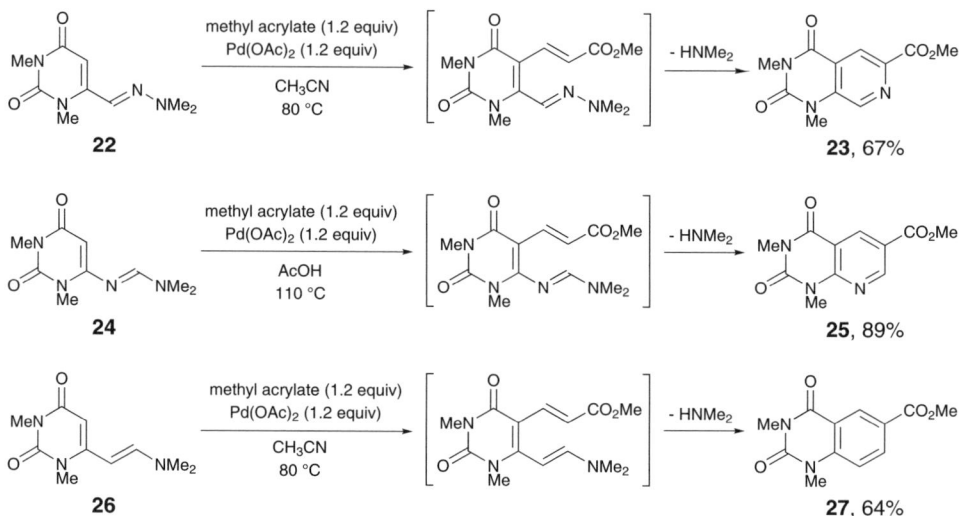

Scheme 9.3

The palladium(II)-assisted alkenylation of aromatic compounds has also been applied to the synthesis of heterocycles. A novel synthesis of pyrido[3,4-*d*] pyrimidines, pyrido[2,3-*d*]pyrimidines and quinazolines was developed by Hirota *et al.* [18] employing the palladium(II)-promoted oxidative coupling of uracil derivatives and alkenes. 1,3-Dimethyluracil-6-carboxaldehyde dimethylhydrazone (**22**), 6-dimethylaminomethylenamino-1,3-dimethyluracil (**24**) and (*E*)-6-(2-dimethylaminovinyl) uracil (**26**) all reacted with methyl acrylate in the presence of stoichiometric Pd(OAc)$_2$, producing pyrido[3,4-*d*]pyrimidine **23**, pyrido[2,3-*d*]pyrimidine **25** and quinazoline **27**, each apparently arising from direct arylation, 6π electrocyclization, and elimination of dimethylamine, in 67%, 89% and 64% yields respectively (Scheme 9.3).

Functionalized benzenes preferentially induced ortho–para substitution with electron-donating groups and meta substitution with electron-withdrawing groups (see above). Additionally, the order of reactivity found with aromatics was similar to that of electrophilic aromatic substitution. These observations implicated an electrophilic metalation of the arene as the key step. Hence, Fujiwara *et al.* [4b] believed that a solvated arylpalladium species is formed from a homogeneous solution of an arene and a palladium(II) salt in a polar solvent via an electrophilic aromatic substitution reaction (Figure 9.2). The alkene then coordinates to the unstable arylpalladium species, followed by an insertion into the aryl–palladium bond. The arylethyl–palladium intermediate then rapidly undergoes β-hydride elimination to form the alkenylated arene and a palladium hydride species, which then presumably decomposes into an acid and free palladium metal. Later on, the formation of the arylpalladium species proposed in this mechanism was confirmed by the isolation of diphenyltripalladium(II) complexes obtained by the C–H activation reaction of benzene with palladium acetate dialkylsulfide systems [19].

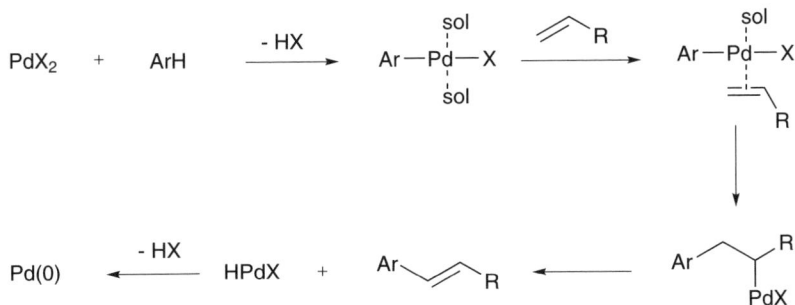

Figure 9.2 The proposed mechanism of the Fujiwara–Moritani reaction.

9.2.2 Catalytic Fujiwara–Moritani Reactions

Based on the proposed mechanism of the stoichiometric alkene arylation, it was anticipated that the same reactions could be achieved using substoichiometric amounts of palladium salts. In the presence of an oxidant, palladium(0) could be reoxidized to palladium(II), which could then reenter the reaction, thus establishing a catalytic cycle for the arene vinylation reactions.

Indeed, Fujiwara and coworkers [4b, 20] discovered that when copper(II) acetate or silver(I) acetate is employed together with oxygen (or air), the palladium-acetate-assisted alkene arylation reaction proceeds catalytically with respect to both palladium and copper (or silver). For example, styrene (**4a**) reacted with benzene (**2a**) in the presence of 10 mol% Pd(OAc)$_2$, 10 mol% Cu(OAc)$_2$ and 50 atm oxygen, producing *trans*-stilbene (**3a**) in 45% yield (Equation (9.7)) [20].

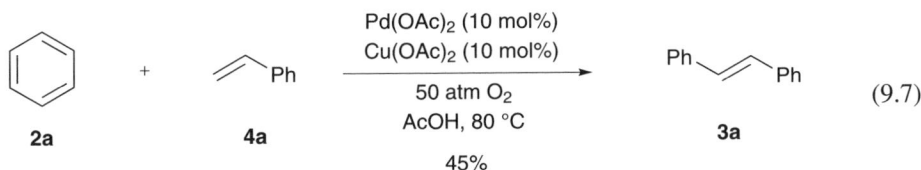

(9.7)

Oxidants other than Cu(OAc)$_2$ and AgOAc have also been examined extensively to reoxidize palladium(0) in the arylation reaction of alkenes. Tsuji and Nagashima [21] reported that *tert*-butyl perbenzoate could be used as the stoichiometric oxidant to facilitate the palladium(II)-catalysed vinylation of benzene. Ishii and coworkers [22] found that the oxidative coupling reaction of substituted benzenes with acrylates could be achieved by the use of a catalytic amount of Pd(OAc)$_2$ combined with molybdovanadophosphoric acid (HPMoV) under atmospheric oxygen as a terminal oxidant.

Fujiwara and coworkers [23] extensively studied the phenylation of ethyl cinnamate (**28**) to produce ethyl 3-phenylcinnamate (**29**) in the presence of catalytic Pd(OAc)$_2$ using a variety of oxidants, such as silver benzoate, manganese dioxide, 30% hydrogen peroxide, *tert*-butyl hydrogen peroxide and the combination of *tert*-butyl hydrogen peroxide and benzoquinone. It was found that the combination of *tert*-butyl hydrogen peroxide and benzoquinone was an effective reoxidation system, with turnover numbers reaching up to 280 (Equation (9.8)).

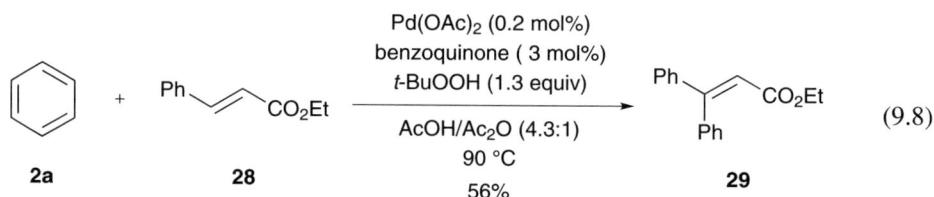

$$\tag{9.8}$$

The first palladium(II)-catalysed C—C bond-forming reaction using oxygen as the sole reoxidant was reported by Shue [24] in 1971. Benzene (**2a**) and styrene (**4a**) were coupled in the presence of Pd(OAc)$_2$ and approximately 20 atm O$_2$ at 100 °C to provide *trans*-stilbene (**3a**). Up to 11 catalytic turnovers were observed under these conditions. More recently, Jacobs and coworkers [25] have described a similar system in the oxidative coupling of benzene derivatives and activated esters. For example, in the presence of Pd(OAc)$_2$ and a cocatalytic amount of benzoic acid under approximately 8 atm O$_2$, anisole (**2d**) was oxidatively coupled with ethyl cinnamate (**28**) to afford a mixture of styrenyl compounds (Equation (9.9)). Notably, the regioselectivity of arylation was highly unselective, suggesting that the standard electrophilic aromatic metalation may not be operative in this case. Under similar conditions, the turnover number and turnover frequency of the catalyst could reach remarkably high values (762 and 73 h^{-1} respectively).

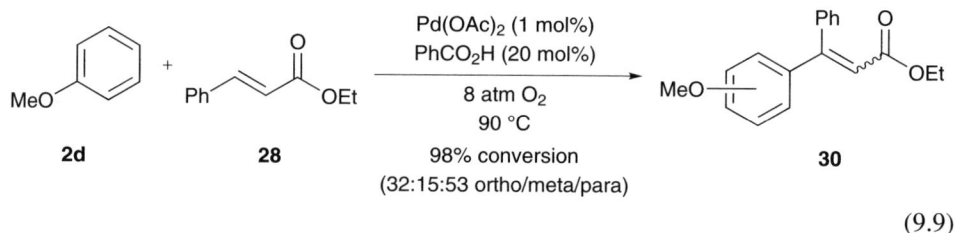

$$\tag{9.9}$$

The palladium(II)-catalysed alkenylation of heterocycles has also been studied extensively. Fujiwara and coworkers [8b, 26] reported that the reactions of furan and thiophene with alkenes such as methyl acrylate and acrylonitrile in the presence of 2 mol% Pd(OAc)$_2$ and 2 equiv of Cu(OAc)$_2$ under atmospheric oxygen or air produced both 2-alkenylated and 2,5-dialkenylated products. This method, however, was not particularly synthetically useful due to the low yields of the reactions (0.3–39%). Tsuji and Nagashima [21] also observed that furans **31a–c** reacted with acrylates **4b** and **4c** to produce monoalkenylated compounds **32a–c**, where the furan has been functionalized solely at the 2-position (or the 5-position if the 2-position is substituted) in moderate to good yields. Even a furan with an electron-withdrawing substituent (2-furaldehyde, **31c**) participated in the oxidative coupling to yield a moderate 34% yield of the arylation product (**32c**, Equation (9.10)).

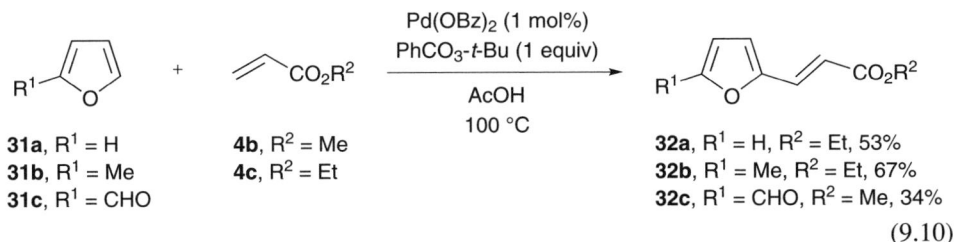

$$\begin{array}{ll}
\textbf{31a, } R^1 = H & \textbf{4b, } R^2 = Me \\
\textbf{31b, } R^1 = Me & \textbf{4c, } R^2 = Et \\
\textbf{31c, } R^1 = CHO &
\end{array}$$

32a, R^1 = H, R^2 = Et, 53%
32b, R^1 = Me, R^2 = Et, 67%
32c, R^1 = CHO, R^2 = Me, 34%

$$(9.10)$$

Palladium(II)-catalysed oxidative couplings involving indoles have been investigated extensively by Itahara and coworkers [9, 10], though these reactions were consistently plagued by low yields. For instance, *N*-2,6-dichlorobenzoylindole (**13**) was oxidatively coupled with methyl acrylate (**4b**) with catalytic Pd(OAc)$_2$ (2 mol%) and a number of stoichiometric oxidants (e.g. AgOAc, Cu(OAc)$_2$, Na$_2$S$_2$O$_8$ and NaNO$_2$) to generate the 3-alkenylated indole [9], but the yield never exceeded 20% in these systems. Alternatively, the oxidative coupling of *N*-tosylindole (**15**) and ethyl acrylate (**4c**) was examined with higher catalyst loadings (10 mol% Pd(OAc)$_2$) [10]. Although the yields of the 3-alkenylated indole product were marginally improved (up to 42% by using 2 equiv Na$_2$S$_2$O$_8$ as the re-oxidant), they were still not particularly useful synthetically. Ricci and coworkers [27] also discovered that substituted indoles can undergo alkenylation reactions at the 3-position of the indole moiety selectively. For example, dimethylcarbamate-substituted indole **33** reacted with methyl acrylate (**4b**) in the presence of a palladium catalyst and benzoquinone, generating the 3-alkenylated indole **34** in 54% yield. Similarly, *N*-benzylindole (**35**) underwent the palladium(II)-catalysed oxidative coupling with methyl acrylate (**4b**), producing 3-alkenylated indole **36** in 95% yield (Scheme 9.4).

Fujiwara and coworkers [23] recently described a single example of a catalytic intermolecular oxidative coupling using the free NH indole nucleus. Under Pd(OAc)$_2$ and a

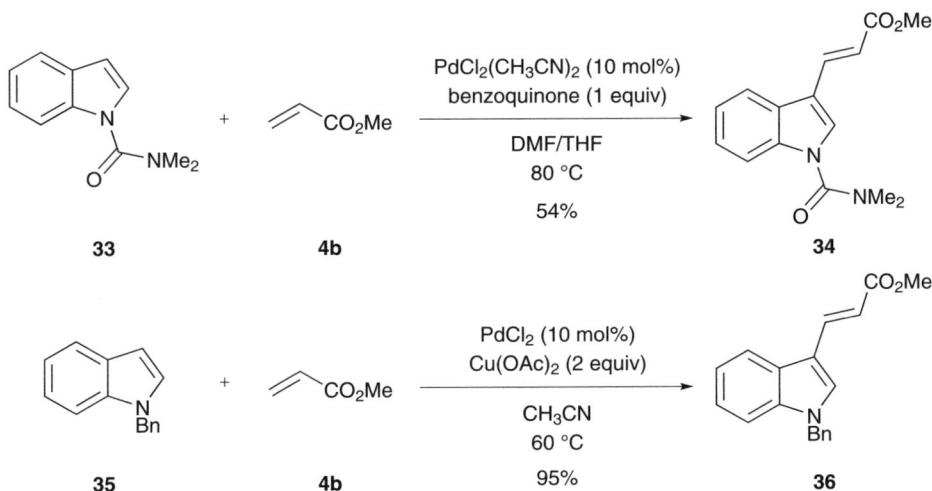

Scheme 9.4

benzoquinone/*tert*-butyl hydrogen peroxide reoxidation system, unsubstituted indole (**37**) was coupled to methyl acrylate (**4b**) to provide methyl 3-indol-3-ylacrylate (**38**) in 52% yield (Equation (9.11)).

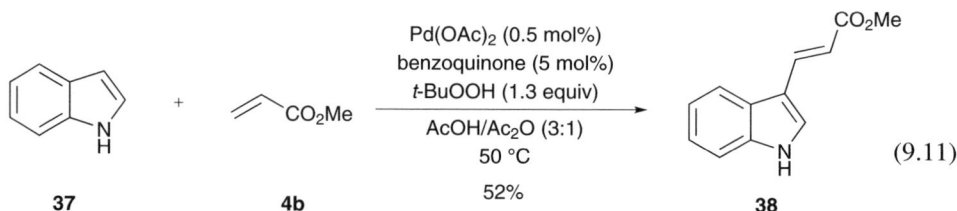

$$(9.11)$$

In all of the aforementioned cases, the palladium(II)-catalysed indole alkenylation occurred preferentially at the 3-position of the indole moiety, which is consistent with the natural reactivity of indole towards electrophilic aromatic substitution reactions. More recently, a palladium(II)-catalysed direct oxidative coupling of indole and alkenes was described by Gaunt and coworkers [28] that exploits a selective, solvent-controlled C—H functionalization of free NH indoles and leads to the substitution of the indole core at either the 2- or the 3-position.

Gaunt and coworkers found that, when the palladium(II)-catalysed coupling reactions of free NH indoles with alkenes were carried out in strongly coordinating solvents such as dimethylsulfoxide (DMSO) and dimethylformamide (DMF) with $Cu(OAc)_2$ as the reoxidant, C(3)-functionalized indoles were produced exclusively. For example, the reaction of indole (**37**) with *n*-butyl acrylate (**4d**) in the presence of 10 mol% $Pd(OAc)_2$ and 1.8 equiv $Cu(OAc)_2$ in DMF/DMSO (10:1) at 70 °C generated exclusively C(3)-functionalized indole **39** in 79% yield (Equation (9.12)). Alternatively, with a weakly coordinating solvent system (3:1 1,4-dioxane/AcOH) and *tert*-butyl perbenzoate as the reoxidant, the reaction provided the C(2)-functionalized indoles as the major or exclusive isomer (7:1 to >19:1 major to minor). For instance, the reaction of indole (**37**) with *n*-butyl acrylate (**4d**) in the presence of 20 mol% $Pd(OAc)_2$ and 0.9 equiv $PhCO_3$-*t*-Bu in 1,4-dioxane/AcOH (3:1) at 70 °C generated C(2)-functionalized indole **40** in 51% yield, along with a small amount (~7%) of the C(3)-alkenylated indole **39** (Equation (9.13)). The yields for the selective C(2)-functionalization of indole were generally lower (34–57%), likely as a result of competitive oxidative decomposition of the starting material and the product under these reaction conditions. Interestingly, the use of *N*-methylindole resulted in no alkenylation at either C(2) or C(3) under these conditions, thus suggesting a potentially crucial role of the free NH indole moiety in these reactions [28].

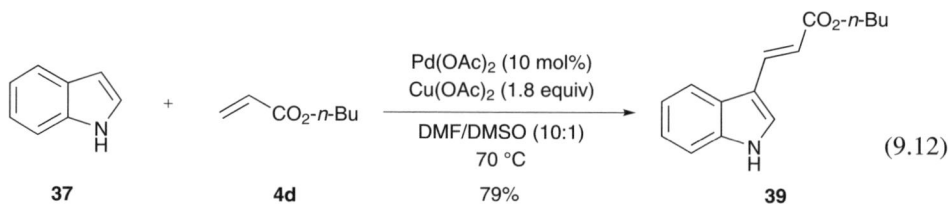

$$(9.12)$$

40, 51%

+

39, 7%

(9.13)

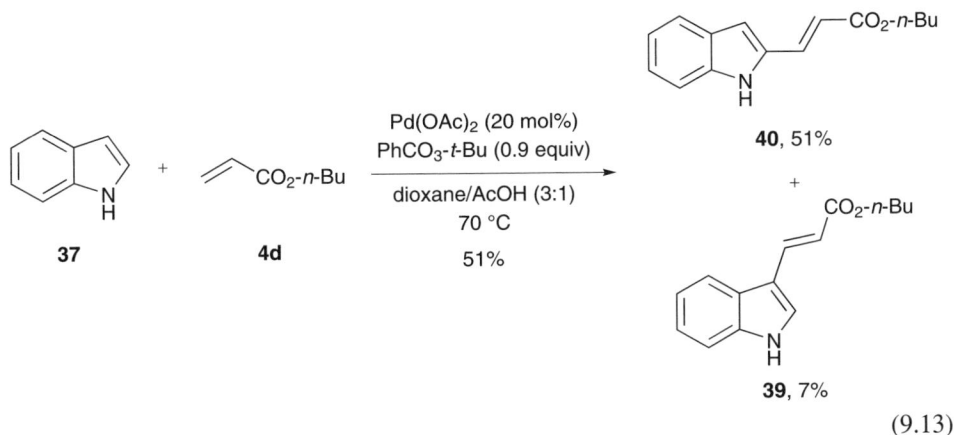

Although much effort has been devoted to the palladium(II)-catalysed C—H functional-ization of indoles, the use of pyrrole in similar processes is surprisingly rare, mainly due to the instability of pyrroles toward acidic and oxidative conditions. Despite the success in their selective alkenylation of indoles, Gaunt and coworkers found that the corresponding pyrrole systems suffered from unselective and multiple alkenylations, as well as signifi-cant polymerization under the conditions employed for selective indole functionalizations. Interestingly, Gaunt and coworkers [29] discovered that the introduction of an electron-withdrawing *N*-protecting group reduced the reactivity of the pyrrole and yielded a more selective process. Accordingly, *N*-Ac, *N*-Ts and *N*-Boc pyrroles **41a–c** reacted with benzyl acrylate (**4e**) to afford only 2-substituted products **42a–c** in good yields in the presence of 10 mol% Pd(OAc)$_2$ and 1.0 equiv PhCO$_3$-*t*-Bu in a dioxane–AcOH–DMSO solvent sys-tem at 35 °C (Scheme 9.5). In contrast, the reaction with *N*-TIPS pyrrole **41d** provided only the 3-alkenylated product **43**. This switch in selectivity is attributed to the sterically demanding nature of the triisopropylsilyl group, which shields the 2-position from reaction with the palladium catalyst, forcing the reactive pyrrole to palladate at C(3). Lastly, the

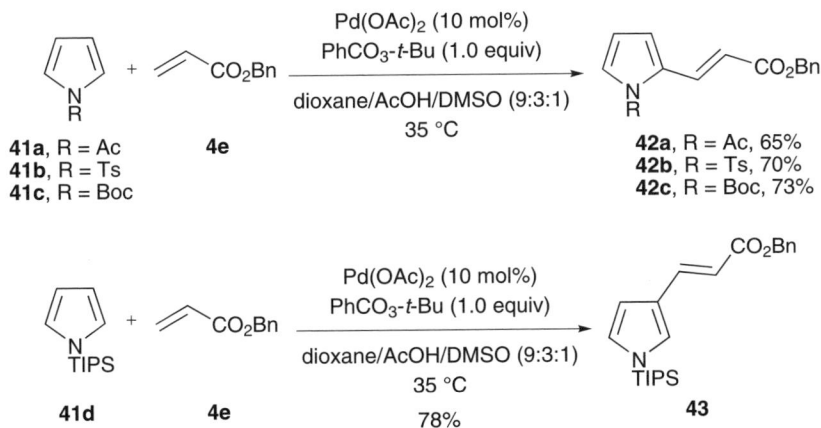

41a, R = Ac   **4e**
41b, R = Ts
41c, R = Boc

42a, R = Ac, 65%
42b, R = Ts, 70%
42c, R = Boc, 73%

41d   **4e**   78%   **43**

Scheme 9.5

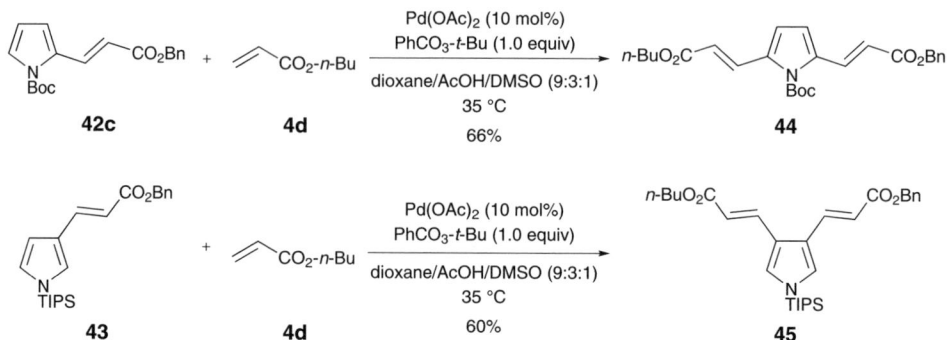

Scheme 9.6

selective vinylation of pyrroles was accomplished in comparable yields by replacing the reoxidant PhCO$_3$-t-Bu with oxygen or air [29].

The substituted pyrrole products can be further differentially elaborated through catalytic regioselective functionalisation. Accordingly, monoalkenylated pyrroles **42c** and **43** underwent selective alkenylation with n-butyl acrylate (**4d**), forming only bisalkenylated pyrroles **44** and **45** respectively in good yields (Scheme 9.6) [29].

The palladium(II)-catalysed direct oxidative coupling reaction was also extended to other heterocycles, such as uracils. Hirota *et al.* [30] reported a simple method for the synthesis of 5-alkenylated uracil derivatives by treatment of 1,3-dimethyluracil (**46**) and 2′,3′-isopropylideneuridine (**48**) with methyl acrylate (**4b**) in the presence of 5 mol% Pd(OAc)$_2$ and 2 equiv *tert*-butyl perbenzoate in refluxing acetonitrile (Scheme 9.7). The reaction successfully generated the corresponding alkenylated heterocycles **47** and **49** in 75% and 46% yields respectively.

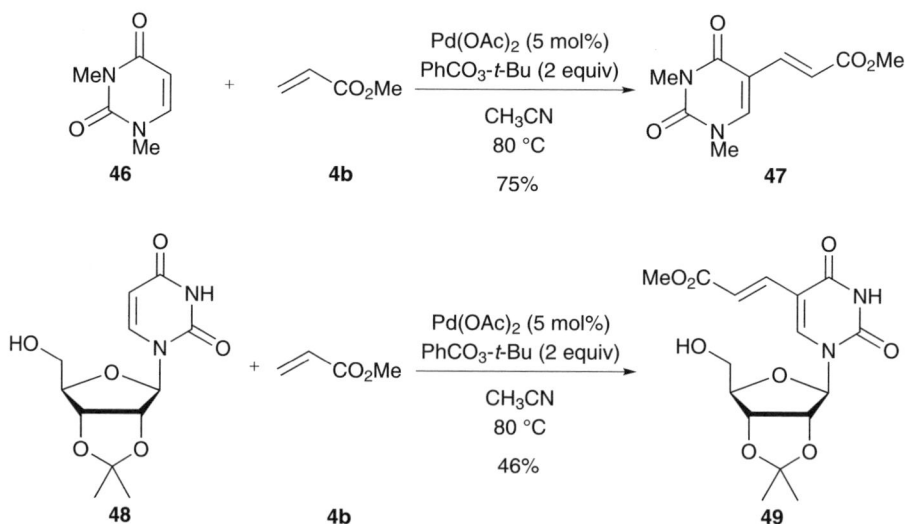

Scheme 9.7

9.2.3 Enantioselective Fujiwara–Moritani Reactions

Compared with the numerous developments of catalytic asymmetric reactions with chiral palladium(0) catalysts [1c,e], catalytic asymmetric reactions by chiral palladium(II) species have so far received only little attention. In fact, the enantioselective Fujiwara–Moritani reaction still remains a significant challenge for organic chemists. Little success has been achieved thus far, presumably because of the inherent nature of the reaction, where styrene-type products absent of chiral centres are typically formed from the *syn-β*-hydride elimination process.

Mikami *et al.* [31] envisioned that an asymmetric Fujiwara–Moritani reaction could be accomplished, however, by using *cyclic* alkenes. In this processs, the endocyclic alkene would insert into the aryl–palladium bond to generate alkyl palladium intermediate **52**, which would then only have one hydrogen available for *syn-β*-hydride elimination. Thus, benzene (**2a**) and cyclohexenecarbonitrile (**50**) were heated at 100 °C in the presence of 10 mol% Pd(OAc)$_2$, 10 mol% chiral sulfonylamino-oxazoline ligand **51**, and *tert*-butyl perbenzoate as the reoxidant, producing phenyl-substituted cyclic alkene **53** in moderate enantiomeric excess (Equation (9.14)). The optimal result was obtained when a chiral ligand with an electron-withdrawing and sterically demanding sulfonyl group was used, which provided **53** in 19% yield and 49% ee. This report represents the first and only example in the literature of an asymmetric Fujiwara–Moritani reaction catalysed by a chiral palladium(II) complex. Although the yields and enantiomeric excesses are modest, this transformation opens a promising area for further discovery and development of enantioselective Fujiwara–Moritani reactions.

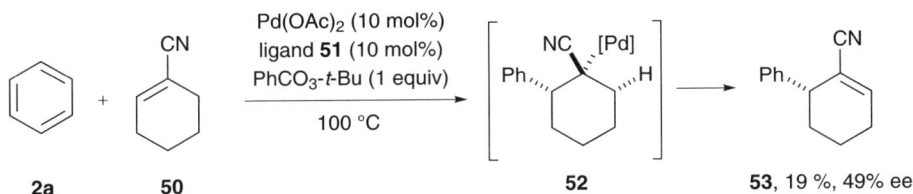

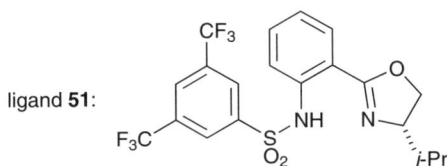

$$(9.14)$$

9.3 Intermolecular Fujiwara–Moritani Reactions via Directing Groups

Unlike the previously discussed Fujiwara–Moritani reactions, which involve the electrophilic palladation of arenes, another category of Fujiwara–Moritani reactions exists that takes advantage of the directing ability of a neighbouring functional group. De Vries and coworkers [32] have reported a selective palladium(II)-catalysed oxidative coupling of anilides with alkenes through C–H bond activation at room temperature. It was found that acetanilide (**54**) reacts with *n*-butyl acrylate (**4d**) in the presence of 2 mol% Pd(OAc)$_2$,

1 equiv benzoquinone as the reoxidant, and 50 mol% *p*-TsOH in toluene/AcOH at room temperature, generating the desired coupling product **55** as a single isomer in 72% yield (Equation (9.15)). This outcome suggests that the acetamide group serves as an *ortho*-directing group, facilitating the selective functionalization at the 2-position by palladium instead of at the 4-position, which would occur if electrophilic aromatic substitution was operative. Similarly, Zaitsev and Daugulis [33] reported a strategy in which palladium(II)-catalysed C—H activation reactions of anilides have been combined with a β-heteroatom elimination process to create a catalytic cycle and achieve the arylation of haloalkenes. For instance, pivaloylanilide (**56**) reacts with methyl *trans*-3-bromoacrylate (**57**) in the presence of 5 mol% PdCl$_2$, 1 equiv AgOTf as an additive in DMF at 90 °C, affording 2-alkenylated product **58** in 85% yield (Equation (9.16)).

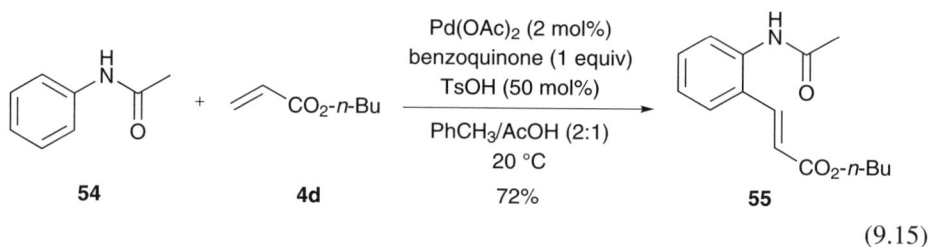

$$(9.15)$$

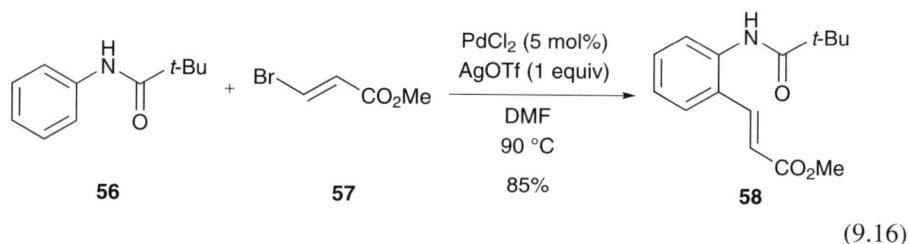

$$(9.16)$$

Literature precedent indicates that electrophilic palladation of indoles usually occurs selectively at the 3-position in the absence of a directing group [9, 10, 23, 27]. Interestingly, *N*-pyridin-2-ylmethylindole (**59**) reacted smoothly with methyl acrylate (**4b**) to give the corresponding 2-indolylethenyl ester (**60**) with complete regiocontrol at the 2-position under conditions where Cu(OAc)$_2$ was present as an oxidant (Equation (9.17)) [27]. This result clearly indicates that the nitrogen atom of the pyridine moiety directs the alkenylation at the 2-position of indole **59**. By comparison, the reaction of *N*-benzylindole (**35**) and methyl acrylate (**4b**) under the same conditions produced only 3-alkenylated indole **36** as the sole coupling product (see above; Scheme 9.4) [27].

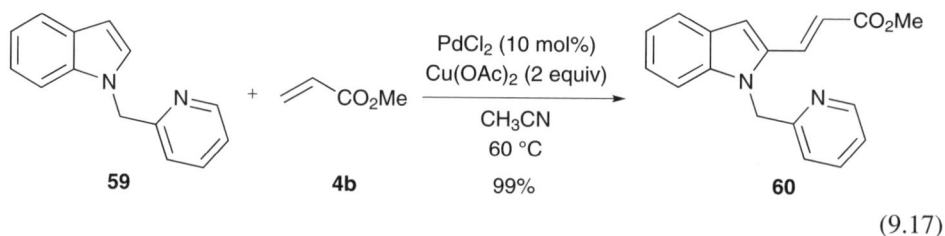

$$(9.17)$$

Neighbouring-group-directed alkenylation of arenes has also been successfully applied to the synthesis of heterocycles by Miura *et al.* [34] using a palladium–copper catalyst system in air. For example, 2-phenylphenol (**61**) readily reacted with ethyl acrylate (**4c**) under these conditions to give 6-substituted-6*H*-dibenzo[*b,d*]pyran **62**, which is believed to be derived from an intramolecular Michael-type addition of the phenol moiety to the double bond in the reaction intermediate (Equation (9.18)). This process has also been extended to *N*-(arylsulfonyl)-2-phenylanilines for the synthesis of dihydrophenanthridine derivatives, and even more remarkably to aromatic carboxylic acids for the preparation of either phthalide and/or isocoumarin derivatives [35]. For example, *N*-(phenylsulfonyl)-2-phenylaniline (**63**) was oxidatively coupled with ethyl acrylate (**4c**), generating dihydrophenanthridine **64** in good yield (Equation (9.19)).

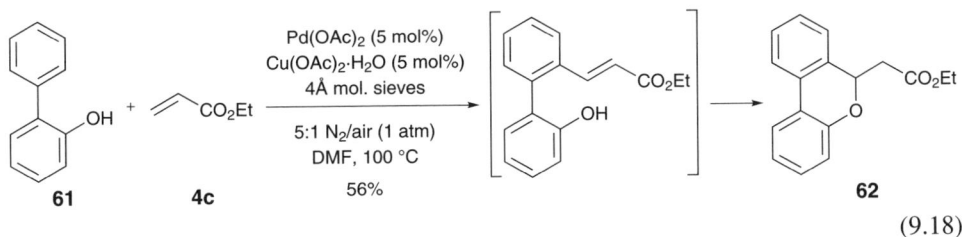

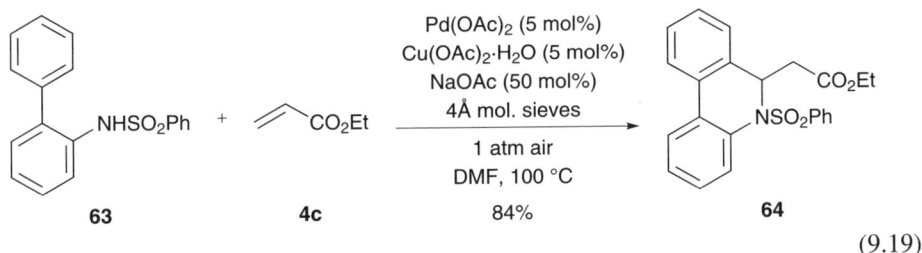

$$(9.18)$$

$$(9.19)$$

Similarly, benzoic acid (**65**) underwent the palladium(II)-catalysed alkenylation and subsequent cyclization to produce phthalide **66**, whereas the reaction between 4-methoxybenzoic acid (**67**) and styrene (**4a**) generated a mixture of isocoumarin **68** and phthalide **69** (Scheme 9.8). In all cases, the substitution was completely regioselective,

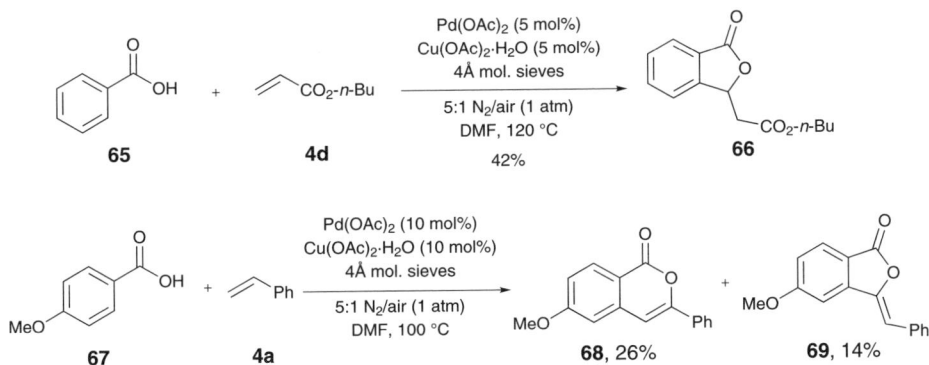

Scheme 9.8

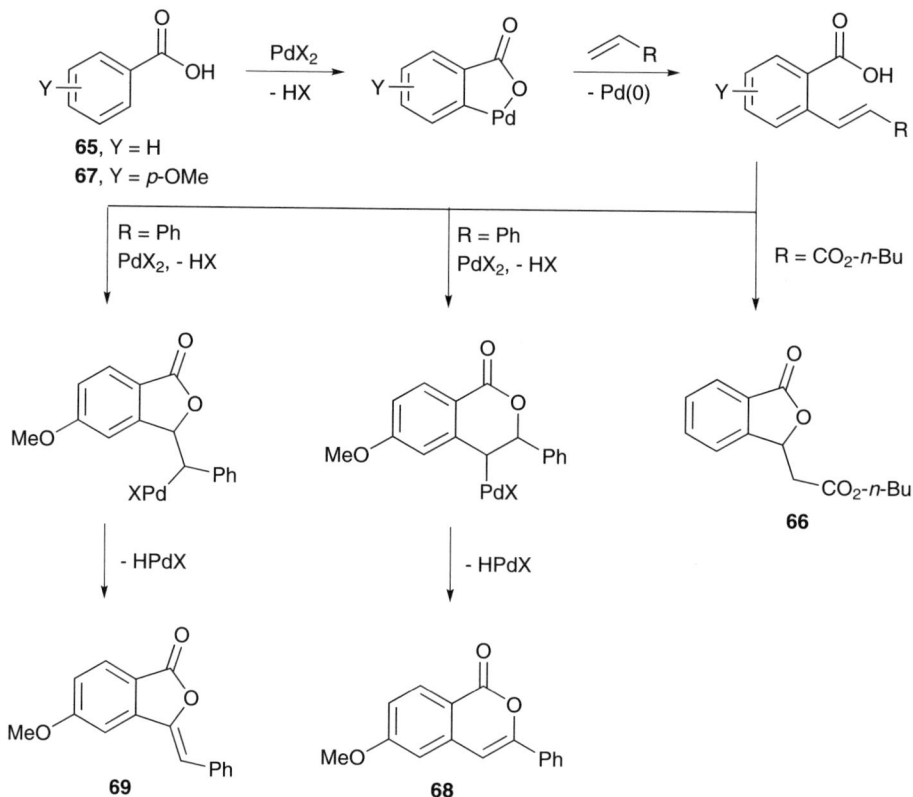

Scheme 9.9

indicative of the important role played by the neighbouring directing group in the initial C–H bond functionalization.

As illustrated in Scheme 9.9, the proposed mechanism of forming phthalides and isocoumarins involves orthopalladation of the carboxylic acid, subsequent alkenylation and nucleophilic cyclization or Wacker-type oxidative cyclization. The observed differences in product distributions from the reactions using *n*-butyl acrylate and styrene suggest that the electronic nature of the alkene plays an important role in product formation, although the exact origin of the difference remains unclear.

9.4 Intramolecular Fujiwara–Moritani Reactions

In addition to the several reports of intermolecular Fujiwara–Moritani reactions, there have been a number of examples of intramolecular reactions, both stoichiometric and catalytic in palladium. In considering an intramolecular oxidative Heck reaction, one can again draw a direct analogy to the classical Heck reaction (Figure 9.3). In the standard Heck reaction, a halogenated arene undergoes an oxidative addition by palladium(0), followed by alkene insertion and β-hydride elimination. In an oxidative version, a C—H bond of

Intramolecular Heck Reaction:

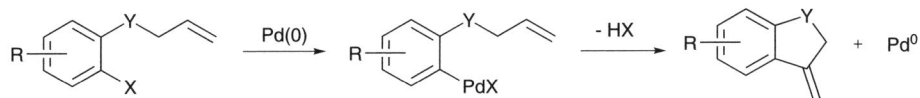

Intramolecular Oxidative Heck Reaction:

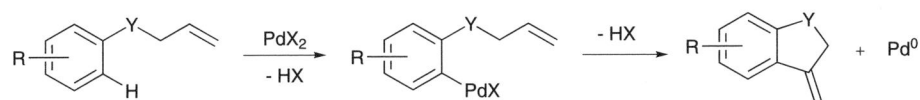

Figure 9.3 *The intramolecular Mizoroki–Heck and oxidative Heck (Fujiwara–Moritani) reactions.*

an arene is directly functionalized by palladium(II), leading to a similar aryl–palladium intermediate. The reaction then proceeds in the same fashion as the Heck reaction. Much like the intermolecular counterpart, this oxidative coupling could again be considered more efficient relative to the classical intramolecular Mizoroki–Heck reaction, circumventing the need for prehalogenation of the arene substrate.

9.4.1 Intramolecular Fujiwara–Moritani Reactions Stoichiometric in Palladium

The first example of an intramolecular Fujiwara–Moritani arylation was reported by Norman and coworkers [36] in 1970. Arenes **70a** and **70b** were treated with a stoichiometric amount of $Pd(OAc)_2$ in AcOH at 80 °C to afford indenes **71a** and **71b** in 54% and 65% yields respectively (Scheme 9.10). The authors proposed two Friedel–Crafts-type

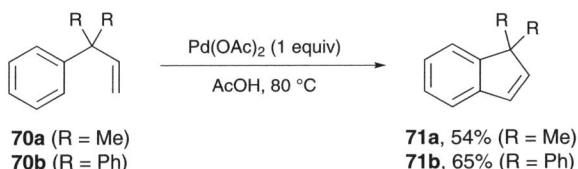

70a (R = Me)
70b (R = Ph)

71a, 54% (R = Me)
71b, 65% (R = Ph)

Mechanism 1

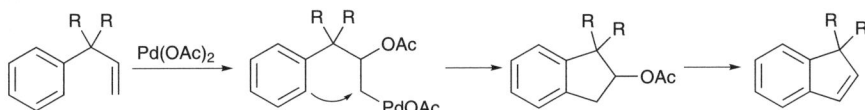

Mechanism 2

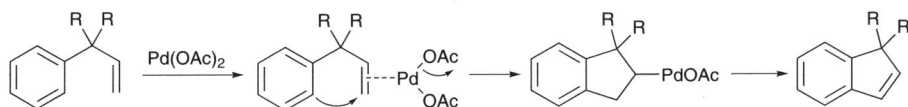

Scheme 9.10

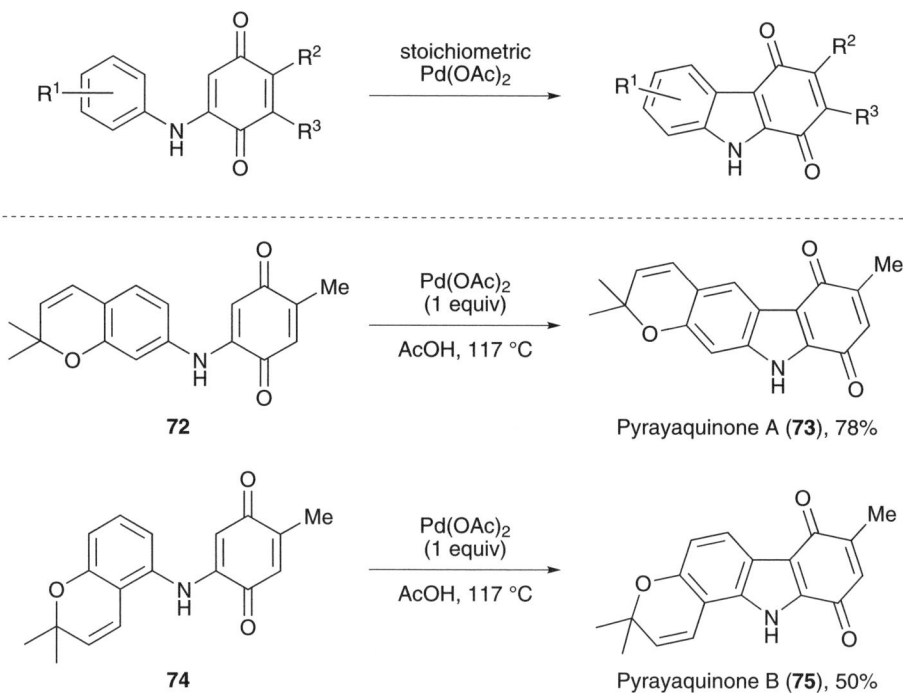

Scheme 9.11

mechanisms for this interesting transformation: (1) acetoxypalladation followed by nu-cleophilic aromatic displacement and subsequent elimination and (2) arene attack on a palladium-coordinated alkene followed by β-hydride elimination. Observations were most consistent with the second mechanism, but the first mechanism could not be ruled out completely. An alternative pathway involving direct palladation of the arene followed by alkene insertion was not considered at the time, but, based on later examples of this type of reaction, is likely the operative mechanism.

Since that initial example of an intramolecular oxidative Heck reaction using sto-ichiometric palladium salts, there have been relatively few applications described in the literature. One notable exception is the use of palladium(II)-mediated cyclizations of anilinoquinones to construct carbazole-1,4-quinones (Scheme 9.11) [37]. Furukawa *et al.* [37a] first described the oxidative cyclization of anilinoquinones **72** and **74** to form pyrayaquinone A (**73**) and pyrayaquinone B (**75**) respectively.

A number of studies on the palladium(II)-mediated oxidative cyclization of anilino-quinones later appeared. Some of the compounds produced via this protocol are depicted in Figure 9.4. Bittner *et al.* [37b] and Furukawa and coworkers [37c] both described the application of the intramolecular cyclization chemistry toward the synthesis of analogues of the carbazole-1,4-quinone alkaloids. Furukawa and coworkers [37c] also reported the synthesis of murrayaquinone A (**79**) using this chemistry. Knölker and O'Sullivan [37d,e] later demonstrated the utility of the palladium(II)-mediated cyclization in the synthesis of **83**, which was initially anticipated to be a prekinamycin analogue precursor. In all

Bittner

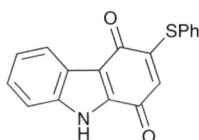 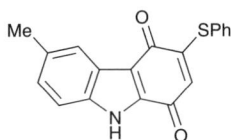 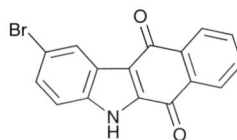

76, 66%   **77**, 94%   **78**, 40%

Furukawa

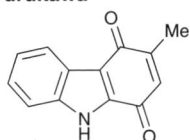 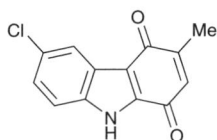 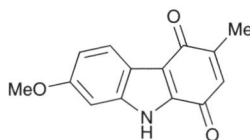 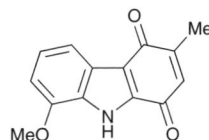

Murrayaquinone A (**79**), 64%   **80**, 46%   **81**, 65%   **82**, 55%

Knölker

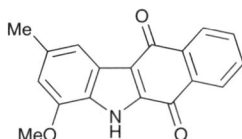

83, 84%

Figure 9.4   *Carbazole-1,4-quinones accessed via intramolecular oxidative Heck cyclizations using stoichiometric palladium.*

cases, stoichiometric amounts of Pd(OAc)$_2$ in refluxing acetic acid were used to effect the oxidative cyclization.

Trost and coworkers [38] described an interesting variant of the Fujiwara–Moritani ox-idative cyclization in his synthesis of ibogamine (Scheme 9.12). Compound **84** was treated with a stoichiometric amount of Pd(CH$_3$CN)$_2$Cl$_2$ and AgBF$_4$ to access alkyl palladium species **85**, which was intercepted by *in situ* reduction with NaBH$_4$ to afford ibogamine (**86**) in modest yield. Deuteration studies helped to elucidate the mechanism of this reaction. By reducing the alkyl palladium intermediate with NaBD$_4$, stereoisomer **87** was obtained exclusively. Since the reduction of an alkylpalladium species proceeds with retention of configuration, this outcome is consistent with a mechanism involving indole palladation followed by alkene insertion. A mechanism involving alkene coordination by palladium, followed by backside nucleophilic attack by the indole moiety, would result in the opposite relative configuration.

Trost *et al.* [39] later reported the formal total synthesis of (±)-catharanthine, applying a similar reaction to indole **88** (Scheme 9.13). Although **89** was obtained in modest yield, it could be readily converted in three steps to alcohol **90**, which Büchi and coworkers [40] have shown can be further transformed to catharanthine. Trost and Fortunak [41] also investigated this reaction by studying structural derivatives without the isoquinuclidine moiety. Indole **92a** cyclized efficiently under carefully optimized conditions (1.1 equiv

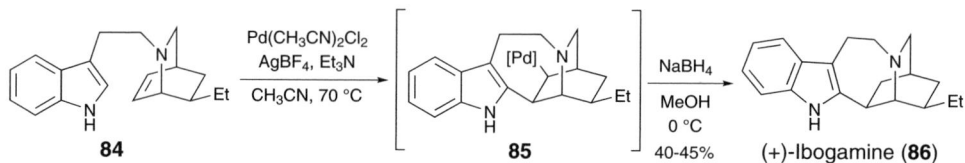

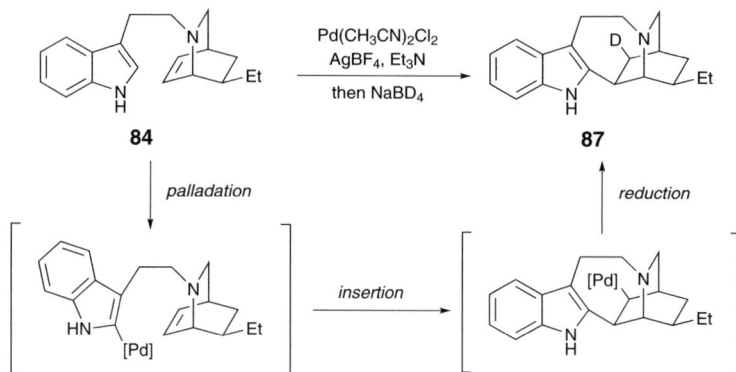

Scheme 9.12

PdCl$_2$, 1.1 equiv AgBF$_4$, room temperature, then NaBH$_4$) to provide **93a**. The *N*-methyl derivative **92b** also cyclized cleanly, ruling out a mechanism involving *N*-palladation followed by rearrangement to a C(2)-palladated species. Indole **94**, with one additional carbon in the linker between the indole and the bicyclooctane moiety, cyclized to form both **95** and **96**. Oxindole **96** likely arose from palladation–cyclization at C(4), followed by air oxidation. Lastly, indole **97**, featuring an exocyclic alkene, underwent the palladium(II)-mediated cyclization to afford compound **98** in good yield.

Williams and coworkers [42] utilized a similar palladium(II)-mediated oxidative cyclization/*in situ* reduction protocol in the total synthesis of (+)-paraherquamide B (**101**, Scheme 9.14). Indole **99** was treated with 1.2 equiv of PdCl$_2$ and 2 equiv of AgBF$_4$ to presumably form an alkylpalladium species, which was reduced with NaBH$_4$ to ultimately provide **100** in good yield. This compound was further transformed to paraherquamide B in six steps. Williams *et al.* [43] later described the total synthesis of paraherquamide A utilizing the palladium(II)-mediated cyclization on a structurally similar compound.

In 2002, Baran and Corey [44] described the total synthesis of (+)-austamide and its naturally occurring relatives (+)-deoxyisoaustamide and (+)-hydratoaustamide (Scheme 9.15). The efficient synthesis of these compounds featured a palladium(II)-mediated cyclization to form an eight-membered ring as the key step. Indole **102** was treated with stoichiometric palladium acetate in a 1 : 1 : 1 mixture of acetic acid, water and tetrahydrofuran (THF) to produce dihydroindoloazocine **103** in 29% yield. A unique mechanism, illustrated in Scheme 9.15, was proposed for this transformation. After palladation of the indole at C(2), 7-*exo* cyclization provides intermediate **108**, which forms cationic intermediate **109** upon acid-mediated heterolysis. This intermediate undergoes a deprotonation and subsequent migration of the indolyl species to afford product **103**. A possible alternative mechanism

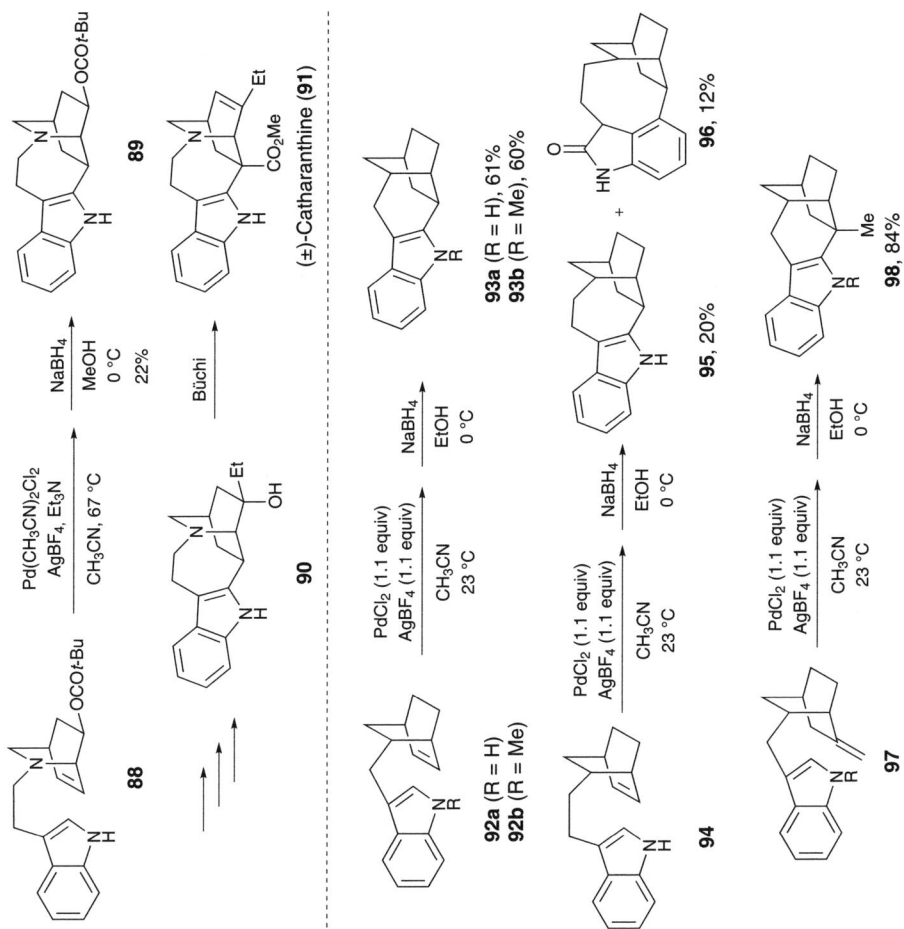

OCO-t-Bu

88

Pd(CH₃CN)₂Cl₂
AgBF₄, Et₃N
CH₃CN, 67 °C

OCO-t-Bu

89

NaBH₄
MeOH
0 °C
22%

Et
OH

90

Büchi

Et
CO₂Me

(±)-Catharanthine (**91**)

92a (R = H)
92b (R = Me)

PdCl₂ (1.1 equiv)
AgBF₄ (1.1 equiv)
CH₃CN
23 °C

NaBH₄
EtOH
0 °C

93a (R = H), 61%
93b (R = Me), 60%

94

PdCl₂ (1.1 equiv)
AgBF₄ (1.1 equiv)
CH₃CN
23 °C

NaBH₄
EtOH
0 °C

95, 20%

+

O
HN

96, 12%

97

PdCl₂ (1.1 equiv)
AgBF₄ (1.1 equiv)
CH₃CN
23 °C

NaBH₄
EtOH
0 °C

Me

98, 84%

Scheme 9.13

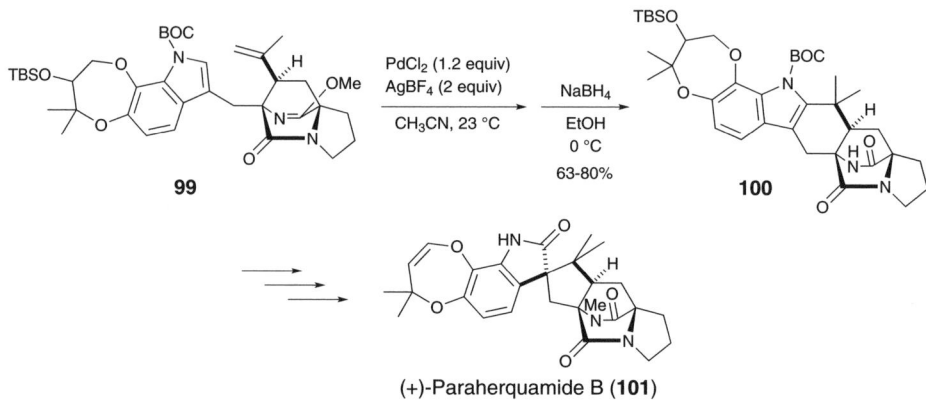

(+)-Paraherquamide B (**101**)

Scheme 9.14

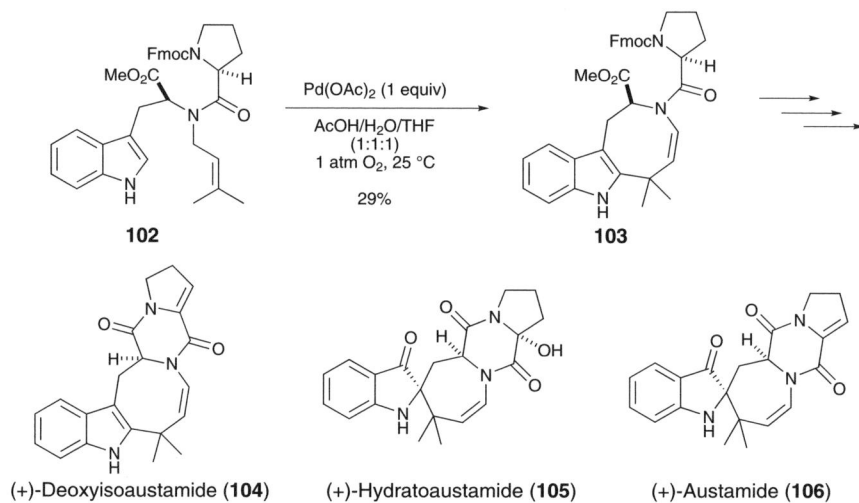

(+)-Deoxyisoaustamide (**104**)   (+)-Hydratoaustamide (**105**)   (+)-Austamide (**106**)

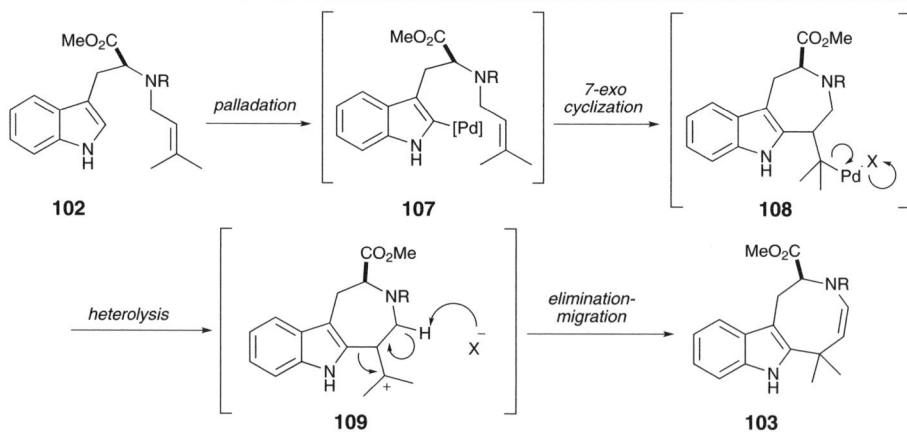

Scheme 9.15

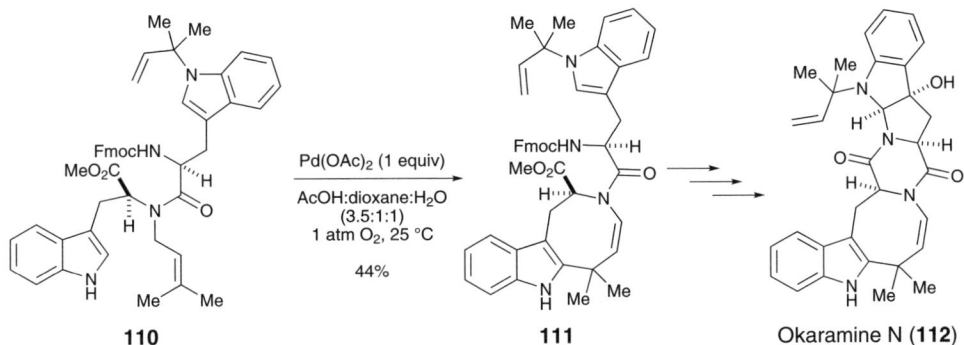

Scheme 9.16

more aligned with the Fujiwara–Moritani reaction involving an 8-*endo* cyclization followed by β-hydride elimination was not addressed but cannot be ruled out for this intriguing transformation.

Corey and coworkers [45] later described the application of this reaction in the total synthesis of okaramine N (Scheme 9.16). Bisindole **110** was oxidatively cyclized using stoichiometric Pd(OAc)$_2$ to form compound **111** in 38% yield. Although the yield was modest, the dihydroindoloazocine product could be elaborated in just two steps to afford okaramine N. This remarkably rapid synthesis highlights the efficacy of the palladium(II)-mediated oxidative Heck reaction to construct complex ring systems that can immediately expedite natural product synthesis.

In 2004, Stoltz and coworkers [46] reported the first total synthesis of dragmacidin F using a palladium(II)-mediated oxidative Heck cyclization as a pivotal transformation (Scheme 9.17). Acyl pyrrole **113**, accessed from (−)-quinic acid in six steps, was treated with 1 equiv Pd(OAc)$_2$ and 2 equiv DMSO in *t*-BuOH/AcOH to afford [3.3.1] bicycle **114** in 74% yield. Interestingly, when the brominated analogue **116** was subjected to a traditional intramolecular Heck cyclization, the reaction was complicated by formation of [3.2.2] bicycle **117**, arising from alkene insertion at the other carbon of the alkene. The highly selective oxidative cyclization was crucial to the success of the synthesis and is demonstrative of how the different catalyst systems for the classical Heck and the oxidative Heck chemistry can have surprising effects on the outcomes of reactions. Efforts to effect this transformation with catalytic amounts of palladium were met with limited success (37% yield with 20 mol% Pd(OAc)$_2$ and 1 atm oxygen as a stoichiometric oxidant).

9.4.2 Intramolecular Fujiwara–Moritani Reactions Catalytic in Palladium

Much like the initial investigations into intramolecular oxidative Heck reactions utilizing stoichiometric palladium salts, much of the early work toward catalytic variants focused on cyclizations of anilinoquinones (Scheme 9.18) [37e, 47]. Knölker and O'Sullivan [37e] reported the first example of an intramolecular oxidative Heck cyclization catalytic in

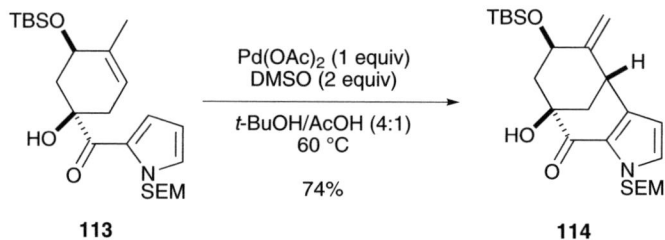

113   **114**

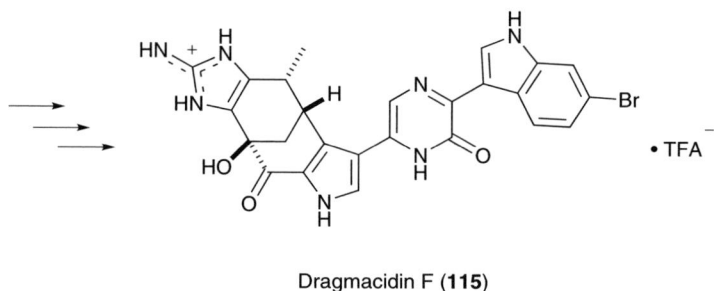

Dragmacidin F (**115**)

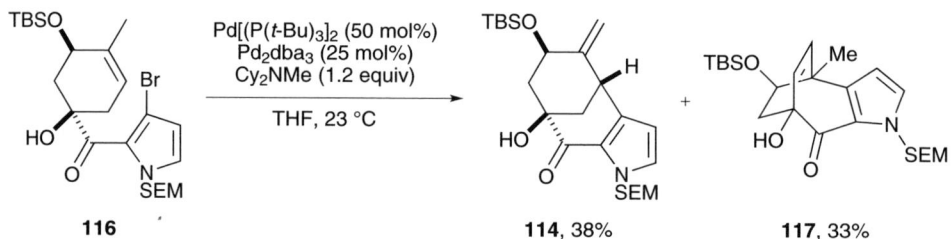

116   **114**, 38%   **117**, 33%

Scheme 9.17

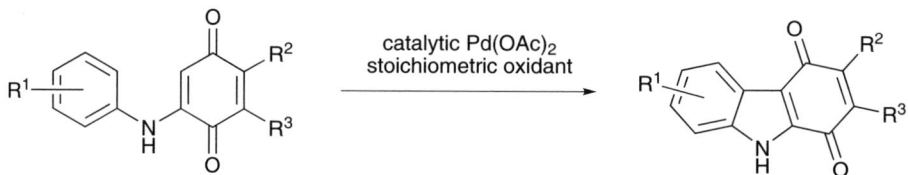

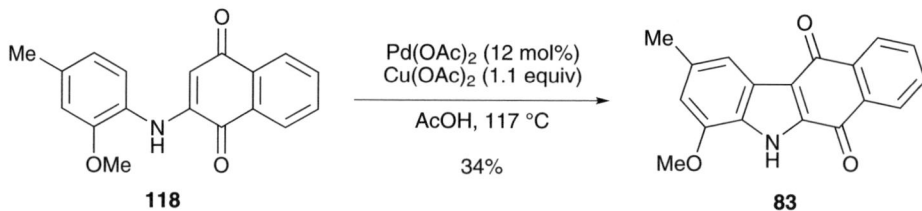

118   **83**

Scheme 9.18

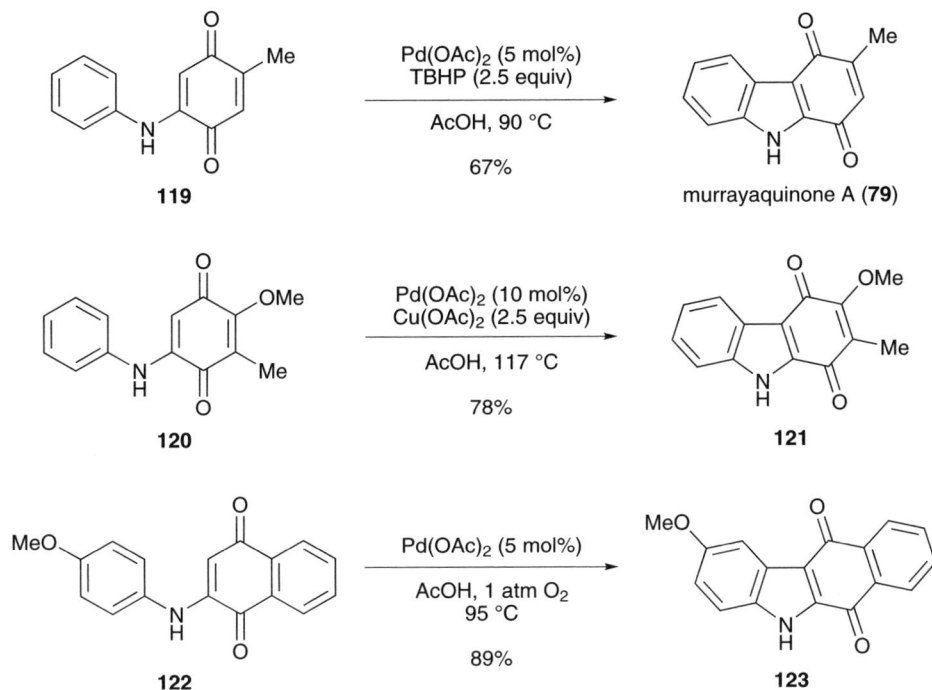

Scheme 9.19

palladium in furthering their efforts toward the syntheses of prekinamycin analogue precursors. Using Cu(OAc)$_2$ as the stoichiometric reoxidant, **118** was cyclized to afford carbazole-1,4-quinone **83**. Although the yield was modest, this result was very significant, as it demonstrated the possibility of an intramolecular oxidative Heck reaction using palladium in substoichiometric amounts.

Since that initial report, a number of cyclizations on similar compounds have been described (Scheme 9.19). Åkermark *et al.* [47a] illustrated that *tert*-butyl hydroperoxide can be used as a stoichiometric reoxidant in these reactions in their syntheses of murrayaquinone A (**79**) and other structural analogues. Knölker *et al.* [47c] demonstrated that the Pd(OAc)$_2$/Cu(OAc)$_2$ oxidation system can be utilized to access carbazole-1,4-quinone **121** in good yield, which was used as an intermediate in the synthesis of carbazoquinocin C. In 1999, Åkermark and coworkers [47d] reported an intriguing example of an intramolecular oxidative Heck reaction of anilinoquinone species **122** using molecular oxygen as the sole stoichiometric oxidant, affording **123** in 89% yield.

In 2003, Ferreira and Stoltz [48] described an aerobic palladium(II)-catalysed intramolecular oxidative Heck reaction to form annulated indoles. The core annulated indole structure is prevalent in a number of biologically active natural products, including paxilline, penitrem A, and yuehchukene (Figure 9.5). Because indoles are susceptible to oxidative decomposition, carefully optimized conditions had to be developed that were compatible with the substrate and product indole species.

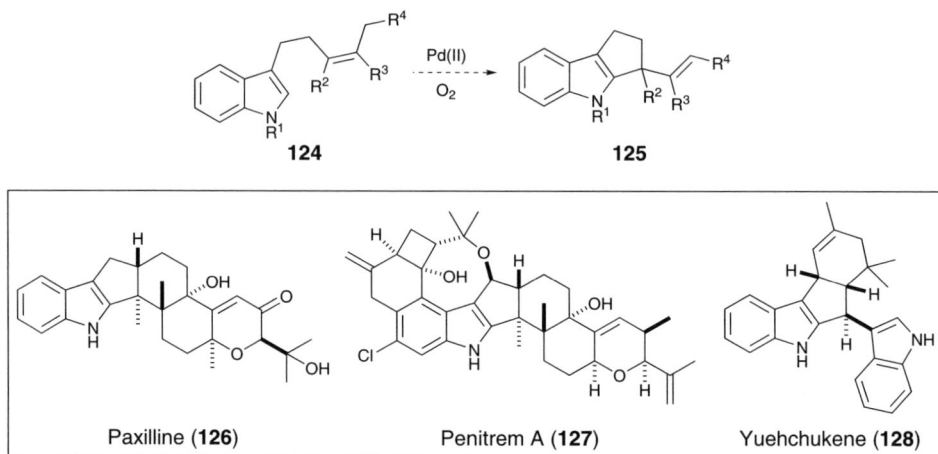

Figure 9.5   *Annulated indoles in natural product architectures.*

As a starting point, indole **129** was subjected to the mild oxidative conditions originally described by Uemura and coworkers [49] for the oxidation of alcohols (catalytic Pd(OAc)$_2$, pyridine (4 equiv relative to palladium), toluene, 1 atm O$_2$, 80 °C). Gratifyingly, oxidative cyclization to annulated indole occurred, albeit with sluggish reactivity. In order to improve upon this reaction, the electronic nature of the pyridine ligand was probed (Table 9.1). Ligands with more electron-donating groups (entries 1 and 2) were found to shut down reactivity completely. Moving to more electron-deficient pyridine ligands (entries 4 and 5),

Table 9.1   *Ligand electronic effects in oxidative indole annulations.*

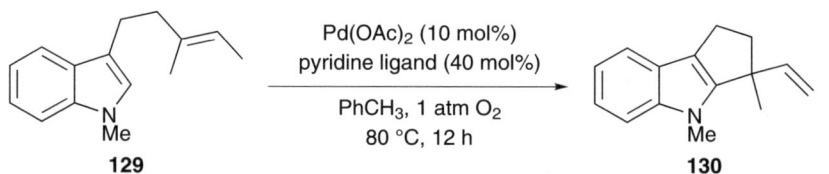

Entry	Pyridine ligand	pK_a (PyrH$^+$)	Conversion (%)
1	4-MeO	6.47	3
2	4-*t*-Bu	5.99	1
3	Unsubstituted	5.25	23
4	4-CO$_2$Et	3.45	52
5	*3-CO$_2$Et*	*3.35*	*76*
6	3-COCH$_3$	3.18	58
7	3-F	2.97	64
8	3-CN	1.39	55
9	3,5-di-Cl	0.90	22

however, boosted the overall reactivity. Further increasing the electron-deficient nature of the pyridine ligand was detrimental to the reaction (entries 6–9). These ligands were likely unable to bind sufficiently to the metal centre, thereby preventing palladium(0) reoxidation. Ultimately, ethyl nicotinate was found to be the optimal ligand for this catalytic palladium system, striking the appropriate electronic balance for overall reactivity.

The choice of solvent also proved to be important in these indole annulations. In standard solvents and in the absence of catalyst, the annulated indole products were susceptible to oxidative decomposition. After a thorough analysis of solvent and catalyst effects, it was found that the products were stabilized in the presence of the Pd(OAc)$_2$/ethyl nicotinate catalyst in a 4:1 mixture of *tert*-amyl alcohol/AcOH as the solvent. Since the oxidative decomposition of indoles is generally initiated by addition at C(3), it was hypothesized that the catalyst and/or AcOH can act as a competitive inhibitor. With this carefully optimized system, the oxidatively annulated indoles could be accessed in good to excellent yields.

A variety of annulated indoles were obtained via this mild palladium oxidation system (Figure 9.6). Five- and six-membered rings were constructed by this method. Indoles with electron-withdrawing and electron-donating groups on the arene were oxidatively cyclized in good yields (**131** and **132**). When a substituent was placed on the tether between the alkene and the indole, diastereoselective cyclization occurred to form **133**. The cyclization can occur not only from C(3) to C(2), but also from N(1) to C(2) (indole **138**) and C(2) to C(3) (indoles **135** and **139**).

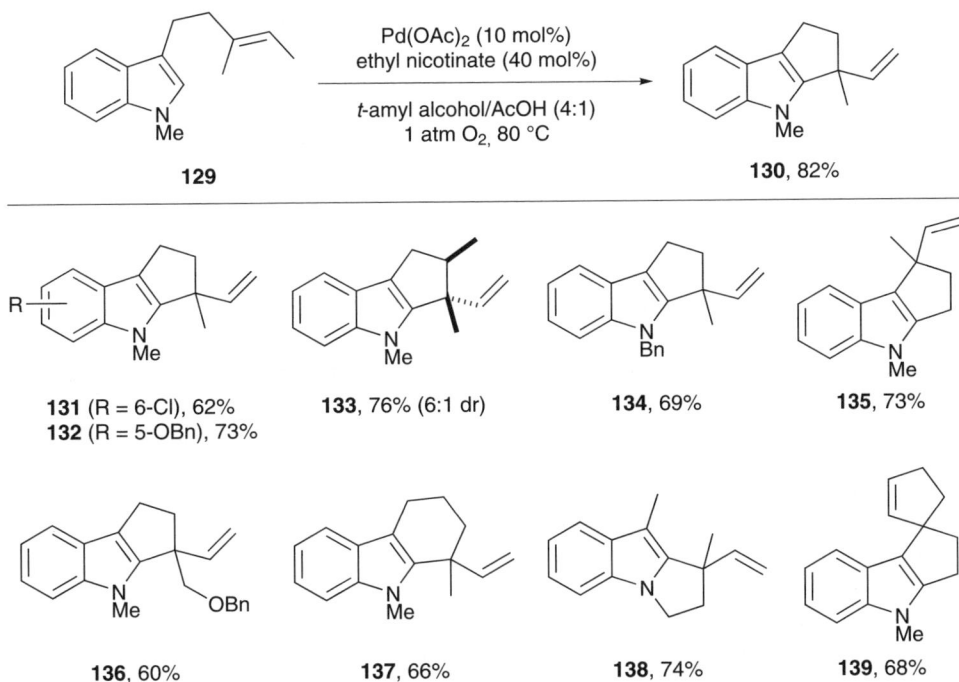

Figure 9.6 *The palladium-catalysed oxidative annulation of indoles.*

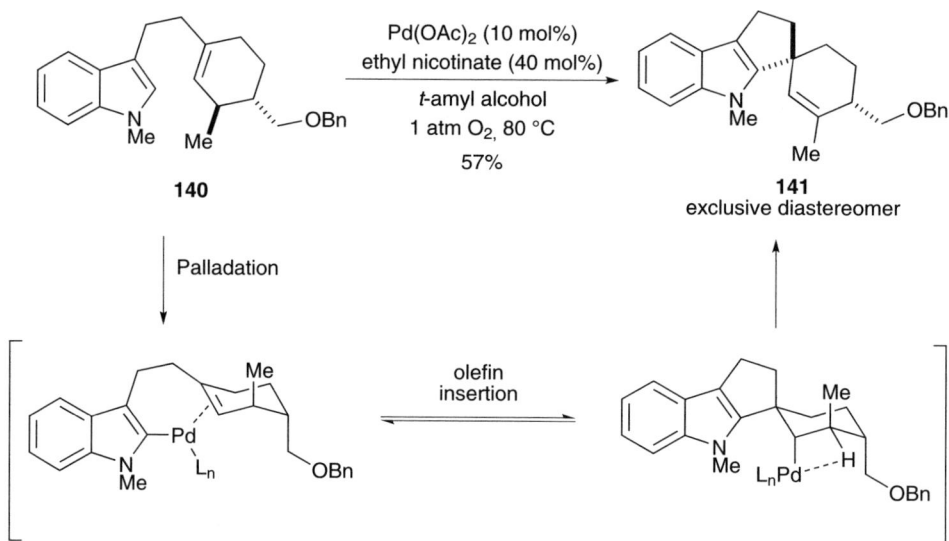

Scheme 9.20

In order to elucidate the mechanism of this reaction, a substrate probe was designed. Diastereomerically pure indole **140** was synthesized and subjected to the aerobic oxidative cyclization (Scheme 9.20). Annulated indole **141** was produced as a single diastereomer. The outcome of this reaction strongly suggested a mechanism involving initial palladation of the indole, followed by alkene insertion and β-hydride elimination (an intramolecular Fujiwara–Moritani reaction). If the reaction proceeded by alkene activation followed by nucleophilic attack of the indole, then the opposite diastereomer would have been observed. This experiment confirmed that an oxidative Heck reaction pathway was operative in this aerobic indole annulation.

Building upon this promising chemistry, Stoltz and coworkers [50] extended the scope of the mild oxidation system to the formation of other cyclic compounds – specifically, benzofurans and dihydrobenzofurans. Aryl allyl ether **142** was subjected to the Pd(OAc)$_2$/ethyl nicotinate catalyst under a variety of oxidants to provide benzofuran **144**. Presumably, this reaction proceeds by initial palladation, followed by alkene insertion and β-hydride elimination to form vinylic intermediate **143**, which then isomerizes to the more thermodynamically stable benzofuran **144**. Although molecular oxygen was a suitable oxidant for this reaction (Table 9.2 entry 1), benzoquinone appeared to be the optimal reoxidant (Table 9.2 entry 2), providing the highest overall yields.

Other parameters of the palladium(II)-catalysed benzofuran synthesis were explored. Ultimately, it was found that the ideal system employed a 2:1 ligand/palladium ratio (10 mol% Pd(OAc)$_2$ and 20 mol% ethyl nicotinate) and a substoichiometric amount (20 mol%) of NaOAc at 100 °C. With this system in hand, a variety of substituted benzofurans were accessed (Figure 9.7). The reaction was limited to electron-rich aryl systems; the palladation event required a sufficiently nucleophilic arene in order to occur. The aryl subunit, however, tolerated an array of alkyl and alkoxy substitution patterns within the

Table 9.2 The palladium-catalysed oxidative cyclization of aryl allyl ethers oxidant studies.

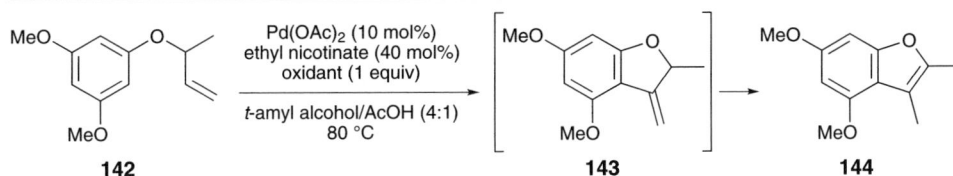

Entry	Oxidant	Yield (%)	Entry	Oxidant	Yield (%)
1	O$_2$ (1 atm)	56	5	Tl(O$_2$CCF$_3$)$_3$	<10
2	Benzoquinone	62	6	K$_2$S$_2$O$_8$	30
3	Cu(OAc)$_2$	31	7	H$_2$NC(S)NH$_2$	<10
4	AgOAc	29	8	PhCO$_3$-t-Bu	42

electronic requirements. The allyl moiety also accommodated several substituents (alkyl, aryl, alkoxy) at both the proximal and distal positions.

If the allyl group of the aryl allyl ether featured a tri- or tetra-substituted alkene, then the thermodynamic isomerization to the aromatic benzofuran could not occur, and dihydroben-zofurans were consequently produced in good to excellent yields (Figure 9.8). As in the benzofuran systems, a host of aryl and alkyl substitution patterns were tolerated in this ox-idative cyclization. A number of polycyclic and highly functionalized dihydrobenzofurans were obtained via this palladium(II)-catalysed oxidation.

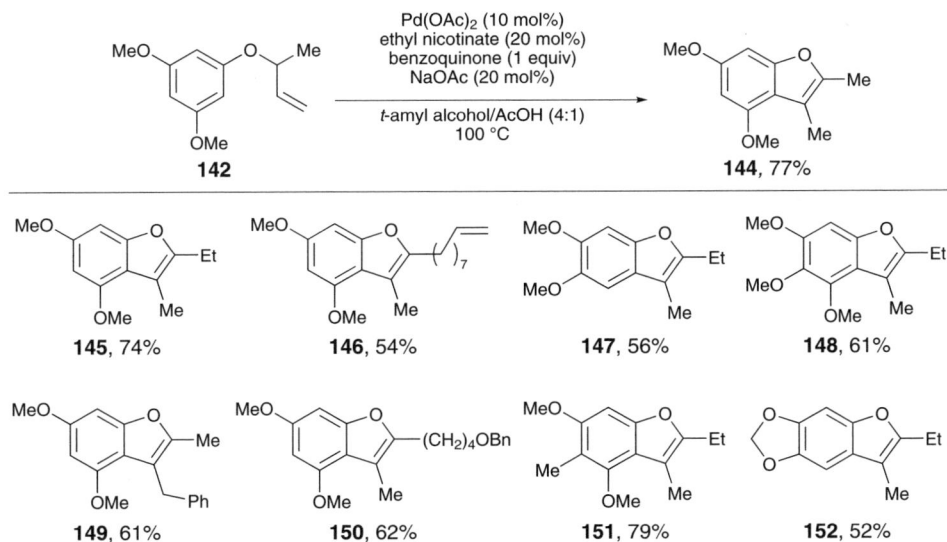

Figure 9.7 The palladium(II)-catalysed oxidative cyclization of aryl allyl ethers to benzofu-rans.

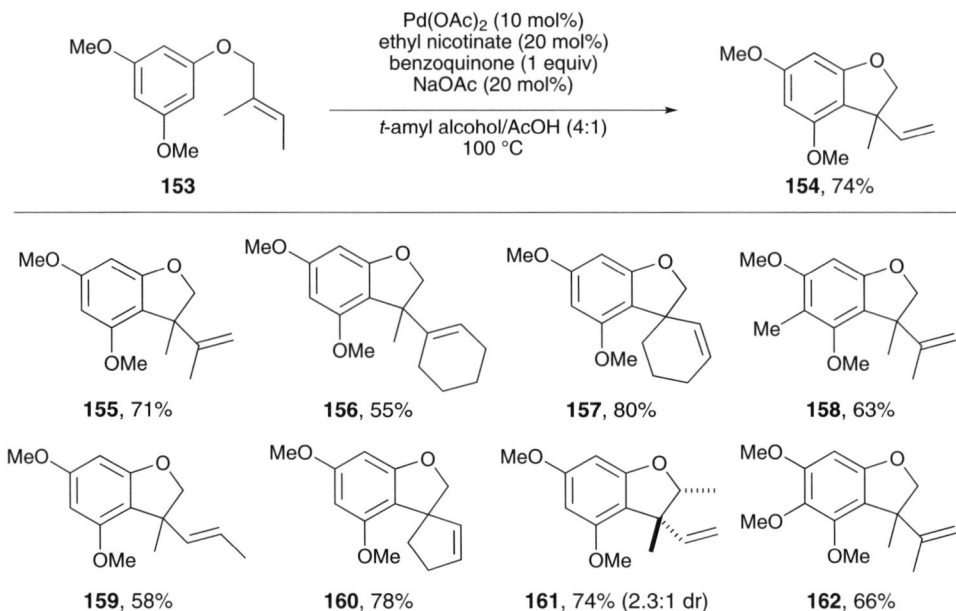

Figure 9.8   *The palladium(II)-catalysed oxidative cyclization of aryl allyl ethers to dihydrobenzofurans.*

As in the palladium(II)-catalysed indole annulation, a diastereomerically pure substrate was designed to help elucidate the mechanism of the benzofuran and dihydrobenzofuran syntheses (Scheme 9.21). When aryl allyl ether **163** was treated with the palladium oxidation catalyst, dihydrobenzofuran **164** was produced as an exclusive diastereomer in 60% yield. This observation confirmed that an oxidative Heck reaction pathway, featuring arene palladation, alkene insertion and β-hydride elimination, was operative.

Beccalli and coworkers [51] described a similar oxidative cyclization of indoles using catalytic palladium and benzoquinone as the stoichiometric reoxidant (Scheme 9.22). By

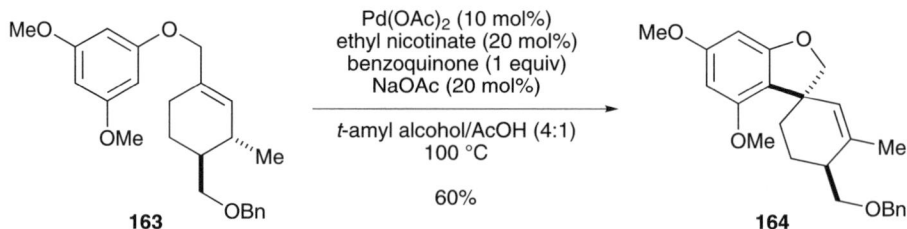

Scheme 9.21

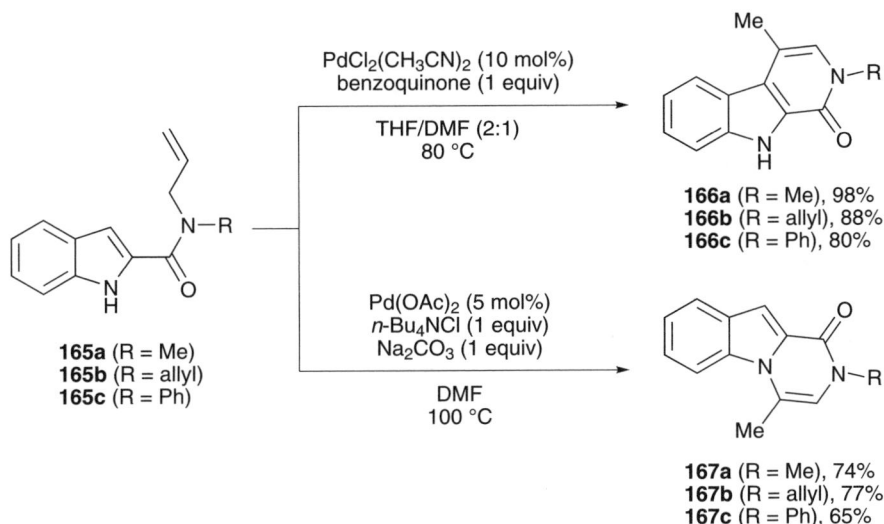

Scheme 9.22

substituting the C(2) position of the indole moiety with an amide tether, β-carbolinones **166a–c** were accessed in good yields. The authors proposed an oxidative Heck reaction mechanism followed by alkene isomerization to the endocyclic alkene to produce the carbolinones. Interestingly, if alternative catalytic palladium conditions were utilized, then an oxidative amination of the tethered alkene occurred to afford pyrazino[1,2-*a*]indole derivatives instead.

This catalytic oxidative annulation process can also be applied to pyrrole systems. Gaunt and coworkers [29] had described the intermolecular selective alkenylation of C(2) or C(3) of pyrroles based on the nitrogen protecting group (Scheme 9.5; see above). This same concept was implemented in an intramolecular sense to achieve selective annulations (Scheme 9.23). When *N*-tosyl pyrrole **168** was subjected to oxidative conditions, the cyclization occurred at C(2) to form pyrrole **169**. Alternatively, when structurally similar *N*-triisopropylsilyl pyrrole **170** was treated with the same palladium(II) system, the functionalization occurred at C(4) to afford pyrrole **171**. This intriguing differentiation based on the choice of protecting group could have broad implications in selective functionalizations of heterocycles.

Youn and Eom [52] recently demonstrated that a palladium(II)-catalysed oxidative cyclization can be employed to synthesize chromene derivatives. Aryl homoallyl ether **172** was subjected to catalytic Pd(CH$_3$CN)$_2$Cl$_2$ (5 mol%), using benzoquinone as the stoichiometric reoxidant, to provide chromene **173** in 72% yield (Scheme 9.24). Although effective for this example, most of the other systems investigated required high catalyst loadings (20–25 mol%) in order to effect reactivity. Regioselectivities in the palladation/cyclization were quite high, however, as illustrated by the cyclization of ether **174** to a 99 : 1 mixture of chromenes **175** and **176**.

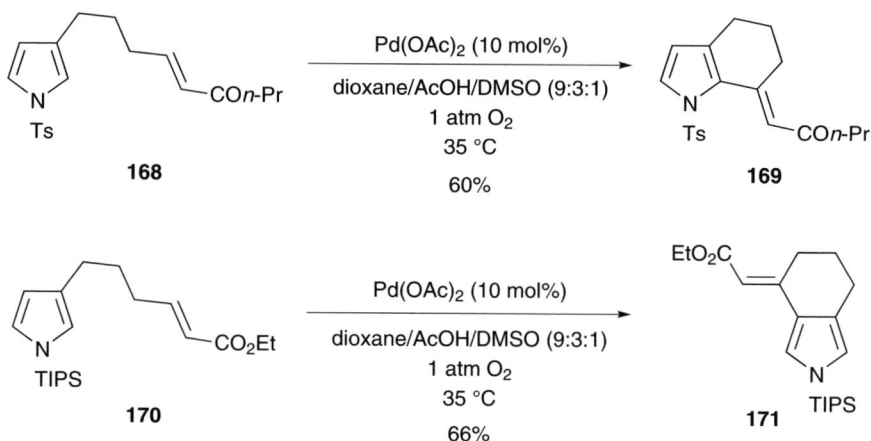

Scheme 9.23

In 2006, Lu and coworkers [53] described a method for the synthesis of carbazoles using the palladium(II)-catalysed oxidative Heck reaction (Scheme 9.25). Indole **177** was subjected to catalytic Pd(OAc)$_2$ and 2.1 equiv of benzoquinone to afford carbazole **178** in 88% yield. Because the alkene is 1,1-disubstituted, a 5-*exo* cyclization mode (as seen in Ferreira and Stoltz's [48] indole annulation) is unproductive, and a 6-*endo* cyclization ultimately occurs. The putative intermediate is then believed to be oxidized to the aromatic carbazole by the excess benzoquinone. When indole **179**, featuring a terminal alkene, was treated with these oxidative conditions, a mixture of products from 6-*endo* cyclization (**180**) and 5-*exo* cyclization (and subsequent isomerization and [3+2] cycloaddition, **181**) was observed. A variety of substituted carbazoles were obtained by this palladium(II)-catalysed oxidative cyclization.

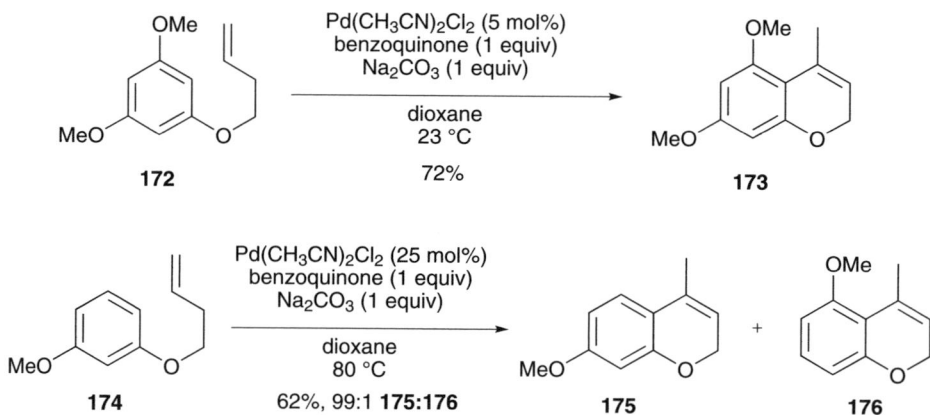

Scheme 9.24

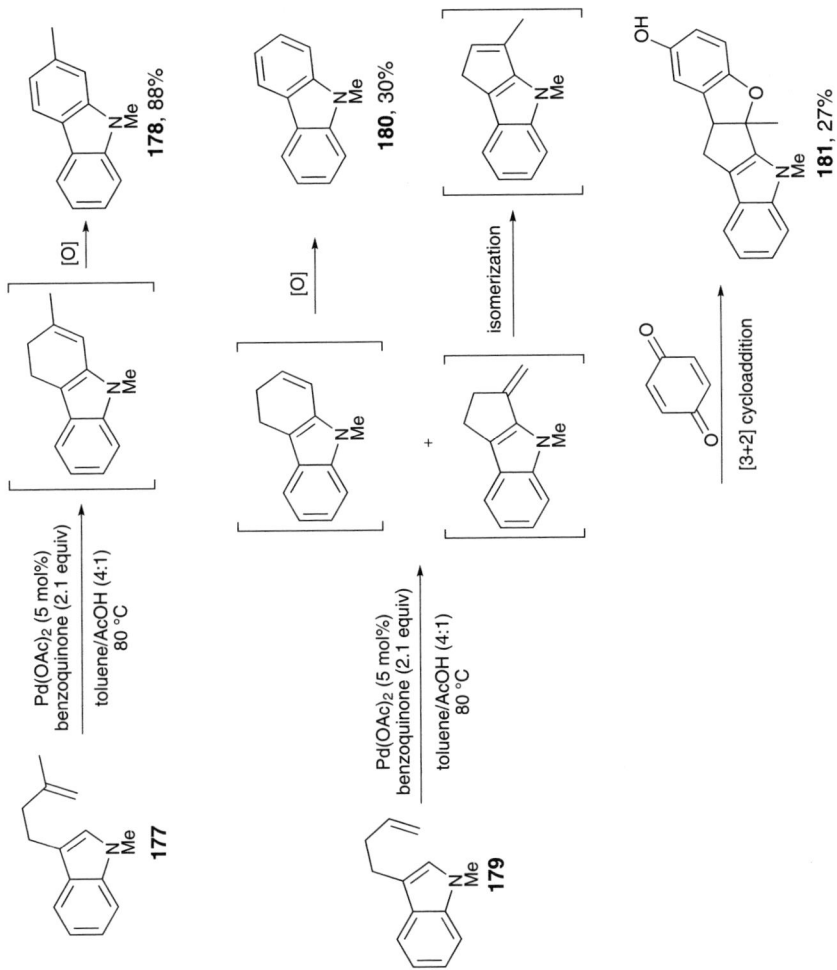

Scheme 9.25

9.5 Summary

The Mizoroki–Heck reaction has proven to be a powerful C—C bond-forming reaction that has seen wide applications in chemical synthesis. Comparably, the oxidative Heck reaction has been less explored, but can clearly be a useful transformation. C—C bonds can be constructed in both an intermolecular and an intramolecular fashion. Recent efforts to perform the reactions with catalytic palladium salts, using inexpensive stoichiometric reoxidants, have met with some success, although there is certainly room for improvement. It is anticipated that, through further understanding of these interesting transformations, the design of new catalysts will lead to unprecedented transformations regarding functional group compatibility, catalyst efficiency and levels of enantioselectivity.

References

1. For recent reviews on the Mizoroki–Heck reaction, see: (a) Alonso, F., Beletskaya, I.P. and Yus, M. (2005) Non-conventional methodologies for transition-metal catalysed carbon–carbon coupling: a critical overview. Part 1: the Heck reaction. *Tetrahedron*, **61**, 11771–835; (b) Bräse, S. and de Meijere, A. (2004) Cross-coupling of organyl halides with alkenes: the Heck reaction, in *Metal-Catalyzed Cross-Coupling Reactions*, Vol. **1**, 2nd edn (eds A. de Meijere and F. Diederich), Wiley–VCH Verlag GmbH, Weinheim, Germany, pp. 217–315; (c) Dounay, A.B. and Overman, L.E. (2003) The asymmetric intramolecular Heck reaction in natural product total synthesis. *Chem. Rev.*, **103**, 2945–63; (d) Link, J.T. (2002) The intramolecular Heck reaction. *Org. React.*, **60**, 157–534; (e) Shibasaki, M. and Miyazaki, F. (2002) Asymmetric Heck reactions, in *Handbook of Organopalladium Chemistry for Organic Synthesis*, Vol. **1** (eds E.-i. Negishi and A. de Meijere), John Wiley & Sons, Inc., New York, pp. 1283–315; (f) Bräse, S. and de Meijere, A. (2002) Intramolecular Heck reaction. Synthesis of carbocycles, in *Handbook of Organopalladium Chemistry for Organic Synthesis*, Vol **1** (eds E.-i. Negishi and A. de Meijere), John Wiley & Sons, Inc., New York, pp. 1223–54; (g) Bräse, S. and de Meijere, A. (2002) Double and multiple Heck reactions, in *Handbook of Organopalladium Chemistry for Organic Synthesis*, Vol **1** (eds E.-i. Negishi and A. de Meijere), John Wiley & Sons, Inc., New York, pp. 1179–208; (h) Larhed, M. and Hallberg, A. (2002) Intermolecular Heck reaction. Scope, mechanism, and other fundamental aspects of the intermolecular Heck reaction, in *Handbook of Organopalladium Chemistry for Organic Synthesis*, Vol. **1** (eds E.-i. Negishi and A. de Meijere), John Wiley & Sons, Inc., New York, pp. 1133–78; (i) Beletskaya, I.P. and Cheprakov, A.V. (2000) The Heck reaction as a sharpening stone of palladium catalysis. *Chem. Rev.*, **100**, 3009–66.
2. For reviews on the Fujiwara–Moritani reaction, see: (a) Fujiwara, Y. (2002) Palladium-promoted alkene–arene coupling via C—H activation, in *Handbook of Organopalladium Chemistry in Organic Synthesis*, Vol. **2** (eds E.-i. Negishi and A. de Meijere), John Wiley & Sons, Inc., New York, pp. 2863–71; (b) Jia, C., Kitamura, T. and Fujiwara, Y. (2001) Catalytic functionalization of arenes and alkanes via C—H bond activation. *Acc. Chem. Res.*, **34**, 633–9; (c) Fujiwara, Y. and Jia, C. (2001) New developments in transition metal-catalyzed synthetic reactions via C—H bond activation. *Pure Appl. Chem.*, **73**, 319–24; (d) Moritani, I. and Fujiwara, Y. (1973) Aromatic substitution of olefins by palladium salts. *Synthesis*, 524–33.
3. (a) Moritani, I. and Fujiwara, Y. (1967) Aromatic substitution of styrene–palladium chloride complex. *Tetrahedron Lett.*, **8**, 1119–22; (b) Fujiwara, Y., Moritani, I. and Matsuda, M. (1968) Aromatic substitution of olefin – III. Reaction of styrene–palladium(II) chloride complex. *Tetrahedron*, **24**, 4819–24.

4. (a) Fujiwara, Y., Moritani, I., Matsuda, M. and Teranishi, S. (1968) Aromatic substitution of styrene–palladium chloride complex. II. Effect of metal acetate. *Tetrahedron Lett.*, **9**, 633–6; (b) Fujiwara, Y., Moritani, I., Danno, S. *et al.* (1969) Aromatic substitution of olefins. VI. Arylation of olefins with palladium(II) acetate. *J. Am. Chem. Soc.*, **91**, 7166–9.

5. (a) Fujiwara, Y., Moritani, I., Asano, R. *et al.* (1969) Aromatic substitution of olefin – VIII. Substituent effects on the reactions of styrene with monosubstituted benzenes in the presence of palladium(II) salts. *Tetrahedron*, **25**, 4815–8; (b) Fujiwara, Y., Asano, R., Moritani, I. and Teranishi, S. (1976) Aromatic substitution of olefins. 25. Reactivity of benzene, naphthalene, ferrocene, and furan toward styrene, and the substituent effect on the reaction of monosubstituted benzenes with styrene. *J. Org. Chem.*, **41**, 1681–3.

6. Danno, S., Moritani, I. and Fujiwara, Y. (1969) Aromatic substitution of olefin – VII. Reactions of lower olefins with benzene by palladium acetate. *Tetrahedron*, **25**, 4809–13.

7. Yamamura, M., Moritani, I., Sonoda, A. *et al.* (1973) Aromatic substitution of olefins. Part XVIII. Reactions of cycloalkenes with benzene. *J. Chem. Soc., Perkin Trans. 1*, 203–5.

8. (a) Asano, R., Moritani, I., Fujiwara, Y. and Teranishi, S. (1973) Aromatic substitution of olefins. XIX. Reaction of five-membered heterocyclic aromatic compounds with styrene. *Bull. Chem. Soc. Jpn.*, **46**, 663–4; (b) Fujiwara, Y., Maruyama, O., Yoshidomi, M. and Taniguchi, H. (1981) Palladium-catalyzed alkenylation of aromatic heterocycles with olefins. Synthesis of functionalized aromatic heterocycles. *J. Org. Chem.*, **46**, 851–5.

9. Itahara, T., Ikeda, M. and Sakakibara, T. (1983) Alkenylation of 1-acylindoles with olefins bearing electron-withdrawing substituents and palladium acetate. *J. Chem. Soc., Perkin Trans. 1*, 1361–3.

10. Itahara, T., Kawasaki, K. and Ouseto, F. (1984) Alkenylation of 1-benzenesulfonylindole with olefins bearing electron-withdrawing substituents. *Synthesis*, 236–7.

11. Yokoyama, Y., Takashima, M., Higaki, C. *et al.* (1993) An interesting chemoselectivity in palladation of aromatic bromides. *Heterocycles*, **36**, 1739–42.

12. Yokoyama, Y., Tsuruta, K. and Murakami, Y. (2002) A study for palladium-catalyzed chemoselective vinylation at C-3 position of 4-bromo-1-tosylindole. *Heterocycles*, **56**, 525–9.

13. Osanai, K., Yokoyama, Y., Kondo, K. and Murakami, Y. (1999) Total synthesis of optically active costaclavine (synthetic studies of indoles and related compounds part 48). *Chem. Pharm. Bull.*, **47**, 1587–90.

14. Yokoyama, Y., Kondo, K., Mitsuhashi, M. and Murakami, Y. (1996) Total synthesis of optically active chanoclavine-I. *Tetrahedron Lett.*, **37**, 9309–12.

15. Hikawa, H., Yokoyama, Y. and Murakami, Y. (2000) A short synthesis of optically active γ,γ-dimethylallyltryptophan (DMAT). *Synthesis*, 214–6.

16. Yokoyama, Y., Matsumoto, T. and Murakami, Y. (1995) Optically active total synthesis of clavicipitic acid. *J. Org. Chem.*, **60**, 1486–7.

17. (a) Asano, R., Moritani, I., Sonoda, A. *et al.* (1971) Aromatic substitution of olefins. Part XVII. Reactions of ferrocene with olefins in the presence of palladium(II) salts. *J. Chem. Soc. C*, 3691–2; (b) Asano, R., Moritani, I., Fujiwara, Y. and Teranishi, S. (1970) Aromatic substitution of olefins. The reaction of ferrocene with styrene in the presence of palladium(II) acetate. *J. Chem. Soc., Chem. Commun.*, 1293.

18. Hirota, K., Kuki, H. and Maki, Y. (1994) Novel synthesis of pyrido[3,4-*d*]pyrimidines, pyrido[2,3-*d*]pyrimidines, and quinazolines via palladium-catalyzed oxidative coupling. *Heterocycles*, **37**, 563–70.

19. Fuchita, Y., Hiraki, K., Kamogawa, Y. *et al.* (1989) Activation of aromatic carbon–hydrogen bonds by palladium(II) acetate–dialkyl sulfide systems. Formation and characterization of novel diphenyltripalladium(II) complexes. *Bull. Chem. Soc. Jpn.*, **62**, 1081–5.

20. Fujiwara, Y., Moritani, I., Matsuda, M. and Teranishi, S. (1968) Aromatic substitution of olefin. IV. Reaction with palladium metal and silver acetate. *Tetrahedron Lett.*, **9**, 3863–5.

21. Tsuji, J. and Nagashima, H. (1984) Palladium-catalyzed oxidative coupling of aromatic compounds with olefins using *t*-butyl perbenzoate as a hydrogen acceptor. *Tetrahedron*, **40**, 2699–702.

22. (a) Yokota, T., Tani, M., Sakaguchi, S. and Ishii, Y. (2003) Direct coupling of benzene with olefin catalyzed by Pd(OAc)$_2$ combined with heteropolyoxometalate under dioxygen. *J. Am. Chem. Soc.*, **125**, 1476–7; (b) Tani, M., Sakaguchi, S. and Ishii, Y. (2004) Pd(OAc)$_2$-catalyzed oxidative coupling reaction of benzenes with olefins in the presence of molybdovanadophosphoric acid under atmospheric dioxygen and air. *J. Org. Chem.*, **69**, 1221–6.

23. Jia, C., Lu, W., Kitamura, T. and Fujiwara, Y. (1999) Highly efficient Pd-catalyzed coupling of arenes with olefins in the presence of *tert*-butyl hydroperoxide as oxidant. *Org. Lett.*, **1**, 2097–100.

24. Shue, R.S. (1971) Catalytic coupling of aromatics and olefins by homogeneous palladium(II) compounds under oxygen. *J. Chem. Soc., Chem. Commun.*, 1510–1.

25. Dams, M., De Vos, D.E., Celen, S. and Jacobs, P.A. (2003) Toward waste-free production of Heck products with a catalytic palladium system under oxygen. *Angew. Chem. Int. Ed.*, **42**, 3512–5.

26. Maruyama, O., Yoshidomi, M., Fujiwara, Y. and Taniguchi, H. (1979) Pd(II)–Cu(II)-catalyzed synthesis of mono- and dialkenyl-substituted five-membered aromatic heterocycles. *Chem. Lett.*, 1229–30.

27. Capito, E., Brown, J.M. and Ricci, A. (2005) Directed palladation: fine tuning permits the catalytic 2-alkenylation of indoles. *Chem. Commun.*, 1854–6.

28. Grimster, N.P., Gauntlett, C., Godfrey, C.R.A. and Gaunt, M.J. (2005) Palladium-catalyzed intermolecular alkenylation of indoles by solvent-controlled regioselective C—H functionalization. *Angew. Chem. Int. Ed.*, **44**, 3125–9.

29. Beck, E.M., Grimster, N.P., Hatley, R. and Gaunt, M.J. (2006) Mild aerobic oxidative palladium(II) catalyzed C—H bond functionalization: regioselective and switchable C—H alkenylation and annulation of pyrroles. *J. Am. Chem. Soc.*, **128**, 2528–9.

30. Hirota, K., Isobe, Y., Kitade, Y. and Maki, Y. (1987) A simple synthesis of 5-(1-alkenyl)uracil derivatives by palladium-catalyzed oxidative coupling of uracils with olefins. *Synthesis*, 495–6.

31. Mikami, K., Hatano, M. and Terada, M. (1999) Catalytic C—H bond activation–asymmetric olefin coupling reaction: the first example of asymmetric Fujiwara–Moritani reaction catalyzed by chiral palladium(II) complexes. *Chem. Lett.*, **28**, 55–6.

32. Boele, M.D.K., van Strijdonck, G.P.F., de Vries, A.H.M. *et al.* (2002) Selective Pd-catalyzed oxidative coupling of anilides with olefins through C—H bond activation at room temperature. *J. Am. Chem. Soc.*, **124**, 1586–7.

33. Zaitsev, V.G. and Daugulis, O. (2005) Catalytic coupling of haloolefins with anilides. *J. Am. Chem. Soc.*, **127**, 4156–7.

34. Miura, M., Tsuda, T., Satoh, T. and Nomura, M. (1997) Palladium-catalyzed oxidative crosscoupling of 2-phenylphenols with alkenes. *Chem. Lett.*, **26**, 1103–4.

35. Miura, M., Tsuda, T., Satoh, T. *et al.* (1998) Oxidative cross-coupling of *N*-(2′-phenylphenyl)benzenesulfonamides or benzoic and naphthoic acids with alkenes using a palladium–copper catalyst system under air. *J. Org. Chem.*, **63**, 5211–5.

36. Bingham, A.J., Dyall, L.K., Norman, R.O.C. and Thomas, C.B. (1970) Reactions of palladium(II) with organic compounds. Part I. Oxidative cyclization of 3-methyl-3-phenylbut-1-ene and 3,3,3-triphenylpropene. *J. Chem. Soc. C*, 1879–83.

37. (a) Furukawa, H., Yogo, M., Ito, C. *et al.* (1985) New carbazolequinones having dimethylpyran ring system, from *Murraya euchrestifolia*. *Chem. Pharm. Bull.*, **33**, 1320–2; (b) Bittner, S., Krief, P. and Massil, T. (1991) Synthesis of carbazoloquinones via direct palladation of 5-anilino-2-phenylthio-1,4-benzoquinones and of 2-anilino-1,4-naphthoquinones. *Synthesis*, 215–6; (c) Yogo, M., Ito, C. and Furukawa, H. (1991) Synthesis of some carbazolequinone alkaloids

and their analogues. Facile palladium-assisted intramolecular ring closure of arylamino-1,4-benzoquinones to carbazole-1,4-quinones. *Chem. Pharm. Bull.*, **39**, 328–34; (d) Knölker, H.-J. and O'Sullivan, N. (1994) Indoloquinones, part 2. Palladium-promoted synthesis of a 7-deoxyprekinamycin isomer. *Tetrahedron Lett.*, **35**, 1695–8; (e) Knölker, H.-J. and O'Sullivan, N. (1994) Indoloquinones-3. Palladium-promoted synthesis of hydroxy-substituted 5-cyano-5*H*-benzo[*b*]carbazole-6,11-diones. *Tetrahedron*, **50**, 10893–908.

38. (a) Trost, B.M., Godleski, S.A. and Genêt, J.P. (1978) A total synthesis of racemic and optically active ibogamine. Utilization and mechanism of a new silver ion assisted palladium catalyzed cyclization. *J. Am. Chem. Soc.*, **100**, 3930–1; (b) For a related example, see Trost, B.M. and Genêt, J.P. (1976) Palladium catalyzed cyclizations to alkaloid skeletons. Facile synthesis of desethylibogamine. *J. Am. Chem. Soc.*, **98**, 8516–7.

39. Trost, B.M., Godleski, S.A. and Belletire, J.L. (1979) Synthesis of (±)-catharanthine via organopalladium chemistry. *J. Org. Chem.*, **44**, 2052–4.

40. (a) Büchi, G., Kulsa, P., Ogasawara, K. and Rosati, R.L. (1970) Syntheses of velbanamine and catharanthine. *J. Am. Chem. Soc.*, **92**, 999–1005; (b) Büchi, G. and Manning, R.E. (1966) Chemical transformations of ibogaine. *J. Am. Chem. Soc.*, **88**, 2532–2355.

41. Trost, B.M. and Fortunak, J.M.D. (1982) Cyclizations initiated by a Pd(2+)–Ag(+) mixed-metal system. *Organometallics*, **1**, 7–13.

42. (a) Cushing, T.D., Sanz-Cervera, J.F. and Williams, R.M. (1993) Stereocontrolled total synthesis of (+)-paraherquamide B. *J. Am. Chem. Soc.*, **115**, 9323–4; (b) Cushing, T.D., Sanz-Cervera, J.F. and Williams, R.M. (1996) Stereocontrolled total synthesis of (+)-paraherquamide B. *J. Am. Chem. Soc.*, **118**, 557–79.

43. Williams, R.M., Cao, J., Tsujishima, H. and Cox, R.J. (2003) Asymmetric, stereocontrolled total synthesis of paraherquamide A. *J. Am. Chem. Soc.*, **125**, 12172–8.

44. Baran, P.S. and Corey, E.J. (2002) A short synthetic route to (+)-austamide, (+)-deoxyisoaustamide, and (+)-hydratoaustamide from a common precursor by a novel palladium-mediated indole → dihydroindoloazocine cyclization. *J. Am. Chem. Soc.*, **124**, 7904–5.

45. Baran, P.S., Guerrero, C.A. and Corey, E.J. (2003) Short, enantioselective total synthesis of okaramine N. *J. Am. Chem. Soc.*, **125**, 5628–9.

46. (a) Garg, N.K., Caspi, D.D. and Stoltz, B.M. (2004) The total synthesis of (+)-dragmacidin F. *J. Am. Chem. Soc.*, **126**, 9552–3; (b) Garg, N.K., Caspi, D.D. and Stoltz, B.M. (2005) Development of an enantiodivergent strategy for the total synthesis of (+)- and (−)-dragmacidin F from a single enantiomer of quinic acid. *J. Am. Chem. Soc.*, **127**, 5970–8.

47. (a) Åkermark, B., Oslob, J.D. and Heuschert, U. (1995) Catalytic oxidative aromatic cyclizations with palladium. *Tetrahedron Lett.*, **36**, 1325–6; (b) Knölker, H.-J. and Fröhner, W. (1998) Palladium-catalyzed total synthesis of the antibiotic carbazole alkaloids carbazomycin G and H. *J. Chem. Soc., Perkin Trans. 1*, 173–5; (c) Knölker, H.-J., Reddy, K.R. and Wagner, A. (1998) Indoloquinones, part 5. Palladium-catalyzed total synthesis of the potent lipid peroxidation inhibitor carbazoquinocin C. *Tetrahedron Lett.*, **39**, 8267–70; (d) Hagelin, H., Oslob, J.D. and Åkermark, B. (1999) Oxygen as oxidant in palladium-catalyzed inter- and intramolecular coupling reactions. *Chem. Eur. J.*, **5**, 2413–6.

48. Ferreira, E.M. and Stoltz, B.M. (2003) Catalytic C–H bond functionalization with palladium(II): aerobic oxidative annulations of indoles. *J. Am. Chem. Soc.*, **125**, 9578–9.

49. Nishimura, T., Onoue, T., Ohe, K. and Uemura, S. (1999) Palladium(II)-catalyzed oxidation of alcohols to aldehydes and ketones by molecular oxygen. *J. Org. Chem.*, **64**, 6750–5.

50. Zhang, H., Ferreira, E.M. and Stoltz, B.M. (2004) Direct oxidative Heck cyclizations: intramolecular Fujiwara–Moritani arylations for the synthesis of functionalized benzofurans and dihydrobenzofurans. *Angew. Chem. Int. Ed.*, **43**, 6144–8.

51. (a) Beccalli, E.M. and Broggini, G. (2003) Uncommon intramolecular palladium-catalyzed cyclization of indole derivatives. *Tetrahedron Lett.*, **44**, 1919–21; (b) Abbiati, G., Beccalli, E.M.,

Broggini, G. and Zoni, C. (2003) Regioselectivity on the palladium-catalyzed intramolecular cyclization of indole derivatives. *J. Org. Chem.*, **68**, 7625–8.

52. Youn, S. W. and Eom, J.I. (2005) Facile construction of the benzofuran and chromene ring systems via Pd(II)-catalyzed oxidative cyclization. *Org. Lett.*, **7**, 3355–8.

53. Kong, A., Han, X. and Lu, X. (2006) Highly efficient construction of benzene ring in carbazoles by palladium-catalyzed *endo*-mode oxidative cyclization of 3-(3′-alkenyl)indoles. *Org. Lett.*, **8**, 1339–42.

10

Mizoroki–Heck Reactions with Metals Other than Palladium

Lutz Ackermann and Robert Born

Institut für Organische und Biomolekulare Chemie, Georg-August-Universität, Tammannstraße 2, D-37077 Göttingen, Germany

10.1 Introduction

The transition-metal-catalysed arylation or alkenylation of alkenes, the Mizoroki–Heck reaction [1, 2], has matured to being one of the most valuable methodologies for the $C(sp^2)$–$C(sp^2)$ bond formation [3, 4]. The most commonly employed, and originally developed, transition metal catalysts for such transformations are based on palladium [5]. However, during recent decades it has been shown that transition metals other than palladium are capable of catalysing such transformations, irrespective of the actual catalyst working mode. While none of them can thus far rival palladium in its synthetic versatility, some features were found complementary to the 'traditional' palladium-catalysed Mizoroki–Heck chemistry. Thus, considering the price for palladium and the often-required coordinating additives, an important asset is found with the use of significantly less expensive catalysts for arylation reactions of alkenes. Furthermore, for specific substrate combinations, transition metals other than palladium can be highly effective catalysts.

Herein, a survey of the literature up to mid 2007 is provided, covering catalytic arylation and alkenylation reactions of alkenes with metals other than palladium. The review summarizes Mizoroki–Heck-type reactions employing organic (pseudo)halides as electrophiles (Scheme 10.1), while oxidative Mizoroki–Heck-type reactions [6] are beyond the scope of this review (Chapters 4 and 9). Valuable transition-metal-catalysed arylation reactions of alkenes employing stoichiometric amounts of organometallic compounds as

The Mizoroki–Heck Reaction   Edited by Martin Oestreich
© 2009 John Wiley & Sons, Ltd

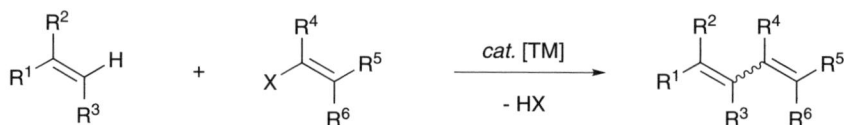

TM = Cu, Ni, Pt, Co, Rh, Ir, Ru.

X = I, Br, Cl, CN, COCl, SO$_2$Cl.

Scheme 10.1 *Transition metal-catalysed Mizoroki–Heck reaction.*

arylating reagents [7–15] or Mizoroki–Heck-type reactions of carbon–heteroatom double bonds [16] will not be covered in any detail. On the contrary, selected metal-catalysed Mizoroki–Heck-type reactions with aliphatic halides as electrophiles set the stage for the development of comparable transformations of aromatic halides and will, therefore, be discussed.

As the detailed mechanism of various metal-catalysed reactions discussed below is often not well understood, a phenomenological presentation is provided, which is categorized according to the transition metal employed.

10.2   Copper-Catalysed Mizoroki–Heck-Type Reactions

Copper salts are well known to mediate or catalyse a variety of valuable C–C and C–heteroatom bond forming reactions with numerous electrophiles [17, 18]. Furthermore, they are generally relatively inexpensive, a beneficial asset compared with the well-established palladium-based catalysts.

Accordingly, catalytic and stoichiometric amounts of cuprous salts were employed for Mizoroki–Heck-type reactions of various conjugated alkenes [19]. Intermolecular catalytic arylations of methyl acrylate (**1**, not shown) and styrene (**2**) were accomplished under 'ligand-free' conditions using CuBr (**3**) or CuI (**4**) as catalyst in *N*-methyl-2-pyrrolidinone (NMP) as solvent; various aryl iodides could be employed (Scheme 10.2). On the contrary, aryl bromides and chlorides, as well as aliphatic halides, were found to be unsuitable substrates. The reactions employing an alkenyl bromide, methylmethacrolein or methyl methacrylate required stoichiometric amounts of copper salts.

Additionally, an intramolecular Mizoroki–Heck-type reaction catalysed by CuI (**4**) was reported (Scheme 10.3) [19].

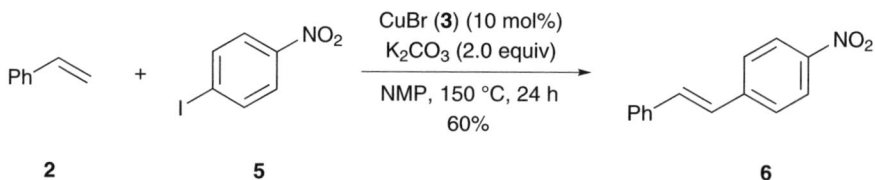

Scheme 10.2 *Copper-catalysed intermolecular arylation of styrene (2).*

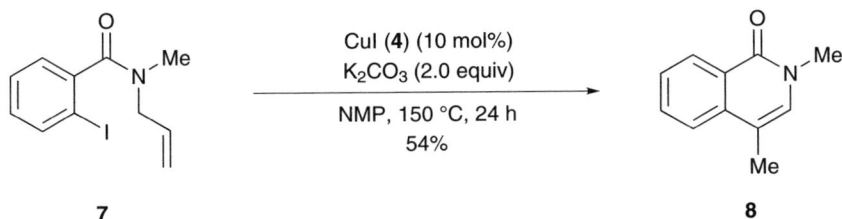

Scheme 10.3 *Copper-catalysed intramolecular Mizoroki–Heck-type reaction.*

While this 'ligand-free' copper-catalysed Mizoroki–Heck-type reaction required relatively high temperatures of 150 °C [19], the use of DABCO (**9**) as ligand allowed for significantly milder reaction conditions [20]. Thereby, satisfying isolated yields were even achieved for *ortho*-substituted electron-rich aryl iodides and alkenyl bromides (Scheme 10.4). However, aryl bromides, particularly electron-rich ones, were converted only sluggishly.

The long reaction times required for the previously reported copper-catalysed Mizoroki–Heck-type reactions could be significantly reduced with microwave irradiation [21] in polyethylene glycol (PEG) as solvent [22]. An induction period [23] was noted, resulting, interestingly, in improved isolated yields for the reused catalytic system in the second cycle (Scheme 10.5) [22]. Various aryl iodides, including *ortho*-substituted electrophiles, were efficiently converted within 30 min. On the contrary, bromo- and chlorobenzene could not be converted under these reaction conditions.

A heterogeneous arylation of ethyl acrylates and styrene (**2**) was successfully catalysed by Cu/Al$_2$O$_3$ (**15**). This catalytic system proved applicable to aryl iodides with both electron-withdrawing and -donating substituents [24]. Subsequently, a silica-supported poly-γ-aminopropylsilane Cu(II) complex was used for Mizoroki–Heck-type reactions of three aryl iodides with methyl acrylate (**1**), acrylic acid (**16**) and styrene (**2**) [23, 25].

More recently, copper bronze (**17**) was shown to generate highly active copper nanocolloids in the ionic liquid tetrabutylammonium bromide (TBAB) as solvent using tetrabutylammonium acetate (TBAA) as base (Scheme 10.6) [26]. These were used as a heterogeneous catalyst, which could be recycled up to 20 times. Interestingly, other commonly used ionic liquids as solvents, such as imidazolium salts, gave no conversion under otherwise identical reaction conditions. As with previously developed copper catalysts, various aryl iodides could be employed successfully. However, even electronically activated aryl bromides provided yields lower than 30%.

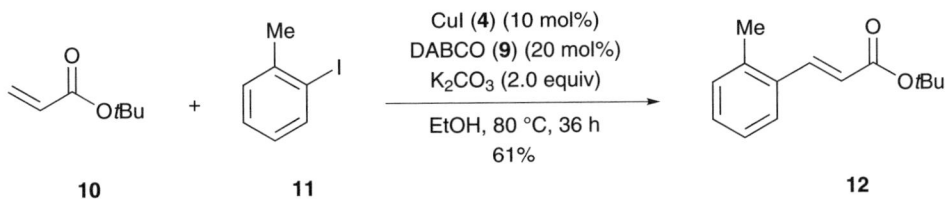

Scheme 10.4 *DABCO (**9**) as ligand for a copper-catalysed Mizoroki–Heck-type reaction.*

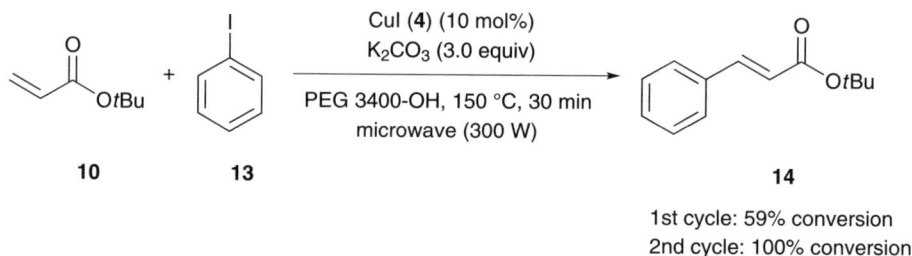

Scheme 10.5   *Microwave-assisted copper-catalysed Mizoroki–Heck-type reaction.*

10.3   Nickel-Catalysed Mizoroki–Heck-Type Reactions

The use of nickel-based catalysts for Mizoroki–Heck-type reactions is very appealing, since nickel complexes are usually rather inexpensive and display an inherent high reactivity towards oxidative addition reactions with various halides as electrophiles. A recent quantum chemical study showed that, for nickel-catalysed Mizoroki–Heck reactions, a mechanism comparable to the one of the corresponding palladium-catalysed transformations is likely to be operative [27]. Interestingly, oxidative addition and alkene insertion are calculated to occur with lower energy barriers when using nickel instead of palladium catalysts in Mizoroki–Heck-type reactions. However, β-hydride elimination and regeneration of the catalytically active species through formal HX elimination was shown to be significantly more difficult with nickel catalysts [27].

An early report on a nickel-catalysed Mizoroki–Heck-type arylation of an alkene was disclosed by Chiusoli *et al.* [28] while studying syntheses of linear, deconjugated unsaturated acids (Scheme 10.7). However, in this report, only one particular alkene, namely potassium salt **20**, was employed. This substrate is supposed to promote the Mizoroki–Heck-type arylation through precoordination to the transition metal catalyst. From a mechanistic viewpoint it is interesting to note that the nickel catalyst did not require the use of a stoichiometric, relatively strong reducing agent (see below).

Generally applicable nickel-catalysed arylation reactions of alkenes were accomplished through the use of relatively strong stoichiometric reducing reagents [29]. Hence, Perichon and coworkers [30] developed an electrochemical, highly selective nickel-catalysed Mizoroki–Heck-type reaction (Scheme 10.8). Importantly, this early report showed that not only aryl iodides and bromides, but also less reactive aryl chlorides could be used for nickel-catalysed Mizoroki–Heck-type reactions.

Scheme 10.6   *Copper-catalysed Mizoroki–Heck-type reaction in an ionic liquid.*

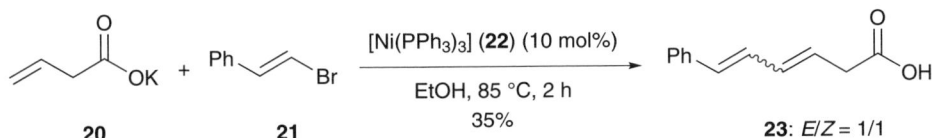

Scheme 10.7   *Nickel-catalysed Mizoroki–Heck-type reaction with salt **20**.*

With stoichiometric amounts of Zn powder (**27**) as reducing agent, [NiCl$_2$(PPh$_3$)$_2$] (**28**) allowed for catalytic arylations of styrene (**2**), and, albeit less selectively, of ethyl acrylate (**29**) [31]. The catalytic system exhibited a relatively broad scope, enabling the use of aryl iodides, bromides and chlorides (Scheme 10.9). When using iodobenzene (**13**) as electrophile, an increase in isolated yield was achieved through the addition of water.

The use of pyridine (**31**) as additive allowed for more selective and efficient nickel-catalysed arylations of styrenes (Scheme 10.10) [32, 33]. Aryl and alkyl bromides gave good yields of isolated products. With respect to the latter, secondary alkyl bromides proved superior to primary ones. However, use of methyl acrylate (**1**) as substrate yielded predominantly products originating from conjugate additions, rather than Mizoroki–Heck-type reactions.

The addition of a strongly reducing agent in nickel-catalysed Mizoroki–Heck-type reactions is not necessary when using highly polar solvents at significantly higher reaction temperatures (Scheme 10.11) [34]. Interestingly, ethyl acrylate (**29**), under optimized reaction conditions, provided the desired Mizoroki–Heck-type substitution products, employing both aryl bromides and iodides with high selectivity. A similar, more recent study probed the nickel/ligand ratio and the effect of H$_2$O as additive (see above) [35].

Stable nickel(0) complexes derived from trisubstituted phosphites, such as [Ni{P(OPh)$_3$}$_4$] (**37**), were employed for inter- and intramolecular arylation reactions of styrene (**2**), ethyl acrylate (**29**) and an alkyne [36]. Both aryl and alkenyl iodides and bromides proved applicable. However, relatively high reaction temperatures and highly polar solvents were found to be mandatory. Interestingly, the nonconjugated alkene cyclooctene (**38**) gave the corresponding 1-arylcyclooctene **39** without migration of the double bond, a valuable feature compared with palladium-catalysed reactions (Scheme 10.12).

Recently, an efficient nickel catalyst for arylation reactions of acrylates was developed by Inamoto *et al.* [37] that relies on the *in-situ* generation of an *N*-heterocyclic carbene

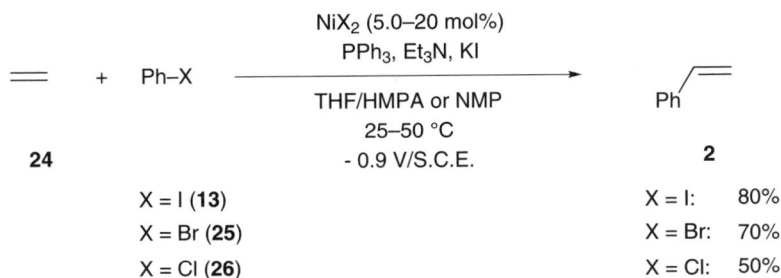

Scheme 10.8   *Nickel-catalysed Mizoroki–Heck-type reaction with electrochemical assistance.*

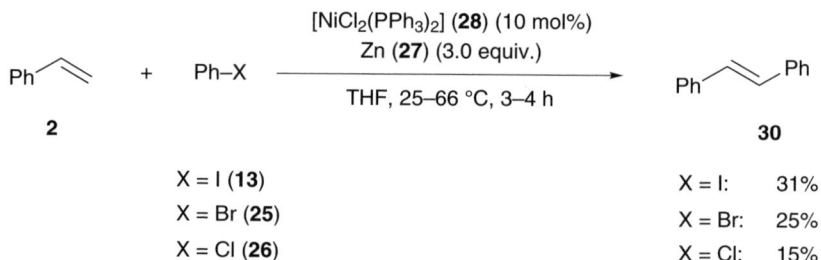

Scheme 10.9 Nickel-catalysed Mizoroki–Heck-type reaction with stoichiometric amounts of Zn (**27**).

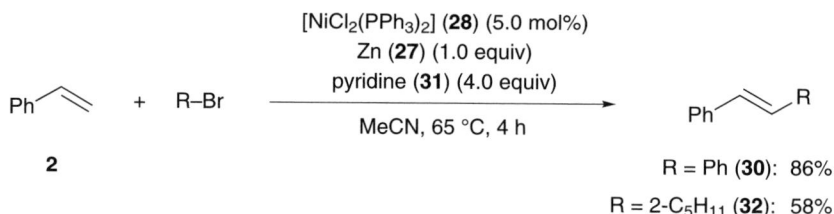

Scheme 10.10 Nickel-catalysed Mizoroki–Heck-type reactions with aryl and alkyl bromides in the presence of pyridine (**31**).

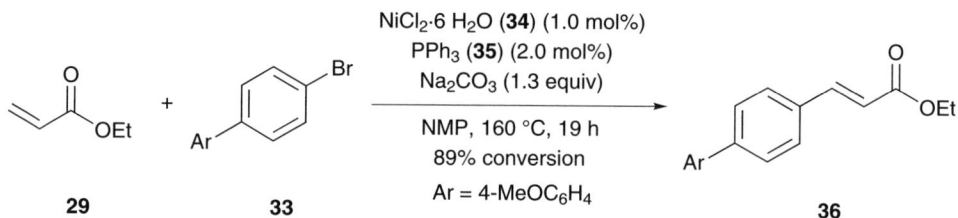

Scheme 10.11 Nickel-catalysed Mizoroki–Heck-type reaction in the absence of a relatively strong reducing agent.

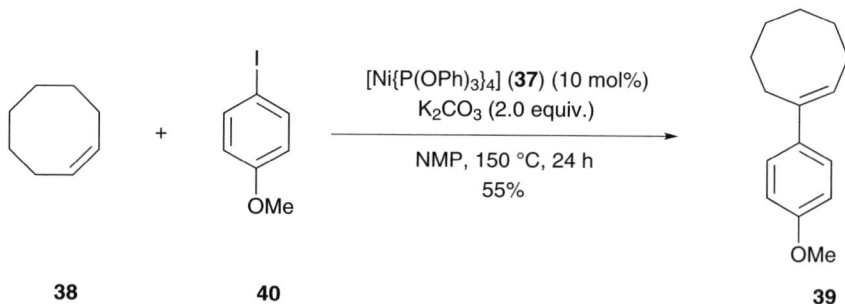

Scheme 10.12 Nickel(0) phosphite complex **37** as catalyst for Mizoroki–Heck-type reactions.

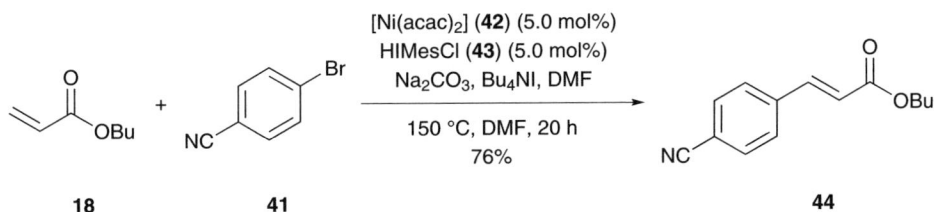

Scheme 10.13 In situ *generated nickel–carbene complex for a Mizoroki–Heck-type reaction with aryl bromide* **41**.

complex. Notably, a relatively strong reducing agent was again not required and relatively mild bases, such as Cs_2CO_3 or, preferably, Na_2CO_3, could be employed. Thus, a number of valuable functional groups were tolerated by the catalytic system. While aryl iodides were efficiently converted, the use of aryl bromides required stoichiometric amounts of Bu_4NI as additive (Scheme 10.13).

For the conversion of more challenging aryl chlorides, a pincer-type nickel(II) complex proved superior (Scheme 10.14) [38]. However, both the *in situ* generated complex and the air-stable well-defined nickel(II) carbene complex **45** required 3 equiv Bu_4NI as additive. Additionally, the protocol proved only highly effective for the conversion of electron-poor aryl chlorides.

Recently, Nakao *et al.* [39] reported an interesting example of a nickel-catalysed Mizoroki–Heck-type reaction employing aryl cyanide **47** as arylating reagent (Scheme 10.15). Although only a single example was disclosed, this report highlights the possibility of employing electrophiles other than halides in nickel-catalysed Mizoroki–Heck-type coupling reactions.

For some applications, the use of homogeneous catalysts can generally cause significant practical problems, since the separation of products from the catalysts, as well as their recyclability, can be rather difficult to achieve. Therefore, Arai and coworkers [40] probed various transition metal salts in a biphasic reaction mixture for the arylation of butyl acrylate (**18**) with iodobenzene (**13**). The biphasic reaction system consisted of ethylene glycol and toluene. Under otherwise identical reaction conditions, the following reactivity trend was

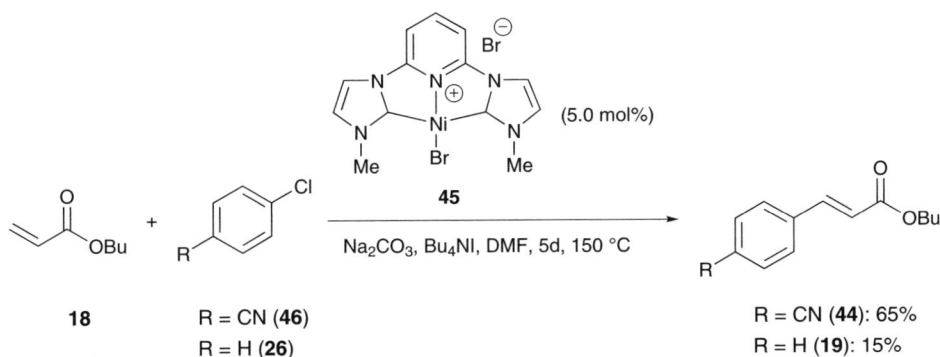

Scheme 10.14 *Pincer-type nickel(II) complex* **45** *for Mizoroki–Heck-type reactions with aryl chlorides.*

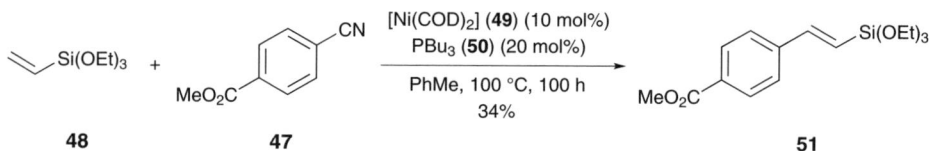

Scheme 10.15   *Nickel-catalysed arylation of alkene **48** with aryl cyanide **47**.*

observed: Pd > Ni > Ru > Co > Pt; however, no desired product was formed with rhodium. Consequently, the scope of the nickel-catalysed reaction was further explored (Scheme 10.16). However, the conversions of iodobenzene (**13**) with two acrylates were still mediocre, and the use of styrene (**2**) led to unsatisfactory results [40].

Arai and coworkers [41] also studied a heterogeneous Ni–TPPTS (TPPTS = triphenylphosphine trisulfonate sodium salt) catalyst immobilized in an ethylene glycol film on silica support for the arylation of butyl acrylate (**18**) with iodobenzene (**13**). The catalyst could be recycled three times without loss of activity, when the catalytic reactions were performed at 140 °C. As for the biphasic system previously reported [40], KOAc as base gave optimal results [41].

Nickel-, cobalt-, copper- and manganese-based heterogeneous catalysts were tested in arylation reactions of styrene (**2**) and various acrylates employing aryl iodides and alkenyl bromides [24]. The most efficient catalysis was again obtained with supported nickel catalysts. However, aryl bromides or chlorides could not be converted with any of the catalysts studied, despite the rather high reaction temperature of 150 °C.

Further, silica-supported poly-γ-aminopropylsilane transition metal complexes derived from Ni(II), Cu(II) and Co(II) salts were probed in arylation reactions of acrylic acid (**16**), methyl acrylate (**1**) and styrene (**2**) using iodobenzene derivatives. The most efficient catalysis was again observed with the nickel-based recyclable catalyst [23, 25].

More recently, nanosized polyaniline/Ni(0) composites were used as heterogeneous catalysts for arylations of acrylates and styrene (**2**) [42]. Efficient catalysis was achieved with aryl iodides, while bromides gave only poor conversions. In contrast to the previously developed nickel-based heterogeneous catalysts, the use of inorganic bases led only to poor yields.

10.4   Platinum-Catalysed Mizoroki–Heck-Type Reactions

An early platinum-catalysed Mizoroki–Heck-type arylation of an alkene with an organic halide was accomplished with [Pt(COD)Cl$_2$] (**54**) as metal precursor and PPh$_3$ (**35**) as ligand (Scheme 10.17) [43]. Relatively high reaction temperatures, an inorganic base and

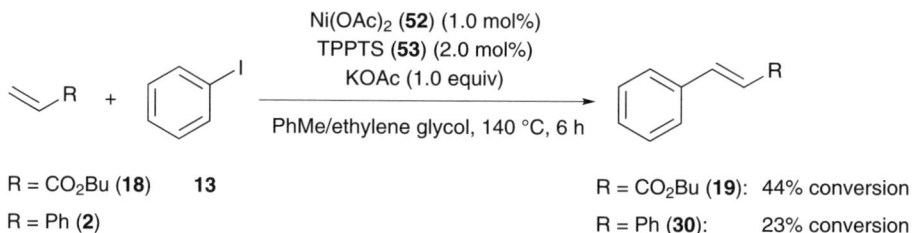

R = CO$_2$Bu (**18**)   **13**   R = CO$_2$Bu (**19**): 44% conversion

R = Ph (**2**)   R = Ph (**30**):   23% conversion

Scheme 10.16   *Nickel-catalysed Mizoroki–Heck-type reactions with TPPTS (**53**) as ligand.*

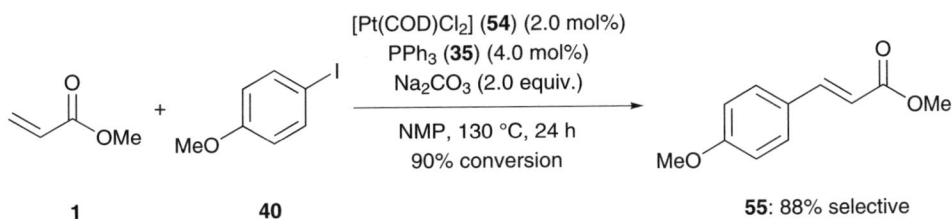

Scheme 10.17 *Homogeneous platinum-catalysed arylation of methyl acrylate (1).*

NMP as highly polar solvent proved essential for effective catalysis. While methyl acrylate (**1**) was converted well, other alkenes, such as styrene (**2**), reacted only sluggishly.

More recently, quinoline–imine platinum complex **56** was reported as heterogeneous catalyst for Mizoroki–Heck-type arylations of methyl acrylate (**1**) with various aryl iodides (Scheme 10.18). However, the use of this modified surface of FSM-16 mesoporous silica-immobilized complex **56** proved significantly less effective than the corresponding palladium or ruthenium catalysts tested in this report [44]. Generally, for both the ruthenium and platinum catalysts, high reaction temperatures and NMP as highly polar solvent were found necessary. Interestingly, NEt$_3$ as base was found superior to inorganic bases.

On the contrary, for *l*ayered *d*ouble-*h*ydroxide-supported nanoplatinum (LDH–Pt(0)) catalyst **57**, highly efficient arylations of styrenes and methyl acrylate (**1**) were accomplished with the inorganic base NaOAc (Scheme 10.19) [45]. As for the previously reported platinum-based catalysts, high reaction temperatures were mandatory. The reusable catalyst was found to be further limited to the use of iodoarenes as electrophiles, since bromoarenes gave no Mizoroki–Heck-type coupling at all.

10.5 Cobalt-Catalysed Mizoroki–Heck-Type Reactions

An organometallic reaction which shows a close similarity to the Mizoroki–Heck reaction is the cobalt-catalysed [46] reaction between alkenes and organic halides. The use of [CoCl(PPh$_3$)$_3$] (**59**) as catalyst for intermolecular arylations of methyl acrylate (**1**) and styrene (**2**) was reported by Iyer [47] (Scheme 10.20). The *para*-substituted aryl iodides could be employed with this homogeneous catalyst, but the more sterically hindered *ortho*-substituted iodoarenes failed to undergo the desired substitution reaction. Aryl bromides and chlorides, as well as alkyl-substituted halides, proved unreactive under these reaction conditions.

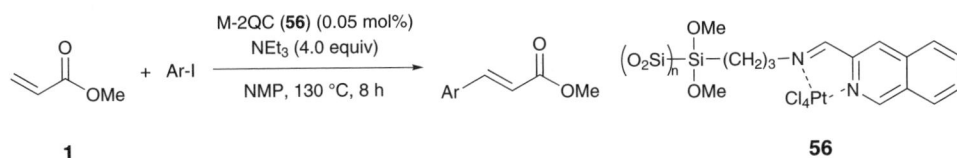

Scheme 10.18 *Heterogeneous platinum-catalysed arylations of methyl acrylate (1) with complex 56.*

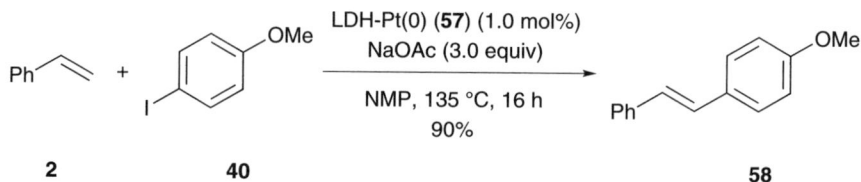

Scheme 10.19 *Platinum-catalysed intermolecular Mizoroki–Heck-type arylation of styrene (2).*

During their comparative study on the use of various transition-metal–TPPTS catalysts under biphasic reaction conditions, Aria and coworkers [40] showed that a cobalt–TPPTS complex could be employed for the arylation of acrylate **18** with iodobenzene (**13**) (Scheme 10.21). However, the selectivity of this transformation is significantly reduced due to a competitive reductive hydrodehalogenation of electrophile **13**.

Cobalt-catalysed electrochemical arylation reactions of acrylates were achieved by Gosmini and coworkers. The presence of 2,2′-bipyridine (Bpy, **63**) was found crucial to reduce the formation of conjugate addition products in this transformation. Notably, this Mizoroki–Heck-type reaction proved applicable to aryl iodides and bromides and to an alkenyl chloride (Scheme 10.22) [48].

A heterogeneous cobalt catalyst was employed for arylations of styrene (**2**) and two acrylates with aryl iodides. Generally, isolated yields were significantly lower than those observed for heterogeneous nickel catalysts [24]. Further, a silica-supported poly-γ-aminopropylsilane cobalt(II) complex was reported as a highly active and stereoselective catalyst for Mizoroki–Heck-type reactions of styrene (**2**) and acrylic acid (**16**) using aryl iodides [23, 25].

Recently, cobalt hollow nanospheres were successfully used in stereoselective Mizoroki–Heck-type reactions of acrylates with aryl bromides and iodides [49]. The recyclable catalyst was most efficient when using NMP and K_2CO_3 as solvent and base respectively (Scheme 10.23).

The palladium-catalysed Mizoroki–Heck reaction is a highly useful tool in synthetic chemistry. Owing to this practical importance, its scope and limitations were thoroughly studied during the recent decades. Thus, it became apparent that aryl and alkenyl (pseudo)halides usually serve well as electrophiles. On the contrary, generally applicable protocols for palladium-catalysed Mizoroki–Heck-type reactions using simple alkyl halides bearing hydrogen(s) in the β-position to the halide have proven elusive. This is largely due

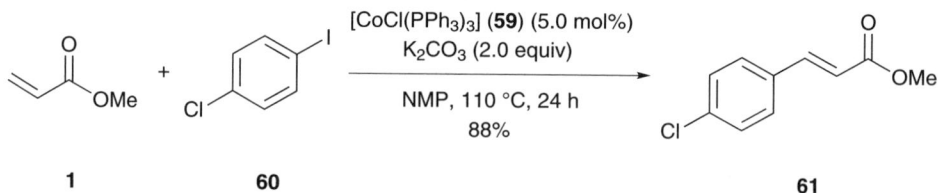

Scheme 10.20 *Cobalt-catalysed intermolecular Mizoroki–Heck-type reaction.*

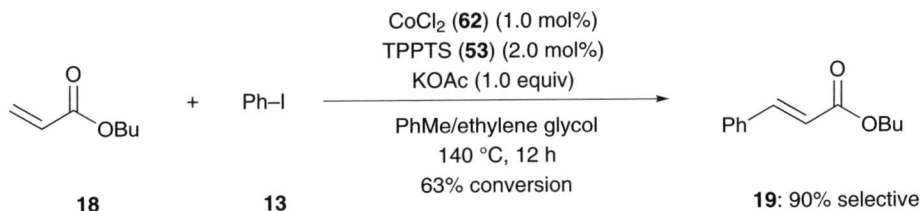

Scheme 10.21 *Cobalt-catalysed Mizoroki–Heck-type arylation under biphasic reaction conditions.*

to the competing β-hydride elimination of the intermediate σ-alkylpalladium(II) species. A valuable alternative is found with complexes of transition metals other than palladium, such as nickel [32] (see above) or titanium [50], for Mizoroki–Heck-type reactions [51].

An early example for cobalt-catalysed Mizoroki–Heck-type reactions with aliphatic halides by Branchaud and Detlefsen showed that an intermolecular substitution of styrene (**2**) could be achieved with [Co(dmgH)₂py] (**70**) (dmgH = dimethylglyoxime monoanion) as catalyst in the presence of visible light. This radical reaction led selectively to the substitution products when using stoichiometric amounts of Zn (**27**) and pyridine (**31**) as additives (Scheme 10.24) [52].

A more efficient and more generally applicable cobalt-catalysed Mizoroki–Heck-type reaction with aliphatic halides was elegantly developed by Oshima and coworkers. A catalytic system comprising CoCl₂ (**62**), 1,6-bis(diphenylphosphino)hexane (dpph; **73**)) and Me₃SiCH₂MgCl (**74**) allowed for intermolecular substitution reactions of alkenes with primary, secondary and tertiary alkyl halides (Scheme 10.25) [51, 53]. The protocol was subsequently applied to a cobalt-catalysed synthesis of homocinnamyl alcohols starting from epoxides and styrene (**2**) [54].

The efficiency of the cobalt-catalysed radical reaction was further used for cyclization reactions of 6-halo-1-hexenes (Scheme 10.26) [55]. Thus, alkyl iodides gave the corresponding exo-methylene products in high yields of isolated products.

The catalytic system proved not only applicable to alkyl halides, but also allowed for the intramolecular conversion of aryl halides. Interestingly, the corresponding Mizoroki–Heck-type cyclization products were formed selectively, without traces of reduced side-products (Scheme 10.27) [55]. Therefore, a radical reaction via a single electron-transfer process was generally disregarded for cobalt-catalysed Mizoroki–Heck-type reactions of aromatic halides. Instead, a mechanism based on oxidative addition to yield an aryl–cobalt complex was suggested [51].

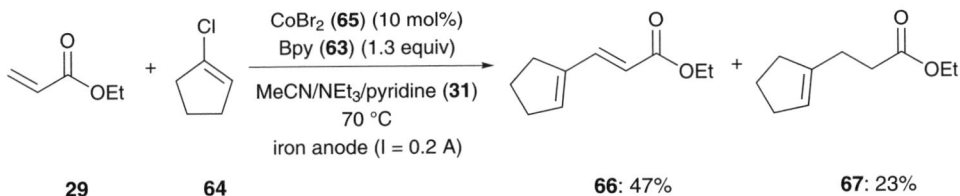

Scheme 10.22 *Electrochemical cobalt-catalysed Mizoroki–Heck-type reaction.*

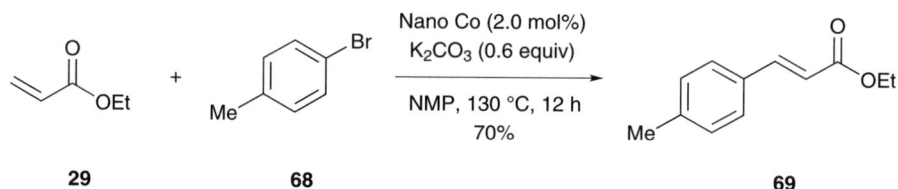

Scheme 10.23   *Cobalt nanospheres in Mizoroki–Heck-type reactions.*

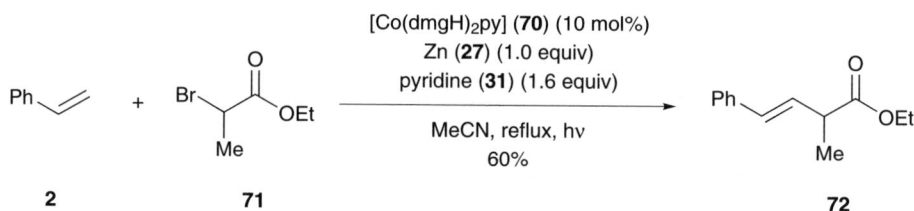

Scheme 10.24   *Cobalt-catalysed photochemical substitution of styrene (2).*

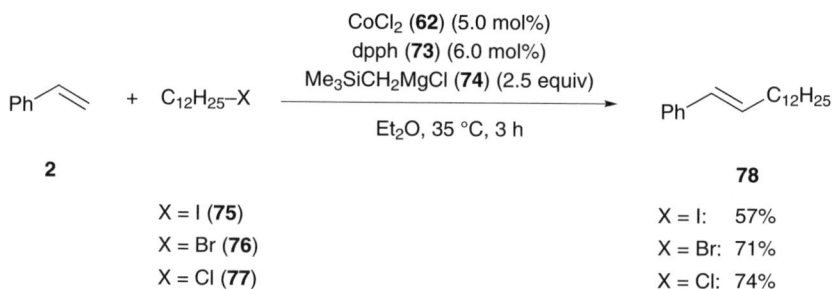

X = I (**75**)   X = I: 57%
X = Br (**76**)   X = Br: 71%
X = Cl (**77**)   X = Cl: 74%

Scheme 10.25   *Cobalt-catalysed intermolecular Mizoroki–Heck-type reaction with alkyl halides.*

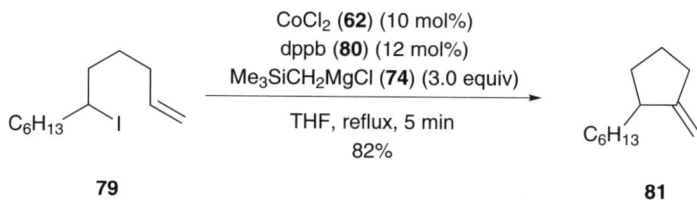

Scheme 10.26   *Intramolecular radical cobalt-catalysed Moziroki–Heck-type reaction.*

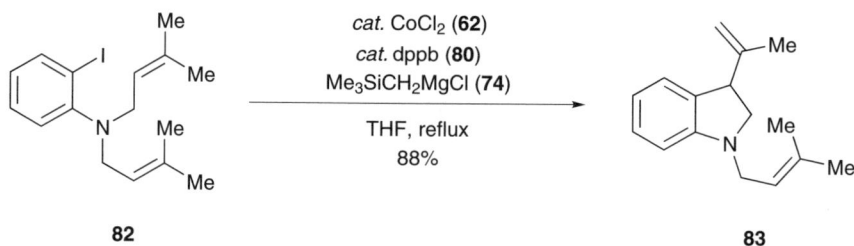

Scheme 10.27   *Intramolecular cobalt-catalysed Mizoroki–Heck-type reaction with an aryl iodide.*

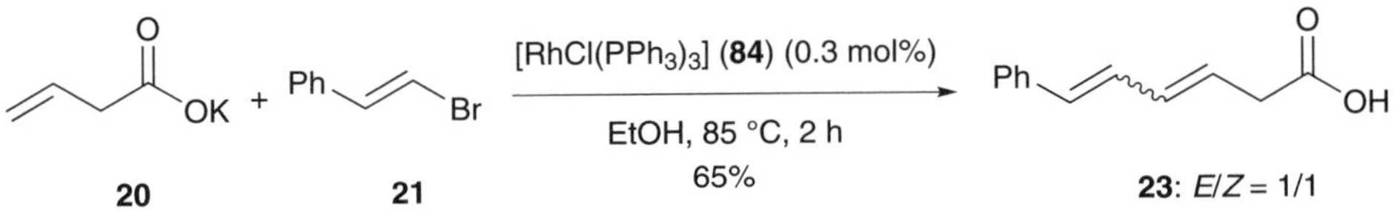

Scheme 10.28   *Rhodium-catalysed alkenylation of potassium 3-butenoate (20).*

10.6 Rhodium-Catalysed Mizoroki–Heck-Type Reactions

An early report on rhodium-catalysed Mizoroki–Heck-type arylations of alkene **20** using two alkenyl bromides was disclosed by Chiusoli *et al.* [28]. Thus, catalytic amounts of [RhCl(PPh$_3$)$_3$] (**84**) allowed for conversion of potassium 3-butenoate (**20**) (Scheme 10.28), a substrate that can potentially direct the rhodium catalyst through coordination of its carboxylate group.

In an intramolecular Mizoroki–Heck-type alkenylation of alkenes, Wilkinson's catalyst [RhCl(PPh$_3$)$_3$] (**84**) proved superior than other rhodium compounds (Scheme 10.29) [56]. Notably, the selectivities of the cyclization of various 2-bromo-1,6-dienes were found to be improved with Wilkinson's catalyst when compared with those observed with palladium complexes. Thereby, the corresponding 1,2-bis(methylene)cyclopentanes, which themselves are valuable substrates for further cycloaddition reactions, such as 1,3-diene **86**, could be isolated in high yields.

In a related intramolecular rhodium-catalysed Mizoroki–Heck-type reaction of an alkene with an aryl iodide, Wilkinson's catalyst (**84**) gave significant amounts of side-products due to isomerization of the resulting double bond. In contrast to the corresponding palladium-catalysed transformation, the presence of Et$_4$NCl had no beneficial influence either on the reactivity or on the selectivity [57].

In a comparative study, Wilkinson's catalyst (**84**) was employed in intermolecular arylation reactions of methyl acrylate (**1**) and styrene (**2**). Rhodium catalyst **84** was reported to be more efficient than a cobalt precursor. Note, however, that the reproducibility of these results was recently questioned [58]. Simple aryl iodides were efficiently converted at 110 °C (Scheme 10.30), but *ortho*-substituted aryl iodides required a higher reaction temperature of 150 °C. Aryl bromides and chlorides, as well as aliphatic halides, could not be converted using rhodium catalyst **84** [47].

As an alternative to the use of aryl halides as electrophiles, Miura and coworkers [59] employed aroyl chlorides in rhodium-catalysed decarbonylative Mizoroki–Heck-type

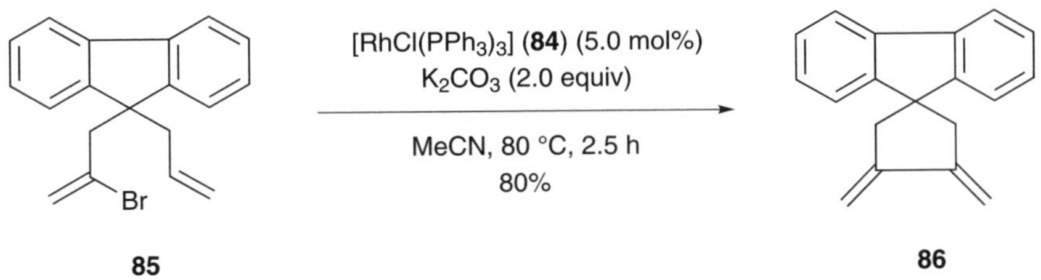

Scheme 10.29   *Rhodium-catalysed intramolecular Mizoroki–Heck-type reaction.*

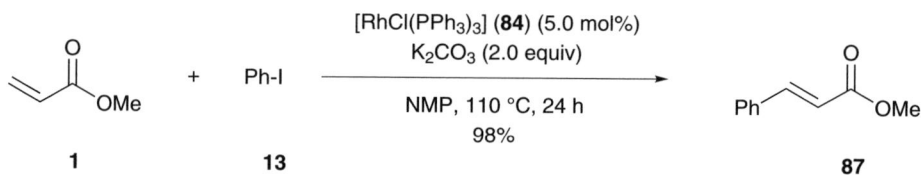

Scheme 10.30 *Rhodium-catalysed intermolecular Mizoroki–Heck-type reaction with Wilkinson's catalyst (**84**).*

reactions (Scheme 10.31). Importantly, these electrophiles allowed for base-free reaction conditions and the sole by-products were gaseous HCl and CO. Rhodium–alkene complexes without phosphine ligands were found most efficient and enabled arylations of styrenes and acrylates [58]. Importantly, a chemoselective functionalization was achieved with substrate **88** (Scheme 10.31).

Dubbaka and Vogel reported on rhodium-catalysed Mizoroki–Heck-type reactions employing arenesulfonyl chlorides as inexpensive electrophiles. Thereby, styrene derivatives were arylated under base-free conditions. Again, rhodium–alkene complex **89** was found to be significantly more active than phosphine-derived rhodium species (Scheme 10.32) [60].

10.7   Iridium-Catalysed Mizoroki–Heck-Type Reactions

As part of comparative studies, Iyer [47] reported the use of Vaska's complex [IrCl(CO)(PPh$_3$)$_2$] (**92**) in intermolecular Mizoroki–Heck-type reactions of methyl acrylate (**1**) and styrene (**2**). Aryl iodides could be used as electrophiles, while bromobenzene, chlorobenzene and aliphatic halides gave no desired product. The catalytic activity was found to be lower than that observed when using Wilkinson's complex [RhCl(PPh$_3$)$_3$] (**84**). Thus, a higher reaction temperature of 150 °C was mostly required. In contrast to the corresponding cobalt-catalysed reaction, however, Vaska's complex (**92**) proved applicable to *ortho*-substituted aryl iodides (Scheme 10.33).

The potential of iridium-catalysed arylations of alkenes with organic halides as electrophiles has thus far not been explored further. However, a notable iridium-catalysed arylation of alkenes with organosilicon reagents was disclosed recently [15].

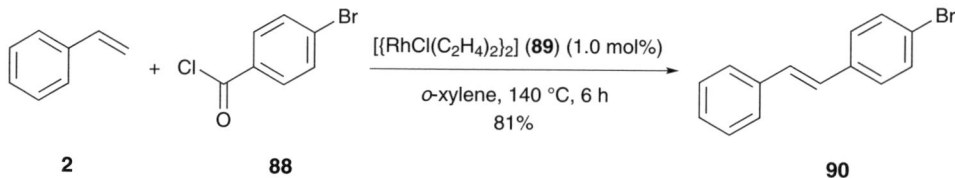

Scheme 10.31 *Rhodium-catalysed decarbonylative Mizoroki–Heck-type reaction.*

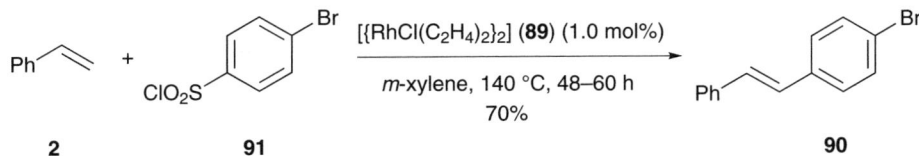

Scheme 10.32 *Rhodium-catalysed Mizoroki–Heck-type reaction with arylsulfonyl chloride 91.*

10.8 Ruthenium-Catalysed Mizoroki–Heck-Type Reactions

In 1985, Kamigata *et al.* [61] reported on ruthenium-catalysed alkenylations of alkenes, employing alkenesulfonyl chlorides. High yields of isolated product were obtained with [RuCl$_2$(PPh$_3$)$_3$] (**94**) as catalyst in benzene as solvent (Scheme 10.34). Importantly, no base was required for the coupling process, for which a radical-based mechanism was proposed.

Subsequently, the ruthenium-catalysed alkenylation of various acrylates was accomplished with alkenyl halides [62]. Most effective catalysis was achieved with [Ru(COD)(COT)] (**98**) as catalyst and NEt$_3$ as base in the absence of additional solvent. Interestingly, both alkenyl bromides and chlorides could be employed as electrophiles (Scheme 10.35). When using an alkenyl chloride, the catalytic activity could be improved through the addition of P(p-C$_6$H$_4$F)$_3$ (**99**) as ligand. The efficiency of this ruthenium catalyst in the alkenylation of β-chlorostyrene (**21**) compared favourably with that observed for either Pd(OAc)$_2$ or Pd(OAc)$_2$/P(o-Tol)$_3$ as catalysts. With respect to the working mode of the catalyst, a radical mechanism was shown to be less likely. Instead, Mitsudo *et al.* [62] proposed an initial oxidative addition of the alkenyl halide to a ruthenium(0) species followed by insertion of the alkene and β-hydride elimination, all in analogy to palladium-catalysed processes.

Arai and coworkers [40] showed that catalytic amounts of RuCl$_3$ (**101**) modified with TPPTS (**53**) allowed for the arylation of butyl acrylate (**18**) with iodobenzene (**13**) in a biphasic system (Scheme 10.36). The catalytic activity of this ruthenium catalyst was found inferior compared with the palladium or nickel catalysts.

Studies on the use of complexes immobilized on quinoline–carboimine-functionalized FSM-16 mesoporous silica for Mizoroki–Heck-type reactions between methyl acrylate (**1**) and various aryl iodides highlighted the superior catalytic activity of a ruthenium(III) catalyst compared with the corresponding platinum(IV) complex [44].

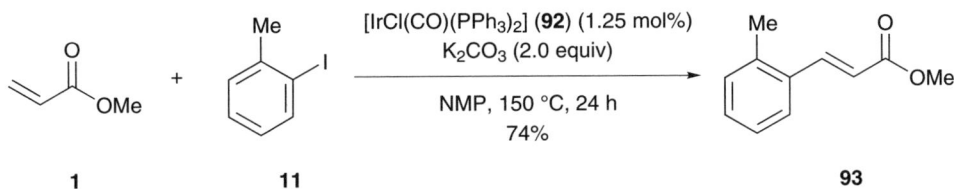

Scheme 10.33 *Intermolecular Mizoroki–Heck-type reaction catalysed by Vaska's complex (**92**).*

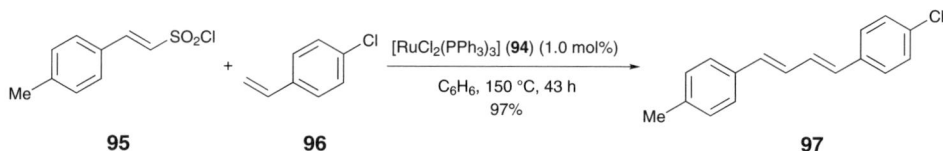

Scheme 10.34   *Ruthenium-catalysed alkenylation with alkenesulfonyl chloride 95.*

Recently, Chang and coworkers [63] developed elegantly a highly efficient ruthenium catalyst for arylations and alkenylations of conjugated alkenes. While studying soluble molecular ruthenium catalysts, they found that these served as precursors to ruthenium(0) colloids. Consequently, Ru/Al$_2$O$_3$ (**102**) was unravelled as a reusable heterogeneous catalyst for Mizoroki–Heck-type reactions with ample scope regarding the conjugated alkene. With respect to the electrophile, the catalyst proved limited to aryl iodides and alkenyl bromides. This allowed for a chemoselective functionalization of electrophile **103** (Scheme 10.37). The high catalytic efficiency is reflected by the comparably low reaction temperature.

Just like all other transition metal catalysts discussed herein, Chang and coworkers's ruthenium catalyst provided good to excellent stereoselectivities in Mizoroki–Heck-type functionalization reactions of alkenes, yielding predominantly or exclusively the corresponding (*E*)-isomer. A stereochemistry that is not only complementary to these catalysts, but also to the commonly employed palladium complexes (Scheme 10.38, (*b*)), was achieved recently [64, 65]. Thus, a ruthenium catalyst allowed for regio- and diastereoselective arylations of 2-pyridyl-substituted alkenes via a C–H bond functionalization process (Scheme 10.38, (*a*)).

Among aryl halides, aryl chlorides are arguably the most attractive class of substrates for coupling reactions, due to their lower price and higher availability. For *traditional* cross-coupling reactions of organometallic species, the development of stabilizing ligands allowed for the use of these readily available inexpensive electrophiles [66]. On the contrary, generally applicable methods for regioselective direct arylation reactions employing aryl chlorides were not available until recently [67–70]. In this context, the authors' research group reported on unprecedented ruthenium-catalysed arylation reactions of alkenes employing aryl chlorides and displaying ample scope [65]. Thus, aryl chlorides with both electron-withdrawing groups and electron-donating groups could be converted in high yields of isolated product. Interestingly, ruthenium precursors in various oxidation states could be employed as catalyst, with ruthenium alkylidene **108** providing the most efficient catalysis. Thereby, even *ortho*-substituted electron-rich aryl chlorides gave high yields of the desired product (Scheme 10.39). Generally, the diastereoselectivity proved complementary to palladium-catalysed Mizoroki–Heck-type reactions of aryl chlorides [71],

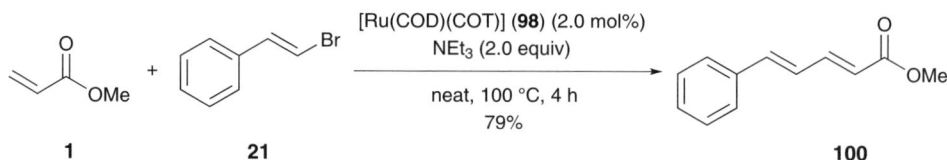

Scheme 10.35   *Ruthenium-catalysed alkenylation of methyl acrylate (1).*

RuCl₃ (**101**) (1.0 mol%)
TPPTS (**53**) (3.0 mol%)
KOAc (1.0 equiv)

toluene / ethylene glycol
140 °C, 12 h
71% conversion

18 **13** **19**: 98.5% selective

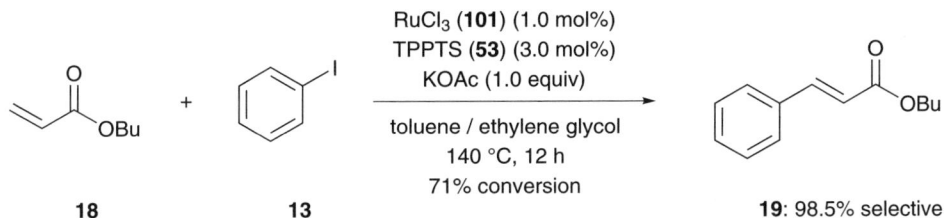

Scheme 10.36 *Ruthenium-catalysed Mizoroki–Heck-type reaction using iodobenzene (**13**).*

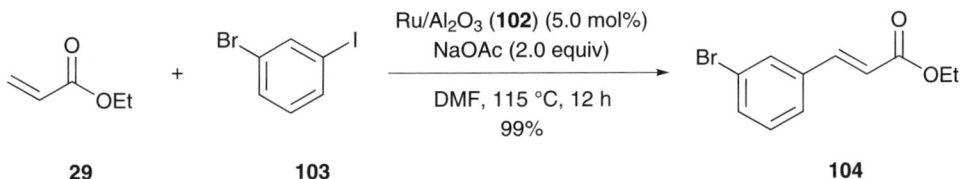

Ru/Al₂O₃ (**102**) (5.0 mol%)
NaOAc (2.0 equiv)

DMF, 115 °C, 12 h
99%

29 **103** **104**

Scheme 10.37 *Ru/Al₂O₃ as catalyst for a chemoselective arylation of alkene **29**.*

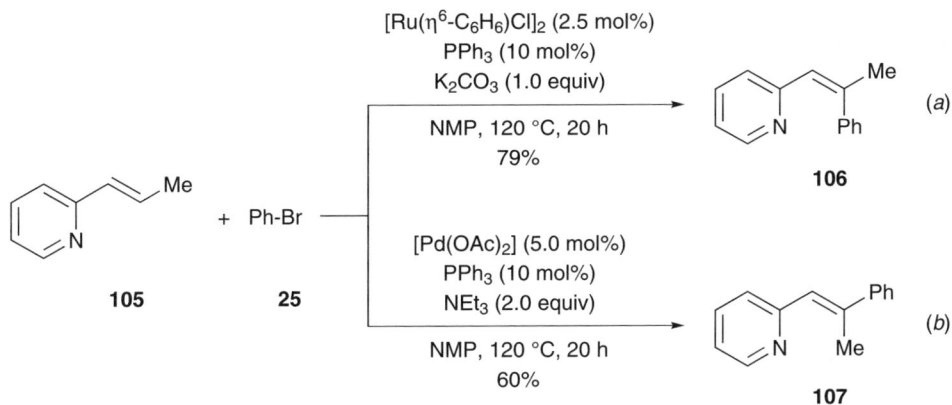

[Ru(η⁶-C₆H₆)Cl]₂ (2.5 mol%)
PPh₃ (10 mol%)
K₂CO₃ (1.0 equiv)

NMP, 120 °C, 20 h
79%

(*a*)

106

+ Ph-Br

[Pd(OAc)₂] (5.0 mol%)
PPh₃ (10 mol%)
NEt₃ (2.0 equiv)

NMP, 120 °C, 20 h
60%

(*b*)

105 **25** **107**

Scheme 10.38 *Ruthenium- versus palladium-catalysed arylation of alkene **105**.*

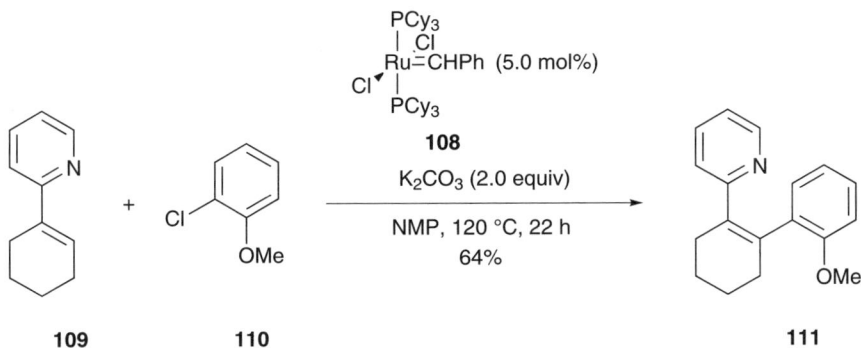

PCy₃
| Cl
Ru=CHPh (5.0 mol%)
Cl |
PCy₃

108

K₂CO₃ (2.0 equiv)

NMP, 120 °C, 22 h
64%

109 **110** **111**

Scheme 10.39 *Ruthenium-catalysed arylation of alkene **109** with aryl chloride **110**.*

providing exclusively the corresponding (Z)-diastereomers [65]. Importantly, the catalytic activity also allowed for reactions to be performed at a lower temperature of 100 °C.

10.9   Conclusions

Various metals other than palladium were employed for catalytic Mizoroki–Heck-type functionalization reactions of alkenes. While none of them can yet rival palladium in its practical diversity, promising alternatives were developed. Thus, relatively inexpensive transition metals, such as copper, cobalt and particularly nickel, exhibited high activities in arylations or alkenylations of alkenes. The use of stabilizing ligands allowed for an increase of efficiency and a broadening of scope, especially with these attractive transition metals. While platinum- and iridium-based catalysts have thus far displayed rather limited scopes, rhodium and particularly ruthenium complexes enabled efficient Mizoroki–Heck-type reactions. Thus, generally applicable protocols for the use of aryl chlorides were revealed. Finally, the use of metals other than palladium for Mizoroki–Heck-type reactions proved in certain applications complementary to the well-established palladium-catalysed methodologies. While cobalt complexes, among others, allowed for the use of alkyl halides with β-hydrogen atoms, ruthenium catalysts enabled arylation reactions with excellent (Z)-diastereoselectivities.

References

1. Mizoroki, T., Mori, K. and Ozaki, A. (1971) Arylation of olefin with aryl iodide catalyzed by palladium. *Bull. Chem. Soc. Jpn.*, **44**, 581.
2. Heck, R.F. and Nolley, J.P. Jr. (1972) Palladium-catalyzed vinylic hydrogen substitution reactions with aryl, benzyl, and styryl halides. *J. Org. Chem.*, **37**, 2320.
3. Beller, M. and Bolm, C. (eds) (2004) *Transition Metals for Organic Synthesis*, 2nd edn, Wiley-VCH Verlag GmbH, Weinheim.
4. De Meijere, A. and Diederich, F. (eds) (2004) *Metal-Catalyzed Cross-Coupling Reactions*, 2nd edn, Wiley-VCH Verlag GmbH, Weinheim.
5. Beletskaya, I.P. and Cheprakov, A.V. (2000) The Heck reaction as a sharpening stone of palladium catalysis. *Chem. Rev.*, **100**, 3009–66.
6. For example: (a) Weissman, H., Song, X. and Milstein, D. (2001) Ru-catalyzed oxidative coupling of arenes with olefins using O_2. *J. Am. Chem. Soc.*, **123**, 337–8; (b) Kakiuchi, F., Sato, T., Yamauchi, M. *et al.* (1999) Ruthenium-catalyzed coupling of aromatic carbon–hydrogen bonds in aromatic imidates with olefins. *Chem. Lett.*, **28**, 19–20.
7. Mori, A., Danda, Y., Fujii, T. *et al.* (2001) Hydroxorhodium complex-catalyzed carbon–carbon bond-forming reactions of silanediols with α,β-unsaturated carbonyl compounds. Mizoroki–Heck-type reactions vs conjugate addition. *J. Am. Chem. Soc.*, **123**, 10774–5.
8. Lautens, M., Roy, A., Fukuoka, K. *et al.* (2001) Rhodium-catalyzed coupling reactions of arylboronic acids to olefins in aqueous media. *J. Am. Chem. Soc.*, **123**, 5358–9.
9. (a) Amengual, R., Michelet, V. and Genêt, J.-P. (2002) New studies of Rh-catalyzed addition of boronic acid under basic conditions in aqueous medium. *Tetrahedron Lett.*, **43**, 5905–8; (b) Martinez, R., Voica, F., Genet, J.-P. and Darses, S. (2007) Base-free Mizoroki–Heck reaction catalyzed by rhodium complexes. *Org. Lett.*, **9**, 3213–6.

10. Zou, G., Wang, Z., Zhu, J. and Tang, J. (2003) Rhodium-catalyzed Heck-type reaction of arylboronic acids with α,β-unsaturated esters: tuning β-hydrogen elimination versus hydrolysis of alkylrhodium species. *Chem. Commun.*, 2438–9.

11. Lautens, M., Mancuso, J. and Grover, H. (2004) Rhodium-catalyzed Heck-type coupling of boronic acids with activated alkenes in an aqueous emulsion. *Synthesis*, 2006–14.

12. De la Herrán, G., Murcia, C. and Csákÿ, A.G. (2005) Rhodium-catalyzed reaction of aryl- and alkenylboronic acids with 2,4-dienoate esters: conjugate addition and Heck reaction products. *Org. Lett.*, **7**, 5629–32.

13. Farrington, E.J., Brown, J.M., Barnard, C.F.J. and Rowsell, E. (2002) Ruthenium-catalyzed oxidative Heck reactions. *Angew. Chem. Int. Ed.*, **41**, 169–71.

14. Farrington, E.J., Barnard, C.F.J., Rowsell, E. and Brown, J.M. (2005) Ruthenium complex-catalysed Heck reactions of areneboronic acids; mechanism, synthesis and halide tolerance. *Adv. Synth. Catal.*, **347**, 185–95.

15. Koike, T., Du, X., Sanada, T. *et al.* (2003) Iridium-catalyzed Mizoroki–Heck-type reaction of organosilicon reagents. *Angew. Chem. Int. Ed.*, **42**, 89–92.

16. Ishiyama, T. and Hartwig, J.F. (2000) A Heck-type reaction involving carbon–heteroatom double bonds. Rhodium(I)-catalyzed coupling of aryl halides with *N*-pyrazyl aldimines. *J. Am. Chem. Soc.*, **122**, 12043–4.

17. Ley, S.V. and Thomas, A.W. (2003) Modern synthetic methods for copper-mediated C(aryl)–O, C(aryl)–N, and C(aryl)–S bond formation. *Angew. Chem. Int. Ed.*, **42**, 5400–49.

18. Kunz, K., Scholz, U. and Ganzer, D. (2003) Renaissance of Ullmann and Goldberg reactions – progress in copper catalyzed C–N-, C–O- and C–S-coupling. *Synlett*, 2428–39.

19. Iyer, S., Ramesh, C., Sarkar, A. and Wadgaonkar, P.P. (1997) The vinylation of aryl and vinyl halides catalyzed by copper salts. *Tetrahedron Lett.*, **38**, 8113–6.

20. Li, J.-H., Wang, D.-P. and Xie, Y.-X. (2005) CuI/Dabco as a highly active catalytic system for the Heck-type reaction. *Tetrahedron Lett.*, **46**, 4941–4.

21. Loupy, A. (ed.) (2002) *Microwaves in Organic Synthesis*, Wiley-VCH Verlag GmbH, Weinheim.

22. Declerck, V., Martinez, J. and Lamaty, F. (2006) Microwave-assisted copper-catalyzed Heck reaction in PEG solvent. *Synlett*, 3029–32.

23. Yang, Y., Zhou, R., Zhao, S. *et al.* (2003) Silica-supported poly-γ-aminopropylsilane Ni^{2+}, Cu^{2+}, Co^{2+} complexes: efficient catalysts for Heck vinylation reaction. *J. Mol. Catal. A*, **192**, 303–6.

24. Iyer, S. and Thakur, V.V. (2000) The novel use of Ni, Co, Cu and Mn heterogenous catalysts for the Heck reaction. *J. Mol. Catal. A*, **157**, 275–8.

25. Yang, Y., Zhou, R., Zhao, S. and Zheng, X. (2003) The vinylation of aryl iodides catalyzed by silica-supported poly-γ-aminopropylsilica Ni, Cu, Co complexes. *Catal. Lett.*, **85**, 87–90.

26. Calò, V., Nacci, A., Monopoli, A. *et al.* (2005) Copper bronze catalyzed Heck reaction in ionic liquids. *Org. Lett.*, **7**, 617–20.

27. Lin, B.-L., Liu, L., Fu, Y. *et al.* (2004) Comparing nickel- and palladium-catalyzed Heck reactions. *Organometallics*, **23**, 2114–23.

28. Chiusoli, G.P., Salerno, G., Giroldini, W. and Pallini, L. (1981) Transition metal-catalyzed synthesis of organic acids by chelation-promoted regioselective double bond insertion. *J. Organomet. Chem.*, **219**, C16–20.

29. Colon, I. (1982) Substituted olefins. US Patent 4334081

30. Rollin, Y., Meyer, G., Troupel, M. *et al.* (1983) Electrocatalytic assistance by nickel complexes of a coupling reaction between alkenes and aromatic halides leading to substituted olefins. *J. Chem. Soc., Chem. Commun.*, 793–4.

31. Boldrini, G.P., Savoia, D., Tagliavini, E. *et al.* (1986) Nickel-catalyzed coupling of activated alkenes with organic halides. *J. Organomet. Chem.*, **301**, C62–4.

32. Lebedev, S.A., Lopatina, V.S., Petrov, E.S. and Beletskaya, I.P. (1988) Condensation of organic bromides with vinyl compounds catalysed by nickel complexes in the presence of zinc. *J. Organomet. Chem.*, **344**, 253–9.

33. Sustmann, R., Hopp, P. and Holl, P. (1989) Reactions of organic halides with olefins under Ni(0)-catalysis. Formal addition of hydrocarbons to CC-double bonds. *Tetrahedron Lett.*, **30**, 689–92.

34. Kelkar, A.A., Hanaoka, T., Kubota, Y. and Sugi, Y. (1994) The vinylation of bromobiphenyls using homogeneous nickel catalysts. *Catal. Lett.*, **29**, 69–75.

35. Ma, S., Wang, H., Gao, K. and Zhao, F. (2006) Nickel complexes catalyzed Heck reaction of iodobenzene and methyl acrylate. *J. Mol. Catal. A: Chem.*, **248**, 17–20.

36. Iyer, S., Ramesh, C. and Ramani, A. (1997) Ni(0) catalyzed reactions of aryl and vinyl halides with alkenes and alkynes. *Tetrahedron Lett.*, **38**, 8533–6.

37. Inamoto, K., Kuroda, J., Danjo, T. and Sakamoto, T. (2005) Highly efficient nickel-catalyzed Heck reaction using Ni(acac)$_2$/*N*-heterocyclic carbene catalyst. *Synlett*, 1624–6.

38. Inamoto, K., Kuroda, J., Hiroya, K. *et al.* (2006) Synthesis and catalytic activity of a pincer-type bis(imidazolin-2-ylidene) nickel(II) complex. *Organometallics*, **25**, 3095–8.

39. Nakao, Y., Yada, A., Satoh, J. *et al.* (2006) Arylcyanation of norbornene and norbornadiene catalyzed by nickel. *Chem. Lett.*, **35**, 790–1.

40. Bhanage, B.M., Zhao, F.G., Shirai, M. and Arai, M. (1998) Comparison of activity and selectivity of various metal–TPPTS complex catalysts in ethylene glycol–toluene biphasic Heck vinylation reactions of iodobenzene. *Tetrahedron Lett.*, **39**, 9509–12.

41. Bhanage, B.M., Zhao, F., Shirai, M. and Arai, M. (1998) Heck reaction using nickel/TPPTS catalyst and inorganic base on supported ethylene glycol phase. *Catal. Lett.*, **54**, 195–8.

42. Houdayer, A., Schneider, R., Billaud, D. *et al.* (2005) New polyaniline/Ni(0) nanocomposites: synthesis, characterization and evaluation of their catalytic activity in Heck couplings. *Synth. Met.*, **151**, 165–74.

43. Kelkar, A.A. (1996) The vinylation of aryl iodides using homogeneous platinum complex catalyst. *Tetrahedron Lett.*, **37**, 8917–20.

44. Horniakova, J., Nakamura, H., Kawase, R. *et al.* (2005) Pyridine-derived ruthenium and platinum complexes immobilized on ordered mesoporous silica as catalysts for Heck vinylation. *J. Mol. Catal. A*, **233**, 49–54.

45. Kantam, M.L., Roy, M., Roy, S. *et al.* (2006) Layered double hydroxide-supported nanoplatinum: an efficient and reusable ligand-free catalyst for Heck and Stille coupling of iodoarenes. *Synlett*, 2266–8.

46. For an early overview on the use of stoichiometric amounts of cobalt complexes, see: Pattenden, G. (1988) Cobalt-mediated radical reactions in organic synthesis. *Chem. Soc. Rev.*, **17**, 361–82.

47. Iyer, S. (1995) The vinylation of aryl iodides catalysed by Co, Rh and Ir complexes. *J. Organomet. Chem.*, **490**, C27–8.

48. Gomes, P., Gosmini, C., Nédélec, J.-Y. and Périchon, J. (2002) Electrochemical vinylation of aryl and vinyl halides with acrylate esters catalyzed by cobalt bromide. *Tetrahedron Lett.*, **43**, 5901–3.

49. Zhou, P., Li, Y., Sun, P. *et al.* (2007) A novel Heck reaction catalyzed by Co hollow nanospheres in ligand-free condition. *Chem. Commun.*, 1418–20.

50. (a) Terao, J. and Kambe, N. (2001) Titanocene-catalyzed reaction of alkenes and dienes with alkyl halides and chlorosilanes. *J. Synth. Org. Chem. Jpn.*, **59**, 1044–51; (b) Terao, J., Watabe, H., Miyamoto, M. and Kambe, N. (2003) Titanocene-catalyzed alkylation of aryl-substituted alkenes with alkyl halides. *Bull. Chem. Soc. Jpn.*, **76**, 2209–14; (c) Terao, J. and Kambe, N. (2006) Transition metal-catalyzed C–C-bond formation reactions using alkyl halides. *Bull. Chem. Soc. Jpn.*, **79**, 663–72.

51. Affo, W., Ohmiya, H., Fujioka, T. *et al.* (2006) Cobalt-catalyzed trimethylsilylmethylmagnesium-promoted radical alkenylation of alkyl halides: a complement to the Heck reaction. *J. Am. Chem. Soc.*, **128**, 8068–77.

52. Branchaud, B.P. and Detlefsen, W.D. (1991) Cobaloxime-catalyzed radical alkyl-styryl cross-couplings. *Tetrahedron Lett.*, **32**, 6273–6.

53. Ikeda, Y., Nakamura, T., Yorimitsu, H. and Oshima, K. (2002) Cobalt-catalyzed Heck-type reaction of alkyl halides with styrenes. *J. Am. Chem. Soc.*, **124**, 6514–5.

54. Ikeda, Y., Yorimitsu, H., Shinokubo, H. and Oshima, K. (2004) Cobalt-mediated Mizoroki–Heck-type reaction of epoxide with styrene. *Adv. Synth. Catal.*, **346**, 1631–4.

55. Fujioka, T., Nakamura, T., Yorimitsu, H. and Oshima, K. (2002) Cobalt-catalyzed intramolecular Heck-type reaction of 6-halo-1-hexene derivatives. *Org. Lett.*, **4**, 2257–9.

56. Grigg, R., Stevenson, P. and Worakun, T. (1988) The regioselectivity of rhodium and palladium-catalysed cyclisations of 2-bromo-1,6- and 1,7-dienes. *Tetrahedron*, **44**, 2033–48.

57. Grigg, R., Sridharan, V., Stevenson, P. *et al.* (1990) The synthesis of fused ring nitrogen heterocycles via regiospecific intramolecular Heck reactions. *Tetrahedron*, **46**, 4003–18.

58. Sugihara, T., Satoh, T., Miura, M. and Nomura, M. (2004) Rhodium-catalyzed coupling reaction of aroyl chlorides with alkenes. *Adv. Synth. Catal.*, **346**, 1765–72.

59. Sugihara, T., Satoh, T., Miura, M. and Nomura, M. (2003) Rhodium-catalyzed Mizoroki-Heck-type arylation of alkenes with aroyl chlorides under phosphane- and base-free conditions. *Angew. Chem. Int. Ed.*, **42**, 4672–4.

60. Dubbaka, S.R. and Vogel, P. (2005) Palladium-catalyzed desulfitative Mizoroki–Heck coupling of sulfonyl chlorides with mono- and disubstituted olefins: rhodium-catalyzed desulfitative Heck-type reactions under phosphine- and base-free conditions. *Chem. Eur. J.*, **11**, 2633–41.

61. Kamigata, N., Ozaki, J.-i., and Kobayashi, M. (1985) Reaction of alkenesulfonyl chlorides with olefins catalyzed by a ruthenium(II) complex. A novel method for synthesis of (*E,E*)-1,4-diaryl-1,3-butadienes. *J. Org. Chem.*, **50**, 5045–50.

62. Mitsudo, T., Takagi, M., Zhang, S.-W., and Watanabe, Y. (1992) Ruthenium complex-catalyzed coupling of vinyl halides with olefins. *J. Organomet. Chem.*, **423**, 405–14.

63. Na, Y., Park, S., Han, S.B. *et al.* (2004) Ruthenium-catalyzed Heck-type olefination and Suzuki coupling reactions: studies on the nature of catalytic species. *J. Am. Chem. Soc.*, **126**, 250–8.

64. Oi, S., Sakai, K. and Inoue, Y. (2005) Ruthenium-catalyzed arylation of 2-alkenylpyridines with aryl bromides: alternative *E,Z*-selectivity to Mizoroki-Heck reaction. *Org. Lett.*, **7**, 4009–11.

65. Ackermann, L., Born, R. and Álvarez Bercedo, P. (2007) Ruthenium(IV) alkylidenes as precatalysts for direct arylations of alkenes with aryl chlorides and an application to sequential catalysis. *Angew. Chem. Int. Ed.*, **46**, 6364–7.

66. Littke, A.F. and Fu, G.C. (2002) Palladium-catalyzed coupling reactions of aryl chlorides. *Angew. Chem. Int. Ed.*, **41**, 4176–211.

67. Ackermann, L. (2005) Phosphine oxides as preligands in ruthenium-catalyzed arylations via C–H-bond functionalization using aryl chlorides. *Org. Lett.*, **7**, 3123–5.

68. Campeau, L.-C., Parisien, M., Jean, A. and Fagnou, K. (2006) Catalytic direct arylation with aryl chlorides, bromides, and iodides: intramolecular studies leading to new intermolecular reactions. *J. Am. Chem. Soc.*, **128**, 581–90.

69. Lafrance, M., Rowley, C.N., Woo, T.K. and Fagnou, K. (2006) Catalytic intermolecular direct arylation of perfluorobenzenes. *J. Am. Chem. Soc.*, **128**, 8754–6.

70. Chiong, H.A. and Daugulis, O. (2007) Palladium-catalyzed arylation of electron-rich heterocycles with aryl chlorides. *Org. Lett.*, **9**, 1449–51.

71. Littke, A.F. and Fu, G.C. (2001) A versatile catalyst for Heck reactions of aryl chlorides and aryl bromides under mild conditions. *J. Am. Chem. Soc.*, **123**, 6989–7000.

11

Ligand Design for Intermolecular Asymmetric Mizoroki–Heck Reactions

Anthony G. Coyne, Martin O. Fitzpatrick and Patrick J. Guiry

Centre for Synthesis and Chemical Biology, UCD School of Chemistry and Chemical Biology, University College Dublin, Belfield, Dublin 4, Ireland

11.1 Introduction

The standard Mizoroki–Heck reaction is the substitution of a vinylic hydrogen by an alkenyl or aryl group catalysed by palladium(0) complexes (Scheme 11.1). Since its discovery in 1968 by Heck [1–3], this elaboration of substituted alkenes by direct C–C bond formation at the vinylic carbon centre has evolved into a synthetic transformation whose potential has only recently been exploited in the key steps of many total syntheses (Chapter 16) [4]. This recent exploitation has been due to a better understanding of the proper choice of reactants, solvent, base, additives, catalyst precursor and ligands necessary for optimal reaction conditions.

This chapter provides a survey of the literature up to the end of 2007 dealing with the asymmetric intermolecular Mizoroki–Heck reaction, specifically focusing on ligand design for this important transformation.

11.2 Mechanism

As for any synthetic transformation, attempts to optimize conditions rely heavily on a well-developed understanding of the reaction mechanism involved. This is particularly true

The Mizoroki–Heck Reaction   Edited by Martin Oestreich
© 2009 John Wiley & Sons, Ltd

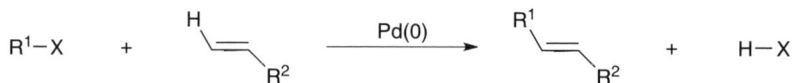

R^1 = aryl, alkenyl; R^2 = aryl, alkyl, alkenyl, OR, CO$_2$R etc.; X = I, Br, Cl, OTf.

Scheme 11.1   *General Mizoroki–Heck reaction.*

for enantioselective catalysis, as attempts to improve enantioselection are difficult if the catalytic pathway is ill-defined [5]. A generally accepted mechanism for the Mizoroki–Heck reaction [6], outlined in Scheme 11.2, will be discussed briefly, as the general features are pertinent to the development of the asymmetric variant.

The first step **A** in the catalytic cycle is the oxidative addition of the aryl halide (RX) to the coordinatively unsaturated 14-electron bis(triphenylphosphine)–palladium(0) complex, generated *in situ*. This affords a σ-arylpalladium(II) complex which associates an alkene in the second step **B**, presumably after dissociation of one of the triphenylphosphine ligands [7]. After the alkene and aryl ligands have adopted the *cis* orientation at palladium(II) necessary for migratory insertion, *syn*-insertion of the alkene into the σ-arylpalladium bond occurs in step **C** to give a σ-alkylpalladium complex. This *syn*-insertion means that internal rotation of the α,β- C–C bond of the resulting σ-alkylpalladium species, step **D**, must occur prior to β-hydride elimination and product decomplexation, step **E**. This gives the arylation product and a hydridopalladium complex from which, in step **F**, the catalytically active palladium(0) complex is regenerated after reductive elimination of hydrogen halide in the presence of base.

At first glance, therefore, the Mizoroki–Heck reaction is not a good candidate for enantioselective catalysis, since the sp^3 centre (labelled* in Scheme 11.2) formed in the migratory *syn*-insertion step is converted back to an sp^2 centre in the achiral product in the β-hydride elimination step. However, closer inspection of these steps suggested that

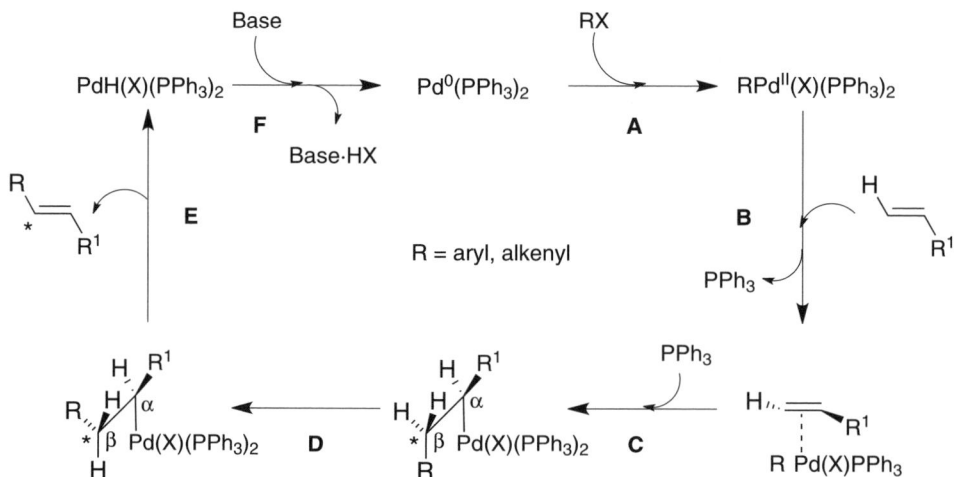

Scheme 11.2   *Accepted catalytic cycle of the Mizoroki–Heck reaction.*

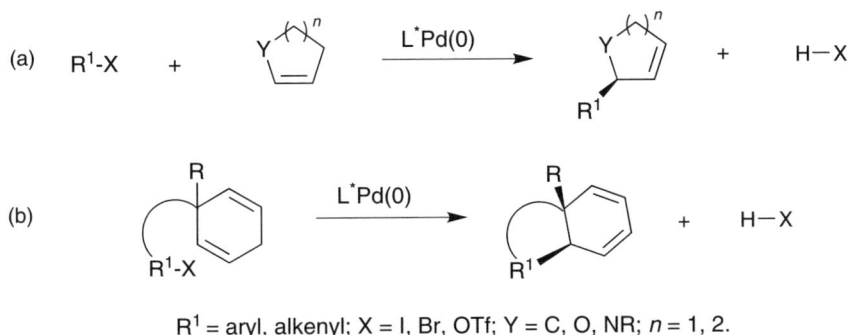

(a) R^1-X + [structure] $\xrightarrow{\text{L}^*\text{Pd(0)}}$ [structure] + H—X

(b) [structure] $\xrightarrow{\text{L}^*\text{Pd(0)}}$ [structure] + H—X

R^1 = aryl, alkenyl; X = I, Br, OTf; Y = C, O, NR; n = 1, 2.

Scheme 11.3 *Inter- and intra-molecular asymmetric Mizoroki–Heck reactions.*

the sp^3 centre formed in the migratory *syn*-insertion step could not be converted back to an sp^2 centre if the hydrogen atom at this sp^3 centre could never be *syn* to palladium; under these circumstances, therefore, the migrating R group would be incorporated onto an sp^3 centre. This situation exists for cyclic alkenes, as the hydrogen at the functionalized sp^3 centre will always be *anti* to palladium and internal rotation of the α,β- C–C bond of the intermediate σ-alkylpalladium species is not possible.

For this reason, two general types of asymmetric Mizoroki–Heck reaction, as illustrated in Scheme 11.3, have been the subject of most investigations:

(a) intermolecular Mizoroki–Heck reactions where the nucleophilic component and the cyclic alkene functionalities are on different molecules;
(b) intramolecular Mizoroki–Heck reactions where the nucleophilic component and cyclic or acyclic alkene functionalities are on the same molecule (one of the possible combinations involving prochiral substrates is shown).

This chapter will focus on the development of the intermolecular asymmetric variant and ligand design for this process. The following sections will be organized according to ligand class.

11.3 P,P Ligands and Derivatives

The first example of the asymmetric intermolecular Mizoroki–Heck reaction was reported by Hayashi and coworkers [8] in 1991. This involved the asymmetric arylation of 2,3-dihydrofuran (**1**) with aryl triflates using a palladium/(R)-BINAP (BINAP = 2,2′-bis(diphenylphosphino)-1,1′-binaphthyl) catalytic system (Scheme 11.4).

The use of phenyl triflate (**2**) was favoured over phenyl iodide, which gave very poor enantioselection, and it was postulated that employing the triflate allows the reaction to proceed by the cationic as opposed to the neutral pathway (Scheme 11.5). The cationic pathway was proposed as early as 1990 and begins with the dissociation of X$^-$ (X = triflate or a halide in the presence of a halide scavenger such as silver or thallium cations) from the σ-arylpalladium(II) complex to generate a tricoordinate 14-electron cationic complex with the accompanying counterion X$^-$ (X = halide or triflate) [9]. Complexation of the alkene

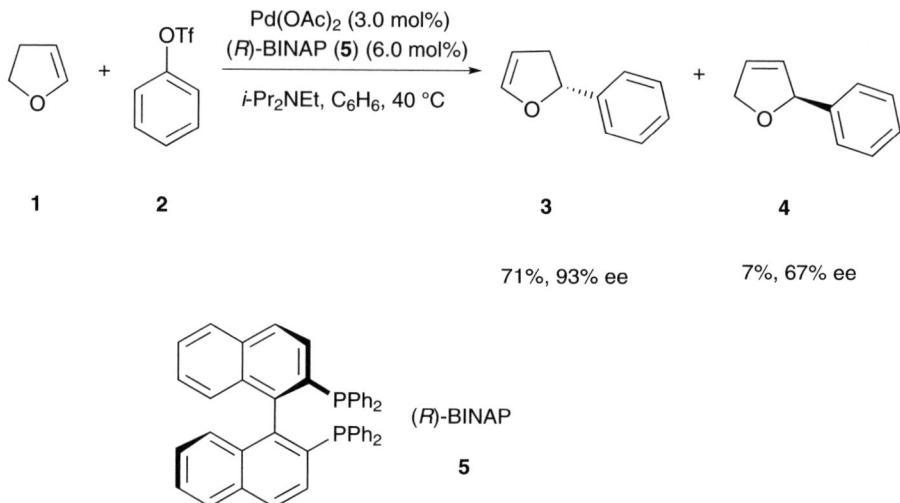

Scheme 11.4   *First catalytic asymmetric intermolecular Mizoroki–Heck reaction.*

into the vacant site gives a 16-electron species and insertion of the alkene into the Pd–R^1 bond followed by reforming the Pd–X bond gives a σ-alkylpalladium complex as desired with the chiral P,P ligand chelated throughout. The other pathway is neutral, with one of the P,P donor atoms dissociated from palladium, resulting in a 14-electron neutral species

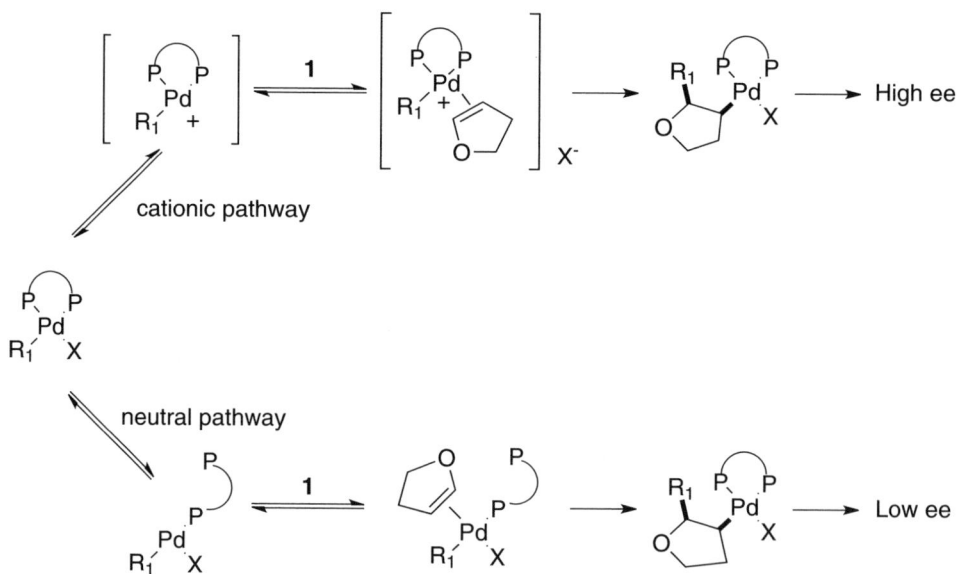

Scheme 11.5   *Cationic versus neutral pathway for the asymmetric intermolecular Mizoroki–Heck reaction.*

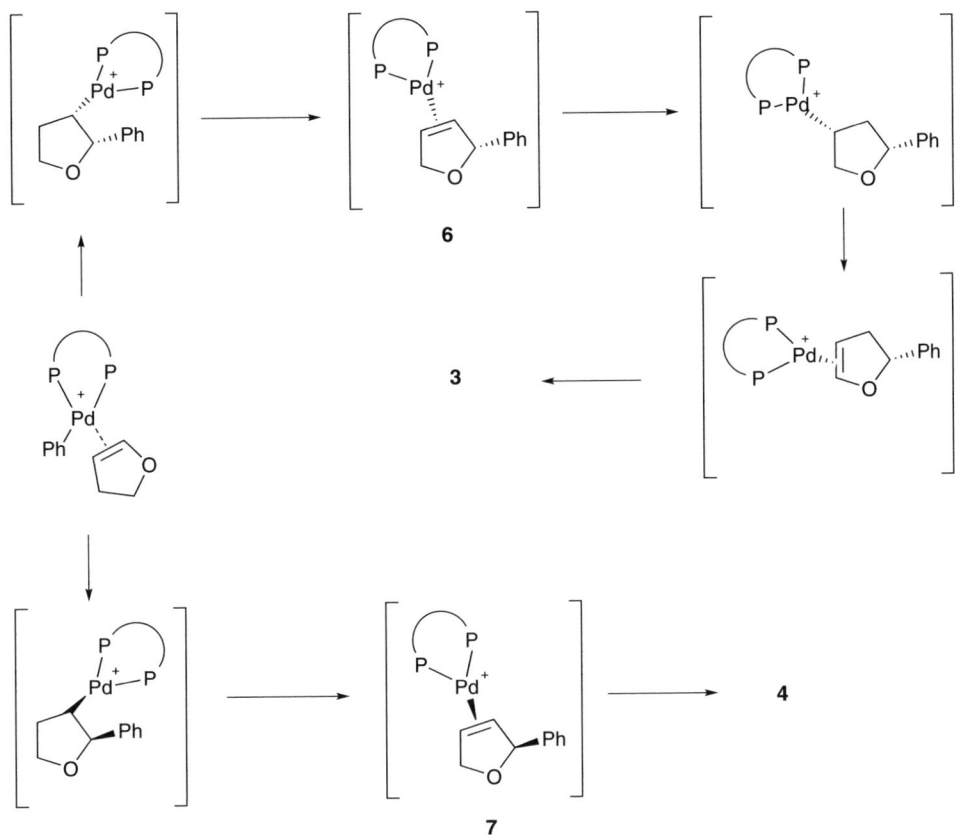

Scheme 11.6 Mechanism of intermolecular asymmetric Mizoroki–Heck reaction of 2,3-dihydrofuran (**1**).

with the resulting low levels of transfer of chiral information. Aryl and alkenyl triflates are generally assumed to follow the cationic pathway [10].

One noteworthy feature of this asymmetric reaction is that the products, **3** and **4**, are of opposite absolute configuration. This can be explained by a kinetic resolution between **6** and **7** (Scheme 11.6). Insertion of the alkene **1** into the Pd–Ph bond followed by β-hydrogen elimination forms hydrido–alkene complex **6** or its diastereomer **7**. It is likely that complex **6** has a preferable structure for further alkene insertion and β-hydrogen elimination processes, giving the major product **3**, while diastereomer **7** readily releases the alkene to give **4**. Brown and coworkers [11–13] reinforced this mechanism by NMR spectroscopy and mass spectrometry.

When alkenyl triflates **8** were used, even higher enantioselectivities were observed, with greater than 96% ee and without the formation of the other isomer (Scheme 11.7) [14].

Shibasaki and coworkers [15] also demonstrated that the reaction can be performed using hypervalent alkene iodinium salts **10** instead of alkenyl triflates, although yields are lower (22%). In this case the 2-alkenyl-2,5-dihydrofuran product **11** is obtained (Scheme 11.8).

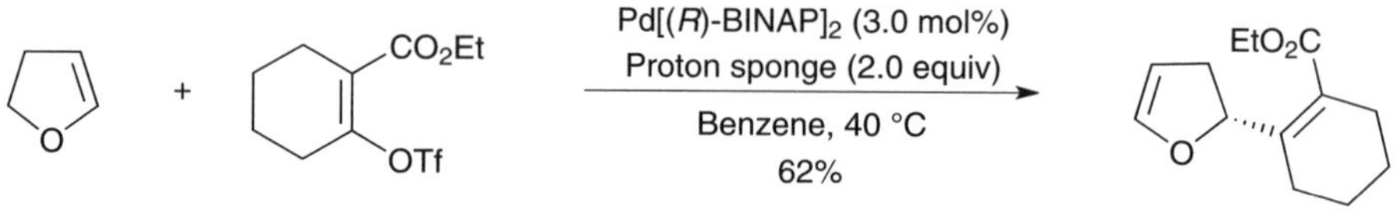

Scheme 11.7 *Intermolecular asymmetric Mizoroki–Heck reaction of alkenyl triflates.*

While 2,3-dihydrofuran (**1**) was the initial test substrate of choice for the intermolecular asymmetric Mizoroki–Heck reaction, the reaction was also applied to 2,3-dihydropyrrole **12**, which shows similar patterns of both regio- and stereoselectivity to 2,3-dihydrofuran (**1**) [16]. The intermolecular Mizoroki–Heck reaction with substituted 2,3-dihydropyrrole **12** and aryl triflates **13** gave mixtures of the 2-aryl-2,3-dihyropyrroles **14** and the 2-aryl-2,5-dihydropyrroles **15**, with the 2,3-product being the major product formed with a 74% ee (Scheme 11.9).

The alkenyl triflate **8** was reacted with the 2,3-dihydropyrrole **12** and the product **16** was formed exclusively in 95% yield and with a greater than 99% ee (Scheme 11.9). Sonesson *et al.* [17] reported the arylation of 2,5-dihydropyrrole **17** with naphthyl triflate **18** using (*R*)-BINAP (**5**) and Pd(OAc)$_2$ to give the 3-naphthyl product **19** in 34% yield and 58% ee (Scheme 11.10).

Shibasaki and coworkers [18] carried out the intermolecular asymmetric Mizoroki–Heck reaction with dihydrodioxepines **20** using the palladium–(*S*)-BINAP catalytic system (Scheme 11.11). The product **21** was obtained in yields up of to 86% and with up to 75% ee. When the aryl group on the triflate **13** was changed, the enantioselectivity was not found to vary appreciably.

Guiry and coworkers [19] reported the preparation and testing of 2,2-dimethyl-2,3-dihydrofuran (**22**) as an interesting substrate for the intermolecular asymmetric Mizoroki–Heck reaction as this substrate only forms the one regioisomeric product **23**. This allows for a true comparative test of a range of palladium complexes, and (*R*)-BINAP

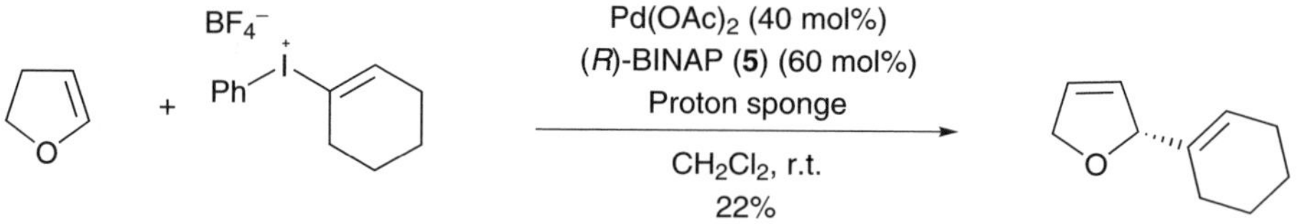

Scheme 11.8 *Intermolecular asymmetric Mizoroki–Heck reaction of hypervalent alkenyliodinium salt **10**.*

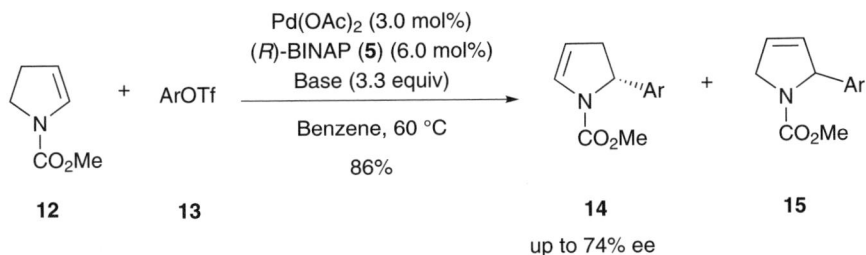

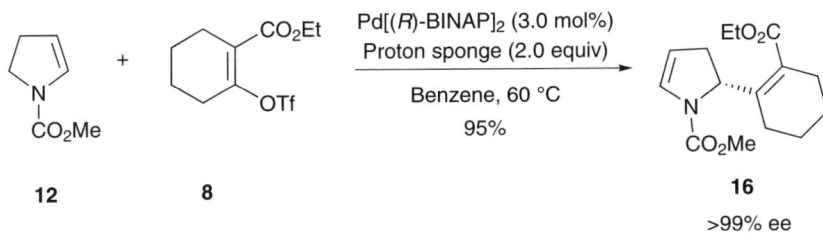

Scheme 11.9 Intermolecular asymmetric Mizoroki–Heck reaction of 2,3-dihydropyrrole **12**.

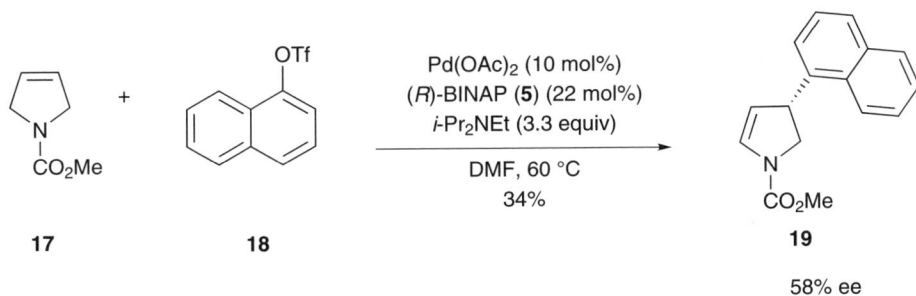

Scheme 11.10 Intermolecular asymmetric Mizoroki–Heck reaction of 2,5-dihydropyrrole **17**.

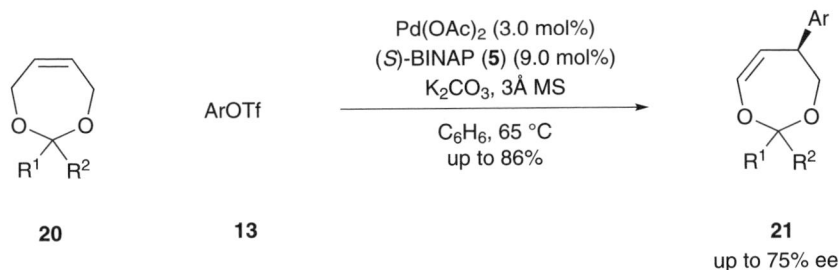

Scheme 11.11 Intermolecular asymmetric Mizoroki–Heck reaction of 4,7-dihydro-1,3-dioxepin **20**.

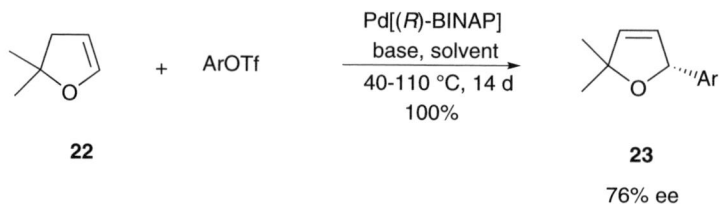

Scheme 11.12 *A new substrate for the intermolecular asymmetric Mizoroki–Heck reaction.*

(**5**) afforded (*R*)-2,2-dimethyl-2,5-dihydrofuran **23** in enantiomeric excesses of up to 76% (Scheme 11.12).

While BINAP (**5**) has been one of the most successful ligands employed in the intermolecular asymmetric Mizoroki–Heck reaction, many other diphosphine ligands have also been synthesized and tested. Sannicolo and coworkers [20] synthesized novel thiophene-derived axially chiral diphosphine ligands **24** and **25** and applied these to the intermolecular Mizoroki–Heck reaction of 2,3-dihydrofuran (**1**) (Scheme 11.13).

The palladium complex of the (*S*)-BITIANP diphosphine ligand (**24**; BITIANP = 2,2'-bis(diphenylphosphino)-3,3'-tetramethyl-3,3'-bibenzo[*b*]thiophene) gave a higher yield of **3** than (*S*)-BINAP (**5**), although the enantiomeric excess was lower at 91%. The similar (*S*)-TetraMeBITIOP ligand (**25**; TetraMeBITIOP = (±)-2,2',5,5'-tetramethyl-4,4'-bis(diphenylphoshino)-3,3'-bithiophene) was found to give a much lower yield of **3** at 58% and a low enantiomeric excess of 20%.

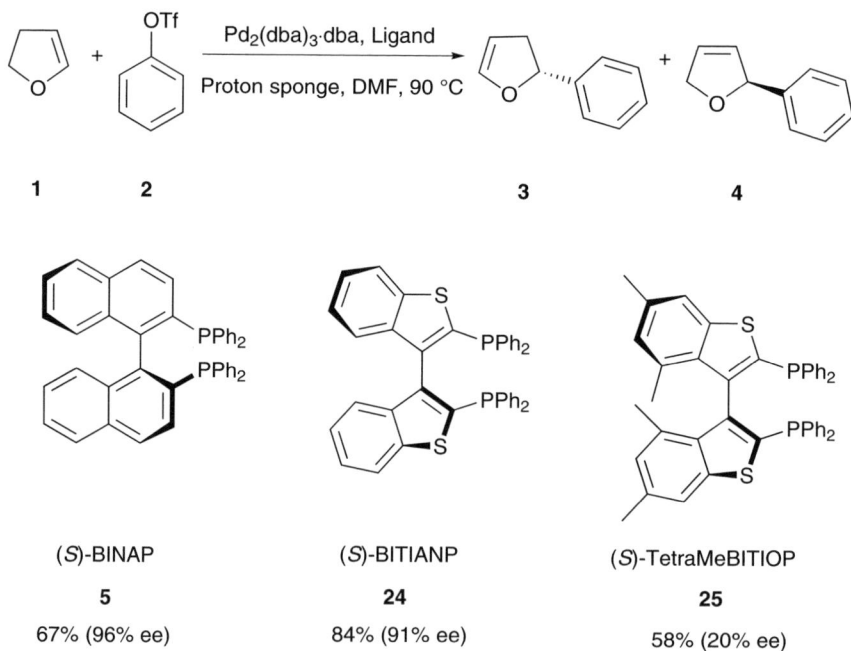

Scheme 11.13 *Intermolecular asymmetric Mizoroki–Heck reaction using thiophene diphosphine ligands.*

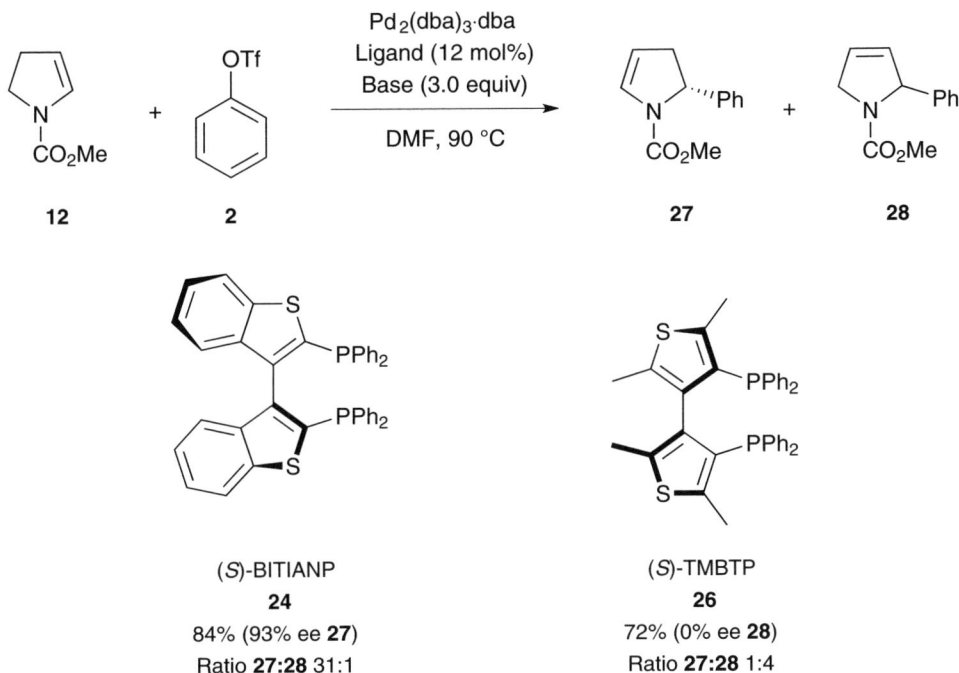

Scheme 11.14 *Intermolecular asymmetric Mizoroki–Heck reaction of 2,3-dihydropyrroles using thiophene-containing diphosphine ligands.*

Tietze and Thede [21] applied (*S*)-BITIANP (**24**) to the intermolecular Mizoroki–Heck reaction of 2,3-dihydropyrrole **12** (Scheme 11.14). They also applied the (*S*)-TMBTP ligand (**26**; TMBTP = 2,2′,5,5′-tetramethyl-3,3′-bis(diphenylphosphine)-4,4′-bithiophene), but (*S*)-BITIANP (**24**) was found to be far superior. The intermolecular Mizoroki–Heck coupling of dihydropyrrole (**12**) and phenyl triflate (**2**) with (*S*)-BITIANP (**24**) as ligand was carried out in high yield (84%) and with high enantioselectivity (93% ee) for the product **27**. The regioselectivity was found to be high (31 : 1), favouring the isomerized product **27**. Lower yields, regioselectivities and no enantioselectivities were observed with ligand **26**. (*S*)-BITIANP (**24**) was further applied to the cyclohexenylation of dihydropyrrole **12** with high levels of enantioselectivity (91% ee). The regioselectivity was found to be lower than that observed with aryl triflates.

Keay and coworkers [22–24] synthesized the novel furan-derived diphosphine ligands, TetFuBINAP (**29**) and BINAPFu (**30**) and applied them to the asymmetric intermolecular Mizoroki–Heck reaction of 2,3-dihydrofuran (**1**) and phenyl triflate (**2**) (Scheme 11.15).

These ligands were found after some optimization studies to give good to excellent enantioselectivities. The TetFuBINAP ligand (**29**), although a less active catalyst than BINAP (**5**), was found to give up to 89% ee. Palladium complexes of BINAPFu (**30**) afforded high enantioselectivities (>97% ee), although the presence of the furan led to a decrease in conversion when compared with BINAP (**5**). The reaction time needed to be increased to 9 days in order to get a conversion similar to that of BINAP (**5**).

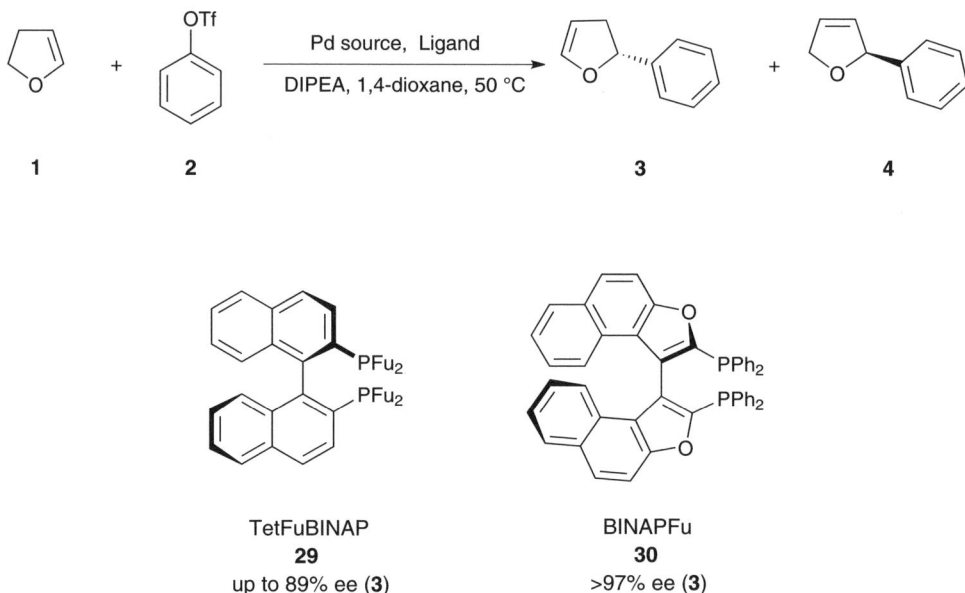

Scheme 11.15 *Intermolecular asymmetric Mizoroki–Heck reaction of 2,3-dihydrofuran (**1**) using furan derived diphosphine ligands.*

Hou and coworkers [25] synthesized novel diphosphine oxazoline ferrocenyl ligands **31** and applied these to the asymmetric intermolecular Mizoroki–Heck reaction of 2,3-dihydrofuran (**1**) with aryl triflates. The ligands can have three possible binding modes with palladium (Scheme 11.16).

From X-ray crystallographic analysis and ^{31}P NMR spectroscopy studies it was found that palladium was chelated in the P,P′ mode. The former technique showed that the P_1–Pd–P_2 bite angle was 99.75°, which was larger than that observed for both 1,1′-bis(diphenylphosphino)ferrocene (dppf) **33** and BINAP **34** palladium complexes (Figure 11.1) [26, 27].

The diphosphine oxazoline ferrocene ligands **38–40** were applied to the asymmetric intermolecular Mizoroki–Heck reaction of 2,3-dihydrofuran (**1**) and cyclohexenyl triflate (**35**) (Scheme 11.17). With 1.5 mol% of catalyst prepared *in situ* the reaction was completed in 36 h, affording the two products **11** and **36**.

The reactions of 2,3-dihydrofuran (**1**) and cyclohexenyl triflate (**35**) when carried out with palladium complexes of ligands **37–39** gave excellent conversions in all cases. The highest enantioselectivity was observed with the 3,5-di(trifluoromethyl)phenyl-substituted phosphine ligand **38** where a 90% ee was obtained. When the substituent was changed to an electron-donating 4-methoxyphenyl, ligand **39**, the enantiomeric excess dropped to 40%. Interestingly, the regioselectivity dropped from 97:3 of **11:36** with ligand **37** to 35:65 with ligand **39**. This dramatic change is presumably due to an electronic effect on the phosphine donor atom.

31

P,N Mode   **P',N Mode**   **P,P' Mode**

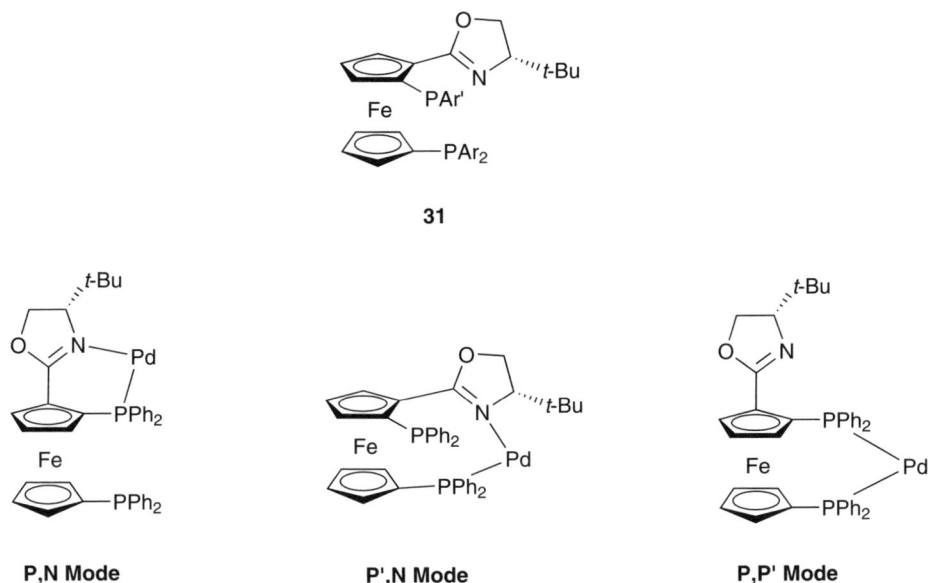

Scheme 11.16 *Possible binding modes of the diphosphine oxazoline ferrocenyl ligands 31.*

Shibasaki and coworkers [28, 29] synthesized the novel arsenic-derived BINAP deriva-tive BINAPAs (**40**). The ligand was tested in the asymmetric intermolecular Mizoroki–Heck reaction of 2,3-dihydrofuran (**1**) with phenyl triflate (**2**) (Scheme 11.18).

The reaction shows that BINAPAs (**40**) shows very different product ratios in comparison with BINAP, resulting in the formation of (*R*)-**3** (29%, 82% ee) and (*R*)-**4** (48%, 45% ee). The product ratio was changed, indicating that dissociation of the BINAPAs–Pd$^+$–H complex from a resulting olefinic double-bond occurs more readily than that of the BINAP–Pd$^+$–H complex, favouring (*R*)-2-phenyl-2,5-dihydrofuran **4** as the major product.

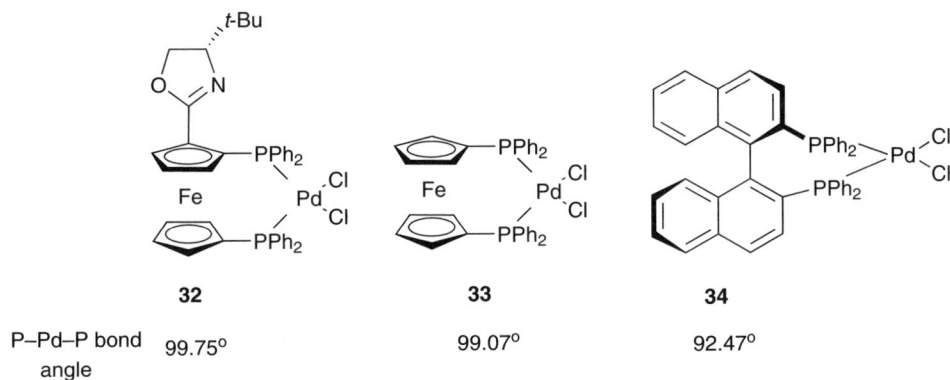

32   **33**   **34**

P–Pd–P bond   99.75°   99.07°   92.47°
angle

Figure 11.1 *Comparison of bite angles of diphosphine ferrocenyl oxazoline with dppf and BINAP.*

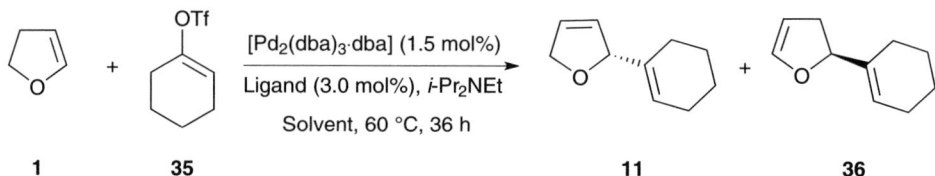

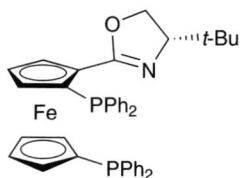 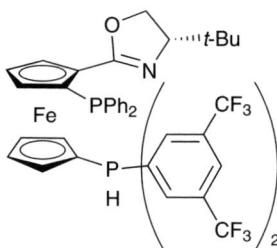 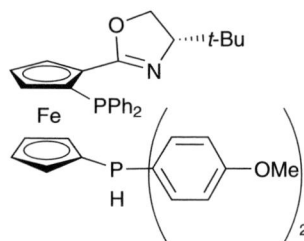

37

99% (80% ee **11**)
Ratio **11:36** 97:3

38

98% (80% ee **11**)
Ratio **11:36** 92:8

39

87% (40% ee **11**)
Ratio **11:36** 35:64

Scheme 11.17 *Asymmetric intermolecular Mizoroki–Heck reaction of 2,3-dihydrofuran (**1**)
with phosphinooxazoline ferrocene ligands.*

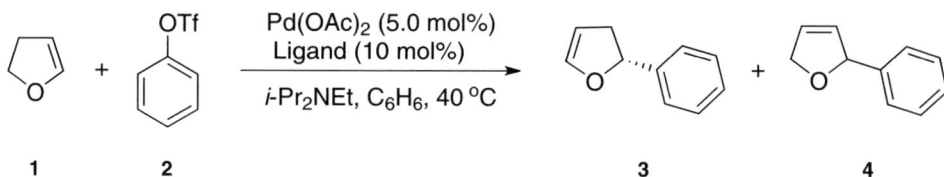

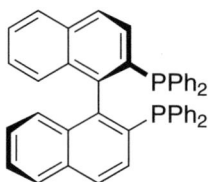 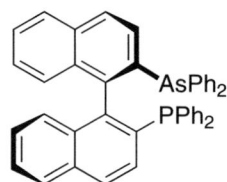

(*R*)-BINAP
5
Yield: 72% (**3**), 4% (**4**)
ee 67% ((*R*)-**3**), 50% ((*S*)-**4**)

(*R*)-BINAPAs
40
Yield: 29% (**3**), 48% (**4**)
ee 82% ((*R*)-**3**), 45% ((*R*)-**4**)

Scheme 11.18 *Intermolecular asymmetric Mizoroki–Heck reaction using BINAP and
BINAPAs ligands.*

11.4   P, N Ligands

To date, the large majority of asymmetric Mizoroki–Heck reactions reported have utilized palladium complexes of BINAP (**5**). However, since their first application to the asymmetric Mizoroki–Heck reaction, P,N ligands have proven successful and have thus received a greater amount of attention recently [30]. The phosphinooxazoline P,N ligands **41–45** developed independently by the groups of Pfaltz [31], Williams [32] and Helmchen [33] have shown dramatic improvement in enantioselectivity in a number of asymmetric transformations, including the intermolecular asymmetric Mizoroki–Heck reaction [34].

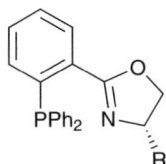

41 - R = *i*-Pr
42 - R = *t*-Bu
43 - R = Ph
44 - R = Bn
45 - R = Me

As an example, when the *t*-Bu ligand **42** was applied in the cyclohexenylation of 2,3-dihydrofuran (**1**) using Pd$_2$(dba)$_3$ (dba = dibenzylideneacetone) and *i*-Pr$_2$NEt as base the reaction proceeded in 92% yield and 99% ee (Scheme 11.19). This result compares favourably with that of BINAP (**5**), which gave a yield of 58% and 87% ee. Perhaps the most surprising result was the low amount of double-bond isomerization observed. In fact, only trace amounts of the isomerized product were formed compared with BINAP (**5**), where it is the more stable, isomerized product **36** that predominates.

Pfaltz and coworkers [34] also found that varying the base had effects on the yield of the cyclohexenylation while enantiomeric excesses remained consistently high. Results of the base screen (Table 11.1) employing 1,8-bis(dimethylamino)naphthalene, 2,2,6,6-tetramethylpiperidine and *N*,*N*-diisopropylethylamine showed that *N*,*N*-diisopropylethylamine was most successful.

Similar results were achieved when phenyl, 1-naphthyl and 1-cyclopentyl triflate were used as reagents. The intermolecular Mizoroki–Heck reaction of 4,7-dihydro-1,3-dioxepin (**46**) with phenyl triflate (**2**) also proceeded with high yield and enantioselectivity (70%, 92% ee) compared with the corresponding Pd(BINAP) complex (84%, 72% ee). Dihydropyran **48** proved to be less reactive than dihydrofuran **1**, but high yield and enantioselectivity were observed at higher temperatures and catalyst loading (78%, 84% ee). As with BINAP (**5**), Guiry and coworkers [19] tested these ligands in the phenylation and

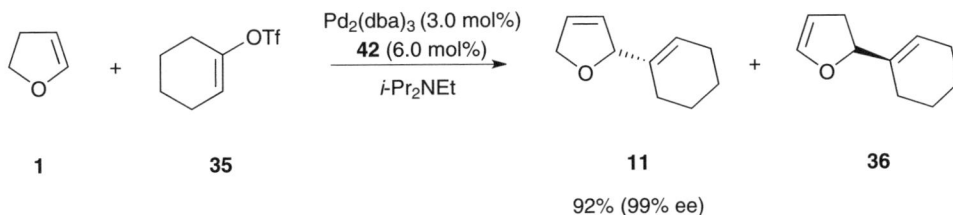

1   **35**   **11**   **36**

92% (99% ee)

Scheme 11.19   *Cyclohexenylation of 2,3-dihydrofuran (**1**) using P,N ligand **42**.*

Table 11.1   *Influence of base on the cyclohexenylation of 2,3-dihydrofuran (1)*

Entry	Base	ee/%	Yield/%
1	1,8-Bis(dimethylamino) naphthalene	98	95
2	2,2,6,6-Tetramethylpiperidine	99	95
3	Triethylamine	>99	78
4	N,N-Diisopropylamine	>99	92
5	N,N-Diisopropylethylamine	99	98
6	Sodium carbonate	98	34
7	Sodium acetate	98	50

cyclohexenylation of 2,2-dimethyl-2,3-dihydrofuran (**22**), with yields of up to 100% and enantiomeric excesses of up to 92% being obtained (Scheme 11.20).

The low amounts of double-bond isomerization observed using these ligands made it possible to consider using substrates such as cyclopentene (**50**), which are converted to a mixture of isomers with palladium complexes of BINAP (**5**) as catalysts. The phenylation of cyclopentene (**50**) proceeded with high regioselectivity for the unisomerized product **51** in 96% yield and 91% ee. The same reaction with BINAP (**5**) gave a mixture of regioisomers **51**, **52** and **53** with low enantioselectivity (Scheme 11.21).

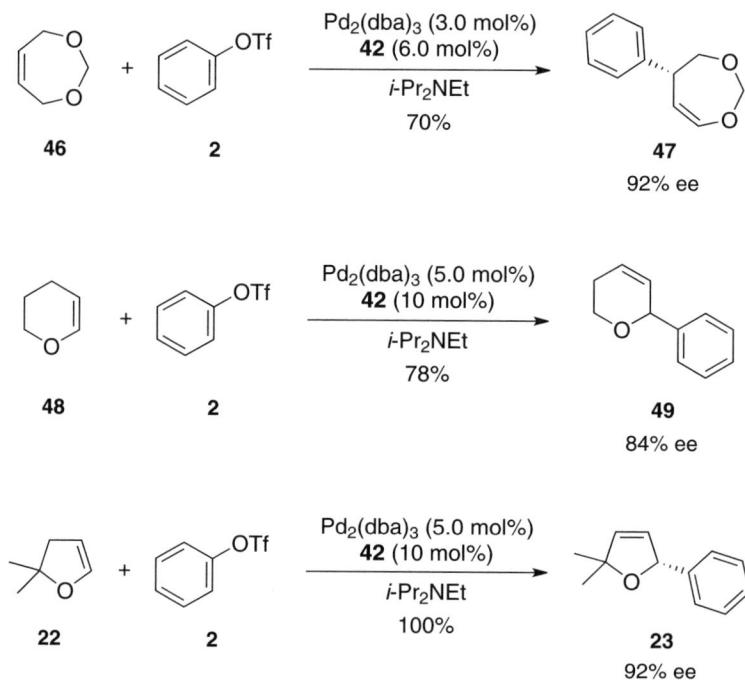

Scheme 11.20   *Selected examples of intermolecular asymmetric Mizoroki–Heck reactions using ligand **42**.*

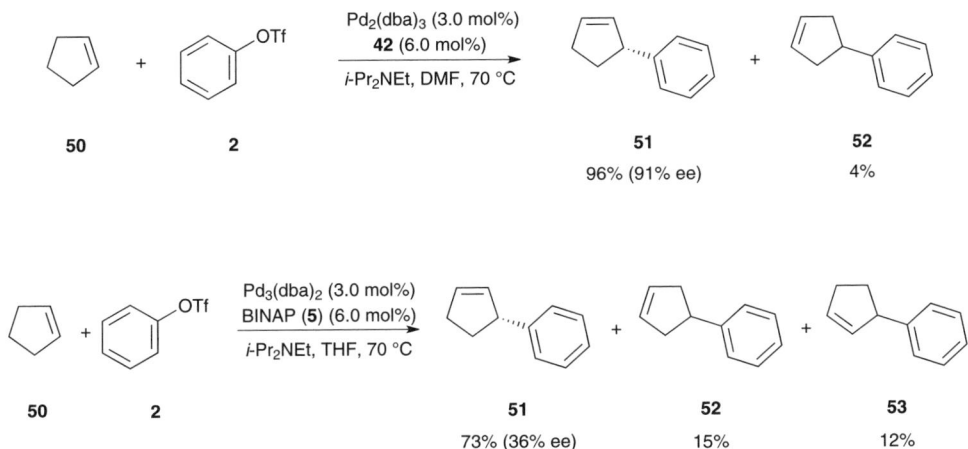

Scheme 11.21   *Intermolecular Mizoroki–Heck reaction with cyclopentene (50) using BINAP (5) and P,N ligand 42.*

Given the success of the PHOX ligands, Hayashi and coworkers [35] and Ikeda and coworkers [36] independently synthesized (*S,S*~P~)- and (*S,R*~P~)-2-[4-(isopropyl)oxazole-2-yl]-2′-diphenylphosphino-1,1′-binapthyl ligands (**54**) and (**55**).

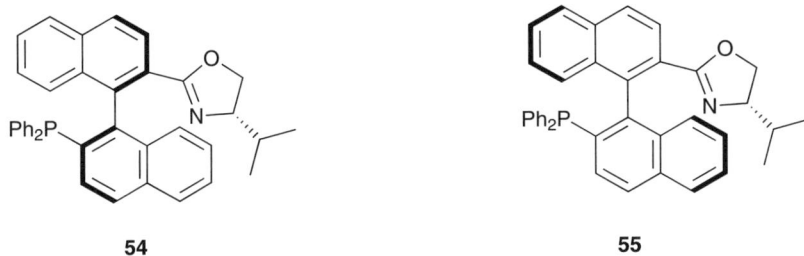

These ligands possess two independent elements of chirality: axial chirality afforded by the binaphthyl and central chirality on the oxazoline unit. Hayashi and coworkers tested these ligands in the intermolecular Mizoroki–Heck reaction of phenyl triflate (**2**) and 2,3-dihydrofuran (**1**). Complete conversions with enantiomeric excesses up to 88% were achieved, whereas none of the double-bond isomer **3** was observed (Scheme 11.22). In addition, the catalyst derived from ligand **55** proved more reactive than that derived from ligand **54**, albeit with slightly lower enantioselectivity. Interestingly, the two ligands **54** and **55** induced opposite configurations in the product, leading to the conclusion that the axial chirality is a more influential factor in the intermolecular Mizoroki–Heck reaction than the central chirality of the oxazoline unit.

Following on from the PHOX ligands **41–45**, Kündig and Meier [37] synthesized the related (diphenylphosphino)benzoxazine ligands **56**. The rationale can be seen in the metal complexes **57** and **58**, where M is the coordinated metal. The substituent on the oxazine ring is closer to the metal centre in **57** than in the analogous oxazoline complex **58**. Thus,

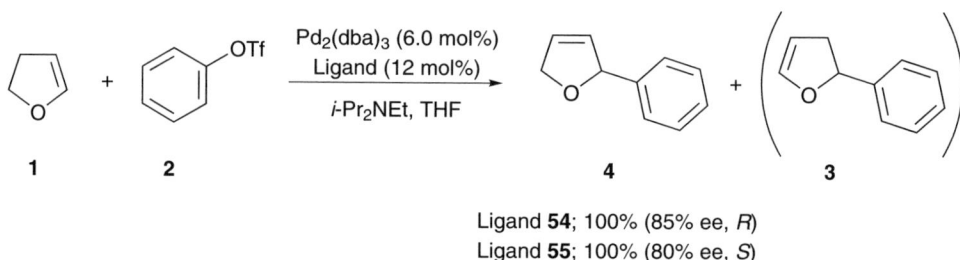

Ligand **54**; 100% (85% ee, *R*)
Ligand **55**; 100% (80% ee, *S*)

Scheme 11.22   *Phenylation of 2,3 dihydrofuran (**1**) using chiral phosphinooxazoline ligands **54** and **55**.*

it was hoped to affect the enantioselective induction to a greater extent. The presence of the fused benzo group ensures that the oxazine ring remains flat (like the oxazoline ring) instead of adopting a chair-like structure.

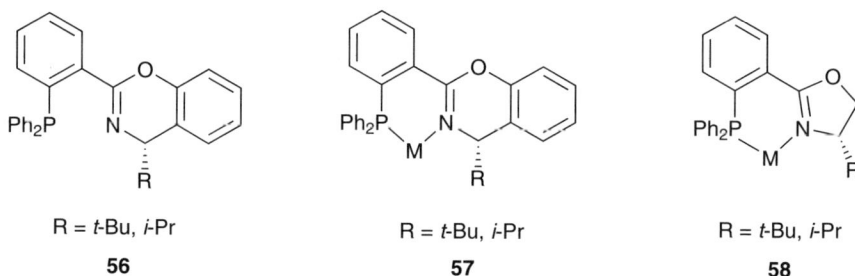

it was hoped to affect the enantioselective induction to a greater extent. The presence of

Again using cyclohexenyl triflate **35** and phenyl triflate (**2**) with 2,3-dihydrofuran (**1**) as test substrates, Kündig and Meier [37] showed that palladium complexes of ligands **56** could induce similar enantioselectivities as the PHOX ligands **41–45**. Yields of 55% and 79%, along with enantiomeric excesses of 94% and 91% for cyclohexenylation and phenylation reactions, respectively, were achieved. Mizoroki–Heck couplings using cyclopentene (**50**) instead of 2,3-dihydrofuran (**1**) afforded lower conversions (25 and 29%) and lower enantiomeric excesses (11 and 84%) than the analogous PHOX ligands **41–45**.

Hashimoto *et al.* [38] reported the synthesis of the P,N oxazoline-containing ligand **59** derived from *cis*-2-amino-3,3-dimethyl-1-indanol and its application to the intermolecular Mizoroki–Heck reaction. Their best reaction conditions for arylation of various cycloalkenes (Scheme 11.23) showed conversions of up to 91% and enantiomeric excesses of up to 98% for the reaction of cyclohexenyl triflate (**35**) with 2,3-dihydrofuran (**1**). Reactions using 4,7-dihydro-1,3-dioxepin (**46**) as the substrate proved less successful, with low conversions of 37% achieved, albeit with high enantiomeric excesses of up to 90%. In all cases, low amounts of isomerization were observed. These results are consistent with, if not better than, many analogous phosphinooxazoline ligands.

Guiry and coworkers [19] reported the application of ferrocene-based P,N ligands **60** and **61** to the Mizoroki–Heck reaction of phenyl triflate (**2**) and cyclohexenyl triflate (**35**)

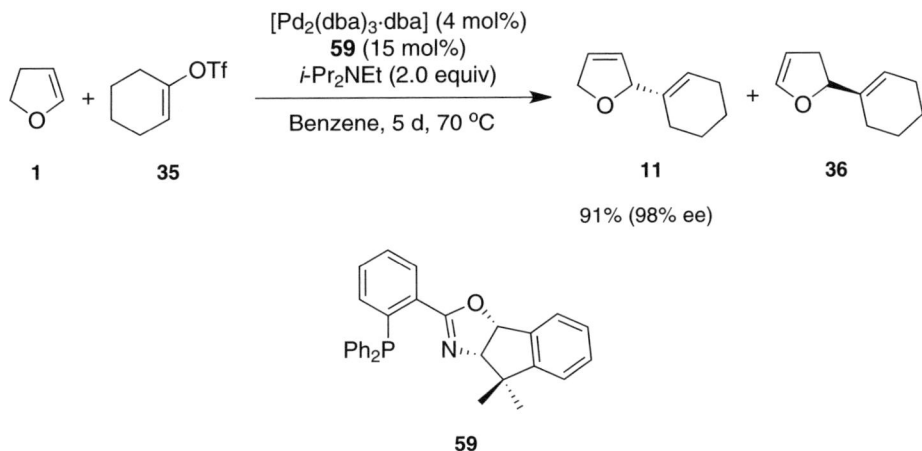

Scheme 11.23 *Intermolecular Mizoroki–Heck reaction of 2,3-dihydrofuran (1) with cyclohexenyl triflate (35) using ligand 59.*

with 2,3-dihydrofuran (**1**). Palladium complexes of ligand **60** gave excellent enantiomeric excesses (up to 99%) albeit with moderate conversions (up to 72%) after long reaction times (14 days). Palladium complexes of ligand **61** gave both low reactivity and selectivity. Using 2,2-dimethyl-2,3-dihydrofuran (**22**) the reaction proceeded in good yield and excellent enantioselectivity (70%, 99% ee) when palladium complexes of ligand **60b** were employed (Scheme 11.24).

The major drawbacks with many Mizoroki–Heck reactions are the long reaction times and high catalyst loadings required to get substantial conversions. Hou and coworkers [39]

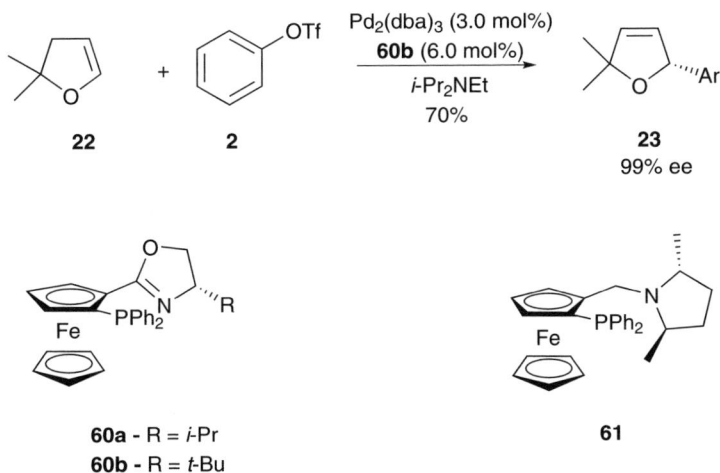

Scheme 11.24 *Intermolecular Mizoroki–Heck reaction using ligands 60 and 61.*

overcame this somewhat when they applied 1,1'-disubstituted and 1,1',2'-substituted-P,N-ferrocene ligands **62–64** to the intermolecular Mizoroki–Heck reaction.

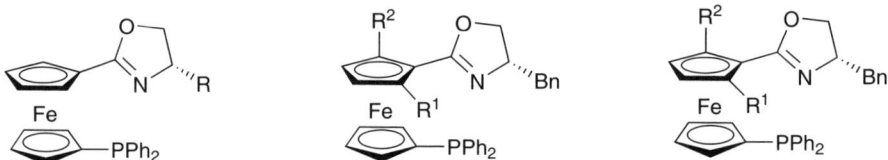

62a - R = *i*-Pr	**63a** - R¹ = TMS, R² = H	**64a** - R¹ = H, R² = TMS
62b - R = Bn	**63b** - R¹ = H, R² = Me	**64b** - R¹ = Me, R² = H
62c - R = *t*-Bu	**63c** - R¹ = H, R² = TMS	
62d - R = Ph		

These ligands were tested in the phenylation of 2,3-dihydrofuran (**1**). The reactions were carried out at 60 °C for 8 h with 6 mol% of ligand. A benzyl substituent on the oxazoline, ligand **62b**, proved to be the most effective, giving rise to 80% conversion and 77% ee. With this in mind, planar chiral ligands **63** and **64** were prepared. A dramatic change in the enantioselectivity of the reaction was observed with ligand **63a** containing a trimethylsilyl group. The enantioselectivity changed from 77% (*R*) to 84% (*S*). They also tested ligand **63b**, which has opposite planar chirality, and this reversed the enantioselectivity, giving 89% (*R*). The same effect in enantioselectivity was observed with ligands **64**, leading to the conclusion that planar chirality plays the dominant role in the selectivity of the intermolecular Mizoroki–Heck reaction. These planar chiral ligands allow moderate reaction conditions and high enantioselectivity, but they also provide access to either configuration of product by changing the planar chirality of the ligand.

Gilbertson *et al.* [40] reported the new class of phosphine–oxazoline ligands **65** and **66**, bearing a chiral phosphanobornadienyl and an oxazoline and their application to the cyclohexenylation of 2,3-dihydrofuran (**1**).

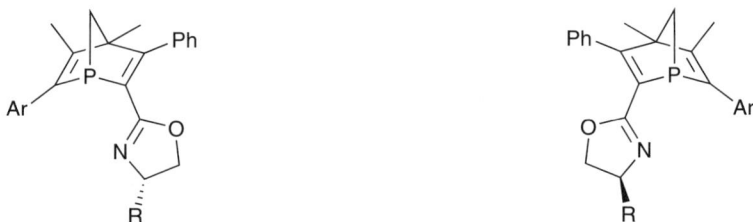

65a - Ar = phenyl, R = *i*-Pr	**66a** - Ar = phenyl, R = *i*-Pr
65b - Ar = phenyl, R = *t*-Bu	**66b** - Ar = phenyl, R = *t*-Bu
65c - Ar = phenanthryl, R = *i*-Pr	**66c** - Ar = phenanthryl, R = *i*-Pr
65d - Ar = anthracyl, R = *i*-Pr	**66d** - Ar = anthracyl, R = *i*-Pr

They observed that the larger the aryl group the faster the reaction proceeded. For example, the palladium complex of **65d** catalysed the reaction of cyclohexenyl triflate **35** with

2,3-dihydrofuran (**1**) to completion in just 20 h at room temperature, whereas the complex of **65a** required 120 h to get the reaction to completion. However, although increasing the size of the substituent increased the reaction rate it had an opposite effect on the enantiose-lectivity. The complex of **65d** gave only 56% ee, whereas the complex of **65a** gave 76% ee. As it was noted previously in oxazoline-containing ligands, increasing the sterics on the oxazoline unit led to an increase in enantioselectivity. Ultimately an enantiomeric excess of 93% was achieved using the palladium complex of ligand **65b**. Again in agreement with many P,N ligands, low amounts (<10%) of isomerized products were observed. It is also worth noting that both ligands **65** and **66** induced the same enantioselectivity in the product, with the (*R*) configuration being the predominant enantiomer. This gave rise to the conclusion that the stereochemistry originates from the chirality on the oxazoline.

The same group also reported similar phosphine–oxazolines **67** and **68** with a norbornyl backbone [41]. The main difference between these ligands and **65** and **66** is the fact that the phosphorus is no longer at the bridgehead position. Preliminary results showed that ligand **67b**, bearing the *t*-Bu substituent on the oxazoline, provided the best results in the reaction of cyclohexenyl triflate (**35**) with 2,3-dihydrofuran (**1**), giving complete conversion with 94% ee after 22 h. Ligands with other substituents on the oxazoline were less reactive and afforded lower enantiomeric excesses. Also, as expected, changing the stereochemistry on the oxazoline (ligand **68**) resulted in the opposite configuration in the product. As before, this shows that the stereochemistry is determined by the substituent on the oxazoline, as opposed to the norbornyl ring. Using these ligands, different coupling partners were tested, achieving excellent enantiomeric excesses in many cases (Scheme 11.25).

Gilbertson *et al.* suggested that although many P,N ligands had proven successful in Mizoroki–Heck transformations many possessed only one chiral centre. Therefore, they prepared proline-based P,N ligands **71** and **72**, which possessed up to three chiral centres [42]. Initial studies provided information that the chirality on the oxazoline was relatively unimportant in determining the enantioselectivity of the reaction. Both ligands, **71** and its diastereomer **72**, gave rise to the same selectivity, giving 80% ee in the reaction of cyclo-hexenyl triflate (**35**) and 2,3-dihydrofuran (**1**). Conversions in most cases were excellent, with up to 99% being achieved. As with their previous reports, they showed that these ligands were applicable to a wide range of Mizoroki–Heck coupling partners. Reaction of phenyl triflate (**2**) and 2,3-dihydrofuran (**1**) gave 95% conversion with 82% ee after 3 days. Phenylation of cyclopentene (**50**) proved slightly more problematic, in that using benzene as solvent afforded low conversion (14%) but good enantiomeric excess (73%), while us-ing dioxane as solvent reversed the situation, giving good conversion (85%) but with low enantiomeric excess (12%). Other alkenyl triflates were coupled with 2,3-dihydrofuran (**1**), resulting in good conversions (up to 95%) with moderate enantiomeric excesses (up to 82%).

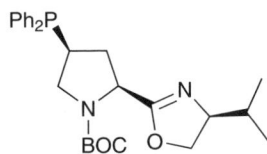

71

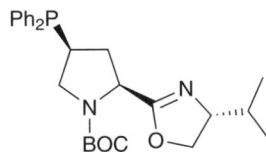

72

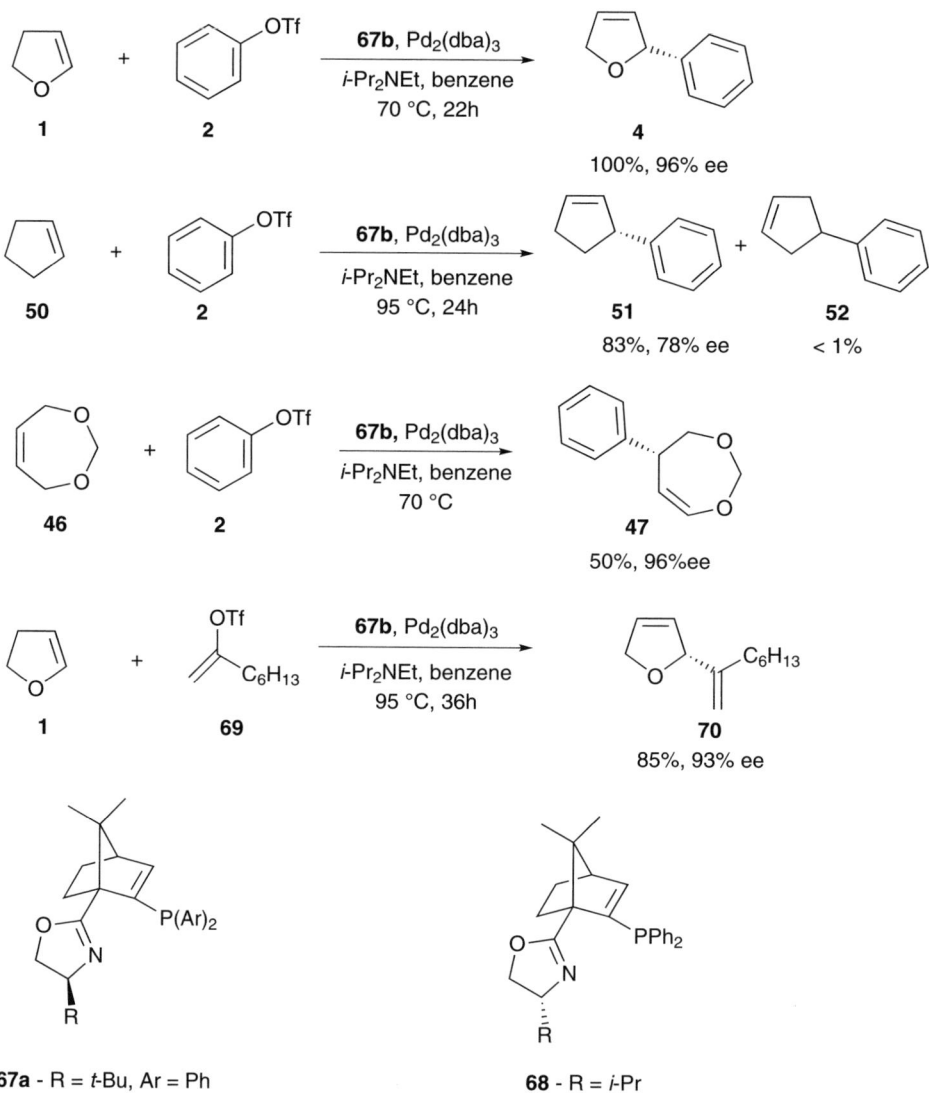

Scheme 11.25 *Various Mizoroki–Heck coupling reactions using Gilbertson's norbornyl P,N ligands **67**.*

Pfaltz and coworkers [43] developed P,N ligands **73** and **74** derived from pyridine and quinoline. They reasoned that oxazoline and pyridine/quinoline ligands, since they have different electronic effects, would induce different patterns of reactivity and selectivity in the asymmetric Mizoroki–Heck reaction. In the reaction of phenyl triflate (**2**) with

2,3-dihydrofuran (**1**), ligand **74b** afforded the best result, with complete conversion after 2 days at 50 °C with an excellent enantiomeric excess of 97%.

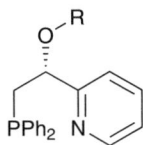

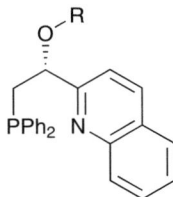

73a - R = Si(*t*-Bu)Me₂
73b - R = Si(*i*-Pr)₃
73c - R = Si(*t*-Bu)Ph₂

74a - R = Si(*t*-Bu)Me₂
74b - R = Si(*i*-Pr)₃
74c - R = Si(*t*-Bu)Ph₂

Guiry and coworkers [44, 45] reported the application of heterocyclic analogues of the PHOX ligands, HetPHOX ligands **75** and **76**, to the phenylation and cyclohexenylation of 2,3-dihydrofuran (**1**). The ligands proved successful in both test reactions, providing enantiomeric excesses and conversions over 90% in many cases. Ligand **78b** provided the best results for both phenylation (97%, 95% ee) and cyclohexenylation (96%, 97% ee) reactions. These results are consistent with the PHOX ligands, where it was found that the ligand with the *t*-Bu substituent on the oxazoline was the most reactive and selective.

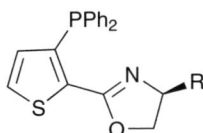

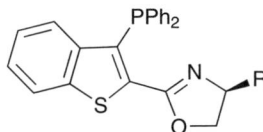

75a - R = *i*-Pr
75b - R = *t*-Bu

76a - R = *i*-Pr
76b - R = *t*-Bu
76c - R = Ph

Zhang and coworkers [46] designed P,N ligands **77** which were also structurally similar to the PHOX ligands and tested them in many asymmetric transformations, including the intermolecular asymmetric Mizoroki–Heck reaction. In the phenylation of 2,3-dihydrofuran (**1**), very little double-bond isomerization was observed. Initially, ligands **77a** and **77d** provided the best conversions (87 and 91%) and enantiomeric excesses (91 and 90%) when the reaction was carried out in benzene. Changing the solvent to tetrahydrofuran afforded higher conversion and enantiomeric excess for ligand **77a** (99%, 94% ee).

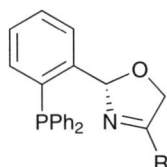

77a - R = *t*-Bu
77b - R = Bn
77c - R = 3,5-di-*t*-Bu-phenyl
77d - R = adamantyl

Ohe and coworkers [47] developed the phosphinite oxazoline ligands **78**, derived from D-glucosamine. They applied these ligands to the Mizoroki–Heck reaction of phenyl triflate

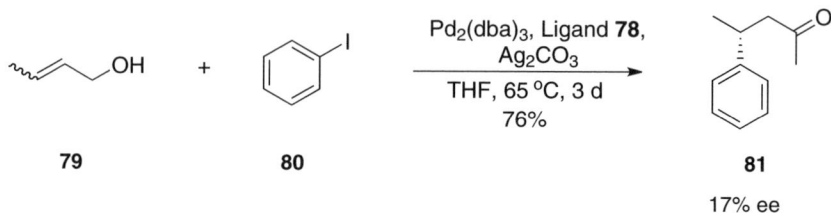

Scheme 11.26   *Phenylation of* cis- *and* trans-*crotyl alcohols* **79**.

(**2**) with 2,3-dihydrofuran (**1**). Ligand **78f** bearing a benzyl substituent on the oxazoline provided the best results for this transformation, giving complete conversion with 96% ee after 1 day.

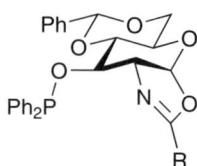

78a - R = Me
78b - R = *i*-Pr
78c - R = *i*-Bu
78d - R = *t*-Bu
78e - R = Ph
78f - R = Bn

Furthermore, these ligands were tested in the phenylation of *cis*- and *trans*-crotyl alcohols **79**. This was the first reported asymmetric Mizoroki–Heck reaction of an acyclic alkene. Using phenyl triflate (**2**) resulted in no conversion after 3 days, whereas the reaction proceeded with good conversions (up to 76%) when phenyl iodide (**80**) was used, albeit with low enantiomeric excess (up to 17%) (Scheme 11.26).

Dieguez and coworkers [48, 49] synthesized similar phosphite oxazoline ligands **82** derived from D-glucosamine. They applied these ligands to many Mizoroki–Heck transformations, including the reaction of phenyl triflate (**2**) with 2,3-dihydrofuran (**1**). Using ligand **82f**, this reaction proceeded to complete conversion with an excellent enantiomeric excess of 99% for the unisomerized product **4**. In all cases, very low amounts of the isomerized product **3** were formed.

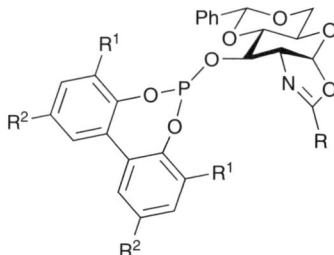

82a - R = Ph, R^1 = R^2 = *t*-Bu
82b - R = *i*-Pr, R^1 = R^2 = *t*-Bu
82c - R = *t*-Bu, R^1= R^2 = *t*-Bu
82d - R = Ph, R^1 = *t*-Bu, R^2 = OMe
82e - R = Ph, R^1 = R^2 = H
82f - R = Ph, R^1 = TMS, R^2 = H

Hou *et al.* [50] prepared phosphinooxazoline ligands **83** where the oxazoline substituent is at the benzylic position. The difference between these ligands and the PHOX ligands **45** is that the chelate formed with palladium is seven membered. In the reaction of phenyl triflate (**2**) with 2,3-dihydrofuran (**1**), ligand **83b** proved the most successful, with high yields (up to 85%) and high enantiomeric excess (up to 94% ee) achieved. Perhaps the most favourable result was the short reaction times (20 h) required using these ligands, compared with the PHOX ligands (up to 4 days).

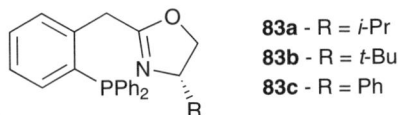

83a - R = *i*-Pr
83b - R = *t*-Bu
83c - R = Ph

11.5   Other Ligand Classes

Although not commonly used in the asymmetric Mizoroki–Heck reaction, N,N donor ligands have proved successful in many asymmetric transformations. Jones and coworkers [52] applied the 'pyox' (pyridyloxazoline) ligands **84** developed by Bolm *et al.* [51] to the intermolecular asymmetric Mizoroki–Heck reaction of phenyl iodide (**80**) and triflate (**2**) with 2,3-dihydrofuran (**1**). Palladium complexes of these ligands were not as reactive or as selective in this transformation as many P,N ligands. Indeed, using phenyl iodide (**80**) as a substrate they observed low conversions of less than 25% along with no enantioselectivity. Using phenyl triflate (**2**), low conversions (<30%) were again observed, albeit with moderate enantioselectivity (up to 60% ee).

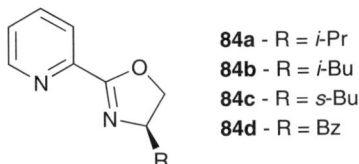

84a - R = *i*-Pr
84b - R = *i*-Bu
84c - R = *s*-Bu
84d - R = Bz

As with N,N ligands, P,S ligands have not received much attention in the area of the Mizoroki–Heck reaction. However, there are a couple of reported examples. Kang *et al.* [53] applied the pseudo-C_2-symmetric P,S-hybrid ferrocenyl ligand **85** to the phenylation of 2,3-dihydrofuran (**1**). However, low conversions (<37%) and low enantiomeric excesses (<34%) were achieved. Molander *et al.* [54] applied cyclopropane-based P,S ligands **86** and **87** to the same reaction. Excellent conversions (up to 99%) and moderate enantiomeric excesses (up to 60%) were obtained.

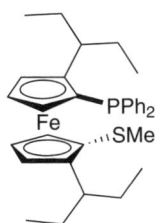

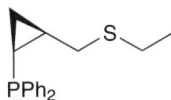

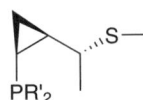

85   **86**   **87**

Dai *et al.* [55] reported the only example of P,O ligands applied to the asymmetric Mizoroki–Heck reaction. The atropisomeric amide-derived ligands **88** were applied to the reaction of phenyl triflate (**2**) and 2,3-dihydrofuran (**1**). Conversions were low (<30%) in all cases, although moderate enantioselectivity was observed (52% ee, ligand **88a**).

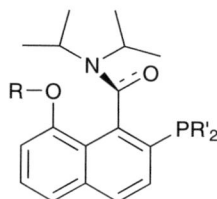

88a - R = Me, R' = Ph
88b - R = Bn, R' = Ph
88c - R = Me, R' = *t*-Bu

11.6 Conclusion

The survey of the asymmetric intermolecular Mizoroki–Heck reaction demonstrates the important improvements that have been made with respect to regioselectivity and enantioselectivity, many of which have come about as a result of rationally designed chiral ligands. Traditionally, as with many asymmetric transformations, BINAP (**5**) and its analogues were the ligands of choice for the Mizoroki–Heck reaction. However, there now exists a large range of ligands that can be used depending on the regioselectivity and enantioselectivity required. Furthermore, these ligands allow the use of previously unreactive 'Mizoroki–Heck substrates'. Simple unfunctionalized cyclic alkenes can now undergo Mizoroki–Heck reactions with excellent regioselectivity and enantioselectivity. This chapter has detailed the different classes of such ligands. First, bisphosphine ligands proved hugely successful in early attempts at asymmetric Mizoroki–Heck reactions. However, important improvements in terms of reactivity, regioselectivity and enantioselectivity have been achieved through the use of heterobidentate P,N ligands. The application of other classes of ligands have only been reported recently, with little evidence that N,N, P,O and P,S donor ligands can compete with the success of many bisphosphine and phosphinooxazoline ligands. Research continues into the development of new and improved ligands for the intermolecular asymmetric Mizoroki–Heck reaction in an attempt to increase substrate scope, enantioselectivities and reactivities. We look forward to future publications on this transformation and its application as key steps in natural product synthesis.

References

1. Heck, R.F. (1968) Arylation, methylation and carboxyalkylation of olefins by Group VIII metal derivatives. *J. Am. Chem. Soc.*, **90**, 5518–22.
2. Mizoroki, T., Mori, K. and Ozaki, A. (1971) Arylation of olefins with aryl iodide catalyzed by palladium. *Bull. Soc. Chem. Jpn.*, **44**, 581–1.
3. Heck, R.F. and Nolley, J.P. Jr (1972) Palladium-catalyzed vinylic hydrogen substitution reactions with aryl, benzyl, and styryl halides. *J. Org. Chem.*, **37**, 2320–2.
4. Dounay, A.B. and Overman, L.E. (2003) The asymmetric Heck reaction in natural products total synthesis. *Chem. Rev.*, **103**, 2945–63.

5. Brown, J.M., Guiry, P.J. and Wienand, A. (1993) Origins of enantioselectivity in catalytic asymmetric synthesis, in *Principles of Molecular Recognition* (eds A.D. Buckingham, A.C. Legon and S.M. Roberts), Chapman and Hall, Glasgow.

6. Hegedus, L.S. (2002) Organopalladium chemistry, in *Organometallics in Synthesis – A Manual* (ed. M. Schlosser), John Wiley & Sons, Ltd, Chichester.

7. Samuel, E.G. and Norton, J.R. (1984) Mechanism of acetylene and olefin insertion into palladium–carbon sigma-bonds. *J. Am. Chem. Soc.*, **106**, 5505–12.

8. Ozawa, F., Kubo, A. and Hayashi, T. (1992) Palladium-catalyzed asymmetric arylation of 2,3-dihydrofuran: 1,8-bis(dimethylamino)naphthalene as an efficient base. *Tetrahedron Lett.*, **33**, 1485–8.

9. Sato, Y., Sodeoka, M. and Shibasaki, M. (1990) On the role of silver salts in asymmetric Heck-type reaction. A greatly improved catalytic asymmetric synthesis of *cis*-decalin derivatives. *Chem. Lett.*, **19**, 1953–4.

10. Dekker, G.P.C.M., Elsevier, C.J., Vrieze, K. and van Leeuwen, P.W.N.M. (1992) Influence of ligands and anions on the rate of carbon monoxide insertion into palladium–methyl bonds in the complexes (P–P)Pd(CH$_3$)Cl and [(P–P)Pd(CH$_3$)(L)]$^+$SO$_3$CF$_3^-$ (P–P = dppe, dppp, dppb, dppf; L = CH$_3$CN, PPh$_3$) . *Organometallics*, **11**, 1598–603.

11. Brown, J.M., Perez-Torrente, J.J., Alcock, N.W. and Clase, H.J. (1995) Stable arylpalladium iodides and reactive arylpalladium trifluoromethanesulfonates in the intramolecular Heck reaction. *Organometallics*, **14**, 207–13.

12. Brown, J.M. and Hii, K.K. (1996) Characterisation of reactive intermediates in palladium-catalysed arylation of methyl acrylate (Heck reaction). *Angew. Chem., Int. Ed. Engl.*, **35**, 657–9.

13. Deeth, R.J., Smith, A. and Brown, J.M. (2004) Electronic control of the regiochemistry in palladium–phosphine catalyzed intermolecular Heck reactions. *J. Am. Chem. Soc.*, **126**, 7144–51.

14. Ozawa, F., Kobatake, Y. and Hayashi, T. (1993) Palladium-catalyzed asymmetric alkenylation of cyclic olefins. *Tetrahedron Lett.*, **34**, 2505–8.

15. Kurihara, Y., Sodeoka, M. and Shibasaki, M. (1994) A novel and efficient coupling reaction of sodium tetraphenylborate with hypervalent iodonium salts. *Chem. Pharm. Bull.*, **42**, 2357–60.

16. Ozawa, F. and Hayashi, T. (1992) Catalytic asymmetric arylation of *N*-substituted 2-pyrrolines with aryl triflates. *J. Organomet. Chem.*, **428**, 267–77.

17. Sonesson, C., Larhed, M., Nyquist, C. and Hallberg, A. (1996) Regiochemical control and suppression of double bond isomerisation in the Heck arylation of 1-(methoxycarbonyl)-2,5-dihydropyrrole. *J. Org. Chem.*, **61**, 4756–63.

18. Koga, Y., Sodeoka, M. and Shibasaki, M. (1994) Palladium-catalyzed asymmetric arylation of 4,7-dihydro-1,3-dioxepin. Catalytic asymmetric synthesis of γ-butyrolactone derivatives. *Tetrahedron Lett.*, **35**, 1227–30.

19. Hennessy, A.J., Malone, Y.M. and Guiry, P.J. (1999) 2,2-Dimethyl-2,3-dihydrofuran, a new substrate for intermolecular asymmetric Heck reactions. *Tetrahedron Lett.*, **40**, 9163–6.

20. Benincori, T., Rizzo, S. and Sannicolo, F. (2002) Free design of chiral diphosphine chelating ligands for stereoselective homogeneous catalysis by assembling five-membered aromatic heterocycles. *J. Heterocyclic Chem.*, **39**, 471–85.

21. Tietze, L.F. and Thede, K. (2000) Highly regio- and enantioselective Heck reactions of *N*-substituted 2-pyrroline with the new chiral ligand BITIANP. *Synlett*, 1470–3.

22. Anderson, N.G., Parvez, M. and Keay, B.A. (2000) Synthesis, resolution, and applications of 2,2′-bis(diphenylphosphino)-3,3′- binaphtho[2,1-*b*]furan. *Org. Lett.*, **2**, 2817–20.

23. Anderson, N.G., McDonald, R. and Keay, B.A. (2001) Synthesis, resolution and application of 2,2′-bis(di-2-furylphosphino)-1,1′-binaphthalene. *Tetrahedron: Asymmetry*, **12**, 263–9.

24. Anderson, N.G., Parvez, M., McDonald, R. and Keay, B.A. (2004) Synthesis, resolution, and application of 2,2′-bis(diphenylphosphino)-3,3′-binaphtho[*b*]furan (BINAPFu). *Can. J. Chem.*, **82**, 145–61.

25. Tu, T., Deng, W.-P., Hou, X.-L. *et al.* (2003) The regioselectivity of the asymmetric intermolecular Heck reaction with planar chiral diphosphine–oxazoline ferrocenyl ligands. *Chem. Eur. J.*, **9**, 3073–81.

26. Hayashi, T., Konishi, M., Kobor, Y. *et al.* (1984) Dichloro[1,1′-bis(diphenylphosphino) ferrocene]palladium(II): an effective catalyst for cross-coupling of secondary and primary alkyl Grignard and alkylzinc reagents with organic halides. *J. Am. Chem. Soc.*, **106**, 158–63.

27. Ozawa, F., Kubo, A., Matsumoto, K. and Hayashi, T. (1993) Palladium-catalyzed asymmetric arylation of 2,3-dihydrofuran with phenyl triflate. A novel asymmetric catalysis involving a kinetic resolution process. *Organometallics*, **12**, 4188–96.

28. Cho, S.-Y. and Shibasaki, M. (1998) Synthesis and evaluation of a new chiral ligand: 2-diphenylarsino-2′-diphenylphosphino-1-1′-binaphthyl (BINAPAs). *Tetrahedron Lett.*, **39**, 1173–6.

29. Kojima, A., Boden, C.D.J. and Shibasaki, M. (1997) Synthesis and evaluation of a new chiral arsine ligand; 2,2′-bis(diphenylarsino)-1,1′-binaphthyl (BINAs). *Tetrahedron Lett.*, **38**, 3459–60.

30. Guiry, P.J. and Saunders, C.P. (2004) The development of bidentate P,N ligands for asymmetric catalysis. *Adv. Synth. Catal.*, **346**, 497–537.

31. Pfaltz, A. and von Matt, P. (1993) Chiral Phosphinoaryldihydrooxazoles as ligands in asymmetric catalysis: Pd-catalysed allylic substitution. *Angew. Chem., Int. Ed. Engl.*, **32**, 566–8.

32. Williams, J.M.J., Dawson, G., Frost, C. and Coote, S. (1993) Asymmetric palladium catalysed allylic substitution using phosphorus containing oxazoline ligands. *Tetrahedron Lett.*, **34**, 3149–50.

33. Helmchen, G. and Sprinz, J. (1993) Phosphinoaryl- and phosphinoalkyloxazolines as new chiral ligands for enantioselective catalysis: very high enantioselectivity in palladium catalysed allylic substitutions. *Tetrahedron Lett.*, **34**, 1769–72.

34. Loiseleur, O., Hayashi, M., Keenan, M. *et al.* (1997) Enantioselective Heck reactions using chiral P,N-ligands. *J. Organomet. Chem.*, **576**, 16–22.

35. Ogasawara, M., Yoshida, K., Kamei, H. *et al.* (1998) Synthesis and application of novel chiral phosphino-oxazoline ligands with 1,1′-binaphthyl skeleton. *Tetrahedron: Asymmetry*, **9**, 1779–87.

36. Imai, Y., Zhang, W., Kida, T. *et al.* (1998) Diphenylphosphinooxazoline ligands with a chiral binaphthyl backbone for Pd-catalyzed allylic alkylation. *Tetrahedron Lett.*, **39**, 4343–6.

37. Kündig, E.P. and Meier, P. (1999) Synthesis of new chiral bidentate (phosphinophenyl)benzoxazine P,N ligands. *Helv. Chim. Acta*, **82**, 1360–70.

38. Hashimoto, Y., Horie, Y., Hayashi, M. and Saigo, K. (2000) An efficient phosphorus-containing oxazoline ligand derived from *cis*-2-amino-3,3-dimethyl-1-indanol: application to the palladium-catalyzed asymmetric Heck reaction. *Tetrahedron: Asymmetry*, **11**, 2205–10.

39. Deng, W.-P., Hou, X.-L., Dai, L.-X. and Dong, X.-W. (2000) Efficient planar chiral 2′-substituted 1,1′-P,N-ferrocene ligands for the asymmetric Heck reaction: control of enantioselectivity and configuration by planar chiral substituent. *Chem. Commun.*, 1483–4.

40. Gilbertson, S.R., Genov, D.G. and Rheingold, A.L. (2000) Synthesis of new bicyclic P–N ligands and their application in asymmetric Pd-catalyzed *p*-allyl alkylation and Heck reaction. *Org. Lett.*, **2**, 2885–8.

41. Gilbertson, S.R. and Fu, Z. (2001) Chiral P,N-ligands based on ketopinic acid in the asymmetric Heck reaction. *Org. Lett.*, **3**, 161–4.

42. Gilbertson, S.R., Xie, D. and Fu, Z. (2001) Proline derived phosphine–oxazoline ligands in the asymmetric Heck reaction. *Tetrahedron Lett.*, **42**, 365–8.

43. Drury, W.J. III, Zimmermann, N., Keenan, M. *et al.* (2004) Synthesis of versatile chiral N,P ligands derived from pyridine and quinoline. *Angew. Chem. Int. Ed.*, **43**, 70–4.

44. Kilroy, T.G., Cozzi, P.G., End, N. and Guiry, P.J. (2004) The application of HETPHOX ligands to the asymmetric intermolecular Heck reaction. *Synlett*, 106–10.

45. Kilroy, T.G., Cozzi, P.G., End, N. and Guiry, P.J. (2004) The application of HETPHOX ligands to the asymmetric intermolecular Heck reaction of 2,3-dihydrofuran and 2,2-disubstituted-2,3-dihydrofurans. *Synthesis*, 1879–88.
46. Liu, D., Dai, Q. and Zhang, X. (2005) A new class of readily available and conformationally rigid phosphino-oxazoline ligands for asymmetric catalysis. *Tetrahedron*, **61**, 6460–71.
47. Yonehara, K., Mori, K., Hashizume, T. *et al.* (2000) Palladium-catalyzed asymmetric inter-molecular arylation of cyclic or acyclic alkenes using phosphinite–oxazoline ligands derived from D-glucosamine. *J. Organomet. Chem.*, **603**, 40–9.
48. Mata, Y., Dieguez, M., Pamies, O. and Claver, C. (2005) Chiral phosphite oxazolines: class of ligands for asymmetric Heck reactions. *Org. Lett.*, **7**, 5597–9.
49. Mata, Y., Pamies, O. and Dieguez, M. (2007) Screening of a modular sugar-based phosphite–oxazoline ligand library in asymmetric Pd-catalyzed Heck reactions. *Chem. Eur. J.*, **13**, 3296–304.
50. Hou, X.-L., Dong, D.X. and Yuan, K. (2004) Synthesis of new chiral benzylically substituted P,N-ligands and their applications in the asymmetric Heck reaction. *Tetrahedron: Asymmetry*, **15**, 2189–91.
51. Bolm, C., Weickhardt, K., Zehnder, M. and Ranff, T. (1991) Synthesis of optically active bis(2-oxazolines): crystal structure of a 1,2-bis(2-oxazolinyl)benzene ZnCl₂ complex. *Chem. Ber.*, **124**, 1173–80.
52. Dodd, D.W., Toews, H.E., Carneiro, F.D.S. *et al.* (2006) Synthesis of metal complexes of peptide-phosphinites: metal complexes in asymmetric Heck reaction. *Inorg. Chim. Acta*, **359**, 3054–65.
53. Kang, J., Lee, J.H. and Im, K.S. (2003) Preparation of pseudo-C_2-symmetric P,S-hybrid ferro-cenyl ligand and its application to some asymmetric reactions. *J. Mol. Cat.*, **196**, 55–63.
54. Molander, G.A., Burke, J.P. and Carroll, P.J. (2004) Synthesis and application of chiral cyclopropane-based ligands in palladium-catalyzed allylic alkylation. *J. Org. Chem.*, **69**, 8062–9.
55. Dai, W.-M., Yeung, K.K.Y. and Wang, Y. (2004) The first example of atropisomeric amide-derived P,O-ligands used for an asymmetric Heck reaction. *Tetrahedron*, **60**, 4425–30.

12

Intramolecular Enantioselective Mizoroki–Heck Reactions

James T. Link and Carol K. Wada

Abbott Laboratories, Department 47F, Building AP10, 100 Abbott Park Road,
Abbott Park, USA

12.1 Introduction

The Mizoroki–Heck reaction is one of the most widely employed C—C bond-forming reactions in organic synthesis, and the intramolecular variant is frequently utilized for small, medium and large ring construction [1]. The intramolecular reaction allows highly substituted alkenes, including tri- and tetra-substituted systems, to participate and often occurs with high regioselectivity and stereoselectivity. The reactions can generally be conducted in the presence of a wide array of functionality, which has made it ideal for multiple applications, particularly complex molecule synthesis, as exemplified by its utilization in natural product syntheses [2]. The reaction remains effective in sterically congested substrates, leading to its use in tertiary and quaternary centre construction [3]. Cascade reactions, also called domino reactions, have further increased the utility of the transformation, allowing for multiple bond formations and the installation of diverse functionality [4]. Over the last 15 years, the enantioselective intramolecular Mizoroki–Heck reaction has undergone substantial development and has emerged as one of the most effective methods to construct cyclic (and polycyclic) systems and quaternary centres enantioselectively [5].

This chapter will introduce novices to the enantioselective intramolecular Mizoroki–Heck reaction by presenting the information necessary to design and execute reactions. To tempt the expert, a summary of recent provocative results, the current status of the field and research opportunities will be highlighted. After a brief introduction, the

The Mizoroki–Heck Reaction   Edited by Martin Oestreich
© 2009 John Wiley & Sons, Ltd

cationic mechanism of the enantioselective intramolecular Mizoroki–Heck reaction with bidentate ligands will be discussed. Rarer enantioselective mechanisms, such as the neutral pathway proceeding via pentacoordinate palladium intermediates, will be mentioned. On the topic of mechanism, recent results with additives and experiments related to the enantioselective step will serve to illustrate some of the current mechanistic research frontiers. Then, the catalyst systems will be covered, focusing on the basic reaction components and the strengths and limitations of the current set of asymmetric ligands. Once the mechanisms and reaction components are understood, the reaction scope will be covered. Currently, the proven scope of the enantioselective reaction is smaller than the intramolecular reaction. The 5-*exo* and 6-*exo* cyclizations are common, and 4-*exo* and 7-*exo* asymmetric reactions have been reported. Many of these reactions can be utilized to create congested quaternary centres. Silane-directed elimination has been used to fashion tertiary centres, and cascade (also called domino) enantioselective reactions have been used to perform several reactions in a one-pot process.

In order for a chemist to take advantage of the asymmetric intramolecular Mizoroki–Heck reaction, developing an understanding of the reaction mechanism is important for target selection, substrate design and catalyst choice [6]. For nongroup-selective (nondesymmetrizing) reactions, β-hydride elimination must take place away from the alkene carbon involved directly in bond formation. Therefore, asymmetric quaternary centre construction (e.g. amide **1** → oxindole **2**) is a frequent application (Scheme 12.1). For a typical substrate, tertiary centre construction is inefficient, since β hydride elimination can destroy

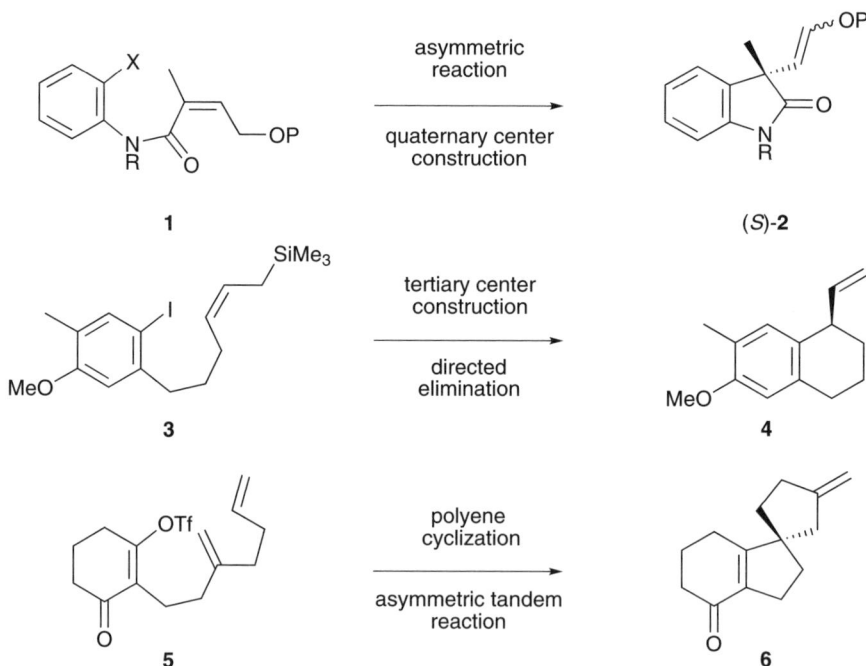

Scheme 12.1   *Representative intramolecular asymmetric Mizoroki–Heck reaction types.*

a newly formed stereocentre. However, directed elimination can be accomplished using silanes (e.g. allyl silane **3** → alkene **4**), allowing the formation of a ring containing a tertiary centre [7]. In addition to β-hydride elimination, other reaction termination manifolds can also be accessed in asymmetric cascade reactions. The σ-alkylpalladium intermediates obtained after migratory insertion can undergo other reactions, including a second alkene insertion reaction such as in the polyene cyclization of triflate **5** to tricycle **6** [8]. These cascade reactions are powerful because they can form multiple stereocentres (enantio- and diastereo-selectively), bonds and/or rings in a single reaction.

12.2 Mechanism

The most common Mizoroki–Heck reaction mechanism is called the neutral mechanism, because its intermediates are uncharged. The catalytic cycle for the neutral manifold of the intramolecular Mizoroki–Heck reaction of alkenyl and aryl halides is shown in Scheme 12.2 [9]. The reaction begins with the generation of the catalytically active palladium(0) species **7** from palladium salts and typically two phosphine ligands. Oxidative addition of halide **8** gives the palladium(II) species **9**. A *syn* migratory insertion to form a C–C bond leads to the σ-alkylpalladium intermediate **10**; *syn* β-hydride elimination rapidly releases the product **11** and the hydridopalladium complex **12**. For tertiary stereocentres, selective elimination of the β′-hydrogen over the β-hydrogen of **11** must occur to maintain the

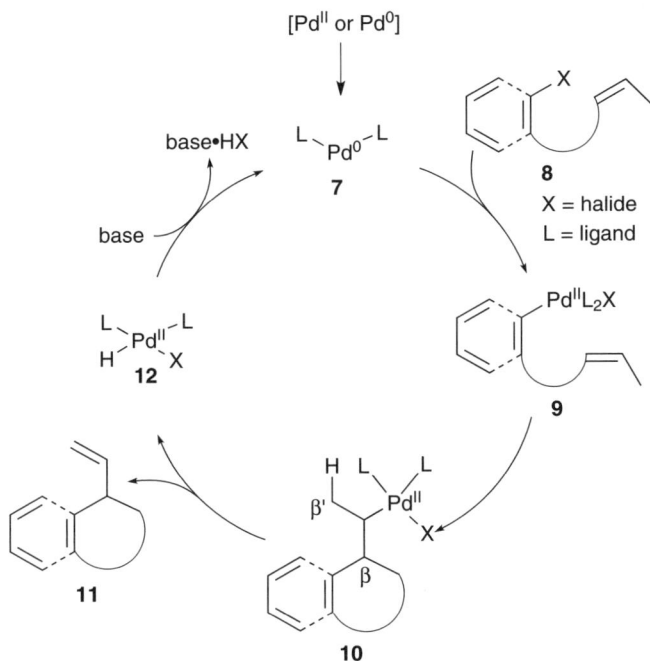

Scheme 12.2 *Neutral mechanism of the intramolecular Mizoroki–Heck reaction.*

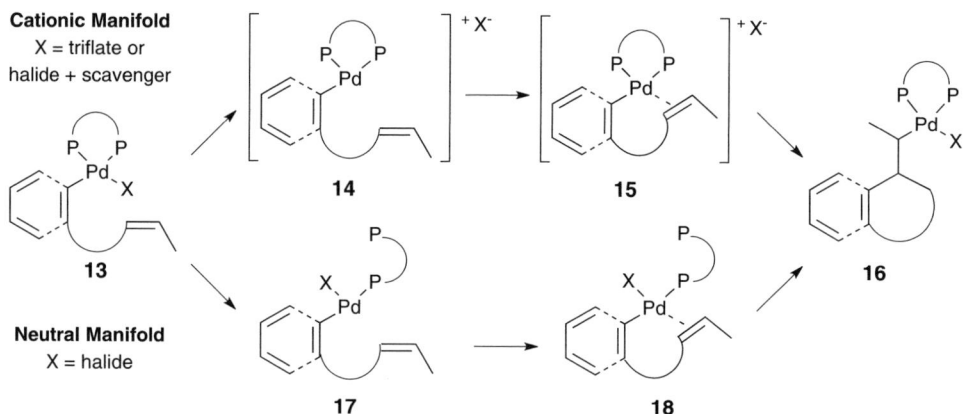

Scheme 12.3   *Cationic and neutral manifolds of the intramolecular Mizoroki–Heck reaction with a bidentate ligand.*

newly formed stereocentre. One equivalent of base then regenerates the catalytically active palladium(0) species **7**.

To rationalize the enantioselective intramolecular Mizoroki–Heck reaction, other mechanistic pathways have been invoked, including the cationic pathway and neutral pentacoordinate intermediates. The different pathways explain the influences that substrate, ligands and additives have upon selectivity. As with many catalytic asymmetric systems, optimization of enantioselectivity is achieved through a combination of experimentation and the application of mechanistic knowledge.

Two mechanistic pathways, termed the cationic and neutral manifolds, account for the differences in reactivity and enantioselectivity observed for alkenyl or aryl triflates and halides (Scheme 12.3). The cationic reaction pathway is invoked for alkenyl and aryl triflates [10]. Following oxidative addition, triflate dissociation (**13**→**14**) leads to the cationic intermediate **14** with a vacant coordination site on palladium. Coordination of the alkene (**14**→**15**) and migratory insertion (**15**→**16**) then occur without dissociation of either arm of the bidentate phosphine ligand. The stability of the bidentate phosphine palladium interaction throughout the catalytic cycle is thought to be responsible for the high enantioselectivity of reactions that utilize this pathway. Frequently, alkene association or migratory insertion is postulated as the stereodifferentiating step in the asymmetric Mizoroki–Heck reaction, and ligand dissociation would lead to diminished enantioselectvity [5, 11]. The most effective enantioselective intramolecular Mizoroki–Heck reactions commonly employ bidentate chiral phosphine ligands, with 2,2'-bis(diphenylphosphino)-1,1'-binaphthyl (BINAP) most often giving the optimal selectivity [1b].

For alkenyl and aryl halides, a neutral mechanistic manifold has been invoked in which one arm of the phosphine ligand must dissociate (**13**→**17**) to create a vacant site on palladium for alkene coordination (**17**→**18**) [6, 11]. The lower enantioselectivities observed for Mizoroki–Heck reactions occurring via the neutral pathway have been attributed to this ligand dissociation. To achieve higher enantioselectivities, the reaction of alkenyl and aryl halides may be directed into the cationic manifold by the addition of the silver or thallium

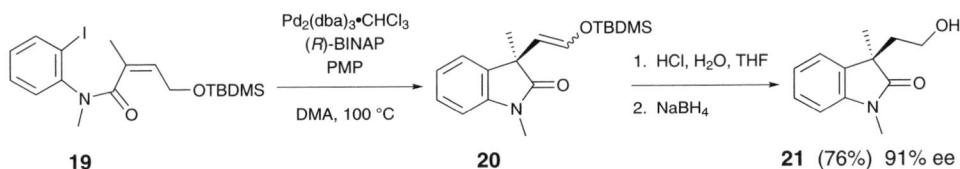

Scheme 12.4 *Enantioselective Mizoroki–Heck reaction of o-iodoanilide.*

salts which act as halide scavengers [12]. Similarly, the addition of tetrabutylammonium halide salts provides access to the neutral manifold for the Mizoroki–Heck reaction of alkenyl and aryl triflates [11, 13]. Thus, by choosing the appropriate reaction conditions, the cationic manifold can be predictably accessed for either aryl triflate or aryl halide substrates. The cationic manifold of the intramolecular Mizoroki–Heck reaction remains the pathway of choice for achieving high enantioselectivities.

Typically, high enantioselectivity cannot be achieved from an aryl iodide in the absence of a halide scavenger. An exception was reported by Overman and Poon [11] and has been found to operate in other systems (Scheme 12.4). The anilino amide **19**, utilizing (*R*)-BINAP as the chiral ligand and 1,2,2,6,6-pentamethylpiperidine (PMP) as the base, cyclized to oxindole **20**. Cleavage of the silyl enol ether and reduction provided the alcohol **21** in good yield and 91% ee. The unusually high selectivity obtained via the neutral pathway led to the proposal of a novel mechanism proceeding through a pentacoordinate palladium intermediate.

The normal neutral pathway (**22**→**24**→**25**→**27**) was ruled out by conducting the re-action with monodentate phosphine BINAP ligand mimics (Scheme 12.5). The products obtained were of low enantiomeric excess relative to reactions employing BINAP. The direct cationic pathway (**24**→**26**) was also eliminated due to the fact that the opposite stereochemistry was obtained under cationic conditions with the addition of silver salts. The switch in stereoselectivity in the presence of silver salts, moreover, indicates that oxidative insertion is not the enantioselective step. β-Hydride elimination was also dis-counted as the enantioselective step due to the influence of the double-bond geometry of the starting material on the enantioselectivity of the cyclization. The proposed enantiose-lective step is the formation of the cationic intermediate **26** by an associative displacement (**24**→**28**→**26**). In the case of square planar palladium(II) complexes, substitution chem-istry can occur through associative processes. Axial coordination of the alkene would form the pentacoordinate palladium(II) complex **28**. Reports of isolated and characterized pentacoordinate palladium(II) species provide support for this proposed intermediate.

12.2.1 New Mechanistic Developments

Much of the recent literature on the mechanism of the enantioselective intramolecular Mizoroki–Heck reaction has focused on the anionic mechanism, *o*-iodoanilide substrates, pathways involving neutral pentacoordinate palladium intermediates and the influence of additives. The new examples and mechanistic findings indicate that the potential may exist to control the stereoselectivity of the intramolecular Mizoroki–Heck reaction through pathways other than the cationic mechanism. However, further research is needed to obtain the level of effectiveness of the traditional cationic pathway.

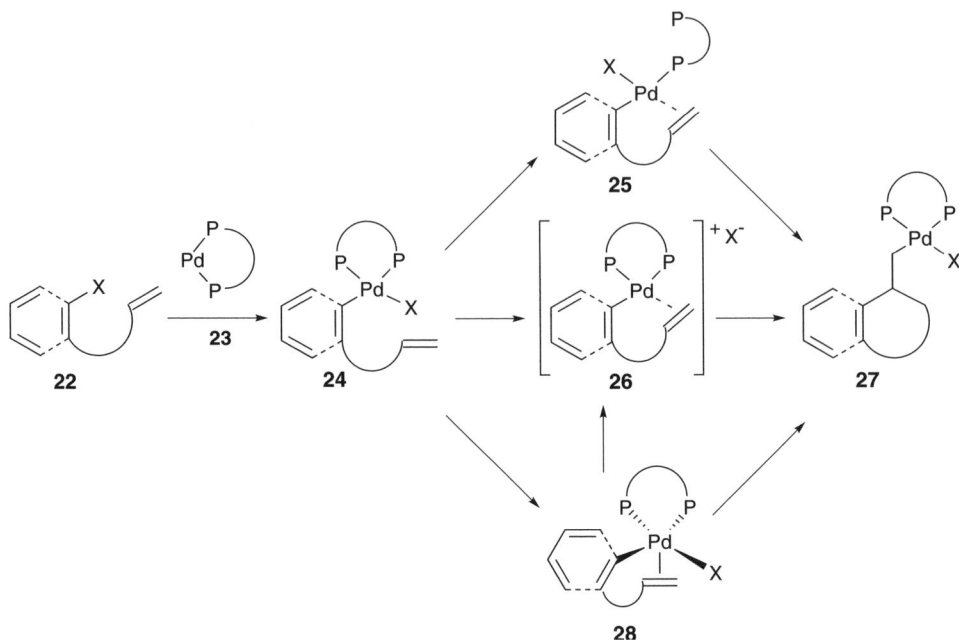

Scheme 12.5   *Potential neutral pathways of the intramolecular Mizoroki–Heck reaction with a bidentate ligand.*

A catalytic cycle arising from the common precatalyst mixture of Pd(OAc)$_2$ and PPh$_3$, termed the anionic pathway, has recently been proposed [14]. This pathway involves anionic palladium(0) and palladium(II) intermediates in which the acetate anion is coordinated with palladium in the catalytically active species persisting after oxidative addition. The anionic pathway has not been invoked or thoroughly explored for enantioselective intramolecular Mizoroki–Heck reactions. However, it may become more significant based on recent studies with Pd(OAc)$_2$ and bidentate phosphine ligands for which the palladium(II) species is only formed in the presence of added acetate ion [15].

The pentacoordinate palladium pathway has been invoked for the highly enantioselective intramolecular Mizoroki–Heck reaction of the cyclohexadienone monoacetal **29a** using the novel monodentate phosphoramidite ligand **30a** (Scheme 12.6) [16]. This is an exciting finding, given that previous monodentate chiral ligands have only achieved low to moderate enantioselectivities [11]. High enantioselectivity typically requires the use of bidentate ligands. Slightly lower enantioselectivity (90% ee) was obtained using the bidentate version of **30a**; when BINAP was used, the product was obtained in low enantioselectivity (0–5% ee) [16a]. To provide evidence as to which mechanistic pathway is involved (cationic or neutral), the triflate analogue **29b** was prepared, which is expected to react via the cationic pathway; the triflate **29b** gave a sluggish reaction, reaching only 25% conversion to tricycle **31** after 2 days in 75% ee. No reaction was observed for **29a** with the addition of Ag$_3$PO$_4$ or Ag$_2$CO$_3$. These results led to the conclusion that the neutral pathway gives the optimal conversion and enantioselectivity for these ligands.

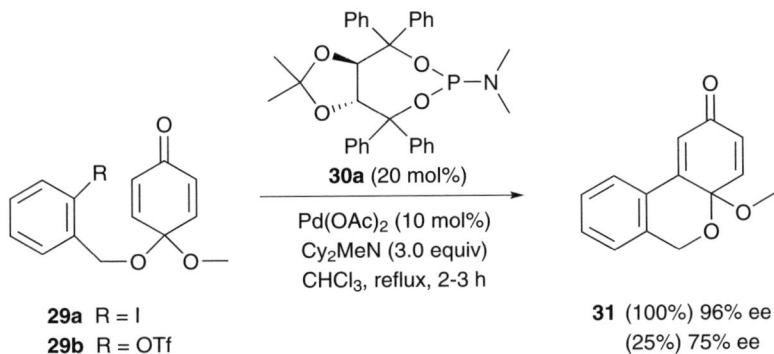

Scheme 12.6 *Enantioselective Mizoroki–Heck reaction with a monodentate phosphoramidite ligand.*

To explore which mechanistic variant of the neutral pathway may be in operation, the ligand **30b** with a pendant amine was prepared (Scheme 12.7) [16b]. This ligand purportedly mimics reaction through the neutral pathway (**32**→**34**→**35**→**37**) with dissociation of the less strongly coordinating pendant amine. A substantial drop in enantioselectivity (26%) and very slow reaction (60% after 48 h) was observed for the reaction of **29a** using ligand **30b**. This is cited as evidence that the reaction proceeds through the pentacoordinate palladium pathway (**32**→**34**→**38**→**36**→**37**). There may also be other mechanistic implications

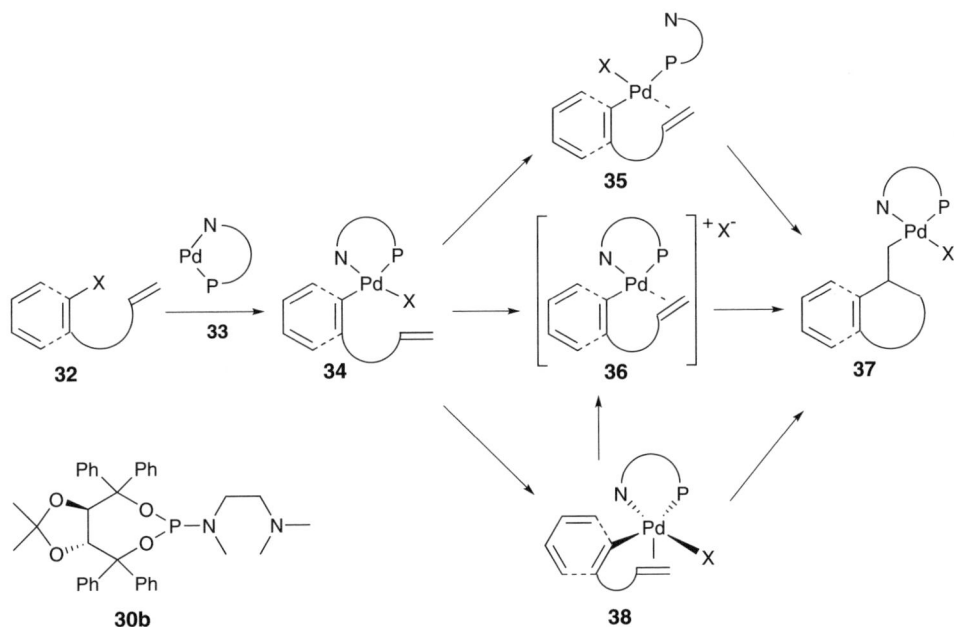

Scheme 12.7 *Pathways accessible to a phosphoramidite ligand with a pendant amine.*

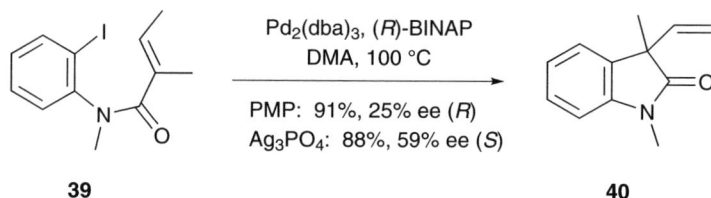

Scheme 12.8   *Mizoroki–Heck reaction of an o-iodoanilide under PMP and Ag₃PO₄ reaction conditions.*

due to the presence of the intramolecular base. Two equivalents of ligand **30a** were found to provide the optimal selectivity. This is significant because several equivalents of monodentate ligands are typically required to form stable palladium(0) complexes and lends support to the idea that the reaction proceeds via the neutral pentacoordinate palladium pathway. Further studies are needed to determine the substrate scope of the monodentate phosphoramidite **31** and whether ligand or substrate governs access to the pentacoordinate palladium pathway.

For both reactions, in which pentacoordinate palladium intermediates have been invoked, the proposed enantioselective step is alkene association. Another interesting observation for the cyclization of *o*-iodoanilides such as **39** is a switch in enantioselectivity triggered by the addition of certain additives (PMP or Ag₃PO₄) while employing the same chiral bisphosphine ligand (Scheme 12.8) [13]. When neutral conditions were employed using PMP as the base (PMP protocol), (*R*)-**40** was obtained in 25% ee. Under cationic conditions in the presence of Ag₃PO₄, the formation of the enantiomer (*S*)-**40** was favoured in 59% ee. Recently, there have been several publications that put forth alternative theories for the stereocontrolling step and provide information on the subtle influence of additives on the cyclization of *o*-iodoanilides.

It has been proposed that oxidative addition, rather than alkene association or migratory insertion, is the enantioselective step in the intramolecular Mizoroki–Heck reaction [17]. The substrates studied are axially chiral *o*-iodoanilides **41** with N–Ar rotational barriers that vary from <20 to 30 kcal mol^{-1} depending on the size of R^1 (Scheme 12.9) [18].

For bis-*ortho*-substituted anilides with a high rotational barrier, such as amide **42**, pure enantiomers (*M*)-(−)-**42** and (*P*)-(+)-**42** can be prepared (Table 12.1). If the intramolecular Mizoroki–Heck reaction of **42** is faster than the rate of racemization by rotation about the N—Ar bond, then reaction of a single enantiomer with an achiral palladium catalyst would produce an enantioenriched product. By employing room-temperature Mizoroki–Heck

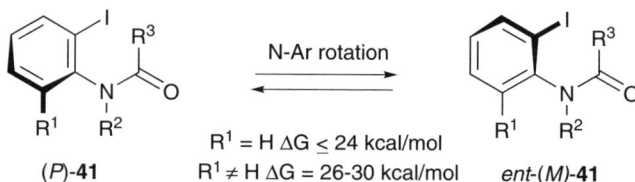

Scheme 12.9   *Axial conformer rotational barriers of an o-iodoanilide.*

Table 12.1 *Intramolecular Mizoroki–Heck reaction of enantioenriched o-iodoanilides*

Pd$_2$(dba)$_3$ (10 mol%)
(t-Bu)$_3$PH•BF$_4$ (40 mol%)

Et$_3$N, PhMe, rt, 24 h

a R = Me
b R = Br

(*M*)-**42a,b**   (*R*)-**43a,b**

	R	Precursor	er **42**	Product	er **43**	Yield (%)	Chirality transfer (%)
1	Me	(*M*)-(−)-**42a**	99.5/0.5	(*R*)-(+)-**43a**	85.5/14.5	95	86
2	Me	(*P*)-(+)-**42a**	98.5/1.5	(*S*)-(−)-**43a**	86.5/13.5	92	88
3	Br	(*M*)-(−)-**42b**	99.5/0.5	(*R*)-(+)-**43b**	89/11	77	89
4	Br	(*P*)-(+)-**42b**	97.5/2.5	(*S*)-(−)-**43b**	89/11	69	91

reaction conditions, pure (*M*)-(−)-**42** and (*P*)-(+)-**42** were cyclized to provide enantioenriched oxindole **43**. These results were interpreted as support for oxidative addition as the stereocontrolling step for this class of asymmetric Mizoroki–Heck reactions. However, the results obtained are likely to be specific to the chosen substrates and closely related examples.

The proposed mechanism is shown in Scheme 12.10. Insertion of the palladium(0) complex into the C–I bond of **42a** or **42b** with retention of axial chirality gives intermediate **44**. This intermediate can still provide either enantiomer of **43**, depending on the facial selectivity of migratory insertion. The diastereotopic alkene faces are accessed by rotation

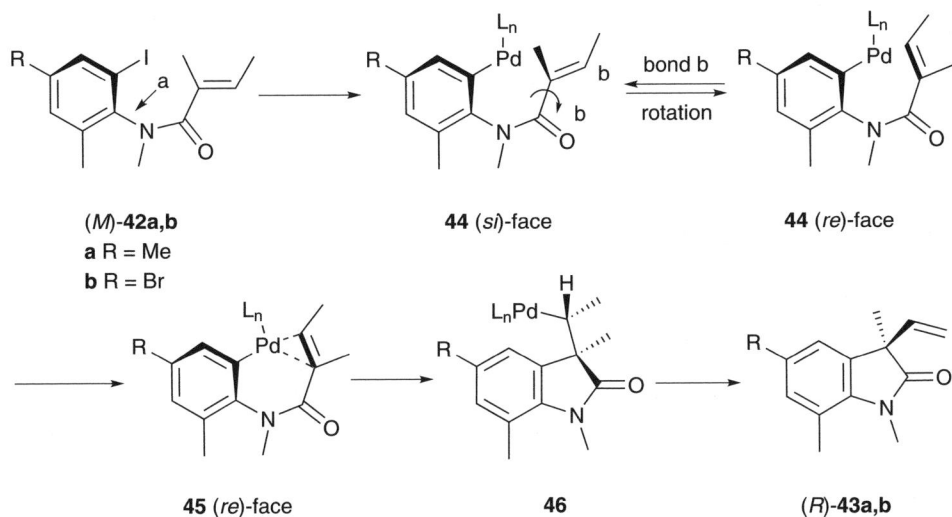

(*M*)-**42a,b**
a R = Me
b R = Br

44 (*si*)-face   44 (*re*)-face

bond b
rotation

45 (*re*)-face   46   (*R*)-**43a,b**

Scheme 12.10 *Intramolecular Mizoroki–Heck reaction mechanism of enantioenriched o-iodoanilides.*

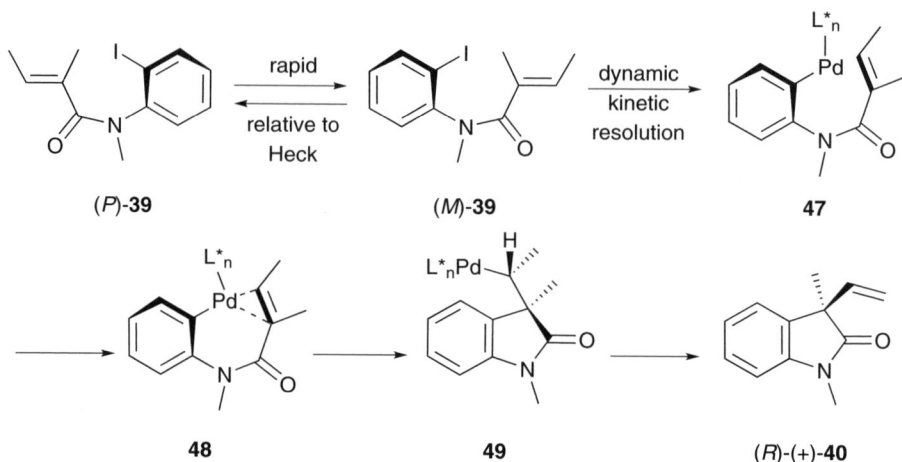

Scheme 12.11 Intramolecular Mizoroki–Heck reaction mechanism of racemic o-iodoanilides.

around the alkene–carbonyl bond (bond b, (si)-**44**→(re)-**44**). These are proposed to be rapidly interconverting, and the geometric requirements for the formation of complex **45** dictate the face of the alkene that reacts (in the rotamer (si)-**44** the β-carbon is inaccessible) and constitutes a transfer of the axial chirality from the oxidative addition.

The authors propose that, for the case of rapidly interconverting racemic o-iodoanilides, oxidative addition of a chiral palladium catalyst to the Ar–I bond constitutes a dynamic kinetic resolution (Scheme 12.11). Theoretically, the oxidative addition using a chiral palladium catalyst would preferentially occur with one enantiomeric conformer of the racemic o-iodoanilide **39** to generate the enantioenriched intermediate **47**. The subsequent insertion step (**47**→**48**→**49**) is expected to be more rapid than N–Ar bond rotation, and the axial chirality of the oxidative addition is relayed to the final product, oxindole **40**.

The suggestion is then made that the stereocontrolling step in asymmetric Mizoroki–Heck reactions is oxidative addition (via dynamic kinetic resolution) rather than alkene association or migratory insertion. The implication is that only substrates capable of a dynamic kinetic resolution may cyclize with high enantioselectivity. This would limit the substrate scope of the asymmetric intramolecular Mizoroki–Heck reaction. While the dynamic kinetic resolution during the oxidative addition may be a component of the overall stereoselectivity, it does not rule out contributions from later events in the mechanistic pathway and does not explain the effect of additives on selectivity. What has been shown is that the axial chirality of the o-iodoanilides (as with any enantioenriched isomer of a chiral precursor) influences the stereochemical outcome of their reactions.

One factor that the mechanistic scheme does not address is the change in stereoselectivity observed for the reaction of o-iodoanilides upon the addition of silver salts (Scheme 12.8). Recently, the influence of additives on the enantioselective intramolecular Mizoroki–Heck reaction of o-iodoanilide **39** was studied in a series of reactions in which reaction times, temperatures, concentration and catalyst/Ag_3PO_4 loading were varied [19]. It was observed that, in the presence of Ag_3PO_4, the enantioselectivity of the product **40** increased

Table 12.2 *Enantioselectivity as a function of reaction conversion*

	Temp. (°C)	Time (min)	Catalyst loading (mol%)	Ag_3PO_4 (equiv)	Conversion (%)	ee **40** (%)
1	60	60	10	5.0/3.0	14	9.0
2	60	240	10	5.0/3.0	39	40
3	80	60	7.5	1.0	51	26
4	80	180	7.5	1.0	100	48

as the reaction progressed (Table 12.2). This suggests that there may be different, or multiple, mechanisms in operation, involving more than simple competition between pro-(*R*) and pro-(*S*) pathways. The results from this and a subsequent study are conflicting and explanations contradictory [19, 20]. An attempt to provide a cohesive summary of the results and proposals within the context of previous mechanistic work on the asymmetric intramolecular Mizoroki–Heck reaction follows.

A rationale for the effect of Ag_3PO_4 on the reaction pathway is illustrated in Scheme 12.12. Variable-temperature NMR experiments and X-ray crystal structures of the starting material revealed several dynamic processes, including *E/Z* isomerization of the amide, rotation about the alkene–carbonyl bond and the presence of pure (*P*) and (*M*) atropoisomers **39** from axial chirality about the N–Ar bond. Under neutral conditions in the presence of PMP and the absence of silver salts (PMP protocol), a dynamic kinetic resolution of the interconverting (*P*) and (*M*) helical *o*-iodoanilides **39** is involved in which oxidative addition to the (*M*)-helix to give (*M*)-**50** is favoured en route to (*R*)-(+)-oxindole **40**. However, kinetic resolution cannot fully account for the variation in enantiomeric excess, as the reaction of **39** progresses in the presence of silver salts. The presence of Ag_3PO_4 is proposed to open up a rapid interconversion between pro-(*R*) and pro-(*S*) oxidative addition species **50**. The pro-(*S*) pathway is now preferred in the C—C bond-forming step. While this Ag_3PO_4-promoted equilibrium explains the change in stereoselectivity in the presence of silver salts, it also does not account for the variation in product enantiopurity as the reaction progresses.

A rationale for the variation in product enantiopurity with the extent of reaction may be gleaned from the reaction of amide **39**, in which the purity of Ag_3PO_4 influenced the stereochemical outcome of the reaction (Table 12.3) [20]. When high-purity Ag_3PO_4 from Strem was used, the product (*R*)-(+)-**40** was obtained. However, when Ag_3PO_4 from Aldrich was employed, the opposite (*S*)-(−) enantiomer was obtained. Upon exposing the Strem Ag_3PO_4 to light, the (*S*)-(−) product was obtained. These results suggest that the pro-(*R*)-(+) selectivity with (*R*)-BINAP is the preferred pathway for both the PMP and

Scheme 12.12   *Influence of Ag₃PO₄ on reaction pathways.*

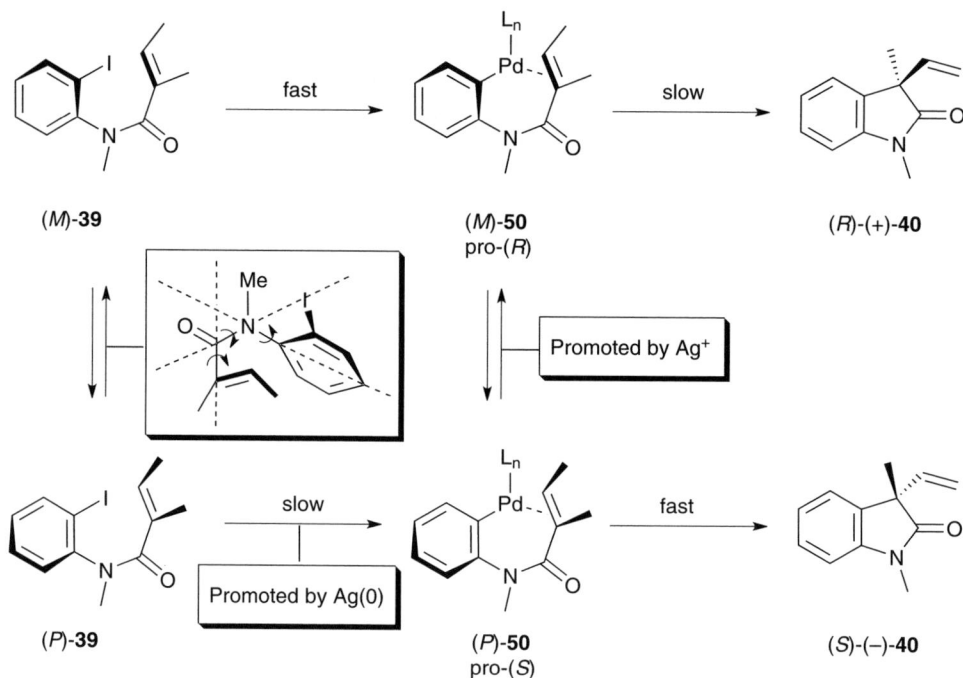

Table 12.3   *Effect of Ag₃PO₄ purity on stereoselectivity*

	Additive	Method[a]	Conversion (%)	Product	ee (%)
1	Strem Ag₃PO₄	A	98	(R)-(+)-**40**	16
2	PMP	A	100	(R)-(+)-**40**	24
3	Aldrich Ag₃PO₄	A	100	(S)-(−)-**40**	43
4	Strem Ag₃PO₄	B	100	(S)-(−)-**40**	33
5	Aldrich Ag₃PO₄	B	80	(S)-(−)-**40**	25
6	PMP	B	100	(R)-(+)-**40**	55

[a] Method A: Anachem SK233 workstation. Method B: conventional Schlenk procedure.

Ag$_3$PO$_4$ protocols, and the presence of Ag(0) impurities promote a competing pro-(*S*) pathway. The opposing influences of Ag(I) and Ag(0) may account for the change in product enantiopurity as the reaction progresses, particularly if Ag(0) is produced during the course of the reaction.

To test further the hypothesis that the presence of Ag(0) promotes the pro-(*S*) pathway, the Mizoroki–Heck reaction of **39** was carried out with *in situ* chemical reduction of Strem Ag$_3$PO$_4$ using formic acid. With 0.33 equiv of formic acid to mimic the effect of the Aldrich Ag$_3$PO$_4$, an improved enantioselectivity of 42% for the (*S*)-(−) enantiomer was observed. However, the addition of 5 equiv of formic acid, which would presumably reduce all Ag(I) to Ag(0), gave (*R*)-(+)-**40** in 24% ee, which is similar to the PMP protocol. These experiments suggest that the presence of both Ag(I) and Ag(0) are required to observe the change in stereoselectivity. Further study is needed to determine the exact effect of Ag(0) on the reaction. A significant finding from these studies is that the neutral and cationic pathways exhibit identical stereocontrol. The authors propose that a common palladium(II) cationic intermediate is involved and that the enantioselective step is the migratory insertion. This common cationic species may be related to intermediate **26** in the pentacoordinate palladium pathway proposed for the reaction of *o*-iodoanilides in the PMP protocol (Scheme 12.5). In the present case, the enantioselective step is the migratory insertion (**26**→**27**), rather than alkene association and formation of the cationic intermediate (**24**→**28**→**26**) proposed previously for the PMP protocol.

12.2.2 Mechanism Summary

These recent mechanistic studies on the enantioselective intramolecular Mizoroki–Heck reaction have revealed complex reaction pathways. Further studies are needed to determine whether monodentate phosphoramidite ligands, neutral pentacoordinate intermediate pathways, or inversion of stereoselectivity through additives have broad and productive value. The cationic manifold with bidentate ligands remains the most reliable method to achieve high enantioselectivity for the intramolecular Mizoroki–Heck reaction. Based on mechanistic insight, the cationic manifold can be routinely accessed through the appropriate choice of substrate and additives. Generally, several combinations of palladium precatalyst, ligand and additives are screened to find the catalyst system that provides the highest enantioselectivity. Future mechanistic studies and improved ligand design have the potential to advance substantially the routine and practical employment of the enantioselective intramolecular Mizoroki–Heck reaction.

12.3 Catalyst Systems

Palladium-based catalyst systems have undergone significant development, enabling and improving a number of interesting transformations like the Mizoroki–Heck reaction [21]. Most Mizoroki–Heck catalyst systems consist of a palladium complex and ligands that are used in conjunction with bases or additives. The choice of components is based on the desired mechanistic pathway (usually cationic), and the following discussion will aid in their selection as well as outline current efforts to discover superior reagents.

A variety of palladium(0) and palladium(II) precatalysts are routinely employed, including Pd$_2$(dba)$_3$ (dba = dibenzylideneacetone), Pd(dba)$_2$, Pd$_2$(dba)$_3$·CHCl$_3$, Pd(OAc)$_2$ and Pd[(allyl)Cl]$_2$. Catalyst loading in the range of 5.0–20 mol% is common. When a palladium(II) precatalyst like Pd(OAc)$_2$ is utilized, phosphines, water or an amine typically reduces the catalyst to the active palladium(0) species [22]. For the asymmetric reaction, the reduction process can consume valuable phosphine ligand, so palladium(0) precatalysts like Pd$_2$(dba)$_3$ are most commonly used.

In terms of ligands, bidentate ligands, usually phosphines, are critical to achieving high enantioselectivity. The groundbreaking initial asymmetric intramolecular Mizoroki–Heck reaction studies were conducted using the chiral bisphosphine BINAP as the ligand [23]. The BINAP ligand continues to be heavily used and is a key starting point to reaction optimization or catalyst benchmarking programmes. A variety of structurally diverse ligands have also been studied; these are a primary focus of current efforts to develop the reaction further and will be covered in detail.

Stoichiometric amounts of base and additives are employed in the reaction. Inorganic bases like potassium carbonate or soluble amine bases like triethylamine are commonly used. The highly hindered base PMP is employed frequently, particularly in reactions designed to proceed via the neutral pathway [24]. Additives can also dramatically affect the asymmetric intramolecular Mizoroki–Heck reaction. Since salts like Ag$_3$PO$_4$ can shift the mechanism to the cationic manifold when used with halides they are used frequently in the asymmetric reaction variant. In many cases, silver and other metal salts increase enantioselectivity, increase reaction rate, reduce alkene isomerization and change regioselectivity [12, 25]. Alternatively, halide salts can be employed to shift a reaction to the neutral or anionic pathway [11]. The anionic pathway has not been thoroughly explored for asymmetric reactions. Alcohols, such as pinacol or t-BuOH, have been employed in asymmetric reactions and may serve to increase catalyst stability by preventing palladium(0) oxidation [26]. Typical reaction solvents are polar and aprotic, like dimethylformamide or N-methylpyrrolidinone, and solvents like dimethoxyethane and toluene are also sometimes utilized. Although the intramolecular Mizoroki–Heck reaction is compatible with polymer supports, the asymmetric reaction has not been reported for substrates attached to them [27].

Catalyst systems can typically be identified that efficiently catalyse most desired reactions in good to high enantioselectivity. Reaction tuning has been facilitated by the introduction of a variety of ligands that can be surveyed. The structural variety of these ligands is highlighted in Schemes 12.13–12.16. The original ligand design is beyond the scope of this chapter, since few were developed specifically for this reaction. Optimization is achieved by studying experimental conditions, and ligand selection is often made based on precedent. Other than BINAP, most ligands have not been tested on a wide range of substrates, thus limiting prediction. Although generally successful, further catalyst innovation is needed.

12.3.1   Ligand Classes

As a ligand class, arylbisphosphines related to BINAP have been the most thoroughly studied and are generally effective (Scheme 12.13). (*R*)-BINAP **51a** and (*R*)-Tol-BINAP **51b** and their enantiomers (e.g. (*S*)-**51a**) are commercially available and are frequently utilized. Substitution of one of the palladium-binding phosphines with arsine

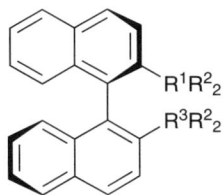 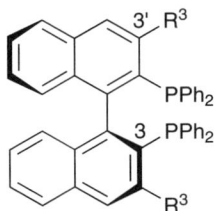

	R¹	R²	R³	Abbreviation
51a	P	Ph	P	(+)-(R)-BINAP
51b	P	tolyl	P	(R)-Tol-BINAP
51c	As	Ph	P	(R)-BINAPAs
51d	As	Ph	As	(R)-BINAs
51e	P	2-furyl	P	TetFuBINAP

51a R³ = H (–)-(S)-BINAP
52a R³ = OMe
52b R³ = O*i*-Pr
52c R³ = OPiv
52d R³ = OBn

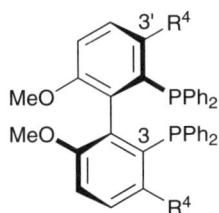

53a R⁴ = H	(R)-MeO-BIPHEP
53b R⁴ = OH	**53g** R⁴ = Ph
53c R⁴ = OMe	**53h** R⁴ = OCO*t*-Bu
53d R⁴ = O*i*-Pr	**53i** R⁴ = Otolyl
53e R⁴ = O*t*-Bu	**53j** R⁴ = OPh
53f R⁴ = *i*-Pr	**53k** R⁴ = furanose

furanose =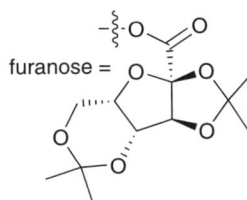

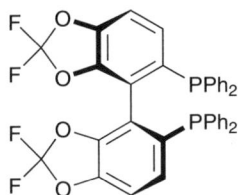 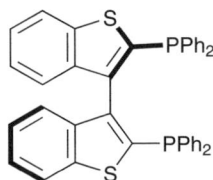 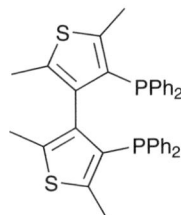

54a (S)-DIFLUORPHOS   **54b** (R)-BITIANP   **55** (+)-TMBTP

Scheme 12.13   *Bisphosphine aryl ligands related to BINAP.*

provides the ligand 2-diphenylarsino-2′-diphenylphosphino-1,1′-binaphthyl (BINAPAs) **51c** [28], or both phosphines give the ligand 2,2′-bis(diphenylarsino)-1,1′binaphthyl (BINAs) **51d** [29]. Furyl phosphines have also been found to be more effective in some palladium-mediated reactions, and the change has been made to the binaphthyl ligand in 2,2′-bis(difurylphosphino)-1,1′binaphthyl (TetFuBINAP) **51e** [30]. Substitution of the position next to the palladium-binding phosphines with an alkoxy substituent in (S)-BINAP **51a** analogues **52a–52c** has also been attempted [31]. Changes to the naphthyl sites have also been studied, as with 2,2′-bis(diphenylphosphino)-1,1′binaphthyl (BIPHEP) ligands like **53a** and some of its substituted analogues **53b–53k** [32], as well as the heterocyclic ligands like 5,5′-bis(diphenylphosphino)-2,2,

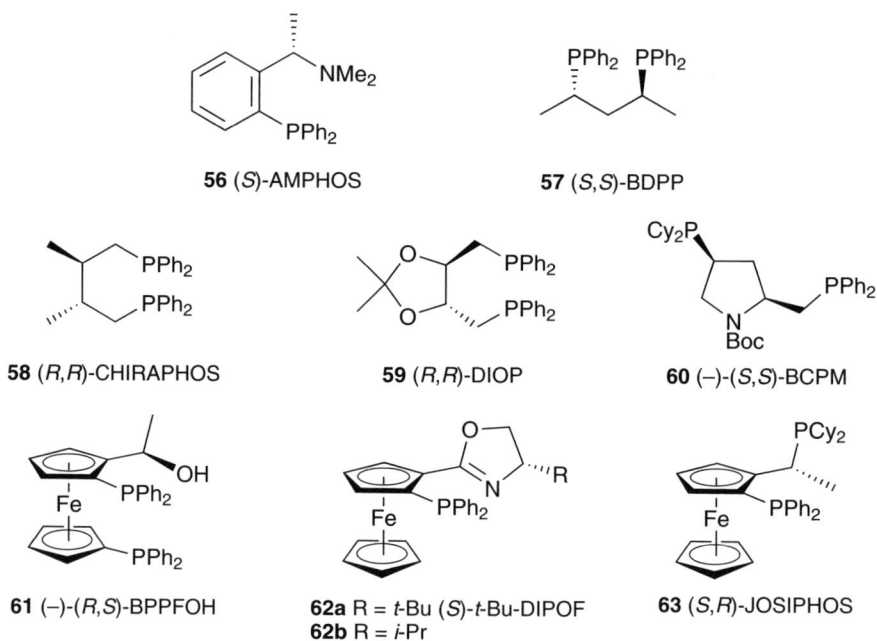

56 (*S*)-AMPHOS

57 (*S,S*)-BDPP

58 (*R,R*)-CHIRAPHOS

59 (*R,R*)-DIOP

60 (–)-(*S,S*)-BCPM

61 (–)-(*R,S*)-BPPFOH

62a R = *t*-Bu (*S*)-*t*-Bu-DIPOF
62b R = *i*-Pr

63 (*S,R*)-JOSIPHOS

Scheme 12.14   *Bisphosphine alkyl and ferrocene ligands.*

2′,2′-tetrafluoro-4,4′-bi-1,3-benzodioxole (DIFLUORPHOS) **54a** [33], benzothiophene ligand 2,2′-bis(diphenylphosphino)-3,3′dibenzo[*b*]thiophene (BITIANP) **54b** [34] and thiophene ligand 4,4′-bis(diphenylphosphino)-2,2′,5,5′-tetramethyl-3,3′-bithiophene (TMBTP) **55** [34]. The optimization of a cascade process that utilizes DIFLUORPHOS **54a** is representative of a typical ligand optimization process and surveys some additional biaryl phosphine ligands not shown herein. Although shown to be more effective than BINAP in some cases, most of the bisphosphine biaryl ligands have not been widely employed or surveyed. The selection of ligands from this subclass remains a primarily experimental exercise for new substrates.

A number of other structurally different bidentate ligands have been used in intramolecular Mizoroki–Heck reactions, including alkyl and ferrocene bidentate ligands. These ligands include the phosphineamine AMPHOS **56** and the diphosphines 2,4-bis(diphenylphosphino)pentane (BDPP) **57**, bis(diphenylphosphino)butane (CHIRAPHOS) **58** [35], 2,3-*O*-isopropylidene-2,3-dihydroxy-1,4-bis(diphenylphosphino) butane (DIOP) **59** [23b] and *N*-(*tert*-butoxycarbonyl)-4-(dicyclohexylphosphino)-2-[(diphenylphosphino)methyl]pyrrolidine (BCPM) **60** [36]. Substituted bidentate ferrocenes have also been studied. In some cases the two palladium-binding groups are attached to different cyclopentadienyl rings, as in (*R*)-α-[(*S*)-1′,2-bisdiphenylphosphino)ferrocenyl]ethyl alcohol (BPPFOH) **61** [37], or the same ring, as in ferrocenyloxazoline DIPOF ligands **62a** and **62b** [38], as well as in the bisphosphine 1-[(*S*)-2-(diphenylphosphino)ferrocenyl]ethyldicyclohexylphosphine (JOSIPHOS) **63** [39]. Although these ligands have not been tested on a diverse substrate set, they have

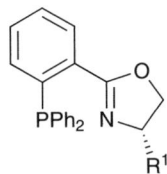

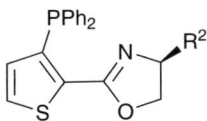

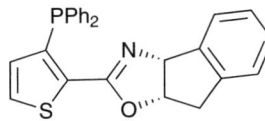

64a R^1 = *t*-Bu (*S*)-*t*-Bu-PHOX

64b R^1 = *i*-Pr

65a R^2 = *i*-Pr

65b R^2 = *t*-Bu

65c R^2 = Ph

66

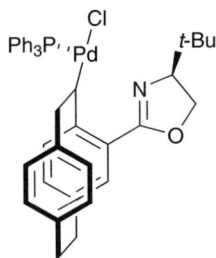

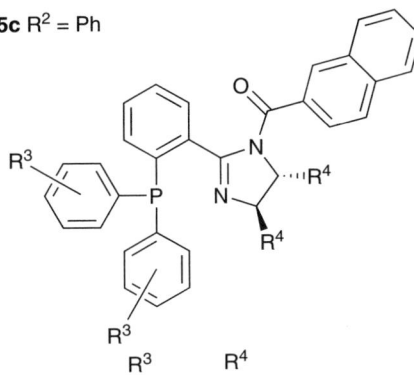

67 KW247

	R^3	R^4
68a	H	(*R,R*)-*p*-CF$_3$C$_6$H$_4$
68b	H	(*S,S*)-3,5-F$_2$C$_6$H$_3$
68c	3,5-F$_2$	(*R,R*)-3,5-F$_2$C$_6$H$_3$
68d	3,5-F$_2$	(*R,R*)-3,5-F$_2$C$_6$H$_3$
68e	3,5-F$_2$	(*S,S*)-Ph
68f	3,5-F$_2$Ph	(*R,R*)-3,5-F$_2$Ph

Scheme 12.15   *Phosphinooxazoline and related ligands.*

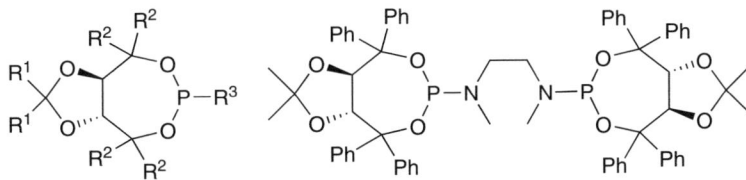

69

	R^1	R^2	R^3
30a	Me	Ph	NMe$_2$
30b	Me	Ph	NMe(CH$_2$)$_2$NMe$_2$
30c	Me	Ph	N-piperidinyl
30d	Me	Ph	N-pyrrolidinyl
30e	Me	2-Naphthyl	NMe$_2$
30f	-(CH$_2$)$_4$-	Ph	NMe$_2$

Scheme 12.16   *Phosphoramidite ligands.*

been shown to offer advantages in specific cases. Currently, there is not a model to determine which substrate will cyclize more effectively with these ligands, so they are employed as part of ligand screening or chosen based on close precedent.

Phosphinooxazolines [40] and phosphinoimidazoline bidentate monophosphine ligands are easily prepared from chiral β-amino alcohols and diamines. A variety of phosphinooxazolines, such as PHOX **64a** and related ligands like **64b** [38], as well as thiophenes **65a–65c** and **66**, have been studied in asymmetric intramolecular Mizoroki–Heck reactions [41]. A more recent report describes KW247 **67** [39]. The phosphinooxazolines are a promising ligand class and should be included in reaction optimization schemes in order to sample the effectiveness of oxazoline monophosphine ligands that may behave differently than bisphosphines. Phosphinoimidazoles have undergone a significant amount of optimization and testing [42]. Amongst the more successful ligands for enantioselective oxindole formation from intramolecular Mizoroki–Heck reactions are phosphinoimidazole ligands **68a–68f**. Determination of the substrate generality of the ligands has not been reported and is worthy of study.

The mono-phosphoramidites **30a–30f** and bis-phosphoramidite **69** are an interesting ligand class that are easily prepared and have been found to be mechanistically unusual [16]. In one class of substrates, high enantioselectivity with the monodentate ligand **30a** was observed via a pentacoordinate intermediate within the neutral pathway (see Section 12.2). Further studies are needed to determine whether the substrate scope of these ligands is broad or narrow.

12.3.2   Catalyst Systems Summary

The development of the enantioselective intramolecular Mizoroki–Heck reaction has stimulated research into new catalyst systems primarily through the exploration of novel asymmetric ligands. The asymmetric intramolecular Mizoroki–Heck reaction places a number of demands upon a catalyst system, including low catalyst loading, high thermal stability, high functional group tolerance, low cost and, for some applications, low toxicity. A robust catalyst system that requires less than 2.0 mol% loading, has high turnover, has moderate air stability, can be employed on a kilogram scale, works in solution and on solid supported substrates, and has wide substrate generality would represent a breakthrough discovery. Efforts to identify catalyst systems with these properties continue.

12.4   Reaction Scope: 4-*exo* and 5-*exo* Enantioselective Cyclizations

The asymmetric intramolecular Mizoroki–Heck reaction is frequently employed to prepare small and medium-sized rings. Typically, 5-*exo* cyclizations are the most facile, leading to more examples of the asymmetric reaction to form this ring size. Cyclizations to form smaller rings are rare, although 4-*exo* asymmetric cyclizations have been reported. For instance, the group-selective cyclization of bis(nonaflate) **70** that also undergoes a intermolecular Mizoroki–Heck reaction with *tert*-butyl acrylate provides the cyclobutane **71** in modest yield and low enantioselectivity (Scheme 12.17) [43]. The formation of cyclopropanes has not yet been reported. The substrates for the 5-*exo* asymmetric cyclizations are commonly designed, or reaction conditions manipulated, so that the cationic Mizoroki–Heck pathway is accessed. For instance, aryl iodide **72** in the presence of a

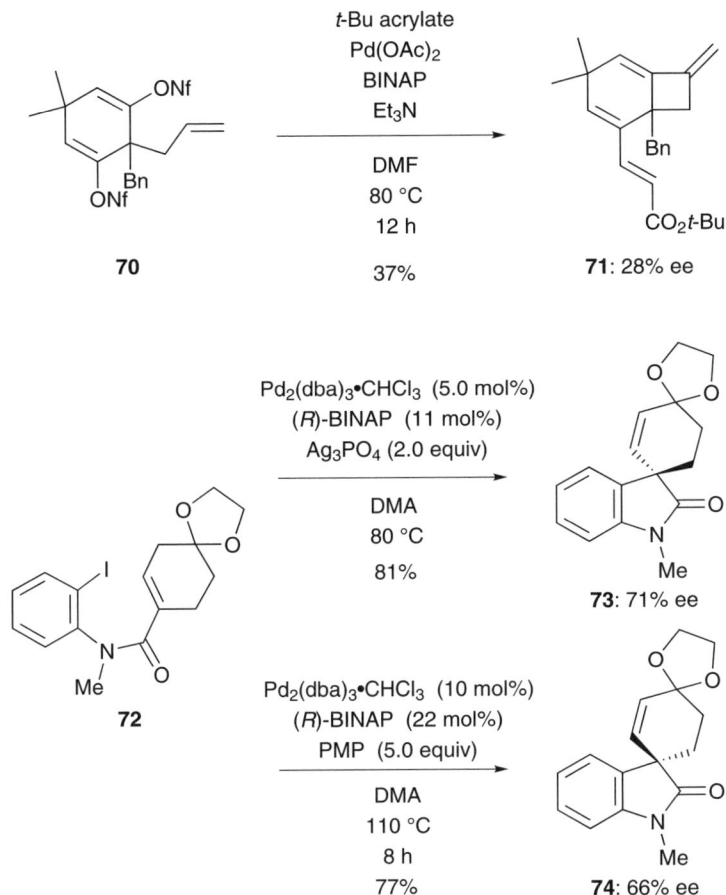

Scheme 12.17   *The 4-exo and 5-exo asymmetric intramolecular Mizoroki–Heck reactions.*

silver salt undergoes 5-*exo* intramolecular Mizoroki–Heck reaction to give oxindole **73** in good yield and enantioselectivity using (*R*)-BINAP as a ligand [24, 44]. This type of reaction shows wide substrate generality and effectively produces quaternary centres. Interestingly, Overman and coworkers [24, 44] have thoroughly studied this type of cyclization and shown that, for some substrates, employing the neutral pathway can yield the enantiomer. For example, treatment of iodide **72** with the same catalyst in the absence of a silver salt provides the enantiomeric oxindole **74** in moderate yield and enantioselectivity. The mechanistic explanation for the reaction outcome involves pentacoordinate palladium intermediates and is described for a different substrate in Scheme 12.5.

12.4.1 Reaction Scope: 6-*exo* Enantioselective Cyclizations

Six-membered rings, particularly those with highly substituted benzylic stereocentres, are readily assembled by the asymmetric intramolecular Mizoroki–Heck reaction. In a seminal example, Shibasaki and coworkers [45] showed that (*E*)-alkenyl aryl triflate **75** can be cyclized in high yield with modest enantioselectivity to bicycle **76** (Scheme 12.18).

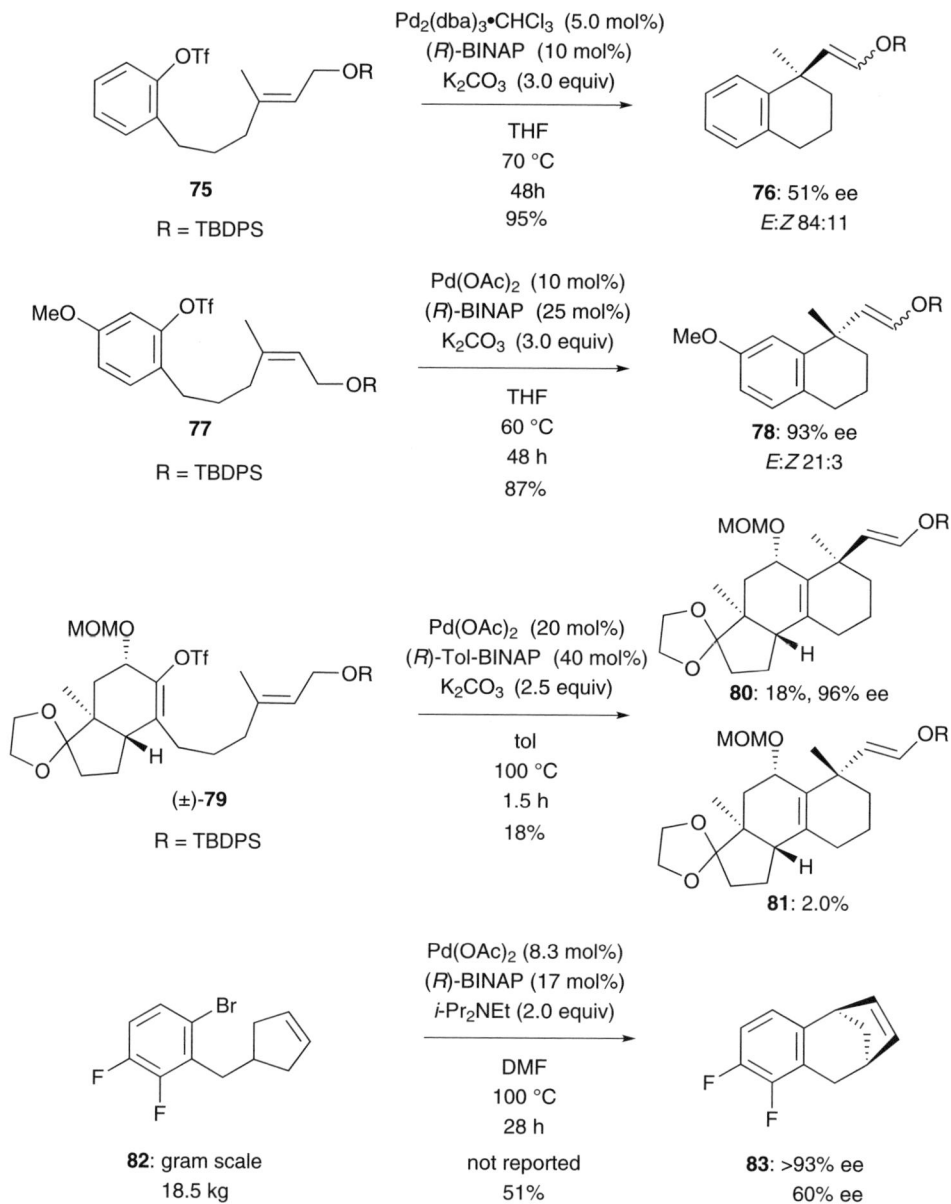

Scheme 12.18   *The 6-exo asymmetric intramolecular Mizoroki–Heck reactions.*

Presumably the reaction occurs via the cationic manifold of the Mizoroki–Heck reaction and a bidentate phosphine is employed. Substrate alkene geometry is critical to the stereochemical outcome and enantioselectivity of the reaction. The (Z)-alkenyl triflate **77** cyclizes to the enantiomeric enol ether **78** in high yield and greater enantioselectivity. The reaction can be utilized to perform a kinetic resolution, although reports are rare. Shibasaki and coworkers [35c] demonstrated that vinyl triflate **79** undergoes reaction to provide the tricyclic intermediate **80** in 18% yield and 96% ee along with diastereomer **81**. Attempts to employ the asymmetric intramolecular Mizoroki–Heck reaction in the scale-up of pharmaceutical intermediates have been made and reveal some of the limitations of the current methodology. Coe and coworkers [46], in their synthesis of a nicotinic receptor probe, reported the gram-scale conversion of aryl bromide **82** to tricycle **83** in excellent enantioselectivity. When scaled up to the kilogram level, a 51% yield of the product was obtained with a modest enantiopurity of 60% ee. Control experiments indicated that oxygen was responsible for the decreased enantioselectivity observed on the large scale. Significant quantities of palladium salt and ligand were also employed. Clearly, the development of oxygen-resistant efficient catalysts is needed to fuel broader industrial use of the reaction.

12.4.2 Reaction Scope: Allylsilane-Terminated Enantioselective Cyclizations

In addition to the possibility of constructing five- and six-membered rings with quaternary centres, medium-sized rings and tertiary centres can also be assembled enantioselectively. Larger ring sizes are possible but have yet to be reported. Allylsilanes can be employed to direct β-hydride elimination, allowing for tertiary centre construction. For instance, the electron-rich aryl iodide **84** with a pendant (Z)-allylsilane can be cyclized to alkene **85** or vinylsilane **86** depending upon the choice of reaction conditions (Scheme 12.19) [47]. Employing Jeffrey conditions primarily leads to β-silyl elimination, giving predominantly alkene **85** (conditions A); adding a silver salt presumably pushes the reaction to the cationic manifold (conditions B) and delivers vinylsilane **86** as the major product. Switching to a bisphosphine and maintaining the cationic reaction conditions shows that the reaction can also be influenced by the catalyst [34]. For instance, (Z)-allylsilane **84** cyclizes to benzazepine **88** in good yield and enantioselectivity, favouring β-silyl elimination. Alternatively, (E)-allylsilane **87** undergoes reaction to provide the benzazepine **89** derived from β-hydride in high enantioselectivity along with bicyclic β-silyl elimination product **88** in modest enantioselectivity. The high enantioselectivity observed for the formation of the tertiary centre within benzazepine **89** is notable, since the reaction termination sequence does not rely upon desilylation.

12.4.3 Reaction Scope: Cascade (or Domino) Asymmetric Intramolecular Mizoroki–Heck Reactions

A powerful feature of the asymmetric intramolecular Mizoroki–Heck reaction is the potential to design substrates that allow the σ-alkylpalladium intermediate resulting from migratory insertion to participate in sequential reactions [4]. Substrates that react with high enantioselectivities to give a σ-alkylpalladium intermediate can be designed to follow reaction pathways other than β-hydride elimination (Scheme 12.20). For substrates such as alkene **90**, designed to lack a β-hydrogen required to carry out termination after the first migratory insertion, the σ-alkylpalladium intermediate **91** can undergo an insertion reaction with another alkene in an inter- (**91**→**92**) or intra-molecular (**91**→**93**) manner to give the cyclic product **92** or polycyclic adduct **93** respectively. The majority of examples are intramolecular and termed polyene cyclization reactions. Capture of the σ-alkylpalladium also takes place with a variety of nucleophiles (**91**→**94**) that are often anionic, neutral, or

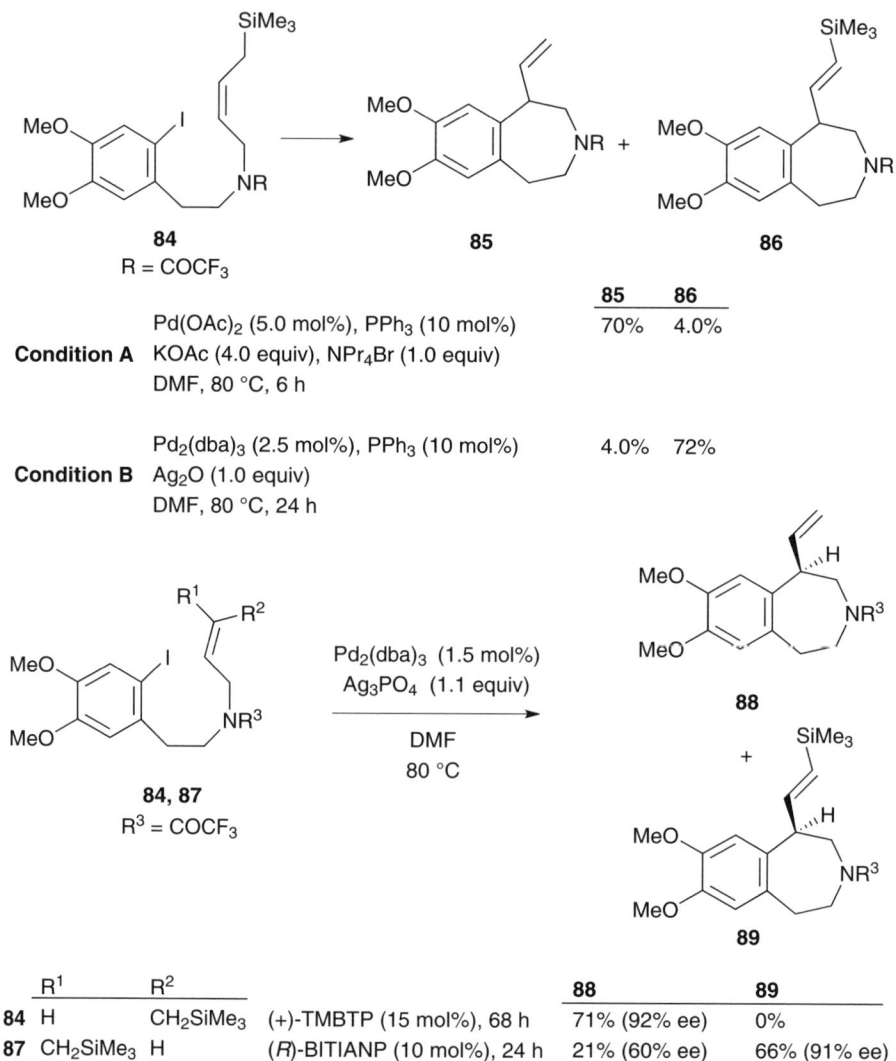

Scheme **12.19** *Allylsilane terminated enantioselective intramolecular Mizoroki–Heck reactions.*

organometallic. Alternatively, carbonylation of the σ-alkylpalladium intermediate **91** leads to acylpalladium **95**, which can be intercepted with nucleophiles to provide functionalized products **96**.

For the enantioselective formation of tertiary stereocentres, endocyclic alkenes or acyclic alkenes with β-hydride-elimination-directing groups, such as allylsilanes, are often employed. In the case of endocyclic alkenes, the migratory insertion intermediate lacks a *syn* β-hydrogen at the newly formed stereocentre and, therefore, cannot undergo termination. For instance, conjugated diene substrates such as **97** lead to a π-allylpalladium

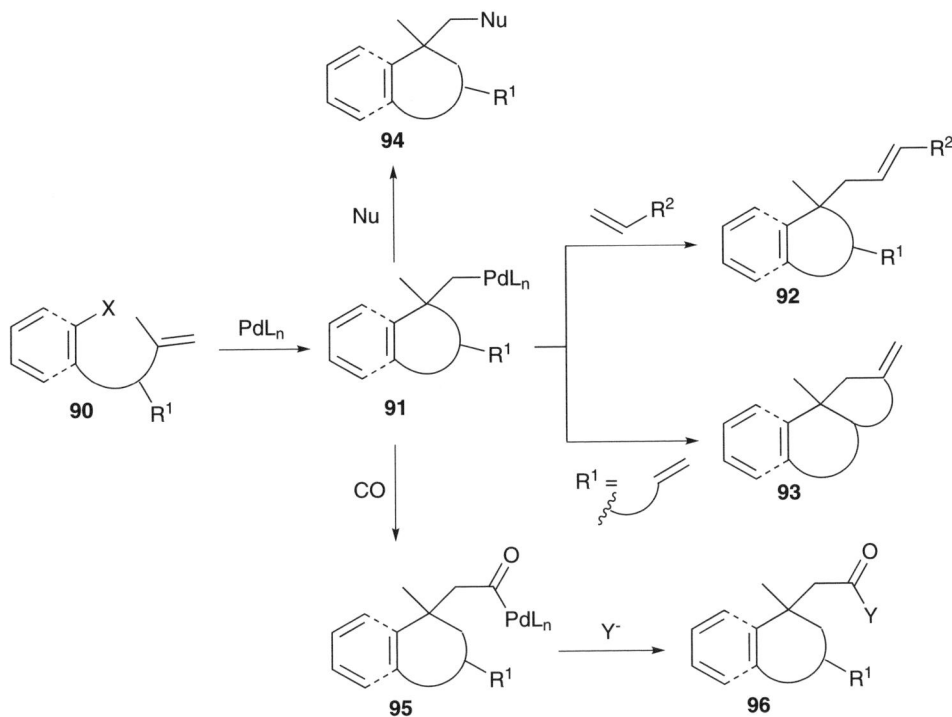

Scheme 12.20 *Cascade reactions via σ-alkylpalladium intermediates.*

intermediate **98** that may be trapped by soft nucleophiles like malonates to form diester **99** (Scheme 12.21) [48]. There are a variety of examples of cascade intramolecular Mizoroki–Heck reactions illustrating transformations for both σ-alkyl- and π-allyl-palladium intermediates to form tertiary or multiple stereocentres. In most cases the reactions are diastereoselective and not enantioselective. Examples of enantioselective cascade intramolecular Mizoroki–Heck reactions to form tertiary or multiple stereocentres have largely been limited to group-selective reactions with cyclic diene substrates that form π-allylpalladium intermediates.

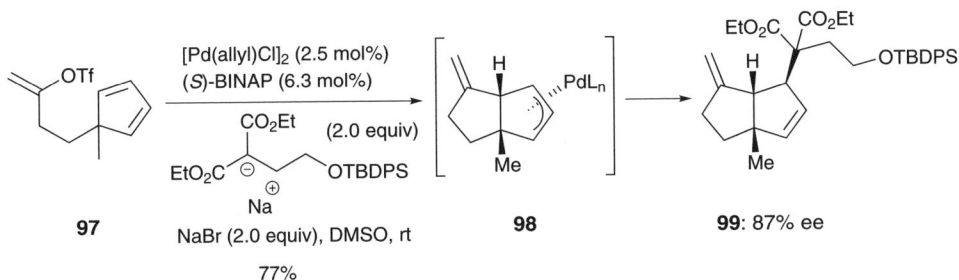

Scheme 12.21 *Cascade reaction via π-allylpalladium intermediates.*

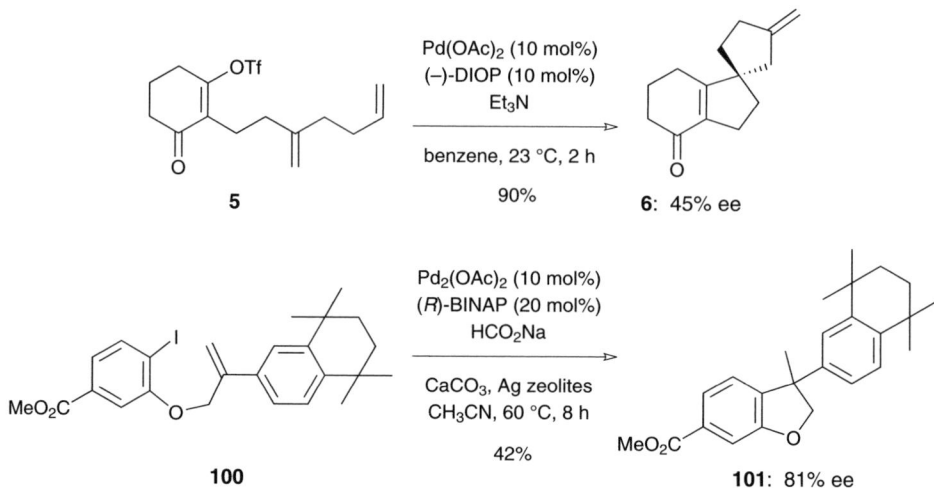

Scheme 12.22   *Polyene cyclization and Mizoroki–Heck–hydride capture cascade reactions.*

For nongroup-selective examples, one of the first reports of an enantioselective intramolecular Mizoroki Heck reaction was a polyene cyclization (Scheme 12.22) [23b]. The trienyl triflate **5** underwent two intramolecular cyclization reactions to give the tricycle **6** in high yield and 45% ee. A cascade intramolecular Mizoroki–Heck–hydride capture sequence was used in the synthesis of retinoid derivatives from aryl iodide **100** to give benzofuran **101** in 80–81% ee [49]. Poor enantioselectivity was observed when 'neutral' reaction conditions were employed.

Another example of an enantioselective polyene cyclization is the formation of tetracycle **103** (Scheme 12.23). Enantioselectivities of the 6-*exo* followed by 6-*endo* cyclization were moderate to good using BINAP. Of note is the substantial increase in enantioselectivity

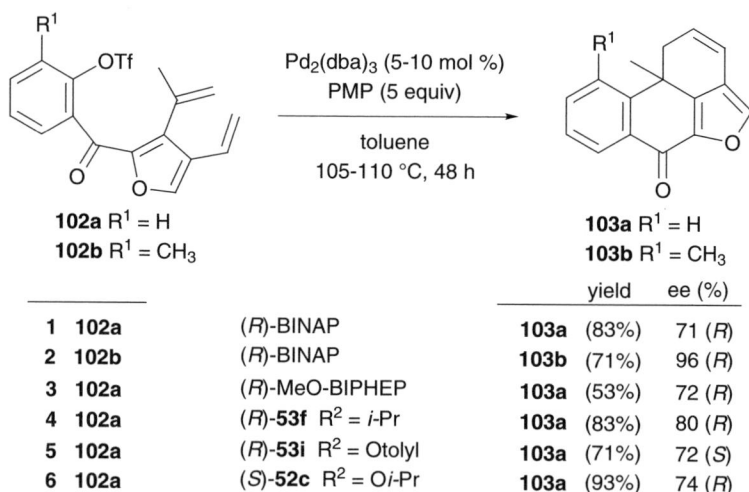

		ligand	yield	ee (%)
1	102a	(*R*)-BINAP	103a (83%)	71 (*R*)
2	102b	(*R*)-BINAP	103b (71%)	96 (*R*)
3	102a	(*R*)-MeO-BIPHEP	103a (53%)	72 (*R*)
4	102a	(*R*)-53f R^2 = *i*-Pr	103a (83%)	80 (*R*)
5	102a	(*R*)-53i R^2 = *O*tolyl	103a (71%)	72 (*S*)
6	102a	(*S*)-52c R^2 = O*i*-Pr	103a (93%)	74 (*R*)

Scheme 12.23   *Polyene cyclizations with ortho-substituted BINAP and BIPHEP ligands.*

MeO [structure **104**]

[Pd(dba)$_2$] (5.0 mol%)
K$_4$[Fe(CN)$_6$] (0.22 equiv)

DMA, 120 °C

MeO [structure **105**] CN

104   **105**

time (h)	ligand (12 mol%)	additive (equiv)	base (1.0 equiv)	yield (%)	ee (%)
5	(*R*)-BINAP	Ag$_3$PO$_4$ (2.0)	Na$_2$CO$_3$	76	24
3	(*R*)-BINAP	PMP (5.0)	-	54	0
3	(*S*)-DIFLUORPHOS	Ag$_3$PO$_4$ (2.0)	K$_2$CO$_3$	78	72 (*S*)

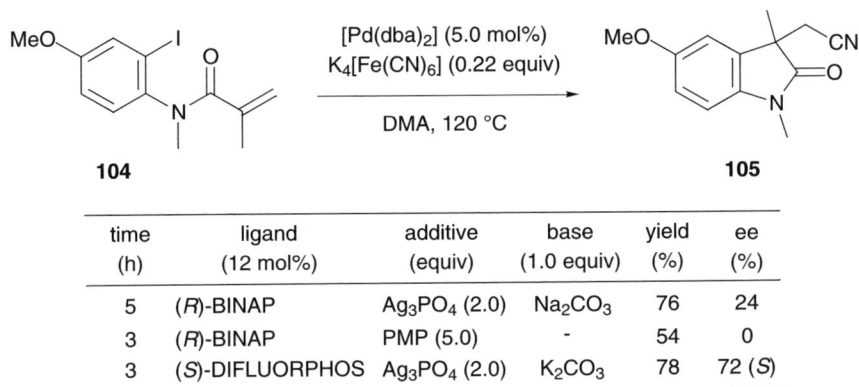

Scheme 12.24 *Enantioselective intramolecular Mizoroki–Heck–cyanation cascade reaction.*

observed for the reaction of the *o*-methyl-substituted triflate **102b** [50]. Semi-empirical calculations indicated that the methyl substituent interacts with the 3′-hydrogen of (*S*)-BINAP after oxidative addition, leading to (*S*)-**103b** [31, 32]. This same interaction is not observed in the intermediate, leading to (*R*)-**103b**. It was proposed that transferring the substitution from the *ortho* position of the substrate to the 3′ position of ligand should give the same improvement in enantioselectivity. Several *ortho*-substituted BINAP and BIPHEP derivatives were synthesized and evaluated. Most substitutions did not lead to an improvement in enantioselectivity, except for the isopropyl-substituted BIPHEP derivative of **53f**. For alkoxy-substituted BINAP and BIPHEP, a fascinating reversal in stereoselectivity was observed for the same ligand axial chirality (compare entry 3 with 5 and entry 1 with 6 in Scheme 12.23). No rationale has yet been offered to account for the change in stereoselectivity.

The first example of an enantioselective intramolecular cascade Mizoroki–Heck–cyanation sequence was recently reported which included the reaction of amide **104** (Scheme 12.24) [33]. The cyanide source employed was potassium ferro(II)cyanide, which has been utilized for the palladium-catalysed cyanation of aryl halides. The proposed reaction pathway for the Mizoroki–Heck–cyanation involves capture of a σ-alkylpalladium intermediate. Previous examples of enantioselective Mizoroki–Heck cyclization–anion capture most often involve trapping of the π-allylpalladium complexes in group-selective reactions. Reaction conditions were surveyed for the Mizoroki–Heck cyanation sequence. It was found that Pd(dba)$_2$ afforded better enantioselectivities than Pd(OAc)$_2$ with Ag$_3$PO$_4$ as the additive. Using PMP under 'neutral' conditions led to racemic product. To improve the enantioselectivity, several bidentate ligands were screened, and the ligand DIFLUORPHOS **54a** was found to give the best enantioselectivity.

Enantioselective intramolecular cascade Mizoroki–Heck reactions have been shown to proceed with moderate to good selectivity via the cationic manifold. There are surprisingly few enantioselective examples, given the wide array of transformations known for σ-alkylpalladium intermediates in racemic or diastereoselective reactions. All of the nongroup-selective, enantioselective, cascade, intramolecular Mizoroki–Heck reactions reported to date involve formation of one quaternary centre. A substantial advance would be to expand the range of transformations available for the σ-alkylpalladium species and allow for the enantioselective construction of tertiary stereocentres. Future innovations in catalyst systems have the potential to realize these advances.

12.5 Conclusion

The asymmetric intramolecular Mizoroki–Heck reaction has undergone remarkable advancement since the first examples were reported by Shibasaki and coworkers [23a] and Overman and coworkers [23b]. Currently, the primary applications of the reaction are in complex molecule synthesis, where it has enabled small and medium ring closures, congested tertiary and quaternary centre construction, and multiple bond formations in one step via cascade/domino reactions. In most cases, the cationic manifold of the Mizoroki–Heck reaction is employed in combination with bidentate asymmetric ligands. More complex mechanistic pathways operate in some cases, and it remains to be determined whether these pathways can be routinely productively engaged. Mechanistic studies to determine the influence of additives, like silver salts and the enantioselective step(s), also promise to enable further improvements in the reaction. BINAP remains the most common ligand choice for the asymmetric intramolecular Mizoroki–Heck reaction. Further reaction optimization is often accomplished through experimentation with other ligands that have been developed, and the continuing exploration of their substrate scope is needed. The discovery and refinement of asymmetric ligand classes holds out promise to improve the reaction further. The scope of the reaction currently covers 5-*exo*, 6-*exo* and some 7-*exo* cyclizations and allows for the formation of congested tertiary centres. Tertiary centres can also be prepared without fearing their destruction by β-hydride elimination by employing an allylsilane terminating moiety. Enantioselective cascade or domino reactions are also effective, but their demonstrated scope is less extensive than for the nonasymmetric intramolecular reaction.

Currently, much of the research in the enantioselective intramolecular Mizoroki–Heck reaction is focused on the development of improved general catalyst/ligand systems that catalyse the reaction of a wide variety of substrates with low loadings and modest air sensitivity. The development would dramatically impact the use, as well as the commercial viability, of the asymmetric intramolecular Mizoroki–Heck reaction. Catalyst advances may ultimately be driven from breakthroughs in a number of areas, including mechanistic studies, catalyst/ligand development and complex molecule synthesis. The advances that have been made in the understanding and execution of the reaction bode well for its continuing development and utilization as an important catalytic asymmetric transformation.

References

1. (a) De Meijere, A. and Meyer, F.E. (1994) Fine feathers make fine birds: the Heck reaction in modern garb. *Angew. Chem., Int. Ed. Engl.*, **33**, 2379–411; (b) Link, J.T. (2002) The intramolecular Heck reaction, in *Organic Reactions*, Vol. **60** (ed. L.E. Overman), John Wiley & Sons, Inc., Hoboken, NJ.
2. (a) Link, J.T. and Overman, L.E. (1998) Intramolecular Heck reactions in natural product chemistry, in *Metal-Catalyzed Cross-Coupling Reactions* (eds F. Diederich and P.J. Stang), Wiley–VCH Verlag GmbH, Weinheim; (b) Link, J.T. (2002) Synthesis of natural products via

carbopalladation, in *Handbook of Organopalladium Chemistry for Organic Synthesis*, Vol. **1** (ed. E. Negishi), John Wiley & Sons, Inc., New York; (c) Dounay, A.B. and Overman, L.E. (2003) The asymmetric intramolecular Heck reaction in natural product total synthesis. *Chem. Rev.*, **103**, 2945–63.

3. (a) Overman, L.E. (1994) Application of intramolecular Heck reactions for forming congested quaternary carbon centers in complex molecule total synthesis. *Pure Appl. Chem.*, **66**, 1423–30; (b) Douglas, C.J. and Overman, L.E. (2004) Catalytic asymmetric synthesis of all-carbon quaternary stereocenters. *Proc. Natl. Acad. Sci.*, **101**, 5363–7; (c) Peterson, E.A. and Overman, L.E. (2004) Contiguous stereogenic quaternary carbons: a daunting challenge in natural products synthesis. *Proc. Natl. Acad. Sci.*, **101**, 11943–8.

4. (a) Negishi, E., Copéret, C., Ma, S. *et al.* (1996) Cyclic carbopalladation. A versatile synthetic methodology for the construction of cyclic organic compounds. *Chem. Rev.*, **96**, 365–93; (b) Grigg, R. and Sridharan, V. (1999) Palladium catalyzed cascade cyclisation–anion capture, relay switches and molecular queues. *J. Organomet. Chem.*, **576**, 65–87; (c) de Meijere, A. and Bräse, S. (1999) Palladium in action: domino coupling and allylic substitution reactions for the efficient construction of complex organic molecules. *J. Organomet. Chem.*, **576**, 88–110; (d) Chapman, C.J. and Frost, C.G. (2007) Tandem and domino catalytic strategies for enantioselective synthesis. *Synthesis*, 1–21.

5. (a) Donde, Y. and Overman, L.E. (2000) Asymmetric intramolecular Heck reactions, in *Catalytic Asymmetric Synthesis* (ed. I. Ojima), Wiley-VCH, Inc., New York; (b) Shibasaki, M. and Miyazaki, F. (2002) Asymmetric Heck reactions, in *Handbook of Organopalladium Chemistry for Organic Synthesis*, Vol **1** (ed. E. Negishi), John Wiley & Sons, Inc., Hoboken, NJ; (c) Shibasaki, M., Vogl, E.M. and Ohshima, T. (2004) Asymmetric Heck reaction. *Adv. Synth. Catal.*, **346**, 1533–52; (d) Tietze, L.F., Ila, H. and Bell, H.P. (2004) Enantioselective palladium-catalyzed transformation. *Chem. Rev.*, **104**, 3453–516.

6. (a) Amatore, C. and Jutand, A. (1999) Mechanistic and kinetic studies of palladium catalytic systems. *J. Organomet. Chem.*, **576**, 254–78; (b) Beletskaya, I.P. and Cheprakov, A.V. (2000) The Heck reaction as a sharpening stone of palladium catalysis. *Chem. Rev.*, **100**, 3009–66.

7. (a) Tietze, L.F. and Raschke, T. (1995) Enantioselective total synthesis of a natural norsesquiterpene of the calamenene group by a silane-terminated intramolecular Heck reaction. *Synlett*, 597–8; (b) Tietze, L.F. and Raschke, T. (1996) Enantioselective total synthesis and absolute configuration of the natural norsesquiterpene 7-demethyl-2-methoxycalamenene by a silane-terminated intramolecular Heck reaction. *Liebigs Ann. Chem.*, 1981–7.

8. (a) Negishi, E. (1992) Zipper-mode cascade carbometallation for construction of polycyclic structures. *Pure Appl. Chem.*, **64**, 323–34; (b) Overman, L.E., Abelman, M.M., Kucera, D.J. *et al.* (1992) Palladium-catalyzed polyene cyclizations. *Pure Appl. Chem.*, **64**, 1813–9.

9. (a) Cabri, W. and Candiani, I. (1995) Recent developments and new perspectives in the Heck reaction. *Acc. Chem. Res.*, **28**, 2–7; (b) Crisp, G.T. (1998) Variations on a theme – recent developments on the mechanism of the Heck reaction and their implications for synthesis. *Chem. Soc. Rev.*, **27**, 427–36.

10. (a) Ozawa, F., Kubo, A. and Hayashi, T. (1991) Catalytic asymmetric arylation of 2,3-dihydrofuran with aryl triflates. *J. Am. Chem. Soc.*, **113**, 1417–9; (b) Cabri, W., Candiani, I., DeBernardinis, S. *et al.* (1991) Heck reaction on anthraquinone derivatives: ligand, solvent and salt effects. *J. Org. Chem.*, **56**, 5796–800.

11. Overman, L.E. and Poon, D.J. (1997) Asymmetric Heck reactions via neutral intermediates: enhanced enantioselectivity with halide additives gives mechanistic insights. *Angew. Chem., Int. Ed. Engl.*, **36**, 518–21.

12. Sato, Y., Sodeoka, M. and Shibasaki, M. (1990) On the role of silver salts in asymmetric Heck-type reaction. A greatly improved catalytic asymmetric synthesis of *cis*-decalin derivatives. *Chem. Lett.*, 1953–4.

13. Ashimori, A., Bachand, B., Calter, M.A. *et al.* (1998) Catalytic asymmetric synthesis of quaternary carbon centers. Exploratory studies of intramolecular Heck reactions of (Z)-α,β-unsaturated anilides and mechanistic investigations of asymmetric Heck reactions proceeding via neutral intermediates. *J. Am. Chem. Soc.*, **120**, 6488–99.

14. Amatore, C., Carre, E., Jutand, A. *et al.* (1995) Evidence for the ligation of palladium(0) complexes by acetate ions: consequences on the mechanism of their oxidative addition with phenyl iodide and PhPd(OAc)(PPh$_3$)$_2$ as intermediate in the Heck reaction. *Organometallics*, **14**, 5605–14.

15. Amatore, C., Jutand, A. and Thuilliez, A. (2001) Formation of palladium(0) complexes from Pd(OAc)$_2$ and a bidentate phosphine ligand (dppp) and their reactivity in oxidative addition. *Organometallics*, **20**, 3241–9.

16. (a) Imbos, R., Minnaard, A.J. and Feringa, B.L. (2002) A highly enantioselective intramolecular Heck reaction with a monodentate ligand. *J. Am. Chem. Soc.*, **124**, 184–5; (b) Imbos, R., Minnaard, A.J. and Feringa, B.L. (2003) Monodentate phosphoramidites; versatile ligands in catalytic asymmetric intramolecular Heck reactions. *Dalton Trans.*, 2017–23.

17. Lapierre, J.B., Geib, S.J. and Curran, D.P. (2007) Low-temperature Heck reactions of axially chiral *o*-iodoacrylanilines occur with chirality transfer: implications for catalytic asymmetric Heck reactions. *J. Am. Chem. Soc.*, **129**, 494–5.

18. Curran, D.P., Hale, G.R., Geib, S.J. *et al.* (1997) Rotational features of carbon–nitrogen bonds in axially chiral *o*-tert-butylanilides and related molecules. Potential substrates for the 'prochiral auxiliary' approach to asymmetric synthesis. *Tetrahedron: Asymmetry*, **8**, 3955–75.

19. McDermott, M.C., Stephenson, G.R., Hughes, D.L. and Walkington, A.J. (2006) Intramolecular asymmetric Heck reactions: evidence for dynamic kinetic resolution effects. *Org. Lett.*, **8**, 2917–20.

20. McDermott, M.C., Stephenson, G.R. and Walkington, A.J. (2007) Silver-sensitive enantioselectivity in asymmetric Heck reactions. *Synlett*, 51–4.

21. Farina, V. (2004) High-turnover palladium catalysts in cross-coupling and Heck chemistry: a critical overview. *Adv. Synth. Catal.*, **346**, 1553–82.

22. (a) Ozawa, F., Kubo, A. and Hayashi, T. (1992) Generation of tertiary phosphine-coordinated Pd(0) species from Pd(OAc)$_2$ in the catalytic Heck reaction. *Chem. Lett.*, 2177–80; (b) Amatore, C., Jutand, A. and M'Barki, M.A. (1992) Evidence of the formation of zerovalent palladium from Pd(OAc)$_2$ and triphenylphosphine. *Organometallics*, **11**, 3009–13; (c) Amatore, C., Carré, E., Jutand, A. and M'Barki, M. (1995) Rates and mechanism of the formation of zerovalent palladium complexes from mixtures of Pd(OAc)$_2$ and tertiary phosphines and their reactivity in oxidative additions. *Organometallics*, **14**, 1818–26.

23. (a) Sato, Y., Sodeoka, M. and Shibasaki, M. (1989) Catalytic asymmetric carbon–carbon bond formation: asymmetric synthesis of *cis*-decalin derivatives by palladium-catalyzed cyclization of prochiral alkenyl iodides. *J. Org. Chem.*, **54**, 4738–39; (b) Carpenter, N.E., Kucera, D.J. and Overman, L.E. (1989) Palladium-catalyzed polyene cyclizations of trienyl triflates. *J. Org. Chem.*, **54**, 5846–8.

24. Ashimori, A. and Overman, L.E. (1992) Catalytic asymmetric synthesis of quaternary carbon centers. Palladium-catalyzed formation of either enantiomer of spirooxindoles and related spirocycles using a single enantiomer of a chiral diphosphine ligand. *J. Org. Chem.*, **57**, 4571–2.

25. (a) Abelman, M.M. and Overman, L.E. (1988) Palladium-catalyzed polyene cyclizations of dienyl aryl iodides. *J. Am. Chem. Soc.*, **110**, 2328–9; (b) Karabelas, K., Westerlund, C. and Hallberg, A. (1985) The effect of added silver nitrate on the palladium-catalyzed arylation of allyltrimethylsilanes. *J. Org. Chem.*, **50**, 3896–900; (c) Grigg, R., Loganathan, V., Santhakumar, V. *et al.* (1991) Suppression of alkene isomerization in products from intramolecular Heck reactions by addition of thallium(I) salts. *Tetrahedron Lett.*, **32**, 687–90.

26. (a) Ohrai, K., Kondo, K., Sodeoka, M. and Shibasaki, M. (1994) Effects of solvents and additives in the asymmetric Heck reaction of alkenyl triflates: catalytic asymmetric synthesis of decalin derivatives and determination of the absolute stereochemistry of (+)-vernolepin. *J. Am. Chem. Soc.*, **116**, 11737–48; (b) Kondo, K., Sodeoka, M., Mori, M. and Shibasaki, M. (1993) Asymmetric Heck reaction: catalytic asymmetric syntheses of bicyclic enones, dienones and the key intermediate for vernolepin. *Synthesis*, 920–30.

27. Bräse, S., Kirchhoff, J.H. and Köbberling, J. (2003) Palladium in catalyzed reactions in solid phase organic synthesis. *Tetrahedron*, **59**, 885–939.

28. Cho, S.Y. and Shibasaki, M. (1998) Synthesis and evaluation of a new chiral ligand: 2-diphenylarsino-2′-diphenylphosphino-1,1′-binaphthyl (BINAPAs). *Tetrahedron Lett.*, **39**, 1773–6.

29. Kojima, A., Boden, C.D.J. and Shibasaki, M. (1997) Synthesis and evaluation of a new chiral arsine ligand: 2,2′-bis(diphenylarsino)-1,1′binaphthyl (BINAs). *Tetrahedron Lett.*, **38**, 3459–60.

30. Andersen, N.G., McDonald, R. and Keay, B.A. (2001) Synthesis, resolution and application of 2,2′-bis(di-2-furylphosphino)-1,1′-binaphthalene. *Tetrahedron: Asymmetry*, **12**, 263–9.

31. Hopkins, J.M., Gorobets, E., Wheatley, B.M.M. *et al.* (2006) Applications of 3,3′disubstituted BINAP derivatives in inter- and intramolecular Heck/Mizoroki reactions. *Synlett*, 3120–4.

32. (a) Gorobets, E., Sun, G.-R., Wheatley, B.M.M. *et al.* (2004) Synthesis, resolution and applications of 3,3′-bis(RO)-MeO–BIPHEP derivatives. *Tetrahedron Lett.*, **45**, 3597–601; (b) Gorobets, E., Wheatley, B.M.M., Hopkins, J.M. *et al.* (2005) Avoiding the classical resolution during the synthesis of MeO–BIPHEP and 3,3′-disubstituted derivatives. *Tetrahedron Lett.*, **46**, 3843–6.

33. Pinto, A., Jia, Y., Neuville, L. and Zhu, J. (2007) Palladium-catalyzed enantioselective domino Heck–cyanation sequence: development and application to the total synthesis of esermethole and physostigmine. *Chem. Eur. J.*, **13**, 961–7.

34. Tietze, L.F., Thede, K., Schimpf, R. and Sannicolò, F. (2000) Enantioselective synthesis of tetrahydroisoquinolines and benzazepines by silane terminated Heck reactions with the chiral ligands (+)-TMBTP and (R)-BITIANP. *Chem. Commun.*, 583–4.

35. (a) Kagechika, K. and Shibasaki, M. (1991) Asymmetric Heck reaction: a catalytic asymmetric synthesis of the key intermediate for $\Delta^{9(12)}$-capnellene-3β,8β,10α-triol and $\Delta^{9(12)}$-capnellene-3β,8β,10α,14-tetrol. *J. Org. Chem.*, **56**, 4093–4; (b) Kagechika, K., Ohshima, T. and Shibasaki, M. (1993) Asymmetric Heck reaction–anion capture process. A catalytic asymmetric synthesis of the key intermediates for the capnellenols. *Tetrahedron*, **49**, 1773–82; (c) Honzawa, S., Mizutani, T. and Shibasaki, M. (1999) Synthetic studies on (+)-wortmannin. An asymmetric construction of an allylic quaternary carbon center by a Heck reaction. *Tetrahedron Lett.*, **40**, 311–4.

36. Kondo, K., Sodeoka, M., Mori, M. and Shibasaki, M. (1993) Asymmetric Heck reaction: catalytic asymmetric syntheses of bicyclic enones, dienones and the key intermediate for vernolepin. *Synthesis*, 920–30.

37. (a) Nukui, S., Sodeoka, M. and Shibasaki, M. (1993) Catalytic asymmetric synthesis of a functionalized indolizidine derivative. A useful intermediate suitable for the synthesis of various glycosidase inhibitors. *Tetrahedron Lett.*, **34**, 4965–8; (b) Sato, Y., Nukui, S., Sodeoka, M. and Shibasaki, M. (1994) Asymmetric Heck reaction of alkenyl iodides in the presence of silver salts. Catalytic asymmetric synthesis of decalin and functionalized indolizidine derivatives. *Tetrahedron*, **50**, 371–82.

38. Kiely, D. and Guiry, P.J. (2003) Palladium complexes of phosphinamine ligands in the intramolecular asymmetric Heck reaction. *J. Organomet. Chem.*, **687**, 545–61.

39. Lormann, M.E.P., Nieger, M. and Bräse, S. (2006) Desymmetrisation of bicyclo[4.4.0]decadienes: a planar-chiral complex proved to be most effective in an asymmetric Heck reaction. *J. Organomet. Chem.*, **691**, 2159–61.

40. Loiseleur, O., Hayashi, M., Keenan, M. *et al.* (1999) Enantioselective Heck reactions using chiral P,N-ligands. *J. Organomet. Chem.*, **576**, 16–22.

41. Fitzpatrick, M.O., Coyne, A.G. and Guiry, P.J. (2006) The application of HETPHOX ligands to the intramolecular asymmetric Heck reaction. *Synlett*, 3150–4.

42. (a) Busacca, C.A., Grossbach, D., So, R.C. *et al.* (2003) Probing electronic effects in the asymmetric Heck reaction with the BIPI ligands. *Org. Lett.*, **5**, 595–8; (b) Busacca, C.A., Grossbach, D., Campbell, S.J. *et al.* (2004) Electron control of chiral quaternary center creation in the intramolecular asymmetric Heck reaction. *J. Org. Chem.*, **69**, 5187–95.

43. Bräse, S. (1999) Synthesis of bis(enolnonaflates) and their 4-*exo*-trig-cyclizations by intramolecular Heck reactions. *Synlett*, 1654–66.

44. (a) Ashimori, A., Bachand, B., Overman, L.E. and Poon, D.J. (1998) Catalytic asymmetric synthesis of quaternary carbon centers. exploratory investigations of intramolecular Heck reactions of (*E*)-α,β-unsaturated 2-haloanilides and analogues to form enantioenriched spirocyclic products. *J. Am. Chem. Soc.*, **120**, 6477–87 [Additions and corrections. *J. Am. Chem. Soc.*, **122**, 192 (2000)]; (b) Dounay, A.B., Hatanaka, K., Kodanko, J.J. *et al.* (2003) Catalytic asymmetric synthesis of quaternary carbons bearing two aryl substituents. Enantioselective synthesis of 3-alkyl-3-aryl oxindoles by catalytic asymmetric intramolecular Heck reactions. *J. Am. Chem. Soc.*, **125**, 6261–71.

45. Takemoto, T., Sodeoka, M., Sasai, H. and Shibasaki, M. (1993) Catalytic asymmetric synthesis of benzylic quaternary carbon centers. An efficient synthesis of (−)-eptazocine. *J. Am. Chem. Soc.*, **115**, 8477–8.

46. Bashore, C.G., Vetelino, M.G., Wirtz, M.C. *et al.* (2006) Enantioselective synthesis of nicotinic receptor probe 7,8-difluoro-1,2,3,4,5,6-hexahydro-1,5-methano-3-benzazocine. *Org. Lett.*, **8**, 5947–50.

47. Tietze, L.F. and Schimpf, R. (1994) Regio- and enantioselective silane-terminated intramolecular Heck reactions. *Angew. Chem., Int. Ed. Engl.*, **33**, 1089–91.

48. (a) Ohshima, T., Kagechika, K., Adachi, M. *et al.* (1996) Asymmetric Heck reaction-carbanion capture process. Catalytic asymmetric total synthesis of (−)-$\Delta^{9(12)}$-capnellene. *J. Am. Chem. Soc.*, **118**, 7108–16; (b) Itano, W., Ohshima, T. and Shibasaki, M. (2006) Synthesis of the tricyclic core of 5α-capnellenols using asymmetric Heck reaction–carbanion capture process. *Synlett*, 3053–6.

49. Diaz, P., Gendre, F., Stella, L. and Charpentier, B. (1998) New synthetic retinoids obtained by palladium-catalyzed tandem cyclization–hydride capture process. *Tetrahedron*, **54**, 4579–90.

50. Lau, S.Y.W. and Keay, B.A. (1999) Remote substituent effects on the enantiomeric excess of intramolecular asymmetric palladium-catalyzed polyene cyclizations. *Synlett*, 605–7.

13

Desymmetrizing Heck Reactions

Masakatsu Shibasaki

Graduate School of Pharmaceutical Sciences, The University of Tokyo, Hongo, Bunkyo-ku, Tokyo, Japan

Takashi Ohshima

Department of Chemistry, Graduate School of Engineering Science, Osaka University, Toyonaka, Osaka, Japan

13.1 Introduction

Because of the usefulness of the Mizoroki–Heck reaction, as described in the preceding chapters, interest in the Mizoroki–Heck reaction has recently increased dramatically. Perhaps the most significant progress to date is the development of enantioselective variants [1]. Given the many reports of chiral phosphine ligands dating from the early 1970s [2], it is, in retrospect, somewhat surprising that the phosphine-mediated Mizoroki–Heck reaction was not subjected to asymmetrization attempts until the late 1980s. This might be rationalized by the fact though that the reaction was usually not used for the construction of stereogenic centres [3] and that, for many years, chelating diphosphines were generally thought to be unsuitable ligands [4]. Reports of successful examples of the asymmetric Mizoroki–Heck reaction were independently published in 1989 by our group [5] and Overman's group [6] and the reaction has since been successfully developed to a point where both tertiary and quaternary centres are accessible. Although the enantioselectivities achieved in these preliminary studies were modest, the great potential of the asymmetric variants of the intramolecular Mizoroki–Heck reaction was demonstrated, leading to later success in this field. Intramolecular asymmetric Mizoroki–Heck reactions might be classified into two types (Figure 13.1): (i) desymmetrization (type i, Shibasaki and coworkers)

The Mizoroki–Heck Reaction   Edited by Martin Oestreich
© 2009 John Wiley & Sons, Ltd

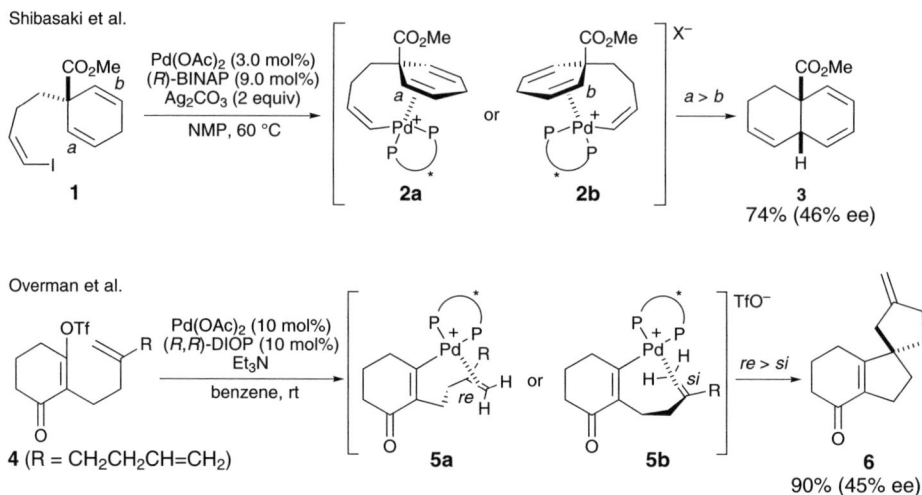

Figure 13.1   *First examples of a asymmetric Mizoroki–Heck reactions.*

and (ii) enantiofacial selection (type ii, Overman and coworkers). For example, the reaction of prochiral substrate **1** to chiral bicyclic compound **3** is classified as a type i reaction (insertion of C=C double bond *a* or *b*) and the reaction of substrate **4** to spiro compound **6** is classified as a type ii reaction (*re*- or *si*-face-selective insertion of C=C double bond). As the latter Mizoroki–Heck reaction (type ii) is covered in Chapter 12, this chapter provides a survey of the former process (type i) up to the end of 2006.

Desymmetrization of a prochiral compound to yield enantiomerically enriched products has proven to be a powerful synthetic tool [7]. In general, to achieve an enantioselective symmetry-breaking synthetic operation, two enantiotopic functional groups must be differentiated. In the desymmetrizing Mizoroki–Heck reaction, this is achieved by using a chiral palladium complex. In most of the reactions reported, prochiral compounds with two alkenyl moieties were used as the substrate. In a single operation, one or, normally, two new stereogenic centres are formed, including otherwise difficult-to-obtain chiral quaternary carbon centres. The following sections are organized according to the carbon skeletons of the substrates.

13.2   Cyclic Dienes as Cyclization Precursors

13.2.1   1,4-Cyclohexadienes

We developed an asymmetric Mizoroki–Heck reaction of the prochiral alkenyl iodides **1**, **7** and **8** into the chiral decalin (bicyclo[4.4.0]decan) system **3**, **9** and **10**, where two vicinal stereocentres (contiguous quaternary and tertiary centres) are set in one step (Figure 13.2) [5]. To achieve high reactivity and enantioselectivity, the choice of solvent, chiral ligand and base was highly important. For example, very low or negligible enantiomeric excesses were obtained using THF, MeCN or DMSO, and *N*-methyl-2-pyrrolidinone (NMP) emerged as the preferred solvent. Similarly, the widely used chiral phosphine ligands 1-*tert*-butoxycarbonyl-4-diphenylphosphino-2-(diphenylphosphinomethyl)pyrrolidine (BPPM)

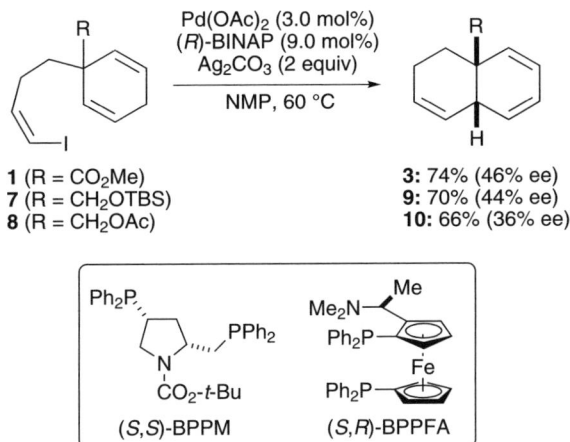

Figure 13.2 *Enantioselective syntheses of cis-decalins using the desymmetrizing Mizoroki–Heck reaction of alkenyl iodides.*

and *N*,*N*-dimethyl-1-[(*R*)-1',2-bis(diphenylphosphino)ferrocenyl]-amine (BPPFA) failed to give significant asymmetric induction, with (*R*)-BINAP (2,2'-bis(diphenylphosphino)-1,1'-binaphthyl) proving to be the ligand of choice, which is consistent with most of the reported examples of the asymmetric Mizoroki–Heck reaction. The nature of the base is also critical to success in this asymmetric reaction [1]; at least 1 equiv of base is required in the Mizoroki–Heck reaction to neutralize the acid (HI in this case) that is produced when the hydridopalladium(II) species is reduced to regenerate the active palladium(0) species. Although the use of silver(I) salt increases the reaction rate in the Mizoroki–Heck reaction [8], it also dramatically enhances enantioselectivity. For example, the use of Et₃N instead of Ag₂CO₃ afforded **3** in low yield with only 0.4% ee [9].

The role of silver(I) salt might be understood as follows [1] (Figure 13.3): (i) without a silver(I) salt the reaction starts with the dissociation of one arm of the bidentate ligand,

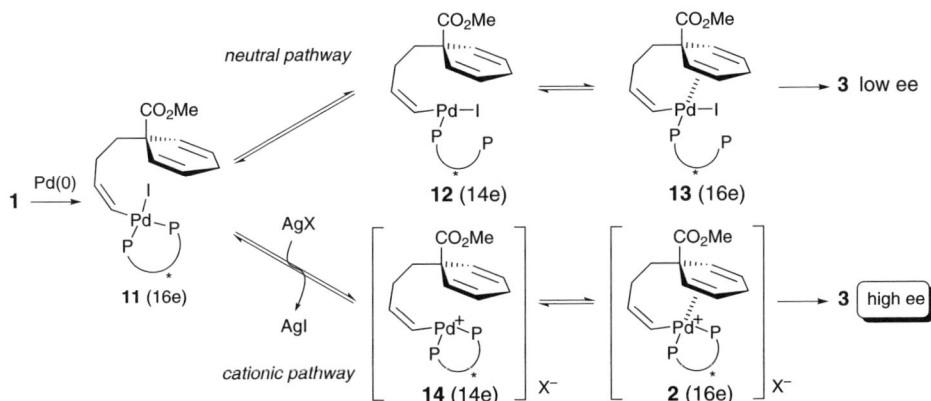

Figure 13.3 *Neutral and cationic pathways for the asymmetric Mizoroki–Heck reaction.*

resulting in the 14-electron neutral species **12**; alkene complexation to the vacant site gives the 16-electron neutral species **13**, and migratory insertion of alkene into the C–Pd bond followed by β-hydride elimination gives **3** (neutral pathway); (ii) with a silver(I) salt, the reaction begins with the counter anion exchange of I^- and X^- at palladium(II) followed by dissociation of X^- from **11** to generate the tri-coordinate 14-electron cationic complex **14** with the accompanying non coordinating or weakly coordinating counterion X^-. Alkene complexation to the vacant site then gives the 16-electron species **2**, which gives **3** with the chiral bidentate ligand remaining fully chelated throughout and, thus, maximizing the asymmetric induction (cationic pathway). Therefore, Ag_2CO_3 acts as both base and halide scavenger.

The modest enantiomeric excesses (33–46% ee) for the conversions shown in Figure 13.2 were greatly improved as a result of studying the effects of varying the anionic component of both the palladium source and, more particularly, the silver salt (Figure 13.4) [9]. The use of a palladium(0) catalyst generated *in situ* from the preformed $PdCl_2–(R)$-BINAP complex [10] and cyclohexene improved the enantiomeric excess relative to the 1:3 $Pd(OAc)_2–(R)$-BINAP transition metal–ligand combination, which was used prereduced in the original work [5]. The use of AgOAc as the silver(I) source reduced the enantiomeric excess to almost zero, indicating that the nucleophilic acetate counterion, which is likely to form a Pd–OAc bond, thereby blocking (for the cationic pathway) the indispensable vacant site at palladium(II) needed for concurrent ligand and alkene coordination. The best silver(I) source in terms of enantiomeric excess was Ag_3PO_4, most likely due to the very low nucleophilicity of $Ag_2PO_4{}^-$, with sparingly soluble $CaCO_3$ being added as the basic component. Under these conditions, **9** was obtained in 67% yield with 80% ee [9]. The introduction of the new ligand 2,2′-bis(diphenylarsino)-1,1′-binaphthyl (BINAs) [11], homologous to BINAP, considerably increased the yield for the conversion of **7** to **9**. After optimization, the product **9** was obtained in 90% yield with 82% ee.

The use of alkenyl triflates **15–18** in place of alkenyl iodides gave better enantioselectivity [12] and allowed for the omission of expensive silver salts and the use of hydrocarbon solvents, such as toluene and benzene (Figure 13.5). Aryl and alkenyl triflates are generally assumed to follow the cationic pathway [13]. In addition, use of the alkenyl triflates

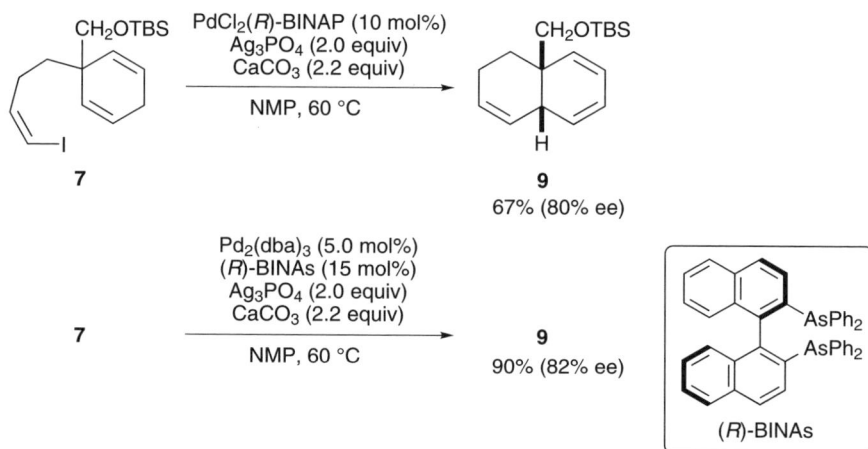

Figure 13.4 *Optimization of the asymmetric Mizoroki–Heck reaction of an alkenyl iodide.*

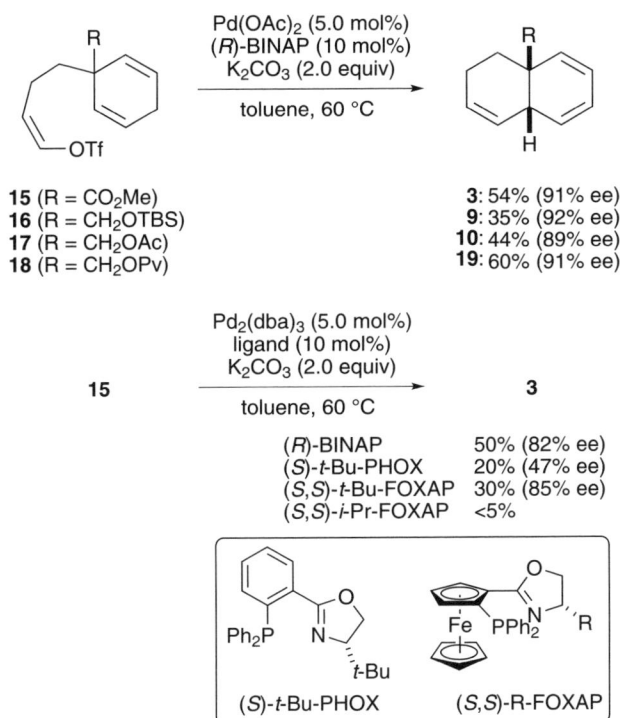

Figure 13.5 *Enantioselective syntheses of cis-decalins using the asymmetric Mizoroki–Heck reaction of alkenyl triflate.*

eliminates the deleterious effects of Pd(OAc)$_2$ on enantiomeric excess observed in NMP. Thus, products **3**, **9**, **10** and **19** were obtained in 35–60% yield with uniformly excellent enantiomeric excesses (89–92%) under the conditions indicated.

The great majority of asymmetric Mizoroki–Heck reactions reported utilized the BINAP ligand system, which was most effective in many cases in which the performance of different ligands was assessed. However, the significant number of exceptions to this rule suggests that experimentation with alternatives will prove worthwhile. The most dramatic recent development has been the introduction by Pfaltz and coworkers [14a–c] of the oxazoline-based P,N-ligands such as PHOX (2-(2-diphenylphosphinophenyl)oxazolines), which produced a distinctly improved enantiomeric excess in intermolecular Mizoroki–Heck reactions. Therefore, Kiely and Guiry [15] examined the applicability of P,N-ligands in the desymmetrization of alkenyl triflate **15**. They used Pd$_2$(dba)$_3$ (dba = dibenzylideneacetone) as a palladium source instead of Pd(OAc)$_2$, which resulted in lower enantioselectivity (82% ee), probably due to the existence of a partial racemic pathway. Although the representative P,N-ligand (*S*)-*t*-Bu-PHOX [14d] proved less effective (20% yield and 47% ee) than BINAP, the ferrocennyloxazoline ligand *t*-Bu-FOXAP [16] slightly improved enantioselectivity to 85% ee. Catalysts prepared from palladium complexes of oxazoline ligands were less reactive than the corresponding BINAP catalyst. The excellent enantioselectivities obtained using those ligands, however, suggests their high potential in the asymmetric Mizoroki–Heck reaction [14d].

A significant extension in the scope of this reaction was highlighted in the synthesis of a range of bicyclic enones and dienones, including a key intermediate **24** in Danishefsky *et al.*'s synthesis [17] of (+)-vernolepin (**25**) (Figure 13.6) [18a]. The asymmetric Mizoroki–Heck reaction was initially the conversion of **20** to the chiral optically active decalin system **23**, via the intermediate **21**. The best solvent for this reaction was 1,2-dichloroethane, with the addition of *t*-BuOH positively affecting the reaction rate and chemical yield without reducing the enantiomeric excess [18b]. Compound **23** was converted to **24** via a nine-step sequence; there was also an alternative approach starting from the more readily available **3** [18c]. Use of the 1,2-dichloroethane–tertiary alcohol solvent system for the conversion of **15** to **3** improved the yield relative to that previously reported; a study of the various tertiary alcohols revealed that pinacol was the most effective, giving **3** in 78% yield with 95% ee. We successfully synthesized (+)-vernolepin (**25**) and were thereby able to determine its absolute configuration.

The general method described above for decalin synthesis was also applied to the synthesis of 6,5-ring systems through the formation of hydrindans (Figure 13.7) [19a].

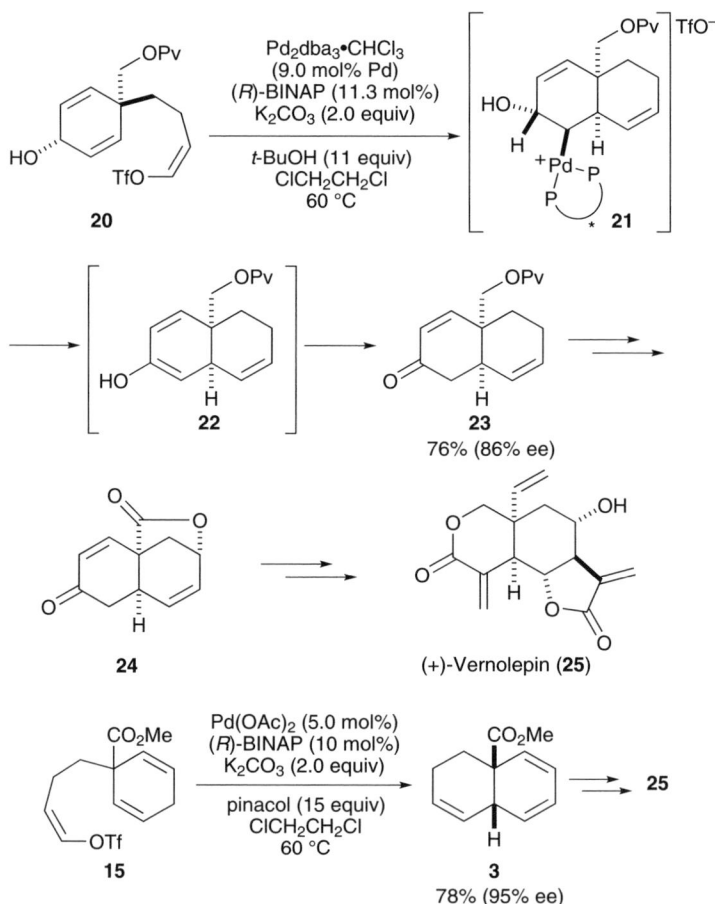

Figure 13.6   *Enantioselective synthesis of (+)-vernolepin.*

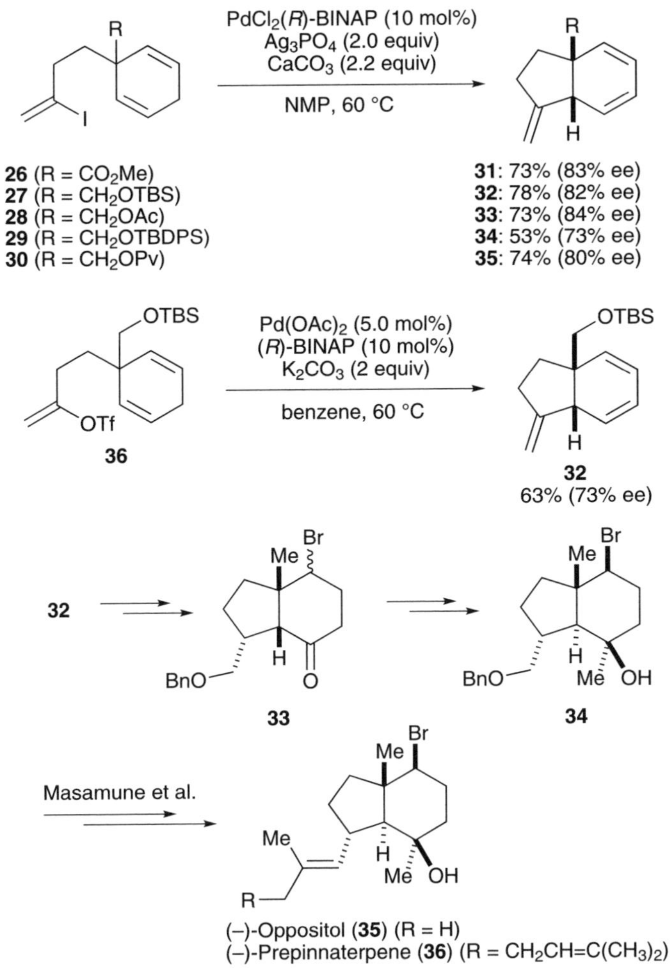

Figure 13.7   *Enantioselective synthesis of* cis-*hydrindans.*

Alkenyl iodides **26–30** were converted to the corresponding *cis*-hydrindans **31–35** by methods similar to those used for decalins, where Ag_3PO_4 was again the most effective silver salt in this conversion. Small increases ($\leq 5\%$) in enantiomeric excess were obtained by prereducing the palladium catalyst *in situ* (**31** : 86% ee, **32** : 86% ee, **33** : 85% ee). The alkenyl triflate **36** gave **32** in slightly lower enantiomeric excess (73% ee) than that observed for the corresponding conversion of alkenyl iodide **27**, with potassium carbonate being the most effective base. The hydrindan **32** obtained was later converted to **34** [19b], which is the key intermediate in the syntheses of (−)-oppositol (**35**) and (−)-prepinnaterpene (**36**) by Masamune and coworkers [20]. The conversion involved oxidation of the diene moiety with singlet oxygen, and is notable for the clean epimerization of the ring junction to give the *trans*-configuration (**33**→**34**), which demonstrates that both *cis*- and *trans*-annulation is respectively directly and indirectly accessible from a single Mizoroki–Heck cyclization.

In 1992, Feringa and coworkers developed an efficient intramolecular asymmetric Mizoroki–Heck reaction of prochiral cyclohexadienones **37** to tricyclic compound **40**. This transformation required inversion of the C2 stereocentre in intermediate **38** by enolization (**38→39**) in order to make a synperiplanar C—H bond available for the subsequent *syn β*-hydride elimination (**39→40**) (Figure 13.8) [21a]. Upon this asymmetric Mizoroki–Heck reaction, the stereogenic centre is not created at the site of the C—C bond formation, but

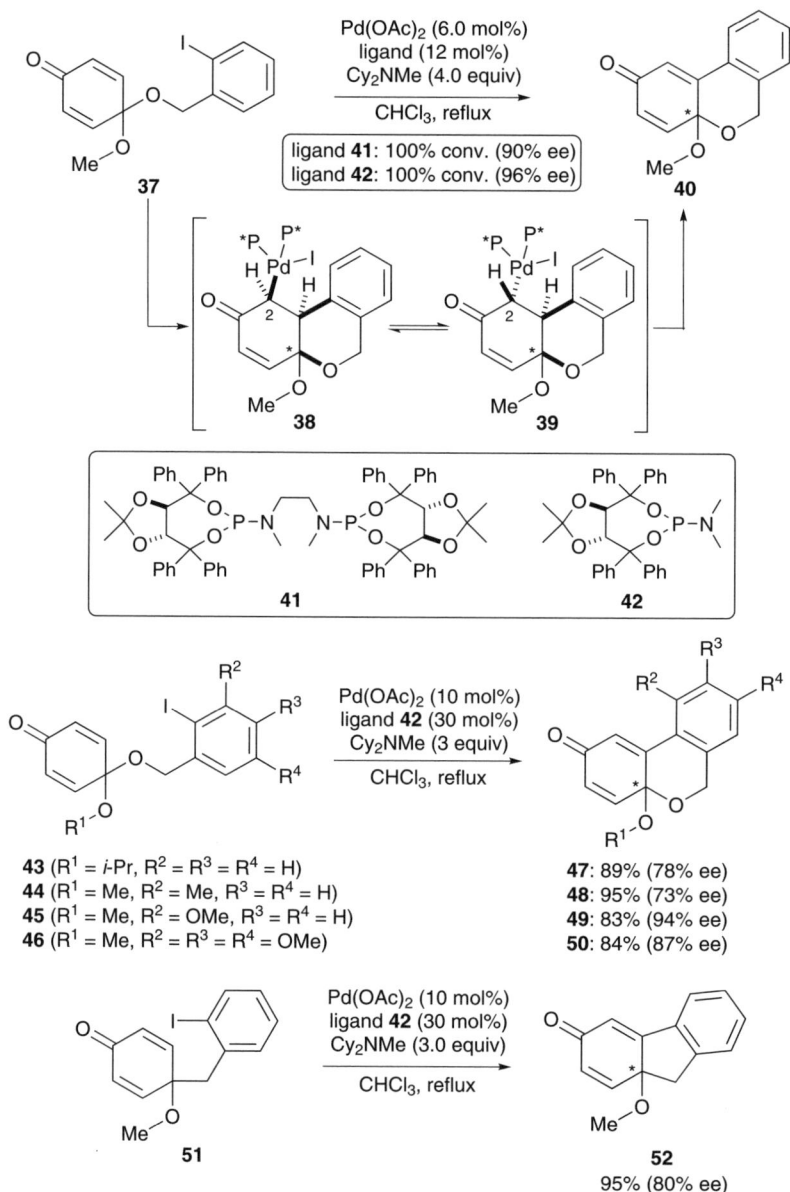

Figure 13.8   *Desymmetrizing Mizoroki–Heck reaction of dienone.*

the substrate is desymmetrized instead. Because phosphoramidites are versatile ligands for a variety of catalytic asymmetric transformations and all successful ligands used for the asymmetric Mizoroki–Heck reaction have so far been bidentate, TADDOL-based bidentate phosphoramidite **41** was first examined. In this reaction system, however, the monodentate ligand **42** gave higher enantiomeric excess than bidentate **41**. To obtain optimal asymmetric induction, 2 equiv of monodentate ligand **42** were necessary. For comparison, BINAP was also tested in this system, affording **40** in 0–50% yield and 0–5% ee. The use of an aryl triflate derivative instead of aryl iodide **37** resulted in a sluggish reaction (25% conversion and 75% ee), and addition of silver(I) salt to the reaction of **37** did not improve the conversion. These facts led to the conclusion that the cationic complex is not reactive in this system and the reaction proceeds through the neutral five-coordinate palladium species! Other dienones **43**–**46**, containing different substituents at the acetal and aromatic rings, gave similar or lower enantioselectivities [21b]. In all cases, monodentate ligand **42** showed slightly higher enantioselectivity than the bidentate ligand **41**. The reaction of dienone **51** formed all-carbon five-membered rings in 95% conversion with 80% ee, suggesting that the acetal group is not essential for high enantiomeric excess and good conversion.

13.2.2   1,3-Cyclopentadienes

The successful execution of asymmetric Mizoroki–Heck reactions for the formation of 6,6- and 6,5-ring systems from prochiral substrates led to a further extension of this method to form 5,5-ring systems, which are the backbone of a large number of natural products. The use of prochiral cyclopentadienyl systems, however, involves the generation of a π-allylpalladium species, which must then be trapped by a suitable nucleophile (Figure 13.9) [22]. The greater reactivity of the 1,3-diene substrate towards the silver salts used in the reactions and the propensity for undesirable side-reactions, such as Diels–Alder cycloadditions, must also be considered. The former problem figures prominently in the first example of an asymmetric Mizoroki–Heck reaction-based diquinane synthesis [23a,b]. Although cyclization of iodide **53** gave the bicyclo[3.3.0]octane **54** in reasonable yield, the observed enantiomeric excess was low ($\sim$20%). We attribute this failure in large part to the instability of **53** in the presence of silver salts, necessitating their omission from the reaction medium and, therefore, forfeiting the beneficial effects noted in an earlier study [9]. The presence of tetrabutylammonium acetate, a source of nucleophilic acetate, is essential, as the reaction does not proceed in its absence. This was the first example of a domino-type asymmetric Mizoroki–Heck reaction (Chapter 8). The problem of low enantiomeric excess was circumvented by using the triflate **55** as a substrate and BINAP as a chiral ligand, which gave the diquinane **57** with 80% ee and 89% yield. Throughout this process, three contiguous stereocentres (one quaternary and two tertiary) were established in one pot. We converted the obtained product **57** to the triquinane **58**, an intermediate in a previously described synthesis of $\Delta^{9(12)}$-capnellenols **59**–**61** [24]. Later, this domino process was further extended to other nucleophiles, such as amines [23b] and carbanions [23c,d]. In the latter transformation, the optical yields obtained were slightly lower (66–80% ee) than acetate (80% ee) and amine (81% ee) capture processes. It appears that counteranion exchange occurs between the hard triflate anion and the soft enolate anion to form a neutral palladium intermediate. The addition of sodium bromide significantly improved the enantiomeric excess of the carbanion capture process (80–94% ee), in which sodium bromide might

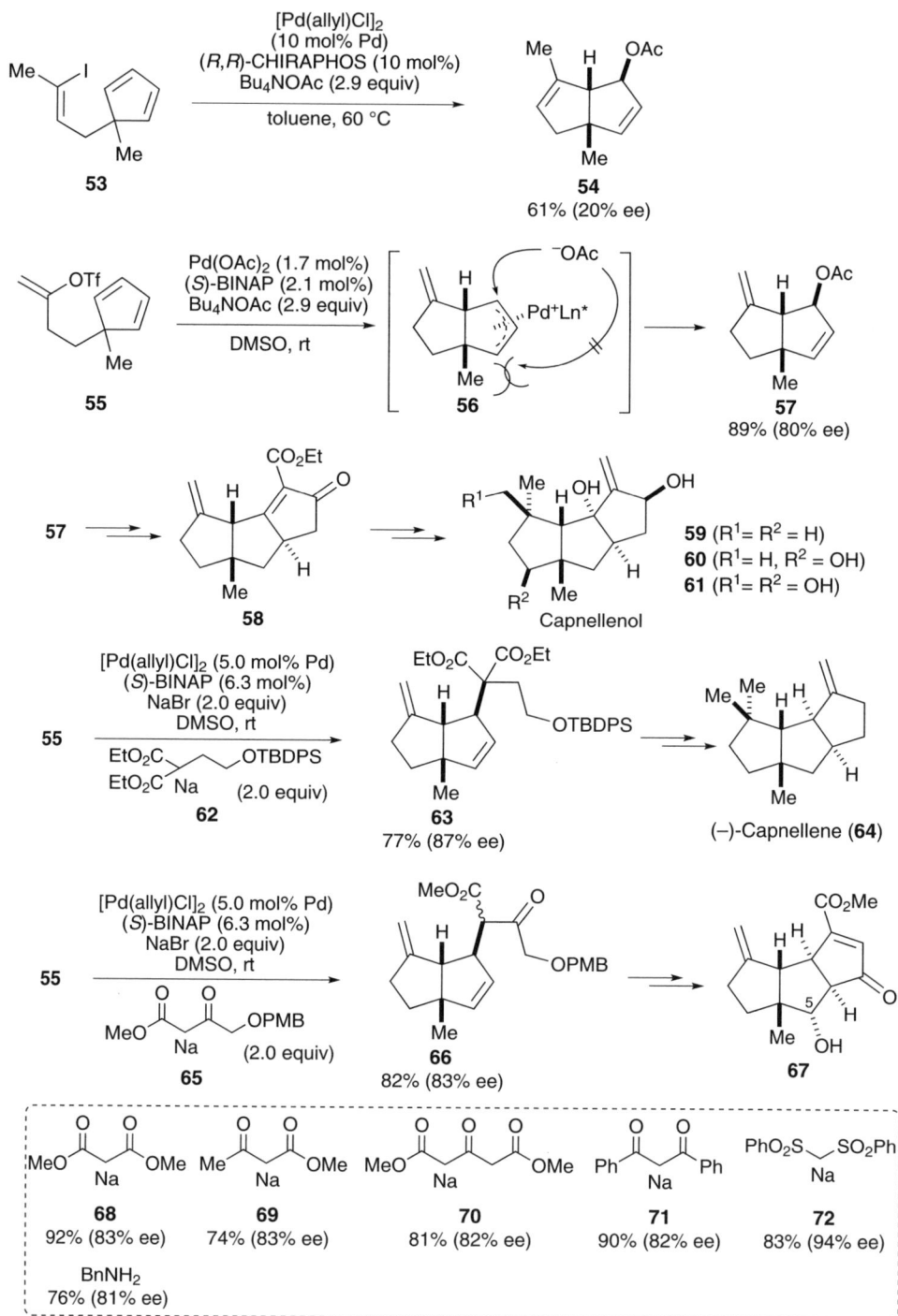

Figure 13.9 Asymmetric Mizoroki–Heck reaction–anion capture process.

prevent counteranion exchange between the triflate anion and the enolate anion by complexing with sodium enolate. Using the carbanion capture process, the first catalytic asymmetric synthesis of $(-)$-$\Delta^{9(12)}$-capnellene (**64**) was achieved [23c]. Furthermore, to synthesize highly functionalized 5α-capnellenols, which have 5α-hydroxyl functionality and exhibit antitumour activities, an asymmetric Mizoroki–Heck–carbanion capture process of triflate **55** with carbanion **65** was performed to give **66** in 83% ee [23d]. Then, product **66** was converted to the tricyclic core **67** of 5α-capnellenols through intramolecular nitrile oxide cycloaddition.

13.3 Bicyclic Diene as Cyclization Precursors

A new entry to the desymmetrizing Mizoroki–Heck reaction was developed by Lautens and Zunic [25] in 2004. They used symmetrical bicyclodienes as substrates, which were synthesized through a diastereoselective double ring-closing metathesis reaction, and the corresponding tetracyclic compounds were obtained with excellent enantioselectivity (up to 99% ee) (Figure 13.10). First, desymmetrization of bicyclo[3.3.0]octadienes was examined. Double ring-closing metathesis of the tetraene **73** with Grubbs' first-generation catalyst [26a] gave the substrate **74** as a sole bicyclic product in 70% yield. The desymmetrizing Mizoroki–Heck reaction of **74** proceeded efficiently in the presence of silver(I) salt to afford tetracyclic compound **75**, in which three contiguous chiral centres were established in a single step. Although the reaction using a tertiary amine base (neutral condition) gave similar levels of enantioselectivity to that using silver(I) salt (cationic condition), conversion rates were low ($\sim$50%). When an aryl triflate derivative was used as a substrate (cationic condition), high enantiomeric excess was again obtained with moderate chemical yield. In this system, tol-BINAP gave the optimal result (87%, 99% ee) and P,N-ligand *t*-Bu-PHOX gave only low chemical yield and selectivity (18%, 29% ee). This system was then applied to the desymmetrization of bicyclo[4.4.0]decadienes. The synthesis of substrate **77** was achieved by ring-opening and double ring-closing metathesis reactions of cyclic triene compound **76**. In the presence of 2 mol% of Grubbs' second-generation catalyst [26b], the metathesis reaction completed in 30 min to give a 4.5 : 1 mixture of bicyclo[4.4.0]decalines *cis*-**77** and *trans*-**77**. The major diastereomer *cis*-**77** was then used for the desymmetrizing Mizoroki–Heck reaction. Different from the reaction using **74**, the reaction using *cis*-**77** gave a mixture of tetracyclic product **78** and **79** in a ratio of 1 : 1.5 with different enantiomeric excess values (**78** : 50% ee, **79** : 35% ee). The latter product would be obtained by alkene isomerization of **78**, including reinsertion of **78** alkene into the Pd–H bond and the following β-hydride elimination either to regenerate **78** or to form the regioisomer **79**. Although Pfaltz and coworkers [27] reported that P,N-ligand *t*-Bu-PHOX prevents alkene isomerization in the intermolecular asymmetric Mizoroki–Heck reaction due to rapid dissociation of the Pd–H species from the product, *t*-Bu-PHOX ligand gave more isomerization product (**78** : **79** $= 1 : 2$) with lower enantiomeric excess. The utilization of Tl$_2$CO$_3$ [28] as a base provided a 1 : 5.6 ratio of **78** : **79** with yields and enantioselectivities similar to those observed with silver(I) salt. Using aryl triflate substrate *cis*-**80**, enantioselectivities of products **78** and **79** were greatly improved to 82% ee and 66% ee respectively, although the yield and regioselectivity were only moderate (60% yield, **78** : **79** $= 1 : 1.6$).

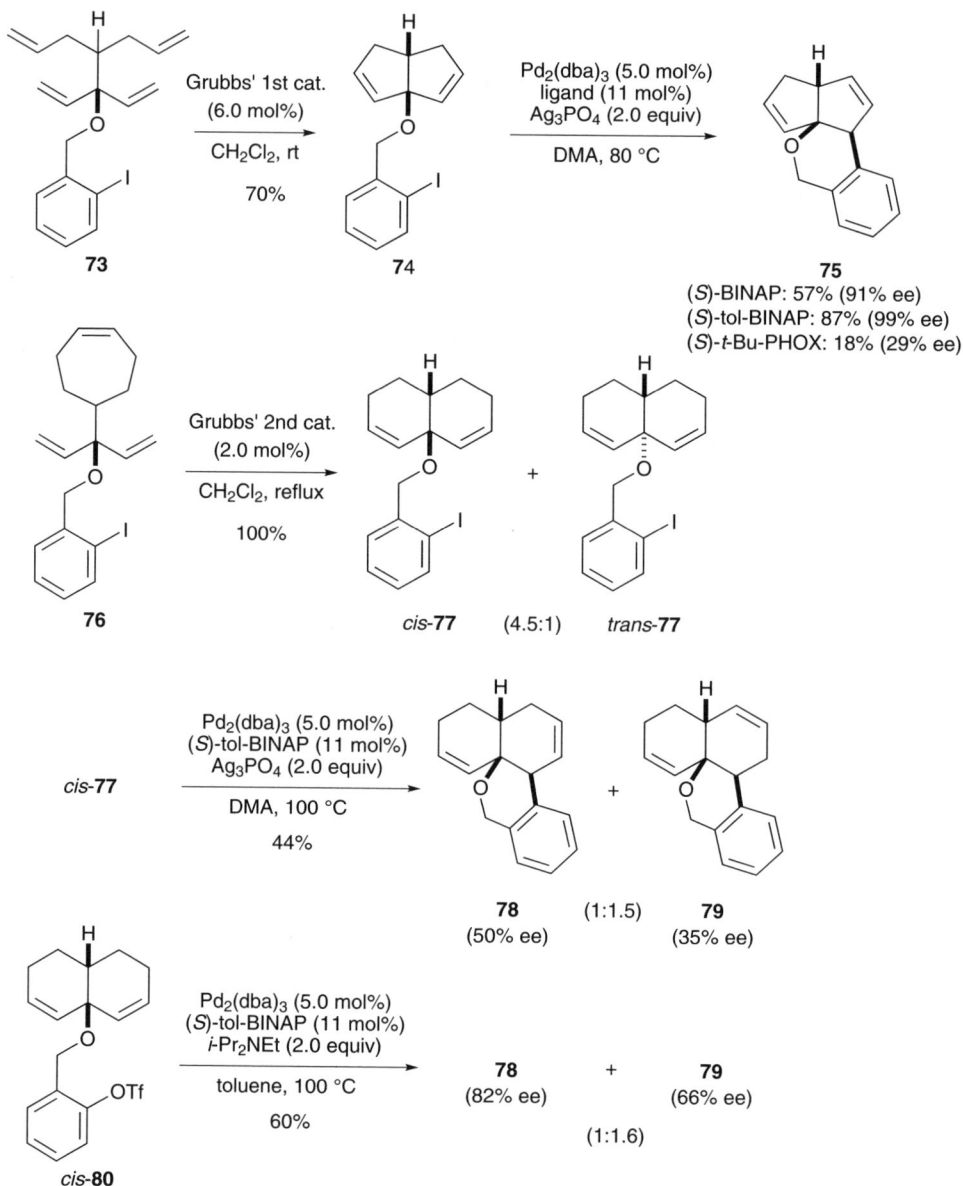

Figure 13.10   *Desymmetrization of bicyclodienes.*

Bräse and coworkers [29] also reported highly enantioselective desymmetrization of *trans*-bicyclo[4.4.0]decadienes **81–84**, which were prepared through Birch reduction of naphthalene, epoxidation and epoxide-opening reaction (Figure 13.11). The position of C=C double bonds also differed from the previous report [25]. Although a Mizoroki–Heck reaction of aryl iodide **81** with the usual chiral ligands such as BINAP, DIOP ((2*R*,3*R*)- or (2*S*,3*S*)-*O*-isopropylidene-2,3-dihydroxy-1,4-bis(diphenylphosphino)butane)

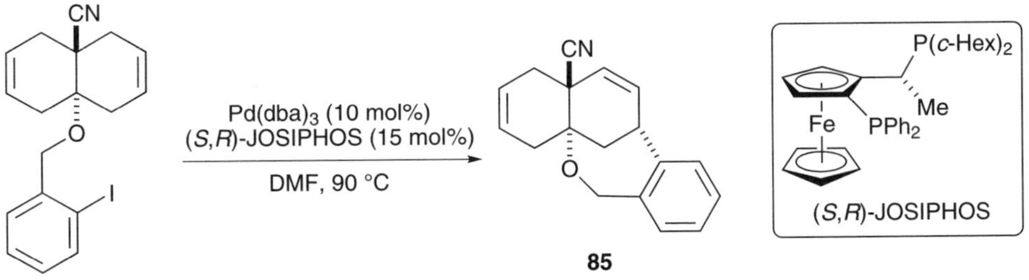

Figure 13.11   Desymmetrization of trans-bicyclo[4.4.0]decadienes.

and *i*-Pr-PHOX resulted in very low enantioselection and/or low conversion, the use of the JOSIPHOS ligand [30] gave the tetracyclic product **85** in 92% yield with high enantiomeric excess (84% ee), forming three new stereocentres in one step. Recrystallization of this product gave optically pure material. The conversion rates were increased by the addition of Ag_2CO_3, but this had negative effects on enantioselectivity. The reaction of aryl bromide **82** gave almost the same enantiomeric excess with decreasing chemical yield due to the formation of the hydrogenated product (X = H). The change of the leaving group from a bromide to a triflate or nonaflate had almost no effect on the stereoinduction of the product. Thus, aryl iodide **81** is the ideal substrate for this system.

13.4   Acyclic Dienes as Cyclization Precursors

The substrates of the desymmetrizing Mizoroki–Heck reaction were restricted to cyclic dienes until Oestreich *et al.* [31a] reported the first desymmetrizing Mizoroki–Heck reaction of acyclic dienes in 2005 (Figure 13.12). Initially, they extensively surveyed reaction conditions using diallyl substrate **86** as a substrate. The enantioselectivities obtained, however, were only moderate (<40% ee). Around the same time, Coogan and Pottenger [32] independently reported a desymmetrizing Mizoroki–Heck reaction of diallyl substrate **87** in which only achiral palladium complexes were utilized. The former group thus considered the introduction of aryl groups at the terminal positions of the allyl moieties as in **88** because such a modification would have two beneficial effects: (i) prevention of post-Mizoroki–Heck C=C double-bond migration on account of the styrene unit and (ii) improvement of enantioselectivity due to increased steric hindrance and potential aryl–aryl interactions. As expected, standard asymmetric Mizoroki–Heck conditions induced exclusive formation of the desired cyclic compound **89** as a single isomer with substantially improved enantioselectivity. After optimization of the reaction conditions, the use of BINAP as a chiral ligand gave the product in 85% yield with 92% ee. Changing the chiral ligand from BINAP to Cl-MeO-BIPHEP [33] improved enantioselectivity to 94% ee. In this system, the free hydroxy group acts as a catalyst-directing group [34]. For example, reactions of silyl ether substrate **90** and deoxygenated substrate **91** resulted in low yields and

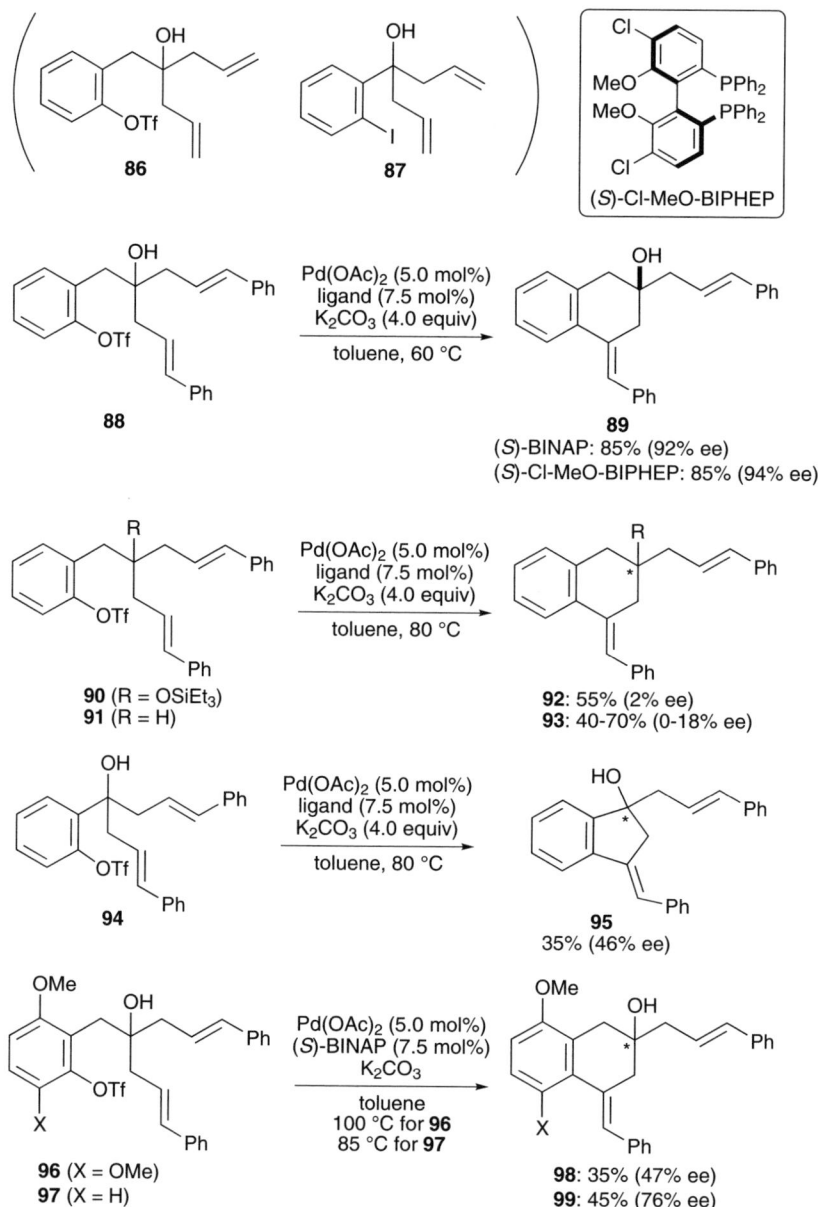

Figure 13.12   *Desymmetrization of diallyl substrates.*

poor enantioselectivities. The position of the unprotected hydroxy group relative to the aryl triflate moiety is also important. Benzyl alcohol derivative **94** gave unsatisfactory results, suggesting that the high enantioselectivity observed for **88** stems from the ideal proximity of the hydroxy group and the palladium centre (six-membered ring coordination). Coordination of the hydroxy group requires a free coordination site at the palladium centre, which is available under cationic conditions. Later, this methodology was applied to the synthesis of an AB synthon of anthracyclines [31b]. Although electronic and steric alteration of **88** was severely detrimental to the level of enantioselection in the cyclization of **96** and **97**, the use of K_2CO_3 as a base and BINAP as a chiral ligand gave enantioselectivities of the products up to 76% ee.

When the arene of the Cp ring is substituted with different groups at the *ortho*- or *meta*-positions, transition metal π-complexes exist in two enantiomeric forms based on a planar chirality. In 1999, Bräse [35] reported an intramolecular Mizoroki–Heck reaction of prochiral arene chromium complexes, although an asymmetric variant was not described (see below). In 2005, for the synthesis of optically active planar-chiral (arene)chromium complexes, Uemura and coworkers [36] utilized the desymmetrizing Mizoroki–Heck reaction of prochiral (arene)chromium complexes (Figure 13.13). Previously, the authors reported the desymmetrization of a prochiral *o*-dichlorobenzene chromium complex by a Suzuki–Miyaura cross-coupling reaction with moderate yield and enantioselectivity, in which discrimination between the two C—Cl bonds at the oxidative addition step was essential [37]. For development of the desymmetrizing Mizoroki–Heck reaction, they surveyed various chiral ligands for the reaction of (2,6-dibutenylchlorobenzene)chromium complex **100**, in which asymmetric induction occurs at the insertion step. Diphosphine ligands, including BINAP, PPFA (2-(1-dimethylaminoethyl)-1-diphenylphosphinoferrocene) and DIOP, as well as TADDOL-based phosphoramidite **41** and **42**, gave products **101** and **102** in a racemic form. On the other hand, 1,1′-bi-2-naphthol (BINOL)-based phosphoramidite **103** afforded the cyclized product **102** in 46% ee. Other P,N-ligands **104–106**, P,O-ligand **107** and bidentate P,N-ligand **108** resulted in lower enantioselectivity. In this system, the addition of silver(I) salt decreased the yield and enantioselectivity. When an amine base was used, the major product was *endo*-cyclic compound **102** along with a small amount of *exo*-cyclic compound **101**. In contrast, when inorganic base K_2CO_3 was used, the *exo*-cyclic compound **101** was obtained as a sole product. After optimization of the reaction conditions, **101** was obtained in 78% yield with 73% ee. The reaction of prochiral chromium complex **109** with a longer side chain also afforded *exo*-cyclic compound **110**; however, in this case, the enantioselectivity decreased to 40% ee. In addition, a domino reaction including the desymmetrizing Mizoroki–Heck reaction and Suzuki–Miyaura cross-coupling reaction was also performed to give the desired product **111–114** as a single diastereomer.

13.5 Desymmetrization of Leaving Groups

Within the scope of the above-described study, Bräse [35] also reported in 1999 an intramolecular Mizoroki–Heck reaction of prochiral arene chromium complexes **115** and **116** using achiral palladium complexes, in which less reactive chloroarenes were effectively activated by chromium tricarbonyl fragments (Figure 13.14). The substrates have two enantiotopic chlorine atoms and, in principle, the chiral palladium complex is able to discriminate one C—Cl bond from the other bond at the oxidative addition step to give optically active compounds – yet only the racemic series was reported. In the same year, the desymmetrizing Mizoroki–Heck reaction of bis(enolnonaflate) was reported by Bräse [38]. First, 4-*exo*-trig cyclization by Mizoroki–Heck reaction of bis(enolnonaflate) **121** was conducted under standard achiral conditions and the highly congested bicyclic compound **123** was isolated in 56% yield as the major product, in which the remaining enolnonaflate

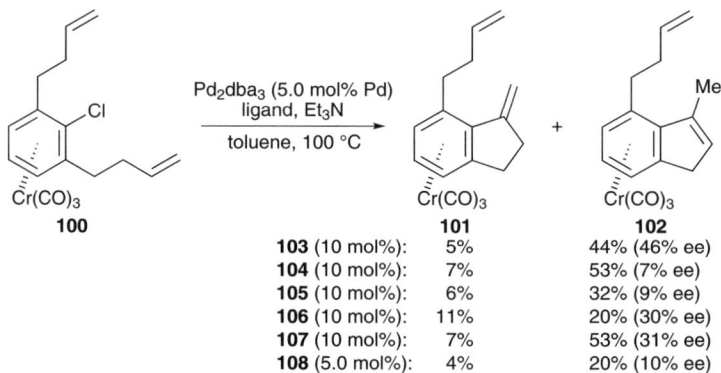

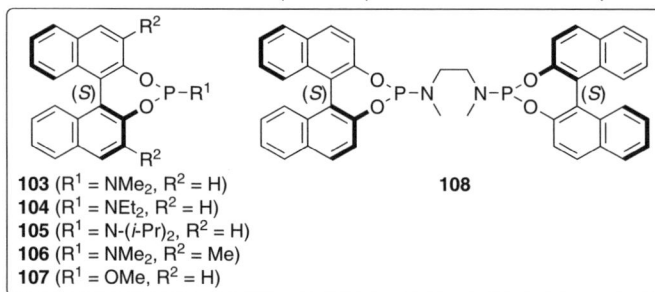

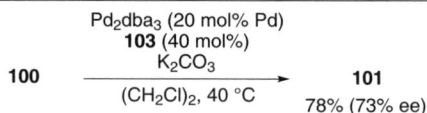

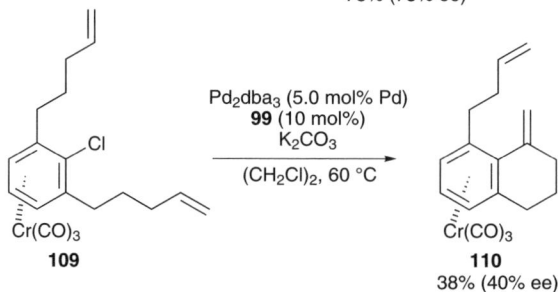

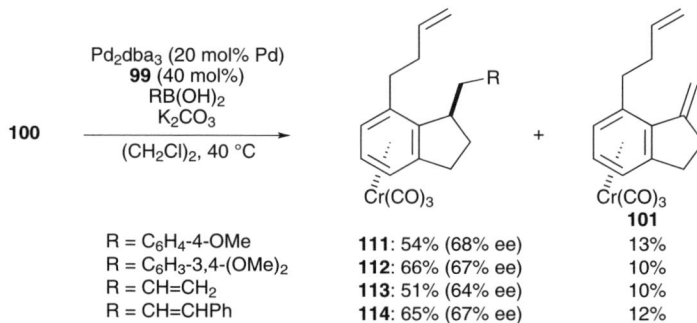

Figure 13.13 *Desymmetrization of (arene)chromium complexes.*

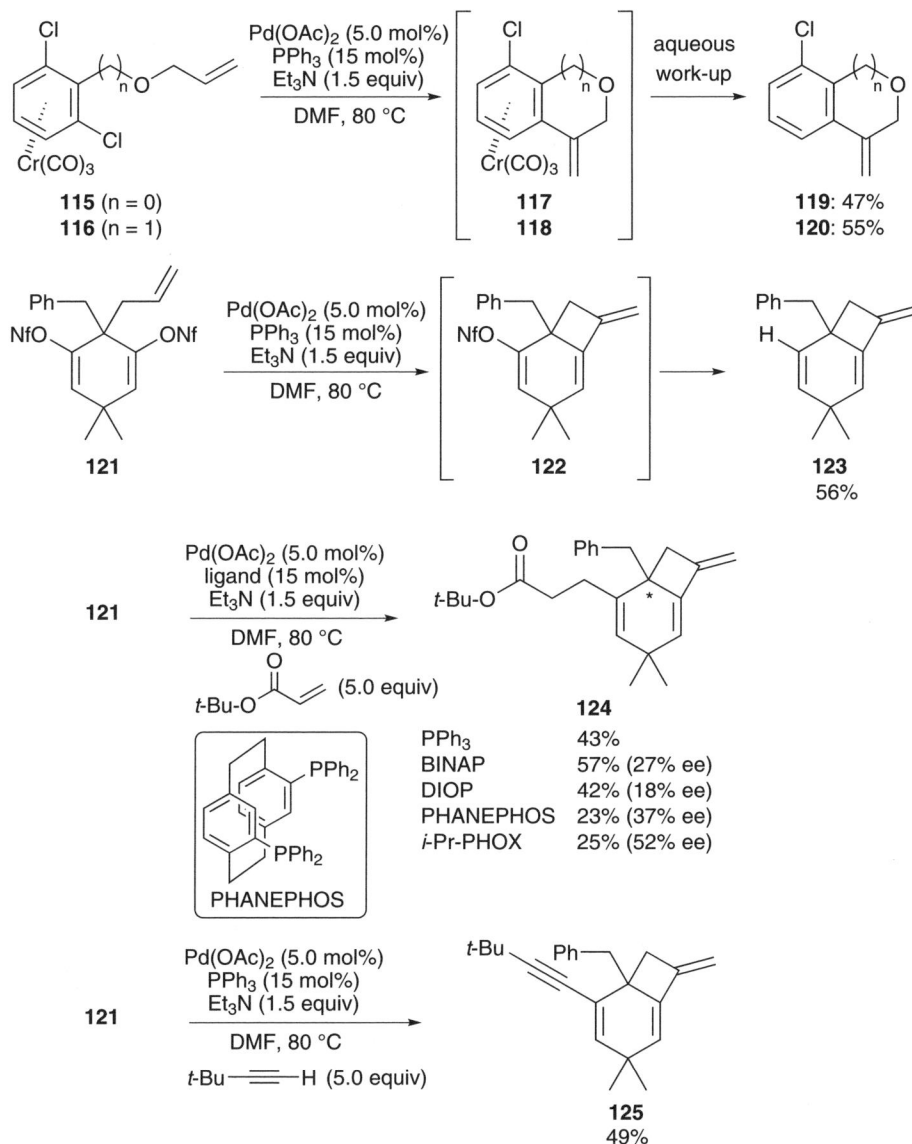

Figure 13.14   *Desymmetrization of bis(enolnonaflate).*

of the cyclized intermediate **122** was reduced to alkene. Next, this process was applied to a domino intramolecular asymmetric Mizoroki–Heck–intermolecular Mizoroki–Heck reaction. In the presence of *tert*-butyl acrylate, the coupling product **124** was obtained in 43% yield. Similarly, a domino intramolecular asymmetric Mizoroki–Heck–Sonogashira coupling reaction also proceeded to give the desired coupling product **125** in 49% yield. To discriminate between the two enolnonaflates, several chiral ligands, including BINAP, DIOP and PHANEPHOS (4,12-bis(diphenylphosphino)-[2.2]-paracyclophane) [39], were examined and P,N-ligand *i*-Pr-PHOX gave the best enantioselectivity (52% ee). Although the level of asymmetric induction was moderate, this result indicates that the concept of two leaving groups in the desymmetrization reaction holds synthetic potential.

13.6 Summary and Conclusions

Since the first desymmetrizing Mizoroki–Heck reaction was reported in the late 1980s, this process has now been developed into a powerful method for the synthesis of optically active cyclic compounds, some of which have found application in natural product synthesis. One or, as in most cases, two new stereogenic centres, including asymmetrically substituted quaternary carbons, were constructed in a single operation with enantiomeric excesses typically greater than 80% and in some cases much higher (up to 99% ee). Under the desymmetrizing strategy, even when the stereogenic centre is not created at the site of C—C bond formation, the substrate can be desymmetrized to give an optically active compound. This strategy has been successfully extended into the so-called domino reaction. Such one-pot multi-bond formations are not only highly elegant, but also reduce the amount of waste produced in a synthesis and thereby conserve resources. Although several elegant desymmetrizing Mizoroki–Heck reactions have been developed, the number of successful substrates is still limited. Therefore, the search for greater substrate diversity remains a challenge for future research.

References

1. For recent reviews of the asymmetric Heck reaction, see: (a) Shibasaki, M., Boden, C.D.J. and Kojima, A. (1997) The asymmetric Heck reaction. *Tetrahedron*, **53**, 7371–95; (b) Shibasaki, M. and Vogl, E.M. (1999) Heck reaction, in *Comprehensive Asymmetric Catalysis*, Vol. **2** (eds E.N. Jacobsen, A. Pfaltz and H. Yamamoto), Springer, Berlin, pp. 457–87; (c) Donde, Y. and Overman, L.E. (2000) Asymmetric intramolecular Heck reactions, in *Catalytic Asymmetric Synthesis* (ed. I. Ojima), Wiley–VCH Verlag GmbH, New York, pp. 675–97; (d) Dounay, A.B. and Overman, L.E. (2003) The asymmetric intramolecular Heck reaction in natural product total synthesis. *Chem. Rev.*, **103**, 2945–64; (e) Shibasaki, M., Vogl, E.M. and Ohshima, T. (2004) Asymmetric Heck reaction. *Adv. Synth. Catal.*, **346**, 1533–52.
2. For a representative review, see: Kagan, H.B., Diter, P., Gref, A., *et al.* (1996) Towards new ferrocenyl ligands for asymmetric catalysis. *Pure Appl. Chem.*, **68**, 29–36, and references cited therein.
3. Heck, R.F. (1985) *Palladium Reagents in Organic Synthesis*, Academic Press, London.
4. Heck, R.F. (1979) Palladium-catalyzed reactions of organic halides with olefins. *Acc. Chem. Res.*, **12**, 146–51.
5. Sato, Y., Sodeoka, M. and Shibasaki, M. (1989) Catalytic asymmetric C—C bond formation: asymmetric synthesis of *cis*-decalin derivatives by palladium-catalyzed cyclization of prochiral alkenyl iodides. *J. Org. Chem.*, **54**, 4738–9.
6. Carpenter, N.E., Kucera, D.J. and Overman, L.E. (1989) Palladium-catalyzed polyene cyclizations of trienyl triflates. *J. Org. Chem.*, **54**, 5846–8.
7. For recent reviews, see: (a) Wills, M.C. (1999) Enantioselective desymmetrisation. *J. Chem. Soc., Perkin Trans. 1*, 1765–84; (b) Studer, A. and Schleth, F. (2005) Desymmetrization and diastereotopic group selection in 1,4-cyclohexadienes. *Synlett*, 3033–41.

8. (a) Karabelas, K., Westerlund, C. and Hallberg, A. (1985) The effect of added silver nitrate on the palladium-catalyzed arylation of allyltrimethylsilanes. *J. Org. Chem.*, **50**, 3896–900; (b) Abelman, M.M., Oh, T. and Overman, L.E. (1987) Intramolecular alkene arylations for rapid assembly of polycyclic systems containing quaternary centers. A new synthesis of spirooxindoles and other fused and bridged ring systems. *J. Org. Chem.*, **52**, 4130–3.
9. Sato, Y., Sodeoka, M. and Shibasaki, M. (1990) *Chem. Lett.*, 1953–4.
10. Hayashi, T., Matsumoto, Y. and Ito, Y. (1988) Palladium-catalyzed asymmetric 1,4-disilylation of α,β-unsaturated ketones: catalytic asymmetric synthesis of β-hydroxy ketones. *J. Am. Chem. Soc.*, **110**, 5579–81.
11. Kojima, A., Boden, C.D.J. and Shibasaki, M. (1997) Synthesis and evaluation of a new chiral arsine ligand; 2,2′-bis(diphenylarsino)-1,1′-binaphthyl (BINAs). *Tetrahedron Lett.*, **38**, 3459–60.
12. Sato, Y., Watanabe, S. and Shibasaki, M. (1992) Further studies on a catalytic asymmetric synthesis of decalin derivatives. *Tetrahedron Lett.*, **35**, 2589–92.
13. Dekker, G.P.C.M., Elsevier, C.J., Vrieze, K. and van Leeuwen, P.W.N.M. (1992) Influence of ligands and anions on the rate of carbon monoxide insertion into palladium–methyl bonds in the complexes (P–P)Pd(CH$_3$)Cl and [(P–P)Pd(CH$_3$)(L)]$^+$SO$_3$CF$_3^-$ (P–P = dppe, dppp, dppb, dppf; L = CH$_3$CN, PPh$_3$) . *Organometallics*, **11**, 1598–603.
14. For representative reviews, see: (a) Pfaltz, A. (1996) Design of chiral ligands for asymmetric catalysis: From C-2-symmetric semicorrins and bisoxazolines to non-symmetric phosphinooxazolines. *Acta Chem. Scand. B*, **50**, 189–94; (b) Loiseleur, O., Hayashi, M., Keenan, M. et al. (1999) Enantioselective Heck reactions using chiral P,N-ligands. *J. Organomet. Chem.*, **576**, 16–22; (c) Helmchen, G. and Pfaltz, A. (2000) Phosphinooxazolines-a new class of versatile, modular P,N-ligands for asymmetric catalysis. *Acc. Chem. Res.*, **33**, 336–45; (d) For a remarkable use of the (S)-*t*-Bu-PHOX ligand in the asymmetric Heck reaction, see: Dounay, A.B., Overman, L.E. and Wrobleski, A.D. (2005) Sequential catalytic asymmetric Heck-iminium ion cyclization: enantioselective total synthesis of the strychnos alkaloid minfiensine. *J. Am. Chem. Soc.*, **127**, 10186–87.
15. (a) Kiely, D. and Guiry, P.J. (2003) A comparison of palladium complexes of BINAP and diphenylphosphinooxazoline ligands in the catalytic asymmetric synthesis of *cis*-decalins. *Tetrahedron Lett.*, **44**, 7377–80; (b) Kiely, D. and Guiry, P.J. (2003) Palladium complexes of phosphinamine ligands in the intramolecular asymmetric Heck reaction. *J. Organomet. Chem.*, **687**, 545–61.
16. (a) Richards, C.J., Damalidis, T., Hibbs, D.E. and Hursthouse, M.B. (1995) Synthesis of 2-[2-(diphenylphosphino)ferrocenyl]oxazoline ligands. *Synlett*, 74–6; (b) Nishibayashi, Y. and Uemura, S. (1995) Asymmetric synthesis and highly diastereoselective *ortho*-lithiation of oxazolinylferrocenes. *Synlett*, 79–81.
17. Danishefsky, S., Schuda, P.F., Kitahara, T. and Etheridge, S.J. (1977) The total synthesis of *dl*-vernolepin and *dl*-vernomenin. *J. Am. Chem. Soc.*, **99**, 6066–75.
18. (a) Kondo, K., Sodeoka, M., Mori, M. and Shibasaki, M. (1993) Asymmetric Heck reaction. A catalytic asymmetric synthesis of the key intermediate for vernolepin. *Tetrahedron Lett.*, **34**, 4219–22; (b) Kondo, K., Sodeoka, M., Mori, M. and Shibasaki, M. (1993) Asymmetric Heck reaction: catalytic asymmetric syntheses of bicyclic enones, dienones and the key intermediate for vernolepin. *Synthesis*, 920–30; (c) Ohrai, K., Kondo, K., Sodeoka, M. and Shibasaki, M. (1994) Effects of solvents and additives in the asymmetric Heck reaction of alkenyl triflates: catalytic asymmetric synthesis of decalin derivatives and determination of the absolute stereochemistry of (+)-vernolepin. *J. Am. Chem. Soc.*, **116**, 11737–48.
19. (a) Sato, Y., Honda, T. and Shibasaki, M. (1992) A catalytic asymmetric synthesis of hydrindans. *Tetrahedron Lett.*, **33**, 2593–6; (b) Sato, Y., Mori, M. and Shibasaki, M. (1995) Asymmetric Heck reaction: a catalytic asymmetric synthesis of the key intermediate of (−)-oppositol and (−)-prepinnaterpene. *Tetrahedron: Asymmetry*, **6**, 757–66.

20. Fukuzawa, A., Sato, H. and Masamune, T. (1987) Synthesis of (±)-prepinnaterpene, a bromoditerpene from the red alga. *Tetrahedron Lett.*, **28**, 4303–6.
21. (a) Imbos, R., Minnaard, A.J. and Feringa, B.L. (2002) A highly enantioselective intramolecular Heck reaction with a monodentate ligand. *J. Am. Chem. Soc.*, **124**, 184–5; (b) Imbos, R., Minnaard, A.J. and Feringa, B.L. (2003) Monodentate phosphoramidites; versatile ligands in catalytic asymmetric intramolecular Heck reactions. *Dalton Trans.*, 2017–23.
22. Nylund, C.S., Klopp, J.M. and Weinreb, S.M. (1994) Consecutive carbon–carbon bond formation via the *p*-allylpalladium variant of the Heck reaction. *Tetrahedron Lett.*, **35**, 4287–90.
23. Kagechika, K. and Shibasaki, M. (1991) Asymmetric Heck reaction: a catalytic asymmetric synthesis of the key intermediate for $\Delta^{9(12)}$-capnellene-3β,8β,10α-triol and $\Delta^{9(12)}$-capnellene-3β,8β,10α,14-tetrol. *J. Org. Chem.*, **56**, 4093–4; (b) Kagechika, K., Ohshima, T. and Shibasaki, M. (1993) Asymmetric Heck reaction–anion capture process. A catalytic asymmetric synthesis of the key intermediates for the capnellenols. *Tetrahedron*, **49**, 1773–82; (c) Ohshima, T., Kagechika, K., Adachi, M., *et al.* (1996) Asymmetric Heck reaction–carbanion capture process. Catalytic asymmetric total synthesis of (−)-$\Delta^{9(12)}$-capnellene. *J. Am. Chem. Soc.*, **118**, 7108–16; (d) Itano, W., Ohshima, T. and Shibasaki, M. (2006) Synthesis of the tricyclic core of 5α-capnellenols using asymmetric Heck reaction–carbanion capture process. *Synlett*, 3053–6.
24. Shibasaki, M., Mase, T. and Ikegami, S. (1986) The first total syntheses of $\Delta^{9(12)}$-capnellene-8β,10α-diol and $\Delta^{9(12)}$-capnellene-3β,8β,10α-triol. *J. Am. Chem. Soc.*, **108**, 2090–1.
25. Lautens, M. and Zunic, V. (2004) Sequential olefin metathhesis–intramolecular asymmetric Heck reactions in the synthesis of polycycles. *Can. J. Chem.* **82**, 399–407.
26. (a) Pchwab, P., France, M.B., Ziller, J.W. and Grubbs, R.H. (1995) A series of well-defined metathesis catalysts-synthesis of [RuCl$_2$(=CHR′)(PR$_3$)$_2$] and its reactions. *Angew. Chem., Int. Ed. Engl.*, **34**, 2039–41; (b) Scholl, M., Ding, S., Lee, C.W. and Grubbs, R.H. (1999) Synthesis and activity of a new generation of ruthenium-based olefin metathesis catalysts coordinated with 1,3-dimesityl-4,5-dihydroimidazol-2-ylidene ligands. *Org. Lett.*, **1**, 953–6.
27. Loiseleur, O., Meier, P. and Pfaltz, A. (1996) Chiral phosphanyldihydrooxazoles in asymmetric catalysis: enantioselective Heck reactions. *Angew. Chem., Int. Ed. Engl.*, **35**, 200–2.
28. Grigg, R., Loganathan, V., Santhamukar, V. *et al.* (1991) Suppression of alkene isomerisation in products from intramolecular Heck reactions by addition of thallium(I) salts. *Tetrahedron Lett.*, **32**, 687–90.
29. Lormann, M.E.P., Nieger, M. and Bräse, S. (2006) Desymmetrisation of bicyclo[4.4.0] decadienes: a planar-chiral complex proved to be most effective in an asymmetric Heck reaction. *J. Organomet. Chem.*, **691**, 2159–61.
30. Togni, A., Breutel, C., Schnyder, A. *et al.* (1994) A novel easily accessible chiral ferrocenyldiphosphine for highly enantioselective hydrogenation, allylic alkylation, and hydroboration reactions. *J. Am. Chem. Soc.*, **116**, 4062–6.
31. (a) Oestreich, M., Sempere-Culler, F. and Machotta, A.B. (2005) Catalytic desymmetrizing intramolecular Heck reaction: evidence for an unusual hydroxy-directed migratory insertion. *Angew. Chem. Int. Ed.*, **44**, 149–52; (b) Oestreich, M., Sempere-Culler, F., and Machotta, A.B. (2006) An enantioselective access to an anthracycline AB synthon by a desymmetrizing Heck cyclization. *Synlett*, 2965–8.
32. Coogan, M.P. and Pottenger, M.J. (2005) Desymmetrisation of a diallyl system by intramolecular Heck reaction. *J. Organomet. Chem.*, **690**, 1409–11.
33. Laue, C., Schröder, G. and Arlt, D. (Bayer AG) (1995) DE-A1 19522293.
34. Oestreich, M. (2005) Neighbouring-group effects in Heck reactions. *Eur. J. Org. Chem.*, 783–92.
35. Bräse, S. (1999) Intramolecular transition-metal catalyzed cyclizations of electron rich chloroarenes. *Tetrahedron Lett.*, **40**, 6757–9.

36. Kamikawa, K., Harada, K. and Uemura, M. (2005) Catalytic asymmetric induction of planar chirality: palladium catalyzed intramolecular Mizoroki–Heck reaction of prochiral (arene)chromium complexes. *Tetrahedron: Asymmetry*, **16**, 1419–23.
37. (a) Uemura, M., Nishimura, H. and Hayashi, T. (1993) Catalytic asymmetric induction of planar chirality by palladium-catalyzed asymmetric cross-coupling of a *meso* (arene)chromium complex. *Tetrahedron Lett.*, **34**, 107–10; (b) Uemura, M., Nishimura, H. and Hayashi, T. (1994) Catalytic asymmetric induction of planar chirality: palladium-catalyzed asymmetric cross-coupling of *meso* tricarbonyl(arene) chromium complexes with alkenyl- and arylboronic acids. *J. Organomet. Chem.*, **473**, 129–37.
38. Bräse S. (1999) Synthesis of bis(enolnonaflates) and their 4-*exo*-trig-cyclizations by intramolecular Heck reactions. *Synlett*, 1654–6.
39. Pye, P.J., Rossen, K., Reamer, R.A. *et al.* (1997) A new planar chiral bisphosphine ligand for asymmetric catalysis: highly enantioselective hydrogenations under mild conditions. *J. Am. Chem. Soc.*, **119**, 6207–8.

14

Combinatorial and Solid-Phase Syntheses

Thierry Muller and Stefan Bräse

Institut für Organische Chemie, Universität Karlsruhe (TH), Fritz-Haber-Weg 6, D-76131 Karlsruhe, Germany

14.1 General Remarks

The development of high-throughput screening assays in crop protection, but mostly in the pharmaceutical sector, engendered a huge demand for small-molecule libraries [1, 2]. Combinatorial chemistry and automated multiparallel synthesis turned out to be the methods of choice to deliver enough material for 'lead generation' and 'lead optimization' in drug discovery. To date, most of the organic chemical libraries have been synthesized using solid-phase methods [3]. As a consequence, since the proper 'invention' of solid-phase organic synthesis by Merrifield [4] in 1963, a lot of effort has been put into the development of versatile methods for the construction of diversified nonoligomeric libraries on solid supports [5]. As for organic chemistry in general, carbon–carbon bond formation is one of the most useful and fundamental reactions during the generation of molecule libraries.

Transition-metal-mediated transformations play an important role in combinatorial synthesis on solid supports because they allow C–C bond formation under mild, neutral conditions and tolerate a broad range of functional groups. In particular, the palladium-mediated cross-coupling reactions, more precisely the Mizoroki–Heck, the Suzuki and the Stille reactions received a lot of attention due to their enormous derivatization potential of functional structures [5]. Although Stille coupling was the first to be explored on and

The Mizoroki–Heck Reaction   Edited by Martin Oestreich
© 2009 John Wiley & Sons, Ltd

adapted to solid phase in the early 1990s, the Mizoroki–Heck reaction is, nevertheless, the prevalent cross-coupling reaction in this area [6–8].

14.2   The Mizoroki–Heck Reaction on Solid Supports

The Mizoroki–Heck reaction is indeed particularly adapted to automated solid-phase synthesis because its substrates are, in general, cheap and easily accessible and it can in most cases be run under aerobic atmosphere [8–12]. Furthermore, addition of aryl or disubstituted alkenyl halides or triflates to unactivated alkenes or alkynes is an excellent way to diversify combinatorial libraries. Thus, the Mizoroki–Heck reaction is nowadays routinely used in solid-phase synthesis for the preparation of relatively complex structures, in inter- and intra-molecular fashion and even in multicomponent reactions [11]. Aside from the general advantages of solid-phase synthesis, namely the possibility to employ excess of reagent, the avoidance of tedious work-up procedures and the noninterference of susceptible functional groups in the building blocks due to a quasi-high-dilution effect, the case of palladium-mediated transformations on solid supports also benefits from the easy removal of the metallic catalyst by washing processes. Moreover, there is ongoing research on new linker systems and solid supports and on novel supported palladium catalysts or ligands with the aim of optimizing these palladium-catalysed combinatorial and solid-phase syntheses to make them even more competitive compared to liquid-phase syntheses.

Throughout this survey, the type of resin used in the different examples will be given. If not stated otherwise, the resin bead insignia symbolizes the terminal part of an aromatic substructure.

14.2.1   Intermolecular Mizoroki–Heck Reactions

Mizoroki–Heck reactions on solid supports can generally be carried out under two different reaction conditions: the standard Mizoroki–Heck conditions ($Pd(OAc)_2$, PPh_3 or $P(o\text{-Tol})_3$, dimethylformamide (DMF), organic base, 80–100 °C, 2–24 h) or the so-called Jeffery conditions ($Pd(OAc)_2$, PPh_3, NBu_4Cl, K_2CO_3, DMF, 20–80 °C), which use an inorganic base and a phase-transfer agent usually leading to a reaction rate increase along with lower reaction temperatures and generally enhanced yields [11].

The Mizoroki–Heck reaction can either be performed on immobilized aryl halides with soluble alkenes [13] or alkynes or with the latter being attached to solid phase and free aryl halides. When performed on the same type of resin and using identical catalyst cocktails, however, it generally seems preferable to use immobilized aryl halides with soluble alkenes rather than doing it the other way around [14]. Both possibilities have been reported and will be discussed below [11].

14.2.1.1   Mizoroki–Heck Reaction Using Immobilized Aryl Halides

Grether and Waldmann [15] developed an enzyme-labile safety catch linker **1** tested in intermolecular Mizoroki–Heck reactions (Scheme 14.1). This linker releases alcohols and amines through enzymatic cleavage of the benzylamide moiety followed by subsequent lactam formation. After a Mizoroki–Heck reaction performed on an immobilized iodoarene

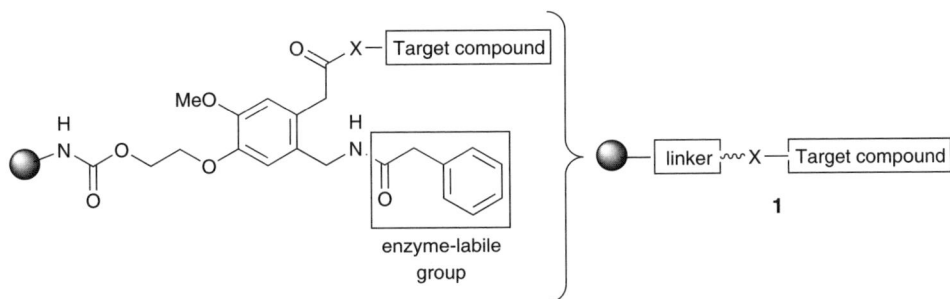

Scheme 14.1 *Polymeric support containing enzyme-labile safety catch linker **1**.*

with *tert*-butyl acrylate, coupling compound **4** was obtained under particularly mild cleavage conditions in good yield (Scheme 14.2).

Brown and coworkers [16] performed Mizoroki–Heck reactions using a novel, readily prepared *tert*-alkoxysiloxane linker system (Scheme 14.3). Besides the fact that small quantities of products were detected in solution under Mizoroki–Heck coupling conditions, the *tert*-alkoxydiphenylsilane linkers showed good stability.

Other examples involve Mizoroki–Heck couplings of aryl bromides using a sulfur linker cleaved under very mild conditions in the presence of samarium iodide [17, 18], the solid-supported hyperbranched polymerization of phenylacetylene derivatives [19] or a

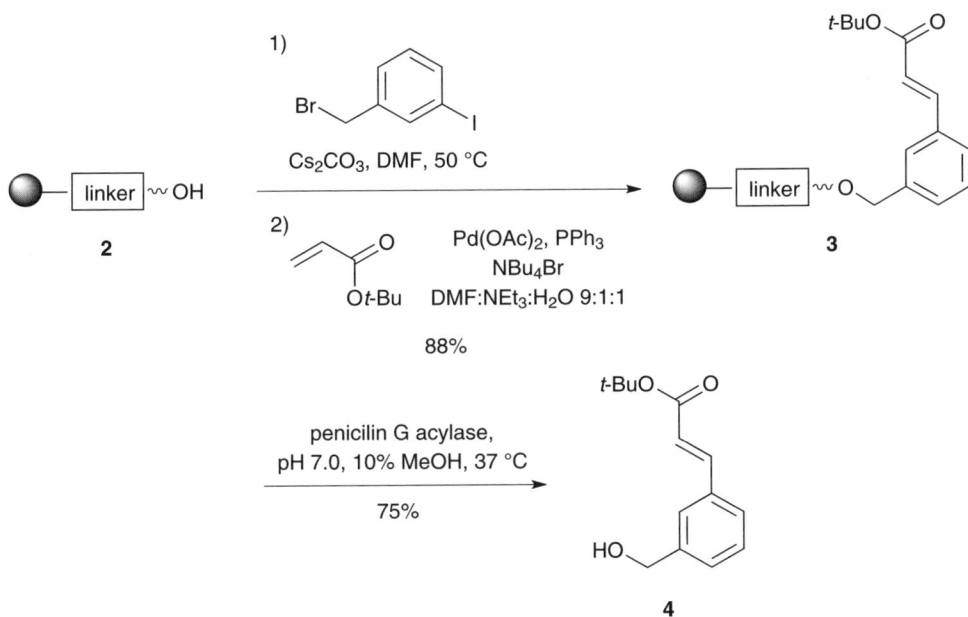

Scheme 14.2 *Mizoroki–Heck reaction of an iodoarene attached by an enzyme-labile safety catch linker.*

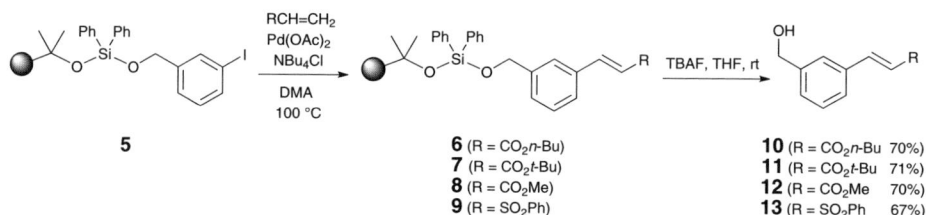

Scheme 14.3   *Mizoroki–Heck reaction using a* tert-*alkoxysiloxane linker system.*

comparative study of microwave-assisted Mizoroki–Heck solid-phase coupling reactions versus conventional Mizoroki–Heck reactions on solid supports [20].

14.2.1.2   Mizoroki–Heck Reaction Using Immobilized Alkenes

Kulkarni and Ganesan [21] reported the synthesis of β-keto esters via sequential Baylis–Hillman and Mizoroki–Heck reactions. The reaction of the Baylis–Hillman product **14** with bromophenol yielded the corresponding Mizoroki–Heck coupling product **15**, which, after cleavage from the resin with concomitant decarboxylation, afforded aryl ketone **16** (Scheme 14.4). A small library of 25 compounds was prepared in this way with overall yields up to 49%.

In general, intermolecular Mizoroki–Heck reactions using immobilized alkenes are not as common as the variant that deals with solid-supported aryl halides, but we will come across another example in Section 14.2.3 on multicomponent Mizoroki–Heck reactions.

14.2.2   Intramolecular Mizoroki–Heck Reactions

Among the previously mentioned advantages of solid-phase synthesis, the pseudodilution effect particularly affects the intramolecular variant of the Mizoroki–Heck reaction, commonly resulting in increased yields and leading to the most complex structures obtained by Mizoroki–Heck reactions on solid phase.

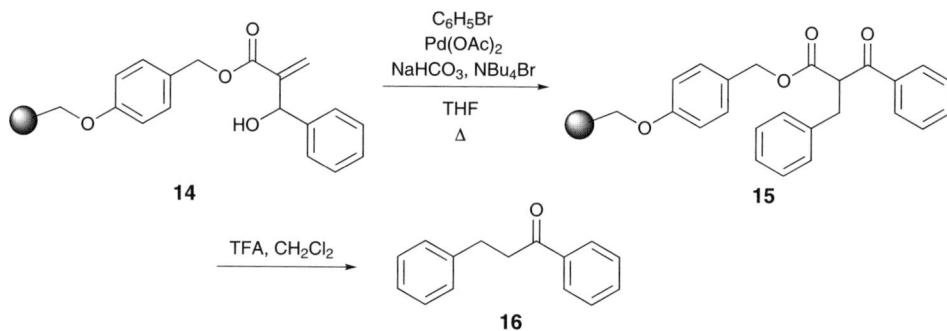

Scheme 14.4   *Solid-phase synthesis of β-keto ester with subsequent cleavage and decarboxylation.*

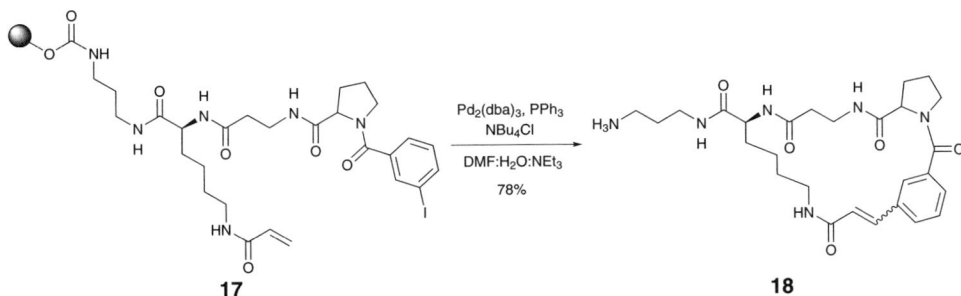

Scheme 14.5 *Macrocyclization via intramolecular Mizoroki–Heck reaction on solid phase.*

An impressive example is the first macrocyclization performed on solid supports by Zhou and coworkers [22] starting from aryliodide **17**, which gave the desired compound **18** under remarkably mild conditions in exceptionally high purity and yield (Scheme 14.5).

Five-membered cycles have also been prepared by intramolecular Mizoroki–Heck reactions on solid phase, as the syntheses of indole and benzofuran derivatives published by Zhang and Maryanoff [23] show (Scheme 14.6).

There are other examples of the concise high-yielding preparations of six-membered rings [24].

Even selective 7-*exo*-cylizations [25] have been reported via intramolecular Mizoroki–Heck reactions on solid supports. One synthesis delivers bicyle **25** in 60% overall yield (Scheme 14.7) [25].

14.2.3 Multicomponent Mizoroki–Heck Reactions

Multicomponent reactions are particularly viable procedures for combinatorial synthesis. The clear advantage of conducting these reactions on solid supports lies in the fact that all nonpolymer-bound components (e.g. excess of reagents) can simply be removed.

An interesting three-component reductive Mizoroki–Heck reaction on solid phase has been undertaken by Koh and Ellman [26] (Scheme 14.8). Whenever resin-bound organopalladium intermediates can be isolated, a split synthesis strategy can be applied, thus allowing a much larger diversity. Reactions of organopalladium intermediate **27** with a range of different nucleophiles lead to molecules bearing two sites of diversity.

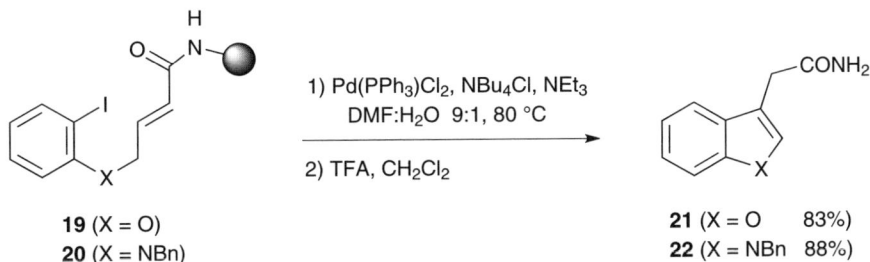

Scheme 14.6 *Synthesis of indole and benzofuran derivatives.*

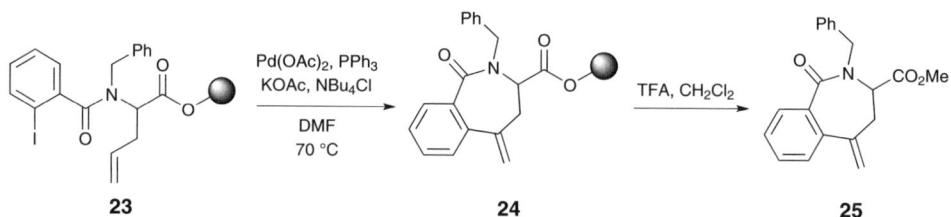

Scheme 14.7   *Concise seven-membered ring formation.*

Another attractive three-component procedure involves the versatile triazene linker T1 and generates spirooctene **30** from a Mizoroki–Heck reaction of immobilized iodoarene **29** with bicyclopropylidene in the presence of an acrylate derivative (Scheme 14.9) [27, 28].

The triazene moiety can be cleaved to diazonium salts which, in turn, act as substrates for Mizoroki–Heck reactions with various alkenes to give spirooctenes **31**. The latter can be obtained without the double bond in the coupled alkene if palladium on charcoal is used instead of palladium acetate. In this case, the same catalyst promotes the Mizoroki–Heck reaction and the subsequent hydrogenation [28].

This cleavage of the triazene linker with subsequent Mizoroki–Heck reaction is interesting in the sense that it adds yet another dimension of diversity by offering an additional possibility of functionalization.

14.3   Outlook and Conclusion

The Mizoroki–Heck reaction is nearly predestinated to be used in solid-phase combinatorial chemistry. The commonly available substrates, the mild reaction conditions and the high yields combined with the ease and rapidity offered by combinatorial and parallel processes are at the origin of a lot of small-molecule libraries.

There are, however, some problems that can be encountered using Mizoroki–Heck coupling reactions on solid supports. Sometimes, when soluble supports are used, the removal of the catalyst and of the excess of reagents may not be that trivial. In order to circumvent this problem, people are constantly looking for solutions. One may be the development of new solid supports [29, 30]; another even more promising approach is the

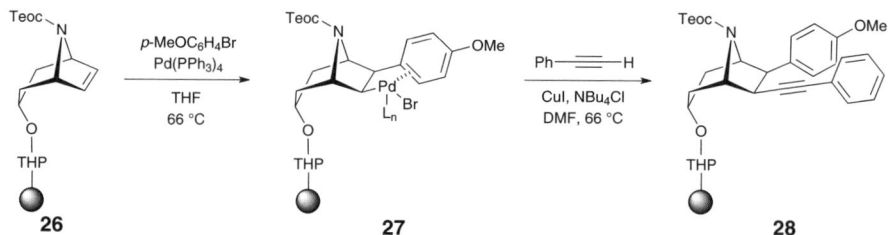

Scheme 14.8   *Carbopalladation of tropane derivative **26**.*

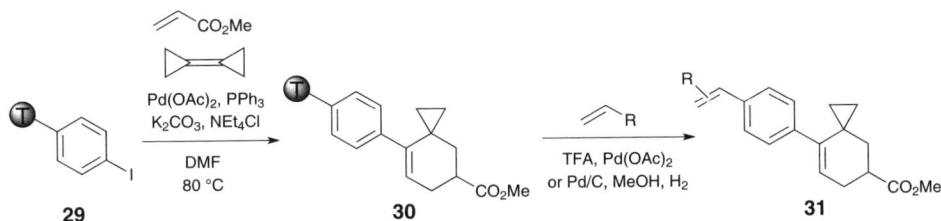

Scheme 14.9 *Versatile three-component reaction.*

use of supported catalysts, ligands or efficient scavenger resins to sequester the catalyst [31–37].

Other major issues are the aspects of stereo- and regio-chemistry. Compared with its liquid-phase variant, where tremendous efforts have been made over recent years to develop chiral ligands and optimize procedures of enantio- and diastereo-selective syntheses of complex molecules, the Mizoroki–Heck reaction on solid supports is still in its infancy. Most libraries, indeed, contain racemates and mixtures of diastereomers [11].

References

1. Thompson, L.A. and Ellman, J.A. (1996) Synthesis and applications of small molecule libraries. *Chem. Rev.*, **96**, 555–600.
2. Ziegert, R.E., Toräng, J., Knepper, K. and Bräse, S. (2005) The recent impact of solid-phase synthesis on medicinally relevant benzoannelated oxygen heterocycles. *J. Comb. Chem.*, **7**, 147–69.
3. Franzén, R. (2000) The Suzuki, the Heck, and the Stille reaction – three versatile methods for the introduction of new C–C bonds on solid support. *Can. J. Chem.*, **78**, 957–62.
4. Merrifield, R.B. (1963) Solid phase peptide synthesis. I. The synthesis of a tetrapeptide. *J. Am. Chem. Soc.*, **85**, 2149–54.
5. Andres, C.J., Whitehouse, D.L. and Deshpande, M.S. (1998) Transititon-metal mediated reactions in combinatorial synthesis. *Curr. Opin. Chem. Biol.*, **2**, 353–62.
6. Berteina, S., Wendeborn, S., Brill, W.K.-D. and De Mesmaeker, A. (1998) Pd-mediated C–C bond formation with olefins and acetylenes on solid support: a scope and limitations study. *Synlett*, 676–8.
7. Wendeborn, S., De Mesmaeker, A., Brill, W.K.-D. and Berteina, S. (2000) Synthesis of diverse and complex molecules on the solid phase. *Acc. Chem. Res.*, **33**, 215–24.
8. Lorsbach, B.A. and Kurth, M.J. (1999) Carbon–carbon bond forming solid-phase reactions. *Chem. Rev.*, **99**, 1549–81.
9. Sammelson, R.E. and Kurth, M.J. (2001) Carbon–carbon bond-forming solid-phase reactions. Part II. *Chem. Rev.*, **101**, 137–202.
10. Burello, E. and Rothenberg, G. (2003) Optimal Heck cross-coupling catalysis: a pseudo-pharmaceutical approach. *Adv. Synth. Catal.*, **345**, 1334–40.
11. Bräse, S., Kirchhoff, J.H. and Köbberling, J. (2003) Palladium-catalysed reactions in solid phase organic synthesis. *Tetrahedron*, **59**, 885–939.
12. Bräse, S. and de Meijere, A. (2004) Cross-coupling of organic halides with alkenes: the Heck reaction, in *Metal-Catalyzed Cross-Coupling Reactions* (eds A. de Meijere and F. Diederich), Wiley–VCH Verlag GmbH, Weinheim, pp. 217–315.

13. Knepper, K., Vanderheiden, S. and Bräse, S. (2006) Nitrogen functionalities in palladium-catalyzed reactions on solid supports: a case study. *Eur. J. Org. Chem.*, 1886–98.
14. Yu, K.-L., Deshpande, M.S. and Vyas, D.M. (1994) Heck reactions in solid phase synthesis. *Tetrahedron Lett.*, **35**, 8919–22.
15. Grether, U. and Waldmann, H. (2001) An enzyme-labile safety catch linker for synthesis on a soluble polymeric support. *Chem. Eur. J.*, **7**, 959–71.
16. Meloni, M.M., White, P.D., Armour, D. and Brown, R.C.D. (2007) Synthesis and applications of *tert*-alkoxysiloxane linkers in solid-phase chemistry. *Tetrahedron*, **63**, 299–311.
17. Turner, K.L., Baker, T.M., Islam, S. *et al.* (2006) Solid-phase approach to tetrahydroquinolones using a sulfur linker cleaved by SmI₂. *Org. Lett.*, **8**, 329–32.
18. McAllister, L.A., Turner, K.L., Brand, S. *et al.* (2006) Solid phase approaches to *N*-heterocycles using sulfur linker cleaved by SmI₂. *J. Org. Chem.*, **71**, 6497–507.
19. Bharathi, P. and Moore, J.S. (1997) Solid-supported hyperbranched polymerization: evidence for self-limited growth. *J. Am. Chem. Soc.*, **119**, 3391–2.
20. Berthault, A., Berteina-Raboin, S., Finaru, A. and Guillaumet, G. (2004) Solid phase synthesis of 2-substituted melatonin derivatives via palladium-mediated coupling reactions using microwave irradiation. *QSAR Comb. Sci.*, **23**, 850–3.
21. Kulkarni, B.A. and Ganesan, A. (1999) Solid-phase synthesis of β-keto esters via sequential Baylis–Hillman and Heck reactions. *J. Comb. Chem.*, **1**, 373–8.
22. Hiroshige, M., Hauske, J.R. and Zhou, P. (1995) Palladium-mediated macrocyclization on solid support and its applications to combinatorial synthesis. *J. Am. Chem. Soc.*, **117**, 11590–91.
23. Zhang, H.-C. and Maryanoff, B.E. (1997) Construction of indole and benzofuran systems on the solid phase via palladium-mediated cyclizations. *J. Org. Chem.*, **62**, 1804–9.
24. Goff, D.A. and Zuckermann, R.N. (1995) Solid-phase synthesis of highly substituted peptoid 1(2*H*)-isoquinolinones. *J. Org. Chem.*, **60**, 5748–9.
25. Bolton, G.L. and Hodges, J.C. (1999) Solid-phase synthesis of substituted benzazepines via intramolecular Heck cyclization. *J. Comb. Chem.*, **1**, 130–3.
26. Koh, J.S. and Ellman, J.A. (1996) Palladium-mediated three component coupling strategy for the solid-phase synthesis of tropane derivatives. *J. Org. Chem.*, **61**, 4494–5.
27. de Meijere, A., Nüske, H., Es-Sayed, M. *et al.* (1999) New efficient multicomponent reactions with C–C coupling for combinatorial application in liquid and on solid phase. *Angew. Chem. Int. Ed.*, **38**, 3669–72.
28. Bräse, S. and Schroen, M. (1999) Efficient cleavage–cross-coupling strategy for solid-phase synthesis – a modular building system for combinatorial chemistry. *Angew. Chem. Int. Ed.*, **38**, 1071–3.
29. Kashiwagi, Y., Chiba, S., Ikezoe, H. *et al.* (2007) Polypyrrole-supported graphite felt for Heck reaction in solid phase synthesis. *Electrochim. Acta*, **52**, 3726–31.
30. Tian, J., Mauer, K., Tesfu, E. and Moeller, K.D. (2005) Building addressable libraries: the use of electrochemistry for spatially isolating a Heck reaction on a chip. *J. Am. Chem. Soc.*, **127**, 1392–3.
31. Molnár, Á., Papp, A., Miklós, K. and Forgo, P. (2003) Organically modified Pd–silica catalysts applied in Heck coupling. *Chem. Commun.*, 2626–7.
32. Chandrasekhar, V. and Athimoolam, A. (2002) New hybrid inorganic–organic polymers as supports for heterogeneous catalysis: a novel Pd(0) metalated cyclophosphazene-containing polymer as an efficient heterogeneous catalyst for the Heck reaction. *Org. Lett.*, **4**, 2113–6.
33. Villemin, D., Jaffrès, P.-A., Nechab, B. and Courivaud, F. (1997) Palladium complexes supported on hybrid organic-inorganic zirconium phosphite: selectivity in the Heck reaction. *Tetrahedron Lett.*, **38**, 6581–4.
34. Daniel, S., Rao, P.P., Nandakumar, M. and Rao, T.P. (2005) Synthesis of heterogeneous phosphine-free palladium-based catalyst assemblies via solid phase extraction and their characterization. *Mat. Chem. Phys.*, **90**, 99–105.

35. Altava, B., Burguete, M.I., García-Verdugo, E. *et al.* (2006) Palladium *N*-methylimidazolium supported complexes as efficient catalysts for the Heck reaction. *Tetrahedron Lett.*, **47**, 2311–4.
36. Poulin, C., Brown, M.A., Wasslen, Y.A. *et al.* (2006) Reactivity of mesoporous palladium yttria-stabilized zirconia for solution phase reactions. *Can. J. Chem.*, **84**, 1520–8.
37. Arisawa, M., Hamada, M., Takamiya, I. *et al.* (2006) Development of a method for preparing a highly reactive and stable, recyclable and environmentally benign organopalladium catalyst supported on sulfur-terminated gallium arsenide(001): a three-component catalyst, {Pd}–S–GaAS(001), and its properties. *Adv. Synth. Catal.*, **348**, 1063–70.

15

Mizoroki–Heck Reactions: Modern Solvent Systems and Reaction Techniques

Werner Bonrath, Ulla Létinois, Thomas Netscher and Jan Schütz

DSM Nutritional Products, PO Box 2676, CH-4002 Basel, Switzerland

15.1 Introduction

The formation of C–C bonds is one of the important fields of synthetic organic chemistry. Modern methods are in the focus of scientists in academia and industry, often being connected to catalysis. Products based on catalysis represent more than 90% of the total amount of new compounds. Presently, palladium-catalysed reactions play a dominant role. For example, Mizoroki–Heck reactions, as well as Stille, Suzuki, Negishi, Kumada and Sonogashira couplings, are in the standard toolbox of organic chemists [1].

Nevertheless, alternative or nontraditional methods are interesting fields for new applications, due to the need for more selective synthetic and environmentally friendly methods, including the aspect of reaction media (solvent) and energy transfer into a reaction system. The use of supercritical fluids (SCFs) or ionic liquids (IL) as process solvents is well established. From the view point of 'green chemistry' or sustainable chemistry, it is SCFs, especially supercritical carbon dioxide ($scCO_2$), that are the main focus [2, 3]. It is currently under discussion whether ionic liquids as alternative process solvents are beneficial. In general, more fundamental problems, like recovery, reuse, purification and ecology (eco-toxicity), have to be solved.

As pointed out, catalytic reactions are important to replace stoichiometric transformations in organic chemistry, where often large amounts of waste are created. In general,

The Mizoroki–Heck Reaction Edited by Martin Oestreich
© 2009 John Wiley & Sons, Ltd

chemical reactions might be classified by the reaction volume [4], atom economy [5] or the *E*-factor [6]. Here, atom economy means the number of atoms of all starting materials which end up in the product [5]. From a more process-oriented view point, the aspects of waste formation, losses of solvents and energy have to be balanced. The *E*-factor, defined as kilograms of by-product per kilogram product, gives distinctive numbers and might be used as a first orientation to assess a reaction [6].

The common feature of Mizoroki–Heck reactions is the palladium(0)-catalysed arylation or alkenylation of alkenes [1]. During recent years, several excellent overviews about this topic have been published [7–12]. The reaction is usually carried out under basic conditions. Stoichiometric amounts of base are needed to neutralize the acid which is formed during the reaction.

One main issue is the waste formation in this type of reaction. Another one is the replacement of aryl halides for cheaper aryl sources. Aryl carboxylic anhydrides could be applied successfully as starting material, thus avoiding salt formation [13].

Furthermore, in Mizoroki–Heck reactions, high catalyst loadings are normally used, corresponding to substrate-to-catalyst ratios (*s*/*c*) of below 50 to 20. This fact prevents most procedures described in literature reports from being applied on a larger scale, since the catalyst is a severe cost driver. In this regard, it is important to note that the *s*/*c* can be reduced to 5000. This was achieved by a fundamental study of the reaction conditions, especially under ligand-free conditions. Also, a decrease in the formation of palladium black could be achieved, which is the main reason for deactivation [14]. A new industrialized product based on Mizoroki–Heck reaction with the advantage of low catalyst loading (*s*/*c* = 5000) is resveratrol, a polyphenol found in red wine grapes and *Polygonum cuspidatum* [15, 16].

The use of nonclassical reaction conditions in Mizoroki–Heck reactions (e.g. ultrasound, microwaves, alternative solvent systems, or combinations of these methods) is known and has been reviewed excellently by Beletskaya and colleagues [9, 12]. Here, we summarize the activities in the field of Mizoroki–Heck reactions from a practical point of view, focusing on preparatively useful laboratory procedures and their possible application to industrial (large-scale) synthetic chemistry.

15.2   Solvent Systems

15.2.1   Ionic Liquids

Ionic liquids are understood as salts, usually organic salts, with a melting point below the boiling point of water [17]. Commonly used ionic liquids have an imidazolium, pyridinium, alkylammonium or alkylphosphonium cation and have been described as liquid compounds with a partially covalent structure. The physical properties are tuneable by variation of the cations and anions (mostly inorganic polyatomic anions, e.g. BF_4^-, PF_6^-, $CF_3SO_3^-$).

Ionic liquids display a limited miscibility with various polar and nonpolar organic substrates, as well as organic and inorganic solvents, and they usually dissolve organometallic catalyst precursors based on rhodium, ruthenium, palladium, nickel, cobalt and iron complexes [18].

The Mizoroki–Heck reaction is usually performed in polar solvents, and salt additives such as tetrabutylammonium chloride have been shown to activate and stabilize the catalytically active palladium species [19]. Furthermore, the reactions in ionic liquids perform differently in terms of thermodynamic and kinetic properties of the reaction system. Additionally, ionic liquids allow a facile recovery of catalyst and substrates, as well as an easy product separation. Here, another beneficial effect might be used by combination of solvent mixtures; for example, of ionic liquids and SCFs. SCFs and ionic liquids have a mixing gap which allows working in two-phase systems, and results in a straightforward phase separation [20].

The first report about Mizoroki–Heck reactions carried out in ionic liquids was published in 1996 by Kaufmann *et al.* [21]. Simple palladium catalyst precursors such as palladium dichloride were dissolved in tetrabutylammonium chloride or bromide without the addition of phosphines as ligands. Bromoarenes could be coupled with olefins in yields over 95% and the product was isolated by simple distillation from the reaction mixture. The catalyst remained in the molten salt and could be reused for further coupling reactions.

Since then, many research groups have described the benefit of ionic liquids as solvents in the Mizoroki–Heck reaction covering a large variety of substrates. Cacchi *et al.* [22] achieved a stereoselective coupling of aryl iodides and methyl cinnamates in a mixture of molten tetrabutylammonium acetate and bromide. Intramolecular Mizoroki–Heck reactions were conducted in 1-*n*-butyl-3-methylimidazolium tetrafluoroborate ([BMIm]BF$_4$) using PdCl$_2$ as a precatalyst. Substituted benzofurans were obtained in satisfactory yields [23]. The ionic liquid containing the palladium catalyst could be reused several times with small decrease in activity.

The use of ionic liquids in Mizoroki–Heck reactions has been described in several reviews [9, 12, 24–26]. Modern trends of Mizoroki–Heck reactions in ionic liquids are the application of new ligand types, like *N*-heterocyclic carbene (NHC) ligands, supported catalysts, combination with ultrasound irradiation, microwave irradiation and multiphase catalysis. The combination of ionic liquids and activation methods is summarized in the representative chapters.

Useful ligands for Mizoroki–Heck reactions stem from the family of imidazolium-based ionic liquids which have a 2-pyridyl residue at C-2 [27]. For example, reactions of such compounds with palladium(II) chloride gave palladium(II) ionic liquid complexes like **1** (Scheme 15.1). Mizoroki–Heck reactions of various aryl iodides or bromides, with acrylic esters or styrene in the presence of these catalysts were achieved in yields up to 94%. The catalyst was recycled efficiently. In most applications the organic compound was

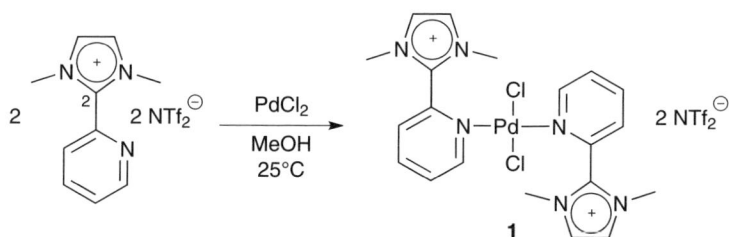

Scheme 15.1 *Synthesis of a pyridyl palladium complex.*

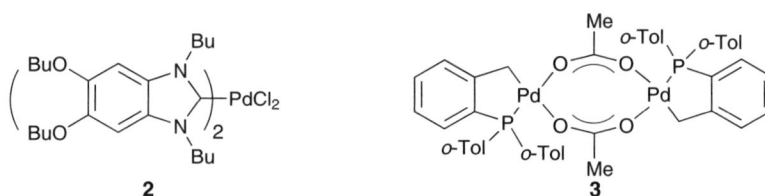

Figure 15.1 *Benzimidazolylidene palladium complex **2** and phospha-palladacycle **3** as Mizoroki–Heck catalysts.*

easily separated from the ionic liquid by phase separation (e.g. decantation) or distillation. Nonvolatile organic compounds can be removed by extraction using another solvent; for example, $scCO_2$ [28].

Another class of useful ligands are NHCs, which are strong σ-donor ligands with negligible π-acceptor abilities. These electronic characteristics are similar to those of electron-rich phosphines. In addition, NHC-based catalyst systems often show superior performance in several aspects. Advantages of these NHC-coordinated catalysts are easy preparation, air and moisture stability, and nontoxicity. Transformations using carbene complexes only require low catalyst loading and, hence, show high catalyst efficiency. This is due to the effect that the carbene ligands do not readily dissociate from the metal centre. Therefore, an excess of ligand is not necessary in order to stabilize the palladium catalyst.

Mizoroki–Heck couplings of bromo- and chloroarenes were successfully achieved by using the electron-rich benzimidazolylidene palladium complex **2** (Figure 15.1), generated *in situ* in tetrabutylammonium bromide as solvent [29]. The coupling of 4-chloroacetophenone with butyl acrylate in the presence of 1 mol% **2** yields 93% 4-acetyl-(*E*)-cinnamic ester in 6 h reaction time. Butyl acrylate was also coupled with 2 equiv of bromobenzene leading to the trisubstituted olefin 3,3-diphenylpropenoic acid butyl ester in 91% yield.

Palladium-containing benzothiazole carbene complexes have been employed for the Mizoroki–Heck reaction in ionic liquids. The catalyst was immobilized in tetrabutylammonium bromide. With an *s/c* of 100, efficient coupling reactions of bromobenzene with *n*-butyl acrylate were conducted in 10 min reaction time at 130 °C [30].

It is not necessary to use preformed and isolated palladium carbene complexes: they can also be generated in situ [31]. When Pd(OAc)$_2$ is dissolved in the presence of a base in [BMIm]Br, styrene was efficiently coupled with iodobenzene affording stilbene (complete conversion and 99% selectivity). The authors were able to isolate the thus-formed palladium carbene complex, derived from the ionic liquid solvent. The counter ion of the ionic liquid also plays an important role. Mizoroki–Heck reactions proceeded much faster in [BMIm]Br than in [BMIm]BF$_4$. In the latter, precipitation of palladium black was encountered. The fact is explained by the necessity of the presence of a halide ion for the stabilization of the carbene–palladium complex [32].

Phospha-palladacycles are also powerful palladium catalysts for the Mizoroki–Heck reaction. Complexes like **3** (Figure 15.1) were successfully tested in Bu$_4$NBr as solvent for the coupling of styrene (**4**) and chloroarenes (e.g. **5**, X = Cl) in the presence of sodium acetate and [AsPh$_4$]Cl, leading to 96% yield ((*E*)-**6** : (*Z*)-**6** : **7** = 95 : 1 : 4) (Scheme 15.2)

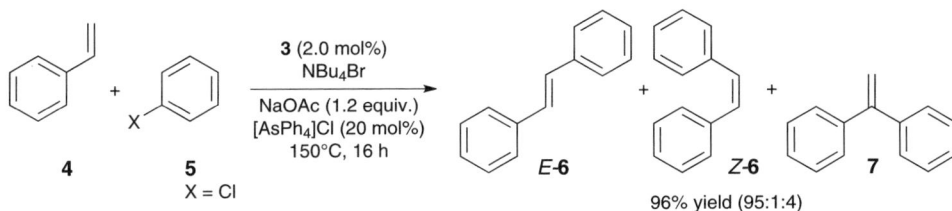

Scheme 15.2 *Palladacycle-catalysed Mizoroki–Heck reaction forming mainly the (E)-configurated product.*

[33]. When recycling the catalyst, the yield of stilbene remained above 95% after at least 12 runs, whereas yields obtained with $PdCl_2$ as catalyst started to decrease after four runs. The main advantage of this reaction protocol is the high selectivity in the predominant formation of the (E)-configured double bond.

A new trend in Mizoroki–Heck reactions is the application of supported palladium catalysts with the aim of easy catalyst recycling and higher selectivity. The application of such catalysts results in a higher regioselectivity, which might be rationalized by the increased steric hindrance of the catalyst at the surface. Immobilization techniques use catalyst on a carrier, catalyst and ionic liquid on a carrier, ionic liquid and ligand on a carrier with and without catalyst, fixation of the base and the starting material.

Palladium acetate immobilized on reversed-phase amorphous silica gel with the aid of an ionic liquid, [BMIm] PF_6, was highly efficient in Mizoroki–Heck type reactions in water under ligand-free conditions. In several runs (up to the sixth reuse) a 95% average yield was obtained with a turnover number (TON, the number of converted substrate molecules per catalytically active centre, $mol \times mol^{-1}$) of 1 600 000 and a turnover frequency (TOF, TON per time unit, $mol \times mol^{-1} \times h^{-1}$) of 71 000 h^{-1} [34].

$Pd(OAc)_2$ in poly(ethylene glycol) supported 3-methylimidazolium chloride, [PEG-MIm]Cl, constitutes a highly efficient and recyclable (five times) catalytic system for the reaction of aryl bromides and activated aryl chlorides in the absence of ligands [35]. For example, *p*-bromoanisole reacted with ethyl acrylate in 10 h reaction time to 96% *p*-methoxy ethylcinnamate (*s/c* = 200).

Palladium(0) nanoparticles of approximately 2 nm diameter, immobilized in [BMIm]PF_6, were also shown to be efficient catalyst precursors for coupling of aryl halides with butyl acrylate [36]. Analysis of the ionic liquid catalytic solution and the organic phase after the catalytic reaction shows the formation of larger nanoparticles and significant (up to 34%) metal leaching from the ionic phase to the organic phase, which strongly suggests that the palladium(0) nanoparticles serve as a reservoir of homogeneous catalytically active species.

The nanoparticles formed *in situ* were investigated in several ionic liquids; for example, *N*-butyronitrile pyridinium triflate and tetrabutylammonium acetate, with acceptable yield (up to 90%) and TOF 20 000–25 000 h^{-1} [37–40].

In situ generation of Pd–NHC complexes in an imidazolium-type ionic liquid matrix and grafting of the catalyst on the surface of silica was reported by Karimi and Enders [41]. Palladium(II) anchored on an ionic liquid matrix, derived from *N*-3-(3-trimethoxysilylpropyl)-3-methyl imidazolium chloride, displays high thermal stability up to 280 °C. Catalyst **8**

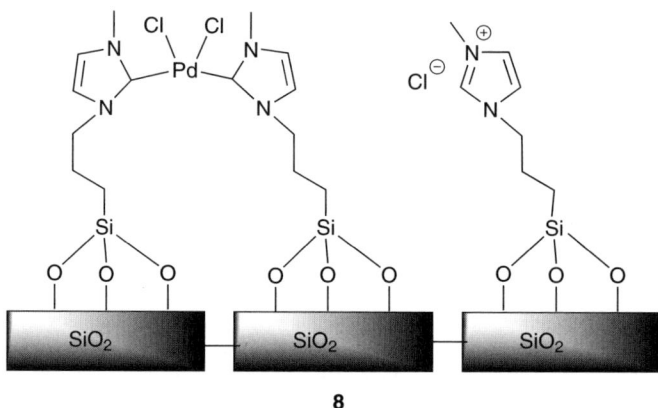

8

Figure 15.2 *Supported ionic liquid and NHC–palladium complex 8.*

(Figure 15.2) showed activity in the reaction of olefins and aryl halides in the presence of triethylamine and/or potassium carbonate in dimethylformamide (DMF) at 100 °C. The selective coupling of bromobenzene and methyl acrylate to (*E*)-methyl cinnamate was successful with a TOF of 420 h^{-1}. The supported NHC–Pd/IL catalyst **8** could be recycled four times with an average yield of 91% and without loss of activity.

Yokoyama and coworkers described the use of palladium catalysts (Pd(NH$_3$)$_4$)Cl$_2$ supported on SiO$_2$ for Mizoroki–Heck reactions in [BMIm]PF$_6$ [43]. In the selective coupling of iodobenzene and ethyl acrylate to (*E*)-ethylcinnamate, the activities of palladium(0) and palladium(II) on SiO$_2$ and on charcoal were compared. Palladium(II) on SiO$_2$ was the preferred catalyst for this coupling reaction. Reuse of the ionic liquid resulted in similar conversion, indicating that the reaction was catalysed by a palladium species dissolved in the ionic liquid.

Palladium acetate on silica remains in a dissolved state in the ionic liquid in the pores of silica when it is prepared by evaporation of tetrahydrofuran from a Pd(OAc)$_2$/SiO$_2$/[BMIm]PF$_6$ mixture [44]. Atomic force microscopy revealed a smooth surface of the silica. Mizoroki–Heck reactions of iodo- and bromobenzene carried out with various acrylates in hydrocarbon solvents resulted in an easy separation of catalyst and product. Leaching of the catalyst into the hydrocarbon solvent occurred to less than 0.24%. The catalyst was successfully recycled and reused six times with conserved activity.

The heterogeneous catalyst Pd/C (10 wt%, 3.0 mol% relative to the olefin) was used in [BMIm]PF$_6$ to couple various iodobenzenes with ethyl acrylate in up to 95% yield [45]. The corresponding bromobenzenes could be coupled with yields of up to 85%. After the reaction, the product was extracted with *n*-hexane. Dissolution of palladium once reported by Earle and coworkers [46] was not observed according to analysis with induced coupled plasma emission spectroscopy.

Palladium nanoparticles supported on chitosan (poly-[β(1–4)-2-amino-2-deoxy-D-glucan]) are very efficient heterogeneous catalysts in the reaction of aryl bromides or activated aryl chlorides in tetrabutylammonium bromide as solvent and tetrabutylammonium acetate as base (reaction time 15 min, in quantitative yield) [47]. The catalyst was

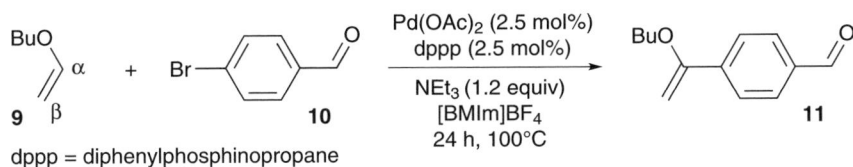

Scheme 15.3 *α-Arylation of electron-rich olefins.*

generated by sacrificial anode electrolysis, performed in an acetonitrile solution of tetra-butylammonium bromide in the presence of chitosan. Palladium in both oxidation states was present on the surface of the nanocomposite, palladium(0) and palladium(II), with the latter prevailing. If iodo derivatives are used, then the catalyst can be reused for several catalytic cycles. The ionic liquid decomposes by Hofmann elimination due to the high temperature needed in the case of bromobenzenes, and the catalyst could not be recycled.

The coupling of electron-rich olefins in conventional solvents often requires additives such as silver triflate or thallium acetate to obtain high α/β selectivity. 1-Butyl-3-methylimidazolium tetrafluoroborate ([BMIm]BF$_4$) is an excellent catalytic system for highly regioselective α-arylation of several electron-rich olefins. Butyl vinyl ether **9** reacted with *p*-bromobenzaldehyde **10** selectively (α/β ratio >99:1) to **11** (Scheme 15.3 [42]). Comparative experiments in conventional solvents (e.g. DMF) resulted in product mixtures with α/β ratios of 46:54 to 69:31. It has been suggested that the ionic pathway is responsible for the unique regiocontrol in the ionic liquid [48–50].

Another approach of regioselective Mizoroki–Heck reactions in ionic liquids is the synthesis of carbonyl compounds [51, 52]. Allylic alcohols (e.g. **12**) react with iodobenzene (**5**, X = I) via an enolic intermediate (e.g. **13**) to the arylated carbonyl product **14** (Scheme 15.4 [52]), which was separated from the reaction mixture by extraction with diethyl ether. However, when choosing the appropriate reaction conditions (solvent and base), the corresponding allylic alcohols were obtained. If Pd(OAc)$_2$ was applied in tetra-butylammonium acetate (as solvent and base) in the coupling of 1-octen-3-ol with bromo- or iodobenzene, then selective formation (96:4) of allylic alcohols was achieved in 94% yield in 30 min reaction time at 70 °C [53].

A key step of the total synthesis of lily-of-the-valley fragrance β-Lilial (3-(4-*tert*-butylphenyl)-2-methylpropanal, **17**) is a Mizoroki–Heck reaction of arene iodide **15** with allyl alcohol **16** leading to the carbonyl product (Scheme 15.5) [51].

Dienol ethers (e.g. **18**) react with several aryl iodides (e.g. **5**, X = I) selectively to (*E*)-1-aryl-4-ethoxybutadienes (e.g. **19**) catalysed by Pd(OAc)$_2$ in presence of sodium acetate in tetrabutylammonium bromide (Scheme 15.6) [54]. Yields up to 80% could be achieved, whereas the (*E*)-configured double bond remains unchanged.

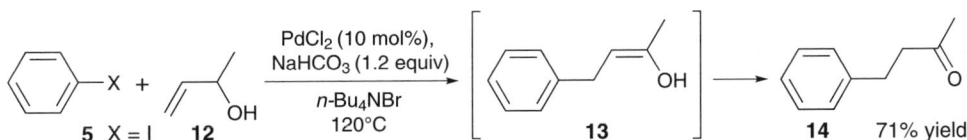

Scheme 15.4 *Mizoroki–Heck reactions of allyl alcohols in ionic liquids.*

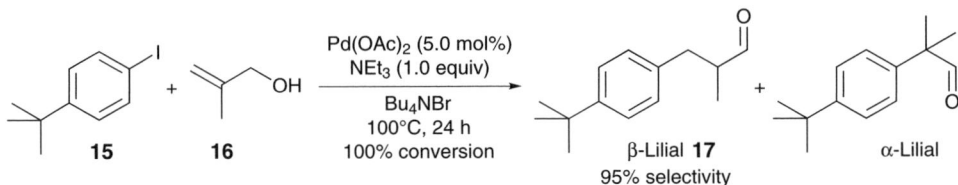

Scheme 15.5   Synthesis of β- and α-Lilial.

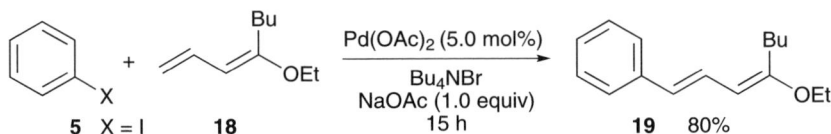

Scheme 15.6   *Mizoroki–Heck reactions of dienol ethers in ionic liquids.*

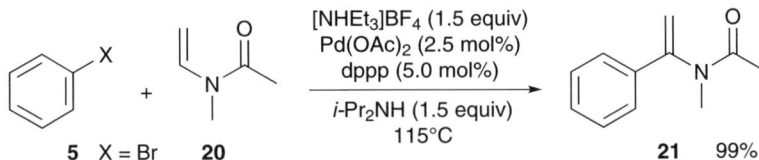

Scheme 15.7   *Mizoroki–Heck reaction of amide **20** in the presence of ammonium salts.*

Aryl chlorides and bromides (e.g. **5**, X = Cl, Br) were regioselectively coupled in the presence of ammonium salts with electron-rich olefins, vinylethers, enamines and amides (e.g. **20** to **21**) (Scheme 15.7) [55, 56]. This behaviour can be explained by hydrogen bond donation of the ammonium salts, which results in a remarkable accelerating effect.

15.2.2   Fluorous Phase

Perfluorinated or very highly fluorinated solvents are commonly called fluorous solvents. They are usually immiscible with many common organic solvents at ambient temperature, although they can become miscible at elevated temperatures. A fluorous solvent was first used in the separation of isotopes in 1940 [57]. For several decades the interest in fluorous systems was limited. Only in 1993 was the first systematic application of perfluorotrialkylamines and perfluoroalkanes as inert solvents published [58]. In the following years the interest grew enormously [59]. With fluorous solvents, it became feasible to perform chemical reactions without the need for volatile or toxic organic solvents. The process becomes less complex by reducing the separation steps. The most remarkable properties of the solvents are their chemical inertness, nonflammability and thermal stability [60]. However, even though fluorous solvents are practically nontoxic, a serious problem is the extremely long lifetime of fluorocarbons in the environment, which is a matter for genuine concern. Perfluoroalkanes, perfluortrialkylamines and perfluorodialkyl ethers are the most common fluorous solvents and they show practically no toxicity [61]. They can be applied together with a fluorous reaction component (e.g. catalyst) with, for instance, a fluorous tag attached to increase its solubility in the fluorous solvent, which has a low miscibility with common organic solvents. Some of the fluorous solvents can become a single phase with an organic solvent at elevated temperatures. Thus, a biphasic system combines the advantages of a one-phase reaction medium with biphasic product separation. The reaction is simply run at elevated temperatures and cooled down for the separation step [62].

There are only very few examples for Mizoroki–Heck reactions in fluorous systems. The catalyst systems, the fluorous or nonfluorous solvent and the additional base are listed in Table 15.1. There are detailed reviews on the Mizoroki–Heck reaction with nonconventional methods that also include fluorous media [12, 63].

Most of the reported work on Mizoroki–Heck reactions in fluorous media applies palladium salts and free ligands. There are only five reports on the usage of preformed fluorous palladium complexes (Table 15.1, entries 1–4) [64–68]. They all used a biphasic system with an organic solvent and a fluorous catalyst that dissolved only at elevated temperatures. Gladysz and coworkers [64, 65] recovered the catalyst **22** in the only patent on Mizoroki–Heck reactions in fluorous media by simple filtration (Table 15.1, entries 1, 2). The first two runs with iodobenzene and methyl acrylate at 100 °C in DMF were almost quantitative after 2 h reaction time (TON: 5251), but after the third run at 100 °C the activity decreased (TON: 2500–2900) and the authors changed the reaction time in the fourth run to 10 h to receive again quantitative conversion and yield (TON: 5251).

The same catalyst was compared with another palladacycle (**23**) at 140 °C and longer reaction times (14–48 h) (Table 15.1, entry 2) [66]. The catalyst was very active in coupling iodobenzene with styrene or methyl acrylate (TON: $(1.3–1.5) \times 10^6$), whereas lower conversions and yields were observed due to catalyst deactivation for the less reactive bromoacetophenone for the coupling with methyl acrylate (TON: $(2.7–3.0) \times 10^5$). The catalytic system could be recycled after the addition of $C_8F_{17}Br$ to give a biphasic mixture. The data obtained by transmission electron microscopy indicates that colloidal palladium nanoparticles were formed as active species for the Mizoroki–Heck reaction.

Another fluorous palladium complex that was applied in a Mizoroki–Heck reaction is the SCS pincer palladium complex **24** (Table 15.1, entry 3) [67]. It was applied under thermal and microwave heating. No fluorous solvent was used and the insoluble catalyst dissolved at the reaction temperature of 140 °C. The catalyst was recovered after 30 to 45 min by solid-phase extraction with a fluorous silica gel. Depending on whether activated or nonactivated substrates were coupled, the yields ranged between 76 and 94%.

The palladium–NHC complex with a fluorous tag on the carbene ligand **25** was used in the Mizoroki–Heck reaction by Ryu and coworkers [68]. Iodobenzene was coupled with acrylic acid at 120 °C for 2 h (Table 15.1, entry 4). The insolubility of the product cinnamic acid in F-626 ($1H,1H,2H,2H$-perfluorooctyl 1,3-dimethylbutyl ether) allowed facile separation by simple filtration. The yield of 93% is comparable to the results obtained with $Pd(OAc)_2$ and the imidazolium salt with a fluorous tag (**26**; Table 15.1, entry 7) together with or without PPh_3. The recovered F-626 phase, containing the palladium catalyst, was reused for five more runs without any detectable loss in catalytic activity. Further iodoaryls were tested with various α,β-unsaturated acids and esters in the presence of *in situ* prepared palladium–fluorous-NHC complex in yields above 90%. In some cases, $P[p\text{-}(C_6H_4)\text{–}CH_2CH_2C_6F_{13}]_3$ (**27**) was used instead of PPh_3 without significant loss of activity (Table 15.1, entry 7).

Table 15.1 Catalysts of Mizoroki–Heck reactions in fluorous solvents.

Entry	Catalyst	Catalyst loading	Base	Solvent[a]	X	R^1	R^2	Ref.
1		$(0.68–1.83) \times 10^{-6}$ mol% 0.02 mol% cat. recycling	NEt$_3$	DMF	Br, I	H, OMe	CO$_2$Me, Ph	[64, 65]
2		$(0.66–1.83) \times 10^{-4}$ mol% 0.02 mol% cat. recycling	NEt$_3$	DMF	Br, I	H, OMe	CO$_2$Me, Ph	[64, 66]
3		2.0–3.0 mol%	NBu$_3$	Dimethylacetamide	Br, I, OTf	H, OMe, CN, COMe, NO$_2$	CO$_2$Me, Ph	[67]

#	Catalyst/ligand	Loading	Amine	Solvent	X	R (aryl)	Product	Ref.
4	Ph$_3$P—Pd—Cl (with imidazolylidene, C$_{10}$F$_{21}$) **25**	2.0 mol%	NPr$_3$	F-626	I	H	CO$_2$H	[68]
5	Perfluorinated polyether-derivatized poly(propylene imine) dendrimers with palladium(0)-nanoparticels	3.0–5.0 mol%	NEt$_3$	Heptane + benzene + FC-75	Br, I	H, NO$_2$	CO$_2$Bu	[69]
6	Pd(OAc)$_2$ and PPh$_3$	2.0 mol%	NPr$_3$	Rf$_6$-DMF **37**	—	H	CO$_2$Bu	[70]
7	Pd(OAc)$_2$, ligands: (imidazolium C$_{10}$F$_{21}$ salt) precursor **26** and PPh$_3$ or P((CH$_2$)$_2$-C$_6$F$_{13}$)$_3$ **27**	2.0 mol%	NPr$_3$	F-626 or Rf$_6$-DMF **37**	—	H, Me, OMe, COMe, F, NO$_2$, Ph	CO$_2$Bu, H	[68, 70]
8	Pd$_2$(dba)$_3$ or Pd(OAc)$_2$, ligands: P(C$_6$H$_4$-R$_f$)$_3$; R$_f$ = OCH$_2$(C$_7$F$_{15}$) **28**, C$_6$F$_{13}$ **29**, O(CH$_2$)$_2$OCH$_2$CF$_2$[CF(CF$_3$)CF$_2$]$_m$(OCF$_2$)$_n$OCF$_3$ m = 3.38; n = 0.11 **30**	1.0 mol%	NEt$_3$	CH$_3$CN + D-100 or solely CH$_3$CN	—	H, NO$_2$, OMe	CO$_2$Me	[71]

(Cont.)

Table 15.1 (Continued)

Entry	Catalyst	Catalyst loading	Base	Solvent[a]	X	R[1]	R[2]	Ref.
9	Pd(OAc)₂, ligand: **29**	10 mol%	Tl₂CO₃	Toluene + hexane + perfluoromethyl-cyclohexane	I, Br	Intramolec.	Intramolec.	[72]
10	Pd(OAc)₂, ligand: R_f = p-(CH₂)₂C₆F₁₃ **31**, m-(CH₂)₂C₆F₁₃ **32**, p-(CH₂)₂C₄F₉ **33**	3.0 mol%	NEt₃	DMSO	OTf	Cyclic vinyl triflates	OBu, N(Me)-COMe	[73]
11	Pd(OAc)₂, ligand: (R)-F₁₃BINAP **34**	3.0 mol%	i-Pr₂NEt	Benzene + FC-72 or solely BFT	OTf	Cl		[74, 75]

12	Pd(OAc)$_2$ ligand: **35**	3.0 mol%	i-Pr$_2$NEt	Toluene + FC-72 or solely C$_6$H$_5$CF$_3$	OTf	H, Cl, OMe, naphthyl	[76]
13	PdCl$_2$ ligand: **36**	4.3–63.5 mol%	NEt$_3$	CH$_3$CN + C$_8$F$_{17}$Br or Galden HT 135	I	H	CO$_2$Et, Ph-Et-acrylate Ph-CN-acrylate	[77]

---

[a] Abbreviations: FC-72: a mixture of perfluorohexanes; F-626: 1H,1H,2H,2H-perfluorooctyl 1,3-dimethylbutyl ether; FC-75: perfluoro-2-butyl-tetrahydrofuran; D-100: mainly n-perfluorooctane; Rf$_6$-DMF: N-(1H,1H,2H,2H,3H,3H-perfluorononanyl)-N-methyl formamide; BFT: benzotrifluoride; Galden HT 135: perfluoropolyether; DMSO: dimethylsulfoxide.

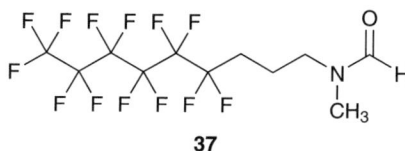

Figure 15.3 *Polyfluorinated-type DMF **37**.*

One report about Mizoroki–Heck coupling with dendritic nanoreactors was published in 2001 [69]. Perfluorinated polyether-derived poly(propylene imine) dendrimers containing palladium(0) nanoparticles were prepared by introducing Pd^{2+} into the interiors of amine-terminated poly(propylene imine) dendrimers, which were previously end-group derivatized with perfluorinated polyether chains (Table 15.1, entry 5). The resulting dendrimer–Pd^{2+} coordination complex was reduced by $NaBH_4$ to yield dendrimer-encapsulated palladium(0) nanoparticles. They could be dissolved in perfluorinated solvents without any agglomeration. Two different dendritic systems were applied at 90 °C for 24 h and the best coupling was observed with iodobenzene at 70% yield. However, bromobenzene gave only 26% yield with the more active dendrimer, and after catalyst recovery (liquid–liquid separation) the activity of the catalysts decreased significantly. It is interesting to note that the usually required addition of a base is not essential for this system due to the interior tertiary amines of the dendrimers.

Polyfluorinated-type DMF (Rf_6-DMF, **37**; Figure 15.3) was tested as a solvent for the Mizoroki–Heck coupling of iodobenzene and butyl acrylate at 120 °C for 2 h [70]. The authors describe the catalysis with the standard system of $Pd(OAc)_2$/PPh_3 as well as with $Pd(OAc)_2$ and a fluorinated imidazolium salt **26**. The catalyst was recovered by addition of cyclohexane and F-72 (perfluorohexanes) and liquid–liquid phase separation. The fluorous catalyst was used in two more catalytic runs without reduction of activity (93, 90 and 98% yields). The standard catalyst system gave a slightly lower yield (87%).

The first time that a Mizoroki–Heck reaction was conducted in a fluorous system was in 1999 when Sinou and coworkers [71] applied $Pd_2(dba)_3$ (dba = dibenzylideneacetone) or $Pd(OAc)_2$ and a perfluorinated phosphine (e.g. **28–30**) in a perfluorinated–nonfluorous solvent mixture in the Mizoroki–Heck reaction of aryl iodides (4 h, 80 °C). In this system, the Mizoroki–Heck product was soluble in acetonitrile and the catalyst was dissolved in the fluorous phase. However, some ligand was lost due to its partial solubility in acetonitrile and some of the palladium was reduced to palladium black. Thus, after each run with the recovered catalyst, lower conversions were obtained.

Ligand **29** (Table 15.1, entry 9) was also used in intramolecular Mizoroki–Heck reactions [72]. In combination with $Pd(OAc)_2$, it catalysed a cascade ring-closing metathesis (RCM)/Mizoroki–Heck reaction. The RCM step was conducted at room temperature on (bromo or iodo) *N*-alkenyl-*N*-allyl-2-halo-benzenesulfonamides and the Mizoroki–Heck reaction was run at 110 °C for 16 h in a perfluorous solvent system. The overall yield with fluorous conditions (0–67%) was significantly lower than a reference system with polymer-bound palladium catalyst (58–80%).

The three fluorous-tagged bidentate phosphine ligands **31–33** (Table 15.1, entry 10) were tested in the Mizoroki–Heck vinylation and standard Mizoroki–Heck reaction [73]. In the

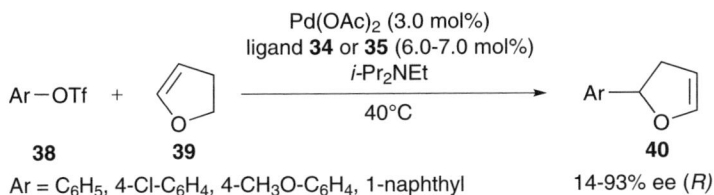

Scheme 15.8 *Asymmetric Mizoroki–Heck reaction of 2,3-dihydrofuran (39) with various aryl triflates 38.*

vinylation, the reaction mixture was conventionally heated at 60 °C for 18 h or microwave heated at 90 °C for 15–20 min in DMSO. There was no difference in yield for both kinds of heating after the respective reaction time and there was only minor influence on yield using the perfluorinated ligands or the nonfluorous ligand.

The ligands, the palladium-complexed ligands, and oxidized ligands could be easily removed from the reaction mixture by solid fluorous phase separation using a methanol/water (9 : 1) eluting system. However, all attempts to reuse the catalyst failed, probably due to degradation of the catalyst. In the coupling between butyl vinyl ether and 1-naphthyl triflate, ligand **33** behaved similarly to its nonfluorous analogue, whereas ligands **31** and **32**, with larger fluorous tails, reduced the reaction rate and selectivity.

Two new fluorous chiral 2,2′-bis(diphenylphosphino)-1,1′-binaphthyl (BINAP; Table 15.1, entries 11 and 12) ligands were synthesized and their efficiency was demonstrated in an asymmetric Mizoroki–Heck reaction (Scheme 15.8) [74–76]. Aryl triflates **38** were coupled with 2,3-dihydrofuran (**39**) to (*R*)-**40** in up to 93% ee. Either pure fluorous solvents or mixtures of fluorous and nonfluorous solvents were applied. Ligand **34** was compared with conventional BINAP, with the fluorous ligand giving lower reaction rates and a similar level of enantioselectivity in 24–77 h reaction time [74, 75]. The catalyst could not be recovered by reverse-phase silica gel chromatography due to its oxidation.

Ligand **35** gave better regioselectivities but worse enantioselectivities than ligand **34**, with reaction times between 72 and 120 h. The reaction mixture was separated after the reaction by liquid–liquid separation. The authors do not report on the recovery of the catalyst system.

Palladium nanoparticles prepared from ligand **36** and $PdCl_2$, using methanol as reducing agent, are soluble in perfluorinated solvents [77, 78]. The Mizoroki–Heck reaction was conducted in five consecutive runs at 80–140 °C for 6–7 h in one case and in all other for 48–168 h. The isolated yields were not proportional to the catalysis runs, but were in general above 50%. However, the 'catalyst' loading was in three out of four examples between 40 and 64 mol% palladium!

15.2.3 Supercritical Fluids

Large-scale applications of SCFs have been well known for a long time. SCFs have been used as solvents for extraction and in chromatography, in the food, pharmaceutical and textile industries, and in some other areas [79–81]. Attractive features for their use as reaction media in preparative and exploratory chemical transformations are the potential for tuning solubilities of reactants (including gases, salts and organometallic complexes) by variation of temperature and pressure, reducing or eliminating interphase transport limitations, stabilizing reactive intermediates and thus influencing reaction rates and selectivities. Furthermore, the use of SCFs delivers new opportunities for recycling of catalysts and reagents, as well as isolation and purification of products (downstream processing) [3, 80, 82].

In particular, with regard to large-scale (industrial) applications [79, 83–85], the aspects of aiming at clean ('green chemistry') processes have to be considered carefully. While, for example, the avoidance of flammable and toxic organic solvents and improvements of the overall transformations represent environmental and safety advantages, the requirement of (expensive and not in every laboratory easily available) specialized pressure equipment and the (often higher) energy consumption necessary for pressure and temperature changes certainly are drawbacks. Conversely, the possibilities of continuous (catalytic) processing [84, 86], performing heterogeneous [87] and multiphase reactions [88], and particle design for manufacturing new materials [89] must also be taken into account for the development of environmentally benign processes.

Of the SCFs available for laboratory use, scCO$_2$ and sub- or super-critical water have been applied in experimental work using Mizoroki–Heck-type reactions [9, 12, 87, 88, 90–92]. That scCO$_2$ is the medium which has attracted most interest so far is mainly due to its unique physical properties, its lack of reactivity in chemical reactions, its low toxicity, and its non-flammability. The critical properties of scCO$_2$ (critical temperature $T_C = 31.0\,°C$, critical pressure $p_C = 73.8$ bar) provide relatively mild conditions for transformations. Supercritical water (scH$_2$O; critical temperature $T_C = 374.2\,°C$, critical pressure $p_C = 220.5$ bar) [81, 82, 93] requires much harsher conditions. Therefore, one major application of scH$_2$O is oxidation as an alternative method for destroying hazardous waste [79]. At the critical point, scH$_2$O has a dielectric constant similar to diethyl ether, and it was also used as a medium for synthetic organic chemistry.

With respect to synthetic (in particular, catalysed) organic reactions [3, 94], it is noteworthy that the first reports on the useful application of scCO$_2$ as a solvent and reactant were described by Noyori and coworkers [95, 96] in 1994 for the efficient ruthenium-catalysed transformations of carbon dioxide to formic acid, dimethyl formamide and methyl formate. During the last decade, SCFs have been applied for a variety of other areas of organic synthesis, including the formation of carbon–carbon bonds [80, 81, 97–99].

The first Mizoroki–Heck reactions under supercritical conditions were described by Parsons and coworkers [100, 101] in 1995. The coupling reaction of benzene derivatives **5** (best results were obtained with iodide **5**, X = I) with styrene (**4**) in overheated (260 °C) and scH$_2$O (400 °C) yielded up to 30% of coupling product **6** (*E/Z* mixture), but also led to hydrogenation and other side reactions (→**7**, **41–46**, Scheme 15.9). By studying the effect of the kind of base and reaction conditions, Ikushima and coworkers [102–104] could improve the yield of stilbenes (*E:Z* = 81 : 19) by performing the coupling without any catalyst and in the presence of the base KOAc to up to 56%. Gron *et al.* [105] found that water density is an important reaction variable, leading to changes in yield and isomeric distribution in the coupling of iodobenzene (**5**, X = I) with cyclohexene.

The group of Tumas published Mizoroki–Heck coupling reactions of iodobenzene (**5**, X = I) with styrene (**4**) (or acrylic ester **47**) in scCO$_2$ with a selectivity of 94–99% by using fluorinated ('CO$_2$-philic') ligands like **48** (Scheme 15.10). These results are comparable to those obtained for transformations in toluene as a solvent [106]. Similar data were

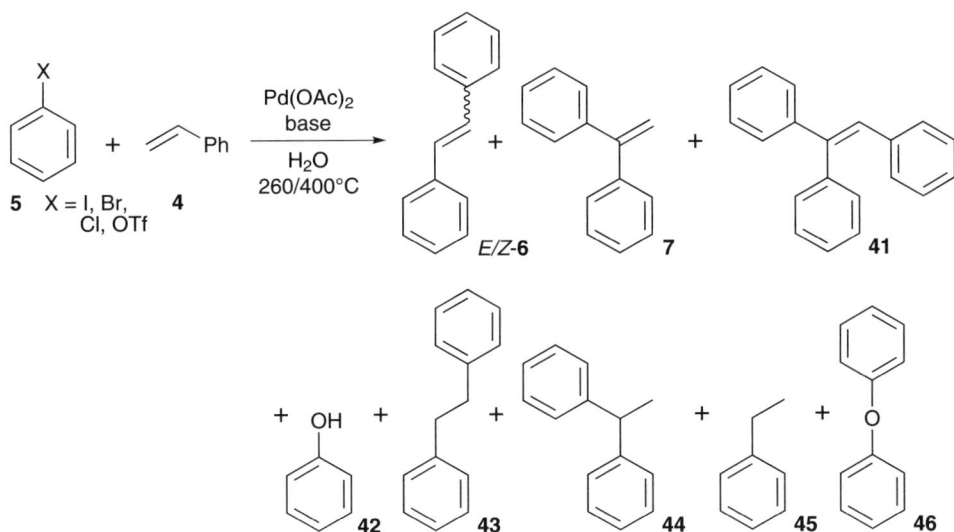

Scheme 15.9 *Mizoroki–Heck coupling reactions in sub- and super-critical water.*

reported by Carroll and Holmes with ligands **49** and **50** [107]. The extent and location of fluorination in phosphine ligands on the activity of catalysts has a significant effect on the activity of catalysts. The results of coupling experiments in scCO$_2$ are comparable to those obtained in conventional polar solvents like ethanol or *N*-methylpyrrolidone (NMP) [108]. Also, commercially available palladium precursors containing fluorinated ligands (trifluoroacetate, F$_6$-acac) showed superior performance over nonfluorinated catalysts [109].

In addition to those improvements by using fluorinated ligands, nonfluorinated systems (palladium acetate with tri-*tert*-butylphosphane) also yield coupling products in up to 98% yield starting from **5** (X = I) and (free and polymer-tethered) acrylates [110]. Further optimization of reagent systems could even suppress unwanted isomerization of double bonds in intramolecular Mizoroki–Heck reactions (Scheme 15.11). Reacting substrates **51a,b** with palladium trifluoroacetate in combination with P(2-furyl)$_3$ as a ligand and Hünig's base delivered high proportions of the exocyclic products **52a,b** (**52** : **53** = 88 : 12 and 83 : 17 at full conversion respectively) in scCO$_2$, whereas endocyclic products **53a,b** are formed predominantly in conventional solvents (toluene, acetonitrile) [111].

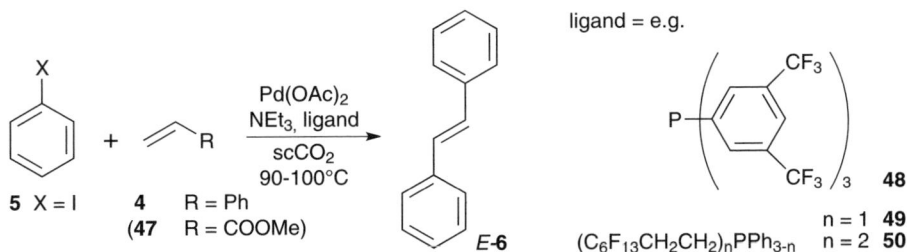

Scheme 15.10 *Mizoroki–Heck coupling reactions in scCO$_2$ with fluorinated ligands.*

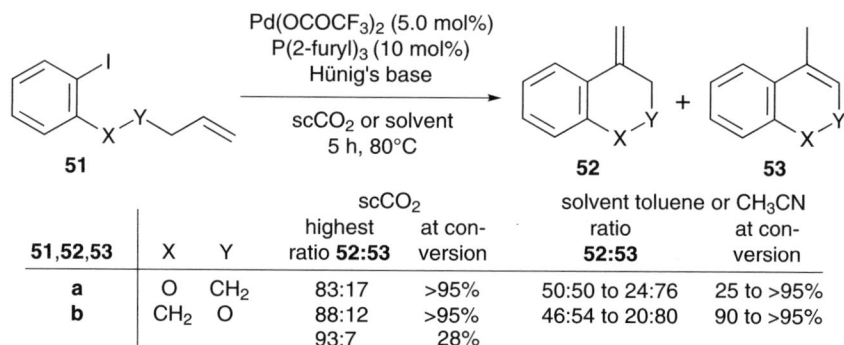

51,52,53	X	Y	scCO$_2$ highest ratio **52:53**	at conversion	solvent toluene or CH$_3$CN ratio **52:53**	at conversion
a	O	CH$_2$	83:17	>95%	50:50 to 24:76	25 to >95%
b	CH$_2$	O	88:12	>95%	46:54 to 20:80	90 to >95%
			93:7	28%		

Scheme 15.11 *Suppression of double-bond isomerization in intramolecular Mizoroki–Heck coupling reactions.*

Another approach to improve conversion and selectivity, as well as separation/isolation of product and catalyst recycling, is to apply multiphase catalysis (the combination of solvents different from SCFs with multiphase catalysis is discussed in Section 15.2.4). The Arai group used scCO$_2$ and the catalyst system Pd(OAc)$_2$/triphenylphosphane trisulfonate sodium salt (TPPTS) with cosolvents [112]. The reaction rates are significantly enhanced when water or ethylene glycol are added, due to increased solubility of the catalyst under the reaction conditions. In the absence of a cosolvent, the system behaves like that of a conventional biphasic toluene–ethylene glycol [113]. Additional advantages of the scCO$_2$–H$_2$O or scCO$_2$–ethylene glycol mixture are simple phase separation for product recovery and catalyst reuse, as well as enhanced stability of the catalyst; that is, no leaching of palladium occurs. In a screening of various tertiary phosphine and phosphite ligands, PdCl$_2$[P(OC$_6$H$_5$)$_3$]$_2$ emerged as the most suitable catalyst precursor for the arylation of ethylene in an scCO$_2$–liquid biphasic system [114]. By variation of the types of ligand (from triphenylphosphane to partly and highly fluorine-substituted ones), the reaction mixture is homogeneous or biphasic depending on the pressure applied [115]. Efficient recycling of the catalyst was also achieved by extraction of the products with scCO$_2$ from an ionic-liquid-based system [32] (see Section 15.2.1).

The concept of heterogenizing catalysts or one of the reactants was followed by various approaches. Cacchi *et al.* [116] used palladium on carbon successfully in scCO$_2$ for the coupling of aryl iodides or vinyl triflates with olefins, although long reaction times (e.g. 60 h at 80 °C and 100 bar) were necessary to get isolated yields of up to 80%. Polyurea-encapsulated PdCl$_2$ is a robust phosphine-free catalyst in scCO$_2$ and in conventional solvents, which can be separated easily by simple filtration and reused [117, 118]. Supported reagents have also been applied successfully for Mizoroki–Heck couplings. An example is the use of commercially available polystyrene-supported amines and phosphine catalysts in the reaction of acrylates with aryl halides [119]. Also, substrates (e.g. acrylates) bound to resins have been used [120].

Recent developments aiming at highly active and versatile catalytic materials for Mizoroki–Heck and Suzuki reactions in various reaction media, including scCO$_2$, are exemplified by dendrimer-encapsulated metal (palladium) nanoparticles [121–123]. Although those systems are not yet fully developed and optimized, they have already delivered promising results in Mizoroki–Heck-type coupling; the TONs of these systems in scCO$_2$ have not reached the level observed in other solvent systems so far. Carbosilane dendrons were prepared for enhancing the solubility of catalysts. In contrast to the catalytically inactive palladium complexes containing PPh$_3$, palladium complexes with functionalized phosphines showed good activity due to increased solubility in scCO$_2$ [124]. Supercritical conditions were also used for the preparation of new materials applicable as catalysts in Mizoroki–Heck chemistry. Aerogel nanocomposites doped with metallic palladium were prepared from silica wet gels or by sol–gel polymerization, followed by ethanol or CO$_2$ supercritical drying [125].

15.2.4 Multiphase Systems

One of the main issues of the Mizoroki–Heck reaction is the recycling of the catalyst. Most of the Mizoroki–Heck catalysts are either deactivated during the course of reaction or they tend to leach from various supports. Furthermore, the salt formation during the reaction may play an important role in the activity of the catalyst [9]. The salt formed during the reaction accumulates in the reaction media and usually leads to degradation of the catalytic system. The environment varies due to the change of the reaction mixture composition, which affords highly flexible catalytic systems. Many approaches for the design of recyclable catalysts have been tested. In multiphasic systems, for instance biphasic liquid–liquid mixtures, the reagents and the coupling product remain in a different phase than the catalyst. Usually, the catalyst-soluble phase is aqueous or protic. Therefore, the catalyst needs to be hydrophilic. This can be achieved by suitably modifying the ligands or by applying aqueous organic solvents or neat water without organic ligands using transition metal salts [8, 126–131].

15.2.4.1 Aqueous Systems without Additional Organic Solvent

There are a few examples for Mizoroki–Heck reactions in neat water with phase-transfer agents [132–138]. Inorganic carbonates were employed as bases under mild reaction conditions. Typical applied phase-transfer catalysts (PTCs) are HDAPS (*N*-hexadecyl-*N*,*N*-dimethyl-3-ammonio-1-propanesulfonate) and PEG (polyethylene glycol). In most cases, Pd(OAc)$_2$ was used in concentrations between 0.7 and 5.0 mol%. Examples for catalyses in neat water without a PTC are rare. They are generally performed at atmospheric pressure at temperatures between 50 and 130 °C [138–151]. The catalyst loading is seldom below 1.0 mol% and, in terms of reactivity, aryl bromides are at least required for the coupling; no example of an aryl chloride has been reported so far. However, Hou *et al.* [141] were able to apply catalyst loadings as low as 0.0025 mol% and Nájera *et al.* [151] used 0.001 mol% palladium catalyst. They both used palladacycles to obtain Mizoroki–Heck products in yields generally higher than 90%.

A variety of bases has been applied. For instance, the ion-exchange resin Amberlite IRA-400 is not only a base, but also a quaternary ammonium salt which, therefore, acts as a PTC (Scheme 15.12) [142]. The product was easily removed from the reaction mixture by extraction with diethyl ether.

Polymer-supported carbine–palladium complexes were used for the Mizoroki–Heck reaction in water [152]. The product could be separated by extraction with diethyl ether. However, the catalytic activity decreased after reuse of the catalyst. The authors claim that

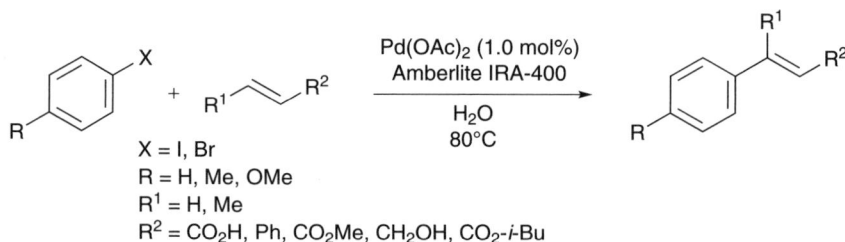

Scheme 15.12   *Mizoroki–Heck reaction in neat water in the presence of basic Amberlite IRA-400.*

this is due to incomplete phase separation rather than catalyst deactivation. In addition, after extraction, a considerable amount of solvent would stay solubilized in the micellar core, which prevents efficient solubilization of the substrate.

Alacid and Nájera [153] reported on supported palladacycles (e.g. **54**, Figure 15.4) as catalysts in the Mizoroki–Heck reaction in neat water at TONs up to 10^5. The reaction was conducted at 120 °C with reaction times between 3 and 30 h, depending on the reactivity of the substrates. The palladium loading was as low as 0.001 mol% and the comparison with DMF as a solvent revealed that activity at this concentration was even higher in water than in DMF. Recycling experiments were performed using the recovered polymer complex in DMF and in water. Although its efficiency was slightly lower than the related dimeric palladacycles, it is easily prepared and reused, at least for eight times, maintaining high activity, especially in water. However, palladium leaching was always detected in the final products.

The coupling of iodo aryls with various olefins was conducted in the presence of palladium chloride (2.0 mol%) at ambient temperature under ultrasonic irradiation in water with yields between 43 and 90% [154]. *In situ* formation of palladium nanoparticles was confirmed by transmission electron microscope and X-ray diffraction analyses, and the palladium nanoparticles could be reused for multiple reactions without significant loss of activity. Biologically relevant stilbene derivates, such as resveratrol, piceatannol and pinosilvine, were efficiently prepared with high regioselectivity and complete stereocontrol (TONs up to 10^4, 0.01 to 0.50 mol% catalyst) [148]. The activity of the oxime-derived palladacycle catalysts was compared in neat water with a DMF–water mixture, with similar

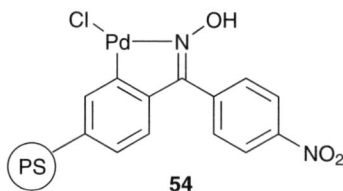

Figure 15.4   *Kaiser oxime-derived palladacycle* **54** *applied in Mizoroki–Heck catalysis in neat water.*

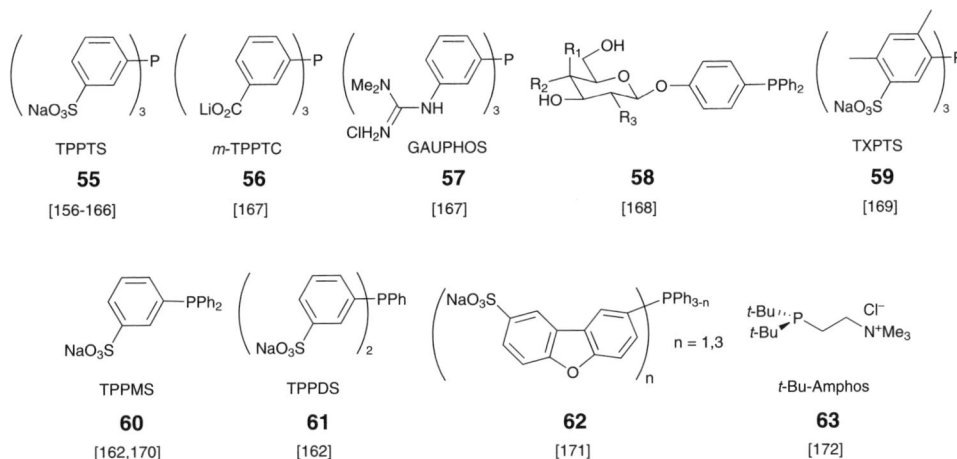

Figure 15.5   *Water-soluble phosphine ligands tested in Mizoroki–Heck reactions under aqueous conditions.*

results for both solvent systems. Superheated and supercritical water was also investigated for the Mizoroki–Heck reaction [100–105, 155] with palladium concentrations between 6 and 10 mol% and is discussed in Section 15.2.3).

15.2.4.2   Aqueous Systems with Additional Organic Solvent

Aqueous organic solvents are the most common system for Mizoroki–Heck reactions in multiphases, often using hydrophilic phosphine ligands. The solvents are usually highly polar; for example, DMF or acetonitrile. The catalysts are easily separated by phase separation. A frequently used ligand for water-soluble complexes is TPPTS. The simple catalyst separation is an important step towards the development of environmentally friendly industrial processes. There is a wide variety of literature on Mizoroki–Heck reactions with TPPTS or related water-soluble phosphine salts **55–63** (Figure 15.5) in aqueous organic solvents [156–172]. Ligand **56** was tested in various Mizoroki–Heck couplings and compared with a few other water-soluble ligands. Pd(OAc)$_2$ was applied in concentrations between 0.1 and 1.0 mol%. The activity is similar to that of **55** and the yield about 20% higher than with the *para*-analogue ligand or with the cationic GUAPHOS (**57** [167]). Beller *et al.* [168] tested triphenylphosphane attached to the residues of D-glucose, D-galactose and D-*N*-acetylglucosamine as ligands. The activity (Pd(OAc)$_2$, 0.10–1.0 mol%) increased compared with TPPTS under similar reaction conditions by about an order of magnitude. Ligand **58** was also compared with **55** and proved to be more active under relatively mild reaction conditions (Pd(OAc)$_2$, 2.5 mol%).

Triphenylphosphane monosulfonate (sodium salt) and triphenylphosphine disulfonate (disodium salt) were compared in one study with TPPTS and the results showed that each ligand together with 1.0 mol% palladium catalyst is similarly effective for the Mizoroki–Heck reaction of iodobenzene with an olefin [162]. Leaching levels are also comparable; the percentaged leaching represents the amount of palladium contamination with respect to the amount of palladium catalyst employed. **62** is inferior to the TPPTS ligand under similar reaction conditions (Pd(OAc)$_2$, 5.0 mol%). An exceptionally high water content is necessary due to the high insolubility of the ligand. Mizoroki–Heck catalysis with **63** resulted in three to four times higher yields than with **55** under similar reaction conditions (Pd(OAc)$_2$, 2.5 mol%). Therefore, the ligand may be applied under significantly milder conditions. Its activity compared with **59** is about the same.

The most common protocol for the Mizoroki–Heck reaction in aqueous systems with additional organic solvent is the application of a palladium salt and a nonwater-soluble phosphine ligand. However, there are only a few reports with an aqueous system in the absence of coordinating organic ligands [156, 173–176]. An NMP–water mixture in the presence of 0.40 mol% Pd(OAc)$_2$ resulted at 80 °C in the Mizoroki–Heck coupling product [174], whereas a water–DMF mixture was applied in the presence of Pd(OAc)$_2$ (3.0 and 10 mol%) or Pd(dba)$_2$ (10 mol%) at 70–100 °C. Only iodo aryls were successfully coupled with olefins. Another approach of the Mizoroki–Heck reaction was in aqueous ethanol with 10 mol% Pd(OAc)$_2$ at only 50 °C [175]. Diazonium salts were coupled in aqueous methanol in the presence of 2.0 mol% Pd(OAc)$_2$ [176]. Other groups applied 1,3-bis(diphenylphosphino)propane (dppp) in DMF–water in the presence of 3.0 mol% Pd(OAc)$_2$ [177], tetraphenylporphyrin (tpp) in DMF–water with 17 mol% Pd(OAc)$_2$ [178], Pd(dba)$_2$ (1.0–5.0 mol%) and P(o-tol)$_3$ in DMF–water [179], a palladacycle (0.001–0.50 mol%) in NMP–water, which even allowed the coupling of chloro aryls [151], or P(p-tol)$_3$ in DMF–water in the presence of Pd(OAc)$_2$ for the synthesis of monomers for coatings of electronic components [180]. The monomers are formed by Mizoroki–Heck reaction of 4-bromobenzocyclobutene with tetramethyldivinylsiloxane. The product is produced by Dow on a scale of several tons per year [83]. The company Albemarle developed a process for the production of Naproxen [181] which includes a Mizoroki–Heck reaction of 2-bromo-6-methoxynaphthalene (**64**) with ethylene (29 bar) to intermediate **66** (Scheme 15.13). The chosen phosphine ligand (**65**) shows a very high activity and, thus, catalyst loadings of 0.03 to 0.05 mol% could be applied. This process is run on a scale of 500 tons per year [83].

An intramolecular Mizoroki–Heck reaction was carried out with Pd(OAc)$_2$ (10 mol%) in the presence of P(o-tol)$_3$ in an acetonitrile–water mixture [182], while another group used the same solvent mixture in an intramolecular Mizoroki–Heck reaction with alkoxy groups as leaving groups [183, 184]. The reaction was conducted at room temperature in the presence of Pd(OAc)$_2$ (10 mol%) and a variety of phosphine ligands: tpp, P(o-tol)$_3$, P(p-tol)$_3$, P(2-furyl)$_3$, diphenylphosphinopropane, diphenylphosphinobutane and diphenylphosphinoferrocene.

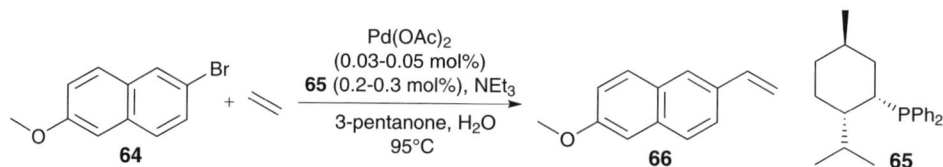

Scheme 15.13 *Synthesis of a precursor to Naproxen.*

15.2.4.3 Nonaqueous Multiphase Systems

Reports about a toluene–ethylene glycol biphasic catalytic system for a Mizoroki–Heck reaction of iodobenzene and butyl acrylate are rare [115, 124]. Potassium acetate was used as a base and various catalyst precursors ($Pd(OAc)_2$, $PdCl_2$, $Ni(OAc)_2$, $Rh(CO)_2acac$, $RhCl_3$, $RuCl_3$, $Pt(COD)Cl_2$, $CoCl_2$, 1.0 mol% each) were applied in the presence of the water-soluble ligand TPPTS. The inorganic base was partly soluble in ethylene glycol and the residual was dispersed as solid granules. The substrates and the product are soluble in toluene; thus, recycling experiments were undertaken by separating the ethylene glycol phase and reusing it along with fresh toluene and potassium acetate. Recycling more than three times was possible by retaining the catalyst's activity and selectivity. The base adduct was separated by precipitation after repeated runs due to saturation of the base.

15.3 Energy Input

New aspects of environmental friendly processes deal with the integrated approach of chemistry and engineering, especially reaction technique. In the main, the focus of these activities is continuous processes and the combination of unit operations. Furthermore, the lowering of energy costs or the application of alternative energy input is of broad interest. The alternative energy input offers new possibilities for a different kind of chemistry resulting from different reaction pathways.

15.3.1 Ultrasound

Ultrasound is defined as sound waves with frequencies in the range from 20 kHz to several megahertz. The term sonochemistry is used for the application of ultrasound to reaction systems with the aim to change reaction pathways. This so-called sonochemical switching was first described by Ando *et al.* [185] (Scheme 15.14).

Fundamental work by Luche resulted in the hypothesis that ultrasound can influence and change reaction pathways in reaction types with single electron transfer [186, 187]. Ultrasound is also believed to influence reaction systems by mechanical effects [187]. An empirical classification of sonochemical reactions is divided into three types of effects: purely chemical effects induced by sonochemical cavitation, hydrodynamic effects (mechanically induced cavitation), and by-passing mass-transport limitation. The latter effects are based on physical rather than chemical phenomena and judged to be 'false' sonochemistry [188]. Nevertheless, these 'false' effects (e.g. emulsification) are often important.

The three types of effect are:

1. homogeneous – proceed via radical or radical ion intermediates;
2. heterogeneous – proceed via ionic intermediates;
3. heterogeneous – proceed via a mixed mechanism, including radical pathways.

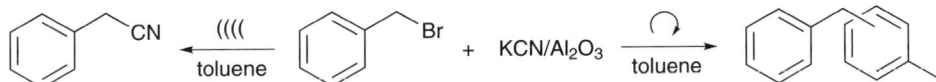

Scheme 15.14 *Sonochemical switching.*

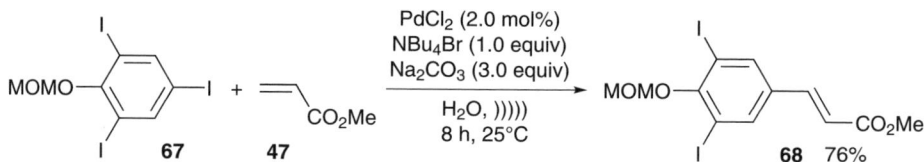

Scheme 15.15 The para-*selective Mizoroki–Heck coupling in water under ultrasound irridiation.*

For the exemplification of the rules given by Luche, see also [189].

In the present work, the expression sonochemistry is used for beneficial chemical and physical effects on synthesis induced by cavitation.

The application of ultrasound in Mizoroki–Heck reactions is not well described. The first application of ultrasound irradiation in Mizoroki–Heck reactions performed in ionic liquids catalysed by palladium–carbene complexes and palladium nanoparticles was published in 2001 [190]. The Pd/C-catalysed reaction of iodobenzene and methyl methacrylate in NMP at room temperature resulted in an increased reaction rate under sonochemical conditions [191]. The catalyst showed high activity under phosphine-free ambient conditions and could be reused. The coupling products could be obtained in up to 70% yield.

Another example of ultrasound-assisted Mizoroki–Heck reactions catalysed by palladium nanoparticles is the coupling under aqueous conditions at ambient temperature. The reaction of iodobenzene and methyl acrylate in water under conventional conditions led to (*E*)-methyl cinnamate in only 10% yield, whereas 86% was obtained under ultrasound irradiation. High *para*-regioselectivity was achieved when starting from di- and tri-iodobenzenes bearing electron-donating groups. Coupling of **67** with acrylate **47** at 25 °C yielded the monosubstituted product **68** in 76% yield (Scheme 15.15 [154]).

Furthermore, the ultrasound- and microwave-promoted reaction of bromonaphthalenes and naphthyl triflate with cyclohexene in ionic liquids catalysed by a palladium complex performed in good yields. Solvent and catalyst was recycled [192].

The beneficial application of ultrasound was also demonstrated in the application of immobilized aryl halides. The air- and water-stable Merryfield-type resins could be reused without loss of activity [193]. A variety of olefins (e.g. styrene) were reacted to the expected products after cleavage from the resin in yields of up to 81% (purity >90%).

Gedanken and coworkers [194] exploited power ultrasound to generate *in situ* amorphous-carbon-activated palladium metallic clusters that proved to be a catalyst for Mizoroki–Heck reactions (without phosphine ligands) of bromobenzene and styrene (yield to an appreciable extent of 30%). The catalyst is stable in most organic solvents, without showing any palladium powder segregation, even after heating them to 400 °C.

Additionally, Mizoroki–Heck reactions were carried out at very low ligand loading or ligand-free with *s/c* of up to 1.0×10^4 (Pd/C, or Pd(OAc)$_2$) under simultaneous ultrasound and microwave irradiation. The reactions result in excellent yields. 4-Iodostyrene, 4-bromoanisole, or 4-bromoacetophenone can be coupled in presence of a cocatalyst (e.g. Wilkinson catalyst, 0.001–0.005%). Under full conversion, the stilbenes could be synthesized in up to 99% yield with an *E/Z* ratio of 5 : 1 to 16 : 1 (Scheme 15.16 [195]).

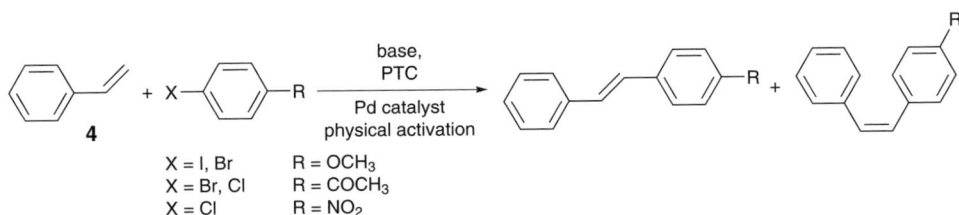

X = I, Br   R = OCH₃
X = Br, Cl   R = COCH₃
X = Cl   R = NO₂

Scheme 15.16 *Microwave- and ultrasound-assisted Mizoroki–Heck reaction.*

15.3.2 Microwaves

Microwaves are electromagnetic waves between infrared and radiofrequency waves. The wavelengths are in the range of 1 cm to 1 m (30 GHz to 300 MHz). To bypass interaction problems with telecommunication, the application of microwaves must be used in defined frequency bands (industrial, scientific and medical frequencies, ISM); see Table 15.2.

Here, it must be emphasized that microwaves do not change the chemistry, because in the temperature range for organic synthesis (200 to 500 K) the energy of microwaves is not high enough to break chemical bonds; see Tables 15.3 and 15.4.

The mechanism by which matter absorbs microwave energy is called dielectric heating [197]. The orientation of dipoles changes with the magnitude and the direction of the electric field. Molecules (with a permanent dipole) are able to align themselves through rotation completely or partly with the direction of the field. In gases and liquids, the molecules cannot follow the inversion of the electric field at an indefinite time (field frequencies 10^6 Hz). This results in dielectric losses and phase shifts. In other words, the field energy is transferred to the medium, meaning that electrical energy is converted into kinetic or thermal energy [198].

The coupling of microwave energy and a medium depends on the dielectric properties of the substance [198]. This can be measured by the dielectric coefficient ε_r, a characteristic for each substance and its state. The dielectric coefficient relates to the capacitance C. The aspects of technical applications of microwaves in chemical reactions and scale-up are discussed in more detail by Nüchter and colleagues [199].

The application of microwaves in Mizoroki–Heck reactions is well documented. There are several overviews of this topic [12, 200–204]. During the last year, new trends in the Mizoroki–Heck reaction conducted under microwave irradiation are the application of solid-supported catalysts and the combination with other activation methods; for example,

Table 15.2 *Microwave frequencies for industrial applications.*

Frequency (MHz)	Wavelength (cm)
433.9	69.14
915	32.75
2450	12.24
5800	5.17
24 120	1.36

Table 15.3 *Chemical bond energies.*

Bond	Energy (eV)	Energy (kJ mol⁻¹)
C–C	3.61	347
C=C	6.35	613
C–OH	3.74	361
C=O	7.71	744
C–H	4.28	413
O–H	4.80	463

ultrasound and ionic liquids. The combination of ultrasound irradiation in microwave-assisted Mizoroki–Heck reactions is discussed in Section 15.3.1.

The Mizoroki–Heck reaction promoted by microwave irradiation was first described by Larhed and Hallberg [205]. For easy catalyst recovery, solid-supported systems were often used. An α-heteroatom-substituted carbonyl linker has been utilized in solid-phase approaches to oxinolines by Pummerer cyclization [206]. The reaction performed with $s/c = 20$ in the presence of phosphine ligands (ratio ligand/palladium-precursor $= 1:1$) in 88–99% yield at 7 h reaction time.

The intermolecular Mizoroki–Heck reaction in macrocyclization of peptides was demonstrated by Byk *et al.* [207]. The reactions were done out in milligram scale with 15–25% yield. The Pd/C-catalysed Mizoroki–Heck reaction of aryl halides and butyl acrylate performed in an excellent manner. The catalyst could be reused five times without loss of activity [208]. Yields up to 90% could be achieved using bromoarenes and a catalyst loading of 1.5 mol% in a few minutes reaction time.

Nanoparticles of palladium supported on resins were used in Mizoroki–Heck reactions of electron-poor olefins under ligand-free conditions. For instance, *p*-bromoanisole proceeded to *p*-methoxy butylcinnamate in 94% yield when the power of microwave irradiation was increased to 375 W (Scheme 15.17) [209].

Microwave irradiation was applied to the synthesis of base-stable ionic liquids [210]. For example, *N*-alkyl-*N*,*N*-dimethylethanolamine salts were prepared and used in several reactions, for instance Mizoroki–Heck reaction, polymerization or condensation reactions, with good yield (95%).

15.4 Additional Systems

Further specialized techniques which can be regarded as belonging to the area of nonclassical methods have already been reviewed by Alonso *et al.* [12]. No such system, however,

Table 15.4 *Energy of microwaves.*

Frequency	Energy (eV)[a]	Energy (kJ mol⁻¹)
MW 0.3 GHz	1.2×10^{-6}	0.00011
MW 2.45 GHz	1.0×10^{-5}	0.00096
MW 300 GHz	1.2×10^{-3}	0.11

[a] 1 eV $= 1.602177 \times 10^{-19}$ J [196].

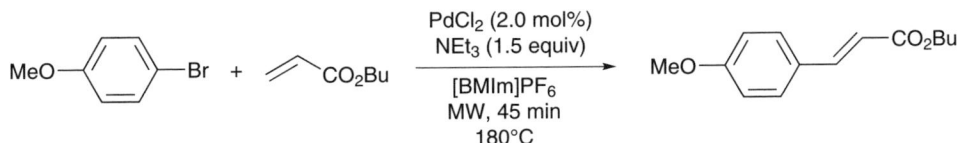

Scheme 15.17 *Mizoroki–Heck reaction under microwave irradiation.*

has so far reached useful levels of practicability. Limited research has been performed on, for example, micellar solutions, electrochemical activation, nanofiltration or ball-milling. Further work is required on these partly rather sophisticated methods. The main limitations of the application of high pressure (up to 20 kbar) are the small volumes and expensive equipment.

In this regard, it is worth mentioning that PEG of molecular weight ≤2000, without any additional solvent, has been used as an efficient reaction medium for the coupling of electron-rich and electron-deficient olefins with aryl bromides, providing stereo- and region-selectivities superior to those observed in conventional and ionic liquid solvent systems. The Mizoroki–Heck reaction of *p*-chlorobromobenzene with *n*-butyl vinyl ether provides the exclusive formation of the (*E*)-configurated olefin in 90% yield.

From the viewpoint of chemical industry, microreactor technology seems promising in principle. For Mizoroki–Heck-type reactions, a few examples have been published. In a continuous-flow reactor loaded with palladium particles, the heterogeneously catalysed coupling of *p*-iodoacetophenone with *n*-butyl acrylate, using DMF as a solvent and tri- ethylamine as a base, could be performed with 78% yield, and successful (seven runs, conversion 71–92% after simple washing) reuse of the catalyst [212]. The enhancement of the reaction rate in a capillary-scale reactor was demonstrated by Wirth and cowork- ers [213]. A 92% yield for the coupling of *p*-nitrobromobenzene with methyl acrylate was claimed in a Japanese patent application using a microreactor packed with a solid palladium catalyst [214]. Ryu and coworkers [215, 216] reacted iodobenzenes with *n*-butyl acrylate in a continuous microflow system with the low-viscosity ionic liquid [BMIm]NTf$_2$ as the re- action medium under homogeneous catalysis; the palladium catalyst could be continuously recycled, and the product was obtained in a yield of 80% in an amount of 10 g h^{-1}. NMP was used as a polar reaction medium for the Pd/C-catalysed transformation of iodobenzene (**5**, X = I) with ethyl acrylate in a modular microreaction system, resulting in values for yield and conversion similar to those obtained in batch-mode experiments [217].

References

1. Carey, F.A. and Sundberg, J.R. (1995) *Organische Chemie*, Wiley–VCH Verlag GmbH, Weinheim.
2. Bonrath, W. and Karge, R. (2002) Application of supercritical fluids in the fine chemical industry, in *High Pressure Chemistry* (eds R. van Eldick and F.G. Klärner), Wiley–VCH Verlag GmbH, Weinheim, pp. 398–421.
3. Jessop, P.G. and Leitner, W. (eds) (1999) *Chemical Synthesis Using Supercritical Fluids*, Wiley–VCH Verlag GmbH, Weinheim.
4. Sheldon, R.A. and van Bekkum, H. (eds) (2001) *Fine Chemicals through Heterogeneous Catalysis*, Wiley–VCH Verlag GmbH, Weinheim.

5. Trost, B.M. (1995) Atom economy – a challenge for organic synthesis: homogeneous catalysis leads the way. *Angew. Chem., Int. Ed. Engl.*, **34**, 259–81.

6. Sheldon, R.A. (1997) Catalysis: the key to waste minimization. *J. Chem. Tech. Biotechnol.*, **68**, 381–8.

7. De Meijere, A. and Meyer, F.E. (1994) Fine feathers make fine birds: the Heck reaction in modern garb. *Angew. Chem., Int. Ed. Engl.*, **33**, 2379–411.

8. Whitcombe, N.J., Hii, K.K. and Gibson, S.E. (2001) Advances in the Heck chemistry of aryl bromides and chlorides. *Tetrahedron*, **57**, 7449–76.

9. Beletskaya, I.P. and Cheprakov, A.V. (2000) The Heck reaction as a sharpening stone of palladium catalysis. *Chem. Rev.*, **100**, 3009–66.

10. Dounay, A.B. and Overman, L.E. (2003) The asymmetric intramolecular Heck reaction in natural product total synthesis. *Chem. Rev.*, **103**, 2945–64.

11. Farina, V. (2004) High-turnover palladium catalysts in cross-coupling and Heck chemistry: a critical overview. *Adv. Synth. Catal.*, **346**, 1553–82.

12. Alonso, F., Beletskaya, I.P. and Yus, M. (2005) Non-conventional methodologies for transition-metal catalyzed carbon–carbon coupling: a critical overview. Part 1: the Heck reaction. *Tetrahedron*, **61**, 11771–835.

13. Stephan, M.S., Teunissen, A.J.J.M., Verzijl, G.K.M. and de Vries, J.G. (1998) Heck reactions without salt formation: aromatic carboxylic anhydrides as arylating agents. *Angew. Chem. Int. Ed.*, **37**, 662–4.

14. Reetz, M.T. and de Vries, J.G. (2004) Ligand-free Heck reactions using low Pd-loading. *Chem. Commun.*, 1559–63.

15. Härter, R., Lemke, U. and Radspieler, A. (2005) Preparation of stilbene derivatives. WO 2005023740.

16. Bonrath, W., Eggersdorfer, M. and Netscher, T. (2007) Catalysis in the industrial preparation of vitamins and nutraceuticals. *Catal. Today*, **121**, 45–57.

17. Wasserscheid, P. and Welton, T. (eds) (2003) in *Ionic Liquids in Synthesis*, Wiley–VCH Verlag GmbH, Weinheim.

18. Seddon, K.R., Stark, A. and Torres, M.J. (2000) Influence of chloride, water, and organic solvents on the physical properties of ionic liquids. *Pure Appl. Chem.*, **72**, 2275–87.

19. (a) Jeffery, T. (1985) Highly stereospecific palladium-catalyzed vinylation of vinylic halides under solid–liquid phase transfer conditions. *Tetrahedron Lett.*, **26**, 2667–70; (b) Herrmann, W.A., Elison, M., Fischer, J. *et al.* (1995) Metal complexes of *N*-heterocyclic carbenes – a new structural principle for catalysts in homogeneous catalysis. *Angew. Chem., Int. Ed. Engl.*, **34**, 2371–4.

20. Dzyuba, S.V. and Bartsch, R.A. (2003) Recent advantages in application of room-temperature ionic liquids/supercritical CO_2 systems. *Angew. Chem. Int. Ed.*, **42**, 148–50.

21. Kaufmann, D.E., Nouroozian, M. and Henze, H. (1996) Molten salts as an efficient medium for palladium-catalyzed C–C coupling reactions. *Synlett*, 1091–2.

22. Cacchi, S., Battistuzzi, G. and Fabrizi, G. (2002) A molten Bu_4NOAc/Bu_4NBr mixture as an efficient medium for the stereoselective synthesis of (*E*)- and (*Z*)-3,3-diarylacrylates. *Synlett*, 439–42.

23. Xie, X., Chen, B., Lu, J. *et al.* (2004) Synthesis of benzofurans in ionic liquid by a $PdCl_2$-catalyzed intramolecular Heck reaction. *Tetrahedron Lett.*, **45**, 6235–7.

24. Calo, V., Nacci, A. and Monopoli, A. (2006) Effects of ionic liquids on Pd-catalysed carbon–carbon bond formation. *Eur. J. Org. Chem.*, 3791–802.

25. Leveque, J.-M. and Cravotto, G. (2006) Microwaves, power ultrasound, and ionic liquids. A new synergy in green organic synthesis. *Chimia*, **60**, 313–20.

26. Imperato, G., König, B. and Chiappe, C. (2007) Ionic green solvents from renewable resources. *Eur. J. Org. Chem.*, 1049–58.

27. Wang, R., Xiao, J., Twamley, B. and Shreeve, J.M. (2007) Efficient Heck reactions catalyzed by a highly recyclable palladium(II) complex of a pyridyl-functionalized imidazolium-based ionic liquid. *Org. Biomol. Chem.*, **5**, 671–8.

28. Yoon, B., Yen, C.H., Mekki, S. *et al.* (2006) Effect of water on the Heck reactions catalyzed by recyclable palladium chloride in ionic liquids coupled with supercritical CO_2 extraction. *Ind. Eng. Chem. Res.*, **45**, 4433–5.

29. Zou, G., Huang, W., Xiao, Y. and Tang, J. (2006) Heck reaction catalysed by palladium supported with an electron-rich benzimidazolylidene generated *in situ*: remarkable ligand electronic effects and controllable mono- and di-arylation. *New J. Chem.*, **30**, 803–9.

30. Calo, V., Nacci, A., Lopez, L. and Mannarini, N. (2000) Heck reaction in ionic liquids catalyzed by a Pd-benzothiazole carbene complex. *Tetrahedron Lett.*, **41**, 8973–6.

31. Xu, L., Chen, W. and Xiao, J. (2000) Heck reaction in ionic liquids and the *in situ* identification of *N*-heterocyclic carbene complexes of palladium. *Organometallics*, **19**, 1123–7.

32. Amatore, C., Azzabi, M. and Jutand, A. (1991) Role and effects of halide ions on the rates and mechanisms of oxidative addition of iodobenzene to low-ligated zerovalent palladium complexes $Pd^0(PPh_3)_2$. *J. Am. Chem. Soc.*, **113**, 8375–84.

33. Böhm, V.P.W. and Herrmann, W.A. (2000) Nonaqueous ionic liquids: superior reaction media for the catalytic Heck-vinylation of chloroarenes. *Chem. Eur. J.*, **6**, 1017–25.

34. Hagiwara, H., Sugawara, Y., Hoshi, T. and Suzuki, T. (2005) Sustainable Mizoroki–Heck reaction in water: remarkably high activity of $Pd(OAc)_2$ immobilized on reversed phase silica gel with the aid of an ionic liquid. *Chem. Commun.*, **23**, 2942–4.

35. Wang, L., Zhang, Y., Xie, C. and Wang, Y. (2005) PEG-supported imidazolium chloride: a highly efficient and reusable reaction medium for the Heck reaction. *Synlett*, 1861–4.

36. Cassol, C.C., Umpierre, A.P., Machado, G. *et al.* (2005) The role of Pd nanoparticles in ionic liquid in the Heck reaction. *J. Am. Chem. Soc.*, **127**, 3298–9.

37. Dubbaka, S.R., Zhao, D., Fei, Z. *et al.* (2006) Palladium-catalyzed desulfitative Mizoroki–Heck coupling reactions of sulfonyl chlorides with olefins in a nitrile-functionalized ionic liquid. *Synlett*, 3155–7.

38. Calo, V., Nacci, A. and Monopoli, A. (2004) Regio- and stereoselective carbon–carbon bond formation in ionic liquids. *J. Mol. Catal. A: Chem.*, **214**, 45–56.

39. Calo, V., Nacci, A., Monopoli, A. *et al.* (2003) Pd nanoparticles catalyzed stereospecific synthesis of β-aryl cinnamic esters in ionic liquids. *J. Org. Chem.*, **68**, 2929–33.

40. Calo, V., Nacci, A., Monopoli, A. *et al.* (2003) Pd nanoparticle catalyzed Heck arylation of 1,1-disubstituted alkenes in ionic liquids. Study on factors affecting the regioselectivity of the coupling process. *Organometallics*, **22**, 4193–7.

41. Karimi, B. and Enders, D. (2006) New *N*-heterocyclic carbene palladium complex/ionic liquid matrix immobilized on silica: application as recoverable catalyst for the Heck reaction. *Org. Lett.*, **8**, 1237–40.

42. Mo, J., Xu, L. and Xiao, J. (2005) Ionic liquid-promoted, highly regioselective Heck arylation of electron-rich olefins by aryl halides. *J. Am. Chem. Soc.*, **127**, 751–60.

43. Okubo, H., Keisukei, S. and Yokoyama, C. (2002) Heck reactions in a non-aqueous ionic liquid using silica supported palladium complex catalysts. *Tetrahedron Lett.*, **43**, 7115–8.

44. Hagiwara, H., Sugawara, Y., Isobe, K. *et al.* (2004) Immobilization of $Pd(OAc)_2$ in ionic liquid on silica: application to sustainable Mizoroki–Heck reaction. *Org. Lett.*, **6**, 2325–8.

45. Hagiwara, H., Shimizu, Y., Hoshi, T. *et al.* (2001) Heterogeneous Heck reaction catalyzed by Pd/C in ionic liquid. *Tetrahedron Lett.*, **42**, 4349–51.

46. Carmichael, A.J., Earle, M.J., Holbrey, J.D. *et al.* (1999) The Heck reaction in ionic liquids: a multiphasic catalyst system. *Org. Lett.*, **1**, 997–1000.

47. Calo, V., Nacci, A., Monopoli, A. *et al.* (2004) Heck reaction catalyzed by nanosized palladium on chitosan in ionic liquids. *Organometallics*, **23**, 5154–8.
48. Pei, W., Mo, J. and Xiao, J. (2005) Highly regioselective Heck reactions of heteroaryl halides with electron-rich olefins in ionic liquid. *J. Organomet. Chem.*, **690**, 3546–51.
49. Mo, J., Liu, S. and Xiao, J. (2005) Palladium-catalyzed regioselective Heck arylation of electron-rich olefins in a molecular solvent-ionic liquid cocktail. *Tetrahedron*, **61**, 9902–7.
50. Sun, L., Pei, W. and Shen, C. (2006) Synthesis of functionalised acetophenone. *J. Chem. Res.*, **6**, 388–9.
51. Forsyth, S.A., Gunaratne, H.Q.N., Hardacre, C. *et al.* (2005) Utilisation of ionic liquid solvents for the synthesis of lily-of-the-valley fragrance {β-Lilial; 3-(4-*t*-butylphenyl)-2-methylpropanal}. *J. Mol. Catal. A: Chem.*, **231**, 61–6.
52. Bouquillon, S., Ganchegui, B., Estrine, B. *et al.* (2001) Heck arylation of allylic alcohols in molten salts. *J. Organomet. Chem.*, **634**, 153–6.
53. Calo, V., Nacci, A., Monopoli, A. and Ferola, V. (2007) Palladium-catalyzed Heck arylations of allyl alcohols in ionic liquids: remarkable base effect on the selectivity. *J. Org. Chem.*, **72**, 2596–601.
54. Beccaria, L., Deagostino, A., Prandi, C. *et al.* (2006) Heck reaction on 1-alkoxy-1,3-dienes in ionic liquids: a superior medium for the regioselective arylation of the conjugated dienic system. *Synlett*, 2989–92.
55. Mo, J., Xu, L., Ruan, J. *et al.* (2006) Regioselective Heck arylation of unsaturated alcohols by palladium catalysis in ionic liquid. *Chem. Commun.*, 3591–3.
56. Mo, J. and Xiao, J. (2006) The Heck reaction of electron-rich olefins with regiocontrol by hydrogen-bond donors. *Angew. Chem. Int. Ed.*, **45**, 4152–7.
57. Brice, T.J. (1950) Chapter 13, in *Fluorine Chemistry*, Vol. **I** (ed. J.H. Simons), Academic Press, New York, p. 615.
58. Zhu, D.-W. (1993) A novel reaction medium: perfluorocarbon fluids. *Synthesis*, 953–4.
59. Gladysz, J.A. and Curran, D.P. (eds) (2002) Fluorous chemistry: from biphasic catalysis to a parallel chemical universe and beyond. *Tetrahedron*, **58**, 3823–5.
60. Howe-Grant, M. (ed.) (1995) *Fluorine Chemistry: A Comprehensive Treatment*, John Wiley & Sons, Inc., New York.
61. Clayton, J.W. Jr (1967) Fluorocarbon toxicity and biological action. *Fluorine Chem. Rev.*, **1**, 197–252.
62. Horváth, I.T. (1998) Fluorous biphase chemistry. *Acc. Chem. Res.*, **31**, 641–50.
63. Bhanage, B.M. and Arai, M. (2001) Catalyst product separation techniques in Heck reaction. *Catal. Rev. Sci. Eng.*, **43**, 315–44.
64. Rocaboy, C. and Gladysz, J.A. (2002) Highly active thermomorphic fluorous palladacycle catalyst precursors for the Heck reaction; evidence for a palladium nanoparticle pathway. *Org. Lett.*, **4**, 1993–6.
65. Gladysz, J.A., Wende, M. and Rocaboy, C. (2003) Fluorinated catalysts or reagents with temperature-dependent solubility. DE 10212424.
66. Rocaboy C. and Gladysz, J.A. (2003) Thermomorphic fluorous imine and thioether palladacycles as precursors for highly active Heck and Suzuki catalysts; evidence for palladium nanoparticle pathways. *New J. Chem.*, **27**, 39–49.
67. Curran, D.P., Fischer, K. and Moura-Letts, G. (2004) A soluble fluorous palladium complex that promotes Heck reactions and can be recovered and reused. *Synlett*, 1379–82.
68. Fukuyama, T., Arai, M., Matsubara, H. and Ryu, I. (2004) Mizoroki–Heck arylation of α,β-unsaturated acids with a hybrid fluorous ether, F-626: facile filtrative separation of products and efficient recycling of a reaction medium containing a catalyst. *J. Org. Chem.*, **69**, 8105–7.
69. Yeung, L.K. and Crooks, R.M. (2001) Heck heterocoupling within a dendritic nanoreactor. *Nano Lett.*, **1**, 14–7.

70. Matsubara, H., Maeda, L. and Ryu, I. (2005) Preparation of fluorous DMF solvents and their use for some Pd-catalyzed cross-coupling reactions. *Chem. Lett.*, **34**, 1548–9.
71. Moineau, J., Pozzi, G., Quici, S. and Sinou, D. (1999) Palladium-catalyzed Heck reaction in perfluorinated solvents. *Tetrahedron Lett.*, **40**, 7683–6.
72. Grigg, R. and York, M. (2000) Bimetallic catalytic cascade ring-closing metathesis–intramolecular Heck reactions using a fluorous biphasic solvent system or a polymer-supported palladium catalyst. *Tetrahedron Lett.*, **41**, 7255–8.
73. Vallin, K.S.A., Zhang, Q., Larhed, M. *et al.* (2003) A new regioselective Heck vinylation with enamides. Synthesis and investigation of fluorous-tagged bidentate ligands for fast separation. *J. Org. Chem.*, **68**, 6639–45.
74. Nakamura, Y., Takeuchi, S., Zhang, S. *et al.* (2002) Preparation of a fluorous chiral BINAP and application to an asymmetric Heck reaction. *Tetrahedron Lett.*, **43**, 3053–6.
75. Nakamura, Y., Takeuchi, S. and Ohgo, Y. (2003) Enantioselective carbon–carbon bond forming reactions using fluorous chiral catalysts. *J. Fluorine Chem.*, **120**, 121–9.
76. Bayardon, J., Cavazzini, M., Maillard, D. *et al.* (2003) Chiral fluorous phosphorus ligands based on the binaphthyl skeleton: synthesis and applications in asymmetric catalysis. *Tetrahedron: Asymmetry*, **14**, 2215–24.
77. Moreno-Manas, M., Pleixats, R. and Villarroya, S. (2001) Fluorous phase soluble palladium nanoparticles as recoverable catalysts for Suzuki cross-coupling and Heck reactions. *Organometallics*, **20**, 4524–8.
78. Moreno-Manas, M., Pleixats, R. and Villarroya, S. (2002) Palladium nanoparticles stabilised by polyfluorinated chains. *Chem. Commun.*, 60–1.
79. Schmieder, H., Dahmen, N., Schön, J. and Wiegand, G. (1997) Industrial and environmental applications of supercritical fluids, in *Chemistry under Extreme or Non-Classical Conditions* (eds R. van Eldick and C.D. Hubbard), John Wiley & Sons, Inc., New York, pp. 273–316.
80. Leitner, W. (2002) Supercritical carbon dioxide as a green reaction medium for catalysis. *Acc. Chem. Res.*, **35**, 746–56.
81. Stewart, I.H., Hutchins, G.J. and Derouane, E.G. (1999) Catalytic reactions at supercritical conditions. *Curr. Top. Catal.*, **2**, 17–38.
82. Dinjus, E., Fornika, R. and Scholz, M. (1997) Organic chemistry in supercritical fluids, in *Chemistry under Extreme or Non-Classical Conditions* (eds R. van Eldick and C.D. Hubbard), John Wiley & Sons, Inc., New York, pp. 219–71.
83. De Vries, J.G. (2001) The Heck reaction in the production of fine chemicals. *Can. J. Chem.*, **79**, 1086–92.
84. Hyde, J.R., Licence, P., Carter, D. and Poliakoff, M. (2001) Continuous catalytic reactions in supercritical fluids. *Appl. Catal. A: General*, **222**, 119–31.
85. Perrut, M. (2000) Supercritical fluid applications: industrial developments and economic issues. *Ind. Eng. Chem. Res.*, **39**, 4531–5.
86. Styring, P., Allen, R.W.K. and Lau, P.L. (2005) Continuous flow catalytic reactions in supercritical carbon dioxide: miniaturisation. 7th World Congress of Chemical Engineering, Glasgow, UK, 10–14 July, Computer Optical Disc, contribution C25–001; *Chem. Abstr.*, **145**, 473392.
87. Amandi, R., Hyde, J. and Poliakoff, M. (2003) Heterogeneous reactions in supercritical carbon dioxide, in *Carbon Dioxide Recovery and Utilization* (ed. M. Aresta), Kluwer Academic Publishers, Dordrecht, Netherlands, pp. 169–79.
88. Jessop, P.G. (2005) SFCs and liquid polymers, in *Multiphase Homogeneous Catalysis*, Vol. **2** (eds B. Cornils, W.A. Herrmann, I.T. Horváth, W. Leitner, S. Mecking, H. Olivier-Bourbigou and D. Vogt), Wiley–VCH Verlag GmbH, Weinheim, pp. 676–88.
89. Jung, J. and Perrut, M. (2001) Particle design using supercritical fluids: Literature and patent survey. *J. Supercrit. Fluids*, **20**, 179–219.

90. Li, J.-H., Jia, L.-Q. and Jiang, H.-F. (2000) Transition metal-catalyzed reactions in supercritical carbon dioxide. *Youji Huaxue, Chinese J. Org. Chem.*, **20**, 293–8; *Chem. Abstr.*, **133**, 90955 (2000); remark: wrong citations of names are given in this review.

91. Ikushima, Y. and Arai, M. (2000) Two-phase catalytic reactions in supercritical carbon dioxide. *Chorinkai Saishin Gijutsu*, **4**, 16–9; *Chem. Abstr.*, **134**, 115662 (2000).

92. Shezad, N., Oakes, R.S., Clifford, A.A. and Rayner, C.M. (2001) Pd-catalyzed coupling reactions in supercritical carbon dioxide. *Chem. Ind.*, **82**, 459–64.

93. Gao, J. (1993) Supercritical hydration of organic compounds. The potential of mean force for benzene dimer in supercritical water. *J. Am. Chem. Soc.*, **115**, 6893–5.

94. Kaupp, G. (1994) Reactions in supercritical carbon dioxide. *Angew. Chem., Int. Ed. Engl.*, **33**, 1452–5.

95. Jessop, P.G., Hsiao, Y., Ikariya, T. and Noyori, R. (1994) Catalytic production of dimethylformamide from supercritical carbon dioxide. *J. Am. Chem. Soc.*, **116**, 8851–2.

96. Jessop, P.G., Hsiao, Y., Ikariya, T. and Noyori, R. (1994) Homogeneous catalytic hydrogenation of supercritical carbon dioxide. *Nature*, **368**, 231–3.

97. Oakes, R.S., Clifford, A.A. and Rayner, C.M. (2001) The use of supercritical fluids in synthetic organic chemistry. *J. Chem. Soc., Perkin Trans. 1*, 917–41.

98. Savage, P.E., Gopalan, S., Mizan, T.I. *et al.* (1995) Reactions at supercritical conditions: applications and fundamentals. *AIChE J.*, **41**, 1723–78.

99. Prajapati, D. and Gohain, M. (2004) Recent advances in the application of supercritical fluids for carbon–carbon bond formation in organic synthesis. *Tetrahedron*, **60**, 815–33.

100. Reardon, P., Metts, S., Crittendon, C. *et al.* (1995) Palladium-catalyzed coupling reactions in superheated water. *Organometallics*, **14**, 3810–6.

101. Diminnie, J., Metts, S. and Parsons, E.J. (1995) *In situ* generation and Heck coupling of alkenes in superheated water. *Organometallics*, **14**, 4023–5.

102. Zhang, R., Zhao, F., Sato, M. and Ikushima, Y. (2003) Noncatalytic Heck coupling reaction using supercritical water. *Chem. Commun.*, 1548–9.

103. Ikushima, Y. (2004) Development of environmentally friendly synthetic process using supercritical water. *Chem. Abstr.*, **142**, 410655.

104. Zhang, R., Sato, O., Zhao, F. *et al.* (2004) Heck coupling reaction of iodobenzene and styrene using supercritical water in the absence of a catalyst. *Chem. Eur. J.*, **10**, 1501–6.

105. Gron, L.U., LaCroix, J.E., Higgins, C.J. *et al.* (2001) Heck reactions in hydrothermal, subcritical water: water density as an important reaction variable. *Tetrahedron Lett.*, **42**, 8555–7.

106. Morita, D.K., Pesiri, D.R., David, S.A. *et al.* (1998) Palladium-catalyzed cross-coupling reactions in supercritical carbon dioxide. *Chem. Commun.*, 1397–8.

107. Carroll, M.A. and Holmes, A.B. (1998) Palladium-catalyzed carbon–carbon bond formation in supercritical carbon dioxide. *Chem. Commun.*, 1395–6.

108. Fujita, S., Yuzawa, K., Bhanage, B.M. *et al.* (2002) Palladium-catalyzed Heck coupling reactions using different fluorinated phosphine ligands in compressed carbon dioxide and conventional organic solvents. *J. Mol. Catal. A: Chem.*, **180**, 35–42.

109. Shezad, N., Oakes, R.S., Clifford, A.A. and Rayner, C.M. (1999) Use of fluorinated palladium sources for efficient Pd-catalyzed coupling reactions in supercritical carbon dioxide. *Tetrahedron Lett.*, **40**, 2221–4.

110. Early, T.R., Gordon, R.S., Carroll, M.A. *et al.* (2001) Palladium-catalyzed cross-coupling reactions in supercritical carbon dioxide. *Chem. Commun.*, 1966–7.

111. Shezad, N., Clifford, A.A. and Rayner, C.M. (2001) Suppression of double bond isomerisation in intramolecular Heck reactions using supercritical carbon dioxide. *Tetrahedron Lett.*, **42**, 323–5.

112. Bhanage, B.M., Ikushima, Y., Shirai, M. and Arai, M. (1999) Heck reactions using water-soluble metal complexes in supercritical carbon dioxide. *Tetrahedron Lett.*, **40**, 6427–30.

113. Bhanage, B.M., Zhao, F.-G., Shirai, M. and Arai, M. (1998) Comparison of activity and selectivity of various metal–TPPTS complex catalysts in ethylene glycol–toluene biphasic Heck vinylation reactions of iodobenzene. *Tetrahedron Lett.*, **39**, 9509–12.

114. Kayaki, Y., Noguchi, Y. and Ikariya, T. (2000) Enhanced product selectivity in the Mizoroki–Heck reaction using a supercritical carbon dioxide–liquid biphasic system. *Chem. Commun.*, 2245–6.

115. Bhanage, B.M., Fujita, S.-i. and Arai, M. (2003) Heck reactions with various types of palladium complex catalysts: application of multiphase catalysis and supercritical carbon dioxide. *J. Organomet. Chem.*, **687**, 211–8.

116. Cacchi, S., Fabrizi, G., Gasparrini, F. and Villani, C. (1999) Carbon–carbon bond forming reactions in supercritical carbon dioxide in the presence of a supported palladium catalyst. *Synlett*, 345–7.

117. Ley, S.V., Ramarao, C., Gordon, R.S. *et al.* (2002) Polyurea-encapsulated palladium(II) acetate: a robust and recyclable catalyst for use in conventional and supercritical media. *Chem. Commun.*, 1134–5.

118. Holmes, A.B., Ley, S.V., Gordon, R.S. *et al.* (2003) The use of microencapsulated transition metal reagents for reactions in supercritical fluids. WO 03/048090 A1.

119. Gordon, R.S. and Holmes, A.B. (2002) Palladium-mediated cross-coupling reactions with supported reagents in supercritical carbon dioxide. *Chem. Commun.*, 640–1.

120. Holmes, A.B., Gordon, R.S. and Early, T.R. (2003) Chemical reactions in compressed carbon dioxide. WO 03/009936 A2.

121. Yeung, L.K., Lee, C.T. Jr, Johnston, K.P. and Crooks, R.M. (2001) Catalysis in supercritical CO_2 using dendrimer-encapsulated palladium nanoparticles. *Chem. Commun.*, 2290–1.

122. Niu, Y. and Crooks, R.M. (2003) Dendrimer-encapsulated metal nanoparticles and their applications to catalysis. *C. R. Chim.*, **6**, 1049–59.

123. Scott, R.W.J., Wilson, O.M. and Crooks, R.M. (2005) Synthesis, characterization, and applications of dendrimer-encapsulated nanoparticles. *J. Phys. Chem. B*, **109**, 692–704.

124. Montilla, F., Galindo, A., Andrés, R. *et al.* (2006) Carbosilane dendrons as solubilizers of metal complexes in supercritical carbon dioxide. *Organometallics*, **25**, 4138–43.

125. Martínez, S., Vallribera, A., Cotet, C.L. *et al.* (2005) Nanosized metallic particles embedded in silica and carbon aerogels as catalysts in the Mizoroki–Heck coupling reaction. *New J. Chem.*, **29**, 1342–5.

126. Herrmann, W.A. and Cornils, B. (eds) (2004) *Aqueous-phase Organometallic Catalysis*, Wiley–VCH Verlag GmbH, Weinheim.

127. Grieco, P.A. (ed.) (1998) *Organic Synthesis in Water*, Blackie Academic and Professional, London.

128. Li, C.-J. and Chan, T.-H. (eds) (1997) *Organic Reactions in Water*, John Wiley & Sons, Inc., New York.

129. Lindstroem, U.M. (2002) Stereoselective organic reactions in water. *Chem. Rev.*, **102**, 2751–71.

130. Li, C.-J. (2005) Organic reactions in aqueous media with a focus on carbon–carbon bond formations. A decade update. *Chem. Rev.*, **105**, 3095–165.

131. Genet, J.-P. and Savignac, M. (1999) Recent developments of palladium(0) catalyzed reactions in aqueous medium. *J. Organomet. Chem.*, **576**, 305–17.

132. Xia, M. and Wang, Y.G. (2001) Polyethylene glycol as support and phase transfer catalyst in aqueous palladium-catalyzed liquid-phase synthesis. *Chin. Chem. Lett.*, **12**, 941–2.

133. Rabeyrin, C. and Sinou, D. (2004) Palladium-catalyzed asymmetric arylation of 2,3-dihydrofuran with aryl triflates in water in the presence of surfactants. *J. Mol. Catal. A: Chem.*, **215**, 89–93.

134. Mukhopadhyay, S., Rothenberg, G., Joshi, A. *et al.* (2002) Heterogeneous palladium-catalyzed Heck reaction of aryl chlorides and styrene in water under mild conditions. *Adv. Synth. Catal.*, **344**, 348–54.

135. Jeffery, T. (1984) Palladium-catalyzed vinylation of organic halides under solid–liquid phase transfer conditions. *J. Chem. Soc., Chem. Commun.*, 1287–9.

136. Cai, M. and Wang, P. (1998) Use of phase-transfer catalysis in palladium-catalyzed Heck carbonylation of aryl halides. *Jiangxi Shifan Daxue Xuebao, Ziran Kexueban*, **22**, 231–4.

137. Li, J.-H., Hu, X.-C., Liang, Y. and Xie, Y.-X. (2006) PEG-400 promoted Pd(OAc)$_2$/DABCO-catalyzed cross-coupling reactions in aqueous media. *Tetrahedron*, **62**, 31–8.

138. Jeffery, T. (1994) Heck-type reactions in water. *Tetrahedron Lett.*, **35**, 3051–4.

139. Zhao, H., Cai, M.-Z. and Peng, C.-Y. (2002) Stereoselective synthesis of (*E*)-cinnamonitriles via Heck arylation of acrylonitrile and aryl iodides in water. *Synth. Commun.*, **32**, 3419–23.

140. Zhao, H., Cai, M.-Z., Peng, C.-Y. and Song, C.-S. (2002) Palladium-catalyzed arylation of butyl acrylate and acrylamide with aryl iodides in water. *J. Chem. Res. (S)*, 28–9.

141. Hou, J.-J., Yang, L.-R., Cui, X.-L. and Wu, Y.-J. (2003) Heck reaction catalyzed by palladacycle in neat water. *Chin. J. Chem.*, **21**, 717–9.

142. Solabannavar, S.B., Desai, U.V. and Mane, R.B. (2002) Heck reaction in aqueous medium using Amberlite IRA-400 (basic). *Green Chem.*, **4**, 347–8.

143. Botella, L. and Nájera, C. (2004) Controlled mono and double Heck reactions in water catalyzed by an oxime-derived palladacycle. *Tetrahedron Lett.*, **45**, 1833–6.

144. Uozumi, Y. and Watanabe, T. (1999) Green catalysis: Hydroxycarbonylation of aryl halides in water catalyzed by an amphiphilic resin-supported phosphine–palladium complex. *J. Org. Chem.*, **64**, 6921–3.

145. Uozumi, Y. and Kimura, T. (2002) Heck reaction in water with amphiphilic resin-supported palladium–phosphine complexes. *Synlett*, 2045–8.

146. Bumagin, N.A., Bykov, V.V., Sukhomlinova, L.I. *et al.* (1995) Palladium-catalyzed arylation of styrene and acrylic acid in water. *J. Organomet. Chem.*, **486**, 259–62.

147. Bumagin, N.A., More, P.G. and Beletskaya, I.P. (1989) Synthesis of substituted cinnamic acids and cinnamonitriles via palladium-catalyzed coupling reactions of aryl halides with acrylic acid and acrylonitrile in aqueous media. *J. Organomet. Chem.*, **371**, 397–401.

148. Botella, L. and Nájera, C. (2004) Synthesis of methylated resveratrol and analogues by Heck reactions in organic and aqueous solvents. *Tetrahedron*, **60**, 5563–70.

149. Yokoyama, Y., Hikawa, H., Mitsuhashi, M. *et al.* (1999) Syntheses without protection: a three-step synthesis of optically active clavicipitic acid by utilizing biomimetic synthesis of 4-bromotryptophan. *Tetrahedron Lett.*, **40**, 7803–6.

150. Basnak, I., Takatori, S. and Walker, R.T. (1997) Palladium-catalyzed carbomethoxyvinylation and thienylation of 5-iodo(bromo)-2,4-dimethoxypyrimidine in water. *Tetrahedron Lett.*, **38**, 4869–72.

151. Nájera, C., Gil-Moltó, J., Karlströem, S. and Falvello, L.R. (2003) Di-2-pyridylmethylamine-based palladium complexes as new catalysts for Heck, Suzuki, and Sonogashira reactions in organic and aqueous solvents. *Org. Lett.*, **5**, 1451–4.

152. Schönfelder, D., Fischer, K., Schmidt, M. *et al.* (2005) Poly(2-oxazoline)s functionalized with palladium carbene complexes: soluble, amphiphilic polymer supports for C–C coupling reactions in water. *Macromolecules*, **38**, 254–62.

153. Alacid, E. and Nájera, C. (2006) Palladated Kaiser oxime resin as precatalyst for the Heck reaction in organic and aqueous media. *Synlett*, 2959–64.

154. Zhang, Z., Zha, Z., Gan, C. *et al.* (2006) Catalysis and regioselectivity of the aqueous Heck reaction by Pd(0) nanoparticles under ultrasonic irradiation. *J. Org. Chem.*, **71**, 4339–42.

155. Gron, L.U. and Tinsley, A.S. (1999) Tailoring aqueous solvents for organic reactions: Heck coupling reactions in high temperature water. *Tetrahedron Lett.*, **40**, 227–30.

156. Williams, D.B.G., Lombard, H. and Holzapfel, C.W. (2001) A comparative study of some Pd-catalysed Heck reactions in polar- and aqueous biphasic media. *Synth. Commun.*, **31**, 2077–81.

157. Genet, J.P., Blart, E. and Savignac, M. (1992) Palladium-catalyzed cross-coupling reactions in a homogeneous aqueous medium. *Synlett*, 715–7.

158. Lemaire-Audoire, S., Savignac, M., Dupuis, C. and Genet, J.-P. (1996) Intramolecular Heck-type reactions in aqueous medium. Dramatic change in regioselectivity. *Tetrahedron Lett.*, **37**, 2003–6.

159. Ha, H.-J., Ahn, Y.-G. and Woo, J.-S. (1998) Ring closure selectivity of the intramolecular Heck reaction. *Bull. Kor. Chem. Soc.*, **19**, 818–9.

160. Choudary, B.M., Lakshmi Kantam, M., Mahender Reddy, N. and Gupta, N.M. (2002) Layered-double-hydroxide-supported Pd(TPPTS)$_2$Cl$_2$: a new heterogeneous catalyst for Heck arylation of olefins. *Catal. Lett.*, **82**, 79–83.

161. Fujita, S.-I., Yoshida, T., Bhanage, B.M. and Arai, M. (2002) Heck reaction with a silica-supported Pd–TPPTS liquid phase catalyst: effects of reaction conditions and various amines on the reaction rate. *J. Mol. Catal. A: Chem.*, **188**, 37–43.

162. Thorpe, T., Brown, S.M., Crosby, J. *et al.* (2000) A practical synthesis of a disulfonated phosphine and its application to biphasic catalysis. *Tetrahedron Lett.*, **41**, 4503–5.

163. Bhanage, B.M., Shirai, M. and Arai, M. (1999) Heterogeneous catalyst system for Heck reaction using supported ethylene glycol phase Pd/TPPTS catalyst with inorganic base. *J. Mol. Catal. A: Chem.*, **145**, 69–74.

164. Yokoyama, Y., Hikawa, H. and Murakami, Y. (1997) Organic synthesis without protecting group total synthesis of optically active clavicipitic acid in aqueous media. *Tennen Yuki Kagobutsu Toronkai Koen Yoshishu*, **39**, 325–30; *Chem. Abstr.*, **131**, 73829 (1999).

165. Mirza, A.R., Anson, M.S., Hellgardt, K. *et al.* (1998) Optimization of palladium-based supported liquid-phase catalysts in the Heck reaction. *Org. Proc. Res. Dev.*, **2**, 325–31.

166. Fujita, S., Yoshida, T., Bhanage, B.M. *et al.* (2002) Palladium-based supported liquid phase catalysts: influence of preparation variables on the activity and enhancement of the activity on recycling in the Heck reaction. *J. Mol. Catal. A: Chem.*, **180**, 277–84.

167. Amengual, R., Genin, E., Michelet, V. *et al.* (2002) Convenient synthesis of new anionic water-soluble phosphanes and applications in inter- and intramolecular Heck reactions. *Adv. Synth. Catal.*, **344**, 393–8.

168. Beller, M., Krauter, J.G.E. and Zapf, A. (1997) Carbohydrate-substituted triarylphosphanes – a new class of ligands for two-phase catalysis. *Angew. Chem., Int. Ed. Engl.*, **36**, 772–4.

169. Moore, L.R. and Shaughnessy, K.H. (2004) Efficient aqueous-phase Heck and Suzuki couplings of aryl bromides using tri(4,6-dimethyl-3-sulfonatophenyl)phosphine trisodium salt (TXPTS). *Org. Lett.*, **6**, 225–8.

170. Casalnuovo, A.L. and Calabrese, J.C. (1990) Palladium-catalyzed alkylations in aqueous media. *J. Am. Chem. Soc.*, **112**, 4324–30.

171. Gelpke, A.E.S., Veerman, J.J.N., Goedheijt, M.S. *et al.* (1999) Synthesis and use of water-soluble sulfonated dibenzofuran-based phosphine ligands. *Tetrahedron*, **55**, 6657–70.

172. DeVasher, R.B., Moore, L.R. and Shaughnessy, K.H. (2004) Aqueous-phase, palladium-catalyzed cross-coupling of aryl bromides under mild conditions, using water-soluble, sterically demanding alkylphosphines. *J. Org. Chem.*, **69**, 7919–27.

173. Nilsson, P., Larhed, M. and Hallberg, A. (2003) A new highly asymmetric chelation-controlled Heck arylation. *J. Am. Chem. Soc.*, **125**, 3430–1.

174. Zhao, F., Shirai, M. and Arai, M. (2000) Palladium-catalyzed homogeneous and heterogeneous Heck reactions in NMP and water-mixed solvents using organic, inorganic and mixed bases. *J. Mol. Catal. A: Chem.*, **154**, 39–44.

175. Zhang, H.C. and Daves, G.D. Jr. (1993) Water facilitation of palladium-mediated coupling reactions. *Organometallics*, **12**, 1499–1500.

176. Sengupta, S. and Bhattacharya, S. (1993) Heck reaction of arenediazonium salts: a palladium-catalyzed reaction in an aqueous medium. *J. Chem. Soc., Perkin Trans. 1*, 1943–4.

177. Vallin, K.S.A., Larhed, M. and Hallberg, A. (2001) Aqueous DMF–potassium carbonate as a substitute for thallium and silver additives in the palladium-catalyzed conversion of aryl bromides to acetyl arenes. *J. Org. Chem.*, **66**, 4340–3.

178. Hiroshige, M., Hauske, J.R. and Zhou, P. (1995) Formation of C–C bond in solid phase synthesis using the Heck reaction. *Tetrahedron Lett.*, **36**, 4567–70.

179. Hayashi, M., Amano, K., Tsukada, K. and Lamberth, C. (1999) Efficient synthesis of unsaturated branched-chain sugars in aqueous media by Heck-type reaction. *J. Chem. Soc., Perkin Trans. 1*, 239–40.

180. DeVries, R.A., Vosejpka, P.C. and Ash, M.L. (1998) Homogeneous palladium catalysed vinylic coupling of aryl bromides used to make benzocyclobutene derivatives. *Chem. Ind.*, **75**, 467–78.

181. Lin, R.W., Herndon, R.C. Jr., Allen, R.H. *et al.* (1998) Preparation of carboxylic compounds and their derivatives. WO 9830529.

182. Rigby, J.H., Hughes, R.C. and Heeg, M.J. (1995) *endo*-Selective cyclization pathways in the intramolecular Heck reaction. *J. Am. Chem. Soc.*, **117**, 7834–5.

183. Nguefack, J.-F., Bolitt, V. and Sinou, D. (1997) Palladium-mediated cyclization on carbohydrate templates. 1. Synthesis of enantiopure bicyclic compounds. *J. Org. Chem.*, **62**, 1341–7.

184. Nguefack, J.-F., Bolitt, V. and Sinou, D. (1997) Palladium-mediated cyclization on carbohydrate templates. 2. Synthesis of enantiopure tricyclic compounds. *J. Org. Chem.*, **62**, 6827–32.

185. Ando, T., Sumi, S., Kawate, T. *et al.* (1984) Sonochemical switching of reaction pathways in solid–liquid two-phase reactions. *J. Chem. Soc., Chem. Commun.*, 439–40.

186. Luche, J.-L., Einhorn, J., Einhorn, C. and Sinisterra-Gago, J.V. (1990) Organic sonochemistry: a new interpretation and its consequences. *Tetrahedron Lett.*, **31**, 4125–8.

187. Luche, J.-L. (ed.) (1998) *Synthetic Organic Sonochemistry*, Plenum Press, New York, p. 53.

188. Luche, J.-L. (1993) Sonochemistry, from experiment to theoretical considerations, in *Advances in Sonochemistry*, Vol. **3** (ed. T.J. Mason), JAI Press, London, p. 85.

189. Mason, T.J. (1997) Ultrasound in synthetic organic chemistry. *Chem. Soc. Rev.*, **26**, 443–51.

190. Deshmukh, R.R., Rajagapol, R. and Srinivasan, R. (2001) Ultrasound promoted C–C bond formation: Heck reaction at ambient conditions in room temperature ionic liquids. *Chem. Commun.*, 1544–5.

191. Ambulgekar, G.V., Bhanage, B.M. and Samant, D.S. (2005) Low temperature recyclable catalyst for Heck reactions using ultrasound. *Tetrahedron Lett.*, **46**, 2483–5.

192. Pei, W. and Shen, C. (2006) Heck arylation of cyclohexene promoted by ultrasound and microwaves in ionic liquids: a novel method of the synthesis of 3-naphthylhexene. *Chin. Chem. Lett.*, **17**, 1534–6.

193. Bräse, S., Enders, D., Köbberling, J. and Avemaria, F. (1998) A surprising solid-phase effect: development of a recyclable 'traceless' linker system for reactions on solid support. *Angew. Chem. Int. Ed.*, **37**, 3413–5.

194. Dhas, N.A., Cohen, H. and Gedanken, A. (1997) *In situ* preparation of amorphous carbon-activated palladium nanoparticles. *J. Phys. Chem. B*, **101**, 6834–8.

195. Palmisano, G., Bonrath, W., Boffa, L. *et al.* (2007) Heck reactions with very low ligandless catalyst load accelerated by microwaves or simultaneous microwaves/ultrasound irradiation. *Adv. Synth. Catal.*, **349**, 2338–44.

196. Atkins, P.W. (1990) *Physical Chemistry*, Oxford University Press, Oxford, p. 938.

197. Mingos, D.M.P. and Baghurst, D.R. (1997) *Microwave Enhanced Chemistry* (eds H.M. Kingston and S.J. Haswell), ACS, Washington, DC, p. 3.

198. Gabriel, C., Gabriel, S., Grant, E.H. *et al.* (1998) Dielectric parameters relevant to microwave dielectric heating. *Chem. Soc. Rev.*, **27**, 213–24.

199. (a) Nüchter, M., Ondruschka, B., Bonrath, W. and Gum, A. (2004) Microwave assisted synthesis-critical technology overview. *Green Chem.*, **6**, 128–41; (b) Nüchter, M., Ondruschka, B., Weiss, D. *et al.* (2005) Contribution of the qualification of technical microwave systems

and to the validation of microwave-assisted reactions and processes. *Chem. Eng. Technol.*, **28**, 871–81.

200. Larhed, M., Moberg, C. and Hallberg, A. (2002) Microwave-accelerated homogeneous catalysis in organic chemistry. *Acc. Chem. Res.*, **35**, 717–27.

201. Getvoldsen, G.S., Elander, N. and Stone-Elander, S.A. (2002) UV monitoring of microwave-heated reactions – a feasibility study. *Chem. Eur. J.*, **8**, 2255–60.

202. Kappe, C.O. (2004) Controlled microwave heating in modern organic synthesis. *Angew. Chem. Int. Ed.*, **43**, 6250–84.

203. Loupy, A. (ed.) (2006) *Microwaves in Organic Synthesis*, 2nd edn, Wiley–VCH Verlag GmbH, Weinheim.

204. Pillai, S.M., Wali, A. and Satish, S. (2003) Heterogeneous palladium catalysts and microwave irradiation in Heck arylation. US 2003100625 A1.

205. Larhed, M. and Hallberg, A. (1996) Microwave-promoted palladium-catalyzed coupling reactions. *J. Org. Chem.*, **61**, 9582–4.

206. McAllister, L.A., Turner, L.K., Brand, S. *et al.* (2006) Solid phase approaches to *N*-heterocycles using a sulfur linker cleavage by SmI_2. *J. Org. Chem.*, **71**, 6497–507.

207. Byk, G., Cohen-Oana, M. and Reichman, D. (2006) Fast and versatile microwave-assisted intramolecular Heck reaction in peptide macrocyclization using microwave energy. *Biopolymers*, **84**, 274–82.

208. Xie, X., Lu, J., Chen, B. *et al.* (2004) Pd/C-catalyzed Heck reaction in ionic liquid accelerated by microwave heating. *Tetrahedron Lett.*, **45**, 809–11.

209. Choudary, B.M., Madhi, S., Chowdari, N.S. *et al.* (2002) Layered double hydroxide supported nanopalladium catalyst for Heck-, Suzuki-, Sonogashira-, and Stille-type coupling reactions of chloroarenes. *J. Am. Chem. Soc.*, **124**, 14127–36.

210. Earle, M.J., Fröhlich, U., Hug, S. *et al.* (2006) Preparation of quaternary ammonium compounds as base stable ionic liquids. WO 2006072785 A2.

211. Chandrasekhar, S., Narsihmulu, C., Sultana, S.S. and Reddy, N.R. (2002) Poly(ethylene glycol) (PEG) as a reusable solvent medium for organic synthesis. Application in the Heck reaction. *Org. Lett.*, **4**, 4399–401.

212. Solodenko, W., Wen, H., Leue, S. *et al.* (2004) Development of a continuous-flow system for catalysis with palladium(0) particles. *Eur. J. Org. Chem.*, 3601–10.

213. Ahmed, B., Barrow, D. and Wirth, T. (2006) Enhancement of reaction rates by segmented fluid flow in capillary scale reactors. *Adv. Synth. Catal.*, **348**, 1043–8.

214. Sato, M., Yoswathananont, N. and Liu, R.-H. (2006) Coupling reactions using palladium catalyst-packed flow reactor, and microreactor for the reactions. JP 2006–193483.

215. Liu, S., Fukuyama, T., Sato, M. and Ryu, I. (2004) [Bmim]NTf$_2$, a low viscosity ionic liquid is a viable reaction medium for Pd-catalyzed cross-coupling reactions. *Synlett*, 1814–6.

216. Liu, S., Fukuyama, T., Sato, M. and Ryu, I. (2004) Continuous microflow synthesis of butyl cinnamate by a Mizoroki–Heck reaction using a low-viscosity ionic liquid as the recycling reaction medium. *Org. Proc. Res. Dev.*, **8**, 477–81.

217. Snyder, D.A., Noti, C., Seeberger, P.H. *et al.* (2005) Modular microreaction system for homogeneously and heterogeneously catalyzed chemical synthesis. *Helv. Chim. Acta*, **88**, 1–9.

16

The Asymmetric Intramolecular Mizoroki–Heck Reaction in Natural Product Total Synthesis

Amy B. Dounay

*Neurosciences Chemistry, Pfizer Global Research and Development, Groton Laboratories USA,
Eastern Point Road, Groton, USA*

Larry E. Overman

*Department of Chemistry, 1102 Natural Sciences II, University of California,
Irvine, Irvine, USA*

16.1 Introduction

Mizoroki and Heck first reported their independent discoveries of palladium(0)-catalysed vinylations of aryl halides over 30 years ago [1]. This transformation, now known as the Mizoroki–Heck reaction, can be broadly defined as the palladium(0)-mediated coupling of an aryl or vinyl halide (or triflate) with an alkene. The potential utility of this transformation in complex molecule synthesis was largely unappreciated for a number of years; however, its application for the construction of complex organic molecules, including natural products, has flourished recently [2].

The lack of efficient methods for enantiocontrolled construction of tertiary and quaternary carbon stereocentres largely inspired the discovery and development of the asymmetric Mizoroki–Heck reaction [3]. Shibasaki and coworkers [4] and Overman and coworkers [5] described the first catalytic asymmetric intramolecular Mizoroki–Heck reactions in 1989. Following these initial reports, research efforts worldwide led to enormous improvements

The Mizoroki–Heck Reaction   Edited by Martin Oestreich
© 2009 John Wiley & Sons, Ltd

in this transformation to the point that it is now a powerful tool for the asymmetric synthesis of complex, polyfunctional molecules [6]. Two approaches for exploiting asymmetric intramolecular Mizoroki–Heck reactions have been developed: desymmetrization of prochiral substrates or direct formation of sp^3 stereocentres. Much of the current insight into how this reaction can be strategically employed in the enantioselective synthesis of chiral cyclic and polycyclic molecules comes from natural products total synthesis endeavours. A range of natural products, including terpenoids, polyketides and alkaloids, have now been prepared in enantioenriched form using catalytic asymmetric Mizoroki–Heck reactions as pivotal steps. This chapter will summarize the application of catalytic asymmetric Mizoroki–Heck cyclizations in natural product total synthesis. Several exploratory studies that provided critical insight into how best to apply the asymmetric Mizoroki–Heck cyclization in target-directed synthesis are also included.

16.2   General Features of the Asymmetric Intramolecular Mizoroki–Heck Reaction

A detailed discussion of the current understanding of the mechanism of the Mizoroki–Heck reaction can be found in earlier chapters of this book and in several excellent reviews [7]. Two mechanistic pathways, typically termed neutral and cationic, have been proposed to account for the differences in reactivity and enantioselectivity observed in asymmetric Mizoroki–Heck cyclizations of unsaturated triflates and halides. These pathways differ in the degree of positive charge and the number of available coordination sites assignable to the palladium(II) intermediates of the catalytic cycle. Because catalytic asymmetric Mizoroki–Heck cyclizations are typically carried out with bidentate ligands, these pathways will be illustrated with a chelating diphosphine ligand.

The cationic variant of the asymmetric intramolecular Mizoroki–Heck reaction is illustrated in Scheme 16.1 [8, 9]. This pathway accounts for the outcome of asymmetric Mizoroki–Heck reactions of unsaturated triflates or unsaturated halides in the presence of silver(I) or thallium(I) additives. Following oxidative addition of the palladium(0) catalyst, either triflate dissociation occurs or halide abstraction is promoted by silver(I) or thallium(I) salts. This event vacates a coordination site, permitting coordination of the

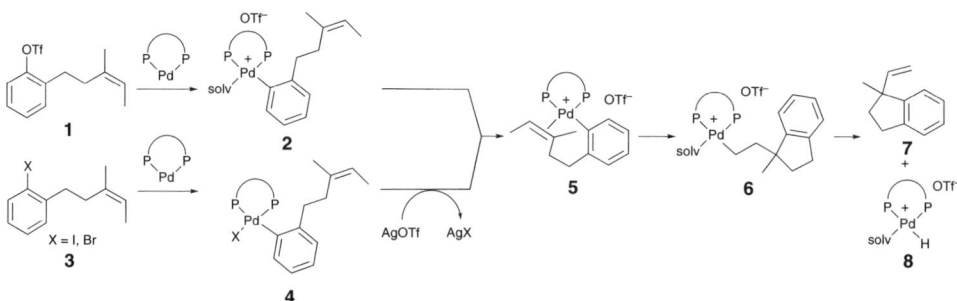

Scheme 16.1

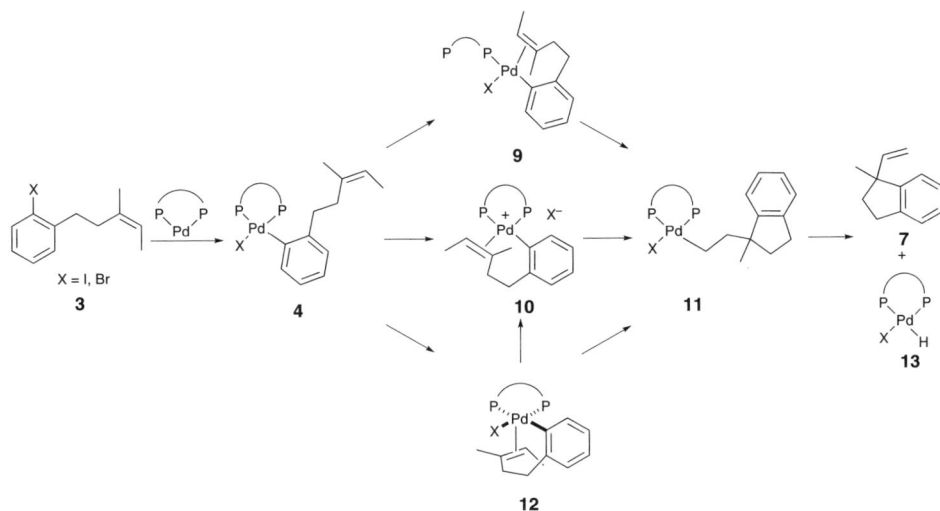

Scheme 16.2

pendant alkene [10, 11]. In the cationic manifold, both phosphorus atoms of the chiral diphosphine ligand remain coordinated to the palladium centre during coordination of the alkene to form cationic complex **5** and migratory insertion to form intermediate **6**. It is generally accepted that either of these steps could be the enantiodifferentiating step of the asymmetric C–C bond formation [12]. Partial dissociation of the phosphine ligand would diminish rigidity of the ligand and could lead to erosion of enantioselectivity. With a vacant coordination site available, ligand dissociation is unfavourable. Thus, the cationic mechanism is consistent with the enhanced enantioselectivity typically achieved by the addition of silver(I) or thallium(I) salts to the asymmetric Mizoroki–Heck reactions of unsaturated halides.

In the absence of additives such as silver(I), the Mizoroki–Heck reaction of vinyl or aryl halides is expected to proceed through a neutral reaction manifold, as depicted for the intramolecular reaction of an aryl halide in Scheme 16.2. Mizoroki–Heck reactions of this type often show low enantioselectivity, which is attributed to the formation of a neutral palladium–alkene complex by partial dissociation of the bidentate diphosphine ligand from the initially formed intermediate (**4→9**). However, as first reported by Ashimori and Overman [13] in 1992, Mizoroki–Heck cyclizations of select aryl halide substrates can proceed with high enantioselectivity in the absence of halide scavengers. With these substrates, monophosphine analogues that mimic the partially dissociated complex **9** led to the formation of Mizoroki–Heck products with low enantiopurity. These results support the conjecture that both phosphines remain bound to palladium during the enantiodifferentiating step in the neutral manifold Mizoroki–Heck reactions [14].

To date, a fully satisfactory explanation of enantioselective Mizoroki–Heck cyclizations that follow a neutral reaction pathway remains elusive. Direct halide ionization and alkene coordination (**4→10**) has been excluded as a possible pathway for this transformation, because different results are obtained when the same substrate is cyclized under cationic or neutral conditions. It is proposed that pentacoordinate complex **12** is an intermediate in

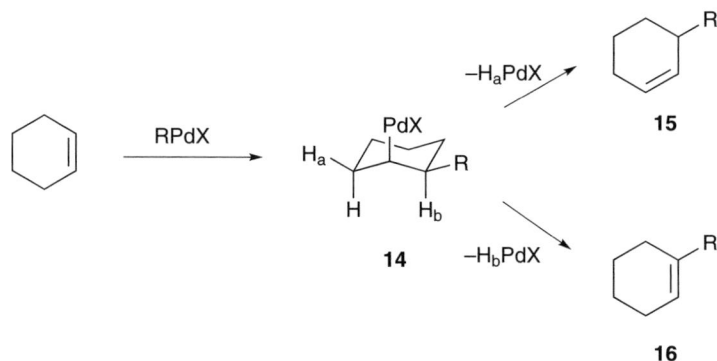

Scheme 16.3

the pathway to generate the cationic alkene complex **10**. The enantioselective step of the neutral pathway likely occurs during formation of this cationic intermediate by associative halide displacement (**4**→**12**→**10**) [14].

For intramolecular Mizoroki–Heck reactions, regiocontrol in the migratory insertion step is largely governed by the size of the ring being formed [2]. Cyclizations in *exo*-mode are highly favoured, with 5- and 6-*exo* cyclizations generally occurring in preference to their 6- and 7-*endo* counterparts [15]. If a tertiary carbon stereocentre is to be formed, then the hydrogen substituent at this stereocentre must not participate in the β-hydride elimination step. The use of cyclic alkenes as Mizoroki–Heck substrates is one design element that prevents the formation of the undesired vinylic substitution product. As illustrated in Scheme 16.3, stereospecific *syn*-addition of an aryl- or alkenyl-palladium species to a cyclic alkene produces a σ-alkylpalladium(II) intermediate (e.g. **14**) bearing a single *syn* β-hydrogen (H_a). The *syn* elimination of this hydrogen provides the allylic product **15**, preserving the newly formed tertiary stereocentre. An alternate approach in which allylsilanes are used to control the β-elimination step in asymmetric Mizoroki–Heck insertions of acyclic substrates has been developed by Tietze and coworkers (see Section 16.3.1) [16].

Another potential complication is the possibility that the hydridopalladium(II) species produced upon β-hydride elimination could re-add to the newly formed double bond. Depending upon the regio- and stereo-chemistry of this hydropalladation step, subsequent β-hydride elimination could regenerate either the initial Mizoroki–Heck product or an alkene regioisomer. The use of low reaction temperatures [17], additives such as silver or thallium salts [18, 19], or coordinating P–N ligands [20] has been shown to minimize alkene isomerization in several cases. Recent reports suggest that the use of aqueous dimethylformamide (DMF)–K_2CO_3 as a reaction medium for the Mizoroki–Heck reactions of aryl bromides may provide an inexpensive and 'green' alternative to the use of silver and thallium additives [21].

A variety of chiral phosphine ligands have been used to promote asymmetric Mizoroki–Heck reactions (**17–26**, Figure 16.1). To date, 2,2'-bis(diphenylphosphino)-1,1'-binaphthyl (BINAP; **17**) has been employed most widely; it is often employed for first testing a new asymmetric Mizoroki–Heck cyclization. However, an empirical approach typically is required for determining the optimal ligand for a given reaction. There are

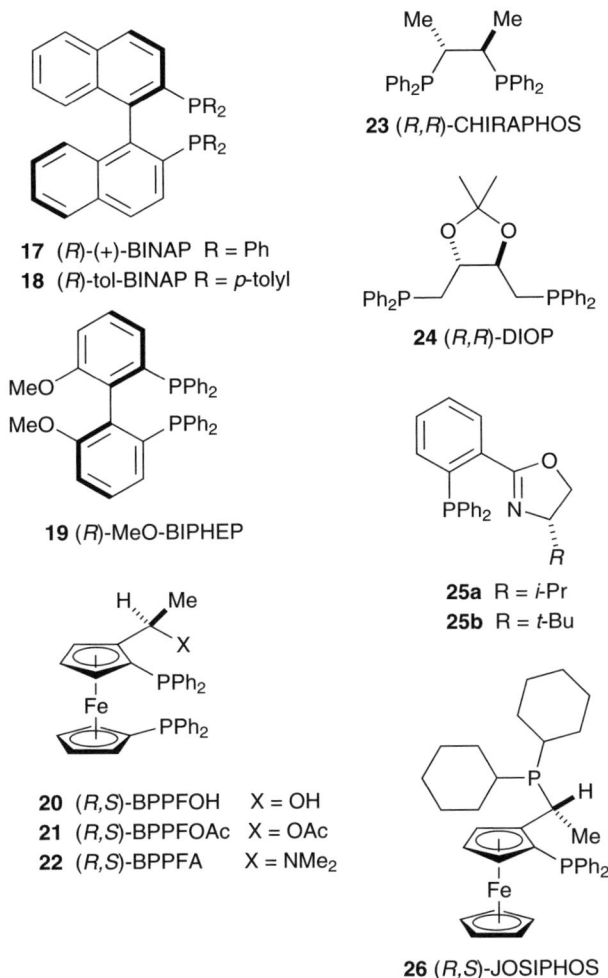

Figure 16.1 *Chiral phosphine ligands promoting asymmetric Mizoroki–Heck reactions.*

numerous examples in the recent literature in which BINAP provides unsatisfactory results, whereas alternate ligands provide superior yields and enantioselectivities [22].

A variety of other reaction variables, such as the solvent, base and various additives, can have profound effects on asymmetric Mizoroki–Heck cyclizations. These factors have been discussed in recent reviews [3]. The reaction temperature can also play an important role. At temperatures above 100 °C, catalyst decomposition can contribute to deterioration of enantioselectivity [23, 24]. Nonetheless, there are numerous examples of successful asymmetric Mizoroki–Heck cyclizations at higher temperature. Carrying out palladium-catalysed reactions in microwave reactors is now common [25], with several examples of microwave-accelerated asymmetric Mizoroki–Heck reactions being reported [26, 27]. Under microwave heating conditions, reaction times can be reduced and rigorous exclusion of oxygen is often not required. This technology was used to promote an asymmetric Mizoroki–Heck cyclization in the recent total synthesis of minfiensine (see Section 16.4.4).

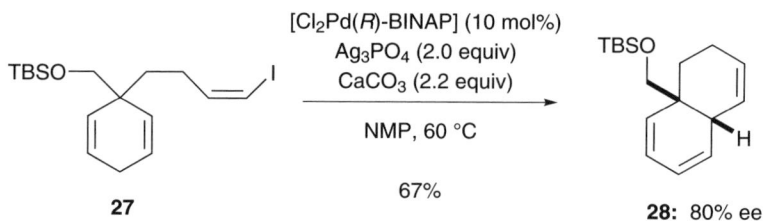

Scheme 16.4

16.3 Natural Product Total Synthesis: Formation of Tertiary Stereocentres

16.3.1 Terpenoids

In their first report of the asymmetric Mizoroki–Heck cyclization of a vinyl iodide substrate, Shibasaki and coworkers [4] found that the enantioselectivities achieved with Ag_2CO_3 were superior to those attained with amine or inorganic bases. In a later study, they demonstrated that enantioselectivities observed in the group-selective cyclization of iodide **27** to form *cis*-decalin triene **28** depended on the counterion of the silver salt, with Ag_3PO_4 providing the best results (Scheme 16.4) [28]. Reactions conducted in polar solvents, such as *N*-methylpyrolidinone (NMP), generally proceeded with higher enantioselectivity than those caried out in nonpolar solvents, such as toluene. The analogous Mizoroki–Heck cyclization of trisubstituted vinyl iodide **29** provided the more highly functionalized *cis*-decalin **30** in up to 87% ee (Scheme 16.5) [29]. Ultimately, the most significant improvement in group-selective asymmetric Mizoroki–Heck cyclizations to form substituted *cis*-decalins was achieved by the use of vinyl triflates rather than vinyl iodides as substrates (Scheme 16.6) [29]. For example, treatment of prochiral triflate **31** with $Pd(OAc)_2$, (*R*)-BINAP and K_2CO_3 in toluene afforded *cis*-decalins **32** in moderate yield and high enantioselectivity (91% ee). Mizoroki–Heck cyclizations of vinyl triflates, rather than vinyl iodides, offer the twofold advantage of improved enantioselectivity without the need for stoichiometric amounts of silver salts.

Further optimization of asymmetric Mizoroki–Heck cyclizations of prochiral vinyl triflates focused on the effects of base, solvent and additives [30, 31]. Tertiary amine bases, such as *i*-Pr_2NEt, gave results inferior to those realized with K_2CO_3. In the absence of additives, reactions in toluene gave superior yields to reactions conducted in 1,2-dichloroethane (DCE) or polar solvents. However, additives such as tertiary alcohols (e.g. pinacol) or KOAc

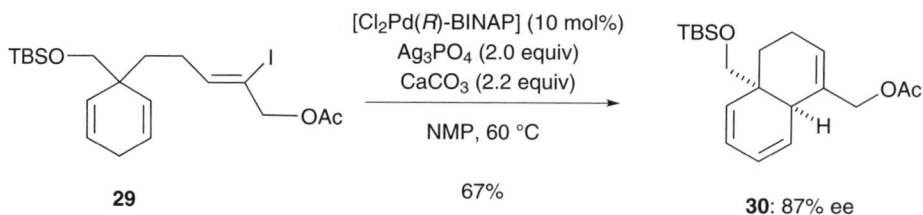

Scheme 16.5

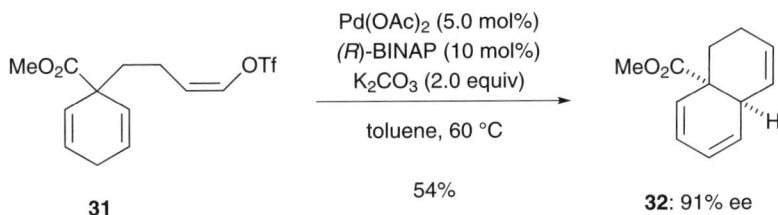

Scheme 16.6

accelerated the reactions in DCE [30, 31]. ^{31}P-NMR studies suggest that these additives prevent the DCE-promoted oxidation of the catalytically active palladium(0) species to a palladium(II) species. Use of $Pd(OAc)_2$, (R)-BINAP, K_2CO_3 and a tertiary alcohol or potassium acetate in DCE for the Mizoroki–Heck cyclization of terminal vinyl triflates provides *cis*-decalin products in high yield and up to 95% ee [31]. Shibasaki and coworkers [30, 31] demonstrated the utility of this catalytic asymmetric synthesis of *cis*-decalins by an enantioselective synthesis of (+)-vernolepin (**33**, Figure 16.2), which is described in Chapter 13.

Recent studies by Kiely and Guiry [32] compared the utility of BINAP and diphenylphosphinooxazoline ligands in the catalytic asymmetric synthesis of *cis*-decalins. Their results suggest that BINAP provides superior conversion to Mizoroki–Heck products, although both ligand systems afford *cis*-decalin products with high enantioselectivities.

By shortening the tether connecting a vinyl iodide or triflate to a cyclohexadiene, tetrahydroindene ring systems can also be formed [33]. Early attempts to prepare an enantioenriched bicyclo[4.3.0]nonane by group-selective Mizoroki–Heck cyclization of terminal vinyl iodide **36** delivered product **37** in good yield, but with no enantioselectivity (Scheme 16.7). Attempts to optimize this reaction proved unsuccessful. However, the Mizoroki–Heck cyclization of closely related vinyl iodide **38** provided *cis*-hydrindane **39** in 78% yield and 82% ee (Scheme 16.8). This result demonstrates that modest structural variation in Mizoroki–Heck substrates can markedly affect enantioselectivity. Furthermore, use of a palladium(0) catalyst generated by the *in situ* reduction of $[Cl_2Pd(R)$-BINAP] with cyclohexene provided tetrahydroindene **39** with a slight improvement in enantioselectivity (86% ee versus 82% ee). Mizoroki–Heck cyclization of vinyl triflate analogue **40** did not require the use of silver salts (Scheme 16.9); however, this reaction delivered **39** in

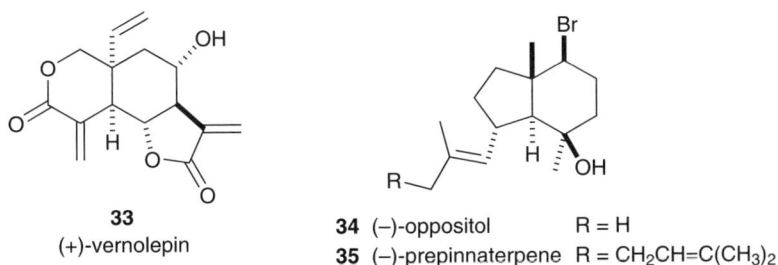

Figure 16.2 *Synthetic targets accessed by Shibasaki and coworkers.*

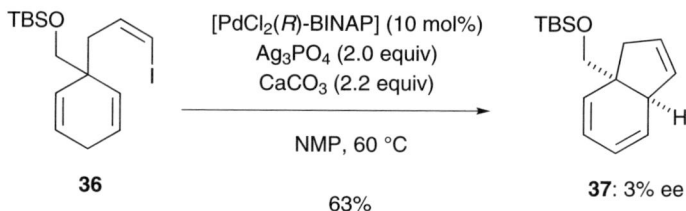

Scheme 16.7

slightly lower yield and enantiopurity. This asymmetric route to tetrahydroindenes was implemented in formal total syntheses of the brominated terpenes (−)-oppositol (**34**) and (−)-prepinnaterpene (**35**, Figure 16.2), as described in Chapter 13 [34].

One limitation of the asymmetric intramolecular Mizoroki–Heck reaction that continues to pose a challenge is the lack of regioselectivity in the β-hydride elimination step when acyclic alkenes are substrates. To address this issue, Tietze and coworkers demonstrated that regioselective elimination to generate products having acyclic alkene side chains can be realized by the use of allylsilanes as terminating units in asymmetric Mizoroki–Heck cyclizations. For example, treatment of aryl iodide **41** with $Pd_2(dba)_3$·$CHCl_3$, (*R*)-BINAP and Ag_3PO_4 in DMF at 80 °C gave vinyl tetralin **42** in 91% yield and 92% ee (Scheme 16.10) [16]. An analogous Mizoroki–Heck precursor lacking the silyl substituent cyclized under identical reaction conditions to afford a mixture of alkene regioisomers. Early studies in this area focused on the use of aryl iodides as Mizoroki–Heck precursors; however, more recent investigations explored the use of vinyl iodides and triflates [35]. The applicability of allylsilane-terminated asymmetric Mizoroki–Heck reactions in natural product synthesis was illustrated by the three-step conversion of product **42** to 7-desmethyl-2-methoxycalamenene (**43**) [16b,c], a natural member of the cadinene sesquiterpenoids. Substituted tetrahydroisoquinolines and tetrahydrobenzo[*d*]azepines have also been synthesized in high enantiopurity using related intramolecular asymmetric Mizoroki–Heck reactions of allylsilanes [16a].

Interception of the alkylpalladium intermediate in a Mizoroki–Heck cyclization can be parlayed to introduce an additional stereocentre and increase molecular complexity. Shibasaki and coworkers insightfully recognized that the linear triquinanes represent a prime opportunity for exploring the feasibility of cascade Mizoroki–Heck processes in the context of natural product synthesis, with their studies leading to the first reported asymmetric Mizoroki–Heck cyclization–η^3-allyl nucleophilic trapping sequence.

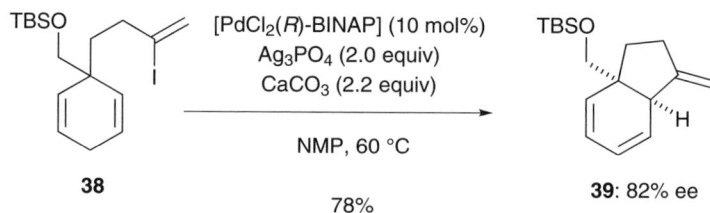

Scheme 16.8

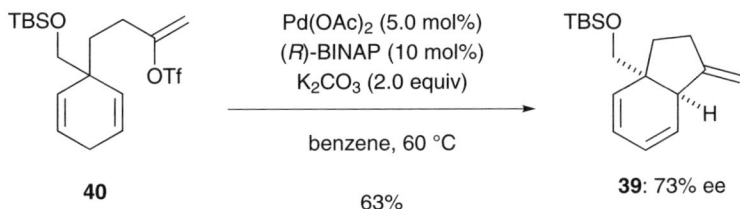

40   63%   **39**: 73% ee

Scheme 16.9

The cascade process was initially explored with prochiral trienyl iodide **44** (Scheme 16.11) [36, 37]. Mizoroki–Heck cyclization of this precursor produced η^3-allylpalladium species **45**, which was trapped by acetate at the least-hindered terminus of the η^3-allyl system to provide *cis*-bicyclo[3.3.0]octadiene **46** in 60% yield, albeit with very low enantioselectivity (20% ee). Attempts to use silver salts as halide scavengers in this reaction led to the decomposition of **44**, presumably resulting from the sensitivity of the cyclopentadienyl moiety. Mizoroki–Heck cyclization of prochiral vinyl triflate **47** with Pd(OAc)$_2$, (*S*)-BINAP and tetrabutylammonium acetate was more productive, giving diquinane product **48** in excellent yield and 80% ee (Scheme 16.12). The corresponding allylic amine **49** was obtained in analogous fashion using benzylamine as the nucleophile [38]. Allylic acetate **48** was elaborated in seven steps to triquinane β-ketoester **50**, an intermediate in Shibasaki and coworkers' [39] earlier total syntheses of $(\pm)$-$\Delta^{9(12)}$-capnellene-$3\beta,8\beta,10\alpha$-triol (**51**) and $(\pm)$-$\Delta^{9(12)}$-capnellene-$3\beta,8\beta,10\alpha,14$-tetraol (**52**).

Shibasaki and coworkers [40] also demonstrated the use of soft carbanionic nucleophiles, initially sodium dimethyl malonate, in cascade asymmetric Mizoroki–Heck cyclization–η^3-allyl trapping sequences. This conversion succeeds with various soft carbanionic nucleophiles to provide functionalized bicyclo[3.3.0]octane derivatives **55** in excellent yields (72–92%) and up to 94% ee (Scheme 16.13). The enantioselectivity of these Mizoroki–Heck reactions is significantly diminished in the absence of NaBr; a speculative rationale to account for the effect of the NaBr additive has been advanced [40].

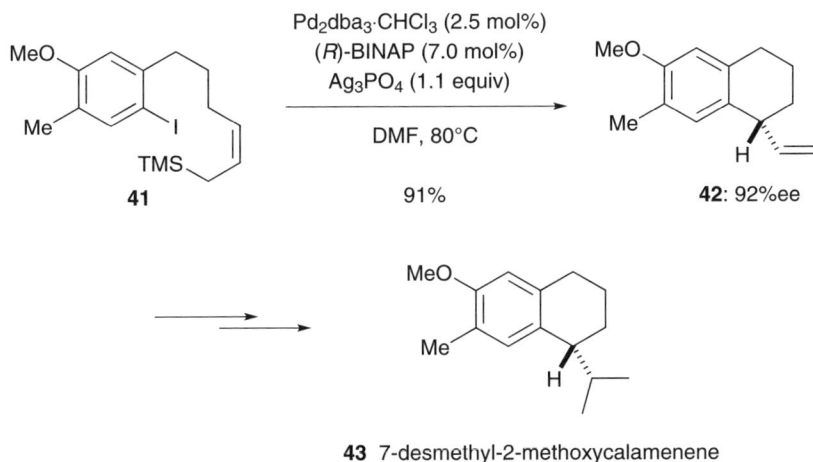

41   91%   **42**: 92%ee

43 7-desmethyl-2-methoxycalamenene

Scheme 16.10

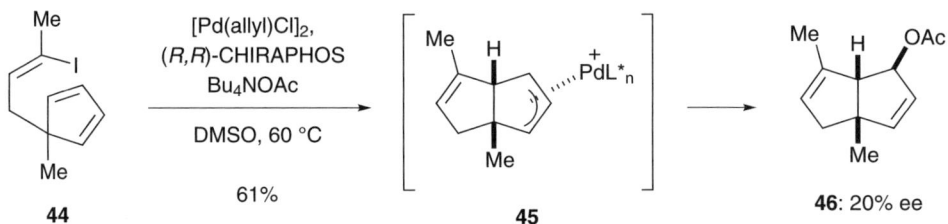

44   61%   **45**   **46**: 20% ee

Scheme 16.11

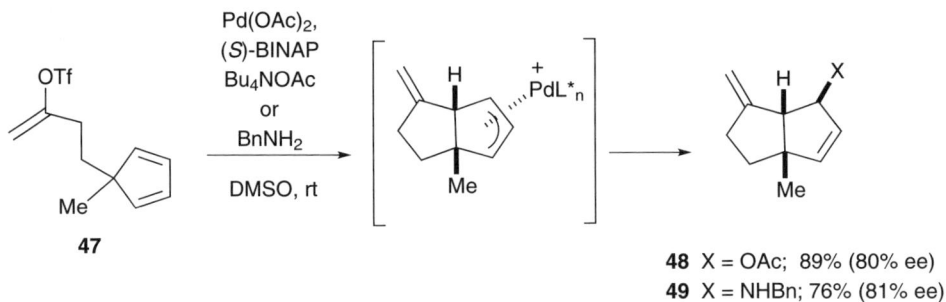

47

48 X = OAc; 89% (80% ee)
49 X = NHBn; 76% (81% ee)

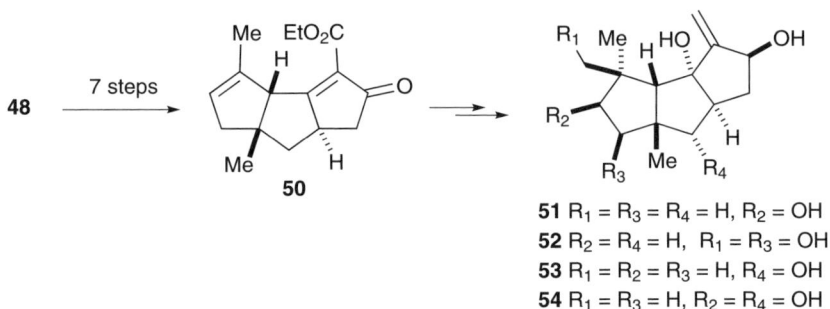

48 —7 steps→ **50**

51 R$_1$ = R$_3$ = R$_4$ = H, R$_2$ = OH
52 R$_2$ = R$_4$ = H, R$_1$ = R$_3$ = OH
53 R$_1$ = R$_2$ = R$_3$ = H, R$_4$ = OH
54 R$_1$ = R$_3$ = H, R$_2$ = R$_4$ = OH

Scheme 16.12

Nu⁻: Na enolate of

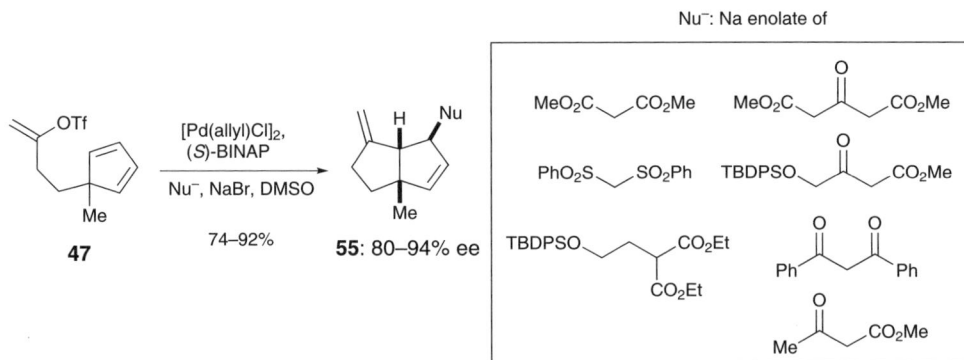

47   74–92%   **55**: 80–94% ee

MeO$_2$C⟋⟍CO$_2$Me   MeO$_2$C⟋⟍⟋CO$_2$Me

PhO$_2$S⟋SO$_2$Ph   TBDPSO⟋⟍CO$_2$Me

TBDPSO⟋⟍CO$_2$Et
⟍CO$_2$Et

Ph⟋⟍Ph

Me⟋⟍CO$_2$Me

Scheme 16.13

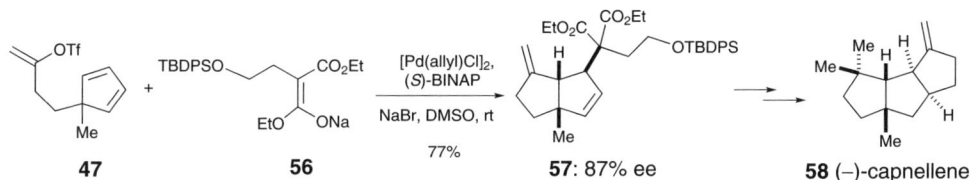

Scheme 16.14

The use of carbanionic nucleophiles in the Mizoroki–Heck cyclization–η^3-allyl nucleophilic trapping sequence allowed for streamlined access to the triquinane core common to various members of the capnellene family of natural products. For example, Shibasaki and coworkers obtained diquinane **57** in 77% yield and 87% ee by Mizoroki–Heck cyclization of trienyl triflate **47** in the presence of malonate nucleophile **56** Scheme 16.14). It is notable that two new C–C bonds and three stereocentres are generated in this reaction. Eleven additional steps were used to convert intermediate **57** to ($\pm$)-$\Delta^{9(12)}$-capnellene (**58**). This first catalytic asymmetric total synthesis ($\pm$)-$\Delta^{9(12)}$-capnellene was achieved in 19 steps and 20% overall yield from commercially available materials. A related approach has recently been employed to prepare intermediates en route to capnellenols **53** and **54** (Scheme 16.12) [41].

16.3.2   Alkaloids

Catalytic asymmetric Mizoroki–Heck cyclizations have been employed to construct a variety of enantioenriched nitrogen heterocycles. In one early example, Shibasaki and coworkers [42] described the asymmetric Mizoroki–Heck cyclization of N-allylpyridones to give simple unsaturated indolizidines in useful enantiopurity. Initial attempts to effect enantioselective cyclization of dihydropyridone vinyl iodide **59** using Pd–BINAP in the presence of Ag$_3$PO$_4$ gave a mixture of unsaturated indolizidine isomers **60** and **61** (Scheme 16.15). The minor isomer **60** could be converted quantitatively to the major α,β-unsaturated lactam product **61** by subsequent reaction of the former with catalytic Pd/C in MeOH.

Despite the demonstrated efficacy of BINAP ligands in related carbocyclic systems, the Pd–BINAP-catalysed Mizoroki–Heck cyclization of dienyl iodide **59** proceeded slowly even at 90 °C, with indolizidine **61** being formed in only 34% ee (Scheme 16.15). A survey of other chiral diphosphine ligands identified (R,S)-BPPFOH (BPP-FOH = 1-[1′,2-bis(diphenylphosphino)ferrocenyl]ethyl alcohol, **20**, Figure 16.1) as the optimal ligand for this transformation [43]. Using (R,S)-BPPFOH, indolizidine diene **61** was produced as the only product in 45% yield and 74% ee. The structurally similar 1-[1′,2-bis(diphenylphosphino)ferrocenyl]ethylamine (BPPFA, **22**) and 1-[1′,2-bis(diphenylphosphino)ferrocenyl]ethyl acetate (BPPFOAc, **23**) ligands (Figure 16.1) imparted somewhat lower enantiocontrol in this transformation. The utility of indolizidine intermediate **61** in natural product synthesis was demonstrated by its conversion to lentiginosine (**62**) and gephyrotoxin 209D (**63**) [44].

Sulikowski and coworkers [45] reported enantioselective Mizoroki–Heck cyclizations of the related N-acyl tetrahydropyridine bromide **64** (Scheme 16.16). In these studies, the selection of solvent proved to be critical. Whereas Mizoroki–Heck cyclization of

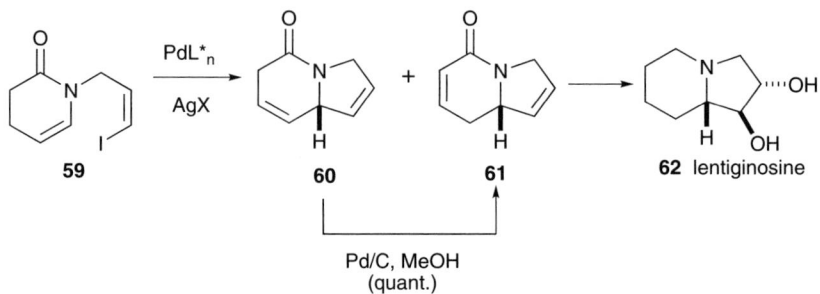

59   **60**   **61**   **62** lentiginosine

Pd/C, MeOH
(quant.)

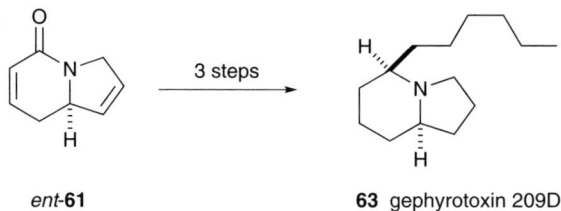

3 steps

ent-**61**   **63** gephyrotoxin 209D

Scheme 16.15

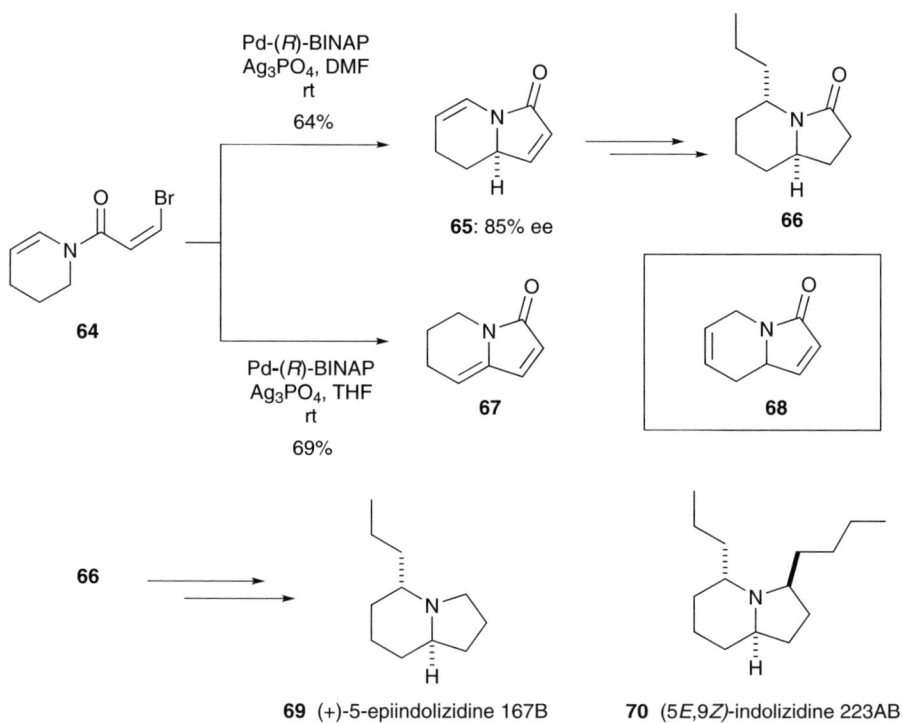

Pd-(*R*)-BINAP
Ag₃PO₄, DMF
rt

64%

65: 85% ee   **66**

64

67   **68**

Pd-(*R*)-BINAP
Ag₃PO₄, THF
rt

69%

66

69 (+)-5-epiindolizidine 167B   **70** (5*E*,9*Z*)-indolizidine 223AB

Scheme 16.16

dienyl bromide **64** using Pd–BINAP and Ag$_3$PO$_4$ in DMF at room temperature delivered enelactam **65** in 85% ee, use of tetrahydrofuran (THF) as solvent yielded only the achiral indolizidinone **67**. In either case, the presumptive initially formed Mizoroki–Heck product, **68**, was not observed. The role of the solvent in dictating the outcome of double-bond migrations to give either indolizidinones **65** or **67** is not clear. Unsaturated indolizidinone **65** was converted in four steps to indolizidinone **66**, a common intermediate in total syntheses of two indolizidine alkaloids: 5-epiindolizidine 167B (**69**) [46] and 5*E*,9*Z*-indolizidine 223AB (**70**) [47].

16.4   Natural Product Total Synthesis: Formation of Quaternary Stereocentres

16.4.1   Terpenoids

Many examples of successful ring construction with concomitant introduction of a quaternary stereocentre followed the initial report by Overman and coworkers [5] of the use of catalytic asymmetric Mizoroki–Heck cyclizations for the enantioselective construction of quaternary carbon stereocentres. In the arena of terpenoid total synthesis, Shibasaki and coworkers [48] studied asymmetric Mizoroki–Heck cyclizations of cyclohexadienyl aryl triflate **71** (Scheme 16.17). Cyclization of this triflate using Pd–BINAP in the presence of K$_2$CO$_3$ provided a mixture of hydrophenanthrenes **72** and **73** in 62% yield. Both 5-*exo* and 6-*exo* cyclizations are generally favourable; but, in this case, only the fused bicyclic products resulting from 6-*exo* cyclization were observed. The authors propose that severe steric repulsion in the 5-*exo* transition state prevents formation of spirocyclic products. Furthermore, the reluctance of tetrasubstituted double bonds to coordinate to palladium(II) would hamper the 5-*exo* pathway. The quantitative conversion of nonconjugated diene **72** to its conjugated isomer **73** was achieved by treating the mixture of isomeric products with catalytic amounts of naphthalene·CrO$_3$. Diene **73** then was transformed over several steps to enone **74**, an intermediate that had been employed previously in total syntheses of the tricyclic diterpenes (±)-kaurene (**75**) [49] and (±)-abietic acid (**76**) [50] (Figure 16.3).

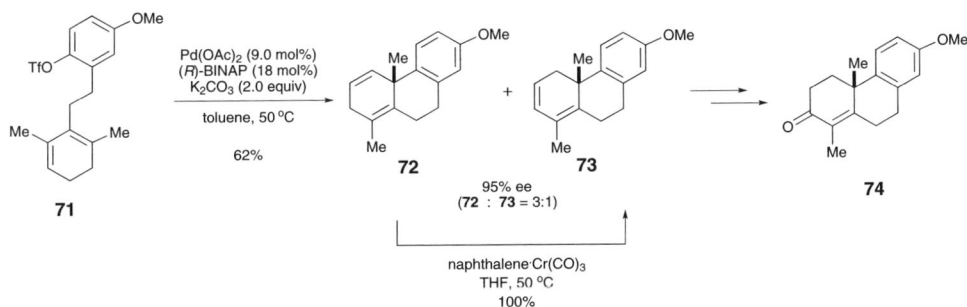

Scheme 16.17

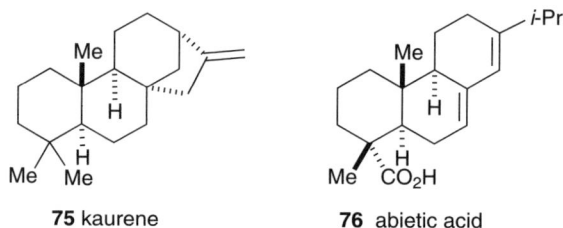

75 kaurene   **76** abietic acid

Figure 16.3   *Tricyclic diterpenes accessible by means of Mizoroki–Heck chemistry.*

16.4.2   Retinoids

Diaz *et al.* [51] explored asymmetric Mizoroki–Heck cyclization/hydride capture reactions (reductive Mizoroki–Heck-type processes) in the assembly of a series of conformationally restricted retinoid analogues bearing benzylic quaternary stereocentres (Scheme 16.18). For example, aryl iodide **77** was allowed to react with Pd(OAc)$_2$, (*R*)-BINAP, Ag-exchanged zeolite, CaCO$_3$ and sodium formate in acetonitrile to produce **79** in moderate yield and good enantioselectivity (81% ee). Cyclization of isomer **78** under identical conditions provided isomeric product **80** in slightly higher yield, albeit with lower enantiomeric purity. Poor stereoinduction was observed when cyclizations of this type were conducted under neutral rather than cationic conditions. Because the σ-bonded palladium intermediates generated upon 5-*exo* cyclization do not have β-hydrogen atoms, a hydride source (HCO$_2$Na) was required to reduce this intermediate and regenerate the palladium(0) catalyst.

16.4.3   Polyketides

Catalytic asymmetric intramolecular Mizoroki–Heck reactions have been employed to synthesize several polyaromatic polyketide natural products. In one such endeavour, the catalytic asymmetric cyclization of alkenyl naphthyl triflate **84** was a central step in total syntheses of halenaquinone (**88**) and halenaquinol (**89**) recorded in the Shibasaki group laboratory (Scheme 16.19) [52]. Naphthyl triflate **84**, which arose from Suzuki coupling of triflate **81** with alkylborane **82**, was cyclized using the conditions that had become nearly

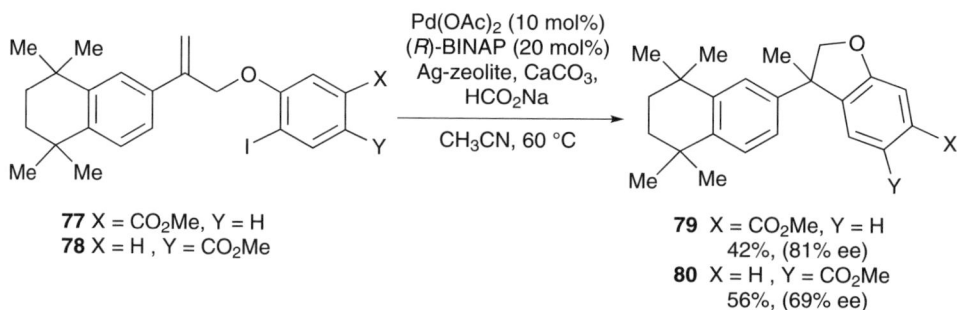

Pd(OAc)$_2$ (10 mol%)
(*R*)-BINAP (20 mol%)
Ag-zeolite, CaCO$_3$,
HCO$_2$Na

CH$_3$CN, 60 °C

77 X = CO$_2$Me, Y = H
78 X = H , Y = CO$_2$Me

79 X = CO$_2$Me, Y = H
        42%, (81% ee)
80 X = H , Y = CO$_2$Me
        56%, (69% ee)

Scheme 16.18

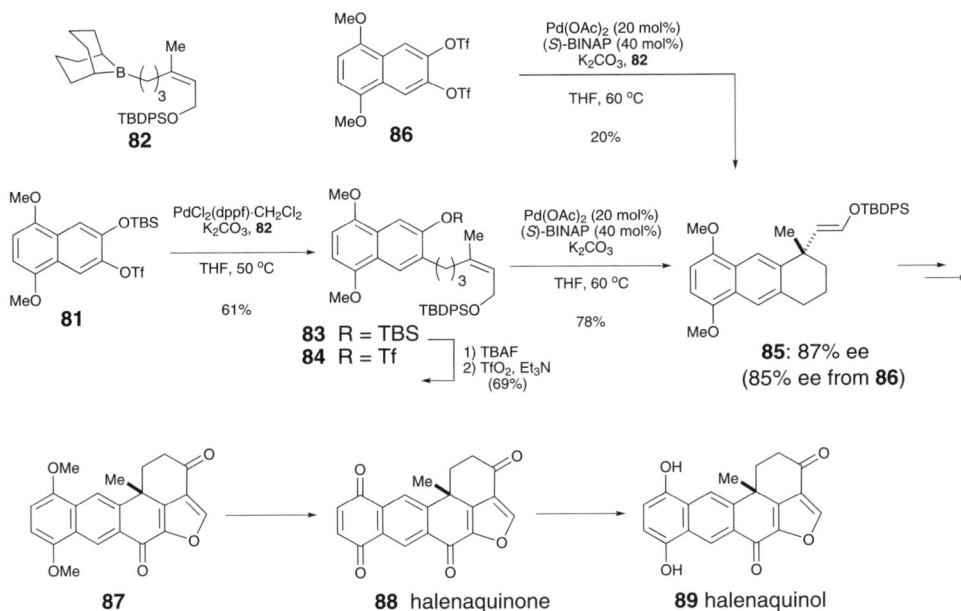

Scheme 16.19

standard for Mizoroki–Heck reactions of aryl triflates (Pd(OAc)$_2$, (S)-BINAP, K$_2$CO$_3$, THF) to provide tetrahydroanthracene **85** in 78% yield and 87% ee.

Alternatively, tetrahydroanthracene **85** was also obtained directly from symmetrical ditriflate **86** by a Suzuki cross-coupling/asymmetric Mizoroki–Heck cascade reaction. In this case, reaction of ditriflate **86** with alkylborane **82**, Pd(OAc)$_2$, (S)-BINAP and K$_2$CO$_3$ in THF at 60 °C directly gave product **85** in high enantiopurity, albeit in 20% yield. Although the yield for this one-pot conversion was poor, this transformation represents a novel application of the asymmetric Mizoroki–Heck cyclization and remains the only reported example of cascade Suzuki cross-coupling/asymmetric Mizoroki–Heck cyclization. A series of additional synthetic transformations was required to convert **85** to pentacyclic intermediate **87**, which had earlier been converted by Harada *et al.* [53] to halenaquinone (**88**) and halenaquinol (**89**).

The identification of novel ways to incorporate an asymmetric intramolecular Mizoroki–Heck reaction as part of a cascade cyclization sequence has led to attractive approaches for assembling complex polycyclic molecules. Keay and coworkers [54] reported the use of a double Mizoroki–Heck cyclization as the pivotal step in the asymmetric total synthesis of xestoquinone (**93**), a reduced congener of halenaquinone (Scheme 16.20). In this step, naphthyl triflate **90** was cyclized with Pd$_2$(dba)$_3$ (dba = dibenzylideneacetone), (S)-BINAP and 1,2,2,6,6-pentamethylpiperidine (PMP) in toluene at 110 °C to give pentacyclic product **92** with impressive efficiency and moderate enantioselectivity. This conversion proceeds by initial asymmetric 6-*exo* Mizoroki–Heck cyclization to form the central six-membered carbocycle and install the benzylic quaternary stereocentre. The first cyclization event is followed by a second Mizoroki–Heck reaction in which neopentyl

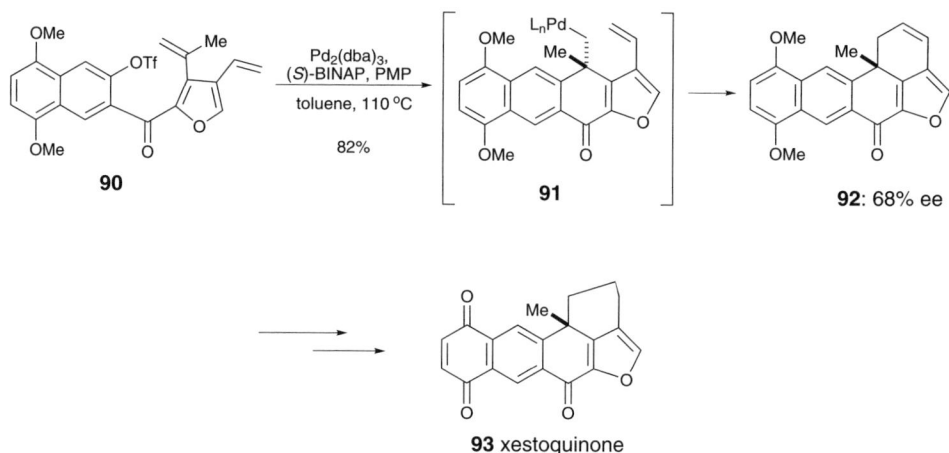

Scheme 16.20

organopalladium intermediate **91** cyclizes upon the pendant alkene to form pentacyclic product **92**. The second insertion occurs in a 6-*endo* sense, because the typically favoured 5-*exo* cyclization would be disfavoured by developing ring strain. As has been observed in other cases, the use of the aryl triflate derivative was critical to the success of this reaction. Attempts to effect the Mizoroki–Heck cyclization of the analogous aryl bromide substrate met with little success, as the desired product was obtained in low enantiopurity (5–13% ee) under either neutral (PMP) or cationic (Ag$_3$PO$_4$) conditions. Only two additional steps were required to complete the first enantioselective synthesis of xestoquinone (**93**).

To extend this asymmetric Mizoroki–Heck cyclization cascade to the synthesis of hale-naquinone, Lau and Keay [55] systematically studied the effects of remote substituents on such reactions. These investigations demonstrated that the presence of an internal methyl substituent on the distal alkene (R^1) of cyclization substrate **94** improved enantioselectivity in forming the benzylic quaternary stereocentre (Scheme 16.21 and Table 16.1, entries 1 and 2). Likewise, the presence of a methyl group *ortho* to the triflate (R^2) resulted in enhanced enantiocontrol (Table 16.1, entry 4). Unfavourable steric interactions apparently exist when both R^1 and R^2 are methyl, as enantioselectivity is significantly lower in the cyclization of **94d** (Table 16.1, entry 5). Surprisingly, the desired reaction pathway is almost completely shut down when substrate **94e**, wherein R^1 = phenyl, is cyclized using (*R*)-BINAP as the ligand (Table 16.1, entry 6), with the major product in this case being tricyclic ketone **96**. The authors suggest that unfavourable interactions between the phenyl substituent and the BINAP ligand disfavour the neopentyl alkylpal-ladium intermediate from coordinating to the styrene side chain, allowing a competitive hydride transfer process to dominate. Use of (*R,R*)-CHIRAPHOS (CHIRAPHOS = 2*R*,3*R*-bis(diphenylphosphino)butane, **23**) as the ligand for this transformation shifts the reaction in favour of formation of the tetracyclic product **95e** (Table 16.1, entry 7). More recent studies suggest that alternate ligands such as 2,2′-bis(diphenylphosphino)-3,3′-binaphtho[2,1-*b*]furan (BINAPFu) [56] and other variably substituted BINAP derivatives offer some advantages for this cascade cyclization reaction [57, 58].

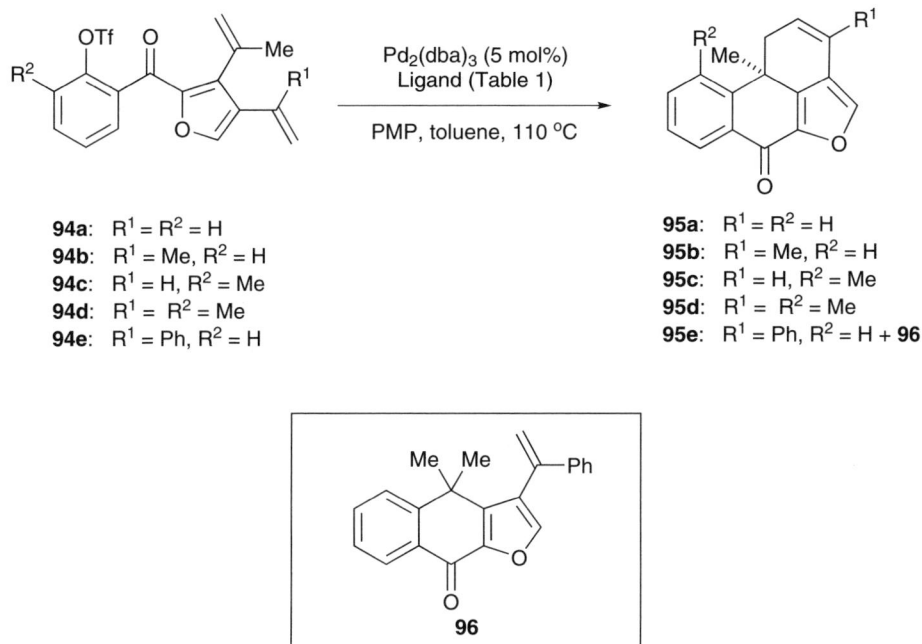

94a: $R^1 = R^2 = H$
94b: $R^1 = Me$, $R^2 = H$
94c: $R^1 = H$, $R^2 = Me$
94d: $R^1 = R^2 = Me$
94e: $R^1 = Ph$, $R^2 = H$

95a: $R^1 = R^2 = H$
95b: $R^1 = Me$, $R^2 = H$
95c: $R^1 = H$, $R^2 = Me$
95d: $R^1 = R^2 = Me$
95e: $R^1 = Ph$, $R^2 = H$ + 96

96

Scheme 16.21

Additional investigations by Keay and coworkers [57] provide insight into the effects of various solvents and bases on Pd–BINAP-catalysed polycyclizations to form tetracyclic products **95a** and **95e**. These studies showed that the use of dioxane as solvent or PMP as base promotes formation of the undesired tricyclic product, because both reagents can serve as hydride donors. Further experiments showed that 1,4-diazabicyclo[2.2.2]octane (DABCO), which is an ineffective hydride donor because of its bridged structure, can be substituted for PMP to reduce formation of by-product **96** without significant loss of enantioselectivity. Likewise, the use of toluene rather than dioxane as the solvent maximizes conversion to the desired tetrayclic products.

Table 16.1   *(Remote) substituent effects in an asymmetric Mizoroki–Heck cyclization cascade*

Entry	Substrate	Ligand	Product	Yield (%)	ee (%)
1	**94a**	(*R*)-BINAP	**95a**	83	71 (*R*)
2	**94b**	(*R*)-BINAP	**95b**	78	90 (*R*)
3	**94b**	(*R,R*)-CHIRAPHOS	**95b**	61	24 (*S*)
4	**94c**	(*R*)-BINAP	**95c**	71	96 (*R*)
5	**94d**	(*R*)-BINAP	**95d**	68	71(*R*)
6	**94e**	(*R*)-BINAP	**95e + 96** (1 : 13)	91	—
7	**94e**	(*R,R*)-CHIRAPHOS	**95e**	66	77 (*S*)

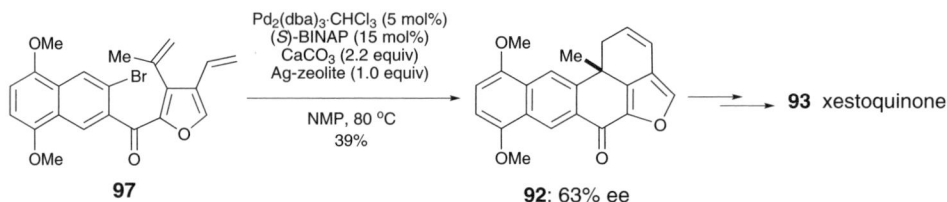

Scheme 16.22

Following the report by Keay and coworkers [54] that naphthyl bromides were poor substrates in the catalytic asymmetric polyene cyclization to form pentacycle **92**, Shibasaki and coworkers [59] revealed reaction conditions that improve somewhat this problematic transformation. For example, reaction of naphthyl bromide **97** with Pd$_2$(dba)$_3$·CHCl$_3$, (S)-BINAP, CaCO$_3$ and Ag-exchanged zeolite (1.0 equiv) in NMP at 80 °C gave product **92** with 63% ee, albeit in only 39% yield (Scheme 16.22). When either 2 or 6 equiv of Ag-exchanged zeolite were used, both the yield and the enantioselectivities were diminished. Furthermore, as in previous examples, higher yields and superior enantioselectivies were achieved using Ag-exchanged zeolite as a halide scavenger rather than Ag$_3$PO$_4$. Pentacyclic intermediate **92** was converted by Shibasaki and coworkers [59] to xestoquinone (**93**) by the sequence developed earlier by Keay and coworkers [54].

We conclude our discussion of aromatic polyketide total synthesis with the preliminary investigations of Shibasaki and coworkers [60, 61] toward a total synthesis of the more elaborate pentacyclic polyketide, wortmannin (**98**, Figure 16.4). As a key step in their proposed synthesis of this natural product, an asymmetric Mizoroki–Heck cyclization of a racemic dienyl hydrindane triflate was envisaged to install the allylic quaternary carbon stereocentre of wortmannin and effect resolution of this intermediate (Scheme 16.23). An initial study of the Mizoroki–Heck cyclization of dienyl triflate **99** with achiral catalysts demonstrated that a bidentate phosphine ligand such as DPPP (1,3-bis(diphenylphosphino)propane) could impart a high level of diastereoselection in forming the new quaternary carbon stereocentre (**100** : **101** = 17 : 1, 90% yield). A series of chiral bidentate phosphine ligands was then surveyed, with the best results being obtained with Pd(OAc)$_2$, (R)-tol-BINAP and K$_2$CO$_3$ in toluene at 100 °C. Using these conditions, triflate **99** gave tetracyclic product **100** with high diastereoselectivity (11 : 1) and enantiomeric purity (96% ee). Although the yield of product **100** was only 20%, this was the first demonstration of kinetic resolution using the asymmetric intramolecular Mizoroki–Heck reaction.

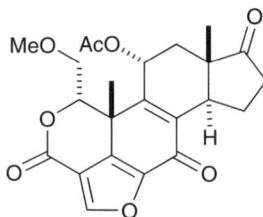

98 wortmannin

Figure 16.4   *A pentacyclic polyketide as a target.*

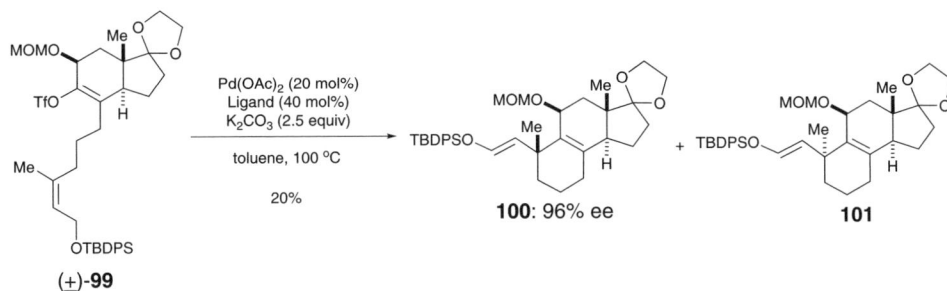

ligand	time (h)	Yield (%) (100:101)	ee of 100 (%)
(R,R)-CHIRAPHOS	20	14 (6:1)	4
(R)-BINAP	2	17 (5:1)	97
(R)-tol-BINAP	1.5	20 (11:1)	96

Scheme 16.23

16.4.4 Alkaloids

Overman and coworkers have extensively developed the utility of asymmetric Mizoroki–Heck reactions for constructing quaternary carbon stereocentres through various alkaloid total synthesis endeavours. A particular focus of these studies has been the catalytic asymmetric synthesis of enantioenriched 3,3-disubstituted oxindoles, because oxindoles of this type are versatile precursors of many natural alkaloids. In an early, unanticipated discovery, it was found that either enantiomer of spirocyclic oxindole **103** could be formed with moderate selectivity using a single enantiomer of a chiral diphosphine ligand (Scheme 16.24) [13]. Specifically, the use of Ag_3PO_4 as an HI scavenger in the cyclization of **102** produced (*S*)-**103** in 71% ee, whereas use of the amine base PMP under otherwise identical reaction conditions yielded (*R*)-**103** in 66% ee.

As a prelude to an asymmetric synthesis of physostigmine, a systematic study of the asymmetric intramolecular Mizoroki–Heck reaction of (*E*)- and (*Z*)-α,β-unsaturated iodoanilides **104** was carried out (Scheme 16.25) [14, 62]. The reaction conditions developed during Overman's previous experiments were employed: (1) silver-promoted cyclizations were conducted with 5 mol% $Pd_2(dba)_3 \cdot CHCl_3$ and 12 mol% (*R*)-BINAP in the presence of 2 equiv of Ag_3PO_4 at 100 °C in *N,N*-dimethylacetamide (DMA) and (2) base-promoted cyclizations were carried out identically in the presence of 4 equiv of PMP,

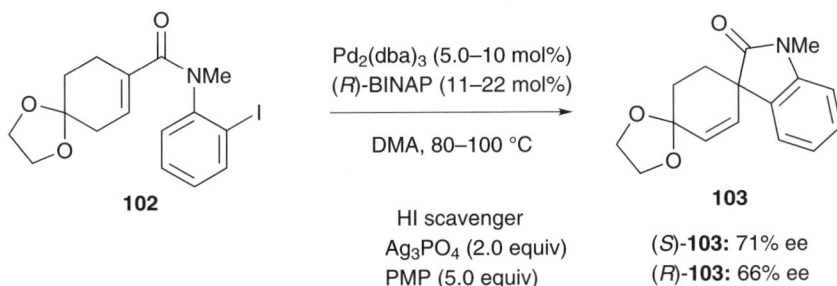

Scheme 16.24

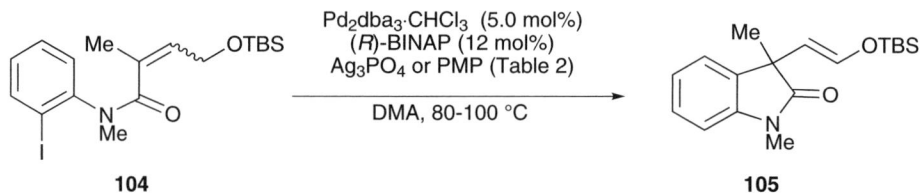

Scheme 16.25

instead of Ag₃PO₄. Mizoroki–Heck cyclization of the *Z* stereoisomer of α,β-unsaturated iodoanilide **104** under PMP-promoted conditions provided the *R* enantiomer of oxindole **105** in excellent yield and 92% ee, whereas the analogous reaction under silver-promoted conditions delivered (*R*)-**105** in lower yield and diminished enantioselectivity (Table 16.2). The *E* stereoisomer of **104** cyclized with lower stereoselectivity under both sets of reaction conditions.

The utility of asymmetric Mizoroki–Heck cyclizations to form 3,3-disubstituted oxindoles was first exemplified by the enantioselective total synthesis of the Calabar alkaloids, (−)-physostigmine (**110**) and (−)-physovenine (**112**) (Schemes 16.26 and 16.27) [63]. Under the conditions discovered during exploratory investigations, the Pd–BINAP-catalysed reaction of (*Z*)-α,β-unsaturated iodoanilide **106** in the presence of PMP occurred smoothly. Subsequent acid hydrolysis of the silyl enol ether intermediate **107** delivered oxindole **108** in 84% overall yield and 95% ee (Scheme 16.26). A single recrystallization of this product provided enantiopure oxindole **108**, which was used to synthesize both (−)-physostigmine (**110**) and (−)-physovenine (**112**). Specifically, condensation of **108** with methylamine followed by reduction of the resultant imine with LiAlH₄ gave (−)-esermethole (**109**), which was converted to enantiopure (−)-physostigmine (**110**) following established procedures [64]. (−)-Physovenine (**112**) was also accessed from (*S*)-oxindole **108** by a three-step sequence (Scheme 16.27). These sequences were the first highly enantioselective catalytic asymmetric routes to Calabar alkaloids.

Using a related strategy, Zhu and coworkers [65] recently described a formal enantioselective total synthesis of (−)-physostigmine (**110**) that featured a novel asymmetric Mizoroki–Heck cyclization/cyanation sequence. Oxindole **114**, an intermediate in an earlier synthesis of physostigmine [66], was prepared from anilide **113** by carrying out the cyclization under cationic Mizoroki–Heck conditions with (*S*)-DIFLUORPHOS as the ligand and potassium ferricyanide as the source of cyanide. Using these conditions, (*S*)-oxindole **114** was formed in 78% yield and 61% ee (Scheme 16.28).

Table 16.2 *Comparison of cationic and neutral pathways in an Overman-type cyclization*

Entry	Substrate	Additive	Yield (%)	ee (%)
1	(*Z*)-**104**	Ag₃PO₄	53	78 (*R*)
2	(*Z*)-**104**	PMP	80	92 (*R*)
3	(*E*)-**104**	Ag₃PO₄	80	45 (*S*)
4	(*E*)-**104**	PMP	85	38 (*R*)

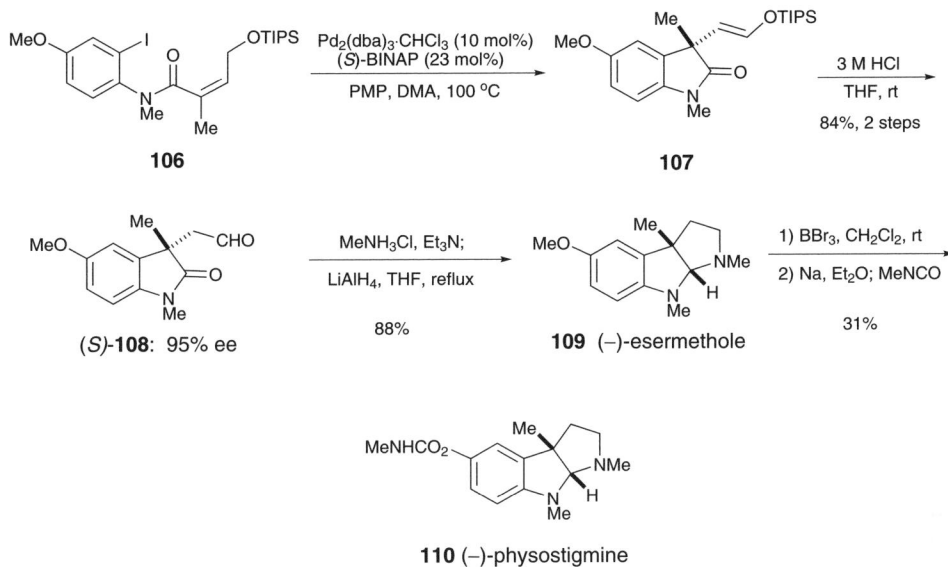

106

Pd$_2$(dba)$_3$·CHCl$_3$ (10 mol%)
(*S*)-BINAP (23 mol%)

PMP, DMA, 100 °C

107

3 M HCl

THF, rt

84%, 2 steps

(*S*)-**108**: 95% ee

MeNH$_3$Cl, Et$_3$N;

LiAlH$_4$, THF, reflux

88%

109 (–)-esermethole

1) BBr$_3$, CH$_2$Cl$_2$, rt

2) Na, Et$_2$O; MeNCO

31%

110 (–)-physostigmine

Scheme 16.26

Recent studies have revealed additional details about the asymmetric Mizoroki–Heck cyclizations of α,β-unsaturated iodoanilides. Stephenson and coworkers [67] examined the cyclization of (*E*)-iodoanilide **115** in the presence of Pd$_2$(dba)$_3$, (*R*)-BINAP and Ag$_3$PO$_4$ and found that the enantiomeric excess of the product changes as the reaction proceeds (Scheme 16.29). In general, the enantiomeric purity of oxindole product **116** was highest at high conversions (86–100%). Furthermore, the authors propose an explanation for the previously described reversals of enantiopreference in cyclizations of (*E*)-iodoanilides using PMP versus Ag$_3$PO$_4$. According to this proposal, both oxidative addition and alkene insertion are important stereocontrolling events.

The potential importance of rotational dynamics about the aryl C–N bond in asymmetric Mizoroki–Heck cyclizations of *o*-iodoacrylanilides was highlighted in a recent study by Curran and coworkers [68]. Because of the presence of two *ortho* substituents, racemic (*E*)-acryl *o*-iodoanilide **117** could be resolved by preparative high-performance liquid chromatography, and the enantiomers were shown to have a barrier for interconversion of 26 kcal mol^{-1} (Scheme 16.30). Although these conformational stereoisomers would racemize readily at the elevated temperatures typically used in asymmetric Mizoroki–Heck

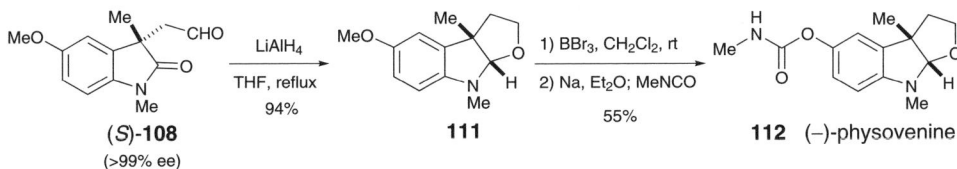

(*S*)-**108**
(>99% ee)

LiAlH$_4$

THF, reflux
94%

111

1) BBr$_3$, CH$_2$Cl$_2$, rt

2) Na, Et$_2$O; MeNCO

55%

112 (–)-physovenine

Scheme 16.27

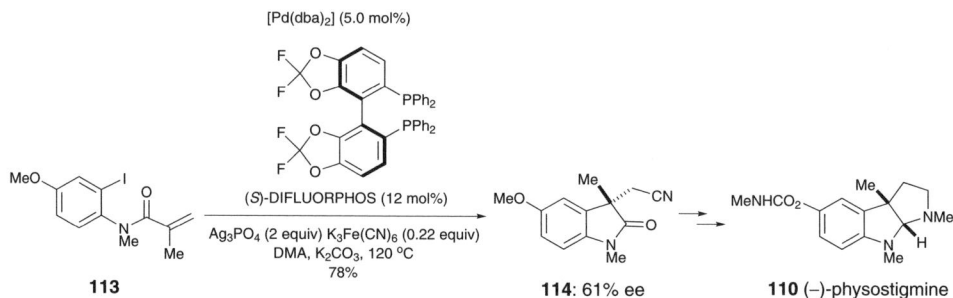

Scheme 16.28

reactions, they were stable at room temperature. Using tri-(*tert*-butyl)phosphine as ligand, the intramolecular Mizoroki–Heck reaction of these stereoisomers could be studied at room temperature. The Mizoroki–Heck cyclizations of these stereoisomers proceeded in excellent yield and with 86–88% transfer of chirality to afford oxindole **118**, demonstrating that axial chirality can be transferred in a Mizoroki–Heck cyclization. The authors propose that the stereocontrolling step in asymmetric Mizoroki–Heck reactions carried out at higher temperature with substrates lacking a second *ortho* substituent likely involves dynamic kinetic resolution in the oxidative addition to the C–I bond, rather than face-selective alkene complexation or insertion.

The first reports by Overman and coworkers [62] of asymmetric Mizoroki–Heck cyclizations of unsaturated iodo- and triflato-anilides focused on constructing oxindoles bearing two alkyl substituents at the C3 quaternary stereocentre. These reports included only a cursory study of the synthesis of oxindoles having an aryl substituent at C3. Efficient access to the latter type of oxindole was required for synthetic entry to a broad range of structurally complex cyclotryptamine alkaloids (Figure 16.5). When applied to α-aryl substrates **119**, the reaction conditions that yielded 3,3-dialkyl-substituted oxindoles with high enantioselectivity from (*Z*)-acrylolyl *o*-iodoanilides produced oxindole **120a** (Scheme 16.31) with low to moderate stereoinduction (19% ee under neutral conditions and 65% ee under cationic conditions). As in numerous previous studies of a diverse range of Mizoroki–Heck substrates, the cyclization to form 3-aryl oxindoles benefited from the switch from aryl iodide to aryl triflate substrates [69]. The asymmetric Mizoroki–Heck cyclization of aryl triflates such as **119** could be realized in useful yields and enantioselectivities in THF using Pd(OAc)$_2$ as a precatalyst, (*R*)-BINAP as a chiral ligand, and PMP as a base. 3,3-Disubstituted oxindoles **120** containing various aromatic or heteroaromatic groups were

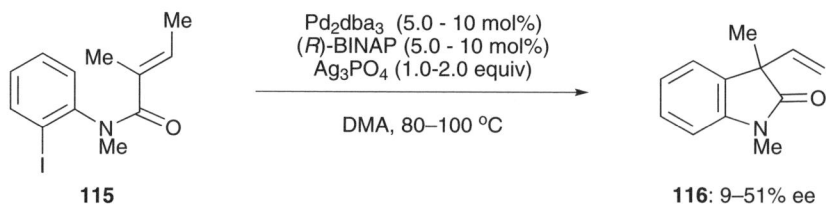

Scheme 16.29

Scheme 16.30

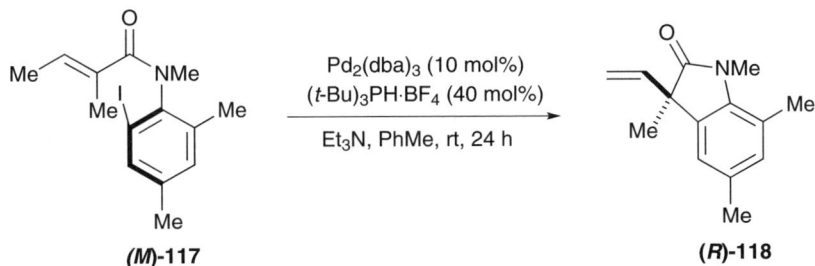

Precursor	er **117**	product	er **118**	Yield (%)	% chirality transfer
(*M*)-(−)-**117**	99.5/0.5	(*R*)-(+)-**118**	85.5/14.5	95	86
(*P*)-(+)-**117**	98.5/1.5	(*S*)-(−)-**118**	86.5/13.5	92	88

121 idiospermuline

122 hodgkinsine

123 quadrigemine C

124 psycholeine

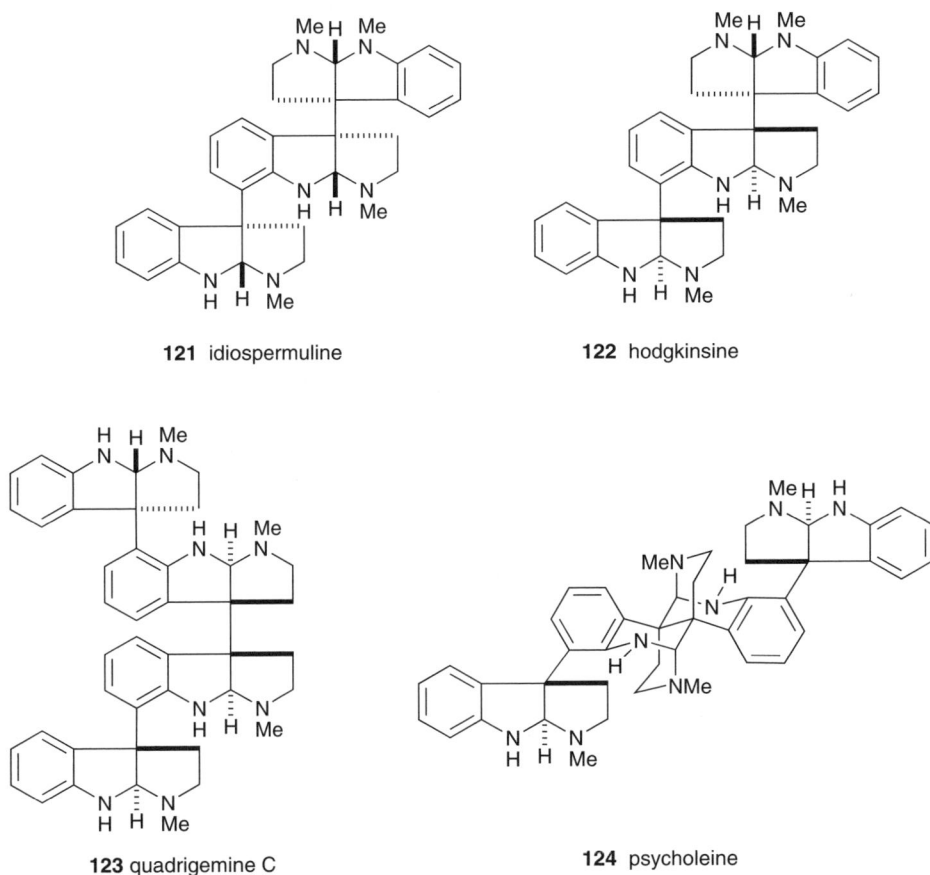

Figure 16.5 *Structurally complex cyclotryptamine alkaloids.*

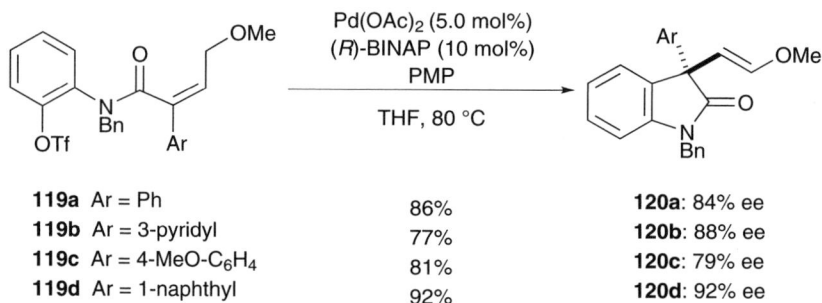

119a Ar = Ph   86%   **120a**: 84% ee
119b Ar = 3-pyridyl   77%   **120b**: 88% ee
119c Ar = 4-MeO-C₆H₄ 81%   **120c**: 79% ee
119d Ar = 1-naphthyl 92%   **120d**: 92% ee

Scheme 16.31

prepared in this way. Modest improvement in enantioselectivity was achieved by using more polar solvents such as DMA or acetonitrile and employing (*R*)-tol-BINAP as the ligand (Scheme 16.32).

These advances in the construction of 3-alkyl-3-aryloxindoles by catalytic asymmetric Mizoroki–Heck cyclizations laid the groundwork for the total synthesis of various members of the cyclotryptamine family of alkaloids (Figure 16.5). Three quite different strategies for exploiting asymmetric Mizoroki–Heck cyclizations to assemble structurally complex target structures were exemplified in these syntheses [70].

In the enantioselective total synthesis of idiospermuline (**121**), Overman and Peterson [71] employed a catalytic asymmetric Mizoroki–Heck cyclization to control diastereoselection in appending the third pyrrolidinoindoline ring of this dodecacyclic alkaloid. The cyclization substrate, enantiopure heptacyclic vinyl triflate **125** (Scheme 16.33), was assembled from isatin by a multistep sequence. The intramolecular Mizoroki–Heck reaction of this intermediate using Pd(OAc)₂ and the achiral diphosphine ligand, 1,1′-bis(diphenylphosphino)ferrocene (dppf), proceeded with modest diastereoselection to afford a mixture of oxindoles **126** and **127** in a 1 : 2 ratio. The reaction conditions developed for the asymmetric construction of relatively simple 3-aryl-3-alkyloxindoles proved to be directly applicable to this more elaborate Mizoroki–Heck precursor. Thus, modulation of this cyclization by ligand control was achieved with the tol-BINAP ligand. When (*R*)-tol-BINAP was used, diastereomer **127** was formed with high diastereoselectivity (*dr* = 18 : 1) in nearly quantitative yield. As expected, diastereoselection was reversed by using (*S*)-tol-BINAP, which provided hexacyclic products **126** and **127** in a 6 : 1 ratio. The diminished diastereoselectivity afforded by the *S* enantiomeric catalyst can be attributed

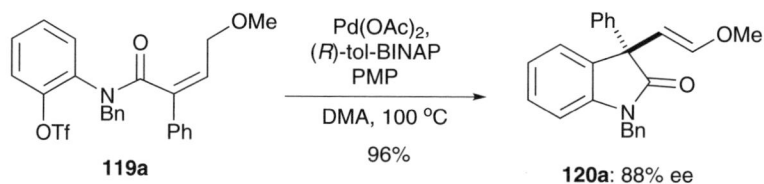

119a   96%   **120a**: 88% ee

Scheme 16.32

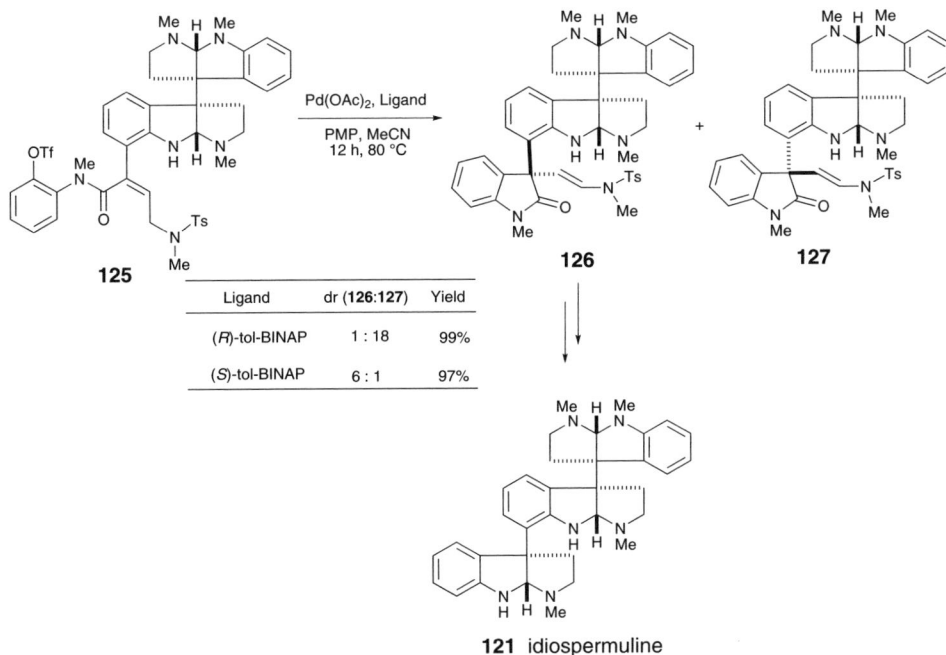

Ligand	dr (126:127)	Yield
(*R*)-tol-BINAP	1 : 18	99%
(*S*)-tol-BINAP	6 : 1	97%

121 idiospermuline

Scheme 16.33

to a mismatch between substrate and ligand control. The asymmetric total synthesis of idiospermuline (**121**) was completed from Mizoroki–Heck product **126** in three additional steps by hydrogenation of the enamide double bond and a one-pot reductive deprotection and cyclization process.

An asymmetric intramolecular Mizoroki–Heck reaction was used in a different fashion in the total syntheses of hodgkinsine (**122**) and hodgkinsine B (**132**) reported by Kodanko and Overman [72] (Scheme 16.34). In this case, Mizoroki–Heck cyclization substrate **129** was assembled as a racemate in three steps from *meso*-chimonanthine (**128**), which itself is available in three steps and 30% overall yield from tryptamine [73]. The asymmetric Mizoroki–Heck cyclization of racemic alkenyl triflate **129** was then carried out with Pd(OAc)$_2$, (*S*)-tol-BINAP and PMP in acetonitrile at 110 °C to give the hexacyclic products **130** and **131** in a 1 : 1 ratio and 93% yield. The enantiomeric purity of these products (**130**: 79% ee; **131**: 83% ee) is consistent with catalyst control in the cyclization of each enantiomer being circa 9 : 1, just slightly less than that observed with much simpler substrates. In two additional steps, Mizoroki–Heck products **130** and **131** were transformed respectively to hodgkinsine (**122**) and hodgkinsine B (**132**), completing the first total syntheses of these nonacyclic alkaloids. The key step in these total syntheses represents the first practical use of the asymmetric Mizoroki–Heck reaction to resolve a racemic substrate, as it delivered products derived from both enantiomers in high yield.

The first example of a double asymmetric Mizoroki–Heck cyclization was reported in the context of the asymmetric total syntheses of quadrigemine C (**123**) and its isomer, psycholeine (**124**) (Scheme 16.35) [74]. In this ambitious approach, the asymmetric

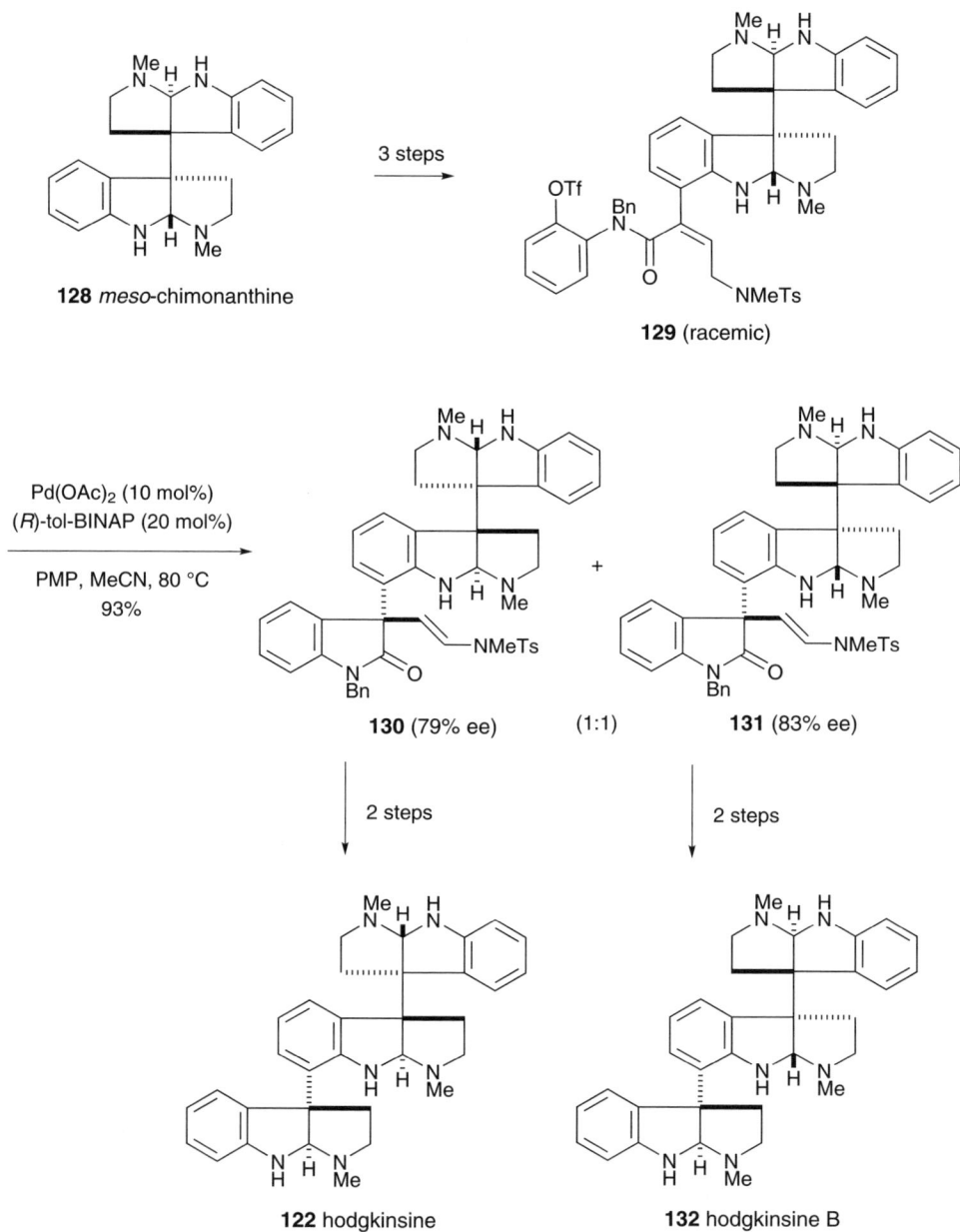

128 *meso*-chimonanthine

3 steps

129 (racemic)

Pd(OAc)$_2$ (10 mol%)
(*R*)-tol-BINAP (20 mol%)

PMP, MeCN, 80 °C
93%

130 (79% ee)   (1:1)   **131** (83% ee)

2 steps   2 steps

122 hodgkinsine   **132** hodgkinsine B

Scheme 16.34

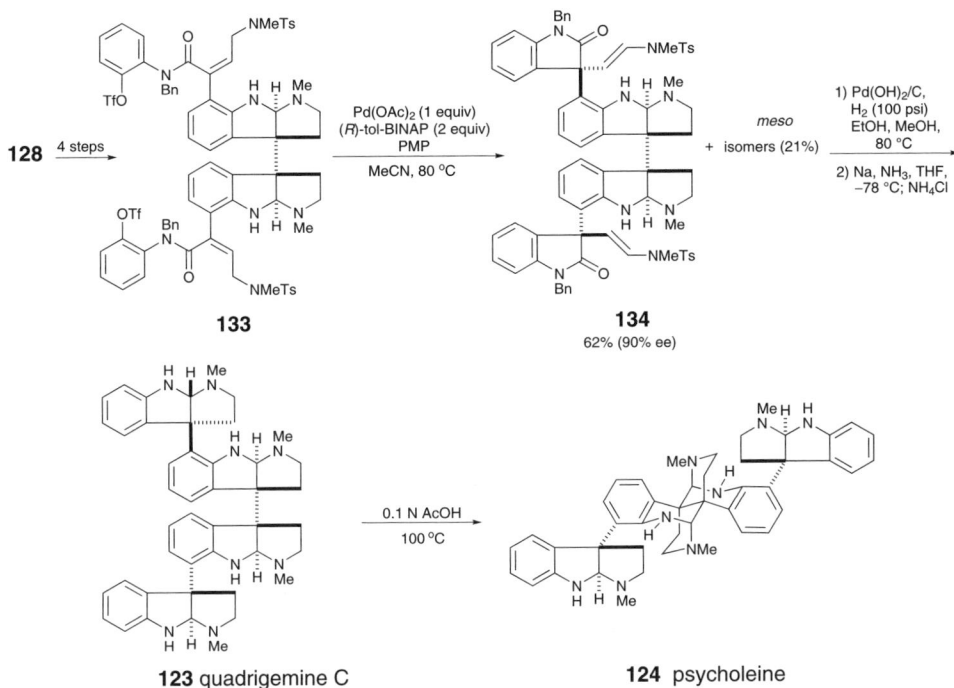

Scheme 16.35

Mizoroki–Heck cyclization is employed to desymmetrize advanced *meso* intermediate **133** and simultaneously install the two peripheral diaryl quaternary stereocentres of these dodecacyclic alkaloids. Starting with *meso*-chimonanthine (**128**), achiral Mizoroki–Heck cyclization substrate **133** was assembled in four efficient steps. The catalyst-controlled double Mizoroki–Heck cyclization of ditriflate **133** was achieved using Pd(OAc)$_2$, (*R*)-tol-BINAP and PMP in acetonitrile at 80 °C to produce decacyclic C_1-symmetric dioxindole **134** in 62% yield and 90% ee. Stereoisomeric *meso* dioxindoles were also formed as minor side-products in this reaction (21% combined yield). The use of (*R*)-BINAP for this transformation resulted in inferior selectivity, providing product **134** in lower yield and only 65% ee. Following the pivotal Mizoroki–Heck cyclization, only two additional steps were required to complete the asymmetric total synthesis of quadrigemine C (**123**), which constituted the first total synthesis of a higher order member of the cyclotryptamine alkaloid family. Subsequent acid-catalysed isomerization of quadrigemine C provided psycholeine (**124**) as well.

This asymmetric total synthesis of quadrigemine C provides a compelling illustration of the power of asymmetric Mizoroki–Heck cyclization strategies in total synthesis. First, the key cyclization of **133**→**134** demonstrates the remarkable functional group tolerance of the Mizoroki–Heck reaction, in this case allowing the enantioselective formation of two C–C bonds in a precursor containing a number of polar and sensitive functional groups. Second, this example, as well as several others discussed in this section, illustrates the

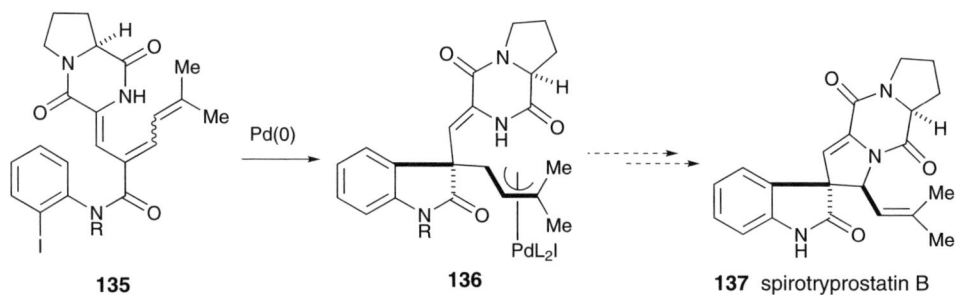

Scheme 16.36

notable ability of intramolecular Mizoroki–Heck reactions to form congested quaternary centres [75]. Third, this asymmetric total synthesis of a dodecacyclic natural product containing eight stereocentres, four of which are quaternary, was realized in only 10 steps from tryptamine.

Overman and Rosen [76] reported a total synthesis of spirotryprostatin B (**137**), a structurally novel diketopiperazine alkaloid, that featured a cascade Mizoroki–Heck cyclization/η^3-allylpalladium capture process and the exploration of a related catalytic asymmetric sequence (Scheme 16.36). The plan was to relay the relative configurations of the quaternary stereocentre and the adjacent tertiary stereocentre in the natural product from the geometry of the trisubstituted alkene in the Mizoroki–Heck cyclization substrate. It was anticipated that the favoured 5-*exo* intramolecular Mizoroki–Heck cyclization of enantiopure triene precursor **135** would generate an η^3-allylpalladium intermediate, with a chiral palladium catalyst controlling the absolute configuration of the initially formed quaternary carbon stereocentre.

The viability of this approach was first investigated with triene **138**, in which the internal alkene possesses a *Z* configuration (Scheme 16.37). Cyclization of iodo triene **138** with Pd$_2$(dba)$_3$·CHCl$_3$, (*S*)-BINAP and PMP in DMA at 100 °C produced a 6 : 1 mixture of spirocycles **139** and **141** in 28% yield. Switching to (*R*)-BINAP for the cyclization of **138**

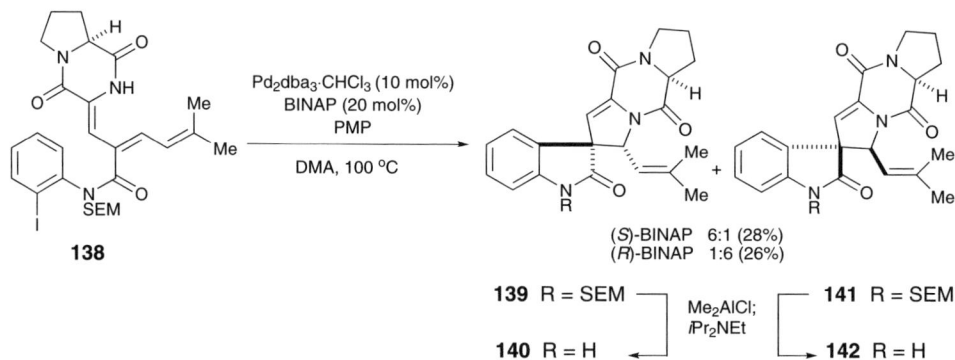

Scheme 16.37

under otherwise identical conditions provided a 1 : 6 mixture of pentacyclic products **139** and **141**. Cleavage of the 2-(trimethylsilyl)ethoxymethyl (SEM) protecting groups from these products produced 18-*epi*-spirotryprostatin B (**140**) and 3-*epi*-spirotryprostatin B (**142**).

The results summarized in Scheme 16.37 demonstrated that (a) the Mizoroki–Heck cyclization takes place with high regioselectivity, (b) the η^3-allylpalladium intermediate is generated and captured with high stereochemical fidelity and (c) the diketopiperazine intercepts this intermediate *anti* to the metal centre. Therefore, these studies suggested that the required configuration of natural spirotryprostatin B (**137**) would be obtained by Pd/(S)-BINAP-catalysed cyclization of a triene analogous to **138** in which the internal alkene possesses the *E* configuration. Unfortunately, attempts to access spirotryprostatin B by this catalytic asymmetric approach were thwarted by isomerization of the internal double-bond of such triene substrates under asymmetric Mizoroki–Heck reaction conditions [76]. The cascade process could be accomplished with an achiral catalyst derived from Pd$_2$(dba)$_3$ and tri(*o*-tolyl)phosphine, and the product of such a Mizoroki–Heck reaction was employed to prepare (−)-spirotryprostatin B (**137**).

A different tandem process was employed by Overman and coworkers [77] to complete the first total synthesis of the structurally unique *Strychnos* alkaloid minfiensine (**148**). In this case, a catalytic asymmetric Mizoroki–Heck cyclization was combined with an iminium ion cyclization to provide rapid access to an enantiopure tetracyclic intermediate (Scheme 16.38). A preliminary survey of reaction conditions demonstrated that the asymmetric Mizoroki–Heck cyclization of **143** could be achieved with several chiral enantiopure ligands. However, the use of BINAP as the ligand resulted in extensive alkene migration to afford the fully conjugated dienyl carbamate. Pfaltz and coworkers [20] have reported the use of chiral (phosphinoaryl)oxazoline ligands for efficient and highly enantioselective intermolecular Mizoroki–Heck reactions of aryl triflates with 2,3-dihydrofuran or cyclopentene, noting that post-insertion double-bond migration is minimized with these

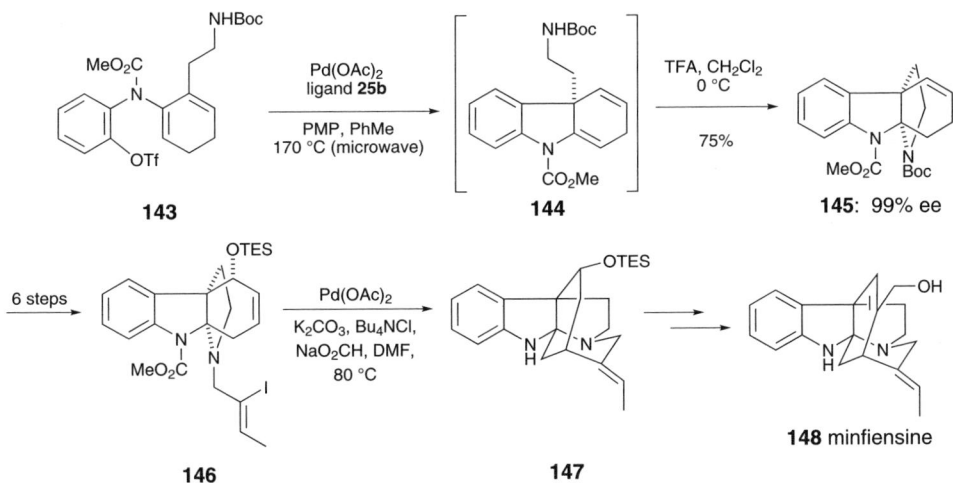

Scheme 16.38

ligands. Ripa and Hallberg [78] observed similar results in their studies of intramolecular asymmetric Mizoroki–Heck arylations of cyclic enamides. Consistent with these precedents, alkene migration was mitigated by the use of (phosphinoaryl)oxazoline **25b**, with the desired cyclization product **144** being obtained in circa 85% yield and 99% ee. One drawback associated with the use of ligand **25b** is the decreased rate of Mizoroki–Heck cyclization in comparison with the corresponding Pd–BINAP-mediated Mizoroki–Heck reactions. Under conventional heating, the Mizoroki–Heck cyclization of **143** with this chiral P–N ligand was sluggish and required more than 70 h at 100 °C to reach completion. Fortunately, the conversion could be accomplished in 30 min at 170 °C in a microwave reactor with no erosion of enantioselectivity. Treatment of crude Mizoroki–Heck product **144** with excess trifluoroacetic acid induced cyclization to furnish (dihydroiminoethano)carbazole **145** in 75% overall yield and 99% ee. In six additional steps, tetracyclic intermediate **145** was converted to vinyl iodide **146**. An intramolecular Mizoroki–Heck cyclization of this intermediate under Jeffery conditions [79], which was terminated by hydride transfer, provided product **147**, thus assembling the full pentacyclic ring system of minfiensine. From this intermediate, the synthesis of minfiensine (**148**) was completed in a straightforward manner in eight additional steps.

16.5   Conclusions and Future Prospects

In the years since the first reports in 1989 [4, 5], the scope of the asymmetric intramolecular Mizoroki–Heck reactions has been substantially increased. This transformation has now been employed as a key strategic step in total syntheses of a wide variety of polycyclic natural products. Among the features that contribute to the broad utility of asymmetric Mizoroki–Heck cyclizations are the high functional group tolerance of palladium(0)-catalysed reactions, the remarkable capacity of this transformation to forge C—C bonds in situations of considerable steric congestion and the ability to orchestrate cascade or tandem processes that form multiple rings.

It is certain that many more applications of catalytic asymmetric intramolecular Mizoroki–Heck reactions will be described in the future. This survey makes apparent the small number of ligands that have been used thus far, with Noyori and coworkers' BINAP ligand being the most widely employed [80]. Two future trends are easy to predict: a larger variety of chiral ligands [56, 58, 81, 82] will be used in asymmetric Mizoroki–Heck processes and a greater variety of cascade processes involving intramolecular Mizoroki–Heck reactions will be developed.

References

1. (a) Mizoroki, T., Mori, K. and Ozaki, A. (1971) Arylation of olefin with aryl iodide catalyzed by palladium. *Bull. Chem. Soc. Jpn.*, **44**, 581; (b) Heck, R.F. and Nolley, J.P. Jr (1972) Palladium-catalyzed vinylic hydrogen substitution reactions with aryl, benzyl, and styrl halides. *J. Org. Chem*, **37**, 2320–2.
2. Selected reviews include: (a) Link, J.T. (2002) The intramolecular Heck reaction, in *Organic Reactions*, Vol. **60** (ed. L.E. Overman), John Wiley & Sons, Inc., Hoboken; (b) de Meijere, A. and Meyer, F.E. (1994) Fine feathers make fine birds: the Heck reaction in modern garb.

Angew. Chem., Int. Ed. Engl., **33**, 2379–411; (c) Bräse, S. and de Meijere, A. (1998) Palladium-catalyzed coupling of organyl halides to alkenes – the Heck reaction, in *Metal-Catalyzed Cross Coupling Reactions* (eds P.J. Stang and F. Diederick), Wiley–VCH Verlag GmbH, Weinheim; (d) Link, J.T. and Overman, L.E. (1998) Intramolecular Heck reactions in natural product chemistry, in *Metal-Catalyzed Cross Coupling Reactions* (eds P.J. Stang and F. Diederich), Wiley–VCH Verlag GmbH, Weinheim.

3. For recent reviews of the asymmetric Heck reaction, see: (a) Shibasaki, M., Vogl, E.M. and Ohshima, T. (2004) Asymmetric Heck Reaction. *Adv. Synth. Catal.*, **346**, 1533–52; (b) Guiry, P.J. and Kiely, D. (2004) The development of the intramolecular asymmetric Heck reaction. *Curr. Org. Chem.*, **8**, 781–94; (c) Tietze, L.F., Ila, H. and Bell, H.P. (2004) Enantioselective palladium-catalyzed transformations. *Chem. Rev.*, **104**, 3453–516; (d) Donde, Y. and Overman, L.E. (2000) Asymmetric intramolecular Heck reactions, in *Catalytic Asymmetric Synthesis* (ed. I. Ojima), Wiley–VCH, New York.

4. Sato, Y., Sodeoka, M. and Shibasaki, M. (1989) Catalytic asymmetric C—C bond formation: asymmetric synthesis of *cis*-decalin derivatives by palladium-catalyzed cyclization of prochiral alkenyl iodides. *J. Org. Chem.*, **54**, 4738–9.

5. Carpenter, N.E., Kucera, D.J. and Overman, L.E. (1989) Palladium-catalyzed polyene cyclizations of trienyl triflates. *J. Org. Chem.*, **54**, 5846–8.

6. For reviews, see: (a) Nicolaou, K.C., Bulger, P.G. and Sarlah, D. (2005) Palladium-catalyzed cross-couplings in total synthesis. *Angew. Chem. Int. Ed.*, **44**, 4442–89; (b) Dounay, A.B. and Overman, L.E. (2003) The asymmetric Heck reaction in natural products total synthesis. *Chem. Rev.*, **103**, 2945–63.

7. For recent reviews of mechanistic studies of the Heck reaction, see: (a) Knowles, J.P. and Whiting, A. (2007) The Heck–Mizoroki cross-coupling reaction: a mechanistic perspective. *Org. Biomol. Chem.*, **5**, 31–44; (b) Jutand, A. (2003) The use of conductivity measurements for the characterization of cationic palladium(II) complexes and for the determination of kinetic and thermodynamic data in palladium-catalyzed reactions. *Eur. J. Inorg. Chem.*, 2017–40; (c) Beletskaya, I.P. and Cheprakov, A.V. (2000) The Heck reaction as a sharpening stone of palladium catalysis. *Chem. Rev.*, **100**, 3009–66; (d) Amatore, C. and Jutand, A.J. (1999) Mechanistic and kinetic studies of palladium catalytic systems. *J. Organomet. Chem.*, **576**, 254–78.

8. Cabri, W., Candiani, I., DeBernardis, S. *et al.* (1991) Heck reaction on anthraquinone derivatives: ligand, solvent and salt effects. *J. Org. Chem.*, **56**, 5796–800.

9. Ozawa, F., Kubo, A. and Hayashi, T. (1991) Catalytic asymmetric arylation of 2,3-dihydrofuran with aryl triflates. *J. Am. Chem. Soc.*, **113**, 1417–9.

10. (a) Karabelas, K., Westerlund, C. and Hallberg, A. (1985) The effect of added silver nitrate on the palladium-catalyzed arylation of allyltrimethylsilanes. *J. Org. Chem.*, **50**, 3896–900; (b) Grigg, R., Loganathan, V., Santhakumar, V. *et al.* (1991) Suppression of alkene isomerization in products from intramolecular Heck reactions by addition of thallium(I) salts. *Tetrahedron Lett.*, **32**, 687–90.

11. Brown, J.M., Pérez-Torrente, J.J., Alcock, N.W. and Clase, H.J. (1995) Stable arylpalladium iodides and reactive arylpalladium trifluoromethanesulfonates in the intramolecular Heck reaction. *Organometallics*, **14**, 207–13.

12. The first irreversible enantiodifferentiating step would determine asymmetric induction: Landis, C.R. and Halpern, J. (1987) Asymmetric hydrogenation of methyl (Z)-α-acetamidocinnamate catalyzed by [1,2-bis(phenyl-*o*-anisoyl)phosphino)ethane]rhodium(I): kinetics, mechanism and origin of enantioselection. *J. Am. Chem. Soc.*, **109**, 1746–54.

13. Ashimori, A. and Overman, L.E. (1992) Catalytic asymmetric synthesis of quaternary carbon centers. Palladium-catalyzed formation of either enantiomer of spirooxindoles and related spirocyclics using a single enantiomer of a chiral diphosphine ligand. *J. Org. Chem.*, **57**, 4571–2.

14. Overman, L.E. and Poon, D.J. (1997) Asymmetric Heck reactions via neutral intermediates: enhanced enantioselectivity with halide additives gives mechanistic insights. *Angew. Chem., Int. Ed. Engl.*, **36**, 518–21.

15. Although 7-*endo* cyclizations are particularly rare, a high-yielding 7-*endo* Heck cyclization was a key step in a recent total synthesis of guanacastepene N: Iimura, S., Overman, L.E., Paulini, R. and Zakarian, A. (2006) Enantioselective total synthesis of guanacastepene N using an uncommon 7-*endo* Heck cyclization as a pivotal step. *J. Am. Chem. Soc.*, **128**, 13095–101.

16. (a) Tietze, L.F. and Schimpf, R. (1994) Regio- and enantioselective silane-terminated intramolecular Heck reactions. *Angew. Chem., Int. Ed. Engl.*, **33**, 1089–91; (b) Tietze, L.F. and Raschke, T. (1995) Enantioselective total synthesis of a natural norsesquiterpene of the calamenene group by a silane-terminated intramolecular Heck reaction. *Synlett*, 597–8; (c) Tietze, L.F. and Raschke, T. (1996) Enantioselective total synthesis and absolute configuration of the natural norsesquiterpene 7-demethyl-2-methoxycalamenene by a silane-terminated intramolecular Heck reaction. *Liebigs Ann.*, 1981–7.

17. Overman, L.E. and Rucker, P.V. (1998) Enantioselective synthesis of cardenolide precursors using an intramolecular Heck reaction. *Tetrahedron Lett.*, **39**, 4643–6.

18. Abelman, M.M., Oh, T. and Overman, L.E. (1987) Intramolecular alkene arylations for rapid assembly of polycyclic systems containing quaternary centers. A new synthesis of spirooxindoles and other fused and bridged ring systems. *J. Org. Chem.*, **52**, 4130–3.

19. Lautens, M. and Zunic, V. (2004) Sequential olefin metathesis – intramolecular asymmetric Heck reactions in the synthesis of polycycles. *Can. J. Chem.*, **82**, 399–407.

20. Loiseleur, O., Meier, P. and Pfaltz, A. (1996) Chiral phosphanyldihydrooxazoles in asymmetric catalysis: enantioselective Heck reactions. *Angew. Chem., Int. Ed. Engl.*, **35**, 200–2.

21. Vallin, K.S.A., Larhed, M. and Hallberg, A. (2001) Aqueous DMF-potassium carbonate as a substitute for thallium and silver additives in the palladium-catalyzed conversion of aryl bromides to acetyl arenes. *J. Org. Chem.*, **66**, 4340–3.

22. For one recent example, see: Lormann, M.E.P., Nieger, M. and Bräse, S. (2006) Desymmetrization of bicyclo[4.4.0]decadienes: a planar-chiral complex proved to be most effective in an asymmetric Heck reaction. *J. Organomet. Chem.*, **691**, 2159–61.

23. Van Leeuwen, P.W.N.M. (2001) Decomposition pathways of homogeneous catalysts. *Appl. Catal., A. Gen.*, **212**, 61–81.

24. McDermott, M.C., Stephenson, G.R., Hughes, D.L. and Walkington, A.J. (2006) Intramolecular asymmetric Heck reactions: evidence for dynamic kinetic resolution effects. *Org. Lett.*, **8**, 2917–20.

25. For a review of microwave accelerated homogeneous catalysis, see: Larhed, M., Moberg, C. and Hallberg, A. (2002) Microwave-accelerated homogeneous catalysis in organic chemistry. *Acc. Chem. Res.*, **35**, 717–27.

26. Nilsson, P., Gold, H., Larhed, M. and Hallberg, A. (2002) Microwave-assisted enantioselective Heck reactions: expediting high reaction speed and preparative convenience. *Synthesis*, 1611–4.

27. Larhed, M., Moberg, C. and Hallberg, A. (2002) Microwave-accelerated homogeneous catalysis in organic chemistry. *Acc. Chem. Res.*, **35**, 717–27.

28. Sato, Y., Sodeoka, M. and Shibasaki, M. (1990) On the role of silver salts in asymmetric Heck-type reaction. A greatly improved catalytic asymmetric synthesis of *cis*-decalin derivatives. *Chem. Lett.*, 1953–4.

29. Sato, Y., Watanabe, S. and Shibasaki, M. (1992) Further studies on a catalytic asymmetric synthesis of decalin derivatives. *Tetrahedron Lett.*, **33**, 2589–92.

30. Kondo, K., Sodeoka, M., Mori, M. and Shibasaki, M. (1993) Asymmetric Heck reaction. A catalytic asymmetric synthesis of the key intermediate for vernolepin. *Tetrahedron Lett.*, **34**, 4219–22.

31. Ohrai, K., Kondo, K., Sodeoka, M. and Shibasaki, M. (1994) Effects of solvents and additives in the asymmetric Heck reaction of alkenyl triflates: catalytic asymmetric synthesis of decalin derivatives and determination of the absolute stereochemistry of (+)-vernolepin. *J. Am. Chem. Soc.*, **116**, 11737–48.

32. Kiely, D. and Guiry, P.J. (2003) A comparison of palladium complexes of BINAP and diphenylphosphinooxazoline ligands in the catalytic asymmetric synthesis of *cis*-decalins. *Tetrahedron Lett.*, **44**, 7377–80.

33. Sato, Y., Honda, T. and Shibasaki, M. (1992) A catalytic asymmetric synthesis of hydrindans. *Tetrahedron Lett.*, **33**, 2593–6.

34. Sato, Y., Mori, M. and Shibasaki, M. (1995) Asymmetric Heck reaction: a catalytic asymmetric synthesis of the key intermediate of (−)-oppositol and (−)-prepinnaterpene. *Tetrahedron: Asymmetry*, **6**, 757–66.

35. Tietze, L.F. and Modi, A. (2000) Regioselective silane-terminated intramolecular Heck reaction with alkenyl triflates and alkenyl iodides. *Eur. J. Org. Chem.*, 1959–64.

36. Kagechika, K. and Shibasaki, M. (1991) Asymmetric Heck reaction: a catalytic asymmetric synthesis of the key intermediate for $\Delta^{9(12)}$-capnellene-3β,8β,10-triol and $\Delta^{9(12)}$-capnellene-3β,8β,10α,14-tetrol. *J. Org. Chem.*, **56**, 4093–4.

37. For previous reports of tandem Heck reaction–anionic capture sequences, see: (a) Grigg, R., Sridharan, V. and Xu, L.–H. (1995) Palladium-catalyzed cyclization-amination of allenes – effect of base on regioselectivity of formation of allylic amines. *J. Chem. Soc., Chem. Commun.*, 1903–4 and references cited therein; (b) Ma, S. and Negishi, E. (1995) Palladium-catalyzed cyclization of ω-haloallenes. A new general route to common, medium, and large ring compounds via cyclic carbopalladation. *J. Am. Chem. Soc.*, **117**, 6345–7 and references cited therein.

38. Kagechika, K., Ohshima, T. and Shibasaki, M. (1993) Asymmetric Heck reaction–anion capture process. A catalytic asymmetric synthesis of the key intermediates for the capnellenols. *Tetrahedron*, **49**, 1773–82.

39. (a) Shibasaki, M., Mase, T. and Ikegami, S. (1986) The first total syntheses of $\Delta^{9(12)}$-capnellene-8β,10α-diol and $\Delta^{9(12)}$-capnellene-3β,8β,10α-triol. *J. Am. Chem. Soc.*, **108**, 2090–1; (b) Mase, T. and Shibasaki, M. (1986) Synthetic studies on the capnellol family: an improved synthesis of $\Delta^{9(12)}$-capnellene-3β,8β,10α-triol and the first total synthesis of $\Delta^{9(12)}$-capnellene-3β,8β,10β,14-tetrol. *Tetrahedron Lett.*, **27**, 5245–8.

40. Ohshima, T., Kagechika, K., Adachi, M. *et al.* (1996) Asymmetric Heck reaction–carbanion capture process. Catalytic asymmetric total synthesis of (−)-$\Delta^{9(12)}$-capnellene. *J. Am. Chem. Soc.*, **118**, 7108–16.

41. Itano, W., Ohshima, T. and Shibasaki, M. (2006) Synthesis of the tricyclic core of 5α-capnellenols using asymmetric Heck reaction–carbanion capture process. *Synlett*, 3053–6.

42. (a) Nukui, S., Sodeoka, M. and Shibasaki, M. (1993) Catalytic asymmetric synthesis of a functionalized indolizidine derivative. A useful intermediate suitable for the synthesis of various glycosidase inhibitors. *Tetrahedron Lett.*, **34**, 4965–8; (b) Sato, Y., Nukui, S., Sodeoka, M. and Shibasaki, M. (1994) Asymmetric Heck reaction of alkenyl iodides in the presence of silver salts. Catalytic asymmetric synthesis of decalin and functionalized indolizidine derivatives. *Tetrahedron*, **50**, 371–82.

43. Hayashi, T., Mise, T. and Kumada, M. (1976) A chiral (hydroxyalkylferrocenyl)phosphine ligand. Highly stereoselective catalytic asymmetric hydrogenation of prochiral carbonyl compounds. *Tetrahedron Lett.*, **48**, 4351–4.

44. Nukui, S., Sodeoka, M., Sasai, H. and Shibasaki, M. (1995) Regio- and stereoselective functionalization of an optically active tetrahydroindolizine derivative. Catalytic asymmetric syntheses of lentiginosine, 1,2-diepilentiginosine, and gephyrotoxin 209D. *J. Org. Chem.*, **60**, 398–404.

45. Kiewel, K., Tallant, M. and Sulikowski, G.A. (2001) Asymmetric Heck cyclization route to indolizidine and azaazulene alkaloids: synthesis of (+)-5-epiindolizidine 167B and indolizidine 223AB. *Tetrahedron Lett.*, **42**, 6621–3.

46. Yoda, H., Katoh, H., Ujihara, Y. and Takabe, K. (2001) SmI2-mediated hetero-coupling reaction of lactams with aldehydes; synthesis of indolizidine alkaloids, (−)-D-coniceine, (+)-5-epiindolizidine 167B and (+)-lentiginosine. *Tetrahedron Lett.*, **42**, 2509–12.

47. Studies by Hart and Tsai have demonstrated that indolizidine **70** is stereoisomeric from the Dendrobatid alkaloid gephyrotoxin-223AB: Hart, D.J. and Tsai, Y.-M. (1982) Stereoselective indolizidine synthesis: preparation of stereoisomers of gephyrotoxin-223AB. *J. Org. Chem.*, **47**, 4403–9.

48. (a) Kondo, K., Sodeoka, M. and Shibasaki, M. (1995) Regioselective olefin insertion in asymmetric Heck reaction. Catalytic asymmetric synthesis of a versatile intermediate for diterpene syntheses. *J. Org. Chem.*, **60**, 4322–3; (b) Kondo, K., Sodeoka, M. and Shibasaki, M. (1995) Catalytic asymmetric synthesis of a versatile intermediate for diterpene syntheses. Regioselective olefin insertion in asymmetric Heck reactions. *Tetrahedron: Asymmetry*, **6**, 2453–64.

49. Bell, R.A., Ireland, R.E. and Partyka, R.A. (1962) Total synthesis of *dl*-kaurene. *J. Org. Chem.*, **27**, 3741–2.

50. Kuehne, M.E. and Nelson, J.A. (1970) Stereospecific syntheses of epimeric diterpenoid. Resin acids through enolate anion reactions. *J. Org. Chem.*, **35**, 161–70.

51. Diaz, P., Gendre, F., Stella, L. and Charpentier, B. (1998) New synthetic retinoids obtained by palladium-catalyzed tandem cyclization-hydride capture process. *Tetrahedron*, **54**, 4579–90.

52. (a) Kojima, A., Takemoto, T., Sodeoka, M. and Shibasaki, M. (1996) Catalytic asymmetric synthesis of halenaquinone and halenaquinol. *J. Org. Chem.*, **61**, 4876–7; (b) Kojima, A., Takemoto, T., Sodeoka, M. and Shibasaki, M. (1998) Catalytic asymmetric synthesis of halenaquinone and halenaquinol. *Synthesis*, 581–9.

53. Harada, N., Sugioka, T., Ando, Y. *et al.* (1988) Total synthesis of (+)-halenaquinol and (+)-halenaquinone. Experimental proof of their absolute stereostructures theoretically determined. *J. Am. Chem. Soc.*, **110**, 8483–7.

54. Maddaford, S.P., Andersen, N.G., Cristofoli, W.A. and Keay, B.A. (1996) Total synthesis of (+)-xestoquinone using an asymmetric palladium-catalyzed polyene cyclization. *J. Am. Chem. Soc.*, **118**, 10766–73.

55. Lau, S.Y.W. and Keay, B.A. (1999) Remote substituent effects on the enantiomeric excess of intramolecular asymmetric palladium-catalyzed polyene cyclizations. *Synlett*, 605–7.

56. Andersen, N.G., Parvez, M. and Keay, B.A. (2000) Synthesis, resolution, and applications of 2,2′-bis(diphenylphosphino)-3,3′-binaphtho[2,1-*b*]furan. *Org. Lett.*, **2**, 2817–20.

57. Lau, S.Y.W., Andersen, N.G. and Keay, B.A. (2001) Optimization of palladium-catalyzed polyene cyclizations: suppression of competing hydride transfer from tertiary amines with DABCO and an unexpected hydride transfer from 1,4-dioxane. *Org. Lett.*, **2**, 181–4.

58. Hopkins, J.M., Gorobets, E., Wheatley, B.M.M. *et al.* (2006) Applications of 3,3′-disubstituted BINAP derivatives in inter- and intramolecular Heck/Mizoroki reactions. *Synlett*, 3120–4.

59. Miyazaki, F., Uotsu, K. and Shibasaki, M. (1998) Silver salt effects on an asymmetric Heck reaction. Catalytic asymmetric total synthesis of (+)-xestoquinone. *Tetrahedron*, **54**, 13073–8.

60. Honzawa, S., Mizutani, T. and Shibasaki, M. (1999) Synthetic studies on (+)-wortmannin. An asymmetric construction of an allylic quaternary carbon center by a Heck reaction. *Tetrahedron Lett.*, **40**, 311–4.

61. A total synthesis of racemic wortmannin has been reported; see: Mizutani, T., Honzawa, S., Tosaki, S.–Y. and Shibasaki, M. (2002) Total synthesis of (±)-wortmannin. *Angew. Chem. Int. Ed.*, **41**, 4680–2. An asymmetric formal synthesis has also been reported; see: Shigehisa, H., Mizutani, T., Tosaki, S.–Y. *et al.* (2005) Formal total synthesis of (+)-wortmannin using catalytic asymmetric intramolecular aldol condensation reaction. *Tetrahedron*, **61**, 5057–65.

62. (a) Ashimori, A., Bachand, B., Overman, L.E. and Poon, D.J. (1998) Catalytic asymmetric synthesis of quaternary carbon centers. Exploratory investigations of intramolecular Heck reactions of (*E*)-α,β-unsaturated 2-haloanilides and analogs to form enantioenriched spirocyclic products. *J. Am. Chem. Soc.*, **120**, 6477–87; (b) Ashimori, A., Bachand, B., Calter, M.A. *et al.* (1998) Catalytic asymmetric synthesis of quaternary carbon centers. Exploratory studies of intramolecular Heck reactions of (*Z*)-α,β-unsaturated anilides and mechanistic investigations of asymmetric Heck reactions proceeding via neutral intermediates. *J. Am. Chem. Soc.*, **120**, 6488–99.

63. (a) Ashimori, A., Matsuura, T., Overman, L.E. and Poon, D.J. (1993) Catalytic asymmetric synthesis of either enantiomer of physostigmine. Formation of quaternary carbon centers with high enantioselection by intramolecular Heck reactions of (*Z*)-2-butenanilides. *J. Org. Chem.*, **58**, 6949–51; (b) Matsuura, T., Overman, L.E. and Poon, D.J. (1998) Catalytic asymmetric synthesis of either enantiomer of the calabar alkaloids physostigmine and physovenine. *J. Am. Chem. Soc.*, **120**, 6500–3.

64. Yu, Q.–S. and Brossi, A. (1988) Practical synthesis of unnatural (+)-physostigmine and carbamate analogs. *Heterocycles*, **27**, 745–50.

65. Pinto, A., Jia, Y., Neuville, L. and Zhu, J. (2007) Palladium-catalyzed enantioselective domino Heck–cyanation sequence: development and application to the total synthesis of esermethole and physostigmine. *Chem. Eur. J.*, **13**, 961–7.

66. Lee, T.B.K. and Wong, G.S.K. (1991) Asymmetric alkylation of oxindoles: an approach to the total synthesis of (−)-physostigmine. *J. Org. Chem.*, **56**, 872–5.

67. McDermott, M.C., Stephenson, G.R., Hughes, D.L. and Walkington, A.J. (2006) Intramolecular asymmetric Heck reactions: evidence for dynamic kinetic resolution effects. *Org. Lett.*, **8**, 2917–20.

68. Lapierre, A.J., Geib, S.J. and Curran, D.P. (2007) Low-temperature Heck reactions of axially chiral *o*-iodoacrylanilides occur with chirality transfer: implications for catalytic asymmetric Heck reactions. *J. Am. Chem. Soc.*, **129**, 494–5.

69. Dounay, A.B., Hatanaka, K., Kodanko, J. *et al.* (2003) Catalytic asymmetric synthesis of quaternary carbons bearing two aryl substituents. Enantioselective synthesis of 3-alkyl-3-aryl oxindoles by catalytic asymmetric intramolecular Heck reactions. *J. Am. Chem. Soc.*, **125**, 6261–71.

70. For a review, see: Steven, A. and Overman, L.E. (2007) Stereocontrolled total synthesis of complex cyclotryptamine alkaloids: development of methods for the stereocontrolled synthesis of quaternary carbons. *Angew. Chem. Int. Ed.* **46**, 5488–508.

71. (a) Overman, L.E. and Peterson, E.A. (2003) Enantioselective total synthesis of the cyclotryptamine alkaloid idiospermuline. *Angew. Chem. Int. Ed.*, **42**, 2525–28; (b) Overman, L.E. and Peterson, E.A. (2003) Enantioselective synthesis of (−)-idiospermuline. *Tetrahedron*, **59**, 6905–19.

72. Kodanko, J.J. and Overman, L.E. (2003) Enantioselective total syntheses of the cyclotryptamine alkaloids hodgkinsine and hodgkinsine B. *Angew. Chem. Int. Ed.*, **42**, 2525–31.

73. Ishikawa, H., Takayama, H. and Aimi, N. (2002) Dimerization of indole derivatives with hypervalent iodines(III): a new entry for the concise total synthesis of *rac*- and *meso*-chimonanthines. *Tetrahedron Lett.*, **43**, 5637–9.

74. Lebsack, A.D., Link, J.T., Overman, L.E. and Stearns, B.A. (2002) Enantioselective total synthesis of quadrigemine C and psycholeine. *J. Am. Chem. Soc.*, **124**, 9008–9.

75. Overman, L.E. (1994) Application of intramolecular Heck reactions for forming congested quaternary carbon centers in complex molecule total synthesis. *Pure Appl. Chem.*, **66**, 1423–30.

76. Overman, L.E. and Rosen, M.D. (2000) Total synthesis of (−)-spirotryprostatin B and three stereoisomers. *Angew. Chem. Int. Ed.*, **39**, 4596–9.

77. Dounay, A.B., Overman, L.E. and Wrobleski, A.D. (2005) Sequential catalytic asymmetric Heck–iminium ion cyclization: enantioselective total synthesis of the *Strychnos* alkaloid minfiensine. *J. Am. Chem. Soc.*, **127**, 10186–7.

78. Ripa, L. and Hallberg, A. (1997) Intramolecular enantioselective palladium-catalyzed Heck arylation of cyclic enamides. *J. Org. Chem.*, **62**, 595–602.
79. (a) Jeffery, T. (1985) Highly stereospecific palladium-catalyzed vinylation of vinylic halides under solid-liquid phase transfer conditions. *Tetrahedron Lett.*, **26**, 2667–70; (b) Jeffery, T. (1996) On the efficiency of tetraalkylammonium salts in Heck type reactions. *Tetrahedron*, **52**, 10113–30.
80. Takaya, H., Mashima, K., Koyano, K. *et al.* (1986) Practical synthesis of (*R*)- or (*S*)-2,2′-bis(diarylphosphino)-1,1′-binaphthyls (BINAPs). *J. Org. Chem.*, **51**, 629–35.
81. For recent reviews of various classes of ligands, see: (a) Flanagan, S.P. and Guiry, P.J. (2006) Substituent electronic effects in chiral ligands for asymmetric catalysis. *J. Organomet. Chem.*, **691**, 2125–54; (b) Shimizu, H., Nagasaki, I. and Saito, T. (2005) Recent advances in biaryl-type bisphosphine ligands. *Tetrahedron*, **61**, 5405–32.
82. For recent reports of the utility of other ligands, see: (a) Fitzpatrick, M.O., Coyne, A.G. and Guiry, P.J. (2006) The application of HETPHOX ligands to the intramolecular asymmetric Heck reaction. *Synlett*, 3150–4; (b) Sulikowski, M., Mercier, F., Ricard, L. and Mathey, F. (2006) C2-BIPNOR: an easily accessible homologue of BIPNOR for asymmetric catalysis. *Organometallics*, **25**, 2585–9; (c) Mata, Y., Dieguez, M., Pamies, O. and Claver, C. (2005) Chiral phosphite-oxazolines: a new class of ligands for asymmetric Heck reactions. *Org. Lett.*, **7**, 5597–9; (d) Busacca, C.A., Grossbach, D., Campbell, S.J. *et al.* (2004) Electronic control of chiral quaternary center creation in the intramolecular asymmetric Heck reaction. *J. Org. Chem.*, **69**, 5187–95; (e) Gibson, S.E., Ibrahim, H., Pasquier, C. and Swamy, V.M. (2004) Novel planar chiral diphosphines and their application in asymmetric hydrogenations and asymmetric Heck reactions. *Tetrahedron: Asymmetry*, **15**, 465–73; (f) Kilroy, T.G., Cozzi, P.G., End, N. and Guiry, P.J. (2004) The application of HETPHOX ligands to the asymmetric intermolecular Heck reaction. *Synlett*, 106–10; (g) Nakamura, Y., Takeuchi, S., Zhang, S. *et al.* (2002) Preparation of a fluorous chiral BINAP and application to an asymmetric Heck reaction. *Tetrahedron Lett.*, **43**, 3053–6; (h) Imbos, R., Minnaard, A.J. and Feringa, B.L. (2002) A highly enantioselective intramolecular Heck reaction with a monodentate ligand. *J. Am. Chem. Soc.*, **124**, 184–5.

Index